Volume 3

Clinical Manifestations of Pollutant Overload

T0248652

William J. Rea, M.D.
Environmental Health Center
Dallas, Texas

CRC Press
Taylor & Francis Group
Boca Raton London New York

CRC Press is an imprint of the
Taylor & Francis Group, an informa business

CRC Press
Taylor & Francis Group
6000 Broken Sound Parkway NW, Suite 300
Boca Raton, FL 33487-2742

© 1996 by Taylor & Francis Group, LLC
CRC Press is an imprint of Taylor & Francis Group, an Informa business

First issued in paperback 2019

No claim to original U.S. Government works

ISBN 13: 978-0-367-44880-6 (pbk)
ISBN 13: 978-0-87371-964-3 (hbk)

Visit the Taylor & Francis Web site at
http://www.taylorandfrancis.com

and the CRC Press Web site at
http://www.crcpress.com

Library of Congress Cataloging-in-Publication Data

Rea, William J.
   Chemical sensitivity.
   Vol. 3 has subtitle: Clinical manifestations of pollutant overload.
      p. cm.
   Includes bibliographical references and index.
   ISBN 0-87371-964-6 (vol. 3)
   1. Environmentally induced diseases. 2. Chemicals—Health aspects.
   3. Toxicology. I. Title.
RB152.R38   1995
616.9'8   92-9436

DEDICATION

To the fathers of modern Clinical Ecology, Drs. Theron Randolph, Larry Dickey, Dor Brown, Herbert Rinkle, Carlton Lee, Russ Williams, Jim Willoughby, Joe Miller, and many more whose names space will not permit but whose precepts inspired our work, which then led to the data presented in this book.

To the chemically sensitive patient, who has often been mistreated and maligned and who has longed not only for help but for understanding from the medical and scientific community.

To the general public, who will benefit from the compilation of all the clinical and scientific data collected in this study.

To my family, especially my wife, Vera, and my children, Joe, Chris, Tim, and Andrea, and loved ones, who have suffered from this problem often to the point of severe frustration and who have provided encouragement and understanding when they were desperately needed.

To the surgical mentors of Dr. Rea, Dr. Tom Shires and Dr. Watts Webb, who taught and allowed him to learn, to innovate, and to provide for patients care that was in their best interest in spite of all adversity and criticism.

FOREWORD

The field of chemical sensitivity has begun to capture the interest and fascination of mainstream physicians, environmental and occupational physicians and scientists, and others interested in the effects of chemicals on health. Sensitivity to chemicals manifests itself in the context of occupational exposure, indoor air, contaminated communities, and in persons with exposures to consumer chemicals and pharmaceuticals/anesthesia. A number of recent volumes on this elusive and difficult problem, written by newcomers to the field, have suggested that chemical sensitivity ought to receive more thorough attention. The National Research Council recently sponsored a workshop on the subject and published a monograph entitled *Multiple Chemical Sensitivities* suggesting future areas for research. The American Occupational and Environmental Health Clinics similarly held a workshop exploring the role of environmental and occupational physicians in recognizing and understanding the problems of patients with nonclassical sensitivities to chemicals.

For the first time, these four volumes on chemical sensitivity place in perspective the long and varied experience of one of the innovative pioneers in the field of clinical ecology, founded by Dr. Theron Randolph. Dr. William Rea provides the much needed clinical perspective derived from observing or treating over 20,000 environmentally sensitive patients. He offers his wide experience concerning the identification, diagnosis, and treatment of chemically sensitive patients. Much of the literature describing the experience of clinicians is dispersed and unfamiliar to most of mainstream medicine. This work provides a unique opportunity to follow the perspective of a major clinician in the field. Many will disagree with his observations, explanations, and conclusions. However, the work represents the coalescence of many patient-years of experience and should be reviewed with an open mind. There is no doubt that our understanding of this difficult area will change over the next few years. These volumes provide a valuable time capsule against which to compare our future observations.

Nicholas A. Ashford
Professor of Technology and Policy
Massachusetts Institute of Technology

ACKNOWLEDGMENTS

Thanks to Christine Bishop and Dr. Hsiech-Chia Liang, whose help in preparing the manuscript and illustrations was invaluable — their efforts were Herculean and the book could not have been completed without them; to Drs. Al Johnson, Jerry Ross, Ralph Smiley, Tom Buckley, Nancy Didriksen, Joel Butler, Irvin Fenyves, John Laseter, and John Pangborn, who supplied cases, data, reports, and criticisms of what should and should not have been done; to Dr. Ya Qin Pan for help in analyzing data; to Drs. Sherry Rogers, Al Lieberman, Bertie Griffiths, and Kalpana Patel, who proofread and helped compile sections of this book; to the staff at the EHC-Dallas for all their support; to the members of the American Academy of Environmental Medicine for their contribution and support to the EHC-Dallas and the American Environmental Health Association, who lent financial support to this seven-year effort; to Drs. Doris Rapp, Theron Randolph, Larry Dickey, John MaClennen, Dor Brown, Carlton Lee, Jim Willoughby, George Kroker, Jean Monro, Jonathan Maberly, Klaus Runow, Colin Little, Marshall Mandell, Joe Krop, Hong yu Zhang, Satoshi Ishikawa, Miki Myata, Joe Miller, and Ronald Finn for advice and for freely exchanging information.

I am especially indebted to Dr. John Pangborn, William B. Jakoby, Andrew L. Reeves, Thad Godish, Steve Levine, Al Levine, Felix Gad Sulman, and Eduardo Gaitan whose research, books, and papers provided an invaluable foundation for the preparation of this text.

William J. Rea

PREFACE

The four volumes comprising *Chemical Sensitivity* are the result of our study of more than 20,000 environmentally sensitive patients under various degrees of environmental control at the Environmental Health Center (EHC)-Dallas. They focus on one aspect of environmental overload, chemical sensitivity. They integrate our experience with the effects of environmental pollutants on known mechanisms of immune and nonimmune detoxication systems, and they emphasize the importance of maintaining a balance between endocrine, immunological, and neurological systems and their nutrient fuels (Figure 1).

The principles of diagnosis and treatment outlined here have been developed from the combined experience of the EHC-Dallas and information accumulated from treatment and study of an estimated 100,000 patients by other environmentally oriented physicians and scientists throughout the world.

The first volume, *Chemical Sensitivity: Principles and Mechanisms*, included chapters that defined the field of chemical sensitivity and identified the basic principles used for its diagnosis and treatment. Chapters on immune and nonimmune mechanisms explained the body's processing of pollutants. The final chapter focused on nutrition, which provides fuel for the endocrine, immunological, and neurological systems and equips the body to respond to pollutant exposure.

The second volume, *Sources of Total Body Load*, focused on the wide range of environmental sources that can contribute to each person's total body load. Detailed attention was given to sources of water pollution and the effects of contaminated water on individual health. Similarly, other sources of pollution including food, outdoor and indoor air, inorganic and organic chemicals, pesticides, formaldehyde, terpenes and terpenoids, the hazards of drugs and medical devices, and factors that compound and, hence, dramatically increase an individual's total body load were also explored.

This third volume, *Clinical Manifestations of Pollutant Overload*, considers the effects of pollutant exposure on the different anatomical systems that may be affected by pollutant exposure. This book includes chapters on the ENT/Upper Respiratory System; the Lower Respiratory System, Chest Wall, and Breast; the Cardiovascular System; the Gastrointestinal System; the

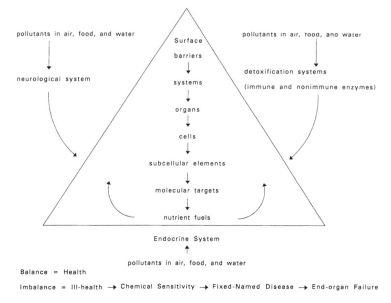

Figure 1. Effects of environment on bodily functions.

Genitourinary System; the Musculoskeletal System; the Endocrine System; Skin; the Nervous System; and the Ophthalmalogical System. It gives special consideration to issues regarding children and their unique susceptibility to chemical/pollutant exposures.

This work is intended for medical students, novice physicians, and practicing clinicians. While the myriad of facts supplied here may initially seem cumbersome, they provide evidence for a new, rational, and exciting way to understand chemical sensitivity, practice medicine, and prevent disease. No longer is symptom suppression and/or intervention after the onset of end-organ damage the focus of healing. Instead, our approach emphasizes the importance of isolating and eradicating root causes of illness before fixed-named disease and permanent damage can occur.

Our study of chemically sensitive individuals has convinced us that following exposure to certain sets of environmental pollution many of these individuals experience a lowered resistance to illness that is followed by periods of vulnerability during which their susceptibility to illness increases. We believe that this is the time when intervention is critical. If, under controlled conditions, these pollutants can be identified along with their route of entry and their effect on individual systems, both intervention and prevention programs can be implemented to combat the threat of chronic illness with multiple symptom manifestation that could lead to fixed-named disease and permanent organ damage.

Our work with the chemically sensitive patient has led to the development of a less-polluted environment, which we use therapeutically to reduce our patients' total pollutant load. We also use this unit to study precisely the effects of environmental pollutants on our individual patients. Use of the less-polluted environment helps us identify specific environmental causes of our patients' symptoms. We can then intervene in the disease process by reducing or eliminating these causative environmental factors (triggers). We can also offset or reverse pollutant damage by bolstering the nutritional state of the patient, thereby fueling and strengthening the systems responsible for protecting the body from pollutant damage.

Attention to the principles (total body load, adaptation, bipolarity, biochemical individuality, spreading and switch phenomenon) that we have developed along with implementation of our method of exposure and challenge followed by diagnosis and treatment has many advantages. Through the identification of causative factors (triggering agents), their elimination or reduction, and the education of patients as to the resourcefulness of good nutrition and the effects of environmental pollutants upon it, physicians can help their patients gain relief from much unnecessary suffering. Patients can achieve improved health, and they can exert control over their life through their own active management of their illness. They can expect a marked decrease in their health care expenses as their chronic symptoms dissipate and end-organ damage or fixed-named disease is avoided.

AUTHOR

 **William J. Rea, M.D., F.A.C.S., F.A.A.E.M.,** is a practicing thoracic and cardiovascular surgeon with an added interest in the environmental aspects of health and disease. Founder of the Environmental Health Center-Dallas, Dr. Rea is currently director of this highly specialized medical facility.

In 1988, Dr. Rea was named to the world's first professorial chair of environmental medicine at the Robens Institute of Industrial and Environmental Health and Safety at the University of Surrey in Guildford, England. Co-author of *Your Home, Your Health and Well-Being*, Dr. Rea has published and peer reviewed more than 100 research papers related to the topics of thoracic and cardiovascular surgery as well as environmental medicine. He is author of several articles for the Correspondence Society of Surgery.

Dr. Rea currently serves on the Board of Directors of the American Academy of Environmental Medicine and on the American Environmental Health Foundation. Previously, he has held the position of chief of surgery at Brookhaven Medical Center and is past president of the American Academy of Environmental Medicine and the Pan American Allergy Society. He has also served on the Science Advisory Board for the U.S. Environmental Protection Agency, on the Research Committee for the American Academy of Otolaryngic Allergy, on the Committee on Aspects of Cardiovascular Endocrine and Autoimmune Diseases of the American College of Allergists, Committee on Immunotoxicology for the Office of Technology Assessment, and on the panel on Chemical Sensitivity of the National Academy of Sciences. Dr. Rea is a fellow of the American College of Surgeons, the American Academy of Environmental Medicine, the American College of Allergists, the American College of Preventive Medicine, the American College of Nutrition and the Royal Society of Medicine.

Born in Jefferson, Ohio, Dr. Rea graduated from Otterbein College in Westerville, Ohio, and Ohio State University College of Medicine in Columbus, Ohio. He then completed a rotating internship at Parkland Memorial hospital in Dallas, Texas. He held a general surgery residency from 1963 to 1967 and a cardiovascular surgery fellowship and residency from 1967 to 1969 with the University of Texas Southwestern Medical School system, which includes Parkland Memorial Hospital, Baylor Medical Center, Veteran's Hospital, and Children's Medical Center.

From 1984 to 1985, Dr. Rea held the position of adjunct professor of environmental sciences and mathematics at the University of Texas, while from 1972 to 1982, he acted as clinical associate professor of thoracic surgery at the University of Texas Southwestern Medical School. He has also served as Chief of Thoracic Surgery at Veteran's Hospital, as adjunct professor of psychology and guest lecturer at the University of North Texas in Denton, Texas. Dr. Rea is currently affiliated with Tri City Health Center of Dallas and Garland Community Hospital in Garland.

Dr. Rea has won several awards for outstanding work in the field of environmental medicine. The Jonathan Forman Gold Medal Award was presented to Dr. Rea for outstanding research in Environmental Medicine by the American Academy of Environmental Medicine. The Herbert Rinkel Award was for outstanding teaching in environmental medicine. The F.A.M.E. Award was for pioneering work in the field of environmental and preventive medicine. The Mountain Valley Water Hall of Fame was for work in the field of study for clean water. Dr. Rea has lectured all over the world, including at the Royal Australian College of Surgery, the Royal College of Medicine, the Royal Society of Medicine, Peking University Medical School, etc.

CONTENTS
CHEMICAL SENSITIVITY
VOLUMES I—IV

Volume I   Principles and Mechanisms

1. Introduction
2. Definition of Chemical Sensitivity
3. Principles of Chemical Sensitivity
4. Nonimmune Mechanisms
5. Pollutant Effects on the Blood and Reticuloendothelial System (Lymphatic and Immune System)
6. Nutritional Status and Pollutant Overload

Volume II   Sources of Total Body Load

7.   Water Pollution
8.   Food and Food Pollution
9.   Outdoor Air Pollution
10. Indoor Air Pollution
11. Inorganic Chemical Pollutants
12. Organic Chemical Pollutants
13. Pesticides
14. Formaldehyde
15. Terpenes and Terpenoids
16. Chemical Hazards in Drugs and Medical Devices
17. Compounding Factors

Volume III   Clinical Manifestations of Pollutant Overload

18. ENT/Upper Respiratory System
19. Pollutant Injury to Lower Respiratory System, Chest Wall, and Breast
20. Cardiovascular System
21. Gastrointestinal System
22. Genitourinary System

23.   Musculoskeletal System
24.   Endocrine System
25.   Integument
26.   Nervous System
27.   Pollutant Injury to the Eye
28.   Children

Volume IV   Tools of Diagnosis and Methods of Treatment

29.   History and Physical
30.   Laboratory
31.   ECU
32.   Avoidance — Air
33.   Avoidance — Water
34.   Avoidance — Food
35.   Heat Depuration and Physical Therapy
36.   Injection Therapy
37.   Nutrition Replacement
38.   Endocrine Treatment
39.   Tolerance Modulators
40.   Behavior Therapy
41.   Surgery
42.   Long-Term Results

CONTENTS

18. ENT/Upper Respiratory System 1105

19. Pollutant Injury to the Lower Respiratory Wall,
    Chest Wall, and Breast 1195

20. Cardiovascular System 1299

21. Gastrointestinal System 1407

22. Genitourinary System 1495

23. Musculoskeletal System 1533

24. Endocrine System 1597

25. Integument 1691

26. Nervous System 1727

27. Pollutant Injury to the Eye 1885

28. Children 1935

Index 2007

18   ENT/Upper Respiratory System

INTRODUCTION

Patients presenting with ear, nose, and throat symptoms are often in the beginning phase of disease, especially chemical sensitivity. For example, vasomotor rhinitis may be one of the early symptoms of more advanced disease such as intractable sinusitis or vasculitis. If the patient's symptoms are the result of environmental exposure, termination of this entity as well as any other environmentally induced ear, nose, and throat diseases by elimination of the pollutant triggers will often prevent advanced end-organ failure some years down the line. Frequently, patients with ear, nose, and throat symptoms can be initially diagnosed and treated by simple measures such as elimination of foods and food additives, nutritional supplementation, and injection therapy. The temptation to treat only symptoms with symptom-suppressing drugs, as is often done for these conditions, and then to allow the patient to go his or her own way is always present. Since many tools including rotary diets, environmental controls, nutrient supplementation, and injection therapy are available to interrupt and prevent an advanced disease process, treatment with symptom-suppressing drugs only should be used as a last resort. Covering symptoms with medication such as antihistamines or steroids and enabling a disease such as rhinitis, sinusitis, or tinnitus to become more complicated should be a procedure of the past. Neglected or improperly treated ear, nose, and throat sensitivities (Ménière's, carotodynia, etc.) result in either irreversible end-organ damage or the spread of sensitivities to other smooth-muscle systems throughout the body. As a result of this spreading, many untreatable inflammatory diseases may occur.

Because symptoms related to ear, nose, and throat dysfunction often signal the onset of chemical sensitivity, it is imperative that the physician when first treating these symptoms consider the possibility of environmental triggers in order to prevent continued deterioration, eventual spreading, and finally onset of end-organ disease. To assist the physician in determining pollutant involvement in patients' ear, nose, and throat symptoms, this chapter examines the particular effects of pollutant injury on the nose and sinuses; the throat, larynx, and neck; and the ear. A variety of fairly common clinical entities are also assessed as manifestations of pollutant damage, and, as such, environmental treatments to halt their spread and severity are further suggested.

Pollutant Injury to the Nose and Sinuses

Following exposure to pollutants, the nose is often the first site of pollutant injury in the chemically sensitive individual. The lymphatic system and nerve supply in the olfactory area, as well as olfaction, can be damaged by pollutants. Each of these is now discussed.

Lymphatic System

The nose has a rich supply of lymphatic vessels located in an anterior and posterior network. Both of these networks can be avenues for pollutant entry. The anterior network made up of lymphatic vessels and nodes is small and drains along the facial vessels to the neck. These vessels serve the most anterior portion of the nose — the vestibule and the preturbinal area. Often, as a result of food sensitivity and early pollutant exposure, chemically sensitive individuals present with these nodes swollen and clearly involved in the onset of their illness.

The posterior network serves the majority of the nasal anatomy, joining three main channels in the posterior area of the nose — the superior, middle, and interior channels. The superior group arises from the middle and superior turbinates and a portion of the nasal wall that passes above the eustachian tube and empties into the retrolaryngeal nodes. The middle group passes below the eustachian tube, drains the inferior turbinate, inferior meatus, and a portion of the floor, and goes to the jugular chain of lymph nodes. The interior group drains from the septum and part of the floor of the nose, passing through the lymph nodes along the internal jugular vein. Pollutants and antigens may enter either, or both, chains and cause local (weeping), regional (swelling), or distal (e.g., vascular spasm and Raynaud's phenomenon) effects throughout the body. Often, the first sign of pollutant exposure that the clinician will observe when these are present is chronic lymph node swelling in the neck. Should these signs occur, a search for triggering agents should be conducted.

Lymph Nodes

The area of the neck most commonly affected by swelling and pain is the lymph nodes. Sometimes submental node swelling is triggered by foods or food additives to which the chemically sensitive patient has developed additional sensitivities. Frequently, environmental exposures induce sore throats and/or tonsillitis. Following exposure, an individual will experience a sudden pain in the neck, and the lymph nodes will swell. In our experience, these nodes will often remain swollen for weeks after the initial exposure, resulting in

increased vulnerability to viruses and bacterias. Often environmental exposures cause swelling in the posterior nodes, and stiff neck with muscle spasm
or headaches occurs. These symptoms are frequently misdiagnosed as tension
headaches. Our work with chemically sensitive patients reveals that these
swellings often persist for weeks after an incitant has been withdrawn. However, conversely, swollen nodes as a result of an acute reaction to pollutant
challenge may be seen, and relief may be obtained with pollutant avoidance.

Nerve Supply

In chemically sensitive individuals, damage to the nerve supply of the nose
and sinuses may result in reduced olfaction or in increased odor sensitivity. The
nerve supply of the nose is innervated by the olfactory nerve located high in
the nasal vault. Autonomic control of nasal physiology, which is usually
imbalanced in the chemically sensitive individual, is mediated primarily by
sympathetic and parasympathetic fibers that reach the nose by way of the
sphenopalatine ganglion. Sensation is from the ethmoid branches of the ophthalmic division of the fifth cranial nerve and the maxillary division of the fifth
nerve (Figure 18.1). All of these nerve endings have been observed to have
been injured in some cases of chemical sensitivity. However, the olfactory and
the autonomic nerves appear to be the most frequently injured.

Olfactory Area and Olfaction

Large individual variations in anatomy characterize the structure of the
olfactory region. These variations are especially pronounced in chemically
sensitive individuals, who may have thick mucus layers (60 micron), but more
commonly have thin layers on nerve endings that are variously located near the
surface and, therefore, have varying degrees of sensitivity. These anatomical
differences contribute to an individual's susceptibility to pollutants so that the
type of toxic (phenol, drugs, etc.) or physical exposure the individual experiences results in damage that is unique to each person.

The olfactory mucosa is pseudostratified columnar epithelium, composed
of bipolar nerve cells, sustentacular supporting cells, and small basal cells, all
of which are involved in the adaptation reaction and may be inadequate in the
chemically sensitive population due to nutrient fuel deficiency.

The olfactory neurons are the only portion of the CNS that reaches the
surface of the body (Figure 18.2). So located, these neurons are vulnerable to
damage during pollutant exposure thus, perhaps, explaining both the involvement of the central and peripheral ANS and the nonautonomic nervous dysfunction often seen in pollutant injury.

The olfactory nerves go through the cribiform plates connecting numerous
nuclei, fasciculi, the limbic system, and other tracts including the hypothalamus. Not infrequently, a chemically sensitive individual presents with complaints of unrelenting pain in this area. Often the pain will fluctuate, depending

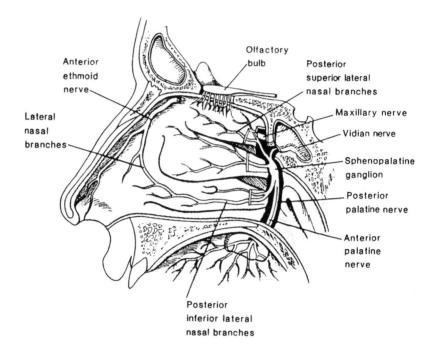

Figure 18.1. Areas of potential damage to the nose and its innervation, emphasizing one aspect of chemical sensitivity. Olfaction is mediated by the olfactory nerve located high in the nasal vault. Autonomic control of nasal physiology is mediated primarily by sympathetic and parasympathetic fibers that reach the nose by way of the sphenopalatine ganglion. Sensation is from the ethmoid branches of the ophthalmic division of the fifth cranial nerve and the maxillary division of the fifth nerve. Pollutant damage may occur at any area. (From Hilger, P. A. 1989. Applied anatomy and physiology of the nose. In *Boies Fundamentals of Otolaryngology*, 6th ed., eds. E. G. Adams, L. R. Boies, Jr., and P. A. Hilyer. 187. Philadelphia: W. B. Saunders. With permission.)

upon the post-pollutant stimuli edema. Often these occur post-trauma with the hypersensitivity being propagated by the pollutant driven reaction.

The central olfactory apparatus is highly complex; thus, there are many areas in this region susceptible to pollutant injury. When the olfactory nerves are injured, their ends regenerate, and the heightened or depressed sense of smell seen in the chemically sensitive patient may result.

Air flow usually curves away from the olfactory clefts, which normally do not reach high enough to receive the full force of an odor. However, in the chemically sensitive individual, odor perception is hyperacute, and all sensory areas appear to respond to stimulation. These responses appear to be due to

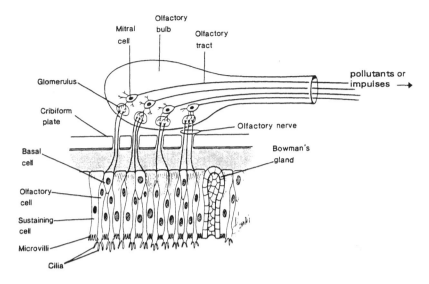

Figure 18.2. The olfactory area, which has a potential of being damaged by pollutants in the chemically sensitive individual. The olfactory area is located high in the vault of the nose. This is the only area of the body in which an extension of the central nervous system comes into contact with the environment. Many cells and parts can be damaged in the chemically sensitive individual, and pollutants or their impulses can go unnoticed. (From Hilger, P. A. 1989. Applied anatomy and physiology of the nose. In *Boies Fundamentals of Otolaryngology*, 6th ed., ed. E. G. Adams, L. R. Boies, Jr., and P. A. Hilyer. 184. Philadelphia: W. B. Saunders. With permission.)

nerve-end damage. A normal sense of smell is in the range of 5×10^{-10} gm/L of air. In the deadapted chemically sensitive individual, however, it is much less. Olfactory sense is quickly exhausted in both the chemically sensitive and the normal individual with adaptation occurring rapidly. The difference is that while adaptation is successful in the normal individual, it breaks down in the pollutant-overloaded chemically sensitive person. The result of this breakdown is triggering of symptoms.

   Odor sensitivity to toxic chemicals, perfumes, fabrics, gasoline, car exhausts, and plastics is one of the most pathognomonic signs of chemical sensitivity. The endings of the olfactory nerves become supersensitive once they are overloaded. This reaction is probably due to local nerve sensitization and/or to direct damage to the olfactory nerve membranes by the toxic chemicals. This overload may often blunt perception due to the masking (adaptation) phenomenon. However, once these patients are deadapted for 3 to 9 days, they have superacute odor perception. The olfactory cilia can slough off and regenerate as their disease clears, and, undoubtedly, regeneration accounts for some lessening of the chemical sensitivity in selected cases. Studies have shown that toluene and xylene travel up the olfactory nerves into the brain of mice who are excessively exposed.[1] Probably these solvents travel similarly in the chemically sensitive human. Direct triggering of the hypothalamus then occurs causing the ANS, the limbic system, the cerebrum, and the vascular tree of the brain to malfunction. The olfactory tract connects with the hypothalamus and the ANS.

Though no dissection studies are possible in live humans, many of our recently cleared chemically sensitive patients describe feeling as if a wave passes over their brain, dulling it, just moments after they perceive a noxious odor. This feeling reflects the direct pollutant triggering that begins in the nose and travels to the CNS in some chemically sensitive patients.

Of chemically sensitive patients, 90% experience odor sensitivity. Rhinorrhea and sneezing, with or without a stuffy nose, are frequently seen following a food or other chemical exposure or even with a physical stimulus such as cold. Often such responses are transient and last less than 30 seconds. They may, however, be long-lived and last 8 to 10 hours, or more. Reaction time is usually determined by the duration and magnitude of exposure as well as previous damage to the nasal mucosa. Cumulative trauma to the nose, directly or as a nasal manifestation of cumulative food or other chemical trauma, is common and usually associated with the appearance of presumably unexplainable manifestations of illness in other organ systems. However, closer scrutiny reveals that, due to the injury, more pollutants become able to enter the body, and damage then can occur in more distal target organs.

Disturbances of the nose and paranasal sinuses in chemically sensitive individuals frequently result from occupational chemical exposure. For example, certain nickel compounds have been shown to be rather powerful carcinogens in animals,[2-4] and an increased risk for the development of nasal carcinoma in nickel workers has been shown in epidemiological studies.[5,6] Histopathological changes of the nasal mucosa in nickel workers have been carefully investigated by Torjussen et al.[7,8] Boysen et al.[9] demonstrated that 12% of process nickel workers and 47% of retired nickel workers exhibited dysplasia of the nasal mucosa. Investigation of the distal organ deposition for nickel was not done in this latter study, but it is interesting to note that nickel is found in blood cells in the highest proportions of all toxic inorganic chemicals in our series of chemically sensitive patients at the EHC-Dallas. The majority of these patients are nickel sensitive. Attention has also been drawn to other noxious agents, such as wood dust,[10-12] coal, coke dust, and leather,[13] that can cause both local and distal disease.

Functional disturbances as well as histopathological changes of the nasal mucosa have been reported in workers exposed to oil mist.[14] Cohr and Stokholm[15] briefly described the irritative effect of toluene on the respiratory tract. These authors evaluated the extent to which alterations of the nasal mucosa were present among workers occupationally exposed to different

Table 18.1.   Changes in Nasal Mucosa from Toxic Solvent Exposure

	A	B	C	D
Exposed Group				
1	X	X	X	
2		X	X	
3				X
4				X
5	X	X		
6		X		
Control Group				
1				X
2	X			
3	X			
4a	X		X	
4b	X			
5a	X			
5b	X	X		

Notes: A = Pseudo-stratified ciliated epithelium present. B = Rather low pseudo-stratified epithlium with no or few cilia or basal cell hyperplasia or goblet cell hyperplasia. C = Squamous metaplasia. D = Stratified, nonkeratinized squamous epithelium.

Source: Irander, K., H. B. Hellquist, C. Edling, and L. M. Ödkvist. 1980. Upper airway problems in industrial workers exposed to oil mist. *Acta Otolaryngol. (Stockh.)* 90:452–459. With permission.

airborne substances such as oil mist, styrene, solvents, and formaldehyde. Ödkvist et al.[16] performed biopsies on 21 workers exposed to oil mist (11 of whom were exposed to styrene, 10 to solvents, 20 to formaldehyde) and 25 who were control patients (Table 18.1). While the controls were normal, they found the majority of exposed workers had an identifiable pathology including tissue change such as loss of cilia, goblet cell hyperplasia, mixed cuboid squamous cell metaplasia, and stratified squamous cell metaplasia. Stratified squamous cell epithelium and squamous cell metaplasia were also seen (Figure 18.3). Ödkvist et al. rarely saw normal ciliated pseudo-stratified epithelium in the exposed group.

Some toxic chemicals have been shown to cause cancer in both man and animals. Some of the chemicals Ödkvist[17] investigated are suspected carcinogens in humans and animals. Formaldehyde, for instance, causes nasal carcinoma in rats and is suspected of causing the same in humans. Other toxic chemicals cause problems in animals. These direct-acting carcinogens,

A.

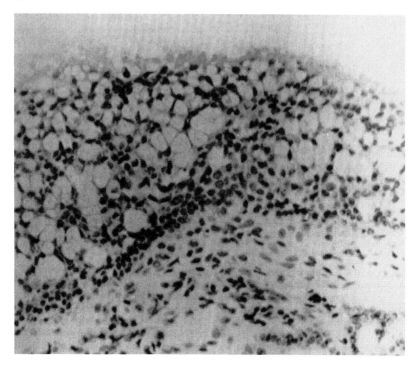

Figure 18.3. (A) Photomicrograph: nasal epithelium with goblet cell hyperplasia, but cilia are present as well (H&E, ×350) — pollutant injury. (B) Photomicrograph: stratified squamous epithelium displaying hyperplasia but no atypia (H&E, ×190) — pollutant injury. (C) Photomicrograph: a very thin nasal mucosa with subepithelial hyalinization (arrow) (H&E, ×150) — pollutant injury. (D) Photomicrograph: nasal mucosa lined with pseudostratified ciliated (arrow) columnar epithelium (totoluidine blue, ×760) — normal. (From Irander, K., H. B. Hellquist, C. H. Edling, and L. M. Ödkvist. 1980. Upper airway problems in industrial workers exposed to oil mist. *Acta Otolaryngol. (Stockh.)* 90:452–459. With permission.)

beta-propiolactone (BPL), methylmethane sulfonate (MMS), and dimethylcarbamyl chloride (DMCC), were evaluated for their carcinogenic potencies in the nasal mucosa of rats and for their ability to bind *in vivo* to rat nasal mucosal DNA. The relative carcinogenic potencies of BPL and MMS corresponded well with their overall levels of binding to nasal mucosal DNA. DMCC, which is the most potent carcinogen of the three compounds, however, produced the lowest level of binding to nasal mucosal DNA. These results indicate that the DNA adducts formed by DMCC in rat nasal mucosa DNA are more readily

B.

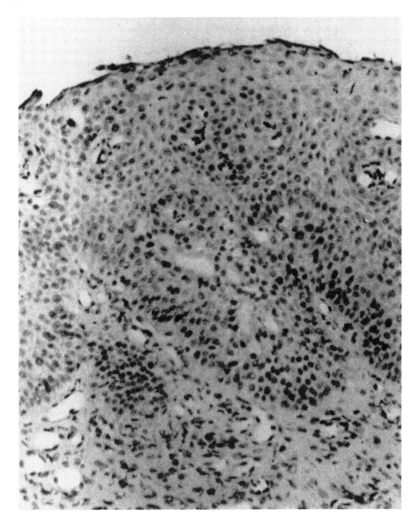

Figure 18.3. (Continued).

expressed as cancer.[17] These same chemicals are known to exacerbate chemical sensitivity.

Epidemiologic studies have shown changes in nasal epithelial histology in individuals chronically exposed to pollutants. For example, an extensive study was performed to compare 1726 occupants of homes that had been insulated with ureaformaldehyde foam insulation (UFFI) with 726 occupants of control homes.[18] The study showed a modest excess of many symptoms among residents of households who were intending to have their UFFI removed. The onset of the symptoms had followed the installation of UFFI. Traditional

C.

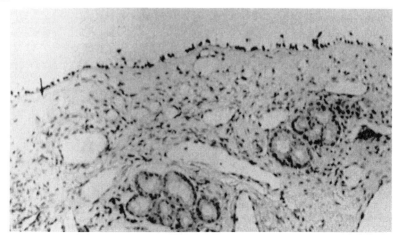

D.

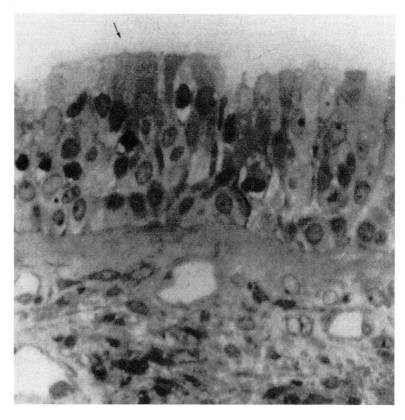

Figure 18.3. (Continued).

measures of response, such as nasal airway resistance, sense of smell, pulmonary function, and patch tests for sensitivity to formaldehyde, failed to demonstrate any specific abnormality. Interestingly, samples of nasal tissue demonstrated a shift from the normal pseudostratified ciliated columnar epithelium to squamous metaplasia just as seen in the series by Ödkvist et al.[16] The epithelial changes were evident only in the noses of residents who were planning to have the UFFI removed from their home. It is possible that the symptom excess and epithelial changes were parallel markers of a toxic effect. One may speculate, however, that the excess symptoms occurred as a result of the epithelial changes.

In another study, nasal specimens obtained from the middle turbinate of formaldehyde-exposed workers showed significant changes in epithelial morphology compared with specimens from control workers. No correlation was found between the histological changes and the duration of exposure, dose of exposure, or smoking habits.[19]

Normally, respiratory epithelium distal to the nasal alae is a pseudostratified columnar epithelium. Nonallergic, asymptomatic individuals may demonstrate mild inflammation of the nasal mucosa on biopsy specimens[20] or increased nasal lavage with neutrophils.[21] Etiologically, the significance of the inflammation in asymptomatic subjects is unknown. Changes in the histology of the airway undoubtedly are associated with changes in epithelial cell biology.

The reports of epithelial changes associated with chronic irritant exposure are provocative, since the effect of metaplasia on the multitude of epithelial functions is unknown. The results of the studies cited below indicate the capacity of a cytokine to induce mucosal symptoms, inflammation, and epithelial changes in the nose of normal humans. Whether this cytokine has relevance in chronic irritation is unknown, but it appears to be relevant in chemically sensitive individuals.

Early studies documented that human leukocyte interferon (HuIFN-*alpha* or HuIFN-*alpha2*), produced by recombinant DNA techniques, was effective in preventing illness or infection after experimental rhinovirus infection.[22,23] While short-term administration of interferon was generally well-tolerated, mild upper respiratory symptoms of nasal stuffiness, sore throat, and epistaxis occurred in five of eight uninfected volunteers.[24] A significant decrease in peripheral blood leukocyte counts was observed in some interferon recipients.[25]

Later studies of chronic intranasal interferon alpha sought to determine the safety of cytokine as a possible prophylactic treatment for rhinovirus infections. Long-term administration was associated with clinically apparent nasal irritation in a minority of recipients and with mucosal histopathologic abnormalities in a majority of recipients.[23] Intranasal interferon (HuIFN-*alpha2*, 8.4×10^6 IU daily and 2.35×10^8 IU over one month) was associated with significantly greater symptoms of nasal burning, sore throat and hoarseness, and less headache in treated subjects as compared with placebo controls.[23] Twenty-three percent of interferon-treated subjects had symptoms or signs of mucosal irritation including blood-tinged mucus, or, on exam, dry mucus membranes, crusts, and friability. Nasal biopsies of subjects treated with interferon demonstrated significantly greater acute or acute and chronic epithelial inflammation (46% vs. 15%). Nineteen percent of interferon recipients had moderate or severe grades of epithelial inflammation, four of whom had microscopic ulcerations on their post-treatment biopsy specimen. In addition, interferon-treated subjects had significantly greater moderate or severe chronic submucosal mononuclear inflammation with localization in the submucosa, around salivary glands, or, in severe cases, spread through the stroma. The authors of these studies noted that abundant mononuclear cell infiltrates have been described as a feature of the early stage of atrophic rhinitis;[26] however, the changes induced by alpha interferon were reversible 8 weeks after cessation of treatment. Hayden et al.[20] also noted that systemic complaints of fever, malaise, and myalgia have been observed with parenteral interferon administration, but not with local administration. Mygind[27] and Poynter[28] noted that the effect of interferon on the mucosa is markedly different from the effect of chronic intranasal corticosteroids in allergic subjects in whom biopsies have not shown inflammation, cellular infiltrates, or atrophy.

POLLUTANT INJURY—CLINICAL EFFECTS ON THE NOSE AND SINUSES

The clinical manifestation of the effects of pollutant exposure on the nose and sinus of the chemically sensitive individual varies. Because these areas are often the first exposed to pollutants and the first to manifest symptoms of pollutant damage, they require close examination. Rhinitis, polyps, and recurrent sinusitis have been associated with allergic and environmental etiologies. They, along with salivary gland malfunction, are among the common entities affecting these areas and signaling early pollutant damage. Therefore, the next few pages are devoted to an independent discussion of these entities.

Rhinitis

The two types of rhinitis, allergic and vasomotor, are lumped together in this discussion. *Allergic rhinitis* is self-evident. It manifests as rhinorrhea, sneezing, and nasal congestion that result from allergies to substances such as pollen and food. It is mediated through the gamma globulin E and is usually easily managed by avoidance and injection therapy. *Vasomotor rhinitis* manifests similarly. It may be triggered by inhalants, foods, and other chemicals. Possibly, vasomotor rhinitis occurs through a mechanism different from that active in allergic rhinitis. Vasomotor rhinitis may be mediated by local and/or central vascular and autonomic dysfunction. Since the simplest of problems can be triggered by biological inhalants, foods, and toxic chemicals, causes of rhinitis should be sought and eliminated. Often, if rhinitis is seasonal, injections for biological inhalants alone will eliminate it. Some complex cases of rhinitis with more severe ear, nose, and throat manifestations will have to be more carefully worked out using environmental control methods. Patients with this type of rhinitis will now be discussed.

Of 100 patients who had advanced environmentally triggered disease involving the neurocardiovascular, respiratory, gastrointestinal, or genitourinary systems and who were seen consecutively in our ECU, 71% reported ear, nose, and throat symptoms prior to the onset of their other disease. Generally, they had vasomotor rhinitis and recurrent sinusitis. This propensity to prior rhinitis and sinusitis suggests that the ear, nose, and throat specialist had a great opportunity to prevent the development of complex disease in these patients had he or she been aware that these illnesses were the result of environmental triggers and had he or she then known how to eliminate them.

Many physicians find themselves frustrated by the lack of an apparent etiology for recurrent sinusitis, vasomotor rhinitis, or Ménière's disease. However, it is becoming clear that the development of an ENT condition that is manifested by altered reactivity has multiple causes, but the actual appearance of the typical manifestations of the sensitivity reaction are initiated by identifiable triggering agents. For example, a patient with recurrent rhinitis may have his or her symptoms triggered by many incitants such as pollens, molds, dusts, foods, formaldehyde, and/or pesticides. To treat such a patient, the physician could use intradermal injections both for diagnosis of the triggering agents and for subsequent treatment. This form of diagnosis and treatment contrasts to the type that finds no etiology and uses antihistamines and steroids for symptom suppression. Repeated use of symptom-suppressing medication eventually results in recurring chronic disease that is very difficult to diagnose and treat. Patients with multiple causes of their intractable ear, nose, and throat symptoms are, at present, best studied by special methods in an ECU. However, less complicated cases can be evaluated in the office.

Polyps

Polyps occur with environmental overload and will decrease if the appropriate triggering agents are found and eliminated or if injection therapy for foods and biological inhalants, including bacteria, are utilized. Among the other causes of rhinitis and polyps are fungi such as aspergillus,[29] mucormycosis,[30] and candida albicans.[31]

Recurrent Sinusitis

Patients seem to be more frequently seeking medical advice for nasal and sinus symptoms related to their jobs. Hellquist et al.[32] studied 10 paint sprayers exposed to solvents, primarily toluene and isobutylacetate, and dust. In addition to the expected tissue alterations, biopsies of the nasal mucosa were all abnormal.

The following case report of a spray painter emphasizes the magnitude of injury sustained by a patient who developed recurrent rhinosinusitis.

> **Case study**. This patient presented at the EHC-Dallas with symptoms of recurrent sinus infections, intractable dizziness with stumbling, and an inability to remember. He was well and active until 2 months prior to being seen at the EHC-Dallas. His history revealed he had worked as a spray painter for years during which time he had used minimal protection. Prior to the onset of the symptoms that brought him to the EHC-Dallas, he had begun working with a new spray compound whereupon he had immediately developed intractable sinusitis and sinus dryness. He stopped work for 2 weeks and improved. He then returned to work and immediately developed more sinusitis followed by dryness accompanied by imbalance, asthma, incapacity, and short-term memory loss. He consulted an astute ear, nose, and throat surgeon who was familiar with environmentally triggered illnesses and who, therefore, referred the patient to the EHC-Dallas.
>
> Intradermal provocative testing revealed sensitivities to natural gas, formaldehyde, cigarette smoke, orris root, molds, pollens, and foods, indicating that sensitivities secondary to the initial toxic exposure had developed.
>
> Laboratory tests showed the following outcomes: autonomic dysfunction measured by the iris corder showed cholinergic effects. Lower extremity neuromotor test showed a decrease in sensitivity indicating a sensory neuropathy. Pulmonary functions showed FVC, 88% of predicted; $FEV_{0.5}$, 51% of predicted; and $FEF_{0.2-1.2}$, 36% of predicted. Blood toxic chemical levels showed 1,1,1-trichloroethane, 7.9 ppb (C <0.5 ppb); toluene, 2.8 ppb (C <0.5 ppb); trimethylbenzene, 1.2 ppb (C <0. 1 ppb); 2 methyl pentane, 45.2 ppb (C <1 ppb); 3 methyl pentane, 9.1 ppb (C <1.0 ppb); and hexane, 5.8 ppb (C <1.0 ppb). Cholesterol levels were 273 mg/dL (C = 120 to 200 mg/dL) and triglycerides 215 mg/dL (C = 30 to 190 mg/dL).
>
> Posturography showed central vestibular sensory and motor defects compatible with toxic exposures, and a SPECT scan showed a neurotoxic pattern in his brain. The patient was placed on a rigid environmental control program with avoidance of air, food, and water pollutants. He was given injections for his secondary sensitivities to biological inhalants and foods. Daily intravenous and then oral nutrient supplementation was performed. Heat depuration and physical therapy were started, and he improved. He is now doing well and being retrained for a new line of work where exposures will be reduced.

At the EHC-Dallas, we have now worked up 81 patients with recurrent sinusitis in the ECU and over 1000 as outpatients. All had failed to respond to previously received standard medical therapy. We, though, were able to define precisely their triggering agents. These patients usually were found to have combinations of biological inhalant, food, and chemical triggers. Bacterial sensitivities were also found. Double-blind, inhaled chemical challenges under environmentally controlled conditions resulted in the reproduction of signs and symptoms by triggering agents such as <0.002 ppm phenol, <0.2 ppm formaldehyde, <0. 5 ppm petroleum derived ethanol, <0.33 ppm chlorine, and a

Table 18.2.   Laboratory Values for Group with Recurrent Sinusitis

Double-blind inhaled challenge	Dose (ppm)	% of Positive results
Phenol	<0.002	87
Formaldehyde	<0.20	80
Ethanol (petroleum derived)	<0.50	79
Chlorine	<0.33	60
Pesticide (2,4-DNP)	<0.0034	50
Saline	—	2
Saline	—	1
Saline	—	2

Notes: Recurrent sinusitis intractable to medical and surgical treatment — 81 patients inhaled double-blind challenges after 4 days deadaptation with the total load decreased in the ECU. Total number of challenges: 640.

Source: EHC-Dallas. 1983–1984.

pesticide <0.0034 ppm 2,4-DNP (Table 18.2). These patients could not be cleared without detailed attention to their food and chemical sensitivities.

In most cases, the etiology of sinusitis can now usually be found. After mechanical problems are ruled out, environmental overload must be suspected and investigated. In our experience, this overload is now the leading etiology of sinusitis. Individuals with sinus infections or recurrent sinus pain should be worked up meticulously to define their triggering agent(s). Laboratory values found in our group of 81 patients are shown in Table 18.3. Laboratory values from another study of 29 chemically sensitive patients whom we treated in 1986 are shown in Table 18.4. The first study indicates that many values are abnormal including some T-lymphocytes. The second study shows a small difference between experimental and control groups.

Salivary Gland Malfunction

In the chemically sensitive individual, a prime target organ in the face and neck for environmental pollutants is the salivary glands. These are known to swell postsurgically or in association with arthritis. However, they also swell independently due to unknown causes. Biopsy of lesions that may appear on the skin concurrently with salivary gland malfunction will show perivascular lymphocytic infiltrates and/or hyperplasia and hypertrophy of the cells similar to what is seen in the chemically sensitive individual. Biopsy of the glands will show a similar pattern. Frequently, the glands themselves do not swell, but, with chemical exposure, the salivary ducts will apparently spasm, stopping the flow of saliva. This reaction occurs early in chemically sensitive individuals who are chemically overloaded. On exposure to any additional pollutants, they rapidly develop a dry mouth. The following case is illustrative of environmental triggering related to malfunction of the salivary glands.

Table 18.3. Sinusitis — 81 Patients, ECU

Patients	Eosinophils 200 to 400/mm³[a]		CH₅₀ THSC 100 ± 20%[a]		T-Lymph 1500 to 2500/mm³[a]	IgG 800 to 1600 mg/dL[a]	IgE 0 to 180 IU/mL[a]	
	Below	Above	Below	Above	Below	Below	Below	Above
81	16	9	4	5	9	1	18	6

[a] Normal range.

Source: EHC-Dallas. 1974–1981.

**Table 18.4.   Immunological Data of Rhino-Sinusitis Patients
and Normals (Mean Values and Differences)**

	Patients (29 persons)	Normals (60 persons)	Difference (patients to normals)	Significance (p)
WBC (#/mm³)	7070 ± 450	7560 ± 220	None	>0.15
L (#/mm³)	2510 ± 170	2770 ± 91	None	>0.08
L %	35.8 ± 1.4	37.3 ± 1.1	None	>0.2
T_{11} (#/mm³)	1800 ± 129	2080 ± 74	Smaller	<0.05
T_{11} %	70.1 ± 2.6	75.2 ± 0.8	Smaller	<0.05
T_4 (#/mm³)	1010 ± 92	1160 ± 43	None	>0.05
T_4 %	40.4 ± 1.6	42.2 ± 0.7	None	>0.1
T_8 (#/mm³)	690 ± 55	740 ± 38	None	>0.2
T_8 %	27.7 ± 2.0	25.8 ± 0.8	None	>0.1
T_4/T_8	1.60 ± 0.11	1.70 ± 0.06	None	>0.2
B (#/mm³)	260 ± 34	270 ± 23	None	>0.3
B %	9.9 ± 1.0	9.4 ± 0.6	None	>0.3

Source: EHC-Dallas. 1987.

Case study. A woman aged 69 years entered the EHC-Dallas with the chief complaint of an inability to talk due to lack of adequate salivation. Her condition had progressed gradually over the previous 10 years to the point that her speech was almost incomprehensible. The parotid glands were swollen from lack of salivation. She was experiencing spontaneous bruising, petechiae, peripheral and periorbital edema, and Raynaud's phenomenon. This patient had received injection hyposensitization therapy using the prick technique from two different allergists for 6 months at a time without success.

Physical examination revealed parotid swelling, extremely dry mouth, and noticeably dry eyes. Gross speech impairment occurred due to the dry mouth. The skin of the extremities exhibited many petechiae and bruises. Mild periorbital and digital edema was present. Her hands and feet were blue.

This patient was placed in the ECU where she fasted for 5 days. All medications were restricted. At the end of this period, she salivated profusely and talked without impediment. Parotid swelling disappeared. Her skin was clear. She underwent food and double-blind inhaled chemical challenge tests (Figure 18.4). Eight foods and three chemicals (natural gas, cigarette smoke, and an insecticide) inhibited salivation (Table 18.5). Eight other foods produced petechiae, bruising, peripheral edema, and cyanosis.

This patient was instructed to clear her home of offending chemicals, not the least of which was cigarette smoke generated by her husband's indoor smoking. She was placed on a rotary diet of chemically less-contaminated foods. She was treated with scheduled neutralizing doses of food injections. She reports continuing improvement over 8 years.

At the EHC-Dallas, we have seen 7 patients with malfunctioning salivary glands. These patients underwent a course of treatment similar to that described

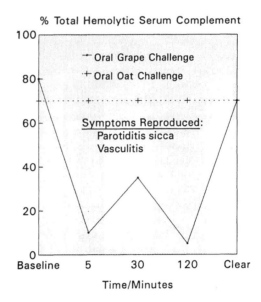

Figure 18.4. 69-year-old white female; oral challenge after 4 days reduction of total body load in the ECU. (From EHC-Dallas. 1980.)

Table 18.5.   Inhaled Double-Blind Challenge after 5 Days Deadaptation in the ECU after Total Load Was Reduced — 69-Year-Old White Female

	Dose (ppm)	Reaction
Insecticide 2,4-DNP	<0.0034	+
Ethanol (petroleum derived)	<0.50	-
Formaldehyde	<0.20	-
Phenol	<0.0050	-
Chlorine	<0.33	-
Saline	ambient	-
Natural gas[a]	open flame	+
Cigarette[a]	open	+

[a] Open challenge

Source: EHC-Dallas, 1980.

in this case report (Table 18.6). Specific triggering agents varied, but all of the patients' symptoms were the result of overexposures. For example, one case involved a beautician whose main sensitivity was to pollutants that emanated from the fumes of hair dressings to which she was exposed occupationally. Another case involved a woman who was the wife of an embalmer. This couple

Table 18.6. Salivary Gland Malfunction

Patient[a] (age, yrs.)	WBC (/mm^3) 4500 to 10,000	EOS (/mm^3) 200 to 400%	CH$_{50}$ THSC (%) 100 ± 20	C$_3$ (mg/dL) 80 to 100	C$_4$ (mg/dL) 15 to 45	T-lymphocytes (E. Rosettes) (/mm^3) 1500 to 2500	IgE (IU/mL) 0 to 180	IgG (mg/dL) 800 to 1600
69	6,000	70	80%	87	32	378	5	1,032
34	4,000	35	94%	80	31	1,038	65	1,115
49	5,800	123	77%	76	44	1,247	25	770
41	11,200	334	52%	66	26	990	120	1,530
55	5,100	53	80%	97	74	460	—	1,540
36	6,900	18	96%	98	38	1,901	190	1,350
58	3,000	35	80%	85	19	350	5	1,950

[a] All patients were white women.

Source: EHC-Dallas. 1983–1984.

lived over the embalming room, and the wife developed sensitivity after experiencing long-term exposure to the formaldehyde that emanated from this area into the room where she slept. These cases illustrate that food and chemical sensitivity should be ruled out before classifying a disease as nonetiologic.

POLLUTANT INJURY TO THE THROAT, NECK, AND LARYNX

Pollutant injury to the throat, neck, and larynx is commonplace in the chemically sensitive individual. As we will show in the following pages, pollutants can trigger numerous forms of dysfunction in these areas. Some of these will be minor, while other kinds can be life threatening.

Throat

In chemically sensitive individuals, recurrent sore throats often occur during pollen and mold seasons. They can also result from excessive exposure to dust. People working around chemicals and those who are chemically sensitive can develop intractable or intermittent sore throats that will stop upon withdrawal of the incitant. Sore throats may occur year round, and, if they are not food triggered, they will usually be chemically or mold triggered. In addition, if allowed to continue, these sore throats will frequently result in bacterial or viral infections. Tobacco is a well-known environmental trigger of pharyngitis. Not so well appreciated triggers are phenol (carbolic acid), formaldehyde, pesticide, and chlorine.

Neck

Commonly seen pollutant injury to the neck involves muscle spasm, the lymphatic system, and carotid artery pain. Since muscle spasm is addressed in Chapter 23, and the lymphatic system is treated in the beginning of this chapter, only carotodynia will be briefly discussed here.

Carotodynia

Carotodynia often occurs with environmental exposure and is frequently involved with neck node enlargement, although direct vessel involvement also occurs. Derrick[33] has described over 100 cases of carotodynia triggered by foods and chemicals. We, too, have confirmed an equal number in our practice. This pain in the carotid artery is probably due to local release of mediators after a food or chemical exposure. In some cases, carotodynia appears to be the result of an early form of vasculitis. In others, it appears to be a self-limiting reaction if the triggering agent is removed (see Chapter 20, section on carotid spasm).

Larynx

Laryngeal dysfunction may occur from pollutant exposure and result in mild to severe edema, spasm, and new tissue growth. This dysfunction can start in infancy and continue through all ages.

Croup

Croup is a rapidly developing infection of the larynx in children. It is mentioned because the local immune dysfunction that allows the hemophilus or staphylococcus to occur may be triggered by biological inhalants, foods, chemicals, and/or nutrient deficiencies.

Leukoplakia

Leukoplakia is known to be caused by cigarette smoking and may have other environmental triggers. (See Chapter 21 for further discussion.)We have seen foods and chemicals trigger this entity.

Hoarseness, Laryngeal Stridor, and Aphonia

Frequently, people with environmental sensitivities develop transient hoarseness. This condition is completely ignored by many patients because of its transient nature. As with odor sensitivity, transient rhinitis, sneezing, or nasal stuffiness, hoarseness is usually an early sign of environmental sensitivity. As environmental sensitivity progresses, laryngeal stridor and spasm due to environmental overload may develop. Also, in many patients, the symptoms of the environmental sensitivities may switch to other organs. Some patients develop spastic aphonia secondary to environmental triggers. We have seen two patients with spastic aphonia whose illness was triggered by exposure to pesticides. Once this illness develops, long-term outlook is guarded unless the individuals are willing to practice strict environmental control in their homes.

Laryngeal Edema

Laryngeal edema may occur de novo. We have studied over 22 patients with chronically recurring acute laryngeal edema (Table 18.7). All had experienced failure with usual medical therapy, including treatment with steroids. All had multi-system complaints in addition to their laryngeal edema. Outside workup tended not to define etiology and produced negative results. However, each patient was placed in the ECU. All were fasted from 3 to 7 days until symptoms lessened. Then double-blind challenges were performed for biological inhalants (intradermal — Table 18.7), foods (oral — Table 18.8), and toxic chemicals (oral, intradermal, and inhaled — Table 18.9).

Table 18.7.   Inhalants Found to Trigger Laryngeal Edema[a] after 4 Days Deadaptation with Total Body Load Decreased

Antigen	Number of patients affected	% of all patients affected
Dust and dust mite	5	25
Molds	9	45
T.O.E.	2	10
Candida	0	0
Cotton linters	3	15
Grasses	2	10
Weeds	2	10
Terpenes (pine, grass, cedar)	7	35
Animal danders	2	10
Total number of subjects sensitive to inhalants	13	65

[a] Inhalant testing was performed by nasal and bronchial challenge and confirmed by intradermal challenge of 0.01 cc dilution: 1/5 to 1/3125. Each subject was tested on all of the inhalants.

Source: EHC-Dallas. 1978–1984.

Table 18.8.   Oral Food Challenge Confirmed by Intradermal Skin Challenge Laryngeal Edema Patients after 4 Days of Deadaptation with Total Body Load Decreased

Food	Number of subjects affected	% of all patients
Meats	12	60
Chicken (3)		
Turkey (3)		
Beef (2)		
Lamb (2)		
Venison (2)		
Grains	7	35
Legumes	6	30
Apples	4	20
Yeast	4	20
Milk	4	20
Seafood	4	20
Nuts	4	20

Source: EHC-Dallas. 1983.

Table 18.9. Double-Blind Inhaled Challenge in ECU after 4 Days Deadaptation with Total Body Load Decreased —15 Minute Exposure; Statistical Analysis of Chemicals Causing Laryngeal Edema

	Gas	Phenol (<0.002 ppm)	Formaldehyde (<0.2 ppm)	Insecticide (<0.0034 ppm)	Alcohol (<0.50 ppm)	Chlorine (<0.33 ppm)	Saline placebo
Number of patients affected	16	18	16	18	17	19	4
Probability (p) of patient being affected	0.08 ± 0.09	0.90 ± 0.07	0.08 ± 0.09	0.09 ± 0.07	0.85 ± 0.08	0.95 ± 0.05	0.20 ± 0.09
Difference between chemical and placebo probabilities	0.60 ± 0.13	0.70 ± 0.11	0.60 ± 0.13	0.70 ± 0.11	0.65 ± 0.12	0.75 ± 0.10	
Level of significance[a]	4.6	6.4	4.6	6.4	5.4	7.5	

[a] Level of significance is expressed by the ratio of the difference (Pc = Chemical/Px = Placebo) to its standard deviation; more than 4 is considered significant.

Source: EHC-Dallas. 1978–1984.

Clearly, faulty antigen recognition at the cell membrane must have occurred. Also, other mechanisms must have been involved since triggering agents were always multiple. The triggers involved biological inhalants, foods, and chemicals. The laryngeal edema of those patients who had inhalant sensitivities was not always triggered with each challenge. Apparently, these sensitivities were in part a function of the patient's total load. When the patient's load increased, the edema occurred as a function of any additional exposures. Also, we observed that the edema of some patients occurred only in response to various combinations such as biological inhalants, foods, and chemicals. It should be emphasized that, in addition to the usual weeds, trees, and grasses, the odors of plants (terpenes) were triggers in 35% of our patients. The edema of 100% of these patients was triggered by foods. Meats, grains, and legumes were the most common triggering agents. Thirteen patients underwent attempts at intradermal neutralization of 500 tested foods that gave some reaction (either laryngeal edema or associated symptoms). Sixty-five (326/500) percent of the food-produced symptoms could be neutralized for a $p<0.001$. The rest of the foods in the diet that triggered the laryngeal edema had to be avoided (Table 18.10). Eighty-nine percent (226/253) of the foods could be neutralized for 11 patients. Of the toxic chemical inhaled double-blind challenge, phenol (<0.002 ppm), petroleum derived ethanol (<0.50 ppm), and insecticides (<0.0034 ppm) were the most prominent triggers of laryngeal edema. In addition, 5 patients experienced difficulty when they were exposed to perfume. Patients who could not be neutralized were unable to go into most public buildings where many of these substances are present. Many of the associated symptoms these patients experienced involved the cardiovascular, extralaryngeal respiratory, gastrointestinal, and genitourinary systems. Symptoms involving the skin also occurred. Laboratory results from these patients are shown in Table 18.11. The following case is illustrative of recurring laryngeal edema.

> **Case study.** A caucasian woman aged 36 years entered the ECU with complaints of "throat closing", seizure-like episodes resulting in partial limb paralysis, and abdominal bloating with severe distention. Her vocal cords were swollen. She had been examined in a physician's office and found to be sensitive to some foods and inhalants. Her clinical course steadily deteriorated in spite of injection therapy for inhalants and limited food avoidance.
>
> Examination on admission to the ECU revealed a well-developed female in moderately acute distress with shortness of breath and stridor. Examination further revealed multiple bruises over the extremities accompanied by petechiae, cyanosis, and mild pitting edema. Laryngoscopy and bronchoscopy done after the patient was clear of symptoms revealed normal cord, larynx, and trachea. Indirect laryngoscopy during her episodes, however, showed swollen vocal cords. Biopsy of the petechiae showed non-necrotizing lymphocytic infiltrates around the vessel wall with edema of associated tissue.
>
> This patient was found to be sensitive to 13 foods and 2 chemicals. Interestingly, five of these foods triggered the laryngeal edema, and nine caused bloating. None brought on the seizure-like episodes. Two chemicals

Table 18.10. Intradermal Neutralization in ECU;
Food Triggered Laryngeal Edema

	Number of patients	Cleared
Short-term (unit)	13	13
Long-term (2 years) with 80–100% reduction	13	13
Patients neutralized (if repeatable)	68% ($p<0.001$)	
Foods neutralized (13 patients): 326/500	65%	
Food neutralized (11 patients): 226/253	89%	

Source: EHC-Dallas. 1978–1986.

Table 18.11. Laryngeal Edema Patients;
Admitting Laboratory Results

	Below normal	Normal	Above normal	Not available
T lymphocytes (normal range: 1000 to 1600/mm³)	7	7	3	3
B lymphocytes (normal range: 400 to 800/mm³)	1	7	9	3
Total complement (normal range: 100 ± 20%)	2	15	1	2
Total eosinophils (normal range: 200 to 400/mm³)	6	11	2	1
CRP	14 neg. 3 pos.			3
IgE	13 below 100 IU/mL; 2 above 100 IU/mL			5
IgG	18 below 1800 mg/dL; 1 above 1800 mg/dL			1

Note: Total patients = 20.

Source: EHC-Dallas. 1978–1984.

triggered both the edema and seizures. From different substances challenged, obvious overlap of symptoms was observed. This patient was advised to clean her house thoroughly and to avoid chemicals as much as possible. She has done well over long-term follow-up.

Patients with recurrent acute laryngeal edema usually have other symptom complexes (Table 18.12). These involve the neurocardiovascular, respiratory, gastrointestinal, and genitourinary systems. If a physician observes multiple symptom complexes, he or she should suspect chemical sensitivity. In our

Table 18.12. Associated Symptoms

System involved		Number of subjects affected
1. Cardiovascular		16
Arrhythmias	Low blood pressure	
Cerebral symptoms	Chest pain	
Hypertension	Headaches	
2. Respiratory		14
Sinus	Bronchitis	
Postnasal drip	Asthma	
3. Gastrointestinal		8
Diarrhea	Gastrointestinal distress	
Abdominal pains	Flatus	
Nausea	Belching	
Mouth cankers		
4. Skin		8
Acne lesions		
5. Genitourinary frequency		6

Note: Subjects displayed associated symptoms in addition to laryngeal edema.
Source: EHC-Dallas. 1978–1984.

series of 22 patients, 20 were able to be worked-up thoroughly, and 21 showed improvement during and after hospitalization (Table 18.13).

POLLUTANT INJURY TO THE EAR

Studies of the external ear show that traumatic irritation, change from the normally acid pH, and environmental changes, especially in combination with increased temperature and humidity, will cause disease processes of the ear. Fungus is particularly common in the external ear.

Diseases of the middle ear and mastoid are common. Next to the common cold, otitis media is the most common problem seen in a pediatrician's office. Inflammation of the middle ear cleft (eustachian tube, middle ear, and mastoid) is especially prevalent in children. Predominant causes are genetic (e.g., congenital stricture of the eustachian tube or congenital narrowing of the middle ear), mechanical (size of canals too small), and environmental (bacterial infection secondary to inhalants, foods, and chemicals). Middle ear disease often presents in a chronic or insidious form, causing hearing loss and drainage.

Several studies now show that the most common cause of otitis is food sensitivity.[34-36] Barotrauma due to diving or flying can be associated with tympanic membrane problems.

Table 18.13.   Reduction in the Frequency of Recurrent
Laryngeal Edema

Patient	Index of reactivity before hospitalization	Index of reactivity after hospitalization	Percentage of improvement
1	10	9	11
2	5	5	0
3	11	4	64
4	10	4	60
5	11	5	55
6	2.5	0	100
7	11	1.5	86
8	11	4	64
9	11	0	100
10	11	0	100
11	11	0	100
12	11	0	100
13	5	0	100
14	12	9	25
15	1.5	0	100
16	2.5	0	100
17	4	3	25
18	11	4	64
19	11	4	64
20	9	0	100
Average	8.6	2.6	71

Code of Frequency of Laryngeal Edema

Frequency	Index of reactivity
8 per day	12
4 per day	11
2 per day	10
1 per day	9
2 per month	5
1 per month	4
2 episodes in last year	1.5
1 episode in last year	0.5

Number improved in ECU: 10 (70–100%)
                                 5 (60–69%)
                                 5 (0–59%)
                                 $p < 0.03$

Source: EHC-Dallas. 1978–1984.

Serous Otitis Media

Serous otitis media is usually caused by inflammation of the eardrum with associated sinusitis and rhinitis and with enlarged adenoids usually triggered by biological inhalants, foods, and chemicals. The chemicals can also trigger autoimmune disease with metabolic dysfunction. Patients should have a work-up for these various pollutants before tubes are inserted. Labrynthine fistula and labrynthinitis can complicate otitis media and should also be searched for.

Facial Nerve Paralysis

Facial nerve paralysis can occur after either a single, or repeated, episode(s) of otitis media, and physicians need to be aware of this possibility. Surgical consultation should be obtained in case decompression is needed to prevent permanent paralysis. Often, triggering due to foods, pollens, viruses, and chemicals can cause or influence facial nerve paralysis. Thus, a meticulous work-up should be instituted when this entity occurs.

Environmental pollutants can affect several nerves such as the facial, trigeminal, lower motor neurons, and glossopharyngeal. Some of the more severe damage caused by these pollutants can be diminished or even eliminated by manipulating and eliminating the pollutants before the individual is ever exposed to them. Pollutant damage to each nerve will be discussed separately.

Bell's Palsy

Bell's palsy is an idiopathic facial paresis of lower motor neuron type that has been attributed to an inflammatory reaction involving the facial nerve near the stylomastoid foramen or in the bony facial canal. A relationship of Bell's palsy to reactivation of herpes simplex virus has recently been suggested, but there is little evidence to support this.

The clinical features of Bell's palsy are characteristic. The facial paresis generally comes on abruptly, but it may worsen over the following day or so. Pain about the ear precedes or accompanies the weakness in many cases but usually lasts for only a few days. The face itself feels stiff and pulled to one side. There may be ipsilateral restriction of eye closure and difficulty with eating and fine facial movements. A disturbance of taste is common, owing to involvement of chorda tympani fibers, and, occasionally, to hyperacusis due to involvement of fibers to the stapedius occurs.

The management of Bell's palsy is controversial. Approximately 60% of cases recover completely without treatment, presumably because the lesion is so mild that it leads merely to a conduction block. Considerable improvement occurs in most other cases and only about 10% of all patients are seriously dissatisfied with the final outcome because of permanent disfigurement or other long-term sequelae. Treatment is unnecessary in most cases, but is indicated for patients in whom an unsatisfactory outcome can be predicted. The best clinical guide to progress is the severity of the palsy during the first few days after presentation. When first seen, patients with clinically complete palsy are less likely to make a full recovery than those with an incomplete one. A poor prognosis for recovery is also associated with advanced age, hyperacusis, and severe initial pain. Electromyography and nerve excitability or conduction studies provide a guide to prognosis but not early enough to aid in the selection of patients for treatment.

The only medical treatment that may influence the outcome of Bell's palsy is administration of corticosteroids, but studies supporting this treatment have been criticized. Many physicians nevertheless routinely prescribe corticosteroids for patients with Bell's palsy who are seen within 5 days of onset. At the EHC-Dallas, we prescribe them only when the palsy is clinically complete or there is severe pain. Treatment with prednisone, 60 or 80 mg daily in divided doses for 4 or 5 days, followed by tapering of the dose over the next 7 to 10 days, is a satisfactory regimen. It is helpful to protect the eye with lubricating drops and a patch if eye closure is not possible.[37] In our experience at the EHC-Dallas, intradermal neutralization with flu vaccine or pork extract will relieve symptoms, but we emphasize that any incitant could cause the problem and the appropriate neutralizing dose of the specific trigger could relieve the palsy.

Trigeminal Nerve Tic Douloureux—Trigeminal or Glossopharyngeal Neuralgia

Trigeminal nerve tic douloureux occasionally afflicts chemically sensitive individuals. It manifests as lancinating pains over one side of the face in part of the sensory area of the fifth or ninth cranial nerves. The pains feel like sudden electric shocks, and they may appear for only a few seconds at a time or they may be continuous. Often they are set off by exceedingly sensitive "trigger areas" on the surface of the face, in the mouth, or in the throat. These pains almost always are caused by a mechanoreceptive stimulus instead of a pain stimulus, for instance, when a patient swallowing a bolus of food triggers the tonsil and initiates a severe lancinating pain in the mandibular portion of the fifth cranial nerve.

Trigeminal neuralgia is extremely difficult to treat. However, there is new hope for patients suffering from it since some can be cleared by finding and eliminating the agents that induce it. In our small series, foods and chemicals were the trigger while biological inhalants seemed to be secondary. The following case is illustrative.

Case study. A 60-year-old white male presented with a 2-year history of severe pain in the right side of his face that had been diagnosed by several physicians as trigeminal neuralgia. The pain was so constant and intense that he had to retire from work. He lived predominantly on pain medication and as a semirecluse. Work-up revealed mild sensitivity to 10 biological inhalants and 15 foods. However, none of these reproduced his symptoms of severe pain. When he was tested for newsprint and pesticides, his neuralgia immediately exacerbated. Within 1 month of avoidance of these two substances, his neuralgia completely subsided for the first time in over 2 years. He not only became pain free; he also stopped using medications. Twice when this patient tried to read the newspaper, he developed slight trigeminal pain. With consistent avoidance of newsprint and pesticides only, he has been well for over 2 years without recurrence.

Hearing Loss

More than 16 million Americans have a significant hearing loss. It is estimated that between 2000 to 4000 infants are born deaf each year, and, of these cases of deafness, only half are believed to have a genetic etiology. The rest may be environmentally induced. Another 15 million people have acquired hearing loss that can be, at least theoretically, prevented or arrested with early identification, since many of these occurrences are the result of potentially preventable disease that is caused by bacterial and viral infection, tumors, trauma, and blood clots as well as noise trauma. Environmental triggers have also been associated with neural hearing loss and conductive hearing loss, both of which will now be discussed.

Neural Hearing Loss

A myriad of toxic agents, including solvents, are known to cause neural hearing loss (Table 18.14). The following case illustrates the involvement of these agents in hearing loss.

Case study. A $5 \frac{3}{4}$ -year-old white female had a history of persistent hearing loss initially noted when she was 2 years of age. At that time, she was found to have a 30 decibel (dB) loss bilaterally. The hearing loss was both sensorineural and conductive in origin. This child had a history of chronic otitis media with bilateral serous effusions. Treatment had included bilateral myringotomy with insertion of drainage tubes in both tympanic membranes, adenoidectomy, and, more recently, mastoidectomy on the right side. She also was treated for allergy with immunotherapy and a limited rotation diet, with eventual resolution of her middle ear disease. Her hearing loss, however, persisted, and, at 3 years of age, she was fitted with a hearing aid. The sensorineural hearing loss slowly progressed over the next 2 years, and a second hearing aid for the right ear was considered. At that time, even with the hearing aid set at maximum amplification, she used lip reading to communicate.

At 9 months of age, this patient was referred to Dr. Douglas Sandberg (University of Miami Clinical Research Unit) for chronic diarrhea. At that time, total serum IgE was 20 U/mL, elevated for her age; sweat chloride concentration, measured twice by iontophoresis, was 17 and 22 mEq/L, within normal limits. Separate oral challenges with cow's milk and peaches

Table 18.14. Some Environmental Agents Known to Cause Ototoxicity

Antibiotics	**Antineoplastics**
Aminoglycosides	Bleomycin
Streptomycin	Nitrogen mustard
Dihydrostreptomycin	*cis*-Platinum
Neomycin	
Gentamicin	**Miscellaneous**
Tobramycin	Pentobarbital
Amikacin	Hexadine
Other Antibiotics	Mandelamine
Vancomycin	Practolol
Erythromycin	
Chloramphenicol	**Chemicals**
Ristocetin	Carbon monoxide
Polymyxin B	Oil of chenopodium
Viomycin	Nicotine
Pharmacetin	Aniline dyes
Colistin	Alcohol
	Potassium bromate
Diuretics	Pesticides
Furosemide	Car exhaust
Ethacrynic acid	
Bumetanide	**Heavy Metals**
Acetazolamide	Mercury
Mannitol	Gold
	Lead
Analgesics and Antipyretics	Arsenic
Salicylates	Mold Toxins
Quinine	
Chloroquine	

Source: Levine, S. C. 1989. Diseases of the inner ear. In *Boies Fundamentals of Otolaryngology*, 6th ed., eds. G. L. Adams, L. R. Boies, Jr., and P. A. Hilger. 132. Philadelphia: W. B. Saunders. With permission.

provoked diarrhea on at least two occasions. Also, intradermal titration skin testing showed positive wheals to milk (1:6250 dilution), with provocation of hives. Banana, oats, and apple also produced positive intradermal wheals. Confirmation of multiple food sensitivities including cow's milk and some fruits resulted in a treatment program of subcutaneous food extract injections with a relatively normal diet. The diarrhea was controlled, and growth and development progressed normally.

This patient was tested intradermally with *D. farinae* (house dust mite) at 14 months of age. She showed a positive wheal at 1:250 dilution. The extract provoked irritability and diaphoresis. IgE radioallergosorbent tests (RAST) were done on serum at 15 months of age. She had negative titers to oak, short ragweed, five molds, *D. farinae*, Hollister-Stier house dust mixture, wheat, egg white, and oats. She had a low level of IgE antibodies to cow's milk. More inhalant skin testing was done at 16 months of age. *Alternaria tenuis* extract produced a positive wheal at 1:250 dilution and caused irritability and aggressiveness. Hollister-Stier house dust mixture produced a positive wheal at 1:250 dilution but provoked no symptoms.

With increasing age, this child had intermittent extreme irritability. Intradermal food testing caused frequent reactions of aggressiveness, hyperactivity, loss of balance, and marked irritability, all of which rapidly disappeared when she was given an appropriate dose of the extracts being tested. These included tests for apple, *Alternaria tenuis*, grass pollen mixture, *D. farinae*, and mold spore mixtures. Her behavior was relatively well controlled with food and inhalant extract subcutaneous injections. Mild diarrhea occasionally recurred but was generally well controlled. Subsequent intradermal testing with Florida tree pollen extracts including Brazilian pepper and Australian pine (*causarina*) also provoked irritability and crying by 18 months of age. Skin testing with a number of food extracts produced positive intradermal wheals and provoked symptoms, including aggressiveness, crying, and marked irritability. Wheat was particularly notable as a producer of these symptoms. However, with continued treatment using subcutaneous extract injections, many extracts no longer provoked overt physical reactions.

At 3 years of age, this patient underwent an audiological evaluation and tympanometry with bilateral drainage tubes present. This evaluation showed a slight bilateral conductive hearing loss, the left ear more than the right, and speech awareness threshold (SAT) of 25 dB and 30 dB for the right and left ears, respectively. Auditory brainstem response (ABR) audiometry obtained at 38 months of age was consistent with a hearing threshold for high frequencies in the range of 40 dB for the right ear.

This patient continued on food and inhalant extract subcutaneous therapy with no diarrhea and fairly good control of behavior and hyperactivity, although frequent retesting was necessary.

Reevaluation of hearing status at $3^{1}/_{2}$ years of age, including pure tone audiometry and ABR, showed bilateral, mild, predominantly sensorineural hearing loss. ABR showed no response below 40 dB for right or left-sided stimulation (Figure 18.5). The latencies of individual components and wave morphology were consistent with a bilateral sensory loss in the range of 40 dB. At that time, bilateral drainage tubes were in place which, by impedance audiometry, were functioning adequately. She was fitted with a hearing aid, and a therapy program for her hearing impairment was instituted. She had persistent articulation difficulties which responded well to therapy during this time.

At 4 years of age, during retesting of components of her food and inhalant extracts, she began to voice complaints of pain in the ear with certain extracts including melaleuca pollen and soy. She had periodic episodes of otitis media requiring antibiotic therapy. However, audiological reevaluation at this time showed only a mild conductive hearing loss superimposed on the sensorineural hypoacusis and no change in this element of her hearing loss. The right drainage tube, however, was not patent at that time.

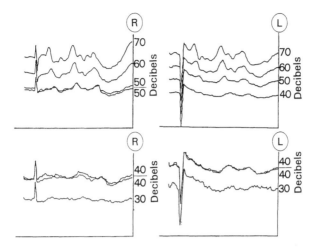

Figure 18.5. Hearing status of white female aged 3¹/₂ years; no response below 40 db. (From Sandberg, D. and J. Freeman. 1983. University of Miami. Personal communication. With permission.)

Tonsillectomy and adenoidectomy were performed in November 1978, and postoperative auditory testing showed improved hearing levels bilaterally with resolution of the conductive element. Otoscopy showed clear drainage tubes bilaterally. Subsequent intradermal testing with other extracts, including extracts of apple, corn, milk, cane sugar, grape, chocolate, potato, banana, grass pollens, cat and dog dander, and epithelium, was noted to provoke ear pain.

Reevaluation at age 4 years, 11 months showed no change in acuity for the right or left ear over the previous 10 months. She had recently had her polyethylene tubes replaced, and her new tubes were functioning well. Subsequent food testing with peanut extract evoked complaints of ear pain. Most of the retesting was done because of periods of irritability and hyperactivity as well as complaints of recurrent abdominal pain. All of these symptoms were clearly related to exposure to specific environmental substances.

Audiological reevaluation at 5 years of age showed some improved hearing acuity bilaterally, tympanometry indicated compliance, and middle ear pressures were within normal limits (Figure 18.6).

Surgical treatment of the right ear for possible occult mastoid disease was carried out in the fall of 1982, with no improvement postoperatively. In fact, there was minor loss of acuity in that ear following surgery. The hearing loss appeared to worsen gradually during the period from age 5 to 5³/₄ years, and, at that time, both ears had a threshold of 50 dB with a slightly higher level in the right ear than in the left.

Because of this persistent hearing loss and the suggestion that this patient would soon require a second hearing aid, she was evaluated for the possibility that the sensorineural hearing loss might be related to sensitivity to environmental factors such as food or inhaled substances. Testing for such sensitivities

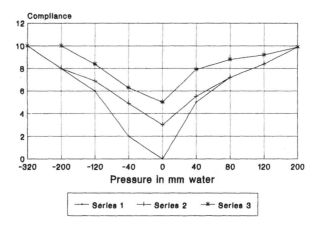

Figure 18.6. Tympanogram of white female aged 5 years. (From EHC-Dallas. 1980.)

was determined to be appropriate. Informed consent was obtained from both parents after an explanation of the procedures to be used and their attendant risks. This discussion included the plan for hospitalization and local study with an understanding that if additional workup were required, she would be transferred to the ECU.

Using ABR audiometry during sleep induced by chloral hydrate, Sandberg and Freeman demonstrated that peanut extract injected intradermally produced a decrease in acuity of approximately 10 dB in the right ear (Figure 18.7A and B). Corn extract caused a similar decrease. In addition, loss of response in the right ear at 60 dB followed injection of peanut extract (1:50 dilution) (Figure 18.8A). Lack of response in the right ear persisted at 60 dB after peanut injection of a 1:250 dilution (Figure 18.8B). Loss of response occurred again following injection of corn extract at 1:6250 and 1:3125 dilutions (Figure 18.8C).

This patient's history included reactions to penicillin (which caused urticaria), erythromycin, Bactrim®, and Septra®. The latter three medications caused urticaria, diarrhea, and hyperactivity.

Following this treatment, the patient was referred to the EHC-Dallas for further evaluation in a more controlled setting (i.e., the ECU). General medical evaluation including complete blood count, erythrocyte sedimentation rate, multiphasic serum screening, and urinalysis were within normal limits, except for serum iron concentration of 230 µg/dL (normal: 40 to 170 µg/dL). C-reactive protein and antinuclear antibodies were negative. Serum total complement and immunoglobulins G, A, and M were within normal limits for age. She had normal absolute numbers of T-lymphocytes at 1627/mm^3 (normal >1000/mm^3). B-lymphocytes were 111/mm^3 (normal >500/mm^3). Modified IgE RAST's demonstrated Class I antibody responses to Russian thistle pollen, *D. farinae*, egg white, milk, potato, wheat barley, rice, soy, and pork. Class II levels were reported for rye. Total serum IgE concentration was 12 U/mL (within normal range for age).

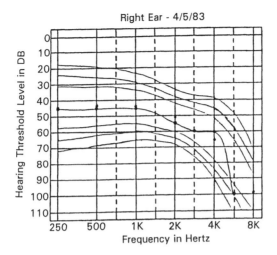

Figure 18.7A. Auditory brainstem response under anesthesia. (From Sandberg, D., H. Goldberg, and J. Freeman. 1983. University of Miami. Personal communication. With permission.)

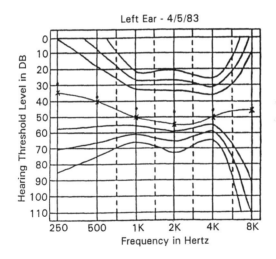

Figure 18.7B. Auditory brainstem response under anesthesia. (From Sandberg, D., H. Goldberg, and J. Freeman. 1983. University of Miami. Personal communication. With permission.)

During an initial period of fasting in the ECU, this patient showed definite improvement in hearing as shown by differences in audiograms done before and after fasting (Figure 18.9). It was determined that use of pure tone audiometry could provide reproducible information, and an audiometer was placed on the ward for continuous use during the period of hospitalization.

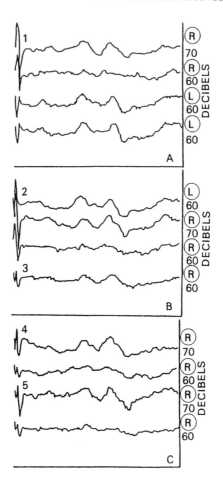

Figure 18.8. **Auditory brainstem responses under anesthesia of various dilutions of intradermal (ID) injections of peanut and corn. (A) 0.05 cc peanut at 1/50 dilution — ID; (B) 0.05 cc peanut at 1/250 dilution — ID; (C) 0.05 cc corn at 1/625 dilution and 1/3125 dilution — ID. (From Sandberg, D., H. Goldberg, and J. Freeman. 1983. University of Miami. Personal communication. With permission.)**

Figure 18.10 shows curves for mean and 1, 1.5, and 2 standard deviations of measurements recorded for each ear for 41 control audiograms done during the hospitalization period. The right ear showed greater hearing loss at higher frequencies than the left. This difference had been present since the surgical procedure performed on the right ear in 1982.

Food testing was done by open challenges using fresh, chemically less-contaminated foods in normal proportions for age. Foods were chosen based on the patient's usual diet. Inhaled challenges were done in a double-blind

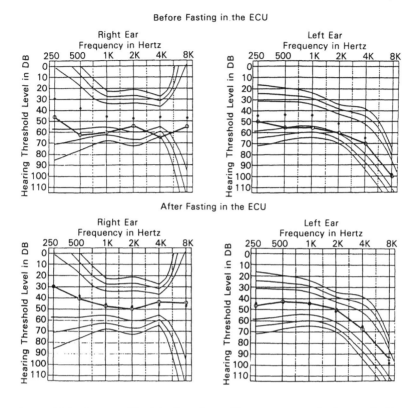

Figure 18.9.   5-year-old white female; auditory responses before and after 4-day fast. (From EHC-Dallas. 1985.)

manner. Phenol was tested in a glass and stainless steel booth with concentration in the booth air of <0. 002 ppm; that test was one of a number of similar double-blind challenges with saline solution used as the control. The saline solution produced no discernable effect on acuity. However, phenol and chalk (calcium carbonate) gave excruciating pain, shutting off the hearing completely (Figure 18.10).

During food testing, repeated audiometry before and after meals demonstrated transiently reduced acuity during specific individual food challenges (Figures 18.11 to 18.13). Figure 18.11 shows that sunflower caused no change, which was characteristic of many individual challenges (Figures 18.12 and 18.13). Beef, green grapes, lamb, oranges, wheat, and dried pineapple caused the greatest changes in the audiogram with acute decreases in hearing threshold as high as 35 dB. Intradermal symptom neutralization of each of these foods was accomplished (Figures 18.12 and 18.13). These injections then allowed the patient to eat the food without further damage. No complaints of vertigo were noted during hospitalization, but the patient may have had tinnitus, according to reports of her mother and nursing personnel.

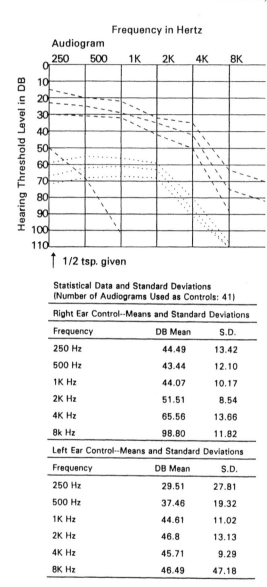

Figure 18.10. 5-year-old white female calcium carbonate (chalk — ½ teaspoonful) challenge. Similar response occurred with phenol <0.0020 ppm challenge. Right ear control; left ear could not be measured. (From EHC-Dallas. 1985.)

Since returning home, she has been able to function well without a hearing aid. She is presently on continued treatment with injections of preservative-free extracts for foods and inhalants and a rotary diversified diet, while completely avoiding those foods found to cause decreased acuity. Her behavior

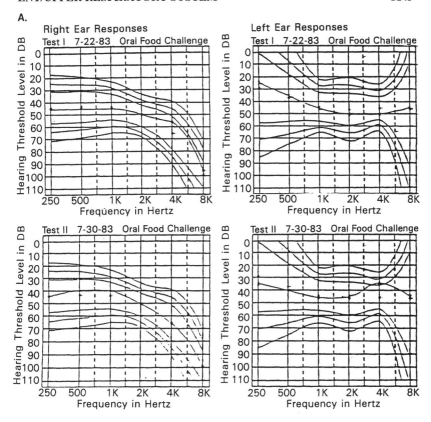

Figure 18.11. Audiogram before and after oral food challenges; 5-year-old white female. (A) Sunflower challenge: no response. (B) Dried pineapple challenges on 7/19/83 and 7/23/83 show reaction in the left ear. Injected neutralizing dose of the food before repeated oral food challenge on 8/2/83 and 8/12/83 prevented reaction. (From EHC-Dallas. 1985.)

has been quite good, and she voices no complaints of abdominal pain, headache, ear pain, etc.

This patient's bedroom was modified to a chemically less-contaminated area to be used as an oasis. An air purifier was installed, and it appears to have been an important adjunct to her treatment. Audiometric evaluation on July 18, 1984, showed no significant change from July 15, 1983, when she was in the ECU. Speech discrimination was reported to be good; impedance was within normal limits for the left ear and consistent with a patent ventilation tube in the right tympanic membrane and with good eustachian tube function on that side. Brain function tests continued to be in the superior range, a marked change from prior to being in the ECU (Table 18.15). Her teacher's report for the year was that she was "reading beautifully with good understanding. She participates in discussions with interest and enthusiasm." After 5 years, follow-up showed she was extremely well. This complex case illustrates

B.

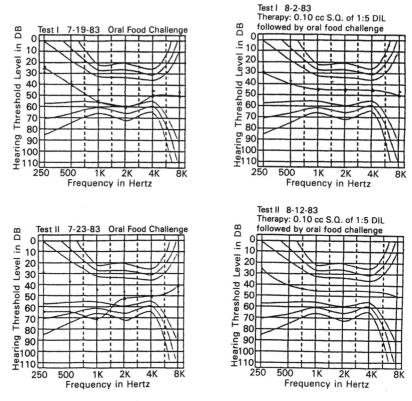

Figure 18.11.   (Continued).

how multiple environmental incitants may affect patients with apparent irreversible hearing problems.

Conductive Hearing Loss

Many environmental causes of conductive hearing loss have been observed. Treatment of problems resulting from sensitivity to inhalants, foods, and chemicals will result in reduction of eustachian tube swelling and fluid in the middle ear. Therefore, before tubes are placed in the ear, workup for inhalants, foods, and chemicals should be done.

Vertigo and Imbalance

According to Rubin et al.,[38] the incidence of dizziness as a presenting complaint in general medical practice is approximately 5%, while in a specialized practice such as otolaryngology or neurology the incidence ranges from 10 to 15%. All physicians should be familiar with the most common causes of dizziness and should be capable of initiating tests to differentiate between

A.

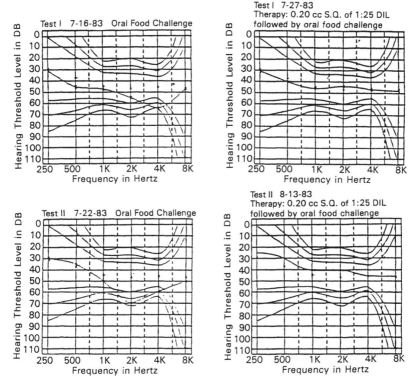

Figure 18.12.   Audiogram before and after oral lamb challenges; 5-year-old white female. (A) Challenges on 7/16/83 and 7/22/83 show reaction in the left ear. Injected neutralizing dose of the food on 7/27/83 and 8/13/83 before repeated oral food challenge prevented a reaction in the left ear. (B) Challenges on 7/16/83 and 7/22/83 show reaction in the right ear. Injected neutralizing dose of the food on 7/27/83 and 8/13/83 before repeated oral food challenge prevented a reaction in the right ear. (From EHC-Dallas. 1985.)

organic vestibular disorders and complaints of imbalance that may be a result of difficulties in other systems. The majority of disorders of the balance mechanism have their origin in the vestibular end-organ (inner ear) or its nerve pathways. In vestibular disorders, the differentiation of peripheral (vestibular end-organ) responses from central (nerve pathways) responses is important since it can help indicate the direction of therapy.

Vertigo and imbalance are common symptoms in chemically sensitive individuals. The CNS and ear are often prime targets of xenobiotics because of their lipophilic nature. The following case exemplifies the seriousness of CNS and ear involvement in chemically sensitive individuals.

B.

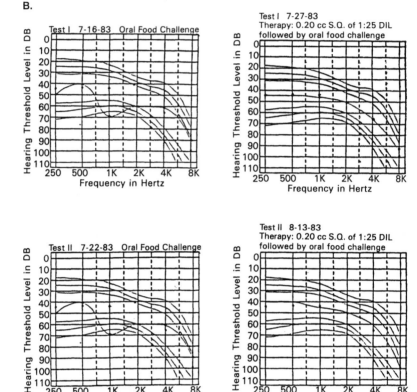

Figure 18.12. (Continued).

Case study. A 48-year-old white male had a new carpet installed in his apartment. He immediately noticed the onset of dizziness, which became true vertigo. He also developed recurrent rhinosinusitis with a loss of smell. He then progressed to short-term memory loss. After being admitted to the ECU, his laboratory workup showed the following results: *Blood minerals* — manganese ND (C = 0.001 to 0.002 ppm), boron 0.56 ppm (C = 0.325 to 0.500 ppm), vanadium ND (C = 0.0005 to 0.005 ppm). *Chemicals* — anion gap 14.1 mEq/L (C = 8.0 to 12.0 mEq/L); albumin/globulin 2.1 (C = 1.1 to 1.9), cholesterol 272 mg/dL (C = 131 to 239 mg/dL), triglycerides 221 mg/dL (C = 0 to 210 mg/dL), calcium 10. 3 mg/dL (C = 8.7 to 10.2 mg/dL). *SPECT brain scan* — Abnormal study showed diffuse defects throughout the cerebral cortex in a pattern associated with neurotoxic substances; after 2 weeks, it showed a dramatic interval improvement, although it did not resolve completely. *Neurologic evaluation* — The ulnar and median sensory values were noted to be diminished; motor conduction velocities, amplitudes, and latencies were essentially normal. *MRI* — No significant abnormalities were demonstrated, with the exception of a prominent amount of signal with the left maxillary sinus and the ethmoid sinus reflecting active sinusitis. *Immune* — T_4, 76% (C = 32 to 56%), T_8, 1909/mm^3 (C = 518 to 1605/mm^3), $T_4/T_8 = 4$

A.

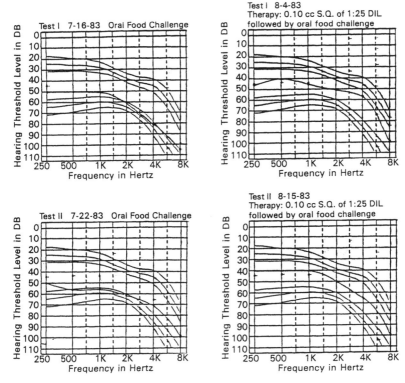

Figure 18.13.   Audiogram before and after oral orange challenges; 5-year-old white female. (A) Challenges on 7/17/83 and 7/24/83 show reaction in the right ear. Injected neutralizing dose of the food on 8/4/83 and 8/15/83 before repeated oral food challenge prevented a reaction in the right ear. (B) Challenges on 8/4/83 and 8/15/83 show reaction in the left ear. Injected neutralizing dose of the food on 8/4/83 and 8/15/83 before repeated oral food challenge prevented a reaction in the left ear. (From EHC-Dallas. 1985.)

(C = 0.9 to 2.9). *Quantitative electroencephalogram* — showed some global increase of theta frequency, both on monopolar and bipolar Z score measures. *Blood toxic chemicals* — toluene, 1.5 ppb (C <0.5 ppb), 1,1,1-trichloroethane, 1.4 ppb (C <0.5 ppb), 2-methylpentane, 13.4 ppb (C <1.0 ppb), 3-methylpentane, 30.4 ppb (C <1.0 ppb), n-hexane, 5.8 ppb (C <1.0 ppb), trimethylbenzenes 1.8 ppb (C <0.1 ppb). *Posturography* — Positive finding on the neurotologic Tandem Romberg was found. Gait was normal. Cranial nerves were intact, except acoustics.

This patient was placed on a rigid avoidance program for pollutants in the air, food, and water. Antigen injections for biological inhalants and foods were administered. Both parenteral and oral nutrient supplementation were

B.

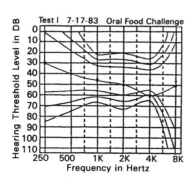

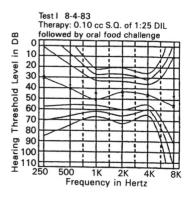

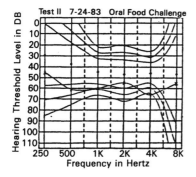

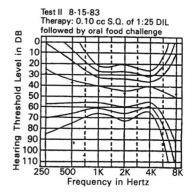

Figure 18.13.   (Continued).

used. Also, this patient underwent heat depuration physical therapy. He has done well at 1-year follow-up without symptoms or medication.

Pathophysiology of Pollutant Injury

Animal studies have shown vestibular disturbances with positional nystagmus induced by hydrocarbon solvents (xylene, styrene, trichloroethylene, methylchloroform).[39-42] The findings of Ödkvist et al.[43] suggest that the mechanism for positional alcohol-induced nystagmus is different from that induced by other solvents. However, both types of mechanisms can produce symptoms in the chemically sensitive population. In another study, Ödkvist et al.[44] found that nystagmus responses in rabbits were inhibited by alcohol and enhanced by alphachloratose and xylene. Further, Tham et al.[45] showed apparent intoxication responses to styrene, while methylchloroform (blood level 125 ppm) and trichloroethylene had no demonstrable influence. Also, they found styrene and toluene caused prolonged rotary nystagmus in 25 rats. This finding exceeded the mean value by 3 SD of the control group. No positional nystagmus was elicited (Table 18.16). Many of these chemicals have been found in the blood

Table 18.15.   5-Year-Old White Female with Hearing Defect

IQ Change

Pre-treatment testing:

WISC-R[a],**	Overall IQ	= 100 average range intelligence
	Performance IQ	= 99 average range
	Verbal IQ	= 99 average range
WAIS-R[b],*	Overall IQ	= 115 high average range intelligence
	Performance IQ	= 112 high average
	Verbal IQ	= 114 high average range

Post-treatment testing

WAIS-R[b],**	Overall IQ	= 117 high average classification
	Performance IQ	= 124 superior classification
	Verbal IQ	= 114 high average range

WAIS-R subtest increase (post-treatment)
Abstract reasoning — increase 1 standard deviation
Comprehension, judgment — increase 2 standard deviations
Attention span (hearing) — increase 1 standard deviation
Perceptual-motor functioning — increase 3 standard deviations

Chemical testing—phenol <0. 0024 ppm
WAIS-R pre-subtest:
   Digit span — attention span above average
   Digit symbol — perceptual-motor functioning — average range Bender
   scores (error scores normal)

WAIS-R post-subtest:
   Borderline range — decrease 2 standard deviations
   Below average — decrease 1 standard deviation
   Bender-Gestalt error scores increase to severe degree
   Cerebral motor dysfunction

[a] WISC-R: Wechsler Intelligence Scale for Children — Revised
[b] WAIS-R: Wechsler Adult Intelligence Scale — Revised

* An increase in standard deviation
** An increase of 2 standard deviations

Source: EHC-Dallas. 1985.

of our chemically sensitive patients seen at the EHC-Dallas. They cause dizziness.

Vertigo occurs with malfunction of four separate and independent systems. First, the vestibular system senses accelerating movement and perceives gravity. Second, proprioceptive cues from joint position senses and muscle

Table 18.16. Mechanisms of Vestibular Disturbances Caused by Industrial Solvents

	Minimal concentration (ppm)	Nystagmus type	Rotatory response
Methanol	900	PAN	inhibited
Ethanol	500	PAN	inhibited
Propanol	900	PAN	inhibited
Xylene	30	PXN	exaggerated
Styrene	40	PXN	paradoxical
Trichloroethylene	30	PXN	insignificant
Methylchloroform	70	PXN	insignificant
Barbiturates	n.d.	0	inhibited
α-Chloralose	n.d.	PXN	exaggerated
Picrotoxin-bicuculline	n.d.	PXN	insignificant

Note: n.d. = not determined.

Source: Ödkvist, L. M., B. Larsby, R. Tham, and G. Aschan. 1979. On the mechanism of vestibular disturbances caused by industrial solvents. *Adv. Oto-Rhino-Laryngol.* 25:167–172. With permission.

tone provide information concerning the relationship of the head to the remainder of the body. Third, vision gives perceptions of position sense, speed, and orientation. Finally, all of these senses are integrated through the brainstem and cerebellum. There are nonvestibular causes of dizziness such as hyperventilation, hypoglycemia, neurotoxins, and vascular insufficiency. In addition, a cervical joint position sense exists. It may be disrupted causing dizziness. Further causes of dizziness are trauma or neoplasm. Metabolic dysfunction such as diabetes, otosclerosis, Paget's disease, thiamine deficiency, and osteoporosis can also cause dizziness.

Once the common causes of vertigo are ruled out, the environmental aspects of this problem should be evaluated. Often patients are exposed to a myriad of incitants that in combination result in excess total pollutant load, and vertigo develops. In one series of 30 patients studied at the EHC-Dallas, multiple incitants were seen to trigger vertigo. Many symptoms were triggered by inhalants and food (Table 18.17). However, toxic chemicals also appeared to play a great role in much of this vertigo. Many investigators, including Ödkvist et al.[46] and ourselves, have seen inner ear, cerebellar, and cerebral involvement, as well as visual changes, from toxic chemicals that cause vertigo (Table 18.18).

CNS disturbances that manifest in ENT symptoms include nystagmus and vestibular and auditory dysfunction.

Organic solvents affect both humans and animals (Table 18.18). Gamberale and Hultengren[47,48] found disturbed psychophysiological functions in workers

Table 18.17. Vertigo

Incitants	Challenge testing (type challenge)	Percent
Inhalants	ID[a]	
Molds		92
Weeds		73
Trees		72
Grasses		76
Terpenes		40
Foods	ID[a]	90
	Oral	36
Corn		83
Cow's milk		73
Egg		73
Brewer's yeast		73
Chicken		73
Baker's yeast		66
Cane sugar		66
Beet		60
Potato		60
Peanut		62
Chemicals	ID[a]	83
	Inhaled[b]	100
Hormones		13

Notes: N = 30 patients. Ages: 8–79 years; M: 20%; F: 80%; average age of onset: 29 years. Associated symptoms: resp. — 73%; CV — 53%; neuropsych — 46%. Blood pesticide — 23%.

[a] Intradermal challenge.
[b] Inhaled challenge was performed after 4 days deadaptation with the chemical load decreased. Double-blind: 1 to 3 saline placebos; formaldehyde <0.20 ppm; phenol <0.0020 ppm; petroleum derived ethanol: <0.50 ppm; chlorine: <0.33 ppm; pesticides 2,4-DNP: <0.0034 ppm.

Source: EHC-Dallas. 1992.

exposed to industrial solvents. Neurasthenia, personality changes, intellectual reduction,[49] vertigo, and nausea[50,51] also may develop in people exposed to solvents. We have found similar dysfunction in our patients (see Chapter 26).

Table 18.18.   Neurotoxic Effects of Hydrocarbons

Reference

Equilibrium Disturbances

Animals

Aschan, G., I. Bunnfors, D. Hydén, B. Larsby, L. M. Ödkvist, and R. Tham. 1977. Xylene exposure: electronystagmographic and gaschromatographic studies in rabbits. *Acta Otolaryngol.* 84: 370.

Larsby, B., L. M. Ödkvist, D. Hydén, and S. R. C. Liedgren. 1976. Disturbances of the vestibular system by toxic agents. *Acta Physiol. Scand. [Supp.]* 440:108.

Larsby, B., R. Tham, L. M. Ödkvist, B. Norlander, D. Hydén, G. Aschan, and A. Rubin. 1978. Exposure of rabbits to methylchloroform. Vestibular disturbances correlated to blood and cerebrospinal fluid levels. *Int. Arch. Occup. Environ. Health* 41:7–15.

Nystagmus

Ödkvist, L. M., B. Larsby, T. Tham, S. R. C. Liedgren, and G. Aschan. 1978. Positional nystagmus elicited by industrial solvents. In *Vestibular Mechanisms in Health and Disease*, ed. J. H. Hood. 188–194. London: Academic Press.

Ödkvist, L. M., B. Larsby, J. M. Fredrickson, and S. R. C. Liedgren. 1980. Vestibular and oculomotor disturbances caused by industrial solvents. *J. Otolaryngol.* 9:53–59.

Ödkvist, L. M., B. Larsby, R. I. Tham, and G. Aschan. 1979. On the mechanism of vestibular disturbances caused by industrial solvents. *Adv. Oto-Rhino-Laryngol.* 25:167–72.

Humans

Tham, R., B. Larsby, L. M. Ödkvist, B. Norlander, D. Hydén, G. Ascham, and Å. Bertler. 1979. The influence of trichlorethylene and related drugs on the vestibular system. *Acta Pharmacol. et Toxicol.* 44:336.

Ödkvist, L. M., B. Larsby, R. Tham, H. Ahlfeldt, B. E. Andersson, and S. R. C. Liedgren. 1982. Vestibulo oculomotor disturbances in humans exposed to styrene. *Acta Otolaryngol.* 94:487–493.

Styrene

Humans (Abnormal)	Götell, P., and B. Lindelöf. 1972. Field studies on human styrene exposure. *Work Environ. Health* 9:76–83.
	Klimkova-Deutschova, E. 1962. Neurologische befunde in der plastikindustrie bei styrol-arbeitern. *Int. Arch. Gewerbepathol. Gewerbehyg.* 19:35–50.
	Klimkova-Deutschova, E., D. Dandova, Z. Salomanova, K. Schwarzova, and O. Titman. 1973. Recent advances concerning the clinical picture of professional styrene exposure. *Cs. Neurol.* 36:20–25.
Reaction Time Changes	Seppäläinene, A. M., and H. Härkönen. 1974. Neurotoxicity of styrene in humans. Presented at the Neurology Congress, Prague, 1974.
	Seppäläinene, A. M., and H. Härkönen. 1976. Neurophysiological findings among workers occupationally exposed to styrene. *Scand. J. Work Environ. Health* 2(3):140–146.
Nerve Conduction Velocity Changes	Lilis, R., W. V. Lorimer, S. Diamond, and I. J. Selikoff. 1978. Neurotoxicity of styrene in production and polymerization workers. *Environ. Res.* 15(1):133–138.
Vestibular Changes	Stewart, R. D., H. C. Dodd, E. D. Baretta, and A. W. Schaffer. 1960. Human exposure to styrene vapor. *Arch. Environ. Health* 16:656.
Ocular-Motor Changes	Ödkvist, L. M., B. Larsby, J. M. Fredrickson, and S. R. C. Liedgren. 1980. Vestibular and oculomotor disturbances caused by industrial solvents. *J. Otolaryngol.* 9:53–59.
	Larsby, B., L. M. Ödkvist, D. Hydén, B. Eriksson, R. Tham, and S. R. C. Liedgren. 1981. Optokinetic disturbances caused by styrene. An experimental study in rabbits. Proceedings of the Neuroequilibriometric Society. New Isenburg, West Germany: Megapress.

Source: Modified by EHC-Dallas from Ödkvist, L. M., B. Larsby, R. Tham, H. Ahlfeldt, B. E. Andersson, and S. R. C. Liedgren. 1982. Vestibulo oculomotor disturbances in humans exposed to styrene. *Acta Otolaryngol.* 94:487–493. With permission.

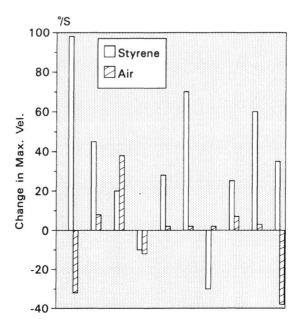

Figure 18.14.   **Vestibulo-oculomotor disturbances in humans exposed to styrene: change in maximum saccade velocity as compared to pre-exposure value for styrene and air. Each pair of columns represents one test subject. (From Ödkvist, L. M., B. Larsby, R. Tham, H. Ahlfeldt, B. E. Andersson, and S. R. C. Leigren. 1982. Vestibulo-oculomotor disturbances in humans exposed to styrene.** *Acta Otolaryngol.* **94:487–493. With permission.)**

Ödkvist et al.[52] investigated the effects of solvents (styrene, trichloroethylene, toluene, and jet fuel) on the vestibule-oculomotor exposure system in healthy volunteers. Speed of the saccade was diminished by trichloroethylene and toluene and increased by styrene (Figure 18.14). Visual suppression and the slow pursuit mechanisms were disturbed by all three of these solvents. These findings are consistent with the theory that cerebellar inhibition of the vestibular oculomotor system is blocked by some organic solvents. Each type of pathology has been seen in some chemically sensitive patients at the EHC-Dallas.

Hydén et al.[53] studied 15 human volunteers exposed to toluene at a concentration comparable to the threshold limit value. Results showed visual suppression. This finding is consistent with his earlier study on styrene. Audiological testing showed five areas of pathology (interrupted speech, cortical responses, brainstem responses, phase audiometry, speech discrimination) that implied disturbances in the central auditory pathways involving polysynaptic pathway functions. These researchers suggested that organic solvents may act on transmitters specialized for specific neural pathways.

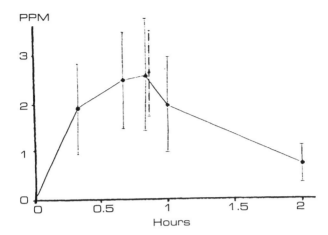

Figure 18.15. Mean venous blood concentrations of styrene in the test persons. Standard deviations are also indicated. The triangle indicates a capillary sample. (From Ödkvist, L. M., B. Larsby, R. Tham, H. Ahlfeldt, B. Andersson, B. Eriksson, and S. R. C. Liedgren. 1982. Vestibulo-oculomotor disturbances in humans exposed to styrene. *Acta Otolaryngol.* 94:487–493. With permission.)

Hogstedt et al.[54] studied 17 men with neuropsychiatric symptoms. Careful analysis indicated diagnoses of psycho-organic syndromes due to solvent exposures. Test batteries showed CNS disturbances, especially of the cerebellum where the visual suppression of vestibular nystagmus is exerted. Ödkvist et al.[55] deduced that solvents preferentially damage the cerebellum. This damage then results in vertigo and balance problems. Our studies with Martinez[56] using computerized posturography emphasize that either inner ear or cerebellar involvement will result in vertigo or imbalance problems.

In our own series of 30 chemically sensitive patients, we found a high incidence of neuropsychiatric symptoms (46%) and an even higher proportion of associated respiratory (73%) and cardiovascular symptoms (53%). Since we performed our studies under environmental control, which Ödkvist did not have available (Figure 18.15), we found a wider spread of triggering agents. These included biological inhalants and foods as well as toxic chemical inhalants. It was quite clear in this series that chemical overload could cause vertigo.

Solvents appear to act in the CNS by blocking GABA.[45] Ödkvist et al.[57] believe that vertigo often can be caused by disturbances in the vascular tree. These disturbances may be caused by changes in cardiac output, embolization, thromboses, and/or vascular spasm. Often the vertigo is accompanied by skin signs of vascular dysfunction, including peripheral spasm, resulting in Raynaud's phenomenon, spontaneous bruises, petechiae, purpura, and vascular chloracne, as shown in many of our studies.[58]

We have seen many patients similar to those described by the European investigators, including Ödkvist. Of our chemically sensitive patients, 56% had xylene in their blood, 8% trichlorethylene, 63% toluene, and 15% styrene. These findings have stimulated us to evaluate patients further who have trouble balancing. Often these patients cannot stand on their toes with their eyes closed and their arms extended. They not only have positive computerized tomography but also have neurotoxic patterns on their SPECT brain scans.

Evaluation of the Patient with Imbalance

According to Rubin and Brookler,[59] the inner ear is a converter (transducer) of mechanical to electrical energy for both hearing and balance functions. This task is accomplished as a result of the presence of chemicals within the perilymph and endolymph. The inner ear functions as an internal body organ, and its efficiency in maintaining normal hearing and balance is chemically dependent upon many systems in the body.

The audiogram is used to evaluate hearing loss. Now, vestibular abnormalities can be evaluated by using electronystagnagram (E.N.G.) with alternative and simultaneous caloric stimulation, harmonic acceleration testing, and posturographic testing. Hearing impairment can be due to endocrine or immune dysfunction or toxic chemical exposure or nutritional deficiency. Clearly, alteration of the chemical make-up can cause dysfunction, which is seen in a subset of chemically sensitive patients.

Patients with imbalance may be accurately evaluated by computerized posturography when using the E.N.G. with caloric testing. According to Rubin,[60] posturography detects abnormalities in 64% of patients with unilateral peripheral vestibular disease and nearly 100% of those with CNS disease. In 80% of the cases, it distinguishes patients with peripheral and central pathology from those with other CNS disease, and in 85% of the cases, it distinguishes correctly between sensory losses and distortions due to trauma, fistula, and positional vertigo.

The posturography test for total equilibrium is performed by a computerized moving platform and moving surrounding system. It is built by Neurocom, Inc., and named Equilibrium Test. The test uses computerized graphics and tracings of the patient's responses to assess systematically motor and sensory deficits affecting posture and equilibrium. It provides quantitative, reproducible, and functional analyses of these entities. It recreates in a clinical setting the type of situation that causes a dysfunction of equilibrium in the patient's daily life. The computerized graphics and tracings of the patient's responses are useful for an accurate diagnosis. They are also of value in determining the benefits of appropriate therapy and in monitoring the results over time.

Posturography is done by using a fixed and swaying platform that is connected to a computer. Patients are attached to sensors that go into the computer. They are also tethered so that they do not fall when the platform sways (Figure 18.16). Orthopedic pathology is ruled out first because abnormalities in the

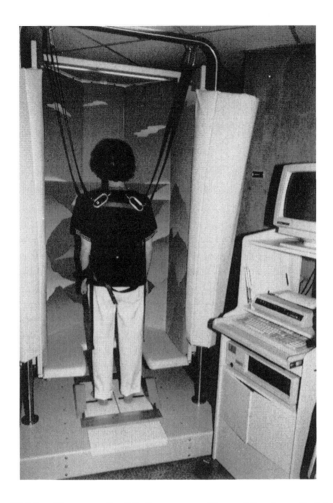

**Figure 18.16.   Computerized balance test. (From Martinez, D. M. 1990.
Personal communication. With permission.)**

musculoskeletal system will render incorrect readings. Measurements will
assess proprioception, visual output, sensory organization, and motor function.

To maintain postural stability while standing, a person must keep his or her
body's center of gravity over the base of support (the feet). To do this, the brain
must first decide how the body is positioned in space. This decision requires
integration of sensory information from three sources: vestibular, visual, and
proprioceptive. Next, the muscles of the body must contract in a coordinated
way to maintain or return the body to an upright position. Normally, this
interaction is an unconscious process.

Balance disorders may be due to a CNS or musculoskeletal disorder
affecting posture, a loss of vestibular function, abnormal interaction between
the three sensory inputs, or any combination of these.

The objectives of the posturography are to quantitate the automatic responses that are used to restore and maintain balance and to isolate the three components — vestibular, visual, and proprioceptive — of the balance system.

The testing procedure of the posturography consists of two main parts — first, evaluation of motor responses to perturbations (rapid forward and backward movements) of the support surface and, second, evaluation of sensory interactions affecting balance. Each will be discussed separately.

Motor Response to Perturbation of the Platform Surface

The motor response test helps evaluate the patient's automatic responses to correct for postural disturbances. Data from this test are particularly useful with patients with CNS or musculoskeletal disorders. During the test, the patient stands quietly on the support surface facing into a visual shield. Brief, but unexpected, forward or backward movements of the support surface cause the patient to sway in the opposite direction. To regain equilibrium, the patient must make a rapid, brief postural adjustment to reposition the body's center of gravity over the feet (Figure 18.17).

Translation of the Support Surface. In Figure 18.17, the solid arrow shows the direction of the support surface movement, while the open arrow shows the direction of the initial body sway motion.

The motor tests consist of six sets of three trials each during which the support is moved. Each translation is adjusted to the patient's height to produce equivalent angular amplitudes of anterior-posterior (AP) sway: (1) small translation forward (0.7° sway); (2) medium translation forward (1.8° sway); (3) large translation forward (3.2° sway); (4) small translation backward (0.7° sway); (5) medium translation backward (1.8° sway); and (6) large translation backward (3.2° sway).

To avoid patient anticipation of the onset of the perturbations, the computer randomizes the timing of the support displacements within each sequence of three trials. The computer records and processes information from the force plates prior to each translation and continues for a total of 2.5 seconds. This data is used to calculate the following parameters (Figure 18.18):

1. **Latency** — the time from the onset of the perturbation to the initiation of a motor response
2. **Symmetry of movement** — the ability of the subject to execute corrective movements by exerting equal effort with the two lower extremities
3. **Strategy of postural movements** — the degree to which the patient uses ankle vs. hip movements to arrest sway

Figure 18.17.   Balance test — rapid brief postural adjustment. (From Martinez, D. M. 1990. Personal communication. With permission.)

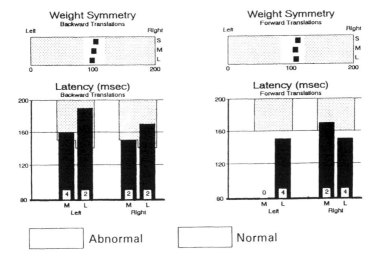

Figure 18.18.   Movement (motor) coordination test. (From NeuroCom International Inc. 1989. Equi Test (RP Version 3.02). With permission.)

4. **Strength of the movement** — the force developed in response to sudden perturbations of the support surface should increase with the amplitude of the perturbation

In addition to measuring the response to translational movements of the support surface, the computer program also measures the response to pitch movements of the support surface. The computer uses the force data to determine:

5. **Adaptation** — the ability to reduce the amplitude of the postural response to repeated trials due to "learning"

The motor test and the calculation of the movement parameters are based on research that shows that normal subjects perform rapid automatic postural movements to correct displacements from normal postural equilibrium within a few seconds or less. When displacements are confined to the forward and backward direction, the force developed is normally exerted symmetrically between the right and left lower extremities. If the support surface is flat and firm, normal subjects tend to keep the hips and knees relatively fixed in extension, and the body is rotated about the ankles. When displacements of the same size are repeated a number of times and then the amplitude is changed, normal subjects can adapt the amplitude of the response within a few trials to produce accurate postural movements.

Depending on the cause of the balance disorder, patients will have abnormal results on different portions of the motor test. For instance, patients with cerebellar disorders affecting postural responses may be unable to grade the strength of the responses to movement of the support surface appropriately, while patients with hemiparesis may have asymmetric responses to support surface perturbations.

Sensory Testing

The sensory organization tests aid in the evaluation of the vestibular input to postural control and the patient's ability to select accurate information when the three sensory inputs conflict. The brain must select the accurate information and ignore that which is inaccurate.

The posturography test exposes the patient to 6-second periods during which the sensory conditions vary. The conditions for each of the six tests can be repeated three times. The support surface and/or the visual surrounding are stationary, or they may be tilted to match the patient's forward and backward sway. These latter conditions are referred to as "sway-referenced support" and "sway-referenced vision". When the support surface moves, the ankle-angle does not change, and the patient does not receive the appropriate proprioceptive cues to maintain balance. Similarly, when the visual surrounding moves parallel to the patient's forward and backward sway, cues such as changes in image size to maintain balance are no longer available to help maintain balance (Figure 18.19).

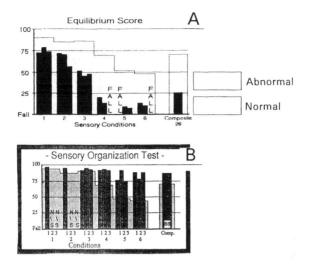

Figure 18.19.   Sensory organization test. (From NeuroCom International Inc. 1989. Equi Test (RP Version 3.02). With permission.)

Test 1. Both the support surface and the visual surrounding are fixed with respect to the patient. The patient can use visual, vestibular, or proprioceptive cues to maintain balance.

Test 2. The support surface is fixed with respect to the patient. The patient's eyes are closed during the test. The patient can use vestibular or proprioceptive cues to maintain balance.

Test 3. The support surface is fixed with respect to the patient, but the visual surrounding moves parallel with the patient's forward and backward sway. Patients can use proprioceptive or vestibular cues to maintain balance, but they must be able to ignore the inappropriate visual inputs.

On the following three tests, the support surface is "sway-referenced" to the patient, and the patient does not get appropriate proprioceptive cues to maintain balance.

Test 4. The patient's eyes are open during the test, and the patient has only vestibular cues to maintain balance.

Test 5. The patient's eyes are closed during the test, and the patient has only vestibular cues to maintain balance.

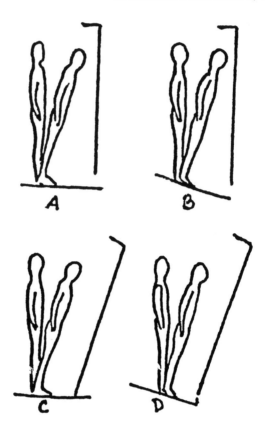

**Figure 18.20.   Different positions during posturography. (A) normal test
conditions; (B) sway-referenced support; (C) sway-referenced vision; (D) a combination of sway-referenced support and vision. (From Martinez, D. M. 1990. Personal
communication. With permission.)**

   Test 6. The visual surrounding is "sway-referenced" to the patient
   during the test, and the patient must ignore inappropriate visual cues
   to maintain balance. The patient can use vestibular cues for balance
   (Figure 18.20).

   At the EHC-Dallas, we use computerized posturography to assess the
contribution of the visual, vestibular, and proprioceptive systems to an
individual's ability to balance and to assess his or her dependence on sensory
inputs from these systems. Patients' performance on these tests helps us
determine their appropriate treatment. We test these patients an additional time
or two during the course of their treatment in order to make appropriate
modifications to their treatment as their status changes. Our experience, as well
as that of investigators at other institutions, reveals that retesting is a valuable

Table 18.19.   Patients with Imbalance after a
Chemical Exposure and Induction of
Chemical Sensitivity — Posturography

| Patients | Abnormal | Normal | After ECU program | |
			Studied	Improved
100	61 (61%)	39 (39%)	21	12 (57%)

Source: EHC-Dallas. 1990.

way to monitor a patient's progress following physical therapy for balance disorders.

In our studies using good environmental control with massive avoidance of pollutants as well as using heat depuration physical therapy in patients with chemically induced balance problems, we have shown reduction not only in the blood levels of solvents but also in cerebral and ENT symptoms.

In a prospective series at the EHC-Dallas, Rea and Martinez[56] used posturography to study 200 chemically sensitive patients who complained of losing their balance after chemical exposure and who had a positive clinical balance study (i.e., they could not stand on their toes with their arms extended and their eyes closed). Eighty-one percent (81%) were abnormal. Most of the abnormalities were in the sensory coordination test. No other etiologic or neurologic abnormalities were found. Table 18.19 shows the first 100 patients when 61% were abnormal. When motor abnormalities were present, they consisted of prolongation of the latency of the neurmuscular response, and this response was thought to be due to peripheral neuritis.

In the abnormal group, the deficiencies were present when proprioceptive and/or visual cues were eliminated (tests 5 and 6). Some patients with an abnormal posturography test were selected to have caloric vestibular tests. In these patients, the responses were normal. This finding suggests that the vestibular abnormalities probably originated in the CNS.

In the group of 61 patients with an abnormal test, 28 (46%) returned for follow-up tests (Table 18.19). Of these patients, 12 (57%) showed improvement on the posturography test. After rigid ECU treatment including pollutant avoidance, heat depuration, injection therapy, and nutritional supplementation, 25% of these showed improvement to normal findings by posturography.

Patients who had abnormal posturography tests (the majority of whom had dysfunction originating in central nerve pathways) and whose clinical examination indicated the need for an MRI of the head usually had negative findings on this exam. This lack of MRI finding may indicate that the toxic chemicals affect neurotransmission rather than produce lytic or demyelinating changes that could be registered by MRI. However, subsequent triple camera SPECT brain scan did show a neurotoxic pattern in each case studied (Figure 18.21). Working at the EHC-Dallas, Pan et al.[61] found a high correlation between blood levels of toxic chemicals and biological inhalant, food, and chemical skin testing/inhalant challenges and balance in 179 patients ranging in age from 8

A.

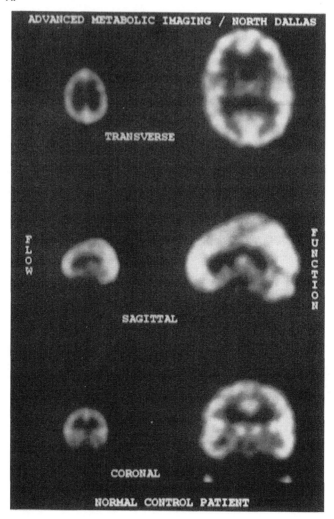

Figure 18.21. Comparison of three-dimensional triple camera brain
SPECT brain scan with posturography: (A) normal SPECT
scan, (B) neurotoxic SPECT scan.

to 81 years (Table 18.20). They also found a high correlation between triple
camera SPECT brain scan, iris corder autonomic nervous system measurements, and computerized balance test (Table 18.21).

Follow-up evaluation of treatment showed that symptom reduction and
leveling of balance paralleled the reduction of the level of solvents in their
blood. As levels initially rose, symptoms increased. As levels decreased,
symptoms dissipated. Biopsies of the dermal vascular lesions showed lymphocytic perivascular infiltrates in these patients.

B.

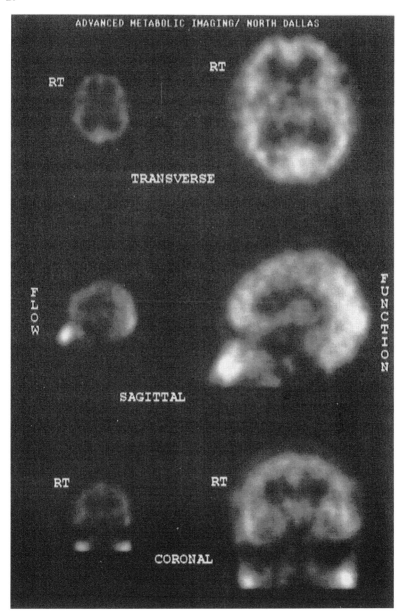

Figure 18.21.   (Continued).

Ménière's Disease

Though probably overlapping with the previous section, Ménière's disease will be discussed separately. Some patients with Ménière's disease appear

Table 18.20. Summary Data of 179 Patients with Balance Test

Sex	Age	Toxic chemical in blood — percent (positive/tested)		
		GVST	HSST	CPST
F: 109 (61%)	8-81 years	93 (113/122)	99 (137/139)	100 (57/57)
M: 70 (39%)	Mean = 43 years			

The highest positive percentage of chemical in blood (four samples of each group)

1,1,1-Trichloroethane	64	3-Methylpentane	88	DDE	98
Toluene	47	2-Methylpentane	84	trans-Nonachlor	39
Xylene	24	n-Hexane	76	HCB	37
Tetrachloroethylene	24	Cyclopentane	12	Oxychlordane	37

Intradermal test				Intradermal test				Chemical inhalation test (Double-blind)	
Chemical	Positive %	(No. pos./ no. tested)	Antigen	Antigen	Positive %	(No. pos./ no. tested)	Chemical	Positive %	(No. pos./ no. tested)
Formaldehyde	96	(110/114)	Dust	Dust	100	(86/86)	Toluene	100	(12/12)
Silicon	83	(5/6)	Mite	Mite	100	(83/83)	Xylene	100	(4/4)
Orris root	82	(98/119)	T.O.E.	T.O.E.	100	(73/73)	Methylethyl ketone	100	(4/4)
Cigarette smoke	77	(96/124)	Weeds	Weeds	100	(78/78)	Diesel fuel	100	(1/1)
Ethanol	66	(79/119)	Grasses	Grasses	100	(77/77)	Ethanol	85	(22/26)
Women's cologne	65	(73/113)	MISC	MISC	100	(2/2)	Pesticides	85	(11/13)
UNL/diesel	63	(70/111)	Hormones	Hormones	100	(28/28)	Chlorine	76	(13/17)
Chlorine-SL	57	(43/76)	Virus	Virus	100	(80/80)	Formaldehyde	75	(18/24)
Fireplace smoke	56	(18/32)	Bacteria	Bacteria	100	(72/72)	1,1,1-Trichlorethane	75	(6/8)
Men's cologne	51	(58/113)	Foods	Foods	100	(105/105)	Phenol	69	(11/16)

Phenol	50	(55/109)	Peptides	100	(1/1)
Propane gas	48	(10/21)	Molds	98	(99/101)
Natural gas	46	(31/67)	Trees	98	(79/81)
Newspaper material	45	(50/111)	Algae	97	(67/69)
			Candida	97	(73/75)
			Terpenes	93	(64/69)
			Smuts	88	(43/49)
			Fabrics	81	(17/21)
			Danders	74	(42/57)
			Cotton	69	(40/58)

Note: T.O.E. = trichophyton oliandomycin epidermatophyton.

Source: Pan, Y., W. J. Rea, and D. M. Martinez. 1994. EHC-Dallas.

Table 18.21. Balance Test of 57 Patients with Both SPECT Scan and Iriscorder

Sex	Age		Balance test		SPECT scan		Iriscorder	
			Abnormal	%	Abnormal	%	Abnormal	%
F = 32	12 to 68 years	Sensory organization test (A)	23	40	57	100	48	84
M = 25	Mean = 44 years	Motor organization test (B)	4	7				
		A & B	22	39				
		Total	49	86				

Source: Pan, Y., W. J. Rea, and D. M. Martinez. 1994. EHC-Dallas.

to be environmentally triggered. Various authors, including Rubin,[62] Boyles,[63] and Derebery,[64] have suggested that some of the etiology of Ménière's disease is sensitivity to pollens, molds, foods, and chemicals.

Viscomi and Bojrab[65] reported that allergy has been cited as a cause of Ménière's disease. A number of observers, beginning with Duke in 1923[66] and in recent times Derlacki,[67] Wilson,[68] Clemis,[69] and Powers[70] have noted that both inhalant and food allergy have been associated with the onset of Ménière's disease. These researchers also described the use of the challenge ingestion feeding test and the provocative food test (PFT) for the diagnosis of food allergy.

In another report, Clemis[71] showed that provocative food testing (PFT) influenced hearing discrimination and nystagmus. He described a patient whose discrimination scores dropped to 16% during a provocative reaction to wheat. During the neutralization period of 1 hour the discrimination improved to the 74% level. In another patient, a 16-year-old boy with severe episodic vertigo, Clemis monitored the provocative food test to milk with electronystagmography. No pretest nystagmus was present. Within minutes of administration of the first injection, this patient developed dizziness and an irregular nystagmus on the right side. These symptoms intensified with the next injection. Within minutes after the third injection, which was the neutralizing dose, the dizziness began to subside, and the recorded nystagmus stopped.

Although the role of food allergy in the causation of Ménière's disease has been emphasized, both inhalant and chemical sensitivity are also important. The diagnosis of inhalant sensitivity can be made with a high degree of accuracy by using a skin endpoint titration (SET), intradermal neutralization, and oral challenge. Chemical sensitivity is evidenced by inhaled, oral, or intradermal challenge.

Viscomi and Bojrab[65] have reported that laboratory studies in the last decade tend to support the concept that sensitivity via an immunological mechanism may affect the inner ear. Harris et al.[72] reviewed a series of studies on the immune function of the endolymphatic sac (ES) and concluded that the inner ear is an immunocompetent site and that inflammation in the ES could impair the function of this organ.

Mogi and Kawaushi[73] reviewed the available data on inner ear immunology. The immunoglobulins in the perilymph (PL) are largely derived from the serum, and the main immunoglobulin is IgG. Direct immunization in the perilymphatic space through an artificial route can evoke local antibody formation. The endolymphatic sac (ES) is the only place in the inner ear that has a network of lymphatic vessels containing lymphocytes, mononuclear phagocytes, plasma cells, and mast cells, strongly suggesting an immune defense system.

Miyamura et al.[74] provoked Type I allergy in guinea pigs who were passively sensitized and then received specific antigen challenge. They found evidence of endolymphatic hydrops and nystagmus within 15 minutes after antigen challenge. They observed negative summating potentials as evidence of hearing loss resulting from endolymphatic hydrops. They concluded that these changes were caused by an antigen-specific immunologic reaction, and the responsible antibody was IgE. As discussed previously in this chapter, direct chemical triggers can also occur.

Viscomi and Bojrab[65] used the intracutaneous provocative food test (IPFT) to diagnose food sensitivity. The IPFT utilizes the intracutaneous injections of progressive five-fold dilutions of a potent food extract (as shown in Chapter 36). The commercially available food extracts used in the Viscomi and Bojrab[65] study were prepared in 1:10 concentrations (corn, soy, wheat, and yeast) and 1:20 concentrations (egg and cow's milk). The test dilutions were prepared by progressive five-fold dilutions.

Viscomi and Bojrab[65] noted that patients reported any change in aural symptoms or provocation of aural fullness, hearing loss, vertigo, or alteration of tinnitus. While positive skin whealing was considered a positive reaction to a particular food, provocation and neutralization of aural symptoms is more specific and relates the food reaction to inner ear disease.

Viscomi and Bojrab[65] showed that in Ménière's disease the provocation of aural symptoms and a positive skin whealing are considered evidence of a positive reaction to the test food. Relief of pretest aural symptoms during the IPFT for a specific food is also considered a positive reaction to that food. However, the limitations to the IPFT in Ménière's disease is that the evidence of provocation, aural symptoms, is dependent on patient observations, which are subjective and raise the possibility of a placebo effect, especially when the provoked symptoms are mild.

Viscomi and Bojrab[65] reported that electrocochleography (ECoG) has proven to be a valuable tool for assessing inner ear function. The early auditory-evoked potentials provide information about the cochlear microphonic (CM), the summating potential (SP), and the whole-nerve action potential (AP). Dauman et al.,[75] Gibson et al.,[76] and others[77-79] have shown that an increase in the low frequency summating potential (SP) is a finding characteristic of Ménière's disease and that SP enlargement is the physiologic manifestation of endolymphatic hydrops. There is agreement that enhancement of the summating potential/action potential (SP/AP) amplitude ratio indicates the presence of endolymphatic hydrops.

Viscomi and Bojrab[65] used the ECoG to monitor the glycerol test. They found that elevation of the SP/AP ratio can be reversed within 1 to 2 hours after ingestion of glycerol. Coats and Alford[80] found this reversible component can be better estimated by transtympanic ECoG monitoring than by behavioral audiometry.

Ferraro et al.[81] have shown that measurements of SP/AP ratio in Ménière's disease were directly influenced by the presence of aural symptoms at the time the examination was being performed. If symptoms were absent at the time of the ECoG study the SP/Ap amplitude ratio was not enlarged. The combination of hearing loss and aural fullness was the strongest predictor of elevation of SP/AP ratio during ECoG. Ferraro et al. concluded that ECoG provided an objective indication of subjective symptoms. Thus, it appeared possible that ECoG could provide an objective method for evaluating the symptom provocation during the PFT. A study incorporating serial ECoG monitoring was proposed to determine whether alterations in SP/AP amplitude ratio would parallel the provocation and neutralization of aural symptoms during the PFT.

Viscomi and Bojrab[65] evaluated hearing for pure tones and speech immediately before the IPFT. The patient's pretest symptoms were reviewed with the patient and recorded. The patient was placed in a comfortable, semirecumbent position for the test. ECoG was then recorded before each IPFT.

The ECoG was performed transtympanically in the involved ear. ECoG potentials were recorded with a transtympanic electrode placed on the promontory, with the reference electrode on the contralateral earlobe and with the ground electrode situated on the forehead. The responses were filtered from 3 to 1500 Hz with an analysis time of 5 milliseconds. Stimuli consisted of 100 microsecond broadband alternating polarity clicks at a rate of 7.1 per second. The stimuli were delivered by an insert transducer at an intensity level of 95 dB HTL, averaged across 200 to 300 presentations.

The patients in Viscomi and Bojrab's study[65] were food tested as described by King et al.[82,83] The foods tested were corn, egg, cow's milk, soy, wheat, and yeast. Three in six food tests were performed at one sitting. Five patients were tested. A trained allergy nurse administered the injections and selected the order in which the foods were tested. The physician monitored and recorded the patients' symptoms, and the physician, patient, and audiologist did not know which food was being tested.

Viscomi and Bojrab[65] included five patients in their preliminary report. Three of the patients had unilateral Ménière's disease, and two had bilateral disease. The diagnosis of Ménière's disease was made on the basis of clinical history and otoneurological evaluation. Two of the patients had prior endolymphatic sac surgery that failed to control their vertigo.

All of the subjects had clinical and skin-test evidence of inhalant allergy. All patients had a previous history of perennial or seasonal rhinitis. Evaluation for food allergy was a standard part of the workup for these patients, some of whom had a history suggestive of adverse reaction to foods.

The results of the IPFT with ECoG monitoring are presented in relation to three parameters. Viscomi and Bojrab[65] recorded the provocation of aural or nasal symptoms or the relief of existing aural or nasal symptoms during the IPFT. The skin-whealing response was recorded. The percentage of real-time increase or decrease in the SP/AP ratio at the maximum point of change was computed in relation to the pretest SP/AP ratio. An increase or decrease of 15% or greater was taken as the standard of meaningful reaction.

Five patients had a total of 27 individual IPFT's performed with ECoG monitoring. Table 18.22 shows representative test results for each patient and all responses in which there was a 15% increase or decrease in the SP/AP ratio.

Of the 27 tests, 8 (29%) showed a real-time increase or decrease greater than 15%, the measure that Arenburg et al.[84] chose as an indicator of sufficient

Table 18.22. Results of IPFTs' Designating Significant Alteration in SP/AP Ratio

| | Patient | | | | | | | | |
| | P.C. | P.G. | | J.H. | M.M. | E.M. | | | |
Food tested	Milk	Corn	Yeast	Wheat	Wheat	Milk	Wheat	Egg	Soy
Aural symptoms	+	+	O	O	++	++	++	++	++
Nasal symptoms	O	O	+	+	O	O	O	O	O
Skin reaction	Pos.	Pos.	Pos.	Pos.	Pos.	Pos.	Pos.	Pos.	Pos.
SP/AP ratio % of change	9%	62%	42%	20%	28%	55%	24%	69%	77%

Notes: + and ++ = degree of severity; O = no symptoms; Pos. = positive.

Source: Viscomi, G. J., and D. I. Bojrab. 1992. Use of eletrocochleography to monitor antigenic challenge in Ménière's disease. Otolaryngol. Head Neck Surg. 107 (6 Pt 1):733–737. With permission.

shift in the SP/AP ratio to be meaningful for his study on surgery of the endolymphatic sac. Six of the eight shifts in SP/AP ratio were increases, and two were decreases. Six of the eight with increases or decreases of 15% in the SP/AP ratio correlated well with symptom provocation in three different patients. In two patients, the rise in the SP/AP ratio was shown to correlate with provocation and the fall in SP/AP ratio with neutralization.

All of the eight tests were associated with positive skin reactions, and 16 (60%) IPFTs were associated with positive skin reactions for the 27 tests. No significant change in the SP/AP ratio occurred in patients who had negative skin reactions, and only two IPFTs developed evidence of nasal provocation. Thus, there was a correlation between the provocation of aural symptoms and a significant change in the SP/AP ratio for 25 of 27 IPFTs (93%).

In discussing their results, Viscomi and Bojrab[65] concluded that presently little is known about serial ECoG response (alteration of the SP/AP ratio) and what constitutes a significant elevation or decrease. In surgical monitoring, Arenburg et al.[84] consider a "real-time" decrease of 15% in the SP/AP ratio as evidence of effective sac opening. Monitoring during the glycerol test also fails to provide specific numerical standards for significant change in the SP/AP ratio. Viscomi and Bojrab[65] decided to apply the standard that Arenburg[84] suggested. If Viscomi and Bojrab[65] reduced the level to 10% change, then two additional reactions with provocation would have been included.

Viscomi and Bojrab[65] emphasized that, while a positive skin response to IPFT is important as a reliable indicator of food allergy, the provocation of symptoms is more specific in conjunction with positive whealing. Thus, in Ménière's disease, Viscomi and Bojrab needed the provocation of aural symptoms to indicate that the food allergy was affecting the inner ear.

Patients who have been previously tested for food allergy by IPFT may not demonstrate the same provocation of aural symptoms when retested. King et al.[82,83] demonstrated that, while the reliability of the skin response and provocation on second and third retests for the same food was statistically reliable, the provocation response was more variable during retest than the skin response. Therefore, if others wish to study the effectiveness of this method of monitoring the effects of food allergy on hearing, we recommend that they choose patients who have not been previously tested.

Hoover[85] presented a series of 27 patients with Ménière's disease, 25 of whom responded to a combination of sublingual therapy for biological inhalants and foods as well as a food elimination diet avoiding the offending foods.

The following case report focuses on a case of Ménière's disease that was treated at the EHC-Dallas. It illustrates the multifactorial etiology of this disease.

Case study. A woman aged 41 years entered the ECU with episodes of dizziness of 5 years duration. She had noticed her condition worsened in the spring and fall. She had fluctuating hearing and pressure in the ears. For the 3 years previous to admission to the ECU, she had experienced the onset of vertigo in early October that became severe by mid-October. Over the year previous to being treated, she had become increasingly incapacitated. She had almost constant ringing in her ears, tended to fall to the left, and had increasing left-sided deafness. Throat swelling inhibited her breathing, and she had severe sinus congestion. Throat swelling had developed every 2 to 3 months during the previous 3 to 4 years. She was treated with cortisone to keep this swelling under control. Injection therapy for pollen, dust, and mold for the previous 5 years had not been successful. Attempts at dietary control had also failed, as had surgical interference with the vestibular apparatus. Audiogram showed a typical Ménière's pattern. Following admission to the ECU, she fasted on a safe water until she reached a basal state with clearing of most of her problems. Challenge with chemically less-contaminated foods produced no symptoms. She was then challenged with commercial food, and after one meal her symptoms were reproduced. All six chemicals on inhalation challenge also reproduced her symptoms. This patient has since removed inciting agents from her environment and is doing quite well.

At the EHC-Dallas, we have seen 20 cases of Ménière's disease that were triggered by biological inhalants, foods, and chemicals (Table 18.23). After meticulous workup, each triggering agent was found and the patient improved. Evidence suggests that autoimmune disease causes Ménière's disease. Since toxic chemicals may trigger autoantibodies, it is reasonable to conclude that exposure to environmental pollutants may well provoke Ménière's disease. Further, the toxic chemicals may directly trigger the nutrient blood vessels and thus cause symptoms of Ménière's with or without producing antibodies. It is becoming well-documented in the literature and our studies that the triggering of autoimmune disease may frequently occur as a result of exposure to chemicals (see Volume I, Chapter 4[86]). Boyles[87] and Derebery and Valenzuela,[88] working independently, have also reported a series of patients with Ménière's disease who were provoked by biological inhalants and foods. Derebery's study results are presented in Table 18.24.

Tinnitus

Noise-induced tinnitus is well-known and discussed in standard ear, nose, and throat books and, therefore, will not be discussed further here.

Various surgeons, including Boyles,[87] Rubin,[89] and Hoover,[90] have reported tinnitus to be triggered by sensitivity to foods. Hoover, for example, studied 57 patients whose symptoms were triggered by foods. She found that the triggering agents were always multiple and involved at least three or more foods. Figure 18.22 shows the variety of offending foods identified in her series.

Non-noise-induced tinnitus is frequently seen in patients with toxic environmental overload. At the EHC-Dallas, we found multiple inhalants, foods, and chemicals were able to reproduce the conditions that allowed tinnitus to occur in 30 sample patients (Table 18.25).

**Table 18.23.   Ménière's Disease Studies in a Controlled
Environment (ECU) after 4 Days Deadaptation
with the Total Load Reduced in the ECU**

	Triggers			
Age	Biological inhalants[a]	Food[b,c]	Toxic chemicals[d,e]	Positive results (years)
35	+	-	+	4
40	+	+	+	3
33	+	+	+	2
37	-	+	+	6
47	-	+	+	10
50	+	+	+	1
56	-	-	+	$1^1/_2$
43	+	-	+	$3^1/_2$
47	-	+	+	$8^1/_2$
54	-	+	+	5
59	-	-	+	6
60	-	-	+	5
38	+	+	+	4
41	+	-	+	3
44	+	+	+	6
57	+	+	+	8
31	-	-	+	7
50	-	-	+	3
55	+	-	+	5
40	-	-	+	1

[a] Intradermal challenge — weeds, trees, grasses.
[b] Intradermal challenge — foods.
[c] Oral challenge — foods.
[d] Inhaled challenge (double-blind) — chlorine, <0.33 ppm; formaldehyde, <0.20 ppm; phenol, <0.005 ppm; pesticide, 2,4-DNP, <0.0034 ppm; ethanol (petroleum derived), <0.50 ppm.
[e] Intradermal challenge — cigarette smoke, newsprint, orris root, jet fuel, car exhaust.

Source: EHC-Dallas. 1992.

Whether inhalant, food, or chemical provocation tests were administered, individual tests sometimes produced tinnitus. Usually resolution of this problem necessitated complete removal of many incitants in addition to injection therapy. An illustrative case report follows.

Case study. A 55-year-old self-employed real estate manager presented at the EHC-Dallas with tinnitus that had come and gone for years for no apparent

Table 18.24. Ménière's Disease—Number and Percentage of Patients Reacting to Biological Inhalants or Foods

Positive Percentage of 87 Patients with Serial Endpoint Titration Intradermal Tests

Antigen	Endpoint Titration				
	1:100	1:500	1:2500	1:12500	1:62500
Dust mite	4	76	10	7	2
Molds	0	90	10	0	0
Alternaria	5	79	12	5	0

Number and Percentage of Patients Reacting to Provocative Food Testing

Food	(N Positive/N Tested)	Percent
Wheat	(66/92)	71.7
Milk	(56/79)	70.9
Corn	(58/90)	64.4
Egg	(55/88)	62.5
Yeast	(32/57)	56.1
Soy	(13/29)	44. 8

Source: Derebery, M. J. and S. Valenzuela. 1992. Ménière's syndrome and allergy. *Otolaryngic Allergy* 25(1):218. With permission.

reason. He also complained of mood swings and personality changes when at airports, where he had to go frequently due to his work situation.

This man experienced seasonal rhinitis, nasal stuffiness, and sneezing, along with hoarseness. All of these symptoms seemed to peak in the spring. He also found that cigarette smoke was much more likely to bother him at the airport. He had some food intolerance with GI upset, and his 22-year-old son had multiple food allergies, along with chemical and inhalant sensitivities.

Upon skin testing, this patient had significant reactions to cow's milk, eggs, brewer's yeast, tea, and jet fuel. The simple elimination of milk from his diet completely stopped his tinnitus. He has since confirmed the direct cause and effect relationship between the two. The reintroduction of milk into his diet consistently reproduces tinnitus. Interestingly, when he was skin tested for jet fuel, he experienced the mood changes that had troubled him while he was at various airports. Since his treatment at the EHC-Dallas, he has maintained good avoidance of his incitants, and in 6 years of follow-up, he has had no recurrence of symptoms.

This case illustrates that some patients may suffer symptoms that are caused or influenced by undiagnosed environmental factors. In these cases, the symptoms may be reduced or eliminated by appropriate environmental inteerovention.

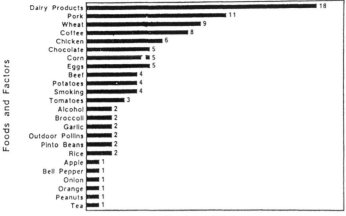

Figure 18.22.   Incidence of foods and factors inducing tinnitus, according
to patients' diaries, following a 4-day diversified rotating
diet for 4 weeks. (From Hoover, S. 1988. Ménière's migraine and allergy. In *Vertigo, Nausea, Tinnitus and
Hypoacusia in Metabolic Disorders*, eds. C.-F. Claussen,
M. V. Kirtane, and K. Schlitter. 296. Elsevier Science Publishers B. V. (Biomedical Division). With permission.)

Headaches

Headaches of the vascular type are discussed in Chapter 20. The ones that
involve the ear, nose, and throat are sinus and apical and posterior headaches.
Also, some headaches can occur from problems involving the temporal
mandibular joint (Table 18.26). These types of headaches can be triggered by
an accumulation of inhalants, foods, and chemicals (Tables 18.27), indicating that triggering is the result of an excessive total load and not necessarily
produced by exposure to an individual incitant. In contrast, however, a study
of 30 patients revealed that 100% had their headaches triggered after a food
or chemical test, indicating that a single exposure is sufficient to trigger a
reaction. In a series of 100 patients at the EHC-Dallas with predominantly
tension and vascular headaches due to inhalant, food, and chemical triggering, 90% cleared and remained clear with either avoidance or a combination
of avoidance and food and inhalant injections and no medications. These
formerly unsalvageable patients now experience a large degree of relief. Our
observations are similar to those reported by Monro[91] who observed a series
of cases of migraine cleared and reproduced by finding the chemical, food,
and inhalant trigger. She also performed a double-blind study on the food
injection therapy, clearly demonstrating that headaches occurred more often

Table 18.25. Tinnitis (30 Patients)

Incitant	Response to intradermal challenge %
Inhalants	
Molds	82
Dander	29
Weeds	41
Trees	76
Grasses	70
Terpenes	30
Foods	93
Cow's milk	82
Brewer's yeast	76
Baker's yeast	52
Cane sugar	52
Corn	52
Wheat	47
Chemicals	52
Includes formaldehyde, phenol, ethanol, perfumes, jet diesel fuel, car exhaust, natural gas, newsprint, cigarette smoke, orris root	

Notes: 6% had low T-cells and 29% had IgE above 150 IU/mL
(C = 10 to 150 IU/mL).
Age: 17–63 years
Mean age: 41 years
F: 70%
M: 30%
Average age of onset: 31 years
Associated symptoms:
NS = 94%
GI: 64%
MS: 52%
Psych: 53%

Source: EHC-Dallas. 1992.

when patients discontinued the food injection than when they took it regularly.[92] Consistent with ours and Monro's findings, Hoover[85] reported on a series of 88 patients with migraines, 87% of whom cleared with sublingual neutralization treatment of biological inhalants and foods and avoidance of the foods that caused their headaches.

Temporal Mandibular Joint

Temporal Mandibular Joint (TMJ) is a common cause of headache and otalgia. This frequently seen disorder has multiple etiologies, including

Table 18.26.   Major Categories of Headaches

I.	Vascular headache of migraine type
	A. Classic migraine
	B. Common migraine
	C. Cluster headache
	D. Hemiplegic and ophthalmologic migraine
	E. Lower half headache
II.	Muscle-contraction headache
III.	Combined headache — vascular and muscular contraction
IV.	Headache of nasal vasomotor reaction
V.	Headache of delusional, conversion, or hypochondriacal states (likely chemical in origin)
VI.	Nonmigraine vascular headaches
VII.	Traction headache
VIII.	Headache due to overt cranial inflammation
IX–XIII.	Headache due to disease of ocular, aural, nasal, sinus, dental, other cranial or neck structures
XIV.	Cranial neuritides
XV.	Cranial neuralgia

Notes: Causes of otalgia are legion and are left at the discretion of the physician.

environmentally triggered muscle spasm. Patients with facial pain or other types of pain are often seen at the EHC-Dallas. Since many patients with TMJ have environmental triggers, they should be worked up accordingly. The environmental medicine specialist should then diagnose and treat temporal mandibular joint involvement in conjunction with a periodontist and an oral surgeon. The following is an example of this environmentally triggered disorder.

> **Case study.** A woman age 30 years presented with severe nausea and vomiting of blood. She reported a continuous low-degree headache. She had bruxism, trismus, and tenderness in the TM joint. In the previous 9 months, her headaches had increased in severity, with pain in back of her eyes and on the right side of her face accompanied by severe pain in the right upper molars. Root canal work was unsuccessful, and the teeth involved were extracted. She continued to have pain in adjoining teeth, which subsequently were extracted. Examination of the teeth showed no pathology. Neurologic evaluation was negative. Her throat swelled frequently, and she had difficulty swallowing. She had frequent earaches, with ringing in her ears, minimal hearing loss, rhinorrhea, nose bleeds, loss of sense of smell, and frequent bleeding of her gums.
>
> This patient was admitted to the ECU, and her symptoms cleared after 7 days of fasting on safe water. She was then challenged with chemically less-contaminated food. Eight foods and five chemicals reproduced her symptoms. The laboratory data are presented in Table 18.28.

**Table 18.27.   Challenge Tests
in 30 Chemically Sensitive
Patients with Headaches**

| Incitants | Challenge tests | |
	Type	Percent
Inhalants	ID[a]	
Mold		96
Smuts		20
Weeds		83
Trees		80
Grasses		83
Terpenes		47
Hormones		23
Foods	ID[a]	100
	Oral	
Chicken		71
Cow's milk		71
Corn		67
Turkey		64
Baker's yeast		60
Bananas		57
Eggs		70
Chemicals	ID[a]	100
	Inhaled[b]	100

Notes:

Headaches (30 patients): sinus, apical posterior, and nonmigraine

Ages: 12–70 years; mean age: 40 years

F: 21; M: 9

Age of onset: under 30 years

Associated symptoms: resp (73%); CV (70%); GI (63%); chlorinated pesticide frequency (7%)

Blood chlorinated pesticide frequency (7% abnormal)

T cells: 14% abnormal

IGE: 92% increased

[a] Intradermal challenge.

[b] Inhaled double-blind challenge — 25 patients, challenge performed after 4 days of deadaptation with the total load reduced in the ECU. Three placebos and formaldehyde at <0.2 ppm, phenol at <0.0020 ppm, petroleum derived ethanol at <0.50 ppm, chlorine at <0.33 ppm, pesticide 2,4-DNP at <0.0034 ppm.

Source: EHC-Dallas. 1992.

Table 18.28.   Laboratory Findings in Patients with TMJ Dysfunction

Laboratory Data

Blood	Value	Blood	Value	Blood	Value	Normal range
HgB (g/dL)	12.4	AB Lymph (E. Rosettes, mm^3)	4046 (1800 ± 200)	IgE (IU/ml)	25	(0–180)
				IgM (mg/dl)	106	(60–280)
HCT (%)	34.4	T Lymph (E. Rosettes, mm^3)	1659	IgG (mg/dl)	740	(800–1600)
				IgA (mg/dl)	200	(90–450)
WBC (/mm^3)	7900	C_3 (mg/dl)	104 (80–120)	IgD (mg/dl)	5.4	
THSC (CH_{50})	107% (90–98%)	C_4 (mg/dl)	58 (20–40)			

Source: EHC-Dallas. 1992.

This patient was discharged on safe water and a rotary diet and given information on how to make her environment more tolerable. She has adhered to this program and has remained almost asymptomatic.

Over the last several years, one of the modalities for treatment of TMJ was to replace the joint with a silicone prostheses. Adverse reactions to these were shown in both animal and human studies. The production or exacerbation of chemical sensitivity was similar to that seen in silicone breast implants or other synthetic implants. Systemic sclerosis with Raynaud's Phenomenon has been observed in some patients studied at the EHC-Dallas who had synthetic implants of their TM joint.

In reviewing the literature on TMJ disc replacement, Henry and Wolford[93] noted that replacement of this disc has been recommended to prevent secondary changes consistent with osteoarthritis within the TMJ that occur as a consequence of disc displacement, perforation, or total discectomy.[94,95] These changes include adhesions, articular surface degeneration, crepitus, and pain.[96] Radiographic changes consisting of condylar sclerosis, flattening, erosion, contour irregularity, lipping, osteophyte formation, and resorption also may be observed.[97] Disc removal and replacement have been performed using autologous,[98-111] allogeneic,[112,113] and alloplastic[114-123] materials in an attempt to promote biological resurfacing, improved function, prevention of adhesions, and elimination of pain.[96]

Both temporary and permanent alloplastic materials have been used to serve as interpositional implants in the TMJ.[114-123] Porous polytetrafluoroethylene (PTFE), reinforced with either vitreous carbon (Proplast I®, Vitek, Inc.) or aluminum oxide (Proplast II, Vitek, Inc.) laminated to smooth, dense PTFE (Teflon®) has been used to reconstruct the TMJ.[124,125] Reconstruction with a PTFE implant may initially give patients acceptable function and relief of symptoms.[126-131]The early reports on these implants were very promising, with 91% of 6182 procedures reported to have produced satisfactory results.[132] Continued radiographic follow-up on patients who were treated with such implants, however, revealed that many of them developed changes in the mandibular condyle.[133,134] Although some patients remained satisfied with their results and function despite these osseous changes,[134,135] other reports in the literature described condylar resorption, severe pain, malocclusion, foreign body giant cell reaction (FBGCR), and headaches associated with PTFE implants.[136-138]

These materials have produced numerous complications, including perforation, fragmentation, unstable occlusion, lymphadenopathy, and severe osteoarthritis that may involve both the mandibular condyle and glenoid fossa with possible perforation into the middle cranial fossa.[135,136,138-146] In addition, chemical sensitivity has occurred.

PTFE implants also have been shown to lose integrity and promote a FBGCR, which continues to increase with time in both mechanically loaded and unloaded animal models.[147-149] The FBGCR associated with previously placed PTFE implants may continue despite meticulous surgical debridement after removal of the implant,[146,150] and the placement of autologous tissues into an environment in which the FBGCR is still occurring can result in significant failure of such reconstructions.[146]

Henry and Wolford[93] reported the outcome of their retrospective study of 107 patients (male, n = 13; female, n = 94) with 163 joints previously treated with Proplast®-Teflon® (PTFE, Vitek, Inc. , Houston, TX) implants. The average time in situ for the PTFE was 59.8 months (range, 2 to 126 months). The average length of follow-up was 84.6 months (range, 59 to 126 months). Only 12% of joints showed no significant osseous changes radiographically. Forty-five patients (42%) continue to have in situ PTFE implants, and 36% of them experience pain that requires medication; 25% have developed an anterior open bite and malocclusion; 9% have limited vertical opening; and 40% are asymptomatic. TMJ reconstruction after PTFE implant failure was performed with five different autologous tissues or a total joint prosthesis. Autologous tissues used to reconstruct the TMJ and the rates of success were as follows: (1) 31% free temporalis fascia and muscle graft with, and 13% without, sagittal split ramus osteotomy; (2) 8% dermis; (3) 25% conchal cartilage; (4) 12% costochondral grafts; and (5) 21% sternoclavicular grafts. The success rate decreased in all autologous tissue groups as the number of TMJ surgeries performed before reconstruction increased. Ankylosis was the most common cause of failure. Results of TMJ reconstruction with a total joint prosthesis were as follows: (1) 88% functional and occlusal stability of total joint prosthesis; (2) level of pain reduction was rated as 46% good, 38% fair, and 16% poor; and (3) an average interincisal opening of 27 mm at 24 months or less and 33 mm at 25 months and beyond. A foreign body giant cell reaction was still present an average of 40 months (range, 32 to 48 months) after implant removal and after 4.5 (2 to 9) additional surgeries. Use of a total joint prosthesis may be indicated to achieve successful reconstruction after PTFE implant failure. (For additional information, see Chapters 19 — section on breast — and 23 and Volume IV, Chapter 41). It is clear that these implants have been disastrous and that chemical sensitivity occurred and is still occurring.

Laryngeal Implants

Reports of sensitivity to Silastic® and Teflon® implants are occurring. Sensitivities include skin symptoms, hoarseness, sinusitis, chronic fatigue, and many others. We have used extracts of silastic and teflon with the intradermal SET and provocation-neutralization method to reproduce patients' symptoms. The following case reported by Captain Darrell H. Hunsacker,[151] Department of Otolaryngology, San Diego Naval Medical Center exemplifies the synthetic-triggered disease that can result from synthetic prosthesis.

Case study. This 43-year-old Caucasian female underwent a Silastic® implant thyroplasty at Tripler Army Hospital, Hawaii, on October 30, 1990, for right vocal cord paralysis secondary to a thyroidectomy 9 months previously for a T-1 (2 cm) follicular carcinoma with capsular invasion. She subsequently received I-131 ablation. No recurrence has appeared, and she had a negative chest X-ray and CT scan recently. She takes Synthroid® 0.15 mgm daily.

   She subsequently developed progressive contact urticaria (immediate dermatographism) throughout her body. When she carried packages, walked barefoot, etc. there was pruritus and urticaria at the points of contact that lasted 10 to 15 minutes. Her voice was fair post-op. Patch testing with the Silastic® by dermatology caused no local reaction but aggravated the contact urticaria elsewhere.

   SET testing using the Silastic® concentrate caused no immediate local reaction. However, immediate itching of her hands and a delayed 10 mm raised reaction for 72 hours at the #1 dilution site did occur. She also had transient reactivation of the contact urticaria, which cleared within 7 days.

   Because of these progressive symptoms supported by the patch testing, the implant was removed on July 13, 1993. The Silastic® implant was replaced with cartilage from the opposite thyroid ala. At this time her voice is better than it has been since the thyroidectomy. She has nearly completely cleared the contact urticaria and has had no pruritus.

   We treated a similar patient who had replacement of one vocal cord with silicon. She developed hoarseness that could be stopped by the neutralizing dose injection of silicon.

Treatment of Environmentally Related ENT Diseases

   The evidence linking environmental factors to the onset of a large portion of ear, nose, and throat disease is compelling. Clinical studies with humans have linked a variety of environmental exposures to the onset of many problems in these areas. As well, studies with animals support clinical observations that olfactory sensitivity in chemically sensitive humans is caused by repeated exposure to pollutants.[152]

   Treatment for patients with chemical sensitivity with ear, nose, and throat involvement best utilizes a holistic approach that extends over a period of years. Although many measures presently used in conventional medical practice to treat these patients with ear, nose, and throat symptoms may be satisfactory on a short-term basis, the long-term outcome of these practices may prove to be detrimental. For example, initial success may be gained with standard injection therapy for pollen, dust, mold, and food sensitivities with patients appearing to improve for months or even several years following treatment. If, however, environmental control principles are violated and injection extracts contaminated with preservatives, this treatment may eventually result in patients' sudden deterioration. These patients may then develop more food sensitivity followed by chemical sensitivity. Although the cause of this spreading phenomenon is uncertain, it appears to result from an overload of synthetic chemicals due to a lack of reduction of the total load of chemicals taken into the body by way of air, food, and water and added to the total load present in injured tissue.

For the chemically sensitive patient with ear, nose, and throat involvement, treatment that fails to reduce the patient's total body load and routinely utilize other environmental-control practices is simply incompatible with long-term wellness. Clearly, the best procedure for dealing with these patients is to search for environmental triggers and treat them accordingly. Treatment aimed at total body load reduction by pollutant avoidance; injection therapy for biological inhalants, foods, and some chemicals; nutritional therapy; heat depuration; and the use of tolerance moderators is no different for these patients than for those suffering from a host of other environmentally-induced illnesses discussed throughout this book (see Volume IV).

REFERENCES

1. Ghantous, H. 1990. Accumulation and turnover of metabolites of toluene and xylene in nasal mucosa and olfactory bulb in the mouse. *Pharmacol. Toxicol.* 66(2):87–92.
2. Gilman, J. P. W. 1962. Metal carcinogenesis. II. A study on the carcinogenic activity of cobalt, copper, iron, and nickel compounds. *Cancer Res.* 22:158.
3. Heath, J. C. and M. R. Daniel. 1964. The production of malignant tumours by nickel in the rat. *Br. J. Cancer* 18:261.
4. Heath, J. C. and M. Webb. 1967. Content and intracellular distribution of the inducing metal in the primary rhabdomyosarcomata induced in the rat by cobalt, nickel and cadmium. *Br. J. Cancer* 21:768.
5. Doll, R., L. G. Morgan, and F. E. Speizer. 1970. Cancers of the lung and nasal sinuses in nickel workers. *Br. J. Cancer* 24:623.
6. Pedersen, E., A. C. Högetveit, and A. Andersen. 1973. Cancer of respiratory organs among workers at a nickel refinery in Norway. *Int. J. Cancer* 12:32.
7. Torjussen, W., L. A. Solberg, and A. Högetveit. 1979. Histopathologic changes of nasal mucosa in nickel workers. A pilot study. *Cancer* 33:963.
8. Torjussen, W., L. A. Solberg, and A. Högetveit. 1979. Histopathological changes of the nasal mucosa in active and retired nickel workers. *Br. J. Cancer* 40:568.
9. Boysen, M., R. Puntervold, B. Schüler, and A. Reith. 1980. The value of scanning electron microscopic identification of surface alterations of nasal mucosa in nickel workers: A correlated light and scanning electron microscopic study. In *Nickel Toxicology*, eds. S. S. Brown and F. W. Sunderman, Jr. 39–42. London: Academic Press.
10. Acheson, E. D., R. H. Cowdell, E. Hadfield, and R. G. Macbeth. 1968. Nasal cancer in woodworkers in the furniture industry. *Br. Med. J.* 2:587.
11. Hadfield, E. H. 1970. A study of adenocarcinoma of the paranasal sinuses in woodworkers in the furniture industry. *Ann. R. Coll. Surg. Engl.* 46:301.
12. Andersen, H. C., J. Solgaard, and I. Andersen. 1976. Nasal cancer and mucus-transport rates in woodworkers. *Acta Otolaryngol. (Stockh)* 82:263.
13. Acheson, E. D., R. H. Cowdell, and E. H. Rang. 1981. Nasal cancer in England and Wales: An occupational survey. *Br. J. Ind. Med.* 38:218.

14. Irander, K., H. B. Hellquist, C. Edling, and L. M. Ödkvist. 1980. Upper airways problems in industrial workers exposed to oil mist. *Acta Otolaryngol. (Stockh.)* 90:452.

15. Cohr, K.-H. and J. Stokholm. 1979. Toluene. A toxicologic review. *Scand. J. Work Environ. Health* 5:71.

16. Ödkvist, L. M., C. Edling, and H. Hellquist. 1985. Influence of vapours on the nasal mucosa among industry workers. *Rhinology* 23:121–127.

17. Ödkvist, L. M. 1988. Personal communication.

18. Broder, I., P. Corey, P. Cole, and M. Lipa. 1988. Comparison of the health of occupants and characteristics of houses among control homes and homes insulated with urea formaldehyde foam. II. Initial health and house variables and exposure-response relationships. *Environ. Res.* 45:156–178.

19. Holmstrom, M., B. Wilhelmsson, H. Hellquist, and G. Rosen. 1989. Histological changes in the nasal mucosa in persons occupationally exposed to formaldehyde alone and in combination with wood dust. *Acta Otolaryngol. (Stockh)* 107:120–129.

20. Hayden, F. G., S. E. Mills, and M. E. Johns. 1983. Human tolerance and histopathologic effects of long-term administration of intranasal interferon-alpha2. *J. Infect. Dis.* 148:914–921.

21. Graham, D., F. Henderson, and D. House. 1988. Neutrophil influx measured in nasal lavages of humans exposed to ozone. *Arch Environ. Health* 43:228–233.

22. Scott, G. M., R. J. Philpotts, J. Wallace, C. L. Gauci, J. Greiner, and D. A. J. Tyrrell. 1982. Prevention of rhinovirus colds by human interferon alpha-2 from *Escherichia coli. Lancet* 2:186–188.

23. Merigan, T. C., S. E. Reed, T. S. Hall, and D. A. J. Tyrrell. 1973. Inhibition of respiratory virus infection by locally applied interferon. *Lancet* 1:563–567.

24. Scott, G. M., R. J. Phillpotts, J. Wallace, D. S. Secher, K. Cantell, and D. A. J. Tyrrell. 1982. Purified interferon as protection against rhinovirus infection. *Br. Med. J.* 284:1822–1825.

25. Hayden, F. G. and J. M. Gwalteny, Jr. 1983. Intranasal interferon-alpha2 for the prevention of rhinovirus infections and illness. *J. Infect. Dis.* 148:543–550.

26. Holopainen, E. 1967. Nasal mucous membrane in atrophic rhinitis with reference to symptom free nasal mucosa: Histology, histochemistry, and exfoliative cytology. *Acta Otolaryngol. [Suppl.] (Stockh.)* 227:7–53.

27. Mygind, N. 1977. Effects of beclomethasone diproprionate aerosol on nasal mucosa. *Br. J. Clin. Pharmacol.* 4(Suppl. 3):295S-301S.

28. Poynter, D. 1977. Beclomethasone diproprionate aerosol and nasal mucosa. *Br. J. Clin. Pharmacol.* 4(Suppl. 3):295S-301S.

29. Muzaffar, M., I. A. Malik, M. Luqman, K. Ullah, and G. Nabi. 1990. Aspergillus granuloma presenting as recurrent nasal polypi. *Trop. Doct.* 20(2):95–96.

30. Kurnatowski, P. 1990. Various forms of mycoses of the oral cavity, pharynx, nasal cavity, paranasal sinuses, larynx and ear. *Otolaryngol. Pol.* 44(2):88–93.

31. Bucur, A. 1967. Mycotic rhinosinusitis in children. *Otorinolaringologie* 12(1):69–72.

32. Hellquist, H., K. Inrander, C. Eding, and L. M. Ödkvist. 1983. Nasal symptoms and histopathology in a group of spray-painters. *Acta Otolaryngol.* 96:495–500.

33. Derrick, C. 1989. Personal communication.

34. Ruokeonen, J., E. Holopainen, T. Palva, and A. Backman. 1981. Secretory otitis media and allergy. *Allergy* 36:59–68.

35. Williams, R. I. 1978. Hypersensitivity problems in otorhinolaryngology. *Ann. Otol. Rhinol. Laryngol.* 87:670–674.

36. Clemis, J. D. 1976. Identification of allergic factors in middle ear effusions. *Ann. Otol. Rhinol. Laryngol.* 23:234–237.

37. Katusic, S. K., C. M. Beard, W. C. Wiederholt, E. J. Bergstralh, and L. T. Kurland. 1986. Incidence, clinical features, and prognosis in Bell's palsy, Rochester, Minnesota. 1968–1982. *Ann. Neurol.* 20(5):622–627.

38. Rubin, W., S. N. Busis, and K. H. Brookler. 1989. Otoneurologic examination. In *Otolaryngology.* 1–47. Philadelphia: J. B. Lippincott Co.

39. Larsby, B., L. M. Ödkvist, D. Hyden, and S. R. C. Liedgren. 1976. Disturbances of the vestibular system by toxic agents. *Acta Phys. Scand. [Suppl.]* 440:108.

40. Larsby, B., R. Tham, L. M. Ödkvist, D. Hyden, I. Dunnfors, and G. Aschan. 1978. Exposure of rabbits to styrene electronstagmographic findings correlated to the styrene level in blood and cerebrospinal fluid. *Scand. J. Work Environ. Health* 4:10.

41. Larsby, B., R. Tham, and L. M. Ödkvist. 1978. Exposure of rabbits to methylcholoroform: Vestibular disturbances correlated to blood and cerebrospinal fluid levels. *Int. Arch. Ocup. Environ. Health* 41:7.

42. Aschan, G., I. Bunnfors, D. Hyden, B. Larsby, L. M. Ödkvist, and R. Tham. 1977. Xylene exposure. Electronystagmographic and gas chromatographic studies in rabbits. *Acta Otolaryngol.* 84:370.

43. Ödkvist, L. M., B. Larsby, R. Tham, and G. Aschan. 1979. On the mechanism of vestibular disturbances caused by industrial solvents. *Adv. Oto-Rhino-Laryngol.* 25:167–172.

44. Ödkvist, L. M., B. Larsby, D. Hyden, G. Aschan, and R. Tham. 1977. Influence of industrial solvents on the balance system: an experimental animal model. *International Symposium on the Control of Air Pollution in the Working Environment.* 88–89. Stockholm.

45. Tham, R., B. Larsby, B. Eriksson, I. Bunnfors, L. M. Ödkvist, and S. R. C. Liedgren. 1982. Electronystagmographic findings in rats exposed to styrene or toluene. *Acta Otolaryngol.* 93:107–112.

46. Ödkvist, L. M., B. Larsby, J. M. Fredrickson, S. R. C. Liedgren, and R. Tham. 1980. Vestibular and oculomotor disturbances caused by industrial solvents. *J. Otolaryngol.* 9(1):53.

47. Gamberale, F. and M. Hultengren. 1972. Toluene exposure. II. Psychophysical findings. *Work Environ. Health* 9:131–139.

48. Gamberale, F. and M. Hultengren. 1983. Methylchloroform exposure. I. Psychophysiological functions. *Work Environ. Health* 10:82.

49. Axelsson, O., M. Hane, and C. Hogstedt. 1976. Neuropsychiatric ill health in workers exposed to solvents — a case-control study. *Lakartidningen* 73:322.

50. Browning, E. 1965. Xylene. In *Toxicity and Metabolism of Industrial Solvents,* ed. E. Browning. 78–89. New York: Elsevier.

51. Goldie, I. 1960. Can xylene (xylol) provide convulsive seizures? *Ind. Med. Surg.* 29:33.

52. Ödkvist, L. M., B. Larsby, R. Tham, H. Ahlfeldt, B. E. Andersson, and S. R. C. Liedgren. 1982. Vestibulo oculomotor disturbances in humans exposed to styrene. *Acta Otolaryngol.* 94:487–493.

53. Hydén, D., B. Larsby, H. Andersson, L. M. Ödkvist, S. R. C. Liedgren, and R. Tham. 1983. Impairment of visuo-vestibular interaction in humans exposed to toluene. *Otorhinolaryngology* 45:262–269.

54. Hogstedt, C., M. Hane, and O. Axelson. 1980. Diagnostic and health care aspects of workers exposed to solvents. In *Developments in Occupational Medicine*, ed. Carl Zenz. 249–258. Chicago: Year Book Medical Publishers, Inc.

55. Ödkvist, L. M., C. H. Edling, L. M. Bergholtz, B. Larsby, and R. Tham. June 11–14, 1984. Vestibulo-oculomotor and audiological disturbances in humans exposed to solvents and jet fuel. Presented in *International Conference on Occupational Health Problems in Aerospace Industry*. 25. Linköping, Sweden.

56. Rea, W. J. and D. M. Martinez. 1995. Use of computerized balance testing in chemically sensitive patients. *Clin. Ecol.* (in press).

57. Ödkvist, L. M., L. M. Bergholtz, H. Åhlfeldt, B. Andersson, C. Edling, and E. Strand. 1982. Otoneurological and audiological findings in workers exposed to industrial solvents. *Acta Otolaryngol. [Suppl.]* 386:249–251.

58. Rea, W. J. 1980. Review of cardiovascular disease in allergy. In *Bi-Annual Review of Allergy,* ed. C. A. Frazier. 282. Springfield, IL: Charles C Thomas.

59. Rubin, W. and K. H. Brookler. 1990. Etiologic diagnosis in neurotologic disease. *Otolaryngol. Head Neck Surg.* 103:693–694.

60. Rubin, W. 1989. Biochemical evaluation of the patient with dizziness. *Seminars in Hearing* 10(2):151–160.

61. Pan, Y., W. J. Rea, and D. M. Martinez. 1994. Unpublished data.

62. Rubin, W. 1990. Personal communication.

63. Boyles, J. H., Jr. 1985. Food allergy diagnosis and treatment. *Otolaryngol. Clin. N. Am.* 18:775–785.

64. Derebery, J. 1990. Personal communication.

65. Viscomi, G. J. and D. I. Bojrab. 1992. The use of electrocochleography to monitor antigenic challenge in Ménière's Disease. *Otolaryngol. Head Neck Surg.* 107(6 Pt1):733–737.

66. Duke, W. W. 1923. Ménière's Disease caused by allergy. *JAMA* 81:2179–2181.

67. Derlacki, E. L. 1968. The medical treatment of Ménière's disease. *Otolaryngol. Clin. N. Am.,* 519–529. Philadelphia: Saunders.

68. Wilson, W. H. 1972. Antigenic excitation in Ménière's disease. *Laryngoscope* 82:1726–1735.

69. Clemis, J. D. 1967. Allergy of the inner ear. *Ann. Allergy* 27:370–378.

70. Powers, W. H. 1971. Allergic phenomena in the inner ear. *Otolaryngol. Clin. N. Am.* 4:557–563. Philadelphia: Saunders.

71. Clemis, J. D. 1972. Allergic cochleovestibular disturbances. *Tran. Am. Acad. Ophthalmol. Otolaryngol.* 1451–1457.

72. Harris, J. P., S. Tomiyama, S. Fukada, and M. Takahashi. 1989. The endolymphatic sac. In *Ménière's Disease*, eds. J. B. Nadol, Jr. Berkeley, CA: Kugler.

73. Mogi, G. and H. Kawaushi. 1987. Immunoglobulin and immune responses in the inner ear. In *Oto-immunology*, Berkeley, CA: Kugler.

74. Miyamura, K., Y. Kanzaki, M. Nagata, and T. Ishikawa. 1987. Provocation of nystagmus and deviation by type I allergy in the inner ear of the guinea pig. *Ann. Allergy* 58:36–40.

75. Dauman, R., J. M. Aran, R. C. de Sauvage, and M. Portmann. 1988. Clinical significance of the summating potential in Ménière's disease. *Am. J. Otolaryngol.* 9:31–38.

76. Gibson, W. P. R., D. A. Moffat, and R. T. Ramsden. 1977. Clinical electrocochleography in the diagnosis and management of Ménière's disorder. *Audiology* 16:389–401.

77. Coats, A. C. 1981. The summating potential and Ménière's disease. *Arch. Otolaryngol.* 107:199.

78. Morrison, A., D. A. Moffat, and A. F. O'Conner. 1980. Clinical usefulness of electrocochleography in Ménière's disease: an analysis of dehydrating agents. *Otolaryngol. Clin. N. Am.* 11:703–721.

79. Goin, D. W., S. J. Staller, D. I. Asher, and R. E. Mischke. 1982. Summating potential in Ménière's disease. *Laryngoscope* 92:1388–1389.

80. Coats, A. C. and B. R. Alford. 1981. Ménière's disease and the summating potential, III: Effect of glycerol administration. *Arch. Otolaryngol.* 107:469.

81. Ferraro, J. A., I. K. Arenberg, and R. S. Hassanein. 1985. Electrocochleography and symptoms of inner ear dysfunction. *Arch. Otolaryngol.* 111:71–74.

82. King, W. P., W. A. Rubin, R. G. Fadal, W. A. Ward, R. J. Trevino, W. B. Pierce, J. A. Steward, and J. H. Boyles. 1988. Provocation-neutralization: a two-part study. Part I. The intracutaneous provocative food test: a multi-center comparison study. *Otolaryngol. Head Neck Surg.* 99(3):263–271.

83. King, W. P., R. G. Fadal, W. A. Ward, R. J. Trevino, W. B. Pierce, J. A. Stewart, and J. H. Boyles. 1988. Provocation-neutralization: a two-part study. Part II. Subcutaneous neutralization therapy: a multi-center study. *Otolaryngol. Head Neck Surg.* 99(3):272–277.

84. Arenburg, I. K., W. P. R. Gibson, and H. K. H. Bohlen. 1989. Improvements in audiometric and electrophysiologic parameters following nondestructive ear surgery utilizing a valved shunt for hydrops and Ménière's disease. In *Ménière's Disease*, ed. J. B. Nadol, Jr. 545–561. Berkeley, CA: Kugler.

85. Hoover, S. 1988. Ménière's migraine and allergy. In *Vertigo, Nausea, Tinnitus and Hypoacusia in Metabolic Disorders*, eds. C.-F. Claussen, M. V. Kirtane, and K. Schlitter. 293–300. New York: Elsevier Science Publishers B. V. (Biomedical Division).

86. Rea, W. J. 1992. *Chemical Sensitivity. Vol. I: Mechanisms of Chemical Sensitivity.* 47. Boca Raton, FL: Lewis Publishers.

87. Boyles, J. 1990. Personal communication.

88. Derebery, M. J. and S. Valenzuela. 1992. Ménière's syndrome and allergy. *Otolaryngic Allergy* 25(1):213–223.

89. Rubin, W. 1990. Personal communication.

90. Hoover, S. 1990. Personal communication.

91. Monro, J. 1988. Personal communication.

92. Monro, J. 1982. Food allergy in migraine. *Proc. Nutr. Soc.* 42:241.

93. Henry, C. H. and L. M. Wolford. 1993. Treatment outcomes for temporomandibular joint reconstruction after Proplast-Teflon implant failure. *J. Oral Maxillofac. Surg.* 51:352–358.

94. Helmy, E. S., R. Bays, and M. Sharawy. 1988. Osteoarthritis of the temporomandibular joint following experimental disc perforation in *Macaca fascicularis*. *J. Oral Maxillofac. Surg.* 46:979.

95. Eriksson, L. and P. Westesson. 1985. Long term evaluation of meniscectomy of the temporomandibular joint. *J. Oral Maxillofac. Surg.* 43:263.

96. Merrill, R. G. 1986. Historical perspectives and comparisons of TMJ surgery for internal derangements and arthropathy. *J. Craniomandib. Pract.* 4:74.

97. Bronstein, S. L. 1987. Retained alloplastic temporomandibular joint disk implants: A retrospective study. *Oral Surg. Oral Med. Oral Pathol.* 64:135.

98. Georgiade, N. 1962. The surgical correction of temporomandibular joint dysfunction by means of autogenous dermal grafts. *Plast. Reconstr. Surg.* 30:68.

99. Narang, R. and R. A. Dixon. 1975. Temporomandibular joint arthroplasty with fascia lata. *Oral Surg. Oral Med. Oral Pathol.* 39:45.

100. Perko, M. 1973. Indikationen und Kontraindikationen fur chirugische Eingriffe am Kiefergelenk. *Schweiz. Mschr. Zahnheilk* 83:73.

101. Markowitz, N. R., T. Patterson, and L. Caputa. 1991. A two-stage procedure for temporomandibular joint disc replacement using free pericranial grafts: a preliminary report. *J. Oral Maxillofac. Surg.* 49:476.

102. Zetz, M. R. and W. B. Irby. 1984. Repair of the adult temporomandibular joint meniscus with an autogenous dermal graft. *J. Oral Maxillofac. Surg.* 42:167.

103. Tucker, M. R., J. R. Jacoway, and R. P. White. 1986. Autogenous dermal grafts for repair of temporomandibular joint disc perforations. *J. Oral Maxillofac. Surg.* 44:781.

104. Feinberg, S. E. and P. E. Larsen. 1989. The use of a pedicled temporalis muscle-pericranial flap for replacement of the TMJ disc: preliminary report. *J. Oral Maxillofac. Surg.* 47:142.

105. Pogrel, M. A. and L. B. Kaban. 1990. The role of a temporalis fascia and muscle flap in temporomandibular joint surgery. *J. Oral Maxillofac. Surg.* 48:14.

106. Herbosa, E. G. and K. S. Rotskoff. 1990. Composite temporalis pedicle flap as an interpositional graft in temporomandibular joint arthroplasty. *J. Oral Maxillofac. Surg.* 48:1049.

107. Greenberg, S. A., J. S. Jacobs, and R. W. Bessette. 1989. Temporomandibular joint dysfunction: evaluation and treatment. *Clin. Plast. Surg.* 16:707.

108. Matukas, V. J. and J. Lachner. 1990. The use of autologous auricular cartilage for temporomandibular joint disc replacement: a preliminary report. *J. Oral Maxillofac. Surg.* 48:348.

109. Tucker, M. R. and I. M. Watzke. 1991. Autogenous auricular cartilage graft for temporomandibular joint repair. *J. Craniomaxillofac. Surg.* 19:108.

110. Longacre, J. J. and R. F. Gelby. 1952. Further observations on the use of autogenous graft in arthroplasty of the temporomandibular joint. *Plast. Reconstr. Surg.* 10:238.

111. Glahn, M. and J. E. Winther. 1967. Metatarsal transplants as replacement for lost mandibular condyle (3 years follow-up). *Scand. J. Plast. Reconstr. Surg.* 1:97.

112. Stringer, D. E., P. J. Boyne, and P. M. Scheer. 1987. Five-year experience using freeze-dried dura mater as a TMJ disc replacement. *J. Oral Maxillofac. Surg. (Supp. Abstr.)* 45: M1.

113. Hartog, J. M., A. B. Slavin, and S. N. Kline. 1990. Reconstruction of the temporomandibular joint with cryopreserved cartilage and freeze-dried dura: a preliminary report. *J. Oral Maxillofac. Surg.* 48:919.

114. Gordon, S. 1958. Surgery of the temporomandibular joint. *Am. J. Surg.* 95:263.

115. Hansen, W. C. and B. W. Deshazo. 1969. Silastic reconstruction of the temporomandibular joint meniscus. *Plast. Reconstr. Surg.* 43:388.

116. Sanders, B., F. A. Brady, and D. Adams. 1977. Silastic cap temporomandibular joint prosthesis. *J. Oral Surg.* 35:933.

117. Howe, D. J. 1979. Preformed silastic temporomandibular joint implant. *J. Oral Surg.* 37:59.

118. Christenson, R. W. 1964. Mandibular joint arthrosis corrected by insertion of a cast vitallium glenoid fossa prosthesis: a new technique. *Oral Surg. Oral Med. Oral Pathol.* 17:712.

119. Cook, H. P. 1972. Teflon implantation in temporomandibular arthroplasty. *Oral Surg. Oral Med. Oral Pathol.* 33:706.

120. House, L. R., D. H. Morgan, W. P. Hall, and S. J. Vamvas. 1984. Temporomandibular joint surgery: results of a 14-year joint implant study. *Laryngoscope* 94:534.

121. Kent, J. N., M. S. Block, C. A. Homsy, J. M. Prewitt, III, and K. Reid. 1986. Experience with a polymer glenoid fossa prosthesis for partial of total temporomandibular joint reconstruction. *J. Oral Maxillofac. Surg.* 44:520.

122. Eriksson, L. and P. L. Westesson. 1986. Deterioration of temporary silicone implant in the temporomandibular joint: a clinical and arthroscopic follow-up study. *Oral Surg. Oral Med. Oral Pathol.* 62:2.

123. Dow Corning Corp. 1985. Silastic temporomandibular joint implant HP (Wilkes design). Midland, MI: Dow Corning Corp. (product information sheets).

124. Homsy, C. A., T. E. Cain, F. B. Kessler, M. S. Anderson, and J. W. King. 1972. Porous implant systems for prosthesis stabilization. *Clin. Orthoped.* 89:220.

125. Homsy, C. A. Jan. 1984. Polymers and metals for alloplastic use. Presented at AAOMS Clinical Congress, San Diego, CA. (abstr)

126. Dusek, J., J. N. Kent, and P. Smith. Sept. 1978. Proplast-Teflon implants for treatment of TMJ degenerative disease. Presented at the AAOMS 60th Annual Meeting, Chicago, IL. (abstr)

127. Malloy, R. B., J. N. Kent, P. Smith, and A. Staples. Sept. 1981. The treatment of TMJ arthrosis with Proplast-Teflon implants. Presented at AAOMS 63rd Annual Meeting. Washington, D.C. (abstr)

128. Gallagher, D. M. and L. M. Wolford. 1982. Comparison of Silastic and Proplast implants in the TMJ after condylectomy for osteoarthritis. *J. Oral Maxillofac. Surg.* 40:627.

129. Carter, J. B., M. S. Goldwasser, and C. O. Taylor. 1983. Meniscectomy in the management of chronic internal derangements of the TMJ. Presented at the AAOMS 65th Annual Meeting. (abstr)

130. Kiersch, T. A. 1984. The use of Proplast-Teflon implants for meniscectomy and disc repair in the temporomandibular joint. Clinical Congress on Reconstruction with Biomaterials. San Diego, CA.

131. Bee, D. E. and D. Zeitler. Sept. 1986. The Proplast-Teflon implant in TMJ reconstruction following meniscectomy. Presented at the AAOMS 68th Annual Meeting, New Orleans, LA. *J. Oral Maxillofac. Surg. (Supp. Abstr.)* 44:M1.

132. Vitek, Inc. 1986. Survey of TMJ surgical results. Houston, TX: Vitek, Inc.

133. McBride, K. L. 1986. Clinical behavior of synthetic meniscus substitutes. Presented at the AAOMS 68th Annual Meeting, New Orleans, LA. *J. Oral Maxillofac. Surg.* 44:M1. (transcript)

134. Florine, B. L., D. Gatto, and M. Wade. 1986. Pre- and postsurgical tomographic analysis of TMJ arthroplasties. Presented at the AAOMS 68th Annual Meeting, New Orleans, LA. *J. Oral Maxillofac. Surg. (Suppl. Abstr.)* 44:M3.

135. Wade, M. L., D. Gatto, and B. Florine. 1986. Assessment of Proplast implants and meniscoplasties as TMJ surgical procedures. Presented at the AAOMS 68th Annual Meeting, New Orleans, LA. *J. Oral Maxillofac. Surg. (Suppl. Abstr.)* 44:M3.

136. Mosby, E. L., J. D. Wagner, and R. C. Robinson. 1988. Assessment of Teflon-Proplast implants in temporomandibular joint reconstruction following meniscectomy. Presented at the AAOMS 70th Annual Meeting, Boston, MA. *J. Oral Maxillofac. Surg. (Suppl. Abstr.)* 46:M3.

137. Homsy, C. A. Oct. 31, 1986. Important information on temporomandibular joint surgery — letter of 1986. Vitek, Inc.

138. Turlington, E. G. and S. R. Welch. Sept. 1986. Foreign-body reaction to Teflon–Proplast fossa implants in TMJ arthroplasty. Presented at the AAOMS 68th Annual Meeting, New Orleans, LA. *J. Oral Maxillofac. Surg. (Suppl. Abstr.)* 44:M3.

139. Wade, M. L., D. J. Gatto, and B. Florine. Sept. 1987. Assessment of discoplasties and disectomies with alloplastic implants as TMJ surgical procedures. Presented at the AAOMS 69th Annual Meeting, Anaheim, CA. *J. Oral Maxillofac. Surg. (Suppl. Abstr.)* 45:M2.

140. Heffez, L., M. F. Mafee, H. Rosenberg, and B. Langer. 1987. CT evaluation of TMJ disc replacement with a Proplast-Teflon laminate. *J. Oral Maxillofac. Surg.* 45:657.

141. Florine, B. L., D. J. Gatto, M. L. Wade, and D. E. Waite. 1988. Tomographic evaluation of temporomandibular joints following discoplasty or placement of polytetrafluorethylene implants. *J. Oral Maxillofac. Surg.* 46:183.

142. Valentine, J. D., Jr., B. E. Reiman, E. A. Beuttenmuller, and M. G. Donovan. 1989. Light and electron microscopic evaluation of Proplast II TMJ disc implants. *J. Oral Maxillofac. Surg.* 47:689.

143. Lagrotteria, L. B., R. Seapino, A. S. Granston, and D. J. Felgenhaue. 1986. Patient with lymphadenopathy following temporomandibular joint arthroplasty with proplast. *J. Craniomandib. Pract.* 4:172.

144. Berarducci, J. P., D. A. Thompson, and R. B. Scheffer. 1990. Perforation into middle cranial fossa as a sequel to use of a Proplast-Teflon implant for temporomandibular joint reconstruction. *J. Oral Maxillofac. Surg.* 48:496.

145. Wagner, J. D. and E. L. Mosby. 1990. Assessment of Proplast-Teflon disc replacements. *J. Oral Maxillofac. Surg.* 48:1140.

146. Yih, W. Y. and R. G. Merrill. 1989. Pathology of alloplastic interpositional implants in the temporomandibular joint. *Oral Maxillofac. Surg. Clin. N. Am.* 1:415.

147. Timmis, D. P., S. B. Aragon, J. E. Van Sickels, T. Aufdemorte. 1986. Comparative study of alloplastic material for temporomandibular joint disc replacement in rabbits. *J. Oral Maxillofac. Surg.* 44:541.

148. Neel, H. B. 1983. Implants of Gore-Tex: comparisons with Teflon-coated polytetrafluoroethylene carbon and porous polyethylene implants. *Arch. Otolaryngol.* 109:427.
149. El-Deeb, M. and R. E. Holmes. 1989. Zygomatic and mandibular augmentation with Proplast porous hydroxyapatite in rhesus monkeys. *J. Oral Maxillofac. Surg.* 47:480.
150. Ryan, D. E. 1989. Alloplastic implants in the temporomandibular joint. *Oral Maxillofac. Surg. Clin. N. Am.* 1:427.
151. Hunsaker, D. H. 1994. Personal communication.
152. Roper, S. D. and J. Atema. 1987. Olfaction and taste IX. *Ann. NY. Acad. Sci.* 510:1–2.

19

Pollutant Injury to the Lower Respiratory Wall, Chest Wall, and Breast

RESPIRATORY TREE—INTRODUCTION

Pollutant injury to the lower respiratory tree, including the trachea, the bronchi, and the alveoli, may be a little different from injury to other organs where direct contact with pollutants does not occur. Because areas of the lower respiratory tree undergo direct contact with pollutants, changes to local membrane barriers are of primary importance. The neurovascular responses, though important, are possibly secondary to the local mucosa membrane changes. The lower respiratory tree is the area where chemical sensitivity is frequently seen following direct exposure to pollutants. In contrast, changes in the chest wall, which does not come into direct contact with pollutants, result from some local, but mostly systemic, irritation.

Principles Involved in Pollutant Damage to the Respiratory Tree in Chemical Sensitivity

Principles involved in pollutant damage to the respiratory tree include total body load, the adaptation phenomenon, bipolarity, and biochemical individuality of response, as well as the spreading phenomenon and switch phenomenon (see Chapter 3 in Volume I). The reason a few of these principles will be reviewed and emphasized in this chapter is that the respiratory mucosa is so often directly exposed to pollutants and, therefore, affords us a unique opportunity to monitor closely its effects and to prevent pollutant-induced end-organ failure. Further, the respiratory tree can function as an early warning system. Symptoms manifesting there indicate that pollutant exposure has occurred, and this warning gives the individual an opportunity to avoid additional exposure with resultant spreading of sensitivities to distal internal end-organs.

Daily, all people encounter pollutant mixtures that go directly to the lungs.[2] Although these exposures are difficult to avoid as they involve fumes

1195

from such commonplace items as combustion products, gasoline, perfumes or colognes, cleaning materials, environmental tobacco smoke, and glues, paints, and finishes used on new furnishings, individual responses to them differ greatly. Many individuals, for instance, recognize that they have experienced a chemical/pollutant exposure, but they do not experience immediate symptoms since their total body load has not been exceeded. They are readily able to adapt to or clear the pollutant from their system. For others, however, the exposures are unpleasant (but only mildly), alter their health (but only temporarily), and may be avoided (when it is convenient to do so). For still others, encounters with various common pollutant mixtures become major life events.[3] For these people, pollutant exposures are uncomfortable, alter perceived health markedly, and must be avoided or neutralized for the individual to maintain a sense of good health. Without appropriate intervention, these individuals will develop chemical sensitivity, and eventually end-organ failure will occur. These various responses seem largely dependent on the functionality of an individual's adaptive and clearing mechanisms as well as his total body load at the time of exposure. For those individuals whose clearing mechanisms predominate, exposure to these pollutant mixtures causes no immediate health problems. If the adaptive mechanisms predominate, however, disease eventually occurs. (The time of onset of disease depends on the individual biochemical response of each person.)

If the clearing mechanisms of the chemically sensitive individual are unable to keep up with the total pollutant load, exposure to pollutants may result in chronic respiratory health problems with chemical sensitivity. Differential biochemical susceptibility of response to chronic pollutant exposure occurs (Chapter 3 in Volume I). The best-known clinical example of differential biochemical individuality (susceptibility) is the varying responses of individuals exposed to tobacco smoke. While a minority of smokers develop lung cancer or chronic obstructive airways disease, others develop cardiovascular or skin disease.[4,5] Exposure to low levels of environmental tobacco smoke is also associated with a range of health hazards[6] including increased respiratory symptoms and illness, decreased lung function in children,[7,8] and increased airflow limitation in adults.[9] Constant exposure to tobacco smoke may result in chemical sensitivity in some patients while others, whose adaptive mechanisms may cover up the acute clinical response that would be seen in chemically sensitive individuals, may develop different health problems such as cardiovascular and skin disease. Therefore, to understand early pollutant-induced injury and to prevent permanent damage, it is essential to first understand the individual biochemical response to exposures to individual pollutants and/or mixtures in the respiratory tree. One must also understand the relationship between acute responsiveness to a pollutant(s) and the risk of chronic disease when the adaptive process is triggered and then breaks down, a process that is often seen in the chemically sensitive individual.

Some investigators seem to recognize that the adaptive process masks sensitivity and deal with it, while others do not. For example, Swedish investigators Berglund and Lindvall[10] recommend that "priority should be given to the protection of the sensitive part of the occupant population" in a building and that "sensory irritants should never exceed the 10th percentile for detection." In contrast, the American Conference of Governmental and Industrial Hygienists, which establishes exposure guidelines for the occupational environment in the U.S., does not attempt to eliminate unpleasant smells and mild mucosal or skin irritation. They appear to be satisfied with an increase in total body load. In fact, their policies encourage some pollution, though probably inadvertently, by failing to correct the conditions that create indoor air pollution. They especially ignore those principles of outgassing and mainly use dilution ventilation as the prevention to indoor air pollution. As a cure-all to produce clean indoor air, this precept is totally unfounded and cannot be supported by any data. This organization seems to recommend diluting indoor air with potentially polluted outside air in order to minimize the indoor pollutants. They fail to advocate elimination of the worst indoor outgassers.

The Federal Hazardous Substances Act of 1967 defined irritants and corrosives as follows: a *corrosive* is any substance that when in contact with living tissue causes destruction of the tissue by chemical action, which is seen to occur with pollutant injury to the lower respiratory tree. In contrast, an *irritant* is defined as any noncorrosive substance that on immediate, prolonged, or repeated contact with normal living tissue induces a local inflammatory reaction[11] or soreness.[12] This type of injury is more often seen in the lower respiratory tree of the chemically sensitive individual. These definitions are consistent with the principles laid out in Chapter 3 in Volume I. The inclusion of "soreness" in the latter definition acknowledges that symptoms are often used to recognize or describe irritation.

The association between a pollutant exposure and the respiratory response of an individual in the adapted state is usually made only when the irritation is severe and the effect immediate (usually indicating that the adaptive mechanism has broken down), the odor is characteristic, or when a recurrent pattern points to the pollutant as the source. For example, an individual working in an environment with low levels of solvents experiences mild symptoms of cough and mucus production while in the workplace. Though irritating, these symptoms are easily ignored because they are intermittent at work and dissipate when the individual is away from his or her work environment. The worker only perceives the relationship between these two events when a large chemical spill (e.g., trichloroethane) occurs and his or her total body load increases to the point that his or her adaptive process breaks down. At this time, the person becomes aware that the chemical has a distinct odor and simultaneous with this discovery realizes that his or her symptoms are recurring and intensifying. Thus, the individual understands that it is this specific pollutant that is triggering his or her lower respiratory symptoms of cough and mucus.

The masking (adaptation) phenomenon is rarely perceived by the clinician, unless he or she has been educated about this process (see Chapter 3 in Volume I). Also, the clinician can easily overlook pollutant damage to the respiratory system because end-organ responses are limited in number and, thus, have to be differentiated through a process of withdrawal and challenge. For example, Bascom[13] has pointed out that nasal irritation may occur as an isolated symptom or with associated drying of the airways, which is often seen in some chemically sensitive patients. However, the symptom "nasal irritation" often occurs in conjunction with rhinitis, symptoms of rhinorrhea, and congestion (see Chapter 18). The symptomatic response to a pollutant may, therefore, resemble the response resulting from other causes. Symptoms of rhinorrhea and congestion may arise not only from exposure to pollutants but also from allergic stimuli (pollen),[14] physical stimuli (cold air),[15] infectious causes (rhinovirus),[16] and psychogenic causes (e.g., rhinorrhea associated with crying). However, in our experience at the EHC-Dallas, pollutant exposure is the main cause of bronchopulmonary symptoms in the chemically sensitive individual. Some opinion suggests that the differences in the pathogenesis of rhinitis response to these aforementioned stimuli compared with the response to pollutants may become apparent only when the biologic response is characterized through the measurement of mediators and cells from lavage samples or analysis of biopsies (Figure 19.1). However, in our experience at the EHC-Dallas, mediator and lavage cell measurements are not as definitive as withdrawal of the pollutant and subsequent inhaled challenge 4 days later. Similar pathological responses and pathology have been observed in various etiologies and may not be definitive in our hands. Inhaled challenge similar to the type described in this book and performed in over 10,000 chemically sensitive patients and using 30,000 double-blind individual challenges at the EHC-Dallas under environmentally controlled conditions has been the single most informative laboratory test in defining the effects of pollutants upon the respiratory tree of the chemically sensitive individual. The results of a small series of 50 patients undergoing 200 individual double-blind inhaled challenges at the EHC-Dallas are shown in Table 19.1. It is clear in these patients that the chemical explored and the patient's individual response to it can be observed.

As evidenced in Table 19.1, individual responses to pollutant exposure can be demonstrated by challenge if the individual's biochemical individuality of response is taken into consideration. For example, one patient may have measurable changes in upper airway function, another in lower airways function, and a third may just have an observable cough. Still other patients may experience various organ-system effects without respiratory involvement, even though the same set of pollutants and same path of exposure are involved. Specific immunologic sensitization does not account for all of the variable responses, since some of these may be the result of direct mucosal triggering by the pollutant while others may be due to nonimmunologic enzymatic detoxification factors.

At the EHC-Dallas, we have extensively studied individuals exposed to chlorine, ethanol, pesticide, formaldehyde, and phenol using double-blind inhaled challenges after a period of deadaptation with the total load decreased. These studies are discussed in Chapter 31, and certainly they add to our present understanding and definition of pollutant-induced illness (Table 19.1). Aside from our work, the best studies on individual responses to outdoor and indoor

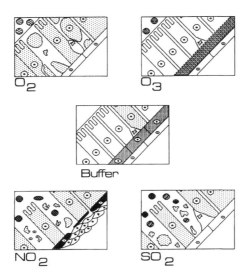

Figure 19.1. Types of pathology seen in pollutant injury.

Table 19.1.   50 Chemically Sensitive Patients (200 Individual
Challenges) Inhaled Double-Blind Challenge after 4 Days of
Deadaptation with Total Load Reduced in the ECU

Chemical	Dose/level (ppm)	Patients with positive reaction peak flow[a] decreased
Ethanol (petroleum derived)	<0.50	13
Phenol	<0.002	10
Chlorine	<0.33	6
Formaldehyde	<0.20	8
2,4-DNP (pesticide)	<0.0034	9
1,1,1-Trichloroethane	(ambient)	4
Placebo—saline		3[b]

[a]   Flow change over 2 standard deviations.

[b]   No patient reacted to more than one placebo.

Source: EHC-Dallas. 1988.

pollutant exposures are those focusing on aldehydes such as formaldehyde,[17-22] acetaldehyde and acrolein,[23,24] the criteria pollutant ozone,[25] sulfur dioxides,[26-28] and nitrogen dioxide.[29] The most commonly studied pollutant mixtures are mainstream and sidestream tobacco smoke.[7,30] Recent studies have addressed the health effects of human exposure to mixtures of volatile organic compounds.[31-33] The mixtures used in these studies were chosen as representative of air samples obtained from new buildings in Denmark. We elaborate on these studies in Chapter 10.[34] Notably, they complement those performed at the EHC-Dallas, emphasizing that combinations of pollutants do trigger symptoms that may or may not be similar.

Our studies, as well as those of others, emphasize that exposure to mixtures may yield substantially different results from exposure to single agents. However, our experience in challenging individuals in the deadapted state with their total load reduced reveals that a reasonably accurate evolution of their problem occurs regardless of the combination of pollutants tested. Studies by Amdur et al.[35] comparing the response in guinea pigs to inhalation of SO_2 alone with their response to inhalation of an SO_2/zinc oxide mixture are an excellent example of the combined challenge. These researchers found that sulfur dioxide caused a brief bronchoconstrictor response, while the addition of zinc oxide (which alone caused no effect) resulted in an extended physiologic response and markedly augmented inflammation and epithelial damage. The mixtures in these studies were carefully developed to mimic products of coal combustion. We have found that often it is dangerous to chemically sensitive individuals to initiate a severe reaction by challenge. Such testing tends to exacerbate their symptoms with long-term responses and slow recovery. Therefore, elements of the combination are often challenged singularly rather than using the combination itself for the challenge test. For example, phenol and chlorine may be challenged separately rather than using hexachlorophenol. However, if the patient has a strong history of response but does not respond to the single incitant, he or she then is challenged with the combination. Patients are always challenged in the deadapted state.

Different responses in chemically sensitive individuals occur for a number of reasons (see Chapter 3 in Volume I). One reason is that individual responsiveness to pollutants has a genetic characteristic that shows variability (the genetic polymorphism — see Chapter 6, section on enzymes[36]). Another reason for different responses is acquired sensitivity that is dependent on nongenetic sensitization in utero or during extrauterine life when the total pollutant load at the time of exposure in relation to the state of nutrition is exceeded. Also, different levels of pollutant stimuli may trigger different degrees of intensity in the same area, thus causing various responses (see Chapter 3 in Volume I, for more information).

Historically, biochemical and biologic variability exists in the responsiveness of groups of healthy adults and chemically sensitive individuals to pollutants. Whether or not the more sensitive of the normals are the ones who develop chemical sensitivity is unclear at this time, but it is highly likely that they do develop sensitivities. Response variability can be measured and, in the case of ozone, it has been shown to persist over time. In a subgroup of chemically sensitive individuals studied at the EHC-Dallas, this variability is the case. More direct evidence of some genetic basis for this variability is available from exposure studies of inbred strains of mice and bacteria. This genetic susceptibility to pollutant exposure appears to be a "time bomb" awaiting the right set of environmental exposures to trigger its expression.

Once stimulated, this genetic susceptibility may result in chemical sensitivity, and, certainly, it can account for some of the variability of response observed in the chemically sensitive population.

Bascom[13] reported variability in individual responses to environmental tobacco smoke in a group of healthy young adults who were being screened for participation in an unrelated study at the University of Maryland Environmental Research Facility. These subjects were asked, incidentally, about their response to environmental tobacco smoke (ETS): 80% of the respondents reported ETS-associated eye irritation, 15% reported nasal irritation, and 33% reported at least one rhinitis symptom (rhinorrhea or congestion) after historical ETS exposure.[37] Speer[38] reported eye irritation in 69% of nonsmokers who were bothered by ETS. Other authors cite ranges of eye irritation of between 47 and 73%.[6]

Epidemiologic data on symptoms occurring in the indoor environment also suggest biologic variability in the occurrence of symptoms related to mucosal irritation. Workers at the U.S. Environmental Protection Agency's administrative buildings in Washington D.C. were asked about the occurrence of mucosal and respiratory symptoms during the year previous to their being questioned. The rating scale was "never", "rarely", "sometimes", "often", and "always". A distribution of responses occurred. The median response for burning eyes and dry throat was "rarely." The median response for headache, runny nose, and dry, itchy, or tearing eyes was between "rarely" and "sometimes". Between 9 and 26% of this population reported these symptoms occurred "often" or "always".[39] A second questionnaire was administered. In general, the high symptom areas continued to report higher symptom rates, while the low symptom areas continued to report lower symptom rates.[40] While the factors responsibile for the different symptom rates are still under investigation, the observation can be made that a range of pollutant-triggered symptoms occurs in a large workforce and that some evidence indicates that the symptom rates are stable over time.

As shown in Volume I, Chapter 3, biologic variability in individuals' responses to ozone is well documented in controlled human exposure studies, even at the current ambient air quality standard of 0.12 parts per million. These responses occurred after challenges in the adapted state and may account for additional variability, which may reflect a faulty adaptive mechanism. Exposure to 6.6 hours of 0.12 ppm ozone under conditions of intermittent moderate exercise caused decrements in FEV_1 in healthy young adults that averaged 13% but ranged up to 48%.[41] Symptoms of cough and chest tightness correlated with the decrement in FEV_1 following ozone exposure.[42] The interindividual variability in acute pulmonary function responsiveness to ozone was highly correlated over time,[43] suggesting stable differences in host susceptibility whether it be actual resistance or adaptability. The relationship between this variable responsiveness and susceptibility to chronic effects such as pulmonary fibrosis is unknown. However, clinical assessment makes clear that those with chemical sensitivity are more obviously susceptible. Further, even if those who eventually develop fibrosis manifest early signs of clinical chemical sensitivity, these may go unrecognized by the patient and physician alike. Much too frequently, the diffuse signs and symptoms of chemical sensitivity are not recognized precisely becaue of their variable and seemingly unrelated manifestation.

Direct evidence of a genetic basis for some variability in responsiveness to ozone has been obtained using studies of inbred strains of mice. However, this information does not exclude a nutritional and acquired basis in another group of patients. Studies by Kleeberger et al.[44] demonstrated strain differences in the inflammatory response to acute ozone exposure. According to Bascom,[13] a single autosomal recessive gene at the inf locus confers susceptibility to acute O_3^--induced influx of neutrophils in mice, but the genetic control of permeability is less clear. The genetics of the physiologic airway response to ozone have not been determined.

According to Bascom,[13] investigation into the genetics of biologic variability in the response to oxidants is also being pursued in strains of bacteria. In these studies, the end-point that is used is cell death, and strains have been identified that are both resistant and susceptible to oxidant-induced cell death. A single gene, the *oxyR* gene, has been identified that positively regulates the expression in at least one hydrogen peroxide-inducible protein. Therefore, *oxyR* mutant alleles alter bacterial sensitivity to oxidants.[45,46] These data and ours confirm that lower respiratory damage in the chemically sensitive population is a function of the total pollutant load, the genetic and acquired biochemical individuality of response, and the state of nutrition at the time of exposure(s) (see Volume I, Chapter 3).

Bascom[37] investigated environmental tobacco smoke (ETS) in healthy young adults. She recruited individuals at either end of the spectrum of responsiveness to ETS. Subjects were asked to rate their history of ETS rhinorrhea and nasal congestion, each on a 0 to 5 scale. An "historical rhinitis index" was calculated as the sum of the two symptoms. Subjects with a score of ≤ 1 were considered ETS-nonsensitive (ETS-NS, N = 11), while subjects with a score of ≥ 3 were considered ETS-sensitive (ETS-S, N = 10). Subjects' atopic status was characterized by skin prick tests to a standard battery of common environmental allergens, and ETS-S subjects were more likely to be atopic. Subjects then underwent exposure in an environmental chamber for 15 minutes each to clean air (2°C, 40% relative humidity) followed by sidestream tobacco smoke with a carbon monoxide concentration of 45 parts per million. Measures of response were symptoms, posterior nasal resistance, and spirometry. The protocol was repeated on a second study day with symptoms and nasal lavage as the measures of response. Nasal lavage fluid was analyzed for the concentration of histamine, albumin, and kinins and for TAME-esterase activity.

Symptom responses on the two study days were significantly correlated. The perception of odor following tobacco smoke exposure and the symptom of eye irritation were the same in the two groups (ETS-S and ETS-NS). Nasal congestion, headache, chest tightness, and cough were symptoms that increased in the ETS-S group, but not in the ETS-NS group. Rhinorrhea and nose–throat irritation increased in both groups but were greater in the ETS-S group.

Baseline nasal resistance was the same in the ETS-S and ETS-NS subjects and did not change with exposure to clean air. Exposure to tobacco smoke resulted in a significant increase in nasal resistance in the ETS-S subjects (4.2 ± 2.2 cm $H_2O/l/sec$, $p<0.001$ vs. pre-exposure), but no change was observed in the ETS-NS subjects. No change occurred in nasal lavage mediators, including histamine, albumin, and kinin concentrations and TAME-esterase activity of ETS-NS. These data provided subjective and objective evidence of a differential upper respiratory response to sidestream tobacco smoke. The nasal lavage data indicated that activation of mast cells was not occurring, since no elevation of histamine was observed. Also, no increase in vascular permeability occurred, as indicated by the lack of an increase in albumin or kinins. Increased TAME-esterase activity in nasal lavage fluid can occur as a result of either the influx of plasma kallikrein or the release of glandular kallikrein or mast cell tryptase.[47] The absence of an increase in TAME-esterase activity in this study suggested that vascular congestion was responsible for the increase in nasal resistance.

Animal studies have indicated that the acute response to tobacco smoke is mediated by activation of capsaicin-sensitive c-fiber neurons. From these findings, Bascom hypothesized that differential responsiveness to environmental tobacco smoke reflected differential responsiveness to c-fiber stimulation. Therefore, she compared the symptomatic and nasal lavage mediator response to intranasal capsaicin in her ETS-S and ETS-NS subjects.[48] She challenged the subjects intranasally using a delivery system that was a nasal speculum in series, a Spinhaler® and a French-Rosenthal Dosimeter®. Subjects inhaled 25 mg lactose powder followed by 25 pg to 25 ng capsaicin diluted in 25 mg lactose. Subjects rated nasal symptoms and underwent nasal lavage with measurement of mediators. ETS-S subjects reported significantly more rhinorrea than ETS-NS subjects (those without a history of ETS-rhinitis). No significant increase occurred in nasal lavage histamine, albumin, or kinins. TAME-esterase activity rose 1000 cpm in 12/21 subjects but was unrelated to ETS-sensitivity. These data suggest that the symptom of ETS-rhinorrhea in humans is related to c-fiber stimulation. The mediator data indicate that this concentration of capsaicin causes symptoms and, in some subjects, glandular stimulation. There was no evidence of increased vascular permeability or mast cell activation.

Two other investigators have quantitated significant increases in nasal secretion induced by much higher doses of capsaicin in subjects with a history of watery vasomotor rhinitis.[49,50] These investigators did not characterize the components of the secretions. Stjarne et al.[49] were able to block the rhinorrhea with a combination of intramuscular and intranasal anticholinergics, indicating a role for cholinergic neurons in the response. These latter studies are included in this chapter rather than Chapter 18 because they have wide application for the tracheal bronchial tree.

The lung is capable of inhaling 3 tons of air per year. In the process, the lung takes into the body approximately 3 gms of toxic environmental substances and probably much more in areas of high pollution. This intake adversely effects chemically sensitive individuals by increasing their total body load. Human and animal studies clearly demonstrate that both natural and manmade environmental chemicals can damage the body's systems, including the enzyme detoxification,[51] bactericidal,[52] and others[53] that are involved in disease resistance (see Volume I, Chapter 4[54]). Breakdown of these systems results in disease states such as bronchitis, bronchiectasis, emphysema, pulmonary fibrosis, asthma, lung failure, and cancer. Clinical evidence indicates that chemical sensitivity often is an early entity in any of these conditions. Substantial evidence has accumulated showing that pollutant penetration or creation of free radicals in mucosa that has been previously damaged may allow for long-term disease states to occur.

The mechanisms in the lung that, when damaged, result in individual sensitivity to pollutant exposure are now being defined, allowing us a better understanding of those chemically sensitive individuals who have respiratory involvement. For example, influenza pneumonia, with or without the use of oxygen or exposure to pesticides, has been shown to change lung sensitivity.[55,56] In one study,[57] the properties of some constitutive and inducible enzyme activities of liver and lung microsomes were determined in B6C3F1 mice pretreated by either intratracheal (i.t.) administration of benzo[a]pyrene (BaP) or polychlorinated biphenyl (PCBs) mixture (Aroclor 1254), or intraperitoneal (i.p.) administration with Aroclor 1254. After i.p. administration of Aroclor 1254, liver cytochrome P-450 content, aryl hydrocarbon hydroxylase (AHH), benzphetamine N-demthylase, and nitroreductase activities increased 2.8-, 2.0-, 2.2-, and 2.0-fold, respectively. Lung cytochrome P-450 content also increased (1.9-fold) after i.p. administration of Arochlor 1254; AHH and nitroreductase activities, however, were not affected, and benzphetamine N-demethylase activity decreased. Aroclor 1254 administered i.t. did not affect liver cytochrome P-450 content. However, AHH and benzphetamine N-demethylase activities decreased 1.4- and 1.2-fold, respectively, and nitroreductase activity increased 1.6-fold. After i.t. adminstration of Aroclor 1254, lung cytochrome P-450 content and AHH activity increased 1.4- and 2.2-fold, respectively. Benzphetamine N-demethylase activity decreased 2.1-fold, and nitroreductase activity was not affected. After i.t. administration of BaP, liver 7-ethoxyresorufin O-deethylase and nitroreductase activities increased 2.2- and 1.5-fold, respectively, and benzphetamine N-demethylase activity decreased 1.3-fold. Lung AHH and 7-ethoxyresorufin O-deethylase activities increased 4.3- and 3.1-fold, respectively, and cytochrome P-450 content, benzphetamine N-demethylase activities decreased 1.4-, 1.2-, and 1.3-fold, respectively, after BaP administration. These data indicate that different cytochrome P-450 isozymes induced in B6C3F1 mice are responsible for monooxygenase and nitroreductase activities and that the route of administration of chemicals is important in the expression of cytochrome P-450 catalyzed activities. These enzymes appear to be the enyzmes of adaptation. Neurogenic change can also produce similar results.[58]

With either large, acute exposures or smaller ones that continue over time, a decreased capacity for destroying toxic chemicals and/or repairing injured tissue occurs. Thus, with either type of pollutant exposure, lung damage with increased sensitivity may result. Conversely, pollutant injury may yield an initial, apparently diminished, clinical sensitivity as a result of an acute toxicological tolerance (adaptation or masking) phenomenon. However, once unmasked, an increased sensitivity declares itself and is obvious. This stage of the disease progression sometimes is necessary for acute survival. However, long-term debits occur with chronic exposure if masking (adaptation) is present, and the end result is lung failure (see Volume I, Chapter 3[1]). This masking phenomenon must be constantly sought, and pollutant triggers must be eliminated in early chemical sensitivity in order to prevent end-stage disease.

As discussed in Chapter 3,[1] Selye[59] originally described this general adaptation phenomenon. Adolph[60] demonstrated its occurrence in specific animals. Randolph[61] outlined specific adaptation in humans. Stokinger and Coffin[62] utilized ozone challenge to demonstrate specific adaptation of the lung in humans (see Volume I, Chapter 3[1]). Other studies have found that, although initial daily exposures to pollutants may decrease pulmonary function from 15 to 20%, by the fourth day subjects recovered to control levels[63-65] (see Volume I, Chapter 3[1]). These studies effectively demonstrated the adaptation phenomenon. Later biopsy studies confirmed the pollutant damage.[66]

Responsiveness to pollutants may be modified by environmental factors. This alteration is seen in the subset of chemically sensitive patients who have respiratory tree involvement. There is solid experimental evidence that the human response to pollutants is not fixed and that the previous exposure to a pollutant will influence the response to subsequent exposure. The key point, however, is that the relationship is not simple. As stated by Storz et al.,[45] "The mechanisms by which cells sense environmental adversity and then transduce the stress signals into a change in action are known for only a limited number of responses." The nature of the modification is influenced by the dose and duration of the original exposure, the dose and duration of the subsequent exposure, and the interval between the initial and subsequent exposures. Examples of these modifications follow.

Studies show that exposure to ozone results in an acute decrease in pulmonary function, which resolves in a matter of hours. Repeat exposure to the same concentration of ozone, 24 hours later, will result in a less significant decrease in pulmonary function, indicating so-called "ozone adaptation" has taken place.[67] If 1 week elapses between two exposures, and the total pollutant load decreases and the patient deadapts, the second exposure will mimic the first, producing a response of the original magnitude.[68] Throughout this book we emphasize the importance of applying these principles of adaptation for accurate diagnosis and treatment of pollutant damage. This "adaptation" does not usually represent the loss of responsiveness through injury but rather an augmentation of a temporary, protective mechanism for acute survival (e.g., antioxidant enzymes increase) that initially effectively reduces the toxic stress on tissue but eventually results in end-organ failure (if pollutant exposure persists and nutrient deficiency with abnormal healing occurs). Demonstration of diversity in the acute response to ozone in an experimental setting, therefore, requires that subjects undergo time delays of several days between ozone exposures, since adaptation will blunt the acute response. The relationship between variability in the acute response to ozone and the occurrence of inflammation or longer-term health risks is unknown.[25] However, what has been shown with ozone, and chemicals generally, is that if the pollutant trigger is not withdrawn, eventual nutrient depletion occurs, and the adaptive response cannot occur. Thus, inflammation results.

In the occupational setting, the initial response to cotton dust has been observed to diminsh with recurrent exposure. The first clinical grade (C1) of byssinosis is tightness of the chest and/or difficulty breathing, which occurs on the first day only of the working week despite continued exposure throughout the work week.[69] The basis for this cross-week effect is unknown; however, endotoxin, the agent in cotton dust suspected to cause byssinosis,[70] clearly is capable of inducing inflammation. Adaptation without symptoms occurs by the second day if the pollutant is not withdrawn; however, inflammation will continue.

Other studies, such as those with capsaicin (active ingredient in red pepper), demonstrate the importance of dose, duration, and sequence and species of exposure on the response to an irritant. Neonatal treatment of rats with capsaicin will permanently induce selective degeneration of primary chemosensitive neurons for the life of the animal.[71] If a similar situation occurred in humans, some chemical sensitivity could occur. Subsequent exposure to capsaicin in these animals failed to elicit a clinical response. Similarly, exposure to an irritant mixture such as tobacco smoke will also fail to elicit edema.[72] Treatment of the adult animal with capsaicin will cause an acute inflammatory response at the site of exposure.[73] Repeated treatment will result in a gradual diminution of the initial inflammatory response and is associated with denervation.[71] This experimental outcome clearly demonstrates the effects of maladaptation when the total pollutant load is not decreased. Capsaicin in the nose of the rat or guinea pig will cause vascular extravasation,[71] while low-dose capsaicin in the nose of humans will cause glandular stimulation without an increase in vascular permeability.[48] The molecular mechanisms associated with these events are not yet understood. However, what has been observed in chemically sensitive patients is that if the early warning neuromechanism is damaged the patient continues to take in the pollutant, which will then damage the target organ in a deeper layer or allow damage to a distal organ to occur.

Adaptation (masking or acute toxicological tolerance) apparently occurs in these situations due to an acceleration of the metabolic function with eventual elimination of Type I columnar epithelial cells by pollutant stress.

Initially, as pollutants come in contact with the mucus, the cellular antipollutant enzymes, secretory IGG, and IGA are induced. For a finite period of time, these substances can bring about physiologic adaptation, which is reversible and not harmful if pollutants are withdrawn. At this point, the individual is well clinically. As injury progresses, actual cell damage occurs. These Type I columnar cells are replaced by Type II ciliated epithelial cells and interstitial and inflammatory cells. Eventually, if pollutant stimuli continues unaltered, the result is scar tissue[74-76] and other end-organ damage. Some pollutants produce proliferation of certain cellular growth, resulting in benign or malignant tumors. Pollutants such as ozone, nitrous oxide, sulfur dioxide, and petroleum derived hydrocarbons help create the free radical superoxide (O_2) or the OH radical.[77] Other free radicals may be generated from lipids and yield peroxides with further lung damage resulting. These free radicals, in turn, may destroy membranes (e.g., mitochondrial for formation and release of ATP). They may also release other toxins from lysosomes or other cellular bodies, such as peroxisomes, that are necessary for optimum function (see Volume I, Chapter 4[54]). Recurrence of this chain of events throughout the lung or in other areas of the body can easily bring about dysfunction and chemical sensitivity. When this process occurs repeatedly over many years, and if the overload of pollutants is not removed, the stage is set for fixed-named disease states.

Pollutant Injury to the Nervous System

Pollutant-induced injury to the nervous system of the respiratory tree can occur in the brain and in the respiratory center as well as in the spinal cord, anterior horn cells, peripheral nerves, and/or the ANS and neuroendocrine receptors. Damage to any of these sites can affect the respiratory response (Figure 19.2). Since nerves are made of fats, they are more prone to damage by solvents, many of which are found in the blood of the chemically sensitive patient. As shown in Volume II, Chapter 13,[78] pollutant injury to the nerves can be extremely selective, as it was in the patient we treated at the EHC-Dallas who had amyolateraltropic disease had developed respiratory failure. Early involvement of the ANS with pollutant injury is now well established (see all chapters in this volume). As can be seen in Table 19.2, dysfunction any place in the neuroresponses of the respiratory system may cause changes in respiration. Also, vascular dysfunction of the respiratory tree causes discomfort and dysfunction in the individual. This early discomfort of pollutant damage to the respiratory tree can be via the nervous system or the local mucosa. Early in the disease process, this injury is often misinterpreted as functional by many clinicians who ignore both the offending pollutants and the potential known neural anatomical pathways for response. If, instead, these pollutants were recognized and the individual tested for sensitivity to them, confirmation of an exposure that was causing illness could result. This misinterpretation leaves the affected individual to proceed on a path that eventually may lead to irreversible

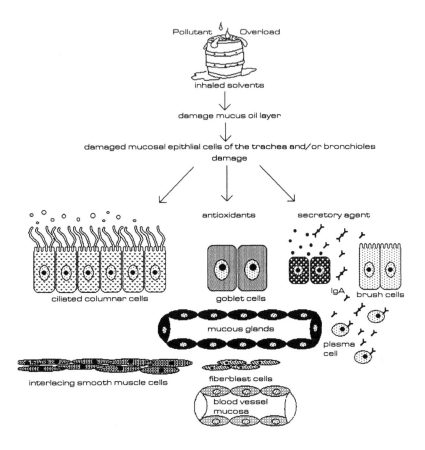

Figure 19.2. Progression of damage caused by unrestrained pollutant exposure.

Table 19.2. Potential Areas of Pollutant Injury to the Neural Responses of the Breathing Mechanism in the Chemically Sensitive

Pneumotactic center	Vagus from brain to deflated lung
Nucleus soliaris	Vagus from brain to inflated lung
Glossopharyngeal IX	Phrenic nerve
Inspiratory center	Intercostal nerve (from brain down cord to
Carotid body	intercostal nerves)
Aortic body	Thalamus

fixed-named disease. Early recognition of pollutant-triggered dysfunction is imperative in prevention of environmentally triggered, end-stage lung or nervous attachment failure.

Pollutant damage may occur in any part of the nervous system of the lung, including the pneumotactic center, nucleus soliaris, glossopharyngeal IX, inspiratory center, carotid body, aortic body, vagus nerve (from the brain to the deflated lung), vagus nerve (from the brain to the inflated lung), phrenic nerve, intercostal nerve (pollutant travels from the brain, down the cord, to the intercostal nerves), and thalmus.

The dorsal motor nucleus of the vagus sends fibers through the medulla oblongata to the vagus and spinal accessory nerves, which end in vagal sympathetic plexuses in the chest and abdomen. The control of respiration in the brain is the solitary nucleus, which also is involved with taste fibers. Damage to the nerves and receptors may cause respiratory dysfunction. Usually the symptom of breathlessness is the main complaint in many early pollutant-stimulated chemically sensitive patients.

At least three stimuli are effective in respiratory control. These stimuli include the following: (1) CO_2 bathing the respiratory center directly (and acting perhaps by influencing the actual reaction of the blood), (2) the chemical composition of the blood acting on the neuroendocrine system, particularly the carotid and aortic bodies, and (3) the degree to which the lung is distended. Excess carbon dioxide or hydrogen ions affect respiration mainly by direct excitatory effects on the respiratory center itself, causing greatly increased strength of both the inspiratory and expiratory signals to the respiratory muscles. The resulting increase in ventilation increases the elimination of carbon dioxide from the blood. This response also removes hydrogen ions from the blood because of decreased blood carbonic acid. Oxygen does not seem to have a direct effect on respiration but instead acts on peripheral chemoreceptors. The motor neurons to the phrenic nerve ($C_4 - C_5$), to the intercostal muscles (in the thoracic segments particularly the upper one half), and presumably to the auxiliary muscles of respiration are under control of an expiratory and inspiratory center in the medulla oblongata. Occasionally, these areas appear to be involved in chemically sensitive individuals. These two centers receive afferent fibers from the pneumotaxic center located in the pons near the hypothalamus and from the solitary nucleus.

The solitary nucleus is under the influence of the messages sent by the chemoreceptor of the carotid body over the glossopharyngeal (IX) and vagus (X) nerves and by the stretch receptors in the lung sent over the vagus. When the lung is collapsed (expiration), impulses traveling from stretch receptors over the vagus to the solitary nucleus have a low frequency. They are picked up by the interior portion of the solitary nucleus and are relayed, probably via the reticulo spinal tract and motor neurons in the spinal cord, to the inspiratory center, which then causes the diaphragm and external intercostal muscles to contract. Expansion of the lung in inspiration stimulates the stretch receptors, and impulses are sent along the vagus at high frequency. They are then picked up by a more superior portion of the solitary nucleus that, in turn, relays impulses to the expiratory center. The latter inhibits the inspiratory muscles and may, in exceptional circumstances, also stimulate expiratory muscles. This mechanism appears sufficient in itself to insure the play of alternate inspiration and expiration. However, it is often observed that if the chemically sensitive individual is pollutant overloaded, he or she will cough when taking in a deep breath. Coughing will then continue on a series of deep inspirations until the patient has been cleared of pollutants or their effects. This response suggests overload of the mechanism. Chemoreceptors and CO_2 provide additional security against overload, perhaps by changing the threshold of response of the neurons composing the respiratory centers. With early pollutant damage to the chemoreceptor or the actual nerve tracts, the nerves of the respiratory centers may cause mild dysfunction, which may be altogether missed by the clinician or misinterpreted as being functional.

The pollutant sensitive sleep apnea patient is a classic example of central respiratory deregulation that may be pollutant triggered with the etiology being missed. Once his or her total pollutant load is removed, the chemically sensitive sleep apnea patient can then breathe well at night without disturbance. At the EHC-Dallas, specific incitants such as onion, milk, soy, formaldehyde, or pesticides have been shown to trigger this type of sleep apnea.

Much evidence is now available suggesting that, in many cases, pollutant injury as seen in chemically sensitive individuals may be due to environmental stimulation or inhibition in both the neuroendocrine and the nerve tissue of the respiratory mechanism. Laboratory evidence supports this point of view. Perhaps the most prevalent cause of respiratory depression and arrest is the overdosage of anesthetics and narcotics. Ether, halothane, cyclopropane, ethylene, chloroform, trichloroethylene, and tetrachloroethylene are found in the blood of chemically sensitive patients and are observed to cause respiratory irregularities. Often these patients complain of shortness of breath and breathlessness.

Another example of pollutant damage resulting in neural center respiratory change is the alteration of the fine nerve structures of the bronchi and lung of rats, which occurred after exposures to the solvents *h*-hexane, butanone, methyl-ethyl-ketone (MEK), or a combination.[2] At the EHC-Dallas, we also have seen pollutant damage that resulted in neural center respiratory change. Of the 200 chemically sensitive patients we studied, 61% had *n*-hexane in their blood. The other 39% had MEK. These patients exhibited signs of neurological and respiratory dysfunction. Not all had lower airway problems, but many complained of breathing difficulties and chest pain, suggesting nerve or neuroendocrine involvement with resultant respiratory deregulation. The following case is an example of chemically triggered nerve dysfunction.

Case study. This 33-year-old occupational therapist presented with laryngeal dysfunction resulting in intermittent hoarseness, hypoventilation to 4 respirations per minute while in episodes of coma, and, at times, difficulty walking. She was found to be sensitive to many foods, inhaled chemicals, and biological inhalants. Pesticides seemed to be her most damaging pollutant, and upon exposure she would go comatose and hypoventilate to 4 breaths per minute. Even though she might remain comatose for 12 to 18 hours, her blood gases always remained normal. Several years into her environmental treatment she developed sudden uncontrollable jerking of her diaphragm and a sudden cessation of her breathing. This condition became quite unbearable in that she could not sleep or get rest. After the pollutant exposure was withdrawn and this patient's microcirculation was opened with a daily oxygen program, her breathing returned to normal. During this time her arterial blood gases remained normal. Clearly, central and peripheral nervous function had been restored.

In another study, Ohnishi et al.[79] reported that rats undergoing single exposure to ethylene oxide (500 parts per million at 6 hours for 3 times per week for 13 weeks) developed ataxia in the hindlegs. Myelinated fibers in hindleg nerves and in the fasciculus gracilis showed axonal degeneration sparing the nerve cell body of the lumbar dorsal root ganglion and myelinated fibers of lumbar dorsal and ventral roots. Clinically, this result was similar to ours, but since biopsies were not done in our patients, we cannot further compare the two series.

Considerable evidence shows that the bronchoconstrictor response to pollutants is mediated in part via cholinergic neurons.[49,80,81] According to both Bascom[13] and our studies at the EHC-Dallas, acquired and genetic differences in cholinergic hyperresponsiveness may, therefore, contribute to variablity in responsiveness to pollutants. Studies by Levitt and Mitzner[82] have demonstrated that a single autosomal recessive gene primarily controls acetylcholine-mediated airway responses in mice. The difference between the responses of different strains of mice can be inhibited by atropine. Studies of methacholine reactivity in 107 pairs of monozygotic and dizygotic twins demonstrated a greater correlation for monozygotic twins ($r_1 = 0.67$) than for dizygotic twins ($r_1 = 0.34$),[83] not only suggesting that genetic factors were important in cholinergic reactivity in humans but also pointing to other factors, such as acquired injury, as the cause of variable responses when genetic time bombs are involved. The allelic variation may be specific to the cholinergic receptor. However, it is also possible that the genetic factors important in the development of cholinergic reactivity in humans are factors that influence the development of airway inflammation in response to common environmental stimuli. Studies at the EHC-Dallas have shown cholinergic and sympatholytic responses to pollutant exposure to be the most frequent in ANS measurements. It is doubtful that these responses are all genetic since many patients were well before the onset of their chemical sensitivity.

A link has also been suggested between cholinergic hyperresponsiveness and familial susceptibility to chronic obstructive lung disease (COPD) (related to chronic exposure to the irritant mixture tobacco smoke). One hypothesis suggests that intrinsic host characteristics, specifically airway responsiveness and atopy, are important determinants of how individuals will respond to subsequent exposures. Bascom[13] suggests, therefore, that the presence of airway hyperrresponsiveness and atopy involve a predisposition to the development of chronic obstructive lung disease.[84,85]

The acute response to inhaled pollutants occurs through stimulation of chemosensitive neurons and the airway epithelium. Both play a key role in modulating the neural response. The variability of this response frequently varies in the chemically sensitive.

According to Bascom,[13] chemosensitive, or c-fiber neurons, are unmyelinated nerve fibers that ramify extensively in the airway mucosa and are approximate to epithelial cells, glands, and blood vessels. These neurons are stimulated by a variety of irritants, including capsaicin. Upon stimulation, antidromic conduction occurs, meaning that the impulse is propagated locally without completing a central reflex arc.[70,86] The c-fiber nerves release biologically active peptides, such as substance P, which may trigger symptoms. Differences among species have been observed in the response to c-fiber stimulation,[87] complicating cross-species extrapolations.

Bascom[13] notes that a series of studies in the early 1980s demonstrated the critical role for c-fiber activation in the inflammatory response of the rodent mucosa to cigarette smoke. These studies took into consideration the fact that neonatal administration of capsaicin permanently depletes c-fiber neurons. Lundblad and Lundberg[88] administered cigarette smoke to guinea pigs and enumerated increased nose wipings as an index of inflammation. They found that capsaicin administered systemically to rats blocked their response, as did local anesthesia. These data showed that the airway inflammatory response to cigarette smoke was mediated via activation of c-fiber neurons. Similar studies were performed in rats, using Evans Blue dye extravasation (indicating increased vascular permeability) as the index of inflammation.[89] Similar data were found for other chemical pollutants including formalin and ether.[72,90] A substance P antagonist also blocked the inflammatory response,[91] indicating that substance P in rodents was the neuropeptide that mediated the neuroinflammatory response (see Chapter 26).

More recently, Dusser et al.[92,93] have demonstrated that the respiratory epithelium modulates the airway response to neuropeptides. These investigators aerosolized substance P in the airways of guinea pigs and measured lung resistance as an index of response. Exposure to the smoke from two cigarettes markedly augmented the bronchoconstrictor response to Substance P. Further investigation demonstrated a parallel depletion of neural endopeptidase, the epithelial enzyme that inactivates substance P. Administration of superoxide dismutase, an antioxidant enzyme, prevented the tobacco smoke-induced augmented response to substance P. These data suggest that the oxidants present in cigarette smoke are responsible for the altered effect similar to those from other pollutants.[94] We often have seen accelerated and blocked responses in the chemically sensitive patient who is challenged with capsaicin containing peppers, solvents, and pesticides. It is clear that pollutants can injure the response nerves, nerve centers, or chemostatic receptors, causing irregular respiration and the excessive bronchial responses that are seen in some chemically sensitive individuals. We have also seen intravenous antioxidants eliminate the reaction once the exposure occurred.

Pathophysiology of Pollutant Injury to the Lower Respiratory Tree

Early pollutant injury to the lung may be extremely subtle and imperceptible because of the massive number of alveoli present. In contrast, pollutant damage to the bronchial mucosa results in irritation with early symptoms. Early pollutant damage to the lower respiratory tree may just disturb function, while a little more advanced exposure will damage just a few alveoli with no apparent immediate change in lung function. It may take 20 to 40 years before end-stage lung damage manifests, since the adaptation phenomenon will help the individual to cover up initial and slowly reoccurring damage until it is too late for reversal. Therefore, in order to prevent permanent bronchial and lung damage, it is extremely important to understand the early changes that result from pollutant injury and to act upon these changes by eliminating the total load of pollutants and treating the resultant underlying nutrient deficits.

Mucosa Damage

The airway epithelium plays a key role in the induction of inflammation of the mucosa following pollutant exposure(s) in the chemically sensitive individual. While the epithelium was once thought to be a passive barrier, its multitude of dynamic functions is now recognized (Figure 19.2). These functions range from mucus production and ciliary transport to the regulation of fluid and electrolytes at the airway surface to the production and release of a host of mediators and cytokines that may induce and amplify inflammatory responses. The epithelium also contains antipollutants such as glutathione, vitamin E, and superoxide dismutase as well as repair nutrients. It is a complex dynamic area that can be damaged in the chemically sensitive individual.

Simple anatomic facts make the bronchial tree more vulnerable to pollutant injury in some chemically sensitive patients. For example, mucus layering on the tracheal bronchial tree is not uniform, and some areas are free of mucus. This lack of uniformity allows for sudden localized pollutant damage in these areas that are not well protected. Here, contact with solvents may easily penetrate the mucosa and cause bronchial irritation and hyperresponsiveness. Secretion of mucus varies from 50 to 150 cc daily and, therefore, is not uniform. It is under the control of the parasympathetic nervous system and the amount of irritation that occurs locally. Faulty production of mucus may make an area more prone to injury. The majority of measurable changes of the ANS are cholinergic and sympatholytic in the chemically sensitive individual, thus resulting in reduced mucus production or bronchial irritability and spasm. However, at times, sympathetic or cholinolytic responses may occur, resulting in reduced mucus buffering capacity. Laryngeal, tracheal, and bronchial clearing are usually under the influence of cilia and mucus whereas the alveolar areas have an added phagocytosis with lymphatic involvement. Any of these processes may be impaired in a subset of chemically sensitive individuals.

The mechanisms by which particulate matter is removed vary according to the site of deposition. The alteration of these mechanisms in a subset of chemically sensitive individuals may allow for their pathological response. If an insoluble particle is deposited in a dead space, it is cleared by the mucociliary escalator. If it is deposited in the lung parenchyma, it is phagocytized by the macrophages, which either migrate to ciliated airways or to the interstitium of the lung and out the lymphatics. As with changes in the mucosa from subtle pollutant damage to the cilia, the macrophages and the lymphatics initially may not be easily measureable, and, therefore, toxic chemicals may cause more severe damage in the chemically sensitive individual before either the chemicals or the damage are detected. Certainly ciliary malfunction has been seen in some chemically sensitive individuals, and impaired phagocytosis is common among those who experience recurrent infections (see Chapter 30).

Pathophysiological responses of the respiratory tract to particles, mists, and gases result in irritation of the airways, bronchoconstriction or damage to parenchyma, or, eventually, occurrence of cancer (in some cases) (Table 19.3). Some parenchymal responses are interstitial fibrosis (asbestos, cobalt, beryllium) or nodular dysfunction (silicosis) or emphysemas (coal, carbon, nitrogen dioxide, chlorine), which can be seen in end-stage chemical sensitivity.

Laboratory findings suggest that initial ciliary damage may be triggered in these exposures. For example, ciliary activity of rat trachea was damaged after exposure to 150 ppm to 1000 ppm of styrene. In control groups without styrene exposure, the ciliary beats were stable within 3 weeks. Changes in the ciliary movements of the trachea of the tested groups exposed to styrene (150 ppm) were slight after 1 week exposure, while there was a drop to 46% in the groups exposed to styrene (1000 ppm). The ciliary movements of the nose decreased by about 40% following exposure to 150 ppm styrene. After a 3-week exposure, the ciliary movements of the trachea fell in 15% of the subjects in the groups exposed to 150 ppm, but in the groups exposed to 1000 ppm, this drop was about 33%. When exposed to 150 ppm styrene, the ciliary movements of the nose dropped to about 55%, and in those rats exposed to 1000 ppm, almost no function was recognized.[95] Five percent of chemically sensitive individuals have styrene in their blood and are made ill by re-exposure, suggesting damage to their respiratory mucosa.

Many environmental factors such as cold and decreased humidity are important factors affecting cilia as are inorganic salts, radioactivity, cigarette smoke, sulfur dioxide, nitrous oxides, ozone, and some congenital structural

Table 19.3. Factors in Pollutants Affecting Pollutant Injury in the Chemically Sensitive

Physical properties

Physical state — particle, mist, fume, vapor, gas

Size and density — determine site of deposition

Shape and penetrability — influence propensity for migration

Solubility — e.g., local reaction insolubility asbestosis; systemic reaction soluble manganese, oxides of nitrogen

Hyperactivity — may increase in size as they go down respiratory tract

Electric charge — influences site of deposition

Chemical properties

pH — excess acidity or alkalinity has toxic effect on cilia cells and enzyme systems; propensity to combine with substances in lung or peripheral tissues; e.g., cyanide and carbon monoxide go systemic; fluorine both local and systemic.

Fibrogenicity — more (asbestos, silica)

less (iron, carbon)

Antigenicity — stimulates antibodies

---

defects of the cilia. Once this ciliary function is disturbed, a snowball effect may occur enabling other less toxic pollutants and foods to trigger reactions. This escalation is often seen in the chemically sensitive population.

Cigarette smoke, exposure to which is totally avoidable, is the most common substance causing pollutant damage. Other common pollutants seen to trigger respiratory chemical sensitivity are inorganic (ozone, nitrous oxide, sulfurs, etc.) and organic chemicals (pesticides, car exhausts, etc.).

Laboratory evidence supports the clinical observations of mucosal changes seen in the chemically sensitive individuals whose respiratory tree is effected by pollutants. For example, animal studies by Leikauf et al.[23] have profiled the inflammatory response following exposure to acrolein (which emanates from new sheetrock), a highly irritating aldehyde. Exposure of guinea pigs for 2 hours to 0.3 to 1.2 ppm acrolein resulted in an acute bronchoconstrictor response with release of prostanoid products thromboxane B_2 and PGF_{2alpha} immediately post-exposure. Later effects included airway hyperresponsivenes (2 to 6 hours later) and airway inflammation (neutrophil influx—24 hours later).[23] These authors concluded that neutrophils were not necessary for bronchial hyperresponsiveness. These investigators subsequently demonstrated that a leukotriene receptor anatagonist (L-649,923) and 5-lipoxygenase inhibitors (L-651,392 and U60,257) diminished acrolein-induced bronchial hyperresponsiveness.[24]

Further studies using the pollutant acrolein have demonstrated that the airway epithelium is a source of at least some of the early mediators demonstrated in the whole animal studies. Exposure of bovine airway epithelium to acrolein *in vitro* results in release of the biologically active arachidonic acid metabolites 12- and 15-HETES, as well as PGE_2, PGF_{2alpha}, and 6-keto-PGF_{1alpha}.[96] When the airway is affected, the chemically sensitive individual's response to pollutants such as acrolein is affected adversely. This clinical pattern often appears to be one of mediator release. The following case is illustrative of the detrimental effects of a toxic exposure on the respiratory mucosa and endothelia.

> **Case study.** A 28-year-old white male was totally well previous to his working in an ice manufacturing plant. There, an ammonia leak occurred. He remained in the contaminated area for several hours during which time he developed a severe cough and shortness of breath. He was seen by his family doctor, who placed him on steroids, without improvement. He was referred to the EHC-Dallas. Seventy-two hours after the exposure, he was admitted to the ECU with a severe cough and wheezing and with a pulmonary flow of 200 cc/min. This patient was immediately placed on bronchial dilators, steroids, and intravenous antibiotics. He was given vigorous respiratory therapy. After 2 weeks of treatment his bronchospasm had subsided, and his flow increased to 400 cc/min. However, each time he ate certain foods his flow would decrease to around 200 cc, lasting for 4 to 6 hours. Intradermal testing to biological inhalants also revealed a decrease in flow to 200 cc with clinical wheezing. This patient had never had these problems from biological inhalants or food previous to his exposure. He was discharged on environmental control, subcutaneous injections of his neutralizing dose of antigens, and nutritional supplementation. His newly acquired chemical sensitivity gradually improved until he developed recurrent bronchial infections. Five episodes occurred over the next few months. Bronchoscopy revealed an inflamed bronchial mucosa. Phagocytic index was 70% normal. He was treated with bacterial vaccines and transfer factor, and gradually the recurrent infections subsided.

The airway epithelium releases mediators in response to oxidants, smoke extracts, foods, biological inhalants, and endotoxin with the response often being accentuated in the subset of chemically sensitive patients who have airway damage. Ozone exposure causes a dose-dependent increase in PGE_2, PGF_{2alpha}, 6 keto PGF_{1alpha}, and leukotriene B_4.[97] Increases in PGF_{2alpha} occurred at a concentration of ozone as low as 0.1 ppm (below the current U.S. ambient air quality standard of 0.12 ppm).[97] Leukotriene B_4 is a potent chemotactic factor for neutrophils and can increase the adhesiveness of leukocytes to endothelial cells.[98] Endotoxin and smoke extracts were shown to release monocyte chemoattractant activity from bovine epithelial cells.[99]

The effect of pollutants on the epithelium has begun to be investigated using other markers of response. For example, heat shock treatment of a variety of cells will induce the selective synthesis of 72kD, 73kKD< 90KD, and 11OkD proteins. Heat shock proteins may be induced *in vitro* in guinea pig tracheal epithelium by exposure to acids and heavy metals, but not ozone.[100,101]

The function of these proteins in the response to pollutants in chemically sensitive individuals is unknown at present.

The human airway epithelial cell supernatants spontaneously release granulocyte-macrophage colony stimulating factor (GM-CSF), which induces leukocyte differentiation and prolongs the survival of eosinophils.[102] Cells regenerating after exposure to isocyanates are reported to release interleukin 1 alpha and beta and interleukin 6.[103] Airway epithelial cells express MHC Class II receptors on their cell surface.[104]

Another cytokine that may function in the response to pollutants is transforming growth factor-beta (TGF-beta). This peptide is chemotactic for neutrophils, although it does not activate them. It recruits and causes both chemotaxis and anchorage-independent growth of fibroblasts.[105] TGF-beta increases epithelial surface density of fibronectin and vitronectin receptors.[106] TGF-beta induces squamous morphology in cultured airway epithelial cells with associated induction of desmoplakin in gene expression.[107] The latter feature is particularly interesting in considering TGF-beta as a potential mediator responsible for the epithelial morphologic changes observed with chronic irritant exposure.

Taken together, these facts suggest that the epithelium has a vital regulatory role in initiating and modulating the response to pollutant exposures in chemically sensitive individuals. Pro-inflammatory mediators released from the epithelium in conjunction with neuropeptides may initiate the response to a pollutant exposure in this population. The epithelium may then become a target site for inflammatory mediators with alteration in one or more of its critical functions resulting in more symptoms in the chemically sensitive individual. Chronic pollutant exposure causes changes both in epithelial histology and in epithelial function.

Numerous studies that also have been associated with chemical sensitivity have implicated toxic chemicals in lung damage. For example, the administration of 1,1-dichloroethylene (1,1-DCE, 125 mg/kg i.p.) to CD-1 mice caused bronchiolar necrosis that was accompanied by substantial covalent binding of radiolabeled compounds and/or metabolites of the lung. Lung injury and covalent binding were not modified by phenobarbital pretreatment. However, 3-methylcholanthrene provided a protective influence, although it failed to alter covalent binding to lung macromolecules.

Methylchloride, which is in the blood of many chemically sensitive patients, appears to be a direct acting genotoxicant. 1,1,1- and 1,1,2-trichloroethane (also in the blood of some chemically sensitive individuals) and 1,1-dichloroethane are all weakly genotoxic. Measurement of DNA repair as unscheduled DNA synthesis (UDS) *in vitro* following exposure *in vivo* in multiple tissues from the same treated animal can provide valuable information relating to the tissue and organ specificity of chemically induced DNA damage.[108] UDS was evaluated in primary cultures of rat tracheal epithelial cells, hepatocytes, and pachytene spermatocytes after these cells were exposed in vitro to methylchloride (MeCl) and after they were isolated from the same treated animal following inhalation exposure *in vivo*.[109] Concentrations of 1 to 10% MeCl *in vitro* induced UDS in hepatocytes and spermatocytes but not in tracheal epithelial cells. Inhalation exposure to MeCl in vivo (3000 to 3500 ppm 6 hours per day for 5 successive days) failed to induce DNA repair in any cell type. *In vivo* exposure to 15,000 ppm for 3 hours also failed to induce UDS in tracheal epithelial cells and spermatocytes, but it did cause a marginal increase in UDS in hepatocytes. Thus, MeCl appears to be a weak direct-acting genotoxicant. While activity could be measured in hepatocytes and spermato- cytes directly *in vitro*, only extremely high concentrations of MeCl elicited a response in the whole animal, and then only in hepatocytes. It is unknown what damage the combination of MeCl and other toxic chemicals found in the chemically sensitive population will cause, but in some cases these substances obviously propagate the chemical sensitivity.

Pulmonary changes from an iatrogenic pollutant (e.g., cyclophosphomide) may be due to the activation of incompetent pulmonary cells with subsequent attraction of systemic inflammatory cells. The extreme in pulmonary pollutant injury is that more squamous cell carcinoma lesions are seen in mustard gas [*bis*-(α-chloroethyl) sulfite] workers. In one study, the bronchial epithelium in stepwise transverse sections was examined histologically in 66 male autopsy cases, composed of the groups of 19 mustard gas (MG) ex-workers with lung cancer, 17 MG ex-workers with nonlung cancer, 10 nonMG lung cancer cases, and 20 nonMG nonlung cancer cases. Foci of moderate or severe atypical cellular lesion or dysplasia or of carcinoma in situ (CIS) in total slides of each group were counted as 146 in 3485, 72 in 2226, 70 in 3797, and 18 in 4611, respectively. The relative frequency of moderate or severe dysplasia and CIS in MG-exposed nonlung cancer cases resembled that found in lung cancer cases of both MG and nonMG exposed. Seven CIS lesions were detected from among all MG-exposed cases, and one CIS was found in a nonMG lung cancer case. Six of eight CIS examples were adjoined by dysplasia. Multivariate analysis revealed a significant correlation between the incidence of atypical lesions and MG exposure, though the incidence of atypical lesions was also influenced significantly by age, smoking, and chronic bronchitis. The inci- dence of atypical lesions was significantly higher in cases of squamous cell lung cancer than those of other histological types, particularly small cell cancer.

During World War I, one of the poison gases used by both sides was *bis*-(α-chloroethyl) sulfite, otherwise known as mustard gas. In addition to the acute effects of exposure to poison gas, chronic respiratory diseases resulted from exposure to this and other gases. The suggestion that there was a possible association between exposure to mustard gas and the development of lung cancer was made by Case and Lea,[110] who studied British soldiers after World War I. It was not, however, until 1959 that Yamada[111] definitely reported the relationship between industrial manufacture of mustard gas and the development of lung cancer in exposed workers. He studied a group of workers who manufactured mustard gas on one of the smaller Japanese islands during World War II, though fortunately the gas was not used in the war. Subsequent investigations have reported on the pathology of tumors caused by exposure to mustard gas.[112] About half of the cases were squamous cell carcinomas, and the latent period was, in most cases, over 20 years. Overall, the death rate among exposed workers from lung cancer was about 16%, compared with less than 1% in a control group. Laboratory models do exist for the study of mustard gas,[113] and it was work with this material that led to the devlelopment of some of the earliest anticancer drugs.[114]

The data on cigarette smoking and lung cancer are well-known. It is clear that long-term smoking causes lung cancers, respiratory failure, emphysema, and cardiovascular disease. The mechanisms of lung damage from smoking are quite varied, but it is clear that benz[a]pyrene is one of the main offenders and will trigger chemical sensitivity.

Examination of the lungs of rabbits exposed 120 times to 3000 ppm of n-hexane, 8 hours per day, 5 days per week, for 24 consecutive weeks revealed three important exposure-related lesions.[115] These consisted of the following: (1) air space enlargement centered on respiratory bronchioles and alveolar ducts (centiacinar emphysema), (2) scattered foci of pulmonary fibrosis, and (3) papillary tumors of nonciliated bronchiolar epithelial cells. n-hexane (10 to 500 ppb) is constantly being isolated from the blood of chemically sensitive patients. In humans, its relationship to lung damage at the ppb level is unclear at this time, due to the lack of longevity studies. However, it is clear that these patients are not functioning well overall, that they respond adversely to hexane and other toxic challenges, and that they get better when the hexane is removed from their blood.

Rabbits given intravenous injected phorbol myrislate (PMN) developed acute and chronic disease.[116] In the acute phase, hemorrhagic pneumonitis with increased lung weight and increased neutrophils and erythrocytes was seen in alveolar fluid. During the interim period (1 to 7 days), interstitial pneumonitis developed, and neutrophils and macrophages were found in the interstial fluid. Fibrosis was evident after chronic doses (repeated for 14 days). If this type of exposure occurs in humans, chemical overload may result in fibrosis in end-stage chemical sensitivity.

One report of a clinical history and the findings of a wedge biopsy specimen of a Vietnam veteran suffering from progressive severe tissue damage of the lung emphasizes how failing to recognize the chemically sensitive individual and neglecting to remove primary and secondary triggering agents can lead to disaster. This patient served as a soldier in defoliated areas of Vietnam for 2 years and developed severe chest pain and dyspnea with chronic postnasal dripping, maxillary sinusitis, and allergic asthmoid bronchitis with pronounced obstructions and eosinophilia. Recurrent onset of symptoms over a period of 10 years led to wedge biopsies of the left upper lobe, right lower lobe, and mediastinal lymph nodes. Histology was consistent with chronic, slightly progressive diffuse alveolar damage including moderate interstitial fibrosis. Total destruction of mediastinal lymph nodes with deposits of amorphous material and foreign body giant cells was noted. Histology findings and clinical course favored hypersensitivity reaction of the lung and congestion of exogenous material in the alveoli probably related to exposure to herbicides.[117] The patient eventually died.

Laboratory evidence shows how failure to reduce total body load may result in adversity. For example, groups of male rats were exposed to acute doses of oxygen, ozone, or paraquat, which produced equivalent mortality (25 to 30%) over a 28-day post-exposure period. The primary response was edema and inflammation with only slight fibrosis. Aerobic pulmonary metabolism was inhibited in lungs of animals exposed to oxygen and ozone as evidenced by decreased oxygen consumption. However, this decrease was transient, and O_2 consumption returned to normal within 24 hours after the rats were removed from the exposure chamber. Conversely, treatment with paraquat caused an immediate, transient stimulation of O_2 consumption. Glucose metabolism was unaltered by the gas exposures and was initially stimulated by paraquat exposure. *In vitro*, only paraquat added to lung slice preparations altered both O_2 consumption and glucose metabolism. Ozone had no effect. Oxygen did not alter O_2 consumption but did cause a slight biphasic response in glucose metabolism. Aerobic metabolism was relatively unchanged by these doses of oxygen and ozone, which resulted in the death of 25 to 30% of all treated animals. Even though paraquat produces similar morphologic changes, it may represent a more severe metabolic insult than "equivalent" doses of oxygen or ozone. Also, if interstitial pulmonary fibrosis is a desired result of experimental exposure, rats may not be a suitable model for oxidant-induced lung injury. Regardless, evidence suggests that changes in metabolism occur in animals exposed to these pollutants. These changes may result from the increased total body load these animals sustain during challenge testing by various levels and repeated doses of single or multiple chemicals, and they may be similar to the changes seen in chemically sensitive patients.

Similar evidence is found in human studies. For example, pneumonitis-fibrosis, which was induced by the treatment with antineoplastic agent(s) and/or irradiation, was encountered in 37 (14.1%) of a total of 515 patients with lung cancer. Of 251 patients who had been treated with bleomycin or pepleomycin alone or in combination with other antineoplastic agent(s) or irradiation, 46 (18.3%) had pneumonitis-fibrosis, and 19 (7.6%) died therefrom.[118]

Trichloroethylene (TRI), often found in the chemically sensitive patient, has been shown to injure bronchial mucosa, produce increased accumulation of pulmonary calcium, lengthen anesthesia recovery time, and decrease pulmonary microsomal cytochrome P-450 content and aryl hydrocarbonase activity in mice.[119] TRI damage to detoxification enzymes (if it occurs in humans) could account for the ability of some individuals with chemical sensitivity to detoxify hydrocarbons entering the body. The result of this overload may be production of a spreading phenomenon whereby the individual becomes more sensitive to other chemicals. We have seen this change repeatedly in chemically sensitive individuals who have not only TRI in their blood but other toxic chemicals as well.

1,2-Dibromoethane (a volatile carcinogen) has been found to bind to the respiratory and gastrointestinal mucosa in mice and rats.[120] This compound is very long-lasting and may take years to degrade. Its presence, which is common, in the chemically sensitive population may continue to aggravate their symptoms.

In the olfactory mucosa, an increased secretion from sustentacular cells and membrane fusion of cilia and sometimes a loosening ciliary membrane has been seen following exposure to toluene and styrene.[121] After single-dose exposures to fenithione, rats have experienced mild inflammation of alveoli including interstitial edema and cellular infiltration.[122] Additionally, increased alveolar macrophages similar to those produced following exposures to areal spraying have been found. Recovery has been seen after 21 to 60 days following exposure.

Humans exposed to paraquat, a toxic chemical, experience symptoms similar to those seen in the previously discussed laboratory animals, often with equally devestating effects. For example, a recent report described a 31-year-old nurseryman who developed progressive respiratory failure following repeated and prolonged paraquat exposures.[123] His condition required a right lung transplantation, which was successful for a period of time, but the transplanted organ subsequently was damaged and failed when lethal levels of paraquat entered his bloodstream from muscle reservoirs after vigorous exercise that resulted in autotransfusions of toxic chemicals from fat and muscle stores. We have seen similar, though not lethal, episodes in chemically sensitive patients who are clearing in the ECU.

Metabolic Damage to the Lower Respiratory System

Some research indicates that metabolic alterations of the lung may be associated with pollutant exposure[124] both in animals and in chemically sensitive patients (Figure 19.3). As a consequence of pollutant exposure, vitamins A,[125] B,[126] B$_1$,[127] C,[128,129] E[130] and minerals, zinc,[131] selenium,[132,133] magnesium,[134] and manganese[135] may be destroyed, or altered, or other nutrient metabolism may be affected. Three mechanisms are associated with vitamin and mineral depletion or excess. The first, overstimulation of the detoxification systems, allows both for overutilization of the available vitamins, minerals, and amino acids and creation of a functional deficiency that results not only in decreased capability of detoxification but also a decrease in the ability to repair damaged tissue. The second is active competition for absorption. Hydralazine, for example, may be absorbed rather than B$_6$. Finally, actual toxic destruction of vitamins and minerals may occur. With cell membrane damage due to pollutants, nutrients may shift into cells or cellular compartments causing

Pollutant Overload

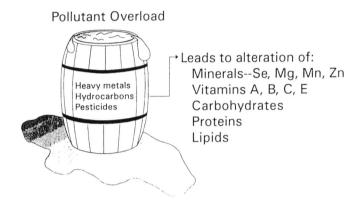

Leads to alteration of:
Minerals--Se, Mg, Mn, Zn
Vitamins A, B, C, E
Carbohydrates
Proteins
Lipids

Heavy metals
Hydrocarbons
Pesticides

Figure 19.3. Pollutant damage to nutrients of the lower respiratory tract.

relative excess. Thus, excess calcium, manganese, etc. may damage the function of that cell.

Generation of free radicals damages mitochondrial membranes. Thus, changes occur in oxygen consumption and ATP metabolism,[136] usually resulting in weakness in the chemically sensitive individual.

Alterations in glucose metabolism may increase CHO metabolism 2 to 3 times with corresponding increases in lactate production.[137-141] For example, exogenous glucose in paraquat-damaged lung slices revealed changes vs. controls. CHO metabolism is frequently altered in chemically sensitive individuals with metabolic and glucose swings causing the whole organism to be unstable.

Glucose changes may be critical after pollutant injury to the lungs.[142,143] Glucose can provide a number of intermediates including energy (ATP) through glycolysis and reducing compounds (NADPH) through the hexose monophosphate shunt (HMP).[144] With adequate oxygen, glucose converts from lactate to pyruvate.[144,145] Pyruvate is then largely oxidized by way of the mitochrondrial tricarboxylic acid cycle.[145] If the mitochondrial membrane is damaged by pollutant-induced superoxides, conversion cannot occur and a concurrent energy loss with increased lactate and resultant damage does occur. The resultant damage is seen in many chemically sensitive individuals. Both nicotinamide adenine dinucleotide (NAD) and nicotine amide adenine dinucleotide phosphate (NADP) increase oxygen consumption and restore energy after pollutant exposure.[144] Thirty percent of the chemically sensitive population are deficient in nicotinamide, and this deficiency decreases their resistance to pollutant exposure.

Changes in DNA, RNA, and protein synthesis occur as a result of pollutant injury (see details in Volume I, Chapter 4[54]).[146] Biological amines and other enzyme systems also are affected by pollutant injury.[146-157] Alterations occur in the monoamine oxidase and pulmonary microsomal mixed function oxidase (cytochrome P-450), aryl hydrocarbon hydroxylase, glucose 6-phosphate dehydrogenase, trypsin, chymotrypsin, and elastase inhibitor activities.[158-162]

Sulfhydral metabolism changes (glutathione) are also involved.[163] Heavy metals (mercury, lead, cadmium) may intervene between the sulfhydral groups thus stopping the regeneration and recycling of glutathione (S-Pb-S), which then leads to abnormalities in the hexose monophosphate shunt and an inability to detoxify.[164] All of these changes are observed in some chemically sensitive patients.

Paraquat-stimulated NADPH depletion in rat lung slices responded to exogenous glucose concentration. Lung slices incubated with 11 mM glucose and 10^{-4} M paraquat had a 40% lower NADPH/NADP$^+$ ratio than did control lung slices. Incubation with no added glucose and 10^{-5} M paraquat caused a 41% decrease in NADPH/NADP$^+$. With paraquat at 10^{-5} M, glucose at 1.1, 5.5, 11, or 22 mM increased NADPH/NADP$^+$ ratios in a concentration-dependent manner until at 22 mM glucose the effect of paraquat was prevented. The sum of NADP + NADPH was only 60% of control with 10^{-5} M paraquat and no glucose. However, with any concentration of glucose from 1.1 to 22 mM, the total was 92% of control. The results of this study indicate that glucose may be beneficial in preventing paraquat-mediated NADPH depletion in rat lung slices.

Activity of the pentose phosphate pathway of glucose metabolism was measured in isolated granular pneumocytes under a variety of metabolic conditions known to alter this pathway in intact lungs. Granular pneumocytes were isolated by trypsinization of rat lungs and maintained in primary culture for 24 hours before use. Cells were incubated for 1 hour at 37°C with 5.5 mM glucose specifically labeled as 1–14C, 6–14C, U-14C, or 5–3H for determination of glucose utilization, pentose cycle activity, and partition of CO_2 production between mitochondrial and pentose pathways. With control cells, total glucose utilization was 111 ± 4.8 nmoles $\times$ hr^{-1} $\times$ (10^6 cells)$^{-1}$ (mean $\pm$ S.E., N = 19), and 2.2% was metabolized by the pentose cycle. Pentose cycle CO_2 production was 7.3 nmoles $\times$ hr^{-1} $\times$ (10^6 cells)$^{-1}$ representing 34% of total CO_2 production. Dinitrophenol (50 μM) stimulated mitochondrial CO_2 production 5-fold but had no effect on the pentose cycle activity 15-fold. Antimycin A (0.4 μg/mL) markedly inhibited both pathways. After a 30-minute preincubation with paraquat (3 mM), the pentose cycle CO_2 production increased to 107 nmoles $\times$ hr^{-1} $\times$ (10^6 cells)$^{-1}$ accounting for 39.6% of glucose utilization and 88.4% of CO_2 production. Mitochondrial CO_2 production was unchanged with paraquat. These studies demonstrate that the pentose cycle in resting granular pneumocytes accounts for a major fraction of the CO_2 production from glucose and that activity of this pathway is regulated by the utilization of cytoplasmic reducing equivalents. Paraquat produces marked stimulation of pentose cycle activity in granular pneumocytes, resulting in maximal utilization of cytoplasmic NADPH.[165]

In the severely damaged chemically sensitive patient, intracellular administration of glucose combined with amino acids, vitamins, and minerals usually helps. In fact, one of our major treatments in the severely pollutant injured chemically sensitive patient with malnutrition is 50% dextrose combined with

8.5% amino acids as well as the other nutrient supplementation (see Chapter 37). Fructose is even more efficacious in some patients, but it is difficult to obtain. At the EHC-Dallas, we are careful to ensure that our patients are not sensitive to the corn residue in glucose or fructose before we administer this treatment.

Administration of benzene-soluble fraction (FAE) and benzene-insoluble fraction (FAR) of fly ash to rats for 3 consecutive days significantly raised cytochrome P-450 levels, aryl hydrocarbon hydroxylase (AHH) activity, and glutathione S-transferase activity in the liver. This treatment also significantly increased pulmonary AHH and glutathione S-transferase activity. Intratracheal administration significantly increased hepatic cytochrome P-450 levels and the activity of glutathione S-transferase. This finding emphasizes the need for adequate nutrient supply in the pollutant overloaded chemically sensitive individual in order to keep up with the increased enzyme activity that results from pollutant exposure.

Cytochrome P-450-dependent monoxygenase systems provide a major pathway in a number of tissues for the oxidative metabolism of many chemicals present in the environment. These enzyme systems are bound to the endoplasmic reticulum and are composed of a flavoprotein (NADPH-cytochrome P-450 reductase; E.C. 1.6.2.4) and a family of hemoproteins called cytochrome P-450. Several forms of the cytochrome that have overlapping, but different, substrate specificities have been purified and characterized. The monooxygenase systems generally produce metabolites that are less toxic than the parent compounds, which include a number of pesticides, drugs, and direct-acting carcinogens and mutagens. However, many procarcinogens, promutagens, and other toxic substances are activated by this metabolic process and are converted from relatively nontoxic compounds to extremely toxic metabolites.[166]

Cytochrome P-450-dependent metabolism is required for the activation of several pulmonary toxins that produce different deleterious effects. For example, activation of 4-methyl-benzaldehyde results in the destruction of pulmonary cytochrome P-450. Activation of benzo[a]pyrene is involved in pulmonary carcinogenesis, and activation of 4-ipomeanol ("lung edema factor" isolated from mold-infected sweet potatoes) leads to pulmonary edema, congestion, and hemorrhage.[166]

Intragastric administration of retinyl palmitate (5000 IU per 100 gm body weight), along with intratracheal FAE and FAR administration, significantly reduced P-450 levels, activity of glutathione S-transferase in liver, and activity of AHH and glutathione S-transferase in the lungs of rats. Intraperitoneal administration of citrate (40 mg per 100 gm body weight) along with FAR significantly reduced FAR-induced increase in hepatic cytochrome P-450 levels and glutathione S-transferase activity. The activity of AHH was not affected by these treatments. Prior administration of the metabolic inhibitors, piperonal butoxide and SKF 525-A, produced differential effects. While piperonal butoxide exacerbated bronchiolar injury by 1,1-DCE, covalent binding remained unaltered.

The presence of lactate dehydrogenase (LDH) activity in the airways was found to be a sensitive indicator of acute toxicity to lung cells. The airway content of LDH increased after bronchopulmonary lavage of Syrian hamsters with increasing amounts of Triton X-100, with a correlation coefficient of 0.98.[167]

The metabolism of benzo[a]pyrene (BaP), which severely exacerbates chemical sensitivity, may be altered by xenobiotic compounds, showing how the spreading phenomenon may occur in chemical sensitivity. BaP detoxification (e.g., 3–OH formation) in rat lung is selectively inhibited by p-xylene, but not ethanol. Ethanol appears to modify the inhibitory effect of p-xylene. p-Xylene was administered by intraperitoneal injection at doses ranging from 0.1 to 1.0 g/kg (1:1 in soybean oil). Ethanol was administered orally at 5 g/kg (40% w/v). Rats given p-xylene, ethanol, or p-xylene and ethanol were sacrificed 1 hour after treatment. Rat tissues were examined at additional time points of 15 minutes, 4 hours, and 24 hours after p-xylene (1 g/kg). 3-Hydroxy-BaP (3–OH) formation was measured fluorometrically as AHH in lung microsomes. p-Xylene (1 g/kg) inhibited the formation of 3–OH BaP 40% at 15 minutes, 27% at 30 minutes, 43% at 1 hour, and 39% at 4 hours after treatment. Inhibition of AHH activity was still present 24 hours after dosing (41%). AHH activity was inhibited 27% and 46% at 0.5 mg/kg and 1.0 mg/kg p-xylene (1 hour), respectively, while the lowest dose (0.1 mg/kg) did not change activity. Analysis of the major metabolites of BaP by high performance liquid chromatography (HPLC) demonstrated that the formation of 3–OH and 4,5-diol BaP was inhibited 32% and 50%, respectively, in lung microsomes prepared 24 hours after a single injection of p-xylene (1 g/kg). None of the other metabolites analyzed were changed by p-xylene. Ethanol had no effect on 3-OH BaP formation during a 1-hour treatment. A combined dose of ethanol and p-xylene moderately inhibited 3–OH BaP formation. These findings suggest that a combination of xenobiotics such as those found in the chemically sensitive population could inhibit detoxification and increase the total body load as is seen in many chemically sensitive patients.

Covalent binding of benzo[a]pyrene to DNA, RNA, and protein of perfused rat lung after intratracheal administration of the substrate was about 50% of that obtained by direct addition of the substrate into the perfusate. Systemic availability of benzo[a]pyrene by absorption from the intratracheal sites was found to be 25%. During single pass perfusion of rat lung as well as in the combined liver-lung recirculating perfusion system, covalent binding after intratracheal exposure to benzo[a]pyrene was reduced to 50 to 75%. These experiments point out the relative importance of systemic vs. inhalation exposure to benzo[a]pyrene in the development of lesions on the lung macromolecules. The elimination capacity of the perfused lung was found to be relatively high as compared to the liver. At constant benzo[a]pyrene infusion by a single pass medium, clearance values of the 5,6-benzoflavone-induced lung with 7.8 mL/min reach almost half of those of a 5,6-benzoflavone-induced liver (15.9 mL/min). The extraction ratios in lung both during bolus administration in the recirculating system are 0.25 and 0.29, respectively, and thus amount to half of those measured in the liver (0.45 and 0.53, respectively). Though the liver is able to exert a protective effect against covalent binding of benzo[a]pyrene in the lung, this protection is not complete, and significant amounts of benzo[a]pyrene are expected to escape metabolic transformation by the liver representing a risk for the lung from the circulatory site in addition to that from the respiratory site.[168] These findings, if applicable to humans, further substantiate the observation that as the total body pollutant load decreases, the ability of chemically sensitive individuals to detoxify increases.

Changes in lipid metabolism may occur with decreases in surfactant activity.[169,170] Dipalmitoyl phosphatidylcholine in the Type II cells appears to increase along with an increase in lipid peroxides. Since many chemically sensitive patients have xylene in their lipid structures and blood, they often are unable to process other toxic chemicals. They then become more sensitive as their total body load increases.

Immune Changes

In addition to metabolic alterations, immune changes as a result of pollutant exposure may occur in the lower respiratory tract of chemically sensitive individuals. In these individuals, exposure to pollutants can alter specific immune responses, a process that has been described in Volume I, Chapter 5,[171] and in Chapter 30. Laboratory data corroborate findings of immune changes. For example, in nonallergic guinea pigs, repeated exposure to high doses of ozone followed by nebulized antigen challenge over 5 days resulted in increased allergic sensitization. The serologic response increased as evidenced by active cutaneous anaphylaxis and increased positive hemagglutination tests. Increased anaphylaxis also occurred by the fifth study day. The effect appeared to be local, since ozone exposure coupled with peritoneal antigen exposure did not increase sensitization.[172] A 30-minute, 8-ppm exposure to ozone caused both accelerated absorption and increased retention of antigen in the lung.[173] Sensitization and anaphylaxis also increased in guinea pigs who were exposed in a similar protocol to NO_2 and sulfur dioxide (330 ppm, 30 minutes).[174] A case of a chemically sensitive physician exposed to NO_2 and subsequently sensitized to antigen with severe anaphylaxis is described in Volume II, Chapter 12.[175]

Recently, Riedel et al.[176] demonstrated that low-to-medium concentrations of sulfur dioxide could facilitate local allergic sensitization in the guinea pig. Guinea pigs were exposed for 8 hours on 5 consecutive days to clean air or 0.1 to 16 ppm sulfur dioxide. On the last 3 days, sulfur dioxide exposure was followed by inhalation of nebulized ovalbumin for 45 minutes. One week later, the sulfur dioxide-exposed animals demonstrated increased ovalbumin-specific antibodies in serum and bronchoalveolar lavage fluid compared with the animals exposed to clean air. Specific bronchoprovocation with ovalbumin resulted in bronchoconstrictor responses in 67% of the sulfur dioxide-exposed animals (SO_2, 0.1 ppm) compared with 7% of the animals exposed to clean air. This animal study is similar to the clinical situation seen in many chemically sensitive patients. These patients report a toxic exposure, and, as they become sicker and are improperly treated, they become sensitive to common foods and biological inhalants to which they are exposed on a daily basis. These secondary triggers then act similarly to the primary chemical trigger causing more bronchospasm.

A similar nonpulmonary study also emphasizes the fact that pollutants will trigger and increase allergic problems. Miyata et al.[177] has shown that O-P pesticides in guinea pigs enhance the onset of allergic conjunctivitis when these animals are exposed to cedar pollen.

These data indicate that pollutants may alter IgE sensitization and make an individual prone to anaphylaxis. The mechanisms by which this augmentation occurs are unknown, but probably it is the result of pollutant injury to the cell membrane that causes altered recognition sites or the creation of haptens (see Volume I, Chapters 4[54] and 5[171]). Studies at the EHC-Dallas on over 100 patients with recurrent anaphylaxis revealed a wide range of sensitivity including sensitivities to biological inhalants (weeds, trees, grasses, molds), foods, and a variety of toxic chemicals. Most of the patients did not clear their recurrent anaphylaxis until their chemical load was reduced by heat depuration/ physical therapy and massive avoidance of pollutants. In addition, some of these patients had alterations of their T-lymphocytes with low suppressor and high helper cells (see Volume I, Chapter 5[171]). Also, most did not improve until they were on a good program of injection therapy for foods and biological inhalants (see Chapter 36). These study results emphasize how the acute effects of chemical overload can result in many secondary sensitivities that can be as severe as the primary expsosure in propagating the anaphylaxis.

Damage to Antibacterial and Antiviral Activity

Many pollutants are known to damage the bacterial and viral killing capacity in the lower respiratory tract (Figure 19.4). Single toluene exposures of 500, 250, 100, and 2.5 ppm, and five 3 hours day exposures of 1.0 ppm significantly decreased pulmonary bactericidal activity in mice.[178] The EHC-Dallas has seen 100 cases of impaired phagocytosis or bactericidal activity in a subset of chemically sensitive patients with recurrent intractable bronchial infections whose blood contained pollutants. As the pollutants left the blood, the antimicrobial activity returned, and the patients' infections subsided (Table 19.4). Exposure to Aroclor 1254[179] also has shown this phenomenon to occur. In addition, the presence of other organochlorine pesticides has been shown to

Pollutant Overload

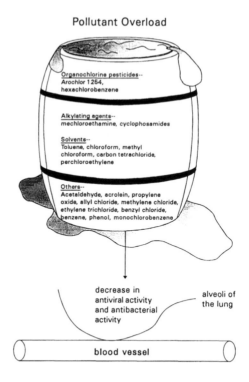

Figure 19.4. Pollutants known to injure the antibacterial and antiviral activity of the lower respiratory tract.

cause defects in phagocytosis and defects in bactericidal and viralcidal capacities of the phagocytes.[180] Correction of these defects in the chemically sensitive patient is difficult, but reduction of total load and occasional use of transfer factor will usually stop the recurrent infections (see Chapter 39).

Other studies report the defects in phagocytosis, such as significant deficits in alveolar macrophage (AM) function, and these deficits have been associated with acute exposure to nitrogen dioxide (NO_2).[181] Also, mice were exposed to formaldehyde and to carbon black and formaldehyde combinations, and increased susceptibility to respiratory infections was quantified by alveolar macrophage-dependent intrapulmonary killing of *Staphylococcus aureus* after an inhalation challenge with bacterium. The salient findings of the bactericidal studies are as follows: 15 ppm formaldehyde impaired the intrapulmonary killing of *S. aureus* when exposure followed the bacterial challenge; 1 ppm formaldehyde impaired the intrapulmonary killing of *S. aureus* when exposure preceded and was continued after the bacterial challenge. CO exposures to target concentrations of 3.5 mg/m³ carbon black and 2.45 ppm formaldehyde or to 10 mg/m³ carbon black and 5 ppm formaldehyde after the bacterial challenge had no effect on the intrapulmonary killing of *S. aureus*. Preexposure for 4 hours per day for 4 days to target concentrations of 3.5 mg/m³ carbon

Table 19.4.   Phagocytic Index (PI) in a Subset (100) of Chemically
Sensitive Patients Who Have Recurrent Infections

Bacteria[a]	PI (%)				
	50–59 (No. of patients)	60–69 (No. of patients)	70–79 (No. of patients)	80–89 (No. of patients)	90–95 (No. of patients)
Staplyloccus Aureus	1	11	44	72	3
Streptococus Epidermidis	0	8	38	61	5
Pseudomonus	0	4	41	59	2
Candida Albicans	0	8	38	55	2

Improvment of Phagocytic Index in Two Chemically Sensitive Patients
Whose Recurrent Infections Stopped with Treatment

Bacteria	Pretreatment		Post-treatment[b]	
	Patient 1	Patient 2	Patient 1	Patient 2
Staphyloccus Aureus	84.5	70.2	95	81.5
Streptococus Epidermidis	84.5	69	95	81.5
Pseudomonus	84.5	69	95	81.5
Candida Albicans	84.5	69	95	81.5

[a] Most had more than one microbe tested.
[b] Individualized: (1) avoidance of air, food, water pollutants; (2) injection therapy for
biological inhalants and foods; and (3) nutrient supplementation.

Source: EHC-Dallas. 1988.

black and 2.5 ppm formaldehyde had no effect on the intrapulmonary killing
of S. aureus when the assay was performed 1 day after the cessation of exposure.[182]

Another study showed that alkylating agents such as mechlorethamine in
rats resulted in decreased clearance of Staphylococcus aureus 502A for 3
weeks. Cyclophosphamide and mechlorethamine both reduced the number of
alveolar macrophages. Alkylating agents have several effects on cellular host
defense responses that could increase either the frequency or the severity of
pulmonary infections.[183] In addition, some of these agents directly injured lung
parenchyma, and they could have effects on intrapulmonary killing processes
independent of any effect on phagocyte number and function.

A murine model was used for staphylococcal clearance to evaluate the
effect of cyclophosphamide and mechlorethamine on intrinsic lung defenses.
Single doses of mechlorethamine (40 micrograms i.v.) or of cyclophosphamide
(150 mg/kg i.p.) reduced peripheral blood neutrophil counts and spleen weights
on the third day after injection. With the exception of neutrophil counts in
mechlorethamine-treated mice, these parameters returned to normal by days 10
to 12. Both drugs reduced the number of alveolar macrophages recoverable by
bronchoalveolar lavage on days 10 to 12, but not day 3. Mechlorethamine

delayed the clearance of *Staphylococcus aureus* 502A from the lung on both days 3 and 12, but cyclophosphamide did not alter clearance on either day 3 or 10. The defect in clearance in mechlorethamine-treated mice resolved by 3 weeks after drug administration. These results demonstrate that alkylating agents do not have uniform effects on antibacterial processes in the murine lung.[183] Since the mechlorethamine effect on staphylococcal elimination appears independent of its effect on macrophage numbers, these results suggest that staphylococcal clearance also depends on nonphagocytic host defense factors. This effect, if applicable to humans, may partially explain why some environmentally susceptible individuals have specific defects for the killing of streptococcus, staphylococcus, candida, and other organisms and yet can kill other microbes without problems.

Several researchers[184] have investigated the potential health hazards of exposure to threshold limit value (TLV) concentrations of acetaldehyde, acrolein, propylene oxide, chloroform, methyl chloroform, carbon tetrachloride, allyl chloride, methylene chloride, ethylene trichloride, perchloroethylene, benzene, phenol, monochlorobenzene, and benzyl chloride compounds. These chemicals may be present in the ambient or workroom atmosphere and often are present in the blood of chemically sensitive individuals. The effects of single and multiple 3-hour inhalation exposures to these chemicals were evaluated in mice by monitoring changes in their susceptibility to experimentally induced streptococcus aerosol infection and pulmonary bactericidal activity to inhaled *Klebsiella pneumonia*. When significant changes in these parameters were found, further exposures were performed at reduced vapor concentrations until the no-measurable-effect level was reached. Multiple exposures on 5 consecutive days were then performed at this concentration. Significant increases in susceptibility to respiratory streptococcus infection were observed after single 3-hour exposures to TLV concentrations of methylene chloride, perchloroethylene, and ethylene trichloride. Tetrachloroethylene (perchloroethylene) is seen in 78% of our patients' blood, possibly explaining why a subset of chemically sensitive indivduals are prone to recurrent respiratory infections.

During the period 1974 to 1987, the frequency of infections in alcoholics discharged from hospitals in Funene showed an increase of 5.53 ± 3.38 ($p = 0.0039$) per 10,000 alcoholics discharged per year. The frequency of discharged alcoholics increased by 3.72 ± 0.60 per 10,000 discharged patients per year. Infections were diagnosed in 0.48% of the discharged alcoholics and in 0.18% of the discharged nonalcoholics ($p<0.00001$). Pneumococcal infections were relatively more frequent in alcoholics ($p<0.05$),[185] thus showing how another toxic chemical will increase the susceptibility to infarctions.

The presence of methylene chloride in 20% of chemically sensitive individuals may, in some instances, contribute to the recurrent infections experienced by a subset of this population. However, not all patients with tetrachloroethylene and methylene chloride in their blood have recurrent infections. Some experience significantly decreased pulmonary bactericidal activity following methylene chloride and (tetra) perchloroethylene exposure.

Clinical Evidence of Pollutant Injury

We have discussed the basic pathophysiology and supporting laboratory data for the subset of chemically sensitive patients who have pulmonary dysfunction. Pollutants cause and/or influence a broad range of symptom complexes and fixed-named diseases including dyspnea and sighing dyspnea, chronic chest wall syndrome, asbestos-mesothlioma, and other asbestos-related disease, bronchitis, asthma, bronchiectasis, cystic fibrosis, allergic alveolitis, granulomatous interstitial disease and parenchymal disease, alveolar capillary block, and acute and chronic respiratory failure. Each has a chemical sensitivity component and will be discussed in the following pages.

Dyspnea and Sighing Dyspnea

Frequently, chemically sensitive patients present with complaints of shortness of breath and breathlessness. The environmentally oriented physician must first discern if a patient's complaints originate in the central or peripheral nervous system or in the bronchopulmonary tree. In our experience, many people with early chemical sensitivity have completely reversible central or peripheral nervous system induced dyspnea. This nervous system induced dyspnea is often missed with the patient being branded a psychoneurotic. Some patients develop severe uncontrollable sighing and breathlessness, which is a clearer clue to respiratory center dysfunction. Randolph[186] originally described a series of cases of sighing dyspnea the cause of which he attributed to environmental incitants. Pulmonary functions are normal in these patients. Usually studies of the breathlessness and sighing dyspnea of chemically sensitive patients at the EHC-Dallas have confirmed Randolph's finding of environmental triggering agents. Some of the patients in our series also developed sleep apnea, which can be entirely cleared with good ecologic management after the environmental triggers are identified.

Chronic Chest Wall Syndrome

Chronic chest wall syndrome is one of the most common clinical syndromes seen in patients with chemical sensitivity. Patients often describe crushing substernal pain, which may or may not radiate down the arms. The patient is always tender in the sternal area and the costal cartilages. Seven hundred patients have been seen at the EHC-Dallas with this problem. One series of 30 representative patients is shown in Table 19.5. Biological inhalants, foods, and chemicals triggered their symptoms.

Table 19.5.   Chest Wall Syndrome — 30 Patients Studied in the ECU after 4 Days Deadaptation with Reduction of Total Body Load

	Type challenge	Percent sensitive
Trees	ID	53
Grasses	ID	87
Weeds	ID	100
Molds	ID	100
Dander	ID	33
Terpenes	ID	40
Smuts	ID	16
Hormones	ID	30
Foods	ID/OC	96
Cow's milk		83
Brewer's yeast		83
Baker's yeast		83
Wheat		66
Sugar cane		63
Beet		56
Corn		50
Beef		50
Toxic Chemicals	INH/ID	93
Formaldehyde (<0.20 ppm)		
Phenol (<0.002 ppm)		
Ethanol (petroleum derived) (<0.05 ppm)		
Chlorine (<0.33 ppm)		
Pesticide (2,4-DNP) (<0.0034 ppm)		
Placebo (spring water)		

Notes: Ages: 20 to 70 years with a mean of 44 years; frequency — 29 years old; 47% of patients; sex — F = 56, M = 44; average age of onset — 27 years old.

Associated symptoms: neurological (76%); respiratory (70%); and cardiovascular (66%). 33% had chlorinated pesticides in their blood. IGE (above control) — 0; T-lymph below 1200 — 43%.

ID = intradermal challenge; OC = oral challenge; and INH = inhaled double-blind

When combination tests were done, they were individual challenges and used as confirmation tests.

Source: EHC-Dallas. 1987.

Asbestosis—Mesothelioma

Asbestosis and malignant changes may be the result of environmental overload. Failure to appreciate the initial exposures and eliminate them may well result in long-term catastrophe. The following case is an example.

Case study. This 83-year-old man consulted an internist complaining of dyspnea on exertion. He reported that except for the relatively recent onset and gradual worsening of his shortness of breath, his health had always been excellent. For more than 30 years, he had worked as a high school physical education instructor and coach. He had never smoked.

Physical examination revealed dry crackles in the lung bases. On pulmonary function tests, all lung volumes were reduced. Chest X-ray revealed irregular interstitial opacities in the lower zones of both lungs and bilateral pleural plaques. He was referred to the author who made the clinical diagnosis of asbestosis.

Only by delving further into the patient's occupational history did this surgeon uncover a key fact: during World War II, 45 years previously, the patient had been employed in a shipyard for several years. His job had involved the spraying of asbestos insulation in a poorly ventilated, confined space, and he did not have adequate respiratory protection. Presenting at the EHC-Dallas, this patient had asbestosis and malignant mesothelioma on biopsy. He also had developed typical chemical sensitivity with a vast array of other environmental triggers such as foods, biological inhalants, and other chemicals that exacerbated his fatigue and pain.

A large number of Americans are at risk of developing asbestosis, a life-threatening disease for which there is no effective therapy. Before enactment of stringent government regulations in the 1970s, millions were employed in workplaces containing high enough levels of airborne asbestos to cause asbestos-related disease.

Asbestos is a fibrous mineral occurring in six major types, all of which are hydrated silicates—chrysotile, crocidolite, amosite, anthophyllite, tremolite, and actinolite. Differences in the chemical structure result in varied physical and biological properties. In the U.S., 95% of commercially used asbestos is chrysotile, the serpentine type, which consists of relatively large, pliable, curly fibrils that tend to occur in bundles. The other forms of asbestos are referred to as the amphibolic group. They have needlelike fibers, which can penetrate more deeply into the lung, in part accounting for their higher pathogenicity.

Today, asbestos has as many as 3000 uses. The construction industry is its major consumer in the U.S. Among products that contain asbestos are cement pipes, flooring, millboard, automobile brake linings and clutch facings, roofing, thermal and electrical insulation, textiles, packing, gaskets, and many others. More than 37,000 people are currently employed in the manufacture of primary asbestos products. Another 300,000 work in secondary asbestos industries. Several million are exposed to asbestos in industries such as brake repair, construction trades, and shipyards.[187]

It has been reported that in the first century A.D. Pliny commented on the sickness of slaves who worked with asbestos. As commercial production began in the late 1800s, a few case reports appeared describing asbestos fiber inhalation in the workplace as injurious to the lung. The first reference to pulmonary fibrosis in an asbestos textile worker appeared in England in 1907. It was not until 1924, however, that Cook described in complete histologic and pathologic detail the presence of "curious bodies" in the lungs of asbestos workers and coined the term "asbestosis".[187]

Asbestosis is currently defined as a diffuse fibrous pneumoconiosis resulting from the inhalation of asbestos particles. The principal lesion is fibrosis caused by fibers that become lodged in the respiratory bronchioles and alveoli. Alveolar macrophages engulf many of the fibers, resulting in cellular death and the release of mediators that appear to induce a chronic alveolitis and ultimately loss of functional alveolar capillary units. Bronchoalveolar lavage of exposed workers with or without radiographically evident asbestosis has revealed an increase in cellularity with an increase in the percentage of neutrophils or lymphocytes, implying a subclinical inflammatory response in the lungs of many individuals.

The pathologic diagnosis of asbestosis is based on the extent and severity of parenchymal fibrosis and on the presence of asbestos bodies (coated asbestos fibers) in the lung parenchyma. Asbestos bodies usually have a beaded surface with clubbed ends. The core is generally a fiber of the amphibolic type. Crysotile asbestos is rarely found in the asbestos-body core. The number of asbestos bodies and the extent of fibrosis tend to correlate poorly.[187]

In most cases, the disease emerges slowly and insidiously. Dyspnea is common, at first occurring on exertion and progressing in severity until it may be present even at rest. Cough appears in the later stages of many cases and is usually dry or produces small amounts of viscid, mucoid sputum. Inspiratory rales sometimes may be heard in the posterior lateral basilar chest. Clubbing of the fingers may be observed in advanced stages.[187] Hemoptysis is not caused by asbestosis, and weight loss is not usual. If either occurs, lung cancer should be suspected.

The diagnosis of asbestosis usually rests on the history of exposure and the presence of parenchymal opacities on the chest radiograph, according to a standard classification system. Although originally designed as an epidemiologic tool, the International Labor Organization (ILO) Classification of the Pneumoconioses has become a benchmark for legal purposes as well. The radiographic appearance of asbestosis is typically irregular opacities in the lower two thirds of both lung fields. Pleural plaques seen along the lateral margins of the thorax and on the diaphragm are highly suggestive of, but not pathognomonic for, asbestos exposure. Progression of radiographic abnormalities is usually observed despite removal of the patient from asbestos exposure.[187]

Pulmonary function in asbestosis is characterized classically by a restrictive impairment with a reduction in forced vital capacity, total lung capacity, functional residual capacity, and residual volume. Gas exchange abnormalities frequently occur as the disease advances. These are characterized by abnormal diffusing capacity, increasing hypoxemia during an exercise test, or resting room air hypoxemia. Obstructive physiology without bronchodilator responsiveness is also a common finding, even in nonsmokers with asbestosis.

Radiographic changes can be seen in the face of normal pulmonary function. Alternately, loss of lung volume and gas exchange may occur before abnormalities appear on X-rays.[187]

Other Asbestos-Related Diseases

In addition to asbestosis, three other major types of diseases have been conclusively related to asbestos exposure. These are benign pleural disorders, lung cancer, and mesothelioma and other malignancies.

Benign pleural disorders include benign pleural effusions, pleural fibrosis, rounded atelectasis, and pleural plaques, with the latter being the most common manifestation of exposure to asbestos. Pleural plaques can be seen with or without asbestosis. Pleural plaques are felt to be markers of exposure and are not premalignant. They can be associated with pulmonary function abnormalities, even in the absence of asbestosis. Asbestos fibers are sometimes identified using electron microscopy within the thickened pleura.

Lung cancer, either squamous cell or adenocarcinoma, is the cancer most frequently associated with asbestos exposure. The tumor often occurs in an area of fibrosis, usually in the lower parts of the lungs. As a rule, there is a dose-response correlation (the greater the dose of asbestos or the longer the exposure period, the higher the risk for developing lung cancer). The average latency period is 20 to 30 years.

The risk for lung cancer is greatly multiplied if the worker exposed to asbestos also smokes, so smoking cessation and regular monitoring of patients with a history of both smoking and significant asbestos exposure are essential so that lung cancer can be abated or detected and treated early.

Mesothelioma, a cancer that arises in the pleura and the peritoneum, is so rare in the general population that it is said to be a "signal tumor" of asbestos exposure. More that 80% of cases occur in persons who have worked with asbestos. Risk may increase by a magnitude of five if exposure begins early in life. The usual latency period is 35 to 40 years. Asbestosis is present in a minority of cases, and there appears to be no relationship with cigarette smoking in the risk for mesothelioma. Most patients survive less than 12 months after diagnosis.

Radiographs of individuals with this disease usually reveal a large pleural effusion, frequently accompanied by a lobulated pleural density that encases a large portion of the lung. The pleural effusion is exudative, and malignant cells can be identified in up to two thirds of patients on thoracentesis. Open pleural biopsy is preferred in order to obtain sufficient tissue for diagnosis. Needle biopsy of the pleura rarely provides enough information. Asbestos bodies are rarely found in the tumor.

Other malignancies with which asbestos has been associated include cancers of the buccal cavity, pharynx, larynx, gastrointestinal tract (stomach and colon), kidney, pancreas, ovary, lymphatic system, and eye. Medical surveillance should therefore include attention to these organ systems in addition to the lungs.[187]

Bronchitis

Environmentally triggered bronchitis can arise from numerous sources creating chemical sensitivity. Workers are exposed in coal mines, steel works, cement works, asbestos factories, ceramics works, heavy engineering trades, and by cigarette smoke. Firefighters with acute smoke inhalation can also develop bronchitis, as can grain, dust, and cotton and hemp, sisal, jute, and flax workers. Any individual with any kind of acute or chronic chemical exposure can develop bronchitis, and it can be propagated for months to years by the secondary sensitization to food and biological inhalants.

Recurrent cough and bronchial infections are commonplace in patients seen at the EHC-Dallas (Table 19.6). Initially, we used bronchoscopy to diagnose all patients fitting this category. However, due to the overwhelming number of patients who have cleared easily with definition and elimination of triggering agents, we have reversed the criteria for the bronchoscopy, reserving it for the recalcitrant. We do check for malignancy when the course of symptoms is prolonged. Overall, adoption of this diagnostic procedure has markedly decreased the cost of medical care in this group of patients.

We have studied 300 chemically sensitive patients with chronic bronchitis. Definition and elimination of triggering agents in these patients is no different from that in any other patients with environmentally triggered problems. These patients are carefully assessed under environmentally controlled conditions after 4 days of deadaptation. We identified the triggers of a sample 30 patients from this population, finding common substances such as food, formaldehyde, phenol, and pesticides (Table 19.6).

In the more recalcitrant cases of bronchitis, detective work is often needed because the triggering agents may be subtle. Careful attention to the bedding including every material and the way it is put together is frequently necessary, since often these patients are worse at night and the triggering agents are in the bedroom. Long-term results in the 300 chemically sensitive patients we monitored who had recurrent bronchitis and for whom triggering agents were identified showed a 95% clearing. This clearing was accomplished without the long-term use of medications. Only avoidance and intradermal injection techniques were used.

Asthma

Responsiveness to pollutants may be modified by inflammatory airway diseases. Asthma is a common inflammatory disease of the respiratory tract that affects between 4 to 8% of the population.[188] Allergic rhinitis is a similar inflammatory disease affecting the upper respiratory tract. Many cases of asthma and allergic rhinitis are triggered by highly specific IgE-mediated allergic responses.[189] The development of specific IgE sensitivity requires a complex process involving antigen processing-presenting cells, T cells, and B cells. The net effect is the production of antigen-specifc IgE, which binds to the

Table 19.6. Bronchitis — 30 Patients Studied in the ECU after 4 Days
Deadaptation with Total Load Reduced

	Type challenge	Percent sensitive
Trees	ID	67
Grasses and weeds	ID	63
Molds	ID	60
Terpenes	ID	30
Foods	ID/OC	
Cow's milk		76
Wheat		70
Corn		66
Yeast		53
Eggs		53
Chicken		50
Toxic Chemicals	ID/INH	66
Formaldehyde (<0.20 ppm)		
Phenol (<0.002 ppm)		
Ethanol (petroleum derived) (<0.50 ppm)		
Chlorine (<0.33 ppm)		
Pesticide (2,4-DNP) (<0.0034 ppm)		

Notes: ID = intradermal challenge; OC = oral challenge; and INH = double-blind
inhaled challenge.

Source: EHC-Dallas. 1992.

surface of mast cells and basophils at the airway surface. On re-exposure, the
sensitizing antigen crosslinks the antigen-specific IgE, activates the mast cell,
and thereby releases an array of mediators from the cell. These mediators,
including histamine and leukotrienes, amplify the biologic response through
the induction of increased vascular permeability and recruitment of inflammatory cells. Recurrent exposure to the same antigen results in an increasing
magnitude of symptoms and inflammation (so-called priming).[190] While the
previous studies describe the experience of many asthmatics seen at the EHC-
Dallas, a larger group of asthmatics have environmental triggers that do not
involve the IgE mechanisms. The asthma experienced by both groups may be
manifestations of chemical sensitivity.

One of the features of asthma is nonspecific bronchial hyperreactivity.[189,191] This nonspecific reactivity is, in fact, a misnomer in our experience
at the EHC-Dallas, since the triggering agents of patients studied in the deadapted
state can almost always be identified. Asthmatics will demonstrate reductions
in lung function at concentrations of substances that will not affect normal
individuals. Hyperreactivity is usually demonstrated in the clinical setting by
inhalation of methacholine or histamine, cold air, exercise, and hypo- or
hyperosmolar aerosols. Cholinergic responsiveness will increase during the

late phase response to antigen challenge testing or following ozone inhalation or viral infection.

Many asthmatics demonstrate an increased responsiveness to low level irritants. Acute decrements in FEV_1 following sulfur dioxide exposure occur at a tenfold lower concentration in asthmatics compared with nonasthmatics.[27] Some, but not all, studies indicate that the response to ozone may be amplified.[192] In contrast, the acute response to formaldehyde is not increased.[20,21] However, these studies were performed on the patients who were in adapted states. Generally, studies at the EHC-Dallas show an increased sensitivity in those chemically sensitive individuals who have been deadapted and are formaldehyde sensitive.

Less documentation in the literature exists on the increased response to mixtures of volatile organic compounds in asthmatics. Burge[193] described increased volatile organic compound (VOC) intolerance in a follow-up study of workers with occupational asthma due to colophony (soldering flux). Shim and Williams,[194] in fact, described the bronchoconstrictor response of a group of asthmatics to inhalation of mixtures of volatile organic compounds. These VOCs included insecticides, perfumes and colognes, household cleaning agents, cigarette smoke, automobile exhaust and gasoline fumes, fresh paint smell, and room deodorant spray.[195] At the EHC-Dallas, these agents have also been shown to trigger chemical sensitivity. A controlled inhaled challenge study of asthmatics has demonstrated variable decrements in pulmonary function after controlled environmental tobacco smoke exposure, but, again, these patients were studied in the adapted state.[188] Those patients we have studied at the EHC-Dallas in the deadapted state demonstrate consistent pulmonary function.

Studies by Weinreich and Undem[196] offer an interesting line of investigation, which may illuminate the interaction between IgE mediated disease and pollutant exposures. These investigators have demonstrated that antigen exposure in previously sensitized guinea pigs will prolong the duration and increase the amplitude of firing of the sympathetic ganglia. Studies at the EHC-Dallas show that a minimum of 70% of chemically sensitive individuals studied have autonomic dysfunction as recorded objectively on the iris corder (see Chapter 27). These studies clearly indicate that nonspecific responsiveness to some pollutants may be increased by the presence of specific immunologic airway diseases. The basis for this increased responsiveness remains an active area of research.

Some groups have deemed acute and chronic asthma to have both an intrinsic and extrinsic etiology. This division is, in our opinion, misleading since all lung problems have an environmental component due to the fact that individuals came into contact with the external environment. Although the etiology of asthma in some patients involves metabolic deficits, most patients with asthma appear to have both environmental and metabolic involvements. Asthma from pollutant injury results from either the pollutant combining with a protein and forming an antigen or from direct bronchial injury that causes hyperreactivity or both.

The environmental triggers of acute and chronic bronchospasm usually can be found if studies are performed under the most precise environmentally controlled conditions with the patient in the deadapted state. Though it is well-known that twitchy airways play a significant part in the diagnosis and treatment of asthma, it has become clear that therapy just for this aspect has not always been productive. In fact, the worldwide mortality for asthma has increased with the use of many medications.[197-199] This is not to say that aminophylline, steroids, and other bronchodilators are useless in the diagnosis and treatment of asthma. However, our data and experience at the EHC-Dallas suggest that more vigorous attempts at defining all the triggering agents and then eliminating exposure to as many as possible for a given asthmatic allows time for the bronchi to heal which, in turn, tends to decrease airway reactivity significantly. Elimination of pollutant exposure almost always reduces the need for medication. In our experience, definition of triggering agents can be so precise as to allow elimination of medication, including bronchodilators and steroids.

At the EHC-Dallas, we have treated 500 asthmatics and over 100 steroid-dependent asthmatics. Over the long term, 3% have remained on steroids at the level of adrenal replacement 5 mg of prednisone every other day. Extreme care must be used when searching for triggering agents in the severe asthmatic, especially those who are cortisone dependent or who have extremely hyperactive airways. We prefer to study most of these patients in the controlled environment of the hospital where all resuscitative equipment is available. We perform initial fasting accompanied by careful monitoring. Once pulmonary flows reach the 250 to 300 range, we can do judicious incitant testing. We perform pulmonary flows pre- and post-challenge on multiple occasions. These flows are used before and after each intradermal skin test as well as with each individual biological inhalant, food, and chemical challenged. The following case study is an example of an asthmatic who not only was sensitive to many environmental incitants but also was violently sensitive to steroids and aminophyllin compounds, including intravenous aminophyllin and steroids.

Case study. This 61-year-old female was admitted to the ECU with the chief complaint of asthma intractable to medications, insomnia, and arthritis. She had sensitivity to all asthma medications. She reported that she had always been a fairly healthy person until March 7, 1986, when she developed wheezing while lying on her couch. She denied exposure to any triggering agent that might have precipitated the asthma attack. Eventually she was taken to the emergency room, where she was treated with cortisone and epinephrine. After 1 month without relief, she began treatment for allergy using the prick method. The allergy testing produced increased wheezing and shortness of breath. She was found to be sensitive to all asthma medications. After testing, her asthma exacerbated severely enough to require hospitalization until she could be stabilized. She was treated in the ICU, and once stabilized, she was referred to the EHC-Dallas.

This patient's physical exam was normal, except for wheezing in both lung fields, with restriction of inspiration. Her EKG showed signs of previous myocardial infarction. Her chest X-ray was interpreted as normal, while pulmonary functions revealed severe obstructive and restrictive disease, with no response to bronchodilators (Table 19.7).

She was placed in the ECU, where she experienced headache as well as persistent asthma. She had exacerbations of her asthma from challenge with an aminophyllin compound so the intravenous aminophyllin had to be stopped. Upon challenge, she proved sensitive to all medication, including steroids, aminophyllin, and all inhalers. Her pulmonary function decreased with each administration.

This patient was placed on nothing orally except spring water for 4 days. At one time, her flow peak was zero for several hours. She finally broke this and improved with fasting. After 4 days, she related that her headaches had improved satisfactorily as had her asthma. Her peak flowmeter was much improved from 50 cc/mm on admission to 340 cc/mm after fasting and time in the controlled environment. Finally, her total pulmonary function was much improved (Table 19.7).

On admission, laboratory studies showed this patient had elevated T_4 lymphocytes at 60% (C = 38 to 52%), elevated T_4/T_8 ratio at 2.4 (C = 2.0 to 2.2), EB virus early antigen at 1:40 and nuclear antigen at 1:5, and anti-VCA antibody at 1:2560. Urine analysis revealed 4+ ketones. Psuedocholinesterase was depressed at 2.1 U/mL (C = 2.4 to 4.9 U/mL), and blood histamine level was elevated at 11.0 µg/dL (C = 3.0 to 9.0 µg/dL).

During her hospitalization in the ECU, this patient underwent extensive intradermal serial dilution injection testing and was found to be quite sensitive to molds, which in turn exacerbated the asthma. The skin testing showed she was sensitive to 21 foods and 6 chemicals, including cigarette smoke, orris root, ethanol, newsprint, perfume, and phenol. She was also sensitive to seven molds, six terpenes, cotton, fluogen, and dust. Double-blind inhaled testing confirmed her chemical sensitivity (Table 19.8).

After this patient was discharged, she did her utmost to minimize exposure in her home. She removed all her curtains, laid ceramic floor tile in the hall and bedrooms, removed the carpet in another bedroom, and removed the gas heat. When she is not in town, she stays at her camp cabin, where there is no carpet. She is feeling quite well. She also is continuing an immunotherapy program and a rotary avoidance diet. She reports that her breathing has improved without medication. She sleeps all night without sleeping pills, and no breathing problems occur during the night. She takes no medication.

In addition to biological inhalants, multiple foods and chemicals often trigger the recurring acute asthma of many patients. The previous case report shows flow changes with intradermal injection neutralization, including food and biological inhalant flows before and after (Table 19.8). If onset can be found early, whether in an adult or child, the etiologies for asthma usually can be demonstrated and eliminated (Figure 19.5). Then the asthma itself can be completely eradicated.

Table 19.7. Admitting Lung Functions in 61-Year-Old Female with Severe Intractible Asthma Who Reacted to All Respiratory Medications

Lung mechanics	Prebronchodilator		Predicted	Postbronchodilator	
	Actual	Percent		Actual	Percent
FVC (liters)	1.00	31	3.16	1.00	31
$FEV_{0.5}$ (liters)	0.36			0.37	
FEV_1 (liters)	1.00	42	2.33	0.55	23
$FEF_{25\%-75\%}$ (L/S)	0.38	14	2.62	0.25	9
$FEV_{0.5}$/FVC (%)		36			37
FEV_1/FVC (%)		100			55
$FEV_{0.5}$/FEV_1 (%)		36			67

Pulmonary Functions After Treatment in the ECU with Total Load Reduced for 3 Weeks; No Bronchodilator or Steroids

FVC (liters)	3.25	99	3.27	3.36	103
FEV_1 (liters)	2.24	93	2.42	2.24	100
$FEF_{0.2-1.2}$ (L/s)	3.00	62	4.84	4.00	83
$FEF_{25\%-75\%}$ (L/s)	1.45	54	2.68	1.30	49
FEV_1/FVC (%)		69			67

Source: EHC-Dallas. 1993.

Table 19.8. The Change of Peak Flow Before and After Intradermal Antigen Injection Testing and Neutralization after 4 Days Deadaptation with Total Load Decreased in the ECU

Antigen dilution[a]	C[b]	0.05/1	0.10/1	0.15/1	0.20/1	0.05/2	0.10/2	0.05/3	0.10/3	0.15/3
					Peak flow					
Dates	380	310	340[c]	330	320		340	380[c]		
Fig	370	325			370[c]	310		300	300	320
Ethanol	350	320	330	310			320			350[c]
Newsprint	380	330	320	345		380[c]	320			310
Formaldehyde	380	150	175	350[c]		150	100			

[a] Intradermal dose.
[b] Before antigen challenge.
[c] Neutralizing dose.

Source: EHC-Dallas. 1985.

In our series of 500 chemically sensitive patients with asthma, the environmental aspects of their illness were discovered to be multiple and varied. A sample series of 30 patients is shown in Table 19.9. A total of 240 double-blind, inhaled challenges for toxic chemicals at ambient doses was performed; 90 placebo challenges, using either normal saline or spring water as blanks, were

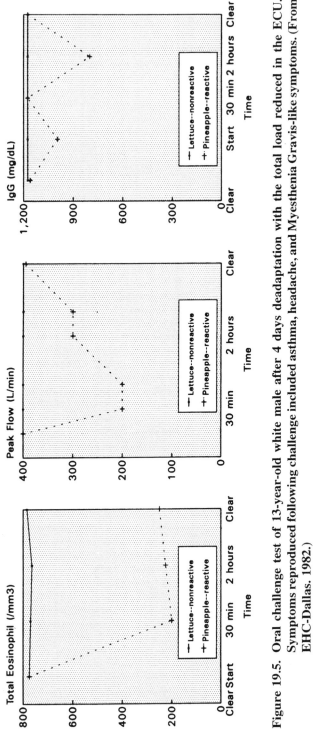

Figure 19.5. Oral challenge test of 13-year-old white male after 4 days deadaptation with the total load reduced in the ECU. Symptoms reproduced following challenge included asthma, headache, and Myesthenia Gravis-like symptoms. (From EHC-Dallas. 1982.)

Table 19.9.   30 Asthma Patients Studied in the ECU after 4 Days of Deadaptation with Reduction of Total Body Load

	Type challenge	Percent sensitive
Trees	ID	53
Grass and Weeds	ID	53
Molds	ID	66
Dander	ID	40
Terpenes	ID	40
Smuts	ID	20
Foods	ID/OC	100
Wheat		79
Cow's milk		72
Eggs		100
Oranges		100
Shrimp		87
Toxic Chemicals	INH/ID	60
Formaldehyde (<0.2 ppm)		14
Phenol (<0.002 ppm)		13
Ethanol (petroleum derived) (<0.50 ppm)		10
Chlorine (<0.33 ppm)		10
Pesticide (2,4-DNP) (<0.0034 ppm)		13
Placebo		<1

Notes: Ages: 13–63 years; mean = 38 years; most common age: 30–59. Sex: F = 53; M = 49. Total challenges (double-blind) = 240; active ingredient = 150; and placebo = 90. Associated symptoms: neurological (60%); cardiovascular (20%). Active ingredients = 81 pos.; placebo = 9 pos.; active ingredients = 69 neg.; placebo = 81 neg. WBC below 4,500: 6%; IgE above 150: 36%; and T lymphocytes: 44% (below 1200/mm^3). ID = intradermal challenge; OC = oral challenge; INH = inhaled challenge (double-blind); when combination tests were done, they were used as confirmation tests.

Source: EHC-Dallas. 1992.

completed. We performed 150 challenges of the active ingredients of toxic chemicals (<0.2 ppm of formaldehyde, <0.33 ppm of chlorine, <0.50 ppm of petroleum derived ethanol, <0.0034 ppm of pesticide, and <0.005 ppm of phenol). Five challenges of active ingredients per patient were performed. Results were judged positive if the patient had observable signs and symptoms in addition to a change in peak flow of 2 standard deviations. All patients were deadapted for at least 4 days with reduction of their total load by using chemically less-contaminated water while fasting in the ECU. Ninety responses to challenge were positive, 81 of which were to the active ingredient and 9 of which were to the placebo (Table 19.9). No patient who reacted to the placebo reacted to more than one of the three placebo challenges. In the opinion of the staff at the EHC-Dallas, all asthmatics should be studied for environmental triggers before being doomed to a life of medication and subsequent chance for increased morbidity and mortality.

Maberly and Anthony[200] studied 19 consecutive asthma patients in a controlled environment similar to the ECU at the EHC-Dallas. He fasted patients for 2 to 6 days using parameters similar to those used for study at the EHC-Dallas. He found that peak expiratory flow fell initially but returned to above the admitting level by a statistically significant range ($p<0.0005$) by the time of discharge (Table 19.10). These deadapted patients were then challenged orally and intradermally with individual incitants. Bronchoconstriction was provoked by both individual foods and chemicals. In the 19 patients with 646 food challenges, Maberly found that peak flow was consistent in some while it fell in others. The peak flow of those whose flow fell returned to normal after intradermal neutralization injections. Follow-up showed that 68% were definitely better and 5% were well or almost well. Symptoms were significantly reduced and use of medication was less at follow-up than at admission ($p<0.01$). Maberly also found that the overall distribution of the maximum change in the PEF after each food challenge in nine patients (with no medication in the ECU) differed positively from the expected best fit normal distribution ($p<0.0001$). The PEF fell after all types of food challenge, but there were differences ($p<0.05$) with more frequent drops (often delayed) in response to challenge with meat, fish, and grains. When the PEF remained low after repeated adverse reactions, recovery most frequently occurred after a fruit meal ($p<0.0001$) (Table 19.11). These findings are similar to ours at the EHC-Dallas.

More than 200 agents have now been associated with occupational asthma. According to Alberts,[201] these agents include high molecular weight substances and low molecular weight agents such as diisocyanates and anhydrides. With some substances, occupational exposure leads to asthma. For example, Alberts found that 5 to 10% of those who reacted to the toluene diisocyanate and 20% of those who reacted to the trimethyl anhydride were previously occupationally exposed, demonstrating some patients can be triggered by toxic chemicals. In some other series, though, the patients are not as well worked-up environmentally as they were in Alberts's study and, therefore, cause and effect has not been as readily identifiable nor has treatment been as successful.

One series of toluene diisocyanate sensitivity in relationship to asthma has been reported that is of great interest to those working in chemical sensitivity;[202] 114 subjects with asthma induced by toluene diisocyanate were identified. Bronchial challenge with exposure of 10 to 25 ppb for 10 to 15 minutes revealed 24 immediate responses, 50 late, and 40 dual responses. At diagnosis, subjects with a dual response showed a longer duration of symptoms and a greater prevalence of airway obstruction. In these subjects, FEV_1 was lower than in subjects with immediate or late reactions to toluene diisocyanate. The percentage of current smokers and ex-smokers was significantly lower in subjects with a late response (26%) than in subjects with immediate or dual responses (56% and 57%, respectively), suggesting that the adaptation mechanism entered into the response. In 27 of the 114 subjects a nonspecific challenge test with methacholine was performed, and subjects with dual responses showed greater nonspecific bronchial hyperresponsiveness than the other groups. This study emphasizes the varied responses seen in this type of chemical sensitivity. Furthermore, these results suggest that a dual response to specific challenge in bronchial asthma related to toluene diisocyanate may be associated with more severe disease than other types of responses, as assessed by duration of symptoms, baseline airway obstruction, and nonspecific bronchial hyperresponsiveness. The high prevalence of nonsmokers and low prevalence of smokers in the group with a late response to specific challenge is so far unexplained but suggests adaptive mechanism involvement.

In another group of toluene diisocyanate (TDI)-induced asthma patients, challenge tests were performed. Twenty patients were examined in order to assess their threshold response to TDI during specific bronchial provocative tests (BPT).[203] Specific bronchial hyperresponsiveness occurred on different days. The threshold response to TDI-induced asthma (low: 0.02 to 0.05 ppm; moderate: 0.1 ppm; high: 0.2 to 0.25 ppm) and the pattern of positive response were evaluated in comparison with some clinical features of the disease. The threshold of airway response to TDI was low in nine, moderate in seven, and high in four patients, again emphasizing the variability seen in chemically sensitive individuals. No evident relationship was observed between the threshold response to TDI and the pattern of positive response to the lower TDI concentration (immediate in five, late in eight, and dual in seven subjects) or other clinical features (duration of asthmatic symptoms, smoking habits, cessation of work, nonspecific bronchial hyperresponsiveness to methacholine). However, 6 out of 9 patients with low threshold had nonspecific bronchial hyperreactivity in comparison with 6 out of 11 patients with moderate or high threshold. In 10 out of 13 patients who performed two positive BPT with different TDI concentrations, the pattern of response was the same both at lower and at higher TDI concentrations. Three subjects who had a late reaction at the lower concentration showed a dual reaction to the higher TDI concentration. A relationship between the degree of the specific bronchial reaction (percent fall in FEV_1 from baseline value) and TDI concentration during BPT was observed for the immediate reaction, but not for the late reaction.

Isolated cases of enzyme and vitamin deficiency in relation to some children's asthma have been reported.[204] Some children have been reported to have their asthma totally cleared by injections of Vitamin B_6, a known cofactor in many enzyme detoxification systems.[205] Treatment with specific nutrients may well be a fruitful option in the next several years and bears investigation. We have quite successfully used both the intravenous and oral nutritional route for the therapy of many chemically sensitive patients, including those with asthma.

Challenge test responses can be plotted in various ways, and both the responses and the methods of plotting them have been discussed throughout this book. Immune and nonimmune changes often can be plotted in addition to pulmonary flow. A challenge for foods is shown in Figure 19.5. Here, marked changes occurred with eosinophils and a biphasic response with gamma globulin.

Table 19.10. Details of the Patients and Their Reactions to Challenge with Inhalant Allergens, Foods and Chemicals, and Their Maximum Changes in PEF

Patient no.	Age	Sex	Asthma group	Asthma severity	Inhalants Wh/PEF	Prick (+ve)	ID (+ve)	Median (EP)	Foods Wh/PEF	EPs No.	Median	Chemicals Wh/PEF	Eps No.	Median	Down	Recovery
1	15	F	Other	Mild	0	Yes	8	11	11(7)	23	2	0	3	2	70	–
2	21	F	A&E	Moderate	0	Yes	11	7	11(5)	32	4	0	13	3	100	130
3	48	F	Asthma	Severe	P	Yes	12	5	23(20)	28	3	3(2)	8	2.5	200	170
4	15	F	A&E	Moderate	5(2)	Yes	10	4.5	11(3)	16	3	2(1)	1	1	100	70
5	15	F	Other	Mild	0	No	9	4	7(2)	17	3	3(0)	5	3	110	–
6	38	F	Asthma	Severe	–	No	12	5	14(10)	36	3	2(1)	9	3	130	60
7	41	F	Other	Moderate	4(0)	Yes	9	5	1(0)	12	5	–	1	(2)	30	30
8	23	F	Asthma	Severe	2(2)	Yes	11	5	*	–	–	0	–	–	*	–
9	51	F	Other	Mild	0	±	9	4	15(8)	31	3	3(0)	8	2.5	100	100
10	54	F	Other	Moderate	–	Yes	11	4	8	28	3	2(0)	8	2	120	100
11	20	F	Asthma	Moderate	–	Yes	10	4	0(1)	2	2	–	1	(2)	40	–
12	43	M	Asthma	Moderate	–	Yes	5	6	16(10)	33	5	3(3)	8	4	110	95
13	36	F	A&E	Moderate	–	Yes	11	4	16(9)	22	3	4(0)	7	4	130	110
14	62	F	Asthma	Moderate	–	±	9	4	9(8)	11	2	0	–	–	100	90
15	39	F	Other	Mild	–	Yes	10	3	15(1)	13	3	0	–	–	90	–
16	15	M	A&E	Moderate	0	Yes	11	6	5(2)	16	3.5	0	12	3	*	90
17	50	F	Other	Mild	–	Yes	12	5	18(1)	12	3	–	–	–	60	–
18	51	F	Asthma	Severe	0	Yes	11	4	17(0)	30	3	4(0)	5	3	130	70

19	54	M	Asthma	Moderate	–	12(4)	Yes	15	195	11	3	8(0)	208(94)	10	2	1(0)	27(7)	2	2.5	70	30
Total		M		M			Yes	11	195	11	3	8(0)	208(94)	10	2	1(0)	27(7)	2	2.5	70	30
Median	39							11		11	4	11(8)		19	3		91		3		
(of those +ve)																	7				

Notes: A&E = asthma and eczema; Wh/PEF = numbers of challenges provoking wheeze or PEF drop ± 30 L/min (>15% of predicted PEF in brackets); ID + ve = numbers of inhalant allergens positive on ID testing; EPs titrated (some for other symptoms), and median titre of EPs; – = not monitored; P = wheeze from prick tests. * = not assessable because of frequent use of solbutamol. NB: Patient 8 used salbutamol throughout and refused EP titration for foods; Patient 10 did not agree to taking post challenge PEFs but wheezed after eight food challenges; Patients 16 and 18 used salbutamol frequently in the first week of challenge.

Source: Maberly, D. J., and H. M. Anthony. 1992. Asthma management in a "clean" environment: 2. Progress and outcome in a cohort of patients. *J. Nutritional Med.* 3:231–248. With permission.

Table 19.11. Number of Challenges with Individual Meats, Fish, Dairy, Grains and Drinks and Changes in the PEF After Challenge in Patients Not Taking Asthma Medication

Food challenge	Fall in the PEF (L/min) after challenge for patient no.:								
	2	4	5	7	9	11	14	17	19
Meats									
Beef	—	—	40	ch	40	rr	R	40	—
Pork	ch	40	—	—	60		60	70	30
Lamb	ch	—	—	—	50	30	—	60	ch
Chicken	50	—	rr	ch	R	—	rr		30
Turkey		70	rr		R	—	50		
Duck					—				
Fish									
Haddock/cod	—	30	60	—	40	40		—	
Sole/plaice	90		—	ch	—		—	R	70
Mackerel						—			—
Salmon	ch	30	—		R	—	80		
Tuna		100			R			—	—
Trout			30	—			R	—	
Prawns			30				60		
Dairy									
Milk	*	—	—	—		rr	—		—
Cheese		rr		—	50	—	—	rr	50
Egg white		30	—	—	40	—	30	R	—
Egg yolk		—	—	—		—	80	—	60
Grains									
Wheat	—	40	—	—	—	—	70	30	50
Corn	—	40	—	—	—	—	—	50	30
Oats	30					—	R	—	
Rice	ch	R		ch	rr	40	90	—	—
Rye	—	—	30						
"Drinks"									
Coffee	40				50	—			—
Tea		rr	—	—	100				
Chocolate		—	—	—	40	R		rr	—
Cane sugar	rr		—		—		R	rr	R
Honey	R								

Notes: No entry, not tested. ch-PEF not fully monitored (less than one pre- and three post challenge readings). – = change ≤20 L/min (or equal rise and fall). R = rise ≥30 L/min. rr = "recovery" rise ≥30 L/min. * = severe reactions to dairy products during previous outpatient testing.

Source: Maberly, D. J., and H. M. Anthony. 1992. Asthma management in a "clean" environment. 1. The effect of challenge with foods and chemicals on the peak flow rate. *J. Nutritional Med.* 3:215–230. With permission.

When the individual was challenged with pineapple, reproduction of flow changes occurred along with clinical asthma. Challenge with a control food, lettuce, resulted in no changes in eosinophils or IGG peak flow. Also asthma was not reproduced.

At the EHC-Dallas, we have studied immune parameters in 27 environmentally sensitive asthmatics, and, as a group, they demonstrated few changes when compared with controls. In one series, 36% of the patients had elevated IGEs. This immune test was the only one that showed any change in this series. Also in this series, complements were no different from the rest of the chemically sensitive, which meant that one third were low. T and B cells showed no significant changes in this group as compared to controls (Table 19.12). In contrast, a later study of a group of 30 patients at the EHC-Dallas showed 49% of the T-lymphocytes were below control levels of 1200 mm^3. Studies of chemically sensitive patients with asthma apparently reflect no different outcomes from those found in chemically sensitive individuals who have not been identified with asthma. In both of these sets of patients, changes in the peripheral immune system do not occur. Rather, changes seem to occur in about 33% of the combined patients.

Bronchiectasis

Injury to the bronchi can be severe enough to cause either saccular dilatation or cylindrical damage to the bronchi. In the past, tuberculosis or whooping cough were the most common causes. Today, however, toxic chemicals with immune suppression are more frequently the cause. Bronchiectasis is not as common as it used to be. A case report in Volume IV, Chapter 41, exemplifies this entity. The patient described there had environmentally induced bronchiectases from exposure to dry cleaning fluids. Her case exemplifies how pollutant exposure can cause bronchial damage. We have seen other cases due to chemical sensitivity and tuberculosis combined. Following is an illustrative case report of a patient with severe chemical sensitivity that induced tuberculosis which, in turn, induced bronchitis.

Case study. In August 1982, a 57-year-old female ophthalmologist began to have acute bronchiectasis, and in December 1982 she had resection of the lingula. The etiology was TBC. She lacked primary peristaltic waves in the esophagus and also had a small schatzski ring. In March 1983, her symptoms of diffculty swallowing and postnasal discharge worsened. In April 1983, she was admitted to the ECU. At that time her chest X-ray showed emphysematous changes with an acute process seen in the left lower lobe, and X-ray of the esophagus showed reflux. The toxic chemical panel of her blood showed benzene, 1 ppb; toluene, 1 ppb; xylene, 0.5 ppb; chloroform, 2 ppb; bromoform, 1 ppb; trichlorethane, 0.3 ppb; beta-BHC, 2.3 ppb; DDE, 0.4 ppb; endrin, 0.8 ppb; and hexachlorbenzene, 0.5 ppb. She had multiple food and chemical sensitivity by challenge. Chemical inhaled challenge showed formaldehyde (<0.33 ppm, 15-minute exposure) caused burning

Table 19.12.   Immunological Data of Asthma Patients and Normals (Mean Values and Differences)

	Patients (n = 27)	Normals (n = 60)	Difference (patients to normal)	Significance
WBC (#/mm³)	7700 ± 360	7560 ± 220	None	>0.3
L (#/mm³)	2610 ± 130	2770 ± 91	None	>0.1
L (%)	35.0 ± 2.0	37.3 ± 1.1	None	>0.1
T_{11} (#/mm³)	1910 ± 102	2080 ± 74	None	>0.05
T_{11} %	73.1 ± 1.8	75.2 ± 0.8	None	>0.1
T_4 (#/mm³)	1110 ± 71	1160 ± 43	None	>0.3
T_4 %	38.2 ± 1.8	42.2 ± 0.7	Smaller	<0.05
T_8 (#/mm³)	770 ± 58	740 ± 38	None	>0.3
T_8 %	29.1 ± 2.4	25.8 ± 0.8	None	>0.05
T_4/T_8	1.55 ± 0.15	1.70 ± 0.06	None	>0.1
B (#/mm³)	320 ± 48	270 ± 23	None	>0.1
B %	12.3 ± 1.8	9.4 ± 0.6	None	>0.05

Source: EHC-Dallas. 1986.

eyes, hoarseness, and symptoms that cleared in 1 to $1\frac{1}{2}$ hours. Phenol (<0.002 ppm) also caused burning eyes, which returned to normal when this patient left the testing booth. Pesticide (<0.0034 ppm, 10-minute exposure) caused stinging eyes and ringing in her ears, the latter of which continued 4 hours after testing was completed. Chlorine (<0.33 ppm, 8-minute exposure) caused burning of the inside of her mouth and hoarseness. These symptoms cleared 1 hour after she left the booth, and she had to use oxygen. Exposure to natural gas (6 minutes) caused ringing in her ears and impaired hearing and headaches. Symptoms lasted 24 hours, and her heart produced two or three extra sytoles per minute. All tested chemicals caused symptoms, but saline placebo caused no reactions.

Laboratory results included the following: Sed rate was low at 4 mm/hr (normal = 10 to 20 mm/hr); CO_2 was elevated at 32. BUN was low at 8. IgG was high at 2240 mg/dL (normal = 800 to 1800 mg/dL). Secretory IgA was high at 36 mg/dL (normal = 3 to 25 mg/dL). WBC was low at $3.2 \times 10^3/mm^3$ (normal = 4 to $11.6 \times 10^3/mm^3$). T cells were low at 38% (normal = 60 to 80%) and 619/mm³ (normal>1000/mm³). B cells were high at 42% (normal = 20 to 60%) and 684/mm³ (normal>500/mm³). C_3, T_4, T_7, and thyroid stimulating hormone were normal.

Discharge diagnosis was as follows: (1) immune suppression character-ized by T-lymphocytic depression; (2) stopic disease; (3) esophageal motility disorder; (4) mild chronic obstructive pulmonary disease and bronchiectasis; (5) pollenosis; and (6) food sensitivities. At discharge, this patient was advised to follow a rotation diet of chemically less-contaminated foods. She was told to drink glass-bottled spring water and to keep a daily diary. She followed a treatment plan of using her food, chemicals, and inhalant antigens.

Following this plan ensured massive avoidance of pollutants in air, food, and water.

In August 1986, this patient's chest X-ray showed that her heart was not widened at the lungs or well-ventilated bilaterally. No infiltrate was seen. There were multiple old healed left rib fractures with resultant deformity including 5th through 9th left ribs posteriorly. WBC, total lymphocytes and the subsets, T_4, T_8, as well as B cells were normal. T_{11} was high at 90% (normal = 73 to 87%). Chromium was low at 0 to 0.8 ppm (normal = 0.119 to 0.209 ppm). Blood toxic chemicals included DDT (1.4 ppb), transnonachlor (0.5 ppb), dieldrin (0.1 ppb), and heptachlor epoxide (1 ppb).

In May 1992 this patient's plasma amino acid analysis showed cystine was low at 0.61 MCM/100 mL (normal = 1.80 to 7.25 MCM/100 mL); phosphoethanolamiine was low at below detection limits (normal = trace, 0.26 MCM/100 mL). Glutamic acid was high at 12.98 MCM/100 mL (normal = 2.30 to 11.00 MCM/100 mL). Hydroxyproline was high at 2.20 MCM/100 mL (normal = nil to 0.65 MCM/100 mL). 1-Methylhistidine was high at 2.64 MCM/100 mL (normal = nil to 1.15 MCM/100 mL). 3-Methylhistidine was high at 0.72 MCM/100 mL (normal = nil to 0.40 MCM/100 mL). Phosphoserine was high at 0.66 MCM/100 mL (normal = 0.28 to 0.65 MCM/100 mL).

Elements in plasma showed zinc was low at 0.63 ppm (normal = 0.745 to 1.45 ppm). Manganese was not detected (normal = 0.001 to 0.002 ppm). Selenium was low at 0.115 ppm (normal = 0.155 to 0.215 ppm). Copper was high at 1.67 ppm (normal = 0.865 to 1.55 ppm). Silicon was low at 0.36 ppm (normal = 0.50 to 200 ppm). The urine amino acid chromatogram showed almost all of the food-protein source amino acids, including the nutritionally essential amino acids, to be low or deficient. Her erythrocyte glutamic oxalo-acetic transaminase (EGOT) was high at 1.53 (normal<1.25 index). Her erythrocyte glutathione reductase (EGR) was high at 1.3 (normal = 0.90 to 1.20 activity coefficient). Toxic chemicals in her blood included the following substances: 2-methylpentane (12.9 ppb), 3-methylpentane (23.4 ppb), n-hexane (3.1 ppb), trimethylbenzenes (5.5 ppb), 1,1,1-trichloroethane (2.2 ppb), HCB (2.1 ppb), oxychlordane (0.3 ppb), and DDT (1.5 ppb).

This patient developed coughing up of sputum in 1993 and was reevaluated. MRI showed a recurrence or missed bronchiectasis in the lower left lobe. She had a resection for this, which stopped her sputum and recurrent infections. Also, in October 1993, her lung biopsy showed a small fragment of fibrous and fibrosclerotic connective tissue. There was no atypia or evidence of malignancy. This patient has done well since cleaning up her home and modifying her diet.

Cystic Fibrosis

Cystic fibrosis is produced by a lethal, recessive gene that occurs in 1 in 2000 Caucasian births. Many organ systems are effected by this disease. The pancreatic acini are replaced by fibrotic tissue, multiple cysts, inspissated mucus, and eventually fat. The clinical manifestations of the consequent pancreatic insufficiency are malnutrition and steatorrhea. The entire gastrointestinal tract may show inspissation so that the diagnosis can be made by rectal biopsy. The liver and gall bladder may be involved. Sweat contains high concentrations of sodium and chloride. The most disabling obstruction affects the lungs with recurrent infections occurring.

The generalized disorder of exocrine dysfunction in patients with cystic fibrosis affects the physical properties of the tracheobronchial mucus and/or the adequacy of mucociliary clearance. This defect causes retention of secretions with partial or complete plugging of the airways. This plugging provides a nidus for implantation and growth of bacteria. Most deaths in patients with cystic fibrosis are now caused by consequences of bronchopulmonary suppuration.

We have treated some patients with cystic fibrosis who developed a secondary food sensitivity along with recurrent infections and chemical sensitivity, apparently due to the malabsorption and modification of the gastrointestinal tract that altered bronchopulmonary response. These patients were helped some by good environmental control to reduce their total load. Rotary diet was implemented to prevent the development of food sensitivity, and varied antigen injections were administered for foods and biological inhalants. In many of the patients that we have treated, we have observed complete cessation of pulmonary infections, and we have observed a decrease in mucus plugging. The following case illustrates these observations.

> **Case study**. This 25-year-old white female with cystic fibrosis was admitted to the EHC-Dallas with complaints of being unable to mobilize secretions, recurrent bronchial infections with recurrent cough, and gastrointestinal upset. She was tested for inhalants, foods, and chemicals. Although she was sensitive to some molds and chemicals, her main sensitivities were to foods (10 of 30 tested). She was treated with a massive avoidance of pollutants in air, food, and water, a rotary diet of chemically less-contaminated foods, avoidance of the foods to which she was most sensitive, injection of inhalants and foods to which she was sensitive, and nutritional replacement. She rapidly lost her cough, inability to mobilize secretions, the recurrent infections, and the gastrointestinal upset. She became totally asymptomatic and has had no infections or antibiotics in over a year following treatment. She is working at her job with no absenteeism.

Allergic Alveolitis

One group of chemically sensitive patients has in common inhalation of antigen material, which results in inflammation affecting the alveolar wall. Pathological changes occur in the bronchioles and intra-alveolar septa, especially in the center of the large lobules. Numerous environmental etiologies for this condition now exist including molds (farmer's lungs) and other organic materials (wood cutter's disease; maple bark disease). Table 19.13 shows various etiologies of this problem.

The classic episode of allergic alveolitis starts with fever, muscular aches, and general malaise, beginning 4 to 8 hours after exposure to an antigen.

Table 19.13.   Types of Allergic Alveolitis

Condition	Antigen
Fungal causes	
Farmer's lung	*Thermophilic actinomycetes*
Air conditioner lung	*Thermophilic actinomycetes*
Bagassosis	*Thermophilic actinomycetes*
Mushroom worker's lung	*Thermophilic actinomycetes*
Maltworker's lung	*Aspergillus clavatus*
Cheesewasher's lung	*Penicillium casei*
Maple bark stripper's disease	*Cryptostroma corticale*
Sequoiiosis	*Aureobasidium pullulans*
Woodworker's disease	*Cryptostroma corticale*
Suberosis	*Penicillium frequentans*
Paprika splitter's lung	Mucor
Dry rot lung	*Merulius lacrymans*
"Dog house disease"	*Aspergillus versicolor*
Lycoperdonosis	Lycoperdon
Animal causes	
Bird breeder's lung	Avian protein, bloom
Rat handler's lung	Rat protein
Wheat weevil disease	Wheat weevil
Furrier's lung	Animal fur
Pituitary snuff taker's lung	Ox and pork protein
Chemical causes	
Isocyanate lung	TDI, MDI, HDT
Pauli's reagent lung	Pauli's reagent
Vineyard sprayer's lung	Bordeaux mixture
Hard metal disease	Cobalt
Cromolyn sodium lung	Cromolyn sodium
Bacterial causes	
Washing powder lung	*Bacillus subtilis* enzymes
B. subtilis alveolitis	*Bacillus subtilis*
B. sereus alveolitis	*Bacillus sereus*
Uncertain causes	
Sauna lung	Lake water (?)
New Guinea lung	Hut thatch (?)
Ramin lung	Ramin wood (?)
Insecticide lung	Pyrethrum (?)

Source: Seaton, A. and W. K. C. Morgan. 1984. Hypersensitivity pneumonitis. In *Occupational Lung Diseases*, 2nd ed., Eds. W. K. C. Morgan and A. Seaton. 587. Philadelphia: W. B. Saunders. With permission.

Occasionally this reaction is preceded, or accompanied, by wheezing or tightness in the chest and a dry cough. Shortness of breath is a feature of more severe attacks. Symptoms usually peak 8 to 12 hours after their onset. Improvement begins at 24 hours after onset.

Lung biopsies are rarely necessary. In our experience, patients will do well without medication if the etiology can be identified, the incitant removed, and intradermal injection neutralization performed. Morris[206] has reported a large series of patients whose grain mold alevolitis was eliminated by injection and sublingual therapy. These patients were sensitive to many antigens, but *aspergillosis* appeared to be the main offender, accompanied by *alternaria* and *cladisporium* as well as the mold mixes. The following cases reported by Morris illustrate the etiology and efficacy of our type treatment.

Case study. A 48-year-old white male farmer had the more typical pulmonary granulomatous form of farmer's lung disease. In March 1964, he complained of sweats, weakness, shortness of breath, and dull substernal pain after being in the barn. His chest X-ray at this time was normal (Figure 19.6). By February 1966, the recurrent symptoms became more severe, and he had cough, chest pain, and shortness of breath. An X-ray taken at this time showed infiltration in both lung fields (Figure 19.7). Potassium iodide was given, and he was instructed to stay out of the barn. In March 1966, he was no better, and treatment was started with triamcinalone 8 mg 4 times daily, which gave considerable improvement. When the dose was decreased to 4 mg per day, he had a recurrence of his previous symptoms. He was treated in the hospital in April 1966 with corticosteroids and general supportive measures, and he showed good improvement. In the summer of 1966, he did quite well on 10 mg of prednisone daily, and he stayed out of the barn completely. In December 1966, he returned to the barn, and the entire disease syndrome recurred, even though he was taking 10 mg of prednisone daily. In January 1967, he showed immediate and delayed skin reaction to alternaria, hormodendrum, cephalosporia, and mold mix A, B and C of Rinkel. The mold mixture A showed the strongest skin reaction, and hay taken from his barn cultured Mucor fungus on Sabouraud's media. The Mucor antigen is in mold mixture A, and he showed skin reaction to a higher dilution of pure Mucor antigen. Treatment was started with subcutaneous antigen to substances to which he had reacted. Corticosteroids were stopped within 1 month, and sublingual antigen was used. From the spring of 1967 to the present, he has been able to work in the barn and has been taking only the ultrasmall dose antigen. His skin reactions have decreased, and antigen has been strengthened accordingly when he is seen each spring and fall. His chest X-ray cleared, and it remains normal (Figure 19.8).

Case study. This 58-year-old white male farmer illustrates the severe acute symptoms that can result during a hypersensitive clinical phase of mold sensitivity. He was seen in April 1968 complaining of chest pain, weakness, and dyspnea. He was admitted to the hospital from April 30 to May 4, 1968, for a medical evaluation. Chest X-rays showed bilateral infiltrate interpreted as representing interstital fibrosis or chronic interstitial pneumonitis. On May 4, he was discharged, and he was told to avoid exposure to mold. On May 6, he went into the barn to give instructions to his men and that evening became much more dyspneic. By the following morning, he was severely dyspneic and cyanotic and was admitted to the hospital, where respiratory acidosis rapidly developed and unconciousness ensued. He was treated with endotracheal intubation, mechanical ventilation, intravenous sodium bicarbonate, corticosteroids, antibiotics, etc. Immediate reactions to several molds were

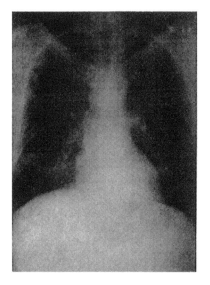 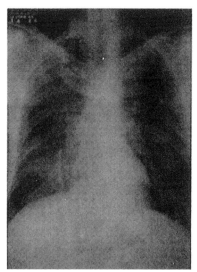

Figure 19.6. X-ray of a 43-year-old white male within normal limits on March 23, 1964, with initial symptoms after working in a moldy barn. (From Morris, D. L. 1970. Treatment of respiratory disease with ultra-small doses of antigens. *Annals Allergy* 23:494–500. With permission.)

Figure 19.7. X-ray same patient (as Figure 19.6), February 12, 1966, with symptoms requiring hospitalization because of continued mold exposure. (From Morris, D. L. 1970. Treatment of respiratory disease with ultra-small doses of antigens. *Annals Allergy* 23:494–500. With permission.)

noted, and he was started on sublingual antigen treatment in the hospital. He has returned to farming and is now able to tolerate the mold and dust.

Granulomatous Interstitial Disease and Parenchymal Disease

Sarcoid-like granulomas with interstitial fibrosis are often seen in chronic lung disease. This condition can be caused by beryllium or, in the case of parenchymal fibrosis, by fibrous hydrated silicates, asbestos, fibrous 2 eolites (erionites), and fibrous clays. Fine fiber forms of tremolites are associated with mesiotheloma.[207,208] Other chemical fibers such as glass, silica, and talc can also cause irritation and fibrosis. At the EHC-Dallas, sarcoid has often been found to be related to mold and perhaps pine terpene sensitivity.

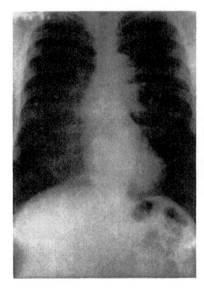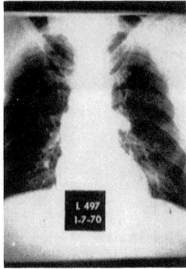

Figure 19.8. X-ray of patient (Figure 19.6) showing (A) clearing (1968) and (B) continued tolerance to mold (1970) with treatment and the continued mold exposure of dairy farming. (From Morris, D. L. 1970. Treatment of respiratory disease with ultra-small doses of antigens. *Annals Allergy* **23:494–500. With permission.)**

The effects of toxic gases and fumes on the lung are local irritation, toxic absorption going to other parts of the body, sensitization, and asphyxiation. Asphyxiation occurs by displacement of oxygen from the inspired air. (Table 19.14) lists those chemicals known to cause respiratory complication. A host of chemicals cause varied reactions.[209-211] "Meat wrapper's" asthma is one of many common reactions from exposure to polyvinyl chloride (PVC) fumes. Patients are seen with reactions similar to those due to exposure to polyvinyl chloride fumes that result from exposure to various other pollutants (Table 19.14).

Alveolar Capillary Block (Figure 19.9)

Pollutant damage may occur to the area of alveolar-vascular membranes. Blocks may be temporary in the case of edema with reversible damage to the alveolar membrane and/or vascular wall and interstitial tissue, or the damage may be permanent due to fibrous scar or hyalin in the interstitial tissue. There appear to be many patients who have reversible edema blocks, which can occur either along or with another pulmonary disease process. The latter case is more common, however, than the former. Mass reduction in total lung pollutant load

Table 19.14. Toxic Gases and Fumes Causing Lung Injury in the Chemically Sensitive

Agent	Principal occupations exposed	Main mechanism of injury	Time-weighted average	IDLH level
Acrolein	Plastic, rubber, textile, resin making	Direct action on mucosa of the eye and respiratory tract, irritant effects	0.1 ppm	5 ppm
Acrylonitrile	Synthetic fiber, acrylic resin, rubber making	Asphyxiant	2 ppm	4 ppm
Ammonia	Fertilizer, refrigerator, explosive production	Direct action on mucosa of the eye and respiratory tract, tracheitis and pulmonary edema	50 ppm	500 ppm
Arsine	Smelting, refining	Systemic effects	0.05 ppm	6 ppm
Cadmium fumes	Ore smelting, alloying, welding	Acute tracheobronchitis, pulmonary edema, emphysema, renal effects	0.1 mg/m^3 (40 μg/m^3)	
Carbon dioxide	Foundry work, mining	Asphyxiant	5000 ppm	50,000 ppm
Carbon disulfide	Degreasing, electroplating, sulfur processing	Systemic effects	20 ppm (1 ppm)	500 ppm
Carbon monoxide	Foundry work, petroleum refining, mining	Asphyxiant	50 ppm (35 ppm)	2500 ppm
Chlorine	Bleaching, disinfectant, and plastic making	Direct action on mucosa of the eyes and respiratory tract, tracheitis and pulmonary edema. Possible chronic effect and airways obstruction?	1 ppm (0.5 ppm)	25 ppm
Copper fumes	Welding	Systemic effects	0.1 mg/m^3	—
Formaldehyde	Disinfectant, embalming fluid use, paper and photography industry	Direct action on mucosa of the eyes and respiratory tract, dermatitis, and asthma?	3 ppm (1 ppm)	100 ppm

Table 19.14 (Continued). Toxic Gases and Fumes Causing Lung Injury in the Chemically Sensitive

Agent	Principal occupations exposed	Main mechanism of injury	Time-weighted average	IDLH level
Hydrogen chloride	Refining, dye making, organic chemical synthesis	Direct action on mucosa of the eyes and lower respiratory tract, tracheobronchitis		
Hydrogen cyanide	Electroplating, fumigant work, steel industry	Systemic effects	10 ppm	50 mg/m^3
Hydrogen fluoride	Etching, petroleum industry, silk working	Direct action on mucosa of the eyes and respiratory tract, tracheitis	3 ppm	20 ppm
Hydrogen sulfide	Natural gas making, paper pulp, sewage treatment, tannery work, oil-well prospecting	Systemic and local effects, pulmonary edema, and asphyxia	20 ppm	300 ppm
Magnesium oxide fumes	Welding, alloy, flare, filament making	Systemic effects	15 mg/m^3	—
Manganese fumes	Foundry work, battery making, permanganate manufacture	Systemic effects, possible predisposition to pneumonia	5 mg/m^3	—
Mercury fumes	Electrolysis	Direct action on mucosa of the eyes, GI tract, and lung. Interstitial pneumonitis, systemic effects	0.1 mg/m^3	28 mg/m^3
Methyl bromide	Fumigating, dye and refrigerant making	Direct action on mucosa of the eyes and respiratory tract	20 ppm	2000 ppm
Natural gas	Mining, petroleum refining, power plant work	Asphyxiant	—	—
Nitrogen	Underwater work, mining	Asphyxiant		

Pollutant	Source/Occupation	Effects		
Nitrogen dioxide	Arc welding, dye and fertilizer making, farming	Irritant to respiratory tract, tracheitis, pulmonary edema, and bronchiolitis obliterans	5 ppm (1 ppm)	50 ppm
Osmium tetroxide fumes	Alloy making, platinum hardening	Direct irritation of respiratory tract	0.002 mg/m^3	1 mg/m^3
Ozone	Arc welding, air, sewage, and water treatment	Direct irritation of respiratory tract	0.1 ppm	10 ppm
Phosgene	Chemical industry, dye and insecticide making	Direct irritation of respiratory tract, pulmonary edema	0.1 ppm	2 ppm
Platinum, soluble salts (mist)	Alloy, mirror making, electroplating, catalyst, and ceramic work	Asthmatic reactions	0.002 mg/m^3	—
Sulfur dioxide	Bleaching, ore smelting, paper manufacture, refrigeration industry	Direct action on the respiratory tract, bronchitis, particularly pulmonary edema	5 ppm	100 ppm
Vanadium pentoxide fumes	Glass, ceramic, alloy making, chemical industry (catalysis)	Direct action on respiratory tract, bronchitis, asthma	0.1 mg/m^3 (0.05 mg/m^3)	70 mg/m^3
Zinc chloride fumes	Dry cell making, soldering, textile finishing	Direct action on respiratory tract, irritant	1 mg/m^3	200 mg/m^3
Zinc oxide fumes	Welding	Systemic effects	5 mg/m^3	—

Source: Seaton, A. and W. K. C. Morgan. 1984. Toxic gases and fumes. In *Occupational Lung Diseases*, 2nd ed., Eds. W. K. C. Morgan and A. Seaton. 611–613. Philadelphia: W. B. Saunders. With permission.

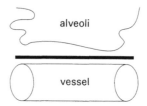

Figure 19.9. Alveolar capillary block seen in a chemically sensitive patient.

will often allow the patient to return to better function. The following is a case report exemplifying the reversible phenomena.

> **Case study**. This 49-year-old male was admitted to the ECU with the chief complaint of severe shortness of breath, coughing up blood, and right shoulder pain.
>
> In 1957, at approximately age 19, he began working as a roughneck in the oil fields. He continued this job for about 6 years, during which time he was exposed to caustic soda and dust. In 1959, he worked on pumps and a compressor, which exposed him to acids and welding gasses. He was exposed to multiple organic fumes, including volatile rocket fuel and Teflon® smoke. He stated that neither respiratory nor ventilation protection was provided for the workers at that time. He had significant chlorine exposure in 1964 and again in 1975. He felt extremely ill after these chlorine exposures and for a while could not eat because of the severe discomfort he experienced.
>
> This patient noted that in 1974, at age 36 years, when he began to work for a chemical company, he had developed shortness of breath while working. At that time, X-rays revealed a shadow on the right lung. His stress tests were considered normal. He took antihistamines, which seemed to improve his symptoms but, in fact, only masked them. In 1981, he developed severe shortness of breath. In 1982, he received allergy testing and began to take allergy shots along with medication. He felt somewhat better. However, he had to discontinue working in September 1982 because of his illness. His condition continued to worsen.
>
> He was in the intensive care unit (ICU) several times before coming to the EHC-Dallas in 1987. He was in the ICU with respiratory distress and hemoptysis for 57 days, including the week before he entered the ECU. He was using oxygen and confined to a wheelchair when he entered the ECU. His physical exam showed an alert 49-year-old male who looked older than his stated age and who had cyanosis with shortness of breath. He was diagnosed with occupational lung disease with bronchial spasm and alveolar capillary block, hemoptysis, and atopic disease (IgE elevated at 464 IU/mL). He was placed in the ECU and fed nothing by mouth except spring water for 4 days. During this time, this patient related that he had less difficulty breathing.
>
> His chest X-ray showed an advanced COPD with pulmonary emphysema, air trapping, and interstitial fibrosis. Pleural and parenchymal scarring was evident, with possible bronchiectatic changes, particularly in the right lower

lung field. His spirometry showed a marked obstructive component with mild response to bronchodilator. However, his respiratory flow was always 450 to 500 cc. His lung scan findings revealed advanced COPD with interstitial fibrosis and bullous emphysema.

His ventilation lung images were abnormal, with both ventilatory and perfusion defects involving the right lung, especially the lower half of the right lobe. The ventilation perfusion match was compatible with airway disease, such as congestive obstructive lung disease.

He had an abnormal EKG which showed a normal sinus rhythm, low voltage QRS compatible with pulmonary disease, and pericardial effusion of normal variant. Nonspecific ST changes on the EKG were apparent also.

His laboratory tests showed depressed RBC at $4.37 \times 10^6/\mu L$ (C = 5.4 ± $7 \times 10^6/\mu L$), elevated iron at 216 µg/dL (C = 50 to 160 µg/dL), elevated MCV at $99.8/mm^3$ (C = 80 to $94/mm^3$), elevated MCH at 34 pg (C = 24 to 31 pg), depressed globulin at 1.9 g/dL (C = 2.0 to 3.5 g/dL), elevated IgE at 464 IU/mL (C = 0 to 126 IU/mL), elevated IgM at 258 mg/dL (C = 45 to 250 mg/dL), and elevated T_4/T_8 ratio at 2.5 (C = 1.0 to 2.2).

Arterial blood gases on March 9, 1987, yielded a pH at 7.44, PaO_2 at 45, $PaCO_2$ at 22, HCO_3 at 18, base excess at 4, O_2 SAT at 83%. After-treatment on March 14 yielded pH at 7.44, $PaCO_2$ at 28, HCO_3 at 20, base excess at 1.6, and O_2 SAT at 92%.

Analysis of his blood toxic chemicals revealed the following results: BHC at 0.4 ppb, DDE at 1.0 ppb, n-hexane at 12.0 ppb, toluene at 0.6 ppb, and 1,1,1-trichloroethane at 0.9 ppb.

His intradermal injection testing showed he had sensitivity to 27 foods; 6 terpenes; cotton; fluogen; MRV; 4 chemicals — orris root, ethanol, phenol, cigarette smoke; 6 molds; lake algae; T.O.E.; candida; dust; dust mite; 2 smuts; histamine; and serotonin (Table 19.15). These findings indicated secondary biological inhalant and food sensitivity resulting from pollutant overload.

This patient dramatically improved in the hospital without medication to the point that he could ride a bicycle for an unlimited time without increasing his hypoxia. He became totally asymptomatic. After treatment, his spirometry findings showed severe obstructive lung disease with improvement in small airways (Table 19.16).

After the time of discharge, he was able to breathe much better than previously. Without medication or redundant shortness of breath, he now rides a bicycle several times a day. Long-term follow-up showed this patient was asymptomatic for $4^1/_2$ years without the use of medication. He did have to build a room in his home that was environmentally controlled and free of pollutants, and he had to stay on a chemically less-contaminated diet and use safer water. He has been free of hemoptysis, infection, and further hospitalizations.

Respiratory Failure — Acute and Chronic

Acute and chronic respiratory failure has many etiologies including shock, viral and bacterial infections, excess oxygen, blood transfusions, pesticides,

Table 19.15.   The Change of Pulse and Peak Flow Before and After
Intradermal Antigen Skin Challenge After 4 Days of Deadaptation
in the ECU with Total Load Reduced

	Pulse/min		Peak flow (L/min)	
Antigen challenged	Before test	After test	Before test	After test
Sweet potato	64	78	460	420
Blackeyed peas	76	84	500	450
Squash	96	116	360	270
Okra	—	—	360	350
Molds	—	—	480	380

Source: EHC-Dallas. 1987.

Table 19.16.   Pulmonary
Function Plot — Before
Treatment vs. After
Treatment in the ECU

	Change (%)
FEV_1 (L)	+11
FEV_3 (L)	+11
PEF (L/S)	+8
$FEF_{25\%-75\%}$ (L/S)	+26
$FEF_{75\%-85\%}$ (L/S)	+22
$FEF_{50\%}$ (L/S)	+17
$FEF_{75\%}$ (L/S)	+66
FEV_1/FVC (%)	+10

Source: EHC-Dallas. 1993.

cystic fibrosis, etc. Even in the chronic patient, attention to the environmental aspects of their illness may prolong life and/or make it more comfortable. Often there is a bronchospastic component that can be eliminated. In addition, edema not only in the bronchial tubes but also in the alveolar capillary membranes occurs. If these responses can be eliminated, the chronic respiratory failure patient can frequently improve. One series at the EHC-Dallas confirmed this point of view (Table 19.17). It is detailed here.

Eight patients ages 18 to 72 years who were admitted to the intensive care unit with acute respiratory failure (ARF) were studied and treated. The methods for determining whether food and/or chemical sensitivity were involved in ARF were careful history of food and chemical sensitivity, blood cytotoxic test of Bryan and Bryan,[212] provocative intradermal testing of Rinkel et al.,[213] Lee et al.,[214] and Miller;[215] skin titration of Rinkel,[216] and total environmental control with withdrawal and direct food challenge of Randolph.[217]

Table 19.17. Respiratory Failure in the
Chemically Sensitive

Age	Sex	Food	Chemicals	Time on respirator (weeks)
43	M	0	+	1
18	F	+	0	1
64	F	0	+	9
66	M	+	+	9
72	M	+	+	10
56	M	+	+	4
56	M	+	+	7
63	M	+	+	7

Source: EHC-Dallas. 1976.

All eight patients were admitted to the intensive care unit with inability to breathe, requiring controlled ventilation by a volume respirator via a tracheostomy or endotracheal tube. Duration of total support by the respirator was from 1 week (18-year-old) to 10 weeks (72-year-old) with an average time for the group of 6 weeks. PaO_2 was 50 mmHg or below with $PaCO_2$ of 60 mmHg or above. Tidal volumes were 200 cc or less, and all patients were clinically dying. Pulses were all above 130, with multiple arrhythmias including multifocal premature ventricular contractions (PVCs) and ventricular tachycardia. All patients initially required controlled ventilation by either respiratory center paralysis with narcotics or muscle paralysis by curare or succinyl choline. All required continuous positive end expiratory pressure breathing (CPPB) of at least 5 cm of H_2O to maintain a PaO_2 of 70 mmHg with a $PaCP_2$ of 50 mmHg or less. All patients were on antibiotics including cephalosporins, garamycin, carbenicillin, and/or kanamycin. Each patient received enough blood to keep hemoglobin above 12 and the hematocrit above 35. Fifty grams of albumin were given daily to treat protein depletion. Intravenous fluids of lactated Ringer's solution and/or 5% dextrose and/or 50% dextrose in protein hydrolysate were given. Total caloric intake was maintained at 1500 calories/24 hours or higher. All were given continuous intravenous vitamins and minerals including potassium, potassium biphosphate, calcium, sodium, and chloride. The dosage of vitamins was up to 3 g/24 hours of C and 1 ampule of B complex, 15,000 units of A, and 1000 units of D. Twenty-five mg of vitamin K in the form of Aqua mephyton were given two times per week, and 2000 units of vitamin B_{12} were injected weekly. For at least 1 week, no food was given by mouth, but all patients were allowed water. During this period of time, tests were made to diagnose food and chemical sensitivities. All eight patients had a strongly positive history compatible with foods and/or chemical sensitivity.

All patients survived, though most of their courses were very stormy. All patients except one required a tracheostomy in order to control the secretions with adequate endotracheal suction. (This study was performed in the early 1970s, and probably today very few patients would have had a tracheostomy since intubation can control the problem). Three patients required chest tubes due to pneuma- and hydrothorax. One had dizziness and hearing loss. Six others required progressive exercise for 1 week to 2 months on an exercycle while still on the respirator in order to develop enough physical strength to breathe spontaneously. With the proper internal and external environmental controls, all patients were able to be off all medications 2 months after discharge.

Two patients had food tests prior to their respiratory failure, but their environment was not well-enough controlled to prevent the respiratory failure. One was a dispatcher for an auto repair garage and had obvious severe chemical sensitivities along with food susceptibility as shown by the cytotoxic test. The second patient had food injections and subsequently oral challenge. She went into acute respiratory failure after a meal of beef and potatoes, which were her two most severe sensitivities. The only patient with pure chemical sensitivity was a spray painter who had a 3-year history of nasal problems of stuffiness and sinus headaches and was found to have no food sensitivities by cytotoxic studies and direct challenge. Previous to admission to the ECU, he went into fulminant pulmonary edema (noncardiac type) after a routine day of spraying using his usual protective mask. The five remaining patients had a combination of food and chemical sensitivities as determined by history, cytotoxic tests, and oral challenge.

After discharge, only one patient was not willing to institute arduous environmental control including a rotary diet using chemically less-contaminated foods and elimination of as many environmental irritants as possible from their homes. The one patient who did not modify his home kept a gas cooking stove, although he had electric heat. He suffered another acute respiratory episode 1 year after his initial failure. He did not have a respiratory arrest this latter time, but he was in the ICU for a week. All other patients were off all medications at this time, except for an 18-year-old who was taking both pollen and food injections. All patients subjectively said they were breathing better than they had in years and objectively appeared so, as evidenced by pulmonary function tests. Also, the only meaningful pulmonary function test (arterial blood gases) that all underwent was done at the time of failure and recovery. PaO_2s were 70 mmHg or above on room air in all patients at the time of discharge. These tests appear to be the only significant pulmonary function tests, since all tested patients were apneic on admission. Four patients have returned to work.

Diagnosis and Treatment

To evaluate the effects of pollutant injury to the lung and bronchial tree in the chemically sensitive patient, many factors need to be considered. Since early pollutant effects often may be subtle and not readily apparent, the physician must carefully observe the patient. He or she should check ventilation to be sure that oxygen is going into the lung, and carbon dioxide is released. The physician should also look for maximum gradients for diffusion. Distribution of inhaled air to separate gas exchange units by the bifurcating tracheal bronchial tree is important and should be observed for abnormalities. Diffusion is the transfer by random molecular motion of gas molecules across the alveolar membranes from the region of high concentration (partial pressure), which is disturbed in some chemically sensitive individuals and should be evaluated. The blood-air barrier of over 90 m² in adults is condensed into a lung volume of 5 liters, made possible by a small radius (150 m²) and a large number (300 million) of alveoli. Perfusion, which is the means by which the outflow of the right ventricle is brought into intimate contact with the capillary bed, should also be observed for malfunction. Often in early pollutant injury, subtle changes will occur in these physiologic functions that may not be measurable by present instruments or, at best, are interpreted as normal variants. The clinician's goal is to find these pollutant-triggered subtle abnormalities in order to eliminate the pollutant triggers and prevent chemical sensitivity.

In addition to central and peripheral nerve triggering, the acute symptomatic response to pollutants primarily involves mucosal surfaces. At the EHC-Dallas, early signs of pollutant exposure have been patients' reports of eye irritation, burning or tearing, nasal irritation, congestion or rhinorrhea (runny nose), throat irritation, hoarseness, cough, chest tightness, wheezing, or shortness of breath.[218] Physiologic changes in response to pollutant exposure in the chemically sensitive individual include an alteration in the pattern of respiration, with a shift to shallower, more rapid breathing. While this change is the classic index of the pollutant response in animal studies, it has generally been ignored as a practical index of mild irritation in the clinical setting. These responses are often seen with early pollutant injury to the chemically sensitive individual, and the astute clinician will be looking out for them. Depending on the severity of the pollutant stimulus, physical examination may show no abnormalities, or it may show erythema, increased secretions, or mucosal ulcerations or alterations of the ANS. Pulmonary examination may be normal or demonstrate wheezing (indicating bronchospasm) or rales (indicating alveolar edema). Other changes that may be apparent include increased airway tone and decreased lung function. According to Bascom,[13] increased airway resistance is the key index of response in the Mead-Amdur guinea pig model, which has been used extensively in the study of the effects of pollutants.[219] Usually pulmonary function measurements are rarely made immediately following a pollutant exposure, although thousands have been made at the EHC-Dallas under controlled inhalation challenge with patients in the deadapted state and confined to an environmentally controlled area. These studies using challenges with formaldehdye (<0.2 ppm), phenol (<0.02 ppm), ethanol (<0.5 ppm), pesticide (<0.0034 ppm), 1,1,1 trichloroethane, and chlorine (<0.33 ppm) showed a decrease in peak flow once the patient became chemically sensitive

(Table 19.8 – Table 19.10). Normal values would be expected from the usual screening blood studies, including a chemistry screen and complete blood count and differential. Delayed effects of pollutant exposures have primarily been recognized from toxic gas inhalations and include airway and alveolar edema and sloughing of the epithelium.[220] Long-term effects have included airway hyperreactivity, bronchiectasis, bronchitis, and pulmonary fibrosis.[218] Greater and lesser responses have been seen in many chemically sensitive patients with respiratory involvement. If pollutant injury has occurred, intradermal, oral, and inhaled challenges under controlled conditions with the total load decreased and the patients in the deadapted state will usually demonstrate changes in some type of pulmonary function.

It is clear that environmentally triggered respiratory disease with early chemical sensitivity can now be diagnosed, reversed, and/or treated by pollutant avoidance in air, food, and water (see Volume IV, Chapters 32 through 34), nutritional manipulation (Volume IV, Chapter 37), and heat depuration/physical therapy (see Volume IV, Chapter 35). The use of the proper controlled environments is necessary for precise diagnosis and treatment (see Volume IV, Chapters 31, 37, and 39).

BREAST

Pollutant injury to the breast is only now beginning to be appreciated. Though the finding of DDT in breast milk led to the banning of this pesticide, most concern was for the effects of this chemical on the baby's and not the mother's health. Recent data about high levels of pesticides found in breast tissue have kindled an awareness of the possible effects of pollutants. This section on the breast presents observations of anatomy and phyiology and the real and prevalent effects of pollutants on the breast tissue. Since the breast is essentially composed of fat globules and lymphatic tissue, it attracts and stores pollutants. Much evidence suggests that this tissue then becomes increasingly vulnerable to illness.

The female breast is an organ subject to many disorders. Although the majority of these disorders relate to benign lesions, the fact that the breast is one of the two most frequent primary sites of cancer in women underscores the importance of determining the nature of the lesion and its potential etiology. Because of its physiologic role in menstruation, pregnancy, lactation, and postmenopausal atrophy, the breast is subject to many alterations in gross and microscopic structure throughout life. The breast is vulnerable to many environmental factors, especially pollutants.

The mammary gland is both ectodermal and mesodermal in origin, and, being an analog of the sweat gland, it is subject to pollutant injury (Figure 19.10).

When one attempts to understand pollutant injury to the breast, he or she must first comprehend the anatomy of the breast. The ducts proceed in retrograde fashion from the nipple. As they radiate peripherally, they subdivide into

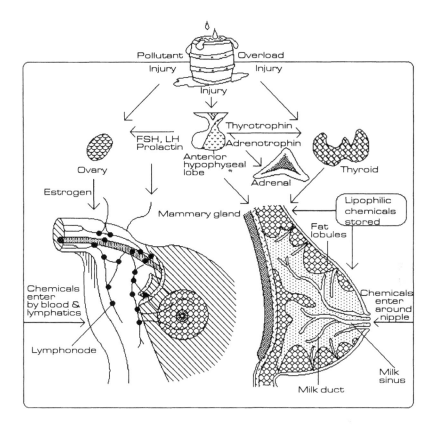

Figure 19.10.   Potential areas for pollutant injury to the breast.

primary, secondary, and tertiary ducts, the last receiving the milk secretions during lactation from the secreting underlying glandular tissue. Fat lobules surround the secretory tissue on most sides and may contain toxic chemicals as was shown in a study on breast milk that contained DDT[221] and our biopsies of patients' breasts containing implants and those of Falck et al.,[222] who found elevated chlorinated hydrocarbons in the breast tissue of women diagnosed with breast cancer (Tables 19.18 and 19.19 and Figure 19.11).

Growth and secretion of the mammary glands are stimulated and controlled by several endocrine factors including those of the ovary, corpus luteum, placenta, and the pars distalis of the hypophysis, thus allowing many avenues for pollutant injury. The luteotrophic hormone prolactin has three effects — luteotrophic, mammotrophic, and lactogenic. The mammae are also influenced by adrenal and thyroid secretions as well as the state of an individual's nutrition. Environmental pollutants are known to imbalance the endocrine system. Study of the environmental influences on the endocrine system reveals the extent of the many and varied effects pollutants may have on the breasts.

Table 19.18.   Pesticide and PCB Concentrations in Mammary Adipose
Tissue (ng/g) from Cases with Breast Cancer (n = 20)
and from Controls with Benign Disease (n = 20)

Pesticides or PCB concentrations	Cases		Controls		
	$\bar{x} \pm SD$	Range	$\bar{x} \pm SD$	Range	p^a
Wet weight basis, corrected for recovery (ng/g)					
HCB	23 ± 8	13–42	20 ± 10	12–54	0.32
HX + OC	116 ± 50	41–203	97 ± 49	25–249	0.22
TN	87 ± 37	24–182	96 ± 80	236–394	0.65[b]
DDE	1877 ± 1283	337–4982	1174 ± 630	237–2246	0.04[b]
DDT	179 ± 135	60–686	14 ± 49	44–248	0.05[b]
PCBs	1669 ± 894	733–4674	1105 ± 424	592–2609	0.02[b]
Lipid basis, uncorrected (ng/g)					
HCB	28 ± 11	16–61	26 ± 11	14–60	0.54
HX + OC	136 ± 52	66–243	121 ± 53	33–278	0.36
TN	103 ± 43	38–197	118 ± 87	53–439	0.49[b]
DDE	2200 ± 1470	425–6398	1487 ± 842	308–3674	0.07[b]
DDT	216 ± 174	72–881	148 ± 75	42–405	0.12[b]
PCBs	1965 ± 927	827–4562	1395 ± 468	823–2875	0.02[b]

Notes: HCB = hexachlorobenzene; HX + OC = heptachlorepoxide and oxychlordane
(the sum); TN = trans-nonachlor; *p,p'*-DDE = *bis*(4-chlorophenyl)-1,1-
dichloroethene; *p,p'*-DDT = *bis*(4-chlorophenyl)-1,1,1-trichloroethane; and
PCBs = polychlorinated biphenyls.

[a] Two-tailed probability for *t*.
[b] For *t* with unequal variances.

Source: Falck, Jr., F., A. Ricci, Jr., M. S. Wolff, J. Godbold, and P. Deckers. 1993.
Pesticides and polychlorinated biphenyl residues in human breast lipids and their
relation to breast cancer. *Environ. Health Monthly* 5(5):143–146. With permission.

Both estrogen and progesterone are needed for full growth and function of
the breast. Pollutant injury to these secretions can cause abnormal development
and function of the breast. Freni-Titulaer et al.[223] demonstrated this process
graphically in his reports of many children who experienced premature breast
growth following exposure to excess estrogen-like compounds that were in
chicken they had ingested that had been contaminated by hormone-treated feed
(see Chapter 24, section on ovaries). We have also seen hypofunction and an
inability to grow normal size breasts in pollutant-injured children.

Of course, since early pollutant injury has been shown to influence the
thalamus, hypothalamus, and pituitary gland, it is easy to see how dysfunction in
this area of the brain could cause mammary dysfunction and poor breast growth.

Hormonal and neural factors lead to lactation. These include prolactin,
adrenotropin, somatotropin, and thyrotropin from the pituitary, estrogen and

Table 19.19.   Logistic Regression Coefficients for Breast Cancer
in Cases and Controls

Residue	Intercept	Age	Age 2	Residue smoking	
DDE	29.052[a]	−1.058[a]	0.00896[a]	0.00085[b]	
	34.967[a]	−1.162[a]	0.00927[a]	0.00122[a]	−2.662[a]
PCB	28.632[a]	−1.059[a]	0.00864[a]	0.0022[a]	
	31.251[a]	−1.088[a]	0.00859[a]	0.00234[a]	−1.6586[b]

Notes: Age is in years; residue in parts per billion; and smoking (ever = 1, never
= 0). Quetelet's index did not achieve statistical significance in these
models. The standardized coefficients were approximately 0.5.

[a] $p<0.05$
[b] $p<0.07$

Source: F. Falck, Jr., A. Ricci, Jr., M. S. Wolff, J. Godbold, and P. Deckers.
1993. Pesticides and polychlorinated biphenyl residues in human breast lipids
and their relation to breast cancer. *Environ. Health Monthly* 5(5):143–146. With
permission.

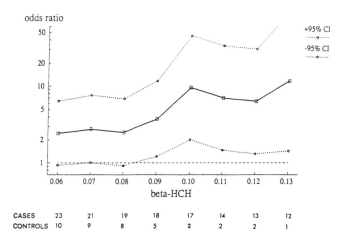

Figure 19.11.   Risk of breast cancer as a function of the concentration of
beta-HCH in breast fat shown on logarithmic scale. Cutoff
point for the age is 50 years. (From Mussalo-Rauhamaa, E.
Häsänen, H. Pyysalo, K. Antervo, R. Kauppila, and P.
Pantzar. 1990. Occurrence of beta-hexachlorocyclohexane
in breast cancer patients. *Cancer* 66(10):2124–2128. With
permission.)

progestrone from the ovary, adrenocortical steroids from the adrenal gland, and
thyroid hormone.

In lactation, the ANS has only vasomotor effects, and it causes the expulsion of a little milk contained in the alveoli and ducts. Pollutants can deregulate the ANS and, consequently, affect these functions (see Chapters 20 and 26).

The components of milk are water, protein, casein, lactalbumin, lactose, fat, vitamins, and minerals as well as numerous antibodies. Pollutant injury can be passed from the mother to the nursing child, as was shown in a series of lactating mothers who had DDT in their milk. This finding led to the banning of DDT in the United States. Any fat soluble or lipophilic chemical can attach to mother's milk. Therefore, conceivably, a child could have a toxic chemical intake each time it nurses. Further, many chemicals such as alcohols, belladonna, opium, antibiotics, salicylates, iodides, and bromides have been shown to be contained in milk. We have seen several children at the EHC-Dallas who were extremely irritable while their mothers were taking aminophylline for their asthma but who would become calm immediately after this medicine was stopped. Australian studies[224] show that mothers who used lanolin that contained pesticide around their nipples transmitted toxic pesticides to their babies. These included lindane and organophosphates.

Mastodynia

Many menstruating females experience mastodynia premenstrually. Some also develop breast swelling and pain with the ingestion of cream. We have observed this phenomenon in 20 environmentally sensitive females. One 41-year-old physician would develop excruciating pain and swelling each time she would ingest cream in any form. This pain and swelling would occur if she drank cream in coffee, tea, or hot water. She was severely chemically sensitive, and 7 years after her initial avoidance treatment, she developed an inflammatory breast cancer that could not be removed. It metastasized, and, within a year, she died.

Toxic chemicals can also lodge in the fat lobules of the breast, causing pain and perhaps tumor.

Inflammation

Pollutant injury may influence inflammatory diseases of the breast. Usually inflammatory lesions result from local trauma from abrasives of the alveolar tissue or around the nipple. This damage then results in bacterial growth with inflammation and abscesses. However, pollutants may also cause the inflammation that results in a sterile or infected abscess. At the EHC-Dallas, one 45-year-old patient who had severe generalized toxic vascular injury provides a graphic example of damage. This patient was undergoing heat chamber depuration for several months due to her chemical overload. She suddenly developed soreness and inflammation of the left breast, which progressed for several days until a hard indurated mass (6 by 10 cm) in the outer quadrant of the left breast occurred. She met all the criteria for surgical drainage and treatment with antibiotics. This patient remembered having had a similar lesion in the same area of the breast several years previously when she had initially been exposed to the pollutants that apparently made her ill. She requested that we not drain the breast or treat it with antibiotics and allow her to have 2 additional days of heat chamber depuration and physical therapy. We agreed with great reluctance. By 24 hours, the lesion had appeared to level off in its growth and inflammation. By 48 hours it had started to shrink, and by 1 week, the mass was gone. In the following 2-week interval no residual mass or infection could be found.

Chronic Cystic Mastitis

A significant part of the female population appears to develop chronic cystic inflammation. Etiologies appear to be multiple. However, Minton et al.[225, 226] has described a series of patients whose cysts and inflammation would subside with avoidance of xanthines found in coffee, tea, and chocolate. This condition would return when these substances were reintroduced. We have observed a similar phenomenon in patients exposed to coffee, tea, and chocolate. However, we carried this idea one step further when we realized that chronic cystic mastitis was influenced by many other environmental incitants. We evaluated a series of 200 patients with chronic cystic mastitis. Three groups emerged. One was influenced by their total body load of pollutants. When the load was up, the mastitis occurred. When the load was down, the mastitis subsided. Another group was influenced by individual food sensitivities in which the breast apparently acted as a target organ. The type food that caused the problem was individual specific. It included milk, corn, chicken, eggs, beef, or other grains. A third group was also influenced by a chemical exposure. This group was dramatic because each time they got into a situation requiring them to breathe a toxic substance, their chronic mastitis was activated. Often, once they were removed from the exposure, the cysts required 2 to 3 weeks before inflammation subsided.

Premalignant Lesions

Premalignant lesions of the breast include lobular carcinoma *in situ* (lobular hyperplasia), intraductal carcinoma, diffuse papillomatosis, and possibly luctal hyperplasia. It is not proven, though it is suspected, that these have environmental influences.

Carcinoma

The etiology of carcinoma of the breast is still elusive. However, present data suggest a complex interaction between hormonal, genetic, nutritional, and environmental factors. It is thought that up to 5 to 7% of breast cancers have a hereditary component. However, the alarming increase in breast cancers in all of the industrialized countries in the world strongly implicates environmental factors. Further, the environmental factors appear to be more important than genetics in most cases, since immigrants from low risk countries approach the risk of breast cancer of the native population of the industrialized country to which they have moved after one generation. It is thought that this increase may be due to the increased consumption of dietary fat.[227] Support for this hypothesis comes from dietary studies[228] that have related dietary factors to estrogen fractions, which are, in turn, likely to be related to the risk of developing breast cancer. Findings suggest that industrialization has added many estrogen-like substances to the environment. These agents include pesticides such as DDT, kepone, heptachlor, and atrazine and toxic chemicals such as PCB, PAH, and polycarbonate.[229]

A second possibility relates to a direct effect of fat on tumor development by way of changes in the lipid content of the cell membrane or the synthesis of prostaglandin. Of course, toxic chemicals, which are known to be mutagenic and carcinogenic, are usually extremely lipophilic and, therefore, can land in the breast, triggering malignancies.

Among postmenopausal women who eat meat (with associated fat), the urinary excretion of estriol and total estrogens was higher than in vegetarians.[230]

Another area of risk may be reproductive factors. In 1713, Ramazzini noted that nuns had more breast cancer than lay women.[231] This finding has been repeatedly documented through the centuries. Subsequent studies show that the risks increase with the age of the mother with her first child and that women who deliver after the age of 30 years have a greater risk than nulliparous women.[232]

It is clear that there are environmental influences upon breast pathology, and these should be considered when evaluating breast problems. At the EHC-Dallas, we have seen all of the forms of pathology in the chemically sensitive patient.

Artificial Breast Implants

Artificial breast implants made of silicone or other forms of plastic have now been performed on approximately one to two million women in the U.S.[233] Reports of problems relating to implantation stretch from local to systemic. The problems have become severe enough to result in an FDA investigation.

We have encountered many patients with synthetic implants who have experienced difficulty from their implants since 1974. Some have had severe chemical sensitivity and had to have them removed. Others have continued to tolerate them marginally with their chemical sensitivity persisting (see Chapter 41).

One-hundred and fifty patients have had their silicone implants removed at the EHC-Dallas over the last 15 years. Fifty have been removed in the past year. Diagnosis was made on a sound history with the patient complaining of chronic fatigue, fibromyalgias, arthralgias, inability to maintain recent memory, and the development of imbalance (especially at night) and clumsiness. Raynaud's phenomenon, petechiae, spontaneous bruising, purpura, and cold sensitivity were the predominant signs. Often there was calcification in the breast or chest (Figure 19.12). Working at the EHC-Dallas, Higuchi[234] studied 45 consecutive white female patients who complained of complications of breast implants (Table 19.20). Of these patients, 24 (49%) had their implants removed, and 2 had them replaced with saline implants; 21 (47%) kept their implants. The signs and symptoms that Higuchi found in this group of patients are listed in Table 19.21. Positive Rhomberg tests were present in the majority (49 of 50) of patients with implants examined at the EHC-Dallas. Nearly 70% of the patients had a positive provocation skin test to silicone or saline implant as well as other toxic chemicals. Their symptoms were reproduced by exposure to these substances (Table 19.22). The majority had neuromuscular atrophy on biopsy of the pectoral and deltoid muscle. All patients had positive skin tests for a battery of chemicals including formaldehyde, phenol, petroleum derived ethanol, newsprint, car exhausts, and perfumes indicating the chemical sensitivity. Three-dimensional SPECT brain scans of 93% were positive (Figure 19.13 and Table 19.23). Iris corder studies indicating autonomic dysfunction were abnormal showing a variety of effects including cholinergic, sympatholytic, cholinergic and sympatholytic, sympathomimetic, cholinolytic and nonspecific changes (Figure 19.14). Blood levels for toxic chemicals revealed elevated toluene, xylenes, 1,1,1-trichloroethanes, 2,3-methyl pentane, and hexanes (Table 19.24). Excision of the implants and the capsules was necessary and revealed that the capsules contained foreign body giant cells including a substance that resembled silicone in the microscopic analysis (Figure 19.15). Regardless of their appearance on mammogram, the implants ranged from intact (Figure 19.16) to the plastic capsule being disintegrated to a jelly-like sticky substance (Figures 19.17 and 19.18). In some cases, biopsy of the breast adjacent to the capsule revealed elevated levels of benzene, toluene, styrene, xylene, etc. (Table 19.25). Two patients had molds in their capsules (Figure 19.19). One mold was black with aspergillosis and the other was candida-like. Some patients have presented at the EHC-Dallas with only the implants removed and the capsules left in place (Figure 19.20). These patients had to have the capsules removed in order to resolve their symptoms. Many patients (26)

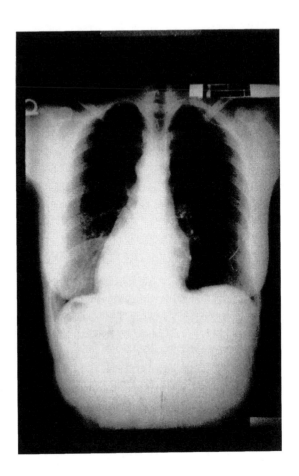

Figure 19.12.   **Calcification of chest x-ray and breast secondary to silicone breast implants. (From EHC-Dallas. 1992.)**

had abnormal immunological findings (Table 19.26). Also, abnormal autoantibodies were found (Table 19.27). The patients were placed on a nutrient regimen of 5 g of vitamin C, 2 multimineral capsules, containing zinc, 30 mg; manganese, 15 μg; chromium, 400 μg; copper, 15 mg; calcium, 250 mg; selenium, 100 μg; magnesium aspartate, 500 μg; glutamic acid, 200 mg; taurine, 2 g; and multiple B vitamins before and after surgery. Many patients cleared their symptoms during the first month after removal of their implants.

Table 19.20.   45 White Female Patients with Breast Implants

Age	Length of time with implant	Present status
Range: 24–65 yrs. Average: 44.8 yrs.	Range: 1–20 yrs. Average: 10.8 yrs.	Implant removed, without replacement: 22 (49%) Implant removed, replaced by saline implant: 2 (4%) Implant kept: 21 (47%)

Source: Higuchi, H. EHC-Dallas. 1994.

Table 19.21.   Symptoms and Signs in 45
Patients with Breast Implants

Affected system	Patient no.	Percent
Musculoskeletal	39	87
Neurological	31	69
Gastrointestinal	12	27
Cardiovascular	11	24
Otologic	9	20
Respiratory	8	18
Dermatological	5	11
Genitourinary	4	9
Ophthalmic	4	9
Endocrinological	3	7
Local breast pain	3	7
Rhomberg	22	49

Source: Higuchi, H. EHC-Dallas. 1994.

However, most had to undergo additional treatment with heat depuration until their chemical sensitivity symptoms cleared. We have observed women developing problems with the saline implants that were placed after the silicone implants were removed.

Burton[235] reports that Gerwashwin at the University of California at Davis found children of patients with implants also to be sick. Similarly, at the EHC-Dallas, we have found that 50% of the children of mothers with implants who

Table 19.22.   Skin Test for Chemicals in Silicone
Breast Implants

Chemical	Total Patients (no.)	Positive (no.)	Patients (%)
Formaldehyde	22	22	100
Jet fuel	3	3	100
Orris root	29	26	90
Cigarette smoke	25	21	84
Propane gas	6	5	83
Ethanol	28	23	82
Men's cologne	25	17	68
UNL/Diesel	28	19	68
Silicone	40	27	68
Phenol	24	16	67
Women's cologne	29	19	66
Natural gas	20	12	60
News material	26	14	54
Chlorine	21	10	48
Fireplace smoke	11	3	27

Source: Higuchi, H. EHC-Dallas. 1994.

we have examined have developed food, biological inhalant, and chemical sensitivities.

In conclusion, the lower respiratory wall, chest wall, and breast may at times be damaged by pollutants. When this injury occurs, diagnosis and treatment is similar to that used with other end-organ involvement in chemical sensitivity. Diagnosis is accomplished by challenge testing. Treatment includes avoidance of pollutants in the air, food, and water and implementation of injection therapy, nutrient supplementation and replacement, heat depuration/physical therapy, and tolerance moderators.

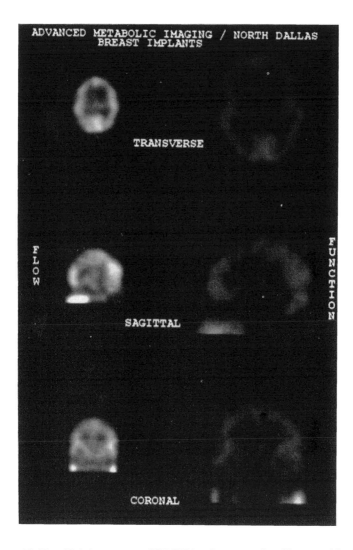

Figure 19.13.   Triple camera SPECT brain scan of a 49-year-old white
female showing a decrease in function in contrast to flow,
salt-and-pepper malformations in the function phase, and
temporal lobe inequity. These are all characteristic of neurotoxicity. Symptoms and signs in the patient included
short-term memory loss, lack of concentration, confusion,
imbalance, and an inability to stand on the toes with the
eyes closed. Of 100 consecutive chemically sensitive patients who underwent triple SPECT brain scan and who
had these signs and symptoms, 98 showed the neurotoxic
patterns. (From EHC-Dallas. 1993.)

Table 19.23.   SPECT Scan with
Neurotoxic Pattern in 29 Silicone
Breast Implant Patients

Abnormal/tested	Abnormal %
27/29	93

Source: Higuchi, H. EHC-Dallas. 1994.

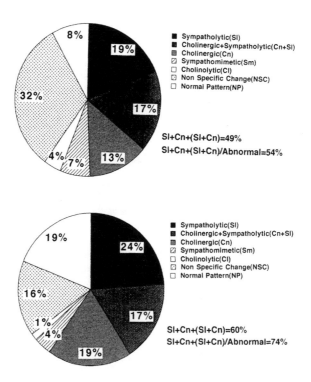

Figure 19.14.   Analysis of ANS. Iris corder changes in 45 chemically
sensitive patients with breast implants; 42 (93%) of these
were abnormal. (From EHC-Dallas. 1993.)

Table 19.24.   Toxic Chemicals in Blood for Silicone
Breast Implant Patients

Chemicals	Total	Over detection limit	
		Positive no.	%
1,1,1-Trichloroethane	36	14	39
Toluene	36	12	33
Benzene	36	12	33
Trimethylbenzenes	36	7	19
Dichloromoethane	36	1	3
3-Methylpentane	35	35	100
2-Methylpentane	35	34	97
n-Hexane	35	28	80
DDE	8	8	100
HCB	8	3	38
Beta-BHC	8	2	25
Oxychlordane	8	2	25
trans-Nonachlor	8	2	25
alpha-BHC	8	1	13
Heptachlor epoxide	8	1	13

Source: Higuchi, H. EHC-Dallas. 1994.

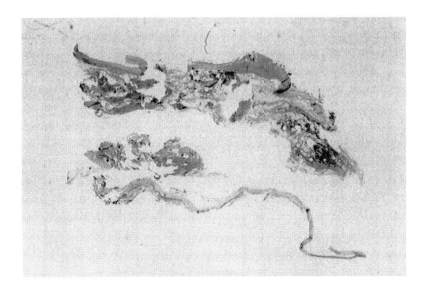

Figure 19.15.   Photomicrograph of breast implant capsule, showing foreign body giant cells containing silicone. (Courtesy of
R. Keene. 1994.)

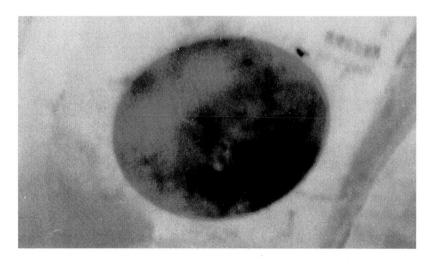

Figure 19.16.   Photograph of intact silicone breast implant. No gross leakage is evident. (From EHC-Dallas. 1994.)

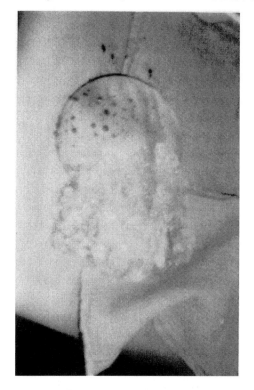

Figure 19.17.   Ruptured silicone breast implant showing jelly-like substance. (From EHC-Dallas. 1994.)

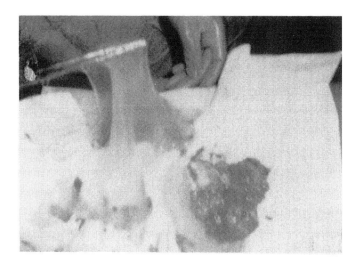

Figure 19.18. Mucous-like formation in a ruptured silicone implant. (From EHC-Dallas. 1993.)

Table 19.25. Toxic Chemicals in Breast Tissue of Silicone Breast Implant Patients

Chemical	Total patients	Over detection limit	
		Patient no.	%
1,1,1-Trichloroethane	10	4	40
Benzene	10	3	30
Toluene	10	2	20
Ethylbenzene	10	1	10
Xylenes	10	1	10
Styrene	10	1	10
Chloroform	10	1	10
n-Pentane	10	1	10

Source: Higuchi, H. EHC-Dallas. 1994.

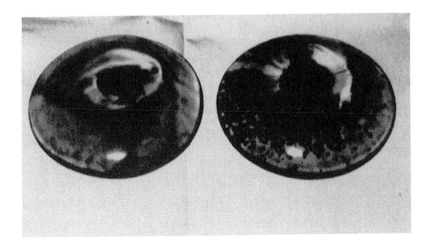

Figure 19.19.   Silicone implants infiltrated with *Aspergillus niger*. These
were removed from a 28-year-old white female exhibiting
symptoms of fibromyalgia, arthralgia, and chronic fatigue.
Skin tests showed sensitivity to the *Aspergillus niger* reproducing these symptoms. (From EHC-Dallas. 1993.)

Table 19.26.   Immunological Findings in Silicone
Breast Implant Patients

Immune parameter	Total patients	Abnormal			
		Patient no.	%	Low	High
WBC	16	1	6	1	
Lymphocytes %	10	1	10	1	
Total T_{11}	10	1	10	1	
T_4	10	2	20		2
T_8	10	2	20	2	
T_4/T_8 ratio	10	4	40		4
$B_4\%$	10	2	20	1	1
Tal/CD26	7	6	86		6
IL2-RI	7	2	29		2

Source: Higuchi, H. EHC-Dallas. 1994.

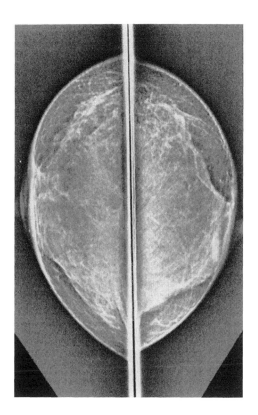

Figure 19.20.   Mammogram illustrating a fibrous capsule containing silicone that was not removed when the silicone implant was
removed. This patient continued to experience symptoms
until the capsule was removed. (From EHC-Dallas. 1993.)

Table 19.27. Autoimmunity in Silicone Breast Implant Patients

	Total patients	Positive Patient no.	%
Anti-smooth muscle ab.	28	20	71
ANA Hep-2 cell	22	12	55
Myelin ab.	22	9	41
Anti-basement membrane ab.	5	2	40
Anti-brushborder ab.	28	9	32
Anti-nuclear ab.	30	9	30
Anti-microsomal ab.	25	7	28
Silicone ab. (IgG)	4	1	25
Anti-thyroglobulin ab.	25	6	24
Anti-recticulin ab.	5	1	20
Anti-parietal cell ab.	28	3	11
Anti-mitochondria ab.	28	0	0

Source: Higuchi, H. EHC-Dallas. 1994.

REFERENCES

1. Rea, W. J. 1992. *Chemical Sensitivity. Vol. I: Mechanisms of Chemical Sensitivity.* 17. Boca Raton, FL: Lewis Publishers.

2. Wallace, L. 1987. The TEAM study: personal exposures to toxic substances in air, drinking water, and breath of 400 residents of New Jersey, North Carolina, and North Dakota. *Environ. Res.* 43:290–307.

3. Cullen, M. R. 1987. The worker with multiple chemical sensitivities: an overview. In *Workers with Multiple Chemical Sensitivities*, ed. M. R. Cullen. 655–662. Philadelphia: Hanley & Belfus.

4. Buist, A. S. 1988. Obstructive diseases: smoking and other risk factors. In *Textbook of Respiratory Medicine*, eds. J. F. Murray and J. A. Nadel. 1001–1029. Philadelphia: W. B. Saunders.

5. Hammond, E. C. and D. Horn. 1958. Smoking and death rates — report on forty-four months of follow-up of 187,783 men. II. Death rates by cause. *JAMA* 166:1294–1408.

6. Eriksen, M. P., C. A. LeMaistre, and G. R. Newell. 1988. Health hazards of passive smoking. *Annu. Rev. Public Health* 9:4–70.

7. Koop, C. E. 1986. The health consequences of involuntary smoking: a report of the Surgeon General. U.S. Department of Health and Human Services, Rockville, MD. 227–260.

8. Ware, J. H., D. W. Dockery, A. Spiroll, F. E. Speizer, and B. G. Ferris, Jr. 1984. Passive smoking, gas cooking, and respiratory health of children living in six cities. *Am. Rev. Respir. Dis.* 129:366–374.

9. Kauffmann, F., J.-F. Tessier, and P. Oriol. 1983. Adult passive smoking in the home environment: a risk factor for chronic airflow limitation. *Am. J. Epidemiol.* 117:269–280.

10. Berglund, B. and T. Lindvall. 1990. Sensory criteria for healthy buildings. The Fifth International Conference on Indoor Air Quality and Climate. 5:65–79.

11. McCutcheon, R. S. 1980. Toxicology and the law. In *Cassaret and Doull's Toxicology: The Basic Science of Poisons*, 2nd ed., eds. J. Doull, C. Klaassen, and M. O. Amdur. 727–733. New York: Macmillan.

12. Gurainik, D. B. 1970. *Webster's New World Dictionary*. 1692. New York: World.

13. Bascom, R. 1991. Differential responsiveness to irritant mixtures: possible mechanisms. Environmental Research Facility, Division of Pulmonary and Critical Care Medicine, Department of Medicine, University of Maryland School of Medicine.

14. Naclerio, R. M., H. L. Meier, A. Kagey-Sobotka, N. F. Adkinson, Jr., D. A. Meyers, P. S. Norman, and L. M. Lichtenstein. 1983. Mediator release after nasal airway challenge with allergen. *Am. Rev. Respir. Dis.* 128:597–602.

15. Togias, A. G., R. M. Naclerio, D. Proud, J. E. Fish, N. Franklin, J. Adkinson, A. Kagey-Sobotka, P. S. Norman, and L. M. Lichtenstein. 1985. Nasal challenge with cold, dry air results in release of inflammatory mediators: possible mast cell involvement. *J. Clin. Invest.* 76:1375–1381.

16. Naclerio, R. M., D. Proud, L. M. Lichtenstein, A. Kagey-Sobotka, J. O. Hendley, J. Sorrentino, and J. M. Gwaltney. 1988. Kinins are generated during experimental rhinovirus colds. *J. Infect. Dis.* 157:133–142.

17. Broder, I., P. Corey, P. Brasher, M. Lipa, and P. Cole. 1988. Comparison of the health of occupants and characteristics of house among control homes and homes insulated with urea formaldehyde foam. III. Health and house variables following remedial work. *Environ. Res.* 45:179–203.

18. Broder, I., P. Corey, P. Cole, and M. Lipa. 1988. Comparison of the health of occupants and characteristics of houses among control homes and homes insulated with urea formaldehyde foam. II. Initial health and house variables and exposure-response relationships. *Environ. Res.* 45:156–178.

19. Broder, I., P. Corey, P. Cole, M. Lipa, S. Mintz, and J. R. Nethercott. 1988. Comparison of health of occupants and characteristics of houses among control homes and homes insulated with urea formaldehyde foam. I. Methodology. *Environ. Res.* 45:141–155.

20. Green, D. J., L. R. Sauder, T. J. Kulle, and R. Bascom. 1987. Acute response to 3.0 ppm formaldehyde in exercising healthy nonsmokers and asthmatics. *Am. Rev. Respir. Dis.* 135:1261.

21. Uba, G. M., D. M. Pachorek, J. M. Bernstein, D. H. M. M. Garabrant, J. R. M. Balmes, W. E. M. M. Wright, and R. B. M. C. Amar. 1989. Prospective study of respiratory effects of formaldehyde among healthy and asthmatic medical students. *Am. J. Ind. Med.* 15(1):91–101.

22. Holmstrom, M., B. Wilhelmsson, H. Hellquist, and G. Rosen. 1989. Histological changes in the nasal mucosa in persons occupationally exposed to formaldehyde alone and in combination with wood dust. *Acta Otolaryngol. (Stockh.)* 107:120–129.

23. Leikauf, G. D., L. M. Leming, J. R. O'Donnell, and C. A. Doupnik. 1989. Bronchial responsiveness and inflammation in guinea pigs exposed to acrolein. *J. Appl. Physiol.* 66:171–178.

24. Leikauf, G. D., C. A. Doupnik, L. M. Leming, and H. E. Wey. 1989. Sulfidopeptide leukotrienes mediate acrolein-induced bronchial hyperresponsiveness. *J. Appl. Physiol.* 66:1838–1845.

25. Lippmann, M. 1989. Health effects of ozone: a critical review. *JAPCA* 39:672–695.

26. Ware, J. H., L. A. Thibodeau, F. E. Speizer, S. Colome, and B. G. Ferris, Jr. 1981. Assessment of the health effects of atmospheric sulfur oxides and particulate matter: evidence from observational studies. *Environ. Health Perspect.* 41:255–276.

27. Sheppard, D., W. S. Wong, C. F. Uehara, J. A. Nadel, and H. A. Boushey. 1980. Lower threshold and greater bronchomotor responsiveness of asthmatic subjects of sulfur dioxide. *Am. Rev. Respir. Dis.* 122:873–878.

28. Sheppard, D. 1988. Sulfur dioxide and asthma — a double-edged sword? *J. Allergy Clin. Immunol.* 82:961–964.

29. Mustafa, M. G. and D. F. Tierney. 1978. State of the art. Biochemical and metabolic changes in the lung with oxygen, ozone, and nitrogen dioxide toxicity. *Am. Rev. Respir. Dis.* 118:1061–1090.

30. U. S. Public Health Service. 1979. Smoking and health. A report of the Surgeon General. DHEW Publication No. (PHS)79–50066, Washington, D.C.: U.S. Government Printing Office.

31. Mølhave, L., B. Bach, and O. F. Pedersen. 1984. Human reactions during controlled exposures to low concentrations of organic gases and vapours known as normal indoor air pollutants. Indoor Air: Proceedings of the Third International Conference on Indoor Air Quality and Climate. 431–436.

32. Mølhave, L. 1990. Volatile organic compounds, indoor air quality and health. Proceedings of the Fifth International Conference on Indoor Air Quality and Climate. 5:15–33.

33. Koren, H. S., D. E. Graham, S. Steingold, and R. Devlin. 1990. The inflammatory response of the human upper airways to volatile organic compounds (VOC). *Am. Rev. Respir. Dis.* 141:A426.

34. Rea, W. J. 1994. *Chemical Sensitivity. Vol. II: Sources of Total Body Load*, 685. Boca Raton, FL: Lewis Publishers.

35. Amdur, M. O., A. F. Sarofim, M. Neville, R. J. Quann, J. F. McCarthy, J. F. Elliott, H. F. Lam, A. E. Rogers, and M. W. Conner. 1986. Coal combustion aerosols and SO_2: an interdisciplinary analysis. *Environ. Sci. Technol.* 20:138–145.

36. Rea, W. J. 1992. *Chemical Sensitivity. Vol. I: Mechanisms of Chemical Sensitivity.* 221. Boca Raton, FL: Lewis Publishers.

37. Bascom, R., T. Kulle, A. Kagey-Sobotka, and D. Proud. 1991. Upper respiratory tract environmental tobacco smoke sensitivity. *Am. Rev. Respir. Dis.* 143:1304–1311.

38. Speer, F. 1968. Tobacco and the non-smoker, a study of subjective symptoms. *Arch. Environ. Health* 16:443–446.

39. Environmental Protection Agency. 1989. Indoor air quality and work environment study. EPA headquarters' buildings. Volume I. Employee survey. 19K-1003, ES-1-C-54. Washington, D.C.: U.S. Government Printing Office.

40. Environmental Protection Agency. 1991. Indoor air quality and work environment study. EPA headquarters' buildings. Volume III. Survey. Supplemental survey results and their relationship to environmental monitoring results. Washington, D.C.: U.S. Government Printing Office.

41. Folinsbee, L. J., W. F. McDonnell, and D.H. Horstman. 1988. Pulmonary function and symptom responses after 6.6 hour exposure to 0.12 ppm ozone with moderate exercise. *JAPCA* 38:28–35.

42. Kulle, T. J., L. R. Sauder, J. R. Hebel, and M. D. Chatham. 1985. Ozone response relationships in healthy nonsmokers. *Am. Rev. Respir. Dis.* 132:36–41.

43. McDonnell, W. F., D. H. Horstman, S. Abdul-Salaam, and D. E. House. 1985. Reproducibility of individual responses to ozone exposure. *Am. Rev. Respir. Dis.* 131:36.

44. Kleeberger, S. R., D. J. Basset, G. J. Jakab, and R. C. Levitt. 1990. A genetic model for evaluation of susceptibility to ozone-induced inflammation. *Am. J. Physiol.* 258:L313-L320.

45. Storz, G., L. A. Tartaglia, and B. N. Ames. 1990. Transcriptional regulator of oxidative stress-inducible genese: direct activation by oxidation. *Science* 248:189–194.

46. Storz, G., L. A. Tartaglia, and B. N. Ames. 1990. The OxyR regulon. *Antonie Van Leeuwenhoek* 58:157–161.

47. Proud, D., A. Togias, R. M. Naclerio, S. A. Crush, P. S. Norman, and L. M. Lichtenstein. 1983. Kinins are generated in vivo following nasal airway challenge of allergic individuals with allergen. *J. Clin. Invest.* 72:1678–1685.

48. Bascom, R., S. Green, D. Green, A. Kagey-Sobotka, and D. Proud. 1988. Effect of intranasal capsaicin on symptoms and mediator release. *Am. Rev. Respir. Dis.* 136:A375.

49. Stjarne, P., L. Lundblad, J. M. Lundberg, and A. Anggard. 1989. Capsaicin and nicotine-sensitive afferent neurones and nasal secretion in healthy human volunteers and in patients with vasomotor rhinitis. *Br. J. Pharmacol.* 1989:693–701.

50. Geppetti, P., B. M. Fusco, S. Marabini, C. A. Maggi, M. Fanciullacci, F. Sicuteri. 1988. Secretion, pain and sneezing induced by the application of capsaicin to the nasal mucosa in man. *Br. J. Pharmacol.* 93:509–514.

51. Testa, B. and P. Jenner. 1981. Inhibitors of cytochrome P-450s and their mechanism of action. *Drug Metab. Rev.* 12:1–117.

52. Weinstein, L. and M. N. Swartz. 1985. Host responses to infection. In *Pathologic Physiology Mechanisms of Disease*, 7th ed., eds. W. A. Sodeman, Jr. and T. M. Sodeman. 548–552. Philadelphia: W. B. Saunders.

53. Reeves, A. L. 1981. *Toxicology: Principles and Practice. Vol. I.* New York: John Wiley & Sons.

54. Rea, W. J. 1992. *Chemical Sensitivity. Vol. I: Mechanisms of Chemical Sensitivity.* 47. Boca Raton, FL: Lewis Publishers.

55. Knight, V. 1980. Influenza. In *Harrison's Principles of Internal Medicine*, 9th ed., eds. K. J. Isselbacher, R. D. Adams, E. Braunwald, R. G. Petersdorf, and J. D. Wilson. 787–788. New York: McGraw-Hill.

56. Hayes, W. J. 1982. *Pesticides Studied in Man.* Baltimore: Williams & Wilkins.

57. Mitchell, C. E., J. P. Fischer, and A. R. Dahl. 1987. Differential induction of cytochrome P-450 catalyzed activities by polychlorinated biphenyls and benzo[a]pyrene in B6C3F1 mouse liver and lung. *Toxicology* 43(3):315–323.

58. Gordon, J., and M. O. Amdur. 1991. Responses of the respiratory system to toxic agents. In *Casarett and Doull's Toxicology: The Basic Science of Poisons*, 4th ed. eds. M. O. Amdur, J. Doull, and C. D. Klaassen. 383. New York: Pergamon Press.

59. Selye, H. 1946. The general adaptation syndrome and the diseases of adaptation. *Allergy* 17:23.

60. Adolph, E. F. 1956. General and specific characteristics of physiological adaptation. *Am. J. Physiol.* 18:184.

61. Randolph, T. G. 1962. *Human Ecology and Susceptibility to the Chemical Environment.* Springfield, IL: Charles Thomas.

62. Stokinger, H. E. and D. L. Coffin. 1968. Biological effects of air pollutants. In *Air Pollution*, ed. A. C. Stern. New York: American Press.

63. National Research Council. 1962. *Atmospheric Studies.* Washington, D.C.: National Academy of Sciences.

64. Bennett, G. 1962. Ozone contamination of high altitude air craft cabins. *Aerospace Med.* 33:969.

65. Stokinger, H. E. 1965. Ozone toxicology: a review of research and industrial experience: 1954–1964. *Arch. Environ. Health* 10:719.

66. U.S. Environmental Protection Agency. 1987. Characterization of HRGC/MS unidentified peaks from the analysis of human adipose tissue. Vol. I. Technical approach. EPA-560/5–87–002A. Office of Toxic Substances. Washington D.C.: U.S. Government Printing Office.

67. Farrell, B. P., H. D. Kerr, T. J. Kulle, L. R. Sauder, and J. L. Young. 1979. Adaptation in human subjects to the effects of inhaled ozone after repeated exposure. *Am. Rev. Respir. Dis.* 119(5):725–730.

68. Kulle, T. J., L. R. Sauder, H. D. Kerr, B. P. Farrell, M. S. Bermel, and D. M. Smith. 1982. Duration of pulmonary function adaptation to ozone in humans. *Am. Ind. Hyg. Assoc. J.* 43(11):832–837.

69. Parkes, W. R. 1982. Occupational asthma (including Byssinosis). In *Occupational Lung Disorders*, 2nd ed., ed. W. R. Parkes, 415–453. London: Butterworths.

70. Rylander, R., P. Haglind, and M. Lundholm. 1985. Endotoxin in cotton dust and respiratory function decrement among cotton workers in an experimental cardroom. *Am. Rev. Respir. Dis.* 131:209–213.

71. Jancso-Gabor, A. and J. Szolcsanyi. 1972. Neurogenic inflammatory responses. *J. Dent. Res.* 51:264–269.

72. Lundberg, J. M. and A. Saria. 1983. Capsaicin-induced desensitization of airway mucosa to cigarette smoke, mechanical and chemical irritants. *Nature* 302:251–253.

73. Lundblad, L., A. Saria, J. M. Lundberg, and A. Anggard. 1983. Increased vascular permeability in rat nasal mucosa induced by substance P and stimulation of capsaicin-sensitive trigeminal neurons. *Acta Otolaryngol.* 96:479–484.

74. Evans, M. J., L. J. Cabral, R. J. Stephens, and G. Freeman. 1973. Renewal of alveolar epithelium in the rat following exposure to NO_2. *Am. J. Pathol.* 70:175–198.

75. Adamson, I. Y. R. and D. H. Bowden. 1974. The type II cell as progenitor of alveolar epithelial regeneration: a cytodynamic study on mice after exposure to oxygen. *Lab. Invest.* 30:35.

76. Evans, M. J., L. J. Cabral, R. J. Stephens, and G. Freeman. 1975. Transformation of alveolar type 2 cell to type 1 cells following exposure to nitrogen dioxide. *Exp. Mol. Pathol.* 22:142.

77. Levine, S. A. and P. M. Kidd. 1985. *Antioxidant Adaptation: Its Role in Free Radical Pathology.* San Leandro, CA: Biocurrents Division, Allergy Research Group.

78. Rea, W. J. 1994. *Chemical sensitivity.* Vol. II: *Sources of Total Body Load*, 837. Boca Raton, FL: Lewis Publishers.

79. Ohnishi, A., N. Inoue, T. Yamamoto, Y. Murai, H. Hori, M. Koga, I. Tanaka, and T. Akiyama. 1985. Ethylene oxide induces central-peripheral distal axonal degeneration of the lumbar primary neurones in rats. *Br. J. Ind. Med.* 42(6):373–379.

80. Beckett, W. S., W. F. McDonnell, D. H. Horstman, and D. E. House. 1985. Role of the parasympathetic nervous system in acute lung response to ozone. *J. Appl. Physiol.* 59:1879–1885.

81. Nadel, J. A., H. Salem, B. Tamplin, and Y. Tokiwa. 1965. Mechanism of bronchoconstriction during inhalation of sulfur dioxide. *J. Appl. Physiol.* 20:164–167.

82. Levitt, R. C. and W. Mitzner. 1988. Expression of airway hyperreactivity to acetycholine as a simple autosomal recessive trait in mice. *FASEB J.* 2(10):2605–2608.

83. Hopp, R. J., A. K. Bewtra, G. D. Watt, N. M. Nair, and R. G. Townley. 1984. Genetic analysis of allergic disease in twins. *J. Allergy Clin. Immunol.* 73:265–270.

84. Sparrow, D., G. O'Connor, and S. T. Weiss. 1988. The relation of airways responsiveness and atopy to the development of chronic obstructive lung disease. *Epidemiol. Rev.* 10:29–47.

85. Barter, C. E. and A. H. Campbell. 1976. Relationship of constitutional factors and cigarette smoking to decrease in 1-second forced expiratory volume. *Am. Rev. Respir. Dis.* 113:305–314.

86. Barnes, P. J., M. G. Belvisi, and D. F. Rogers. 1990. Modulation of neurogenic inflammation: novel approaches to inflammatory disease. *Trends Pharmacol. Sci.* 11:185–189.

87. Lundberg, J. M., L. Lundblad, C. R. Martling, A. Saria, P. Stjarne, and A. Anggard. 1987. Coexistence of multiple peptides and classic transmitters in airway neurons: functional and pathophysiologic aspects. *Am. Rev. Respir. Dis.* 136:S16–S22.

88. Lundblad, L. and J. M. Lundberg. 1984. Capsaicin sensitive sensory neurons mediate the response to nasal irritation induced by the vapour phase of cigarette smoke. *Toxicology* 33:1–7.

89. Lundberg, J. M., A. Saria, and E. R. Martking. 1983. Capsaicin pretreatment abolishes cigarette smoke-induced oedema in rat tracheo-bronchial mucosa. *Eur. J. Pharmacol.* 86:317–318.

90. Lundblad, L., J. M. Lundberg, and A. Anggard. 1984. Local and systemic capsaicin pretreatment inhibits sneezing and the increase in nasal vascular permeability induced by certain chemical irritants. *Naunyn-Schmiedeberg's Arch. Pharmacol.* 326:254–261.

91. Lundberg, J. M., L. Lundblad, A. Saria, and A. Anggard. 1984. Inhibition of cigarette smoke induced oedema in the nasal mucosa by capsaicin pretreatment and a substance P antagonist. *Naunyn-Schmiedeberg's Arch. Pharmacol.* 326:181–185.

92. Dusser, D. J., E. Umeno, P. D. Graf, T. Djokic, D. B. Borson, and J. A. Nadel. 1988. Airway neutral endopeptidase-like enzyme modulates tachykinin-induced bronchoconstriction in vivo. *J. Appl. Physiol.* 65:2585–2591.

93. Dusser, D. J., T. D. Djokic, D. B. Borson, and J. A. Nadel. 1989. Cigarette smoke induces bronchoconstrictor hyperresponsiveness to substance P and inactivates airway neutral endopeptidase in the guinea pig. Possible role of free radicals. *J. Clin. Invest.* 84:900–906.

94. Nakayama, T., M. Kodama, and C. Nagata. 1984. Generation of hydrogen peroxide and superoxide anion radical from cigarette smoke. *Gann* 75:95–98.

95. Ohashi, Y., Y. Nakai, H. Harada, S. Horiguchi, and K. Teramoto. 1982. Studies of industrial styrene poisoning (part XIII): experimental studies about the damage on the mucosal membrane of respiratory tracts of rats exposed to styrene — contemporary observations of the changes in ciliary functions and fine structures. *Jpn. J. Ind. Health* 24:360–366.

96. Doupnik, C. A. and G. D. Leikauf. 1990. Acrolein stimulates eicosanoid release from bovine airway epithelial cells. *Am. J. Physiol.* 259:L222–229.

97. Leikauf, G. D., K. E. Driscoll, and H. E. Wey. 1988. Ozone-induced augmentation of eicosanoid metabolism in epithelial cells from bovine trachea. *Am. Rev. Respir. Dis.* 137:435–442.

98. Lewis, R. A., K. F. Austen, and R. J. Soberman. 1990. Leukotrienes and other products of the 5-lipoxygenase pathway. Biochemistry and relation to pathobiology in human diseases. *N. Engl. J. Med.* 323:645–655.

99. Koyama, S., S. I. Rennard, and R. A. Robbins. 1990. Monocyte migration: one mechanism for the expansion of the mononuclear phagocyte population in the inflamed airway. *Am. Rev. Respir. Dis.* 141:A680.

100. Palmer, E. and D. Sheppard. 1990. Effects of surface acidification on tracheal epithelial cells. *Am. Rev. Respir. Dis.* 141:A100.

101. Cohen, D. S., W. J. Welch, and D. Sheppard. 1990. Ambient concentrations of ozone cause cytotoxicity and suppression of protein synthesis in cultured alveolar macrophages and epithelial cells without inducing stress protein synthesis. *Am. Rev. Respir. Dis.* 141:A101.

102. Cox, G., T. Ohtoshi, J. Gauldie, J. Denburg, and M. Jordana. 1990. Human bronchial epithelial cells cause differentiation of HL-60 cells and prolong the *in vitro* survival of eosinophils. *Am. Rev. Respir. Dis.* 141:A108.

103. Mattoli, S., S. Miante, F. Calabro, M. Mezzetti, A. Fasoli, and L. Allegra. 1990. Bronchial epithelial cells exposed to isocyanates potentiate activation and proliferation of T cells. *Am. J. Physiol.* 259:L320-L327.

104. Spurzem, J. R., O. Sacco, G. A. Rossi, and S. I. Rennard. 1990. MHC Class II expression by bronchial epithelial cells is modulated by lumphokines and corticosteroids. *Am. Rev. Respir. Dis.* 141:A681.

105. Lee, T. C., L. I. Gold, B. Cronstein, and J. Reibman. 1990. Transforming growth factor-beta (TGF-beta): a potent neutrophil chemoattractant. *Am. Rev. Respir. Dis.* 141:A681.

106. Spurzem, J. R., O. Sacco, and S. I. Rennard. 1990. The expression of fibronectin and vitronectin receptors is regulated on bronchial epithelial cells. *Am. Rev. Respir. Dis.* 141:A705.

107. Yoshida, M., D. Romberger, M. Illig, H. Takizawa, S. I. Rennard, and J. D. Beckman. 1990. Transforming growth factor-beta 1 increases desmoplakin transcript levels in bronchial epithelial cells. *Am. Rev. Respir. Dis.* 141:A699.

108. Kleihues, P., K. Patzchke, and G. Doerjer. 1982. DNA modification and repair in the experimental induction of nervous system tumors by chemical carcinogens. *Ann. N. Y. Acad. Sci.* 381:290–303.

109. Working, P. K., D. J. Doolittle, T. Smith-Oliver, R. D. White, and B. E. Butterworth. 1986. Unscheduled DNA synthesis in rat tracheal epithelial cells, hepatocytes and spermatocytes following exposure to methyl chloride in vitro and in vivo. *Mutat. Res.* 162(2):219–224.

110. Case, R. A. M. and A. J. Lea. 1955. Mustard gas poisoning, chronic bronchitis, and lung cancer. *Br. J. Prev. Soc. Med.* 9:62–72.

111. Yamada, A. 1959. Patho-anatomical studies on respiratory cancers developed in workers with occupational exposure to mustard gas. *Hiroshima Med. J.* 7:719–761.

112. Wada, S. 1963. Neoplasms of the respiratory tract among poison gas workers. *Hiroshima Igaku* 16:728–745.

113. Boyland, E. and S. Horning. 1953. The induction of tumors with nitrogen mustards. *Br. J. Cancer* 3:118–123.

114. Frank, A. L. 1982. The epidemiology and etiology of lung cancer. *Clin. Chest Med.* 3(2):219–228.

115. Lungarella, G., I. Barni-Comparini, and L. Fonzi. 1984. Pulmonary changes induced in rabbits by long-term exposure to *n*-hexane. *Arch. Toxicol.* 55(4):224–228.

116. Taylor, R. G., C. E. McCall, R. S. Thrall, R. D. Woodruff, and J. T. O'Flaherty. 1988. Histopathologic features of phorbol myristate acetate-induced lung injury. *Lab. Invest.* 52(1):61–70.

117. Matthew, H., A. Logan, M. F. A. Woodruff, and B. Heard. 1968. Paraquate poisoning–lung transplantation. *Br. Med. J.* 3:759–763.

118. Oizumi, K., Y. Nakai, and K. Konno. 1984. Antineoplastic drug-induced pneumonitis-fibrosis. *Gan No Rinsho-Japanese J. Cancer Clin.* 30(Suppl. 9):1217–1224.

119. Forkert, P. G., P. L. Sylvestre, and J. S. Poland. Lung injury induced by trichloroethylene. *Toxicology* 35(2):143–160.

120. Kowalski, B., E. B. Brittebo, and J. Brandt. 1985. Epithelial binding of 1,2,-dibromoethane in the respiratory and upper alimentary tracts of mice and rats. *Cancer Res.* 45(6):2616–2625.

121. Ekblom, A., A. Flock, P. Hansson, and D. Ottoson. 1984. Ultrastructural and electrophysiological changes in the olfactory epithelium following exposure to organic solvents. *Acta Otolaryngol.* 98(3–4):351–361.

122. Chevalier, G., J. P. H'enin, H. Vannier, C. Canevet, M. G. C'ote, and L. LeBouffant. 1984. Pulmonary toxicity of aerosolized oil-formulated benitrothion in rats. *Toxicol. Appl. Pharmacol.* 76(2):349–355.

123. Toronto Lung Transplant Group, Toronto, Canada. 1985. Personal communication.

124. Lee, D. H. K. 1972. *Environmental Factors in Respiratory Disease.* New York: Academic Press.

125. Seshadry, S. P. and J. Ganguly. 1961. Studies on vitamin A enterase 5: a comparative study of vitamin A enterase and cholesterol esterase of rat and chicken liver. *Biochem. J.* 80:397–406.

126. Goldstein, B. D., M. R. Levine, R. Cuzzi-Spada, R. Cardenas, R. D. Buckley, and O. J. Balcham. 1972. P-aminobenzoic acid as a protective agent in ozone toxicity. *Arch. Environ. Health* 24:243–247.

127. Innami, S., A. Nakamura, and M. Miyazaki. 1977. Effects of vitamin A on polychlorinated biphenyls (PCB). *Fukuoka Acta Med.* 68(3):12–16.

128. Forsman, S. and K. O. Frykholm. 1947. Benzene poisoning. I. Examination of workers exposed to benzene with reference to the presence of extersulfate, muconic acid, urochrome A and polychenols in the urine together with vitamin C deficiency prophylactic measures. *Acta Med. Scand.* 128:256.

129. Gabovich, R. D. and P. N. Maistruk. 1963. On the therapeutic and prophylactic diet in the fluorine manufacturing industry. *Vopuesy Pitanua* 22:323–338.

130. Goldstein, B. D., O. S. Balchum, N. R. Demopoulues, and N. B. Duke. 1968. Electron paramagnetic resonance spectroscopy. Free radical signals associated with ozonization of linoleic acid. *Arch. Environ. Health* 17:46–49.

131. Chrapil, M., S. L. Elias, J. N. Ryan, and C. F. Zuloski. Considerations on the biological effects of zinc. In *International Review of Neurobiology (Suppl. 1)*, ed. C. C. Pfeifer. 115–173. New York: Academic Press.

132. LeVander, O. A., V. C. Morris, D. J. Higgs, and R. S. Ferretti. 1975. Lead poisoning in vitamin E deficient rats. *J. Nutr.* 105:1481.

133. Dam, H., G. K. Nielsen, I. Prange, and E. Sondergaard. 1958. Influence of linoleic and linolenic acids on symptoms of vitamin E deficiency in chicks. *Nature (London)* 182:802–803.

134. West, G. B. 1964. The influence of diet on the toxicity of acetylsalicylic acid. *J. Pharm. Pharmacol.* 16:788–793.

135. Molokhia, M. M. and H. Smith. 1967. Trace elements in the lung. *Arch. Evniron. Health* 15:745–750.

136. Mustafa, M. G. and D. F. Tierney. 1978. Biochemical and metabolic changes in the lung with oxygen, ozone, and nitrogen dioxide toxicity. *Am. Rev. Resp. Dis.* 118:6.

137. Chvapil, M. and Y. M. Peng. 1975. Oxygen and lung fibrosis. *Arch. Environ. Health* 30:528–532.

138. Ospital, J. J., N. Elsayed, S. D. Hacker, M. G. Mustafa, and D. F. Tierney. 1970. Altered glucose metabolism in lungs of rats exposed to nitrogen dioxide. *Am. Rev. Respir. Dis.* 113:107.

139. Currie, W. D., P. C. Pratt, and A. P. Sanders. 1974. Hyperoxia and lung metabolism. *Chest* 66 *(Suppl.)*:105.

140. Cavanaugh, P. J. and B. Woodhall. 1966. Effects of hyperbaric oxygenation on metabolism. I. ATP concentration in rat brain, liver, and kidney. *Proc. Soc. Exp. Biol. Med.* 121:23.

141. Sanders, A. P. and J. N. Hall. 1966. Effects of hyperbaric oxygenation on metabolism. II. Oxidative phosphorylation in rat brain, liver, and kidney. *Proc. Soc. Exp. Biol. Med.* 121:34.

142. Tierney, D. F. 1971. Lactate metabolism in rat lung tissue. *Arch. Intern. Med.* 127:858.

143. Young, S. L. and J. H. Knelson. 1973. Increased glucose uptake by rat lung with onset of edema. *Physiologist* 16:494.

144. Mustafa, M. G. and D. F. Tierney. 1978. Biochemical and metabolic changes in the lung with oxygen, ozone, and nitrogen dioxide toxicity. *Am. Rev. Resp. Dis.* 118(6):1061–1090.

145. Beekman, D. L. and N. S. Weiss. 1969. Hyperoxia compared to surfactant washout on pulmonary compliance in rats. *J. Appl. Physiol.* 26:700-709.

146. Crapo, J. D., K. Sjostiom, and R. T. Drew. 1978. Tolerance and cross-tolerance using NO_2 and O_2. II. Pulmonary morphology and morphometry. *J. Appl. Physiol.* 44:370.

147. Said, S. I. 1968. The lung as a metabolic organ. *N. Engl. J. Med.* 279:1330.
148. Bekhle, Y. S. and J. R. Vane. 1974. Pharmacokinetic functions of the pulmonary circulation. *Physiol. Rev.* 54:1007.
149. Junod, A. 1976. Uptake, release and metabolism of drugs in the lung. *Pharmacol. Ther.* (B) 2:511.
150. Vane, J. R. and Y. S. Bakhle, eds. 1977. *Metabolic functions of the lung.* New York: Marcel Dekker.
151. Bend, J. R., G. E. R. Hook, R. E. Easterling, T. E. Gram, and J. R. Fouts. 1972. A comparative study of the hepatic and pulmonary mital-function oxidase systems in the rabbit. *J. Pharmacol. Exp. Ther.* 183:206.
152. Hook, G. E. R. and J. R. Bend. 1976. Pulmonary metabolism of xenobiotics. *Life Sci.* 18:279.
153. Reid, W. D., K. F. Tlett, J. M. Glick, and G. Krishna. 1973. Metabolism and binding of aromatic hydrocarbons in the lung. *Respir. Dis.* 107:539.
154. Gram, T. E. 1973. Comparative aspects of mixed-function oxidation by lung and liver of rabbits. *Drug Metab. Rev.* 2:1.
155. Grover, P. L., A. Hewer, and P. Sims. 1974. Metabolism of polycyclic hydrocarbons by rat-lung preparations. *Biochem. Pharmacol.* 23:323.
156. Block, E. R. and A. B. Fisher. 1977. Depression of serotonin clearance by rat lungs during oxygen exposure. *J. Appl. Physiol.* 42:33.
157. Clark, J. M. and C. J. Lambersten. 1971. Pulmonary oxygen toxicity: a review. *Pharmacol. Rev.* 23:37.
158. Palmer, N. S., D. N. Swanson, and D. L. Coffin. 1971. Effect of ozone on benzopyrene hydroxylase activity in the Syrian golden hamster. *Cancer Res.* 31:730.
159. Goldstein, B. D., S. Soloma, R. S. Pasternack, and D. R. Bickers. 1975. Decrease in rabbit lung microsomal cytochrome P-450 levels following ozone exposure. *Res. Commun. Chem. Pathol. Pharmacol.* 10:759.
160. DeLucia, A. J., P. M. Hogue, M. G. Mustafa, and C. E. Cross. 1972. Ozone interaction with rodent lung. I. Effects of sulfhydryls and sulfhydryl-containing enzyme activities. *J. Lab. Clin. Med.* 80:559.
161. Mustafa, M. G., A. D. Hacker, J. J. Ospital, M. Z. Hussain, and S. D. Lee. 1977. Biochemical effects of environmental oxidant pollutants in animal lungs. In *Biochemical Effects of Environmental Pollutants*, ed. S. D. Lee. 59. Ann Arbor, MI: Ann Arbor Science Publishers.
162. Kleinerman, J. 1971. Emphysema and bronchial epithelial hyperplasia in hamsters exposed to long term NO_2. *Am. J. Pathol.* 62:93a.
163. Mannervik, B. 1980. Mercaptans. In *Metabolic Basis of Detoxication: Metabolism of Functional Groups*, ed. W. B. Jakoby. 189. New York: Academic Press.
164. Rea, W. J. 1992. *Chemical Sensitivity. Vol. I: Mechanisms of Chemical Sensitivity.* 110. Boca Raton, FL: Lewis Publishers.
165. Fisher, A. B. and J. Reicherter. 1984. Pentose pathway of glucose metabolism in isolated granular pneumocytes. Metabolic regulation and stimulation by paraquat. *Biochem. Pharmacol.* 33(8):1349–1353.
166. Serabjit-Singh, C. J., C. R. Wolf, C. G. Plopper, and R. M. Philpot. 1980. Cytochrome P-450: localization in rabbit lung. *Science* 207(28):1469–1470.
167. Henderson, R. F., E. G. Damon, and T. R. Henderson. 1978. Early damage indicators in the lung. I. Lactate dehydrogenase activity in the airways. *Toxicol. Appl. Pharmacol.* 44:291–297.

168. Foth, H., M. Molliere, R. Kahl, E. Jahnchen, and G. F. Kahl. 1984. Covalent binding of benzo[a]pyrene in perfused rat lung following systemic and intratracheal administration. *Drug Metab. Disposition* 12(6):760–766.

169. Levine, S. A. and P. M. Kidd. 1985. *Antioxidant Adaptation: Its Role in Free Radical Pathology.* 110–111. San Leandro, CA: Biocurrents Division, Allergy Research Group.

170. Borden, D. G. and J. P. Wyatt. 1970. Lung injury and repair: a contemporary view. In *Pathology Annual*, Vol. 5, ed. S. C. Sommers. 279. New York: Appleton.

171. Rea, W. J. 1992. *Chemical Sensitivity. Vol. I: Mechanisms of Chemical Sensitivity.* 155. Boca Raton, FL: Lewis Publishers.

172. Matsumura, Y. 1970. The effects of ozone, nitrogen dioxide, and sulfur dioxide on the experimentally induced allergic respiratory disorder in guinea pigs. II. The effects of ozone on the absorption and the retention of antigen in the lung. *Am. Rev. Respir. Dis.* 102:438–443.

173. Matsumura, Y. 1970. The effects of ozone, nitrogen dioxide, and sulfur dioxide on the experimentally induced allergic respiratory disorder in guinea pigs. III. The effect of the occurrence of dyspneic attacks. *Am. Rev. Respir. Dis.* 102:444–447.

174. Matsumura, Y. 1970. The effects of ozone, nitrogen dioxide, and sulfur dioxide on the experimentally induced allergic respiratory disorder in guinea pigs. I. The effect on sensitization with albumin through the airway. *Am. Rev. Respir. Dis.* 102:430–437.

175. Rea, W. J. 1994. *Chemical Sensitivity. Vol. II: Sources of Total Body Load.* 765. Boca Raton, FL: Lewis Publishers.

176. Riedel, F., M. Krämer, C. Scheibenbogen, and C. H. Rieger. 1988. Effects of SO_2 exposure on allergic sensitization in the guinea pig. *J. Allergy Clin. Immunol.* 82(4):527–534.

177. Miyata, M., T. Namba, and S. Ishikawa. Feb. 22–25, 1990. Experimental allergic conjunctivitis due to environmental chemicals. Presented at the 8th Annual International Symposium on Man and His Environment in Health and Disease. Dallas, TX: The American Environmental Health Foundation.

178. Alexandersson, R., G. Hedenstierna, E. Randma, G. Rosen, A. Swenson, and G. Tornling. 1985. Symptoms and lung function in low-exposure to TDI by polyurethane foam manufacturing. *Int. Arch. Occup. Environ. Health* 55(2):149–157.

179. Shigematsu, N., S. Ishmaru, R. Saito, T. Ikeda, K. Matsaba, K. Sugiyams, and Y. Masuda. 1978. Respiratory involvement in PCB poisoning. *Environ. Res.* 16:92–100.

180. Aripdzharov, T. M. 1973. Effect of the pesticides Anthio and Milbex on the immunological reactivity and certain autoimmunological reactivity and certain autoimmune processes of the body. *Gigiena i Sanitariya* 7:39–42e (Russian).

181. Robison, T. W. and H. J. Forman. 1993. Dual effect of nitrogen dioxide on rat alveolar macrophage arachidonate metabolism. *Experimental Lung Res.* 19(1):21–36.

182. Jakab, G. J., T. H. Risby, and D. R. Henemway. 1992. Use of physical chemistry and in vivo exposure to investigate the toxicity of formaldehyde bound to carbonaceous particles in the murine lung. *Res. Rep. Health Eff. Inst.* 53:1–39.

183. Nugent, K. M. and J. M. Onofrio. 1987. Effect of alkylating agents on the clearance of *Staphylococcus aureus* from murine lungs. *J. Leukocyte Biol.* 41(1):78–82.

184. Proctor, N. H. and J. P. Hughes. 1978. *Chemical Hazards of the Workplace.* Philadelphia: J. B. Lippincott.

185. Siboni, A., F. Solander, and O. Sondergaard. 1989. Serious infections in alcoholics. I. Bacteremia, lobar pneumococcal pneumonia and pneumococcal meningitis in alcoholics 1974–1987. *Ugeskrift for Laeger* 151(6):374–376.

186. Randolph, T. G. 1956. O tratamento da asma brônquica na infância. *Separata de resenha clinico-cientifica* 25(7):3–7.

187. Newman, L. S. and C. S. Rose. 1989. Occupational asbestosis and related diseases. *Medical/Scientific Update.* 8(8):28–33.

188. Smith, J. M. 1988. Epidemiology and natural history of asthma, allergic rhinitis, and atopic dermatitis (eczema). In *Allergy Principles and Practice*, eds. E. Middleton, Jr., C. E. Reed, E. F. Ellis, N. F. Adkinson, Jr., J. W. Yunginger. 891–929. St. Louis: C. V. Mosby.

189. Reed, C. E. and R. G. Townley. 1978. Asthma: classification and pathogenesis. In *Allergy: Principles and Practice*, eds. E. Middleton, Jr., C. E. Reed, and E. F. Ellis. 659–677. St. Louis: C. V. Mosby.

190. Pipkorn, U., D. Proud, L. M. Lichtenstein, R. P. Schleimer, S. P. Peters, N. F. Adkinson, Jr., A. Kagey-Sobotka, P. S. Norman, and R. M. Naclerio. 1987. Effect of short-term systemic glucocorticoid treatment on human nasal mediator release after antigen challenge. *J. Clin. Invest.* 80:957–961.

191. Boushey, H. A., M. J. Holtzman, J. R. Sheller, and J. A. Nadel. 1980. State of the art. Bronchial hyperreactivity. *Am. Rev. Respir. Dis.* 121:389–413.

192. Kreit, J. W., K. B. Gross, T. B. Moore, T. J. Lorenzen, J. D'Arcy, and W. L. Eschenbacher. 1989. Ozone-induced changes in pulmonary function and bronchial responsiveness in asthmatics. *J. Appl. Physiol.* 66:217–222.

193. Burge, P. S. 1982. Occupational asthma in electronics workers caused by colophony fumes: follow-up of affected workers. *Thorax* 37:348–353.

194. Shim, C. and M. H. Williams. 1986. Effect of odors in asthma. *Amer. J. Med.* 80:18–22.

195. Stankus, R. P., P. K. Menon, R. J. Rando, H. Glindmeyer, J. E. Salvaggio, and S. B. Lehrer. Cigarette smoke-sensitive asthma: challenge studies. *J. Allergy Clin. Immunol.* 82:331–338.

196. Weinreich, D. and B. J. Undem. 1987. Immunological regulation of synaptic transmission in isolated guinea pig autonomic ganglia. *J. Clin. Invest.* 79:1529–1532.

197. Spitzer, W. O., S. Suissa, P. Ernst, R. I. Horwitz, B. Habbick, D. Cockcroft, J.-F. Boivin, M. McNutt, A. S. Buist, and A. S. Rebuck. 1992. The use of β-agonists and the risk of death and near death from asthma. *N. Engl. J. Med.* 326:501–506.

198. Mellis, C. M. and P. D. Phelan. 1977. Asthma deaths in children—a continuing problem. *Thorax* 32:29–34.

199. Grainger, J., K. Woodman, N. Pearce, J. Crane, C. Burgess, A. Keane, and R. Beasley. 1991. Prescribed fenoterol and death from asthma in New Zealand 1981–1987: a further case-control study. *Thorax* 46:105–111.

200. Maberly, D. J. and H. M. Anthony. 1992. Asthma management in a "clean" environment. 2. Progress and outcome in a cohort of patients. *J. Nutr. Med.* 3:231–248.

201. Alberts, W. M. 1993. A guide to common causes, and when to suspect them: is your patient's asthma occupationally linked? *J. Respir. Dis.* 14(4):564–578.

202. Paggiaro, P. L., A. Innocenti, E. Bacci, O. Rossi, and D. Talini. 1986. Specific bronchial reactivity to toluene diisocyanate: relationship with baseline clinical findings. *Thorax* 41:279–282.

203. Paggiaro, P. L., L. Lastrucci, F. Paroli, O. Rossi, E. Bacci, and D. Talini. 1986. Specific bronchial reactivity to toluene diisocyanate: dose-response relationship. *Respiration* 50(3):167–173.

204. Hall, M. A., H. Thom, and G. Russell. 1981. Erythrocyte aspartate aminotransferase activity in asthmatic and nonasthmatic children and its enhancement by vitamin B_6. *Ann. Allergy* 47:464–466.

205. Collipp, P. J., S. Goldzier, N. Weiss, Y. Soleymani, and R. Snyder. 1975. Pridoxine treatment of childhood bronchial asthma. *Ann. Allergy* 35:93–97.

206. Morris, D. L. 1988. Treatment of respiratory disease with ultra-small doses of antigens. *Ann. Allergy* 28: 494–500.

207. Wagner, J. C., M. Chamberlain, R. C. Brown, G. Berry, F. D. Pooley, R. Davies, and D. M. Griffiths. 1982. Biological effects of tremolite. *Br. J. Cancer* 45:35.

208. Gibbs, A. R., R. M. E. Seal, and J. C. Wagner. 1984. Pathological reactions of the lung to dust. In *Occupational Lung Diseases*, 2nd ed., eds. W. K. C. Morgan and A. Seaton. 139. Philadelphia: W. B. Saunders.

209. Seaton, A. and W. K. C. Morgan. 1984. Toxic gases and fumes. In *Occupational Lung Diseases*, 2nd ed., eds. W. K. C. Morgan and A. Seaton. 609–639. Philadelphia: W. B. Saunders.

210. Polakoff, P. L., N. L. Lapp, and R. Reger. 1975. Polyvinyl chloride pyrolysis products. *Arch. Environ. Health* 30:269.

211. Jones, R. N. and H. Weill. 1977. Respiratory health and polyvinyl chloride fumes. *JAMA* 237:1826.

212. Bryan, W. T. K. and M. P. Bryan. 1972. Allergy in otolaryngology. In *Otolaryngology.* Vol. 5, ed. W. H. Maloney. 230. New York: Harper & Row.

213. Rinkel, H. J., C. Lee, D. W. Brown, J. W. Willoughby, and J. M. Williams. 1964. The diagnosis of food allergy. *Arch. Otolaryngol.* 79:71.

214. Lee, C. H., R. I. Williams, and E. L. Binkley, Jr. 1969. Provocative inhalant testing and treatment. *Arch. Otolaryngol.* 90:113.

215. Miller, J. B. 1972. *Food Allergy: Provocative Testing and Injection Therapy.* Springfield, IL: Charles C Thomas.

216. Rinkel, H. J. 1949. Inhalant allergy. I. The whealing response of the skin to serial dilution testing. *Ann. Allergy* 7:625.

217. Randolph, T. G. 1962. *Human Ecology and Susceptibility to the Chemical Environment.* Springfield: Charles C Thomas.

218. Menzel, D. B. and R. O. McClellan. 1980. Toxic responses of the respiratory system. In *Cassaret and Doull's Toxicology: The Basic Science of Poisons*, 2nd ed., eds. J. Doull, C. Klaassen, and M. O. Amdur. 246–274. New York: Macmillan.

219. Amdur, M. O. and J. Mead. 1958. Mechanics of respiration in unanesthetized guinea pigs. *Am. J. Physiol.* 192:364–368.

220. Haponik, E. and W. Summer. 1981. Inhalation of irritant gases. In *Occupation Lung Diseases* Vol. 1, eds. S. M. Brooks, J. Lockey, and P. Harber. 273–287. Philadelphia: W. B. Saunders.

221. Mehta, R. K. 1972. Public health hazards of DDT excretion in milk. *Indian J. Public Health* 16(3):107–111.

222. Falck, F., Jr., A. Ricci, Jr., M. S. Wolff, J. Godbold, and P. Deckers. 1993. Pesticides and polychlorinated biphenyl residues in human breast lipids and their relation to breast cancer. *Environ. Health Mon.* 5(5):143–146.

223. Frim-Titulaer, L. W., J. F. Cordero, L. Haddock, G. Lebrón, R. Martinez, and J. L. Mills. 1986. Premature thelarche in Puerto Rico: a search for environmental factors. *AJDC* 140:1263–1267.

224. Gilpin, A. 1978. *Air Pollution*, 2nd ed. Lucia, Queensland: University of Queensland Press.

225. Minton, J. P. 1988. Dietary factors in benign breast disease. *Cancer Bull.* 40:44–50.

226. Minton, J. P. and H. Abou-Issa. 1986. Fibrocystic breast disease — its management and explanation. In *Advances in Breast Surgery*, 15–24. University of Minnesota Medical School, Department of Surgery and Continuing Medical Education. Chicago: Year Book Medical Publishers.

227. Mettelin, C. 1984. Diet and the epidemiology of breast cancer. *Cancer* 53:605–671.

228. Cooperman, A. M. and R. Hermann. 1984. Breast cancer: an overview. *Surg. Clin. N. Am.* 64(6):1031–1038.

229. Raloff, J. 1993. Ecocancers: Do environmental factors underlie a breast cancer epidemic? *Science News* 144:10–12.

230. Armstrong, B. K., J. B. Brown, H. T. Clarke, D. K. Crooke, R. Hähnel, J. R. Masarei, and T. Ratajczak. 1981. Diet and reproductive hormones: a study of vegetarian and nonvegetarian postmenopausal women. *JNCI* 67(4):761–767.

231. Thomas, D. B. and A. M. Lilienfeld. 1976. Geographic reproductive and sociobiological factors in risk factors in breast cancer. In *Risk Factors in Breast Cancer*, ed. B. A. Stoll. Chicago: Year Book Medical Publishers.

232. Macmahon, B., P. Cole, and J. Brown. 1973. Etiology of human breast cancer: a review. *J. Nat. Cancer Inst.* 50(1):21–42.

233. Shows, A. R. and W. Schubert. 1992. Silicone breast implants and immune disease. *Ann. Plast. Surg.* 28(5):491–499.

234. Higuchi, H. 1994. Unpublished data at the EHC-Dallas.

235. Burton, T. M. 1993. Breast implants raise more safety questions: research links silicone version to new diseases. *The Wall Street Journal* Feb. 2: p. 131.

20   Cardiovascular System

INTRODUCTION

In the chemically sensitive patient, vascular injury associated with chemical insults manifests in a variety of signs and symptoms. The clinical picture is mainly dependent upon the types of vessels involved (vein, capillary, or large or small artery), the numbers of vessels involved, the intensity and duration of the pollutant stimuli present, and the strength of the nutrient-derived vessel repair mechanisms. The spectrum of injury can range from a very mild localized edema to a major end-organ failure with loss of limb or life to any problems in between. Problems may develop over a long period of time or within a relatively short period. Most situations of vascular malfunction that bring the patient to a physician are acute, although the events leading up to the acute distress may have begun at a much earlier time. Exposures to toxic substances may permit vascular malfunction, often with resultant chemical sensitivity. These exposures often lead to spastic or deregulated blood vessels associated with many different target organs. In the chemically sensitive individual, the results of this dysfunction may at times account for transient organ ischemia giving a variety of clinical entities such as transient cerebral ischemia, angina with or without infarction, Raynaud's phenomenon, leg and renal ischemia with hypertension, migraine headaches, or severe myalgia with loss of energy. Additional cumulative or more severe stimulants may result in more severely inflamed vessels. In the chemically sensitive individual, inflammation usually can be reversed when the incitants are withdrawn. However, continued release of very vasoactive substances or prolonged exposures will eventually bring on a fixed or irreversible disease state if the incitants are not withdrawn. Cardiovascular ailments and environmental substances have been clearly linked in the literature (Table 20.1).[1-6]

To understand the effects of pollutants on the vascular system, underlying mechanisms (see Volume I, Chapter 4[7]) and their effect on the regulatory receptor of the neurovascular tree must also be understood.

Table 20.1.   Reports ᴏf Environmentally Triggered
Vascular Dysfunction

Author	Date of report	Type of vascular dysfunction reported
		Foods
Hare[64]	1905	Vascular responses
Schofield[189]	1908	Angioedema
Connors[171]	1920	Phlebitis
Rowe[190]	1931	Phlebitis and vascular dysfunction
Rea [67,147]	1976	Phlebitis
	1981	Phlebitis
Harkavy[15]	1939	Arrhythmias
Rea[66]	1977	Arrhythmias
Levi[191]	1979	Arrhythmias
		Chemicals
Randolph[146]	1962	Arrhythmias
Rea[66]	1977	Arrhythmias
Taylor[6]	1970	Death
Spizer[5]	1975	Arrhythmias
Nour-Eldon[3]	1970	Phenol affinity
Yevick[59]	1975	Fibrotic lesions
Rea[150]	1992	Hypertension
		Pollens and molds
Theorell[68]	1976	Hypersensitivity vasculitis
		Biological inhalants, foods, chemicals
Rea[141]	1975	Large vessel vasculitis
Rea[66]	1977	Small vessel vasculitis
Balazs[192]	1981	Animal heart lesions
Balazs[192]	1981	Environmental triggering of the vascular tree

Source: EHC-Dallas. 1994.

POLLUTANT INJURY TO THE AUTONOMIC
NERVOUS SYSTEM

As discussed throughout this book, pollutant injury will cause dysfunction of the autonomic nervous system (ANS) going to the heart and blood vessels.

The effects of toxic chemicals upon the heart manifest as change in rate, rhythm, ventricular function, and oxygen exchange. The basic heart rate and rhythm are functions of the intrinsic electrical activity. Normally, the pacemaker cells with the most rapid intrinsic rate are located in the sinoatrial node. Regulation of heart rate is usually via the ANS.

The cardiovascular system has a rich ANS supply (Figures 20.1 and 20.2). To understand early pollutant injury to this system, one must also understand the anatomy of the autonomic as well as the neuroendocrine system-(see Chapter 24). Since the neuroendocrine system has numerous baroreceptors and chemoreceptors that are ultimately related to the vasculature of the heart and other blood vessels, it is necessary to be familiar with the aforementioned locations and physiological responses of these systems. In addition, the paraganglia end-organ response through its neuroendocrine cells and release of vasoactive amines and neurotransmitters has vast implications for the nature of the response of the cardiovascular system to pollutants. These substances may activate or alter vascular response. The final outcome of pollutant injury to the ANS is malfunction of the cardiovascular system. However, in other cases, occlusive disease occurs either due to scarring or arteriosclerosis, which also may be pollutant and dietary triggered. This dysfunctional response may be localized in one region of the body, or it may be generalized. In the chemically sensitive individual, for example, one path of response to pollutant stimuli may end in cerebral arterial malfunction with headaches that then become general-ized brain dysfunction with short-term memory loss and finally extend to frank hemiplegia. Another regional response might be coronary spasm with resultant angina pectoris, while another might be renal arterial spasm with changes in angiotensin and, thus, blood pressure. Eventually, as damage progresses, es-sential hypertension, as seen in a subset of chemically sensitive patients, can result. On the other hand, if generalized stimulation of the ANS occurs, deregulation of the entire vascular tree with inappropriate vasodilation and constriction of the skin and other internal organ vessels will be seen. This dysfunction occurs in virtually every chemically sensitive patient.

Pollutant damage to the blood supply to the skin results in vessel deregu-lation that manifests as changing peripheral skin colors from blanched white to severe red to deep blue. Even perhaps the yellow color of the "chemical yellows" is due to this deregulation. In addition, blood pressure (see Chapter 25), brain function (see Chapter 26), breathing (see Chapter 19), and visual acuity (see Chapter 27) may fluctuate.

In the chemically sensitive individual, pollutants have various routes of entry and stimulation. When peripheral cutaneous or gastrointestinal noxious stimuli are first detected, there is a retrograde impulse to the dorsal nerve root ganglia through the afferent fibers of peripheral nerves (slow C or rapid delta A) or the gastrointestinal plexus. If the stimulus enters through the nose, it travels up to the hypothalamus and then to the ANS. If it goes to the lung, it travels via the bloodstream to the hypothalamus and then down the autonomic nerves. The sensory neurotransmitter, substance P, is then released causing immediate vasodilation and increased permeability of the microcirculation in

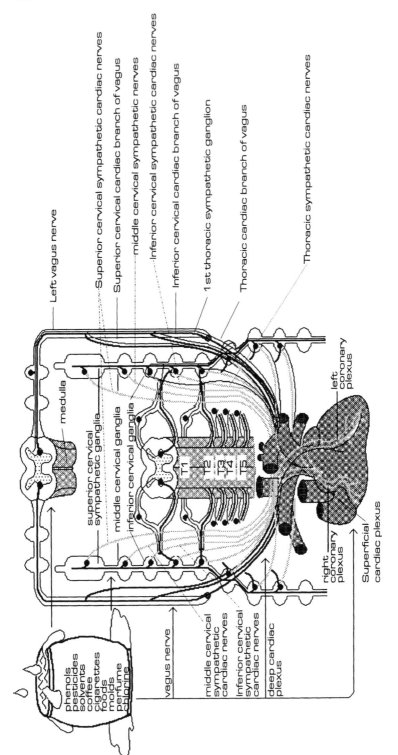

Figure 20.1. Areas of potential pollutant injury of the neural innervation of the heart.

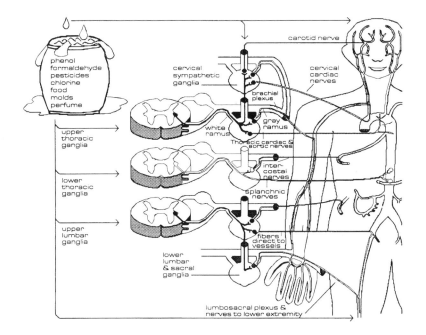

Figure 20.2. Areas of potential pollutant injury to the neurovascular system.

the area of the nerve and activating the non–IgE-mediated release of histamine via the mast cells. This permeability may result in localized edema. In addition, the release of leukotactic factors and leukotrienes is stimulated. Somatostatin is released in other cells of the dorsal root but can also be released from the CNS and the pancreas. The relationship between somatostatin and substance P is complicated, and both have effects on other cell interactions. For more detailed information, see Chapter 26.

To provide the clinician with a better understanding of the physiology of pollutant damage, various regions that are affected are discussed separately. These comprise the brain and head including carotid sinus reflex, heart and mediastinum including coronary spasm, abdomen and renal vessels, and periperal vessels.

Brain and Head

Pollutant injury to the brain and head can result in a very complicated response in the chemically sensitive individual. There are several reasons for this. The inflow track of the hypothalamus is connected to the olfactory nerve, either passing through or bypassing the limbic system, with the outflow to the cerebrum and the rest of the body. The autonomic ganglia, which are located in the cephalic region, are connected with the brainstem through cranial nerve fibers. Fibers that arise in the cervical sympathetic trunk ganglia also extend into the plexus associated with the thoracic and the abdominal viscera through the vagus nerve. Furthermore, paraganglia consisting of neuroendocrine cells are found in the area of the jugullo tympanic membrane, angle of the mandible, intercarotid areas (carotid bodies), interthyroid, laryngeal, aortico-pulmonary area of the superior mediastinum, and the coronary-interatrial at the base of the heart. Upon contact with pollutant stimuli, these paraganglia secrete numerous substances such as monoamines (dopamine, epi- and norepinephrine, and serotonin), oligopeptides, and enzymes. They also contain immunoreactive substances. The carotid body responds to varying levels of chemicals such as carbon monoxide, while the aortic bodies respond to pressure changes. Chemically sensitive individuals are super-sensitive to pressure and ionic changes, perhaps through damage to these neuroendocrine bodies. With either a predominant sympathetic or parasympathetic stimulation, cerebral vessels can malfunction, resulting in brain dysfunction or headaches. In most of the chemically sensitive patients treated at the EHC-Dallas who have brain involvements, triple camera SPECT brain scans have shown a diffuse path and flow pattern, the latter of which is often decreased, but sometimes increased. These scans show typical neurotoxic patterns with altered flow, distribution, and function, usually of the "salt and pepper" type where patchy function is present. Frequently, these patterns reflect a discrepancy between flow and function patterns (Figure 20.3). If the changes from mediator release and immune receptors are added to total body load, the end-organ responses can become much more complicated. We have seen several cases of environmentally triggered hemiplegia from carotid malfunction with resultant unilateral spasm. These spasms could be turned on and off with pollutant withdrawal and subsequent re-exposure. Figure 20.4 is a SPECT scan showing normal and dysfunctional flow and distribution.

Heart and Mediastinum

The heart and mediastinum have a complex arrangement of autonomic innervation as well as neuroendocrine cells (Figure 20.1). Because of this anatomy, pollutant-induced stimulation or deregulation of these areas may occur via many avenues. The result of either stimulation or deregulation may be arrhythmias as well as changes in vessel size.

The intrinsic rate of the sinoatrial node is from 100 to 120 per minute, which is slowed by parasympathetic stimulation of the vagus nerve to 70 per minute. The parasympathetic nerve humor acetylcholine decreases the slope of diastolic depolarization in the SA node to accomplish this breaking effect. Of course, in some chemically sensitive patients, acetylcholine can be disturbed by organophosphate insecticides, causing rate malfunction. Conversely, norepinephrine, which may be released by pollutant trigger of the sympathetic end-organ mediator, increases the slope of diastolic depolarization causing an increase in heart rate as a result of sympathetic stimulation. Impaired methionine

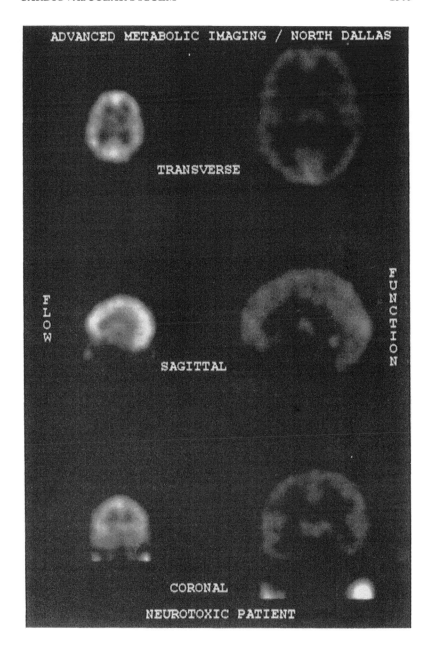

Figure 20.3. SPECT brain scan showing discrepancy between flow and function patterns in a neurotoxic chemically sensitive patient. (From T. Simon and D. Hickey. Metabolic Imaging, 1993. Personal communication.)

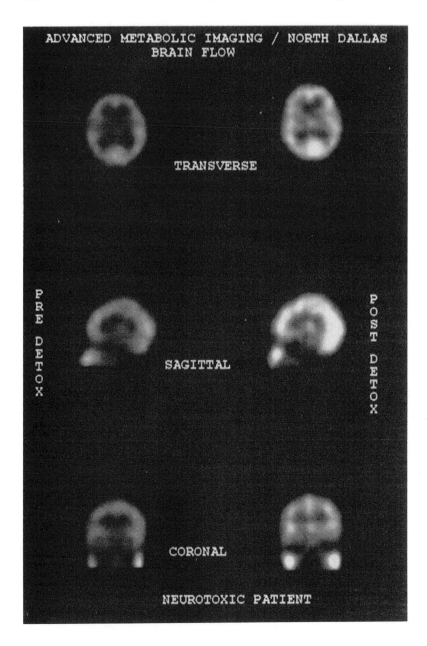

Figure 20.4. Triple camera SPECT brain scan of cerebral blood flow before and after detoxification. (From T. Simon and D. Hickey. Metabolic Imaging, 1993. Personal communication.)

metabolism is the primary conjugation malfunction currently seen in the chemically sensitive patient. Impaired methylation prevents norepinephrine from being metabolized, thus high and prolonged levels result in some chemically sensitive patients, explaining their tachycardia when it occurs. Usually these patients also experience palpitations. As the patient is deadapted with the total pollutant load decreasing, the pulses slow, probably as a result of decreased norepinephrine output (see Chapter 31). Measurements of catecholamine levels in the ECU have shown their reduction with fasting.

Electrical stimulation and pollutant-mediated vasoconstriction effectively limit the metabolic vasodilation by the sympathetic fibers that innervate the heart and coronaries, resulting in tachycardia, increased myocardial constriction, and augmented myocardial oxygen consumption. This increased myocardial metabolism leads to coronary vasodilation by local mechanisms of increased CO_2 and decreased O_2, and, therefore, increases O_2 demand. Paradoxically, there is a concomitant adrenergic coronary vasoconstriction effect that limits the metabolic vasodilation and leads to an augmented myocardial AVO_2 difference.

Although the net effect of sympathetic activation is almost always an increase in coronary blood flow, the effects of adrenergic vasoconstriction retard metabolic vasodilation 20 to 30%.[8] This slow down may aggravate symptoms in the already damaged chemically sensitive individual.

The cardiac chronotropic and inotropic effects of sympathetic activation are mediated via beta receptors. They are blunted by β-receptor-blocking agents. Therefore, it is possible to unmask sympathetic coronary vasoconstriction by prior beta receptor blockade (since the vasoconstriction is mediated by the receptor).[9] Pollutant stimuli can result in beta blocks as well as sympathetic stimulation producing the aforementioned affects in some chemically sensitive individuals.

After the β-receptor blockade, the usual response to sympathetic activation is a net decrease in coronary flow, which is unmasked in vasoconstriction. This response explains why the chemically sensitive patient does not do well unless the triggering agent is removed. Drug treatment may afford a temporary respite but, if exposure to the pollutants causing the disease process continues, adrenergic stimulation also continues. Then dysfunction of the coronary flow occurs and persists. Norepinephrine administration will duplicate this response.[10]

In unanesthetized dogs, resting coronary flow is normal, and sympathetic stimulated (by 6- OH dopamine) regions of the same heart show a 60% greater flow in those areas of the heart than in nonstimulated regions. When animals are in the true resting basal state with few circulating catecholamines, both the coronary and renal arteries show little or no sympathetic affect (vasoconstriction). Pollutant injury can occur to the point of eliminating sympathetic effect. For example, if enough is given, phenol will cause sympathectomy. In chemically sensitive individuals who have lots of phenols in their blood, though an insufficient quantity to cause a total sympathectomy, sympathetic dysfunction may occur because of the chronic stimulation of the sympathetic fibers going to the heart. This malfunction then results in problems related to both local cardiac dysfunction and generalized vascular dysfunction. Many chemically sensitive patients complain of palpitations and have hyperdynamic type hearts. These facts are exemplified in the case of a 41-year-old white female whose alterations are shown in Figure 20.5. As can be seen from the results of this challenge test in this chemically sensitive patient, an inhaled dose of phenol caused peripheral vascular and coronary spasm as well as total complement change.

It should be emphasized that many chemically sensitive patients frequently complain of chest pains and recurrent palpitations. These symptoms appear to be triggered by pollutant injury of the neuroendocrine system and are often associated with magnesium deficiency, which is difficult to diagnose other than by reproduction of symptoms with challenge tests performed under environmentally controlled conditions after the patient's total load is decreased.

Carotid Sinus Reflex

Lowering carotid sinus pressure also leads to tachycardia with peripheral vasoconstriction and increased aortic blood pressure. The tachycardia results from parasympathetic inhibition and sympathetic activation, whereas the peripheral vasoconstriction is mediated via the sympathetic nervous system. Although there is a net increase in coronary blood flow during baroreceptor reflex that is elicited by carotid sinus hypotension, a concomitant andrenoreceptor mediated vasoconstriction competes with the metabolic vasodilation that results from the tachycardia and augmented left ventricular afterload.[8,11,12] Exaggerated responses of this mechanism have been seen in the pollutant deregulated chemically sensitive patient.

The coronary vasoconstrictor component of the reflex is paradoxical since the coronary blood flow must increase to provide oxygen required by the tachycardia and is augmented by the left ventricular afterload. This adrenergic vasoconstriction may help preserve blood flow to the inner layers of the left ventricle. Adrenergic activity of the heart is increased by exercise, anger, cold, and pollutant stimulation.[13] The following case is illustrative of autonomic dysfunction and dysfunction of the carotid sinus reflex influencing the heart (Table 20.2).

> **Case study**. This 60-year-old white male developed drop attacks. Cardiac monitoring revealed that he would develop sudden bradycardia, asystole, then tachycardia. This process was triggered by perfumes and other environmental incitants and parallelled his drop attacks. This chemically sensitive patient's development of cerebral and cardiovascular symptoms indicated he had a very sensitive carotid sinus reflex. When it was stimulated by perfume or

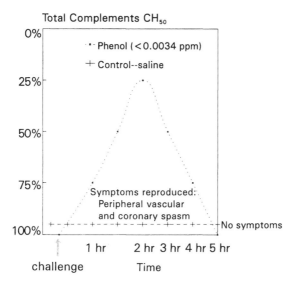

Figure 20.5. 41-year-old, white female. Inhaled double-blind challenge after 4 days of deadaptation with the total load reduced in the ECU. Symptoms reproduced with challenge included vascular and coronary spasm. (From EHC-Dallas. 1979.)

Table 20.2. 60-Year-Old White Male — Double-Blind Inhaled Challenge Reproducing Drop Attacks after 5 Days Deadaptation in the ECU with Total Load Reduced

Incitant	Dose ppm	Result
Perfume	ambient	severe drop attack
Ethanol petroleum derived	<0.50	transient drop attack
Insecticide (2,4-DNP)	<0.0034	—
Formaldehyde	<0.20	—
Phenol	<0.0024	—
Chlorine	<0.33	—
Saline 1	ambient	—
Saline 2	ambient	—

Note: Biological Inhalants (intradermal challenge confirmed by inhaled challenge): molds, dust. Foods (oral challenge confirmed by intradermal): wheat, corn, cane sugar, beet. Chemicals (inhaled challenge confirmed by intradermal): perfume, ethanol, insecticide (2,4-DNP).

Source: EHC-Dallas. 1976.

other toxic substances or when the carotid sinus was touched, he would have a sudden drop attack. This period of unconsciousness and hypotension would last until the pollutants were removed and the patient was given the neutralizing dose of the appropriate antigen. This patient required a pacemaker as well as antigen injections, rotary diet, and good environmental control in order to stop his carotid reflex and cardiac problems.

Coronary Spasm

Coronary spasm implies some pathological changes in the aforementioned mechanism that lead to prolonged vascular constriction. There may be pathological vasoconstrictor agents of environmental origin, such as the mycotoxin, ergot and toxic chemicals and toxins from other molds, food and chemical stimuli, as well as pollutant-derived, altered receptor function or augmented smooth muscle responsiveness. The following case is an example.

> **Case study**. This 41-year-old white female nurse entered the ECU with the chief complaint of recurring chest and neck pain (Table 20.3). For the previous 20 years she had recurrent sinusitis that had been treated symptomatically with medication that covered her pollutant-stimulated injury. Two years prior to admission, this patient had developed severe spasm of her peripheral leg vessels (Figure 20.6). She was treated symptomatically with antispasmodic medications until she developed elevated liver enzymes. Six months prior to ECU admission, this patient developed severe, crushing chest pain that radiated down her left arm and was accompanied by S-T elevation. Her symptoms were refractory to all medications, and her situation became more complicated as she rapidly developed sensitivity to them. Coronary angiograms showed a spastic left anterior descending coronary artery (Figure 20.7).
>
> When she was confined to the ECU, she was taken off all medication. Five days later she was asymptomatic with good peripheral pulses. The S-T depression and coronary pain were gone. Controlled challenge tests revealed that a variety of chemicals such as the pap smear fixative, iodinated instrument wash, petroleum-derived ethanol, pesticide, formaldehyde, phenol, and many other chemicals triggered her spasm and her spontaneous bruising (Figure 20.8). Photomicrograph of the spontaneous bruising showed lymphocytic infiltrate of the precapillary arterial (Figure 20.9). Immune changes were also observed. T lymphocyte depression was seen after inhaled challenges (Figure 20.10). In addition, molds, cotton linters, and turkey triggered her problem. She lived near a cotton gin, and each harvest season the emanations from the gin triggered symptoms. This patient had trouble with recurrent coronary spasm until she was able to develop a controlled home environment. She experienced several readmissions initially due to unavoidable exposures. Eventually, with good environmental control, rotary diet, injection therapy for foods and inhalants, and transfer factor (2 units weekly), this patient has been able to remain well. She has not been hospitalized in the last 15 years for this problem and has been extremely active without medication.

Table 20.3. 41-Year-Old White Female — Fasted and Challenged in an Environmentally Controlled Area of a Hospital after 5 Days Deadaptation with the Total Load Decreased

Diagnosis:	Coronary spasm, peripheral vascular spasm, gastrocnemius spasm, spontaneous bruising, severe tetany		

Laboratory:		Patient	Control
(Biopsy–perivascular	Sedimentation rate	32–55	10 ± 10 mm/hr
lymphocytic vasculitis)	Total complement CH_{50}	138%	$100 \pm 20\%$
	T-lymphocyte absolute	762 (46%)	$1500 \pm 500/mm^3$ $(70 \pm 5\%)$

Triggering agent:	Inhalants — molds, cotton linter, terpenes Foods — turkey Chemicals — formaldehyde, phenol

Sequence of challenge reaction progression:	Immediate — nausea and vomiting 5 minutes — loss of peripheral pulse 10 minutes — severe tetany 30 minutes — change in T-lymphocytes 12 hours — spontaneous

Discharge status:	Absent tetany, coronary and peripheral artery spasm. No treatment with medications — only inhalants and food injection therapy and environmental control.

Follow-up:	Well and active life, except for rehospitalization for acute coronary spasm after massive chemical exposure; on food and inhalant injections and transfer factor. 15 year follow-up — doing well without medication

Source: EHC-Dallas. 1977.

A second case emphasizes the localized type of ventricular arrhythmia of endocardial vessel vasospasm.

Case study. This 33-year-old white female (Table 20.4) presented with the chief complaint of recurrent, incapacitating, ventricular arrhythmias characterized by bigeminy, multifocal PVCs, and ventricular tachycardia. Her

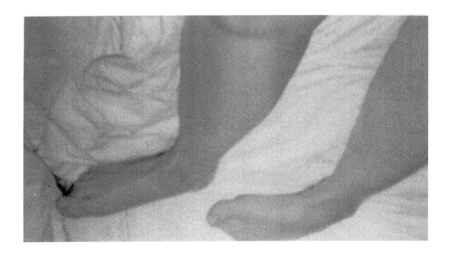

Figure 20.6. 41-year-old, white female. Tetany after double-blind inhaled challenge with phenol (<0.002 ppm) after 4 days deadaptation in the ECU with total load reduced. (From EHC-Dallas. 1979.)

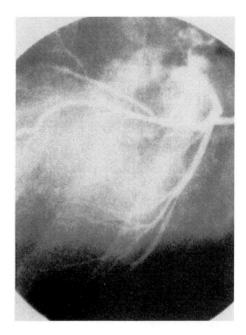

Figure 20.7. 41-year-old, white female. Spasm of left anterior descending coronary artery after double-blind challenge with phenol (<0.002 ppm) after 4 days deadaptation in the ECU with total load reduced. (From EHC-Dallas. 1979.)

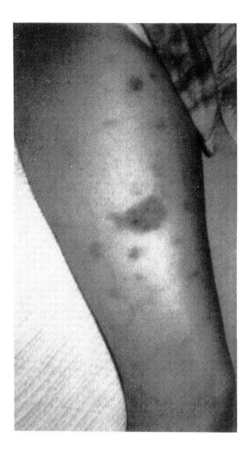

Figure 20.8. 41-year-old, white female. Spontaneous bruising after double-blind inhaled challenge with phenol (<0.002 ppm) after 4 days deadaptation in the ECU with total load reduced. (From EHC-Dallas. 1979.)

environmental history was significant in that she had had thrombophlebitis and pulmonary embolism, asthma, recurrent headaches, and diarrhea. The patient was sensitive to most antiarrhythmic medications. She underwent coronary angiograms, which showed normal coronary arteries. She was placed in the ECU and deadapted for 4 days. Her arrhythmias cleared without medications and then reappeared on challenge. The patient has done well over the past 10 years utilizing an avoidance program and a home oasis.

Abdomen and Renal Vessels

Pollutant injury can be generalized in its symptomatology or be isolated to the abdomen and renal vessels of the body in the chemically sensitive individual. Stimulation by pollutants may result in varied responses because of the

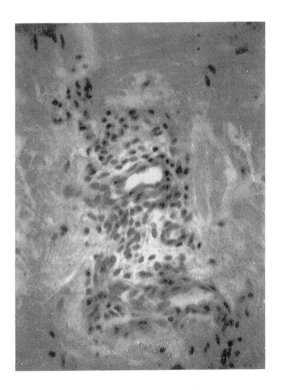

Figure 20.9. 41-year-old, white female. Photomicrograph of small vessel biopsy following double-blind inhaled challenge with phenol (<0.002 ppm) after 4 days of deadaptation in the ECU with total load reduced. Photomicrograph shows the following: (1) precapillary arteriole; (2) monocellular infiltrate; (3) leak or rupture; (4) hemorrhage. (From EHC-Dallas. 1979.)

vessels of the abdominal viscera that are involved as well as the deviation from a strictly segmental arrangement of the parasympathetic nervous system in this area, particularly in the lumbar region. In addition, there is a vast array of neuroendocrine cells that may respond to ANS malfunction as well as to pollutant stimuli in this area. These include the intravagal ganglion nodosum of the nasopharynx and angle of the mandible as well as the end organs of peripheral distribution of the inferior vagus nerve and the aortic-sympathetic cells of the paravertebral intrathoracic and paravertebral areas. Also included are the organ of Zucker-Kandl at the bifurcation of the aorta, the visceral autonomic paraganglia in the gastrointestinal system, porta hepatis, genital tract, urinary bladder, and cauda equina regions.

Pollutant stimuli resulting in autonomic abdominal dysfunction in the chemically sensitive individual can cause a myriad of responses from vascular

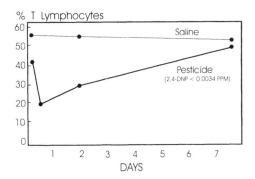

Figure 20.10.   41-year-old, white female. T-cell change after double-blind,
inhaled pesticide (2,4 DNP <0.0034 ppm) challenge. Signs
and symptoms reproduced after 4 days of deadaptation in
the ECU with the total load reduced were nausea and
vomiting within 5 minutes, cough, shortness of breath, loss
of peripheral pulses, and coronary spasm within 20 minutes. Spontaneous bruising within 12 hours. (From EHC-
Dallas. 1979.)

Table 20.4.   33-Year-Old White Female — Challenge after a Minimum
of 4 Days Deadaptation in ECU with Total Load Reduced —
EHC-Dallas

Signs
Oral challenge (less chemically contaminated food)—confirmed by intradermal challenge:

Shrimp	Recurrent asthma
Beef	Diarrhea
Pork	Spontaneous bruising
Wheat, oats, rye	Phlebitis

Oral challenge with commercial food—previously tested safe in its less
chemically contaminated form:

1 meal	Intractable ventricular arrhythmia— refractory to drugs
	Intractable ventricular arrhythmias— refractory to drugs

Inhaled fumes (double-blind):
Natural gas—ambient dose
Phenol (<0.002) ppm

Source: EHC-Dallas. 1978.

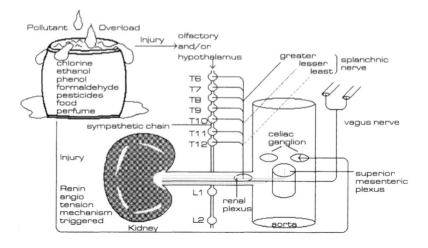

Figure 20.11.   Pollutant injury to the renal autonomic nervous system.

spasm with abdominal cramps, uterine cramps, nausea, and loss of appetite to renal vessel spasm with resultant essential hypertension (Figure 20.11). The case history of a 33-year-old white female who presented to the EHC-Dallas illustrates this point (Figure 20.12).

> **Case study**. This patient had severe symptoms including claudication when she walked 10 feet. She experienced severe abdominal cramps with nausea, bloating, and gas. Aortogram revealed a 90% distal aortic occlusion which was repaired by the author by surgical removal of the fibrous nonarteriosclerotic stenosis. After repair, she had 4 plus peripheral pulses including dorsalis pedis and posterior tibial, but, upon pollutant exposure, she would develop abdominal cramps and peripheral vascular spasm. Her legs would become cyanotic. If she were allowed to go long enough with her exposure, she would develop claudication as severe as that experienced prior to her aortic occlusion repair. The agents that triggerd her symptoms were defined, and she markedly improved with avoidance. Last follow-up at 15 years showed no recurrence of occlusions. She also had improvement of her peripheral and abdominal arterial spasm. She experienced no more abdominal pain (Figure 20.13).

Peripheral Vessels

Pollutant injury to the extremities may result in deregulation of the vessels initially until true inflammation occurs. It is well known that sympathetic dystrophy can occur after trauma, and it also apparently occurs after pollutant injury. Vascular spasm appears to be one of the early clinical injuries that are apparent with pollutant damage (Figure 20.13). In addition, venous and capillary deregulation due to pollutants may occur with resultant

33-year-old white female housewife
1. Peripheral arterial spasm with myalgia.
2. Superfatigue, depression.
3. Increased odor sensitivity.
4. Pustular eruptions angioma.
5. Aortic occlusion--non-arteriosclerotic.
6. Multiple abdominal cramps.
7. Nausea and loss of appetite.
8. Foods--multiple oral confirmed by intradermal challenges.
9. Inhaled double-blind chemical--multiple.

Figure 20.12.   33-year-old white female. Symptoms reproduced: edema, headache, peripheral artery spasm, and peripheral cyanosis. (From EHC-Dallas. 1978.)

cold sensitivity, area blanching, redness, or cyanosis; petechiae; spontaneous bruising; and purpura. Innervation of the peripheral vessels is not as complex as that of the chest and viscera (Figure 20.2). However, pollutant deregulation can result in muscle pain, aches, and cramps as well as fascial swelling. This deregulation will often confuse the picture of pollutant injury to the vasculature.

POLLUTANT INJURY TO METABOLISM OF THE BLOOD VESSEL

Metabolism of the blood vessel wall may be easily disturbed by pollutants. Vascular deregulation will then occur. Antioxidant changes are common with alterations in superoxide dismutase, glutathione peroxidase, and catalase occurring when free radicals are created from pollutant exposure. With excess pollutant exposure, local vessel wall irritation will occur resulting in energy loss due to mitochondrial damage and edema. Repair mechanisms may be retarded resulting in aberrant or retarded repair.

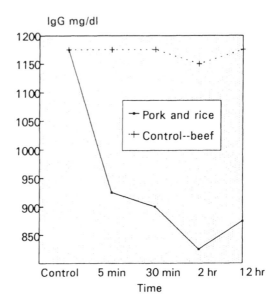

Figure 20.13.   35-year-old, white female. Pork and rice oral challenge resulted in symptom reproduction with pork and rice, but not the control food, beef. (From EHC-Dallas. 1979.)

Effects of Pollutants on Vessel Walls[14]

Physical factors as well as various substances such as toxic chemicals, food and its additives, water pollutants, and minerals can disrupt the vessel metabolism.[14] Each will be discussed separately.

Physical Factors

Physical factors such as heat, cold, weather changes, and light and various cycles such as weather, pollutant, and the individual's intrinsic cycle have for centuries been known to influence health. Advances in modern technology have increased the number of environmental factors able to influence health such as noise and electric and electromagnetic fields (Figure 20.14). Both the induction of disease and the maintenance of cardiovascular health can be regarded as functions of how ably an individual's immune and biochemical detoxification mechanisms can process the total load of pollutants to which he or she is exposed.

Agents Affecting the Cardiovascular System—Toxic Chemicals

While chemical incitants may trigger a maladaptive response(s) in virtually any of the smooth muscle systems, the responses in the cardiovascular system appear to be the most susceptible.[14,15] Recent literature has verified

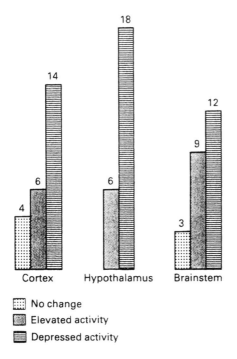

Figure 20.14.   Relation of EEG response from the cortex, hypothalamus, and brainstem due to exposure at 3GHz. The numbers indicate rabbits with a given response. (From Rea, W. J., and O. D. Brown. 1987. Cardiovascular disease in response to chemicals and foods. In *Food Allergy*, eds. J. Brostoff and S. J. Challacombe. 740. London: Ballière Tindall. With permission.)

previous findings regarding the harmful effects of a variety of chemicals on the cardiovascular system (Table 20.5). Most toxic chemical substances are not entirely exhaled but are usually transformed and eliminated in the urine and feces after being affected by the cytochrome P-450 system in the liver or stored in some lipid membrane. The effects of toxic chemicals upon the heart are changes in the rate, rhythm, ventricular function, and balance between the supply of oxygen and demand of oxygen by the heart. New changes seen in the peripheral vessels are constriction, dilatation, and vessel wall leaks. Some pollutants such as anesthetics produce a dose-related decrease in the intrinsic rate of discharge on the sino atrial node.[16] Others such as diethyl ether increase the pacemaker discharge.[17] Most also decrease automatically in latent pacemaker tissue. Two of the anesthetics that cause changes are chloroform and trichloroethylene, both of which are found in the blood of many chemically sensitive patients (40%). Not all chemically sensitive patients with these two anesthetics in their blood have arrhythmias. However, they all have vascular

Table 20.5.   Chemical Agents Affecting the Cardiovascular System

Fluorocarbons[6,15]	Glycerine[193]	Formaldehyde[4]	Chlorophenothane
Phenol[3]	Toluene[4]	Chlorine[1]	(DDT)[4]
Petroleum alcohol[59]	Hydralazine[194]	Cigarette smoke[2]	Turpentine[4]

Source: Rea, W. J. and O. D. Brown. 1987. Cardiovascular disease in response to chemicals and foods. In *Food Allergy and Intolerance*, eds. J. Brostoff and S. J. Challacombe. 738. London: Baillière Tindall. With permission.

deregulation, and arrhythmias are more easily triggered with overexposure to pollutants in those patients containing chemicals than in those who do not have them.

Many other anesthetics such as diethyl ether, cyclopropane, and fluroxene and nitrous oxide have been found to suppress myocardial function.[18-21] The pollutant-depressed chemically sensitive individual often responds adversely not only to nitrous oxide (see Volume II, Chapter 11) but also to these other pollutants. Myocardial perfusion and oxygenation have also been found to be disturbed by some of these pollutants.[18,19,22-25] Ethanol in doses that are mildly intoxicating has been shown to affect left ventricular function adversely.[26-34] These observations are very important in chemically sensitive individuals who have an increased total body load that may break down the xenobiotics in the body into alcohols as part of their detoxification process. Changes in salts in the muscle are of major importance in the direct influences of ethanol administered acutely to cardiac cells. The action necessary to reducing the force of a contraction may well entail altered transcellular calcium movement with ion[35] fluxes at the microsomal level.[36] Changes similar to these fluxes are seen in the pollutant damaged chemically sensitive patient (see Volume I, Chapter 6[37] and Chapter 30). Also, lipid transport in the myocardium is altered[27] in the acutely ethanol exposed patient. Any xenobiotic that is broken down to alcohol can trigger a similar problem (see Volume I, Chapter 4[7]).

Drug induced cardiomyopathy can occur from isoproterenol.[38] As shown later in this chapter, apparently so can cardiopathy that is not drug induced. For example, coffee has been associated with cardiac myopathy in some patients. In addition to isoproterenol,[39] the pollutants hydralazine and miroxidol[40] are also known to cause cardiac sensitization.

The role of electrolyte derangements and electrolyte hormone interrelationships in the pathogenesis of myocardial injury and necrosis was for many years one of the focal points of investigations carried out by Lehr et al.[41-47] They showed that severe ventricular arrhythmia could be induced by α-carnitine and suppressed by the administration of $MgSO_4$, $MgCl$, or Mg gluconate.[48] These findings are similar to ours in the pollutant-injured, chemically sensitive patients who responded to magnesium supplementation (see Volume I, Chapter 6[49] and Chapter 37). It has also been shown that alpha and beta adrenergic amines such as phenylephrine, isoproterenol or epinephrine can cause cardiac necrosis. Beta blockers can also be cardiotoxic, as can excess digitalis. Cardiovascular reactions may result from exposure to many other drugs such as neuroleptics (i.e., phenothiazines), tricyclic antidepressants, control depressants (i.e., phenobarbital), anticonvulsants (i.e., hydantoins), MAO inhibitors, and CNS stimulants (i.e., amphetamines).[50] The majority of chemically sensitive individuals react adversely to these drugs. Some antibiotics such as neomycin, gentomycin, penicillin, tetracycline, chloramphenicol and streptomycin can cause cardiac effects in susceptible individuals. Many antineoplastic drugs such as the anthracyclines cause cardiac toxicity.[51]

Water

Water usually contains minerals, organic chemicals, particulate matter, and radiation, all of which can trigger the cardiovascular system (see Volume II, Chapter 7[52]). Chemically contaminated water is a major component of the total environmental load and may influence the cardiovascular system of the chemically sensitive individual. Most public water systems are full of organic and inorganic chemicals that may increase the chemical body burden of some individuals several fold. These chemicals include trihalomethanes,[53] pesticides,[54-57] formaldehyde,[58] solvents,[58] oils,[59] heavy metals,[58] and other metals such as copper.[58] Public drinking water has been described by Laseter et al.[60] and others[61] as containing most of the contents of an organic chemical laboratory (see Volume II, Chapter 7[52]).

In developed nations, the prevalence of many chronic diseases, particularly cardiovascular diseases, can be associated with various water characteristics related to hardness. Those diseases involved include coronary heart disease, hypertension, and stroke. The theorized protective agents found in hard water are calcium, magnesium, vanadium, lithium, chromium, and manganese[58] (see Volume II, Chapter 7[52]). Recent studies report fewer heart attacks, less coronary disease, and lower mortality rates in patients with existing cardiovascular disease in areas with hard water.[62]

Suspected harmful agents include the metals cadmium, lead, copper, and zinc, which tend to be found in higher concentrations in soft water as a result of its relative corrosiveness. Barium, frequently found in the blood of the chemically sensitive patient, can bring about an increase in blood pressure and even cause death.[63]

Food

The cardiovascular health of the chemically sensitive individual can be greatly affected by the quality of their diet. Particularly, three aspects of the foods an individual eats should be evaluated by the clinician treating the chemically sensitive individual with cardiovascular involvement. These include the food sensitivity(ies) of the individual, the food additives (natural and man-made toxins) that may have been applied to the foods, and the nutritive quality of the food itself.

Food Sensitivity

Involvement of the cardiovascular system in food sensitivities has been reported by a number of researchers. It was first shown by Hare[64] in 1905. He recognized tachycardia and bradycardia in patients following ingestion of some foods to which they were sensitive. Since then a variety of cardiovascular disease states have been reported to be related to food, including increased heart rates, angina pectoris, arrhythmias, myocardial infarction, extrasystoles, and atrial and ventricular fibrillation.[65] Other studies have found phlebitis and vasculitis to be triggered by foods.[66-68] Reports on the effects of fat and arteriosclerotic cardiovascular disease are legion. Various mineral deficiencies have been shown to influence the cardiovascular system (see Volume I, Chapter 6[49]).

Pollutant injury occurs as pollutants enter the body where they may create free radicals. These may be superoxides, singlet oxygens, hydroxy radicals, and lipid peroxides and may damage mitochondrial and cellular membranes (see Volume I, Chapter 6,[69] section on enzymes). This production of free radicals results in dysfunction of the vessel directly as well as through the ANS, endocrine, and immune system. This injury, according to Zeek,[70] results initially in vascular dysfunction followed by leaking of fluid, which then results in edema. As the inflammation progresses, red cells extravasate followed by larger holes in the vessel wall with resultant leaking of the white cells. Any type of white cell can leak resulting in eosinophilia, lymphophilia, or polymorphonuclear response and inflammation. Finally, clots occur, and, if the vessel is small enough, the area will survive. If it is larger so that an extremity or region without enough collateral is involved, tissue necrosis occurs (Figure 20.15).

Attempts at healing may occur in various ways leading to granulomatous or fibrous scar formation. Triggering agents may be infectious (bacteria, virus, fungi, parasitic), chemical (sulphur dioxide, phenol hydrocarbon),[71,72] nutritional,[72] or traumatic (physical environmental agents).

Food Additives

Cobalt and brominated vegetable oils are food additives that have been shown to trigger cardiomyopathy.[73-76] In 1965 and 1966 fulminating cases of cardiomyopathy were observed in heavy beer drinkers in Quebec City, Omaha, and Louvan.[73-76] Dingle and co-workers[77] have shown that cobalt inhibits the oxidation of pyruvate and α-keto-glutaric by mitochondrial suppression of rat liver and thigh muscle suspensions. Eventually, the cardiomyopathy occurs.

Brominated vegetable oil is used to adjust the essential flavoring oils of certain citrus flavored beverages. This element produces inflammatory cells and then fatty degeneration of the myocardium occurs.

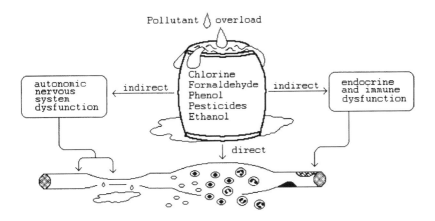

Figure 20.15.   Mechanisms of pollutant injury to the blood vessels.

Nutrient Quality

The nutrient quality of foods is extremely important in its effects upon the cardiovascular tree in the chemically sensitive individual. Foods with a high magnesium content are some of the most important due to the dampening effect of magnesium on arrhythmias and myocardial infarction. High potassium foods such as bananas are important in myocardial stabilization (see Volume II, Chapter 8[78]). In addition, the antioxidant effects of the vitamins E and C are important.

Effects of Pollutants on the Clotting System

Countless medical and surgical procedures have been associated with chemically triggered reactions. Nickel sensitivity has been reported secondary to the use of skin clips[79] and in a patient with a nickel steel heart valve[80] who developed valve thrombosis. All synthetic heart valves and artificial hearts, kidneys, and lungs are known to be able to trigger the clotting mechanism. Hemolysis has been associated with necrotizing dermatitis, and this process can occur after exposure to the epoxy resin in needles and polyvinylchloride tubing.

Effects of Pollutants on Cells

The effects of industrial chemicals on the heart may be both acute and chronic. Upon acute exposure in one series of 249 patients who were taken to the hospital after inhaling toxic concentrations of carbon monoxide or other types of irritant gases, 28% had sinus arrhythmia; 20% had tachycardia; 17% had ST segment depression and 12% had abnormal impulse conduction. Bradycardia, ventricular extrasystole, and auricular fibrillation were observed in some cases.[81] Several industrial and ambient chemicals to which the chemically sensitive individual reacts and which are known to have cardiac effects are carbon disulfide (see Volume II, Chapter 11,[82] sections on carbon monoxide, lead, and other heavy metals), halogenated alkenes, and organic nitrates (see Volume II, Chapter 12[83]).

Carbon disulfide exposure inhibits dopamine-β-hydroxylase, the enzyme that converts dopamine to noradrenaline.[84-86] Exposure to carbon disulfide increases the incidence of coronary heart disease and hypertension.[87-93] Microaneurysms occur in some people.[94]

Concentrates of carbon monoxide (9 to 10% C.O.H.G.) produce arteriosclerosis in rabbits.[95] There are suggestions the carbon monoxide helps produce arteriosclerosis in man. (Smokers have increased carbon monoxide.) Certainly carbon monoxide produces vascular changes in the chemically sensitive patient.

The low molecular weight halogenated alkanes have widespread industrial use as solvents, refrigerants, and fire extinguishing agents. They can cause cardiac arrest, arrhythmias, and sensitization of the myocardium to sympatho-adrenal discharge or exogenous catecholamine.[96] Younger persons are more prone to sudden death than older ones.[97-100]

While sublethal effects of fluorocarbons and chlorocarbons such as trichloroethylene are usually reversed in an occasional exposure, prolonged cellular changes may lead to degenerative cardiac lesions as seen in some chemically sensitive patients. The following case is illustrative.

> **Case study.** This 33-year-old white female entered the ECU with chest pain and an inability to walk. Her past history had been one of severe leg cramps with generalized myalgia for at least 3 years duration. She had run an elevated creatine phosphokinase (C.P.K.) in the 3000 IU/L plus range (normal = 33 to 165 IU/L). Cytotoxic drugs given by her rheumatologist had been of little help. She was placed in the ECU, and her chest pain and ventricular arrhythmias subsided in 72 hours. In addition to a sudden reduction in her total body load, she was treated with 15 g of vitamin C, 20 mEq of $MgSO_4$, 2 g of taurine, and 600 mg of glutathione, all given intravenously daily. Her history revealed she had worked around toxic dump sites, and her blood chemical analysis revealed the presence of several chlorinated pesticides including DDT, DDE, chlordane, and dieldrin. Challenge tests reproduced her symptoms. It was clear that the chlorinated hydrocarbons had triggered her myalgia, myositis, and cardiac arrhythmias.

Chemical pollutants can have wide-ranging effects on the cardiovascular system. Lead, for example, may cause gangrene of the fingers, angina pectoris, hypertension, chronic nephritis and arteriosclerosis. It also appears to disturb catecholamine metabolism and glutathione replenishing. Mercury and cadmium seem to affect these processes similarly.[101] Cadmium can suppress the mononuclear phagocyte system.[102] Being potent coronary dilators that increase coronary flow, organic nitrites such as nitroglycerin and nitroglycol also effect blood vessels.[103] Some chemicals like dichlorodiphenyltrichloroethane (DDT) suppress the mast cells so thoroughly that anaphylaxis is less likely to occur.[104,105] However, cell membrane disturbance occurs due to injury of the lipid content. Ozone can cause lipid plastic perithyroiditis with leukocytic infiltration and capillary proliferation.[106] Chloracne perivasculitis lesions have been produced in monkeys fed a pesticide, Arochlor 1248 (polychlorinated biphenyl). Mycotoxins, especially the ergot type, can cause severe vasospasm and even gangrene (see Volume II, Chapter 17[107]). Cocaine is well known to cause cardiac arrest.

Some substances such as phenol have a particular affinity for the cardiovascular system.[108] Yevick[59] demonstrated cardiovascular changes in sea animals exposed to oil spills.

Types of Mechanisms in Vascular Damage

The mechanisms for vascular damage will be similar to those of other body cells described in Volume I, Chapters 4[7] and 5.[109] They will be only briefly discussed here.

Nonimmune Mechanisms

Nonimmune triggering of the vessel wall may occur as a result of pollutant exposure. Complements may be triggered directly via the alternative pathway by molds, foods, or toxic chemicals.[110] Mediators like kinins and prostaglandins may also be directly triggered. Interestingly, in addition to causing allergic responses, pollens have been shown to contain toxic substances that will trigger hemolysis and other responses.[111] Vasoactive peptides may be released from the intestine from food triggers.

Immune Mediated Pollutant Changes

The intrinsic mechanisms by which blood vessels are damaged can be mediated either via the immune system or via the nonimmune biological detoxification system. The immune hypersensitivity responses in the vessel wall can be any of four types and frequently can be a combination of types.

Type I hypersensitivity is mediated through the IgE mechanism on the vessel wall. The classic examples of vascular change are angioedema, urticaria, and anaphylaxis due to sensitivity to pollen, dust, mold, or food.[65]

Type II cytotoxic damage may occur with direct injury to the cell. A clinical example of this is seen with exposures to mercury,[112] although this exposure might be directly toxic rather than antibody mediated.

Type III immune complex syndromes include lupus vasculitis. Numerous chemicals including procaineamide[113] and chlorothiazide[59] are known to trigger the autoantibody reactions of lupus, but they also effect the myocardium and electrolyte contents. Other chemicals such as vinyl chloride[114] will produce microaneurysms.

Type IV cell-mediated immunity occurs with sensitization and stimulation of T lymphocytes. Numerous chemicals such as phenol, pesticides, organohalides, and some metals will also alter immune responses possibly triggering lymphokines and producing a Type IV reaction.[115] Clinical examples are polyarteritis nodosa, hypersensitivity angiitis, Henoch-Schonlein purpura, and Wegener's granulomatosis.[66] (See Volume I, Chapter 5,[109] for more details.)

Microcirculation

Alteration of the harmony of the microcirculation with its coordinated function appears to be the key to the pathological malfunction at the tissue level in the chemically sensitive. Numerous studies have shown that tissue ischemia may occur chronically, altering the local ph from the resultant hypoxia. Kinzel[116] reported that Lohman discovered in 1930 that energy rich phosphates, especially ATP, are not limited by the intake of food as much as they are by oxygen transport to the body tissue. Lack of energy is one of the cornerstones of symptomatology in the chemically sensitive individual. Damaged or absent mitochondria have been observed on human muscle biopsy in the chemically sensitive patient by our group and in animal studies after pollutant exposures.[117] According to von Ardenne,[118] chronic O_2 deficiency occurs in the local tissue level not only with aging but also with a variety of environmental stimuli such as viral infection, cigarette smoke, and cold (Figure 20.16). Hauss[119] has further listed sclerogenic noxae, which will reduce microcirculation oxygen (Table 20.6). This deficiency is a combined or individual effect of critically reduced O_2 transport to the tissues (O_2 offer to the tissues), critically reduced O_2 utilization by the tissue, and critically increased O_2 requirements of the tissue. von Ardenne[120] has shown that with various pollutants, or with aging, there is swelling of the pericytes and endothelial cells in the distal capillary venule due to chronic hypoxia (Figure 20.17A and B). This swelling results in a lowered venous PO_2 extraction and, thus, less oxygen being transported to local tissue with a loss in cell energy. The Na^+/K^+ ATPase, which regulates the sodium pump, is inactivated, and osmoregulation of the cell is damaged, causing edema with its accumulated hydrated sodium ions. Edema is one of the prime signs of chemical sensitivity.

von Ardenne[121] further showed that, once this chronic mild hypoxia occurs, there is a generalized switch of the microcirculation throughout the body (allowing for end-organ variability) to a decreased energy mode (Figure 20.18). Using his multi-step O_2 therapy, he can reverse the switch to an energy producing mode with improved health.[122] He further showed that swelling also occurs in the alveoli of the lung and that there is an increase in O_2 partial pressure in erythrocytes during passage through the lung capillaries. This partial pressure will be reduced with chronic pericytes and endothelial swelling

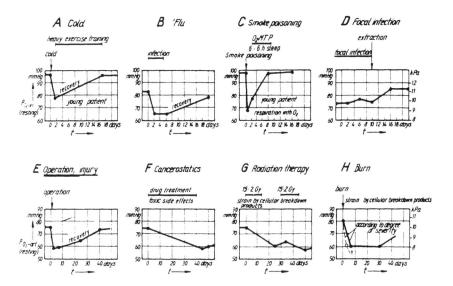

Figure 20.16. Examples of the lowering of the arterial resting PO_2 by stressful processes of infectious, toxic, and quasi-toxic stimuli. (From von Ardenne, M. 1990. *Oxygen Multistep Therapy*. 45. New York: Georg Thieme Verlag Stuttgart. With permission.)

and a vicious, downhill cycle of hypoxia and degeneration will occur. This poor O_2 supply leads, via the triggered narrowing of the vessel diameter, to a drop in the microcirculation with more local O_2 deficiency. Blood viscosity increases, lowering local tissue pH to the 6.7 to 6.3 range. The drop in pH is caused by a stronger or weaker transition to fermentation metabolism in the tissue surrounding the capillaries and by the hampering of the drainage of the lactic acid formed. In this process, pH levels less than 6.6 occur at the venous end of the capillaries, leading, after a latent period, to hemostasis. This pH reduction causes a rearrangement in the architecture of the biomembranes,[123] both in the blood cells and the inner surface of the endothelium, which impedes microcirculation. Flexibility is critically reduced in the red cells and leukocytes.[124-127] This loss of flexibility leads to a slowing down of capillary flow and a jamming of the red cells with the leukocytes. A prestage of capillary damage looms.[128] The subsequent process of the pressing of leukocytes against the venule wall in the capillary and postcapillary transition zone by accumulated (stiffened) red cells that overtake the leukocytes has been investigated.[129]

Apparently fractions of a second (at pH>6.5) in which the blood cells pass along the venous part of a capillary are sufficient to trigger the alteration of the membrane structure and the formation of pores. The alteration of structure and pores is reversible with adequate O_2 and pH. This fact means that a grossly impeded microcirculation can be brought back to normal flow by increasing

Table 20.6.   List of Sclerogenous Noxae.[a]

Sclerogenous noxa	Disturbing factors
1 Lack of oxygen[b]	Hypobaric chamber with P_{O2} values <150 mmHg
2 Hypotension	Blood pressure amplitude too low (e.g., <25 mmHg)
3 Hypertension[c]	Reduced kidney circulation; hypertension (systolic blood pressure >200 mmHg)
4 Operations	Anesthesia, etc.
5 Toxins	Endotoxins, staphylococcus toxin, diphtheria toxin
6 Infections	*Pasteurella multocida*, staphylococci
7 Nicotine	Inhalation of cigarette smoke
8 Foreign protein	Albumin injection
9 Allergic reactions	Serum shock, Arthus' phenomenon
10 Foreign substances	Croton oil, plastic products
11 Overexertion	Excessive movement in the treadmill
12 Mechanical stress	Laceration of muscle, dilatation of the aorta
13 Interbrain stimulation	Thalamus stimulation by means of electric current
14 Emotional stimuli	Binding, restricted movement
15 Noise	Roaring, uniform random noise
16 Hormones	Adrenaline, testosterone, thyroxine, cortisone, glucagon
17 Influences of diet	Sclerogenous diets

Note: Disturbing factors that, according to W. H. Hauss,[119] regularly trigger unspecific mesenchyme reaction, detected by the increased rate of ^{35}S-sulfate incorporation into the sulfomucopolysaccharides of the ground substance of connective tissue in various organs (animal experiments).

[a] Most of these noxae deteriorate the oxygen status.

[b] Subnormal values of $P_{O2\text{-}art}$ or η; locally reduced O_2 delivery to the vessel wall; mechanical irritation at ramifications in the arterial system.

[c] See Figure 39 from von Ardenne, M. 1990. *Oxygen Multistep Therapy*, 46. New York: Georg Thieme Verlag Stuttgart.

Source: von Ardenne, M. 1990. *Oxygen Multistep Therapy*, 45. New York: Georg Thieme Verlag Stuttgart. With permission.

the pH (as is done in alkylization therapy for the chemically sensitive patient) of the venous capillary blood (e.g., by means of the quick elimination of O_2 deficiency, which suppresses glycolysis or the rapid administration of g-strophanthin, which then re-normalizes the pH of the mycocardia) (Figures 20.19, 20.20, and 20.21). It should be emphasized that the individual cell needs for maintenance of its structure and its readiness and ability to perform its

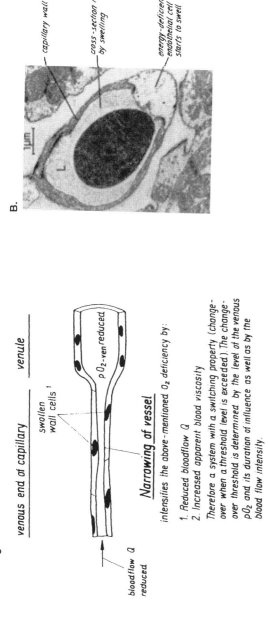

A.

Primary O_2 deficiency causes an O_2 (energy) deficit in the wall cells at the venous capillary end, which then swell and thereby lead to a narrowing of the vessel with reduced bloodflow.

venous end of capillary   venule

swollen wall cells [1]

pO_2-ven reduced

bloodflow Q reduced

Narrowing of vessel

intensifies the above-mentioned O_2 deficiency by:

1. Reduced bloodflow Q
2. Increased apparent blood viscosity

Therefore a system with a switching property (change-over when a threshold level is exceeded). The change-over threshold is determined by the level of the venous pO_2 and its duration of influence as well as by the blood flow intensity.

B.

capillary wall

cross-section reduced by swelling

energy-deficient endothelial cell starts to swell

1 µm

L

Figure 20.17. (A) Ideas for the triggering of the switching mechanism of the blood microcirculation at the venous end of the capillaries in O_2 deficiency. *Note:* H_2O flows into the cells as a consequence of the failure of the K^+/Na^+ pump, which requires a great deal of energy. (B) The elementary process of the switching and regulating mechanism of the blood microcirculation at the venous end of the capillaries in the electronmicroscopic picture. It was discovered that these swellings of the endothelial cells, if not yet at too advanced a stage, can be lastingly reduced by O_2MT procedures. Blood microcirculation strengthened in all perfused capillaries of the organism, weakened in distress. L = capillary lumen; E = erythrocyte. (From von Ardenne, M. 1990. *Oxygen Multistep Therapy.* 7. New York: Georg Thieme Verlag Stuttgart. With permission.)

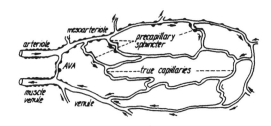

Figure 20.18.   Structure of the terminal network with its resistance vessel in the skeletal muscle. The smooth vessel musculature, which, as a contractile element, controls the vasomotor activity, is shown in black. (From von Ardenne, M. 1990. *Oxygen Multistep Therapy*. 148. New York: Georg Thieme Verlag Stuttgart. With permission.)

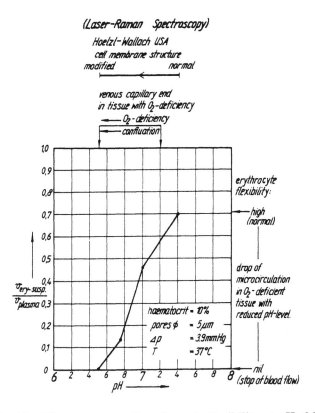

Figure 20.19.   Disappearance of erythrocyte flexibility at pH≈6.5. Measurements of the relative flow rate of a red cell suspension as compared to plasma through pores with 5 μm Ø. (From von Ardenne, M. 1990. *Oxygen Multistep Therapy*. 75. New York: Georg Thieme Verlag Stuttgart. With permission.)

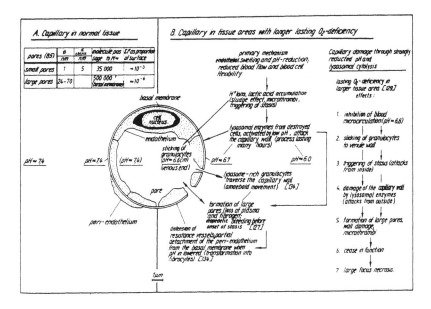

Figure 20.20. Cross-section of a capillary (A) and postulated processes in capillary damage in a long-term O₂ deficient and, hence, overacidified tissue area (B). Formation of large pores, evasation of plasma, red cell aggregation. (From von Ardenne, M. 1990. *Oxygen Multistep Therapy.* 75. New York: Georg Thieme Verlag Stuttgart. With permission.)

function, a certain energy supply that it mainly obtains from the oxidative catabolism of nutrients when the O₂ offer is sufficient.

When oxygen is lacking, the required energy can be gained by anaerobic glycolysis. Under *in vivo* conditions, the oxidative breakdown of 1 mole of glucose delivers 689 K calories (2855 KS) and is connected to the formation of 38 moles of ATP. With 270 to 380 K calories (1130 to 1590 KS), these represent the physiologically utilizable free energy that corresponds to between 39 and 55%. By comparison, the transformation of 1 mole of glucose into 2 moles of lactate is connected with a reaction energy of only 47 K calories (196 KS) by which 2 moles of ATP are formed. These findings correspond to only 16 to 20 K calories (67 to 84 KS) and an energy yield of 3.4 to 4.3%. Thus, the aerobic decomposition of glucose supplies a total amount of energy that is 15 times higher than anaerobic glycolysis.[130,131] It is, therefore, not surprising that the end-product of the glycolysis, lactic acid, still contains a high amount of energy.

Special relationships exist in the heart inasmuch as the heart muscle, which is sufficiently supplied with oxygen, covers more than 50% of its energy requirements from the oxidation of fatty acids. Glucose contributes 18% to the energy gain; lactate, 17%; pyruvate, 1%; and ketone bodies, 5%.[132] In conditions

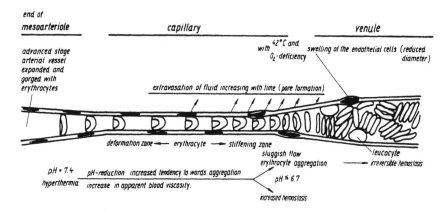

Figure 20.21.   Schematically presented ideas about the triggering of the
mechanism of blood microcirculation inhibition in longer-
lasting strong tissue overacidification (and hyperthermia).
Examples: myocardial infarct, inflammation, gangrene.
Manipulated selective hemostasis in cancer multistep oxygen therapy (CMT) with two-step regional hyperthemia
(selectotherm process). (From von Ardenne, M. 1990.
Oxygen Multistep Therapy. 79. New York: Georg Thieme
Verlag Stuttgart. With permission.)

of hypoxia as seen in the chemically sensitive individual, the breakdown of
fatty acids is reduced, and glucose uptake and breakdown increase, as do
glycogen reserves. Lactate then accumulates to a greater extent,[133] and overacidification occurs. The joint use of the lactic acid as an energy substrate in
the myocardium (and perhaps in the brain with a combination of glycolysis to
metabolism of approximately 19%) can be strengthened by the use of rapid
acting g-strophanthum. In increasing O_2 deficiency the cell is forced to cover
its energy deficit by glycolysis. Due to the energy yield of fermentation
metabolism being only 7% compared with respiration glycolysis, O_2 increases
very quickly below a certain PO_2 level. A sudden change in cell metabolism
then follows with severe energy loss, as seen in chemically sensitive individuals. von Ardenne's[118] model of O_2 deficiency states agrees with our observations of the chemically sensitive individual who responds to alkylization after
a pollutant or allergic insult.

von Ardenne[134] has shown that lactic acid formation in the space of the
narrow glycolytic zone of a single capillary area is not enough to lead to a
considerable reduction of the mean pH level in the interstitium (and at the
venous end of the capillary wall). Only when glycolytic zones of many
intracapillary areas come together (V>0.2 mm³) can the reduction of the mean
pH in the tissue exceed 0.5 pH units. pH reduction of more than 1 unit in the
tissue and severe pH reductions also at the venous end of the capillary wells,

which finally trigger the blood microcirculation inhibition and capillary damage, do not occur until volume V of the O_2-deficient area exceeds 10 mm^3. The consequences of an O_2 deficit in various organs and tissues in the chemically sensitive individual are often determined by the relation presented here. Thus, the connection between the size of the volume affected by O_2 deficiency and the level of pH reduction that occurs is one of the elementary pathophysiological bases of O_2 deficiency diseases and conditions. It can now be understood why critical consequences (e.g., large-area necrosis after a myocardial infarction or tissue degradation in inflammatory processes such as seen in the chemically sensitive individual) only occur when the affected tissue area exceeds a certain volume (e.g., V≥10 mm^3).

O_2 deficiency is the main sclerogenous noxae that triggers the nonspecific mesenchyme reaction discovered and described by Hauss.[119,135] The mesenchyme reaction instigates the process of arteriosclerosis. The reactivity of cells included under the triggering of mesenchyme cells (e.g., endothelial cells of the vascular intima, fibroblasts, pericytes, smooth muscle cells of the vascular media, etc.) exceeds all other cells. von Ardenne's studies in the rat show that 90 minutes after the impact of strong sclerogenous noxae the nonspecific mesenchyme reaction occurs (Figure 20.22).[136] The extent of the triggered incorporation of the mesenchyme into the vessel wall and the quantity of the initially gradually gathering deposition products depends upon the strength and duration of the sclerogenic irritation. Thus, these determine whether or not the occurring change in structure should not be deemed as pathologic. The structural changes in the vessel wall usually regress after the sclerogenous influence has faded away (reversible phase). If, however, the sclerogenous irritation (i.e., toxic chemical) or O_2 deficiency lasts for a long time (reduced clearance capacity due to energy deficiency) and particularly if such long-lasting conditions frequently reoccur, then the nondecomposable deposition products from the metabolism (collagen fiber bundles, calcification, etc.) accumulate. The damage to the vessel wall caused by these breakdown products no longer recedes (irreversible phase) as the increase of the diffusion length in both directions of the vessel wall (radial, central) begins definitively to limit the metabolism. A final stabilization of the hypertrophic structure thus occurs (Figure 20.23A and B).

O_2 deficiency as seen in many conditions, as well as chemical sensitivity, triggers the spontaneous lysosome formation once glycolysis occurs. Few lysosomes are found in normal heart muscle,[137] while there is a very high lysosome content in the myocaradial cells of hypertrophied cardiac muscle. Once the low pH, which is the result of the low PO_2, extends over an area of 10 mm,^{3} lysosome formation increases. By lysosome formation and by the sudden change to fermentation metabolism with the strong pH reduction, nature seems to be preparing for the dissolution (digestion) of the cell material when cell death becomes apparent as a result of a longer duration of O_2 deficit. If larger cell complexes are drawn into the deficiency situation, the reduction

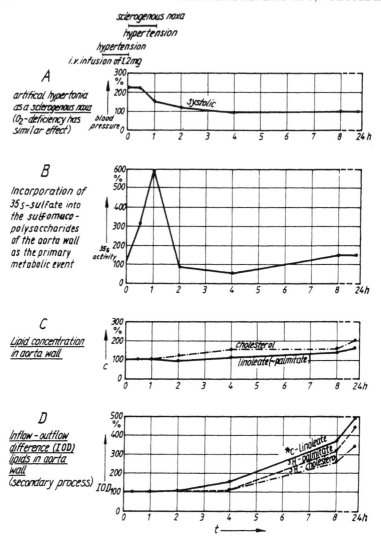

Figure 20.22. Measurement of rats to show the fast onset of the unspecific mesenchyme reaction (B) after the raising of a sclerogenous noxa (A). The deposition of lipids into the vessel wall only occurs about 6 hours after the incorporation of mesenchyme (C) and (D). (From von Ardenne, M. 1990. *Oxygen Multistep Therapy*, 112. New York: Georg Thieme Verlag Stuttgart. With permission.)

can reach levels of pH = 6.0 to 6.3, at which the membranes of the lysosomes become permeable, the lysosomal enzymes are released, and a strong activation of enzymes occurs. These vasoactive substances trigger the inflammation and further hypoxia of the microcirculation resulting in more vascular

A. and B.

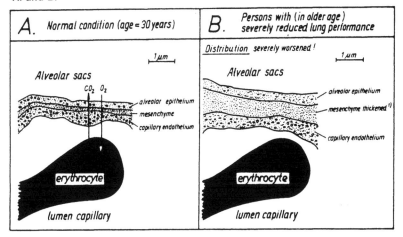

C.

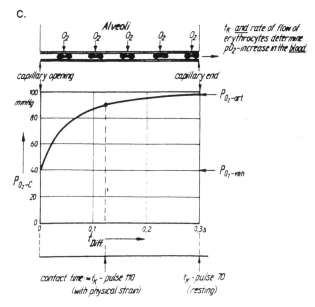

Figure 20.23.   (A) and (B) Diffusion paths for O_2 and CO_2 in the exchange of gases in the lungs. (C) Increase in O_2 partial pressure in erythrocytes during passage through the lung capillaries, represented for the diffusion contact times $t_K = 0.3$ s (resting conditions, pulse 70; strain conditions, pulse 110); 1 mmHg = 1.33•10² Pa. (From von Ardenne, M. 1990. *Oxygen Multistep Therapy*. 40. New York: Georg Thieme Verlag Stuttgart. With permission.)

dysfunction. The final stage is the dissolution of the cells (necrosis). Several cases of chemical sensitivity with microcirculation dysfunction are presented in this chapter as well as in Volume II, Chapter 13.[138]

ENDOCRINE EFFECTS IN THE VASCULAR SYSTEM

Estrogen has long been known to have a mild dampening effect on the cardiovascular system. The late onset of arteriosclerosis in normal females with early onset of arteriosclerosis with early menopause is commonly observed. A study by Couch and Wortman[139] supports this observation in that they found a significantly greater number of occurrences of migraine in pathologically anovulatory females (polycystic ovary, galactorrhea, amenorrhea) compared with pregnant women or women taking the contraceptive pill. It was suggested that this increase might also be due to hypothalamic problems. Excess estrogen has been shown to have an adverse effect on vessel walls giving rise to venous inflammation, which results in thrombophlebitis and pulmonary emboli. Debled[140] has demonstrated that androgen supplementation will decrease cholesterol and retard arteriosclerosis.

POLLUTANT INJURY—CLINICAL SYNDROMES

The clinical syndromes produced by pollutants are myriad. Some are named and others are not. Discussed separately in the remainder of this chapter are vasculitis, small vessel vasculitis, large vessel vasculitis, carotid spasm, Raynaud's disease, hypertension, hypersensitive vasculitis, periarteritis nodosa, Wegener's granulomatosis, rheumatoid vasculitis, eosinophilic vasculitis, recurrent phlebitis, and cardiac arrhythmias and dysfunction.

Vasculitis

Many necrotizing and nonnecrotizing vasculitides are directly caused or closely associated with immunopathogenic mechanisms and environmental triggers. It was originally thought that the mechanism most commonly associated with vasculitis is that of the deposition of circulating immune complexes in the blood vessel walls. However, this association appears to be limited since complexes are not always necessary for inflammation of vessels to occur. In fact, a larger percentage of blood vessel deregulation in the chemically sensitive patient appears to occur through nonimmune mechanisms (see Volume I, Chapter 4[7]). Usually pollutants triggering vasoactive amines are involved.

Small Vessel Vasculitis

At the EHC-Dallas, Rea et al.[141] described a group of chemically sensitive patients with multisystem involvement distinguished by a wide variety of

symptoms. All evidenced frequent peripheral vasospasms, spontaneous cutaneous bruising, and/or petechiae, and/or purpura, and/or acneiform lesions, and
peripheral edema (inflamed leaking blood vessels) (Figures 20.24 and 20.25).
Their vascular responses apparently mirrored their internal pathology. Some
cases of vasculitis were triggered by pollens, dusts, and/or molds, while others
resulted from foods tested by intradermal or oral challenge and/or by chemicals
injected or challenged by inhalation or ingestion. Different susceptibilities
were found in each patient, but each produced a sequential progression of
symptoms of color change of the skin, especially of the hands, feet, nose, and

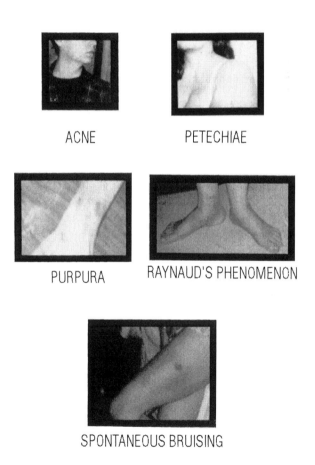

Figure 20.24.   Pollutant injury causes a syndrome of the various vascular
phenomena including acne, petechiae, purpura, Raynaud's
phenomena, spontaneous bruising, and peripheral edema.
(From EHC-Dallas. 1978.)

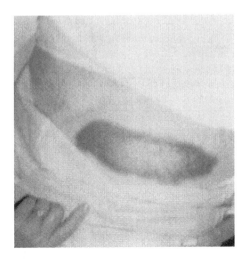

Figure 20.25. **45-year-old, white female. Pollutant injury after a double-blind inhalant challenge — phenol (0.002 ppm) after 4 days deadaptation with total load decreased in the ECU. Symptoms reproduced by challenge included spontaneous bruising, cold susceptibility, and blue hands, feet and nose. (From EHC-Dallas. 1988.)**

face, followed by pulse alteration, periorbital and peripheral edema, petechiae, and/or spontaneous bruising and/or purpura. Biopsies have shown perivascular lymphocytic infiltrates (Figure 20.9).

This initial study was followed by a second study in a similar group of chemically sensitive patients as compared to 60 normal controls. Immune parameters in the form of T and B lymphocytes and subsets revealed significantly lower suppressor cells than the control group and a significantly larger T_4/T_8 1.70 ± 0.06 $p<0.0001$ level (Table 20.7).

T_{11} or total T lymphocytes were less in the chemically sensitive vascular group to the <0.005 level as were total lymphocytes in the group.

A third group of patients with vascular dysfunction was studied. There were over 300 cases in this group. A prospective study revealed multiple triggering agents. There were 4800 challenges, 14,400 intradermal confirmations after positive oral challenge of foods, and over 800 individual double-blind inhaled challenges in this group of patients. One hundred of the group with 684 separate, individual challenges are shown in Table 20.8. The most common foods reproducing symptoms were wheat, corn, beef, pork, chicken, eggs, milk, and sugar. The most common chemicals were petroleum derived ethanol, formaldehyde, phenol, insecticide, and chlorine. It is clear from these studies that small vessel vascular disease predominates in pollutant injury.

Large Vessel Vasculitis

Large vessel involvement associated with environmental sensitivities has been shown by various researchers.[142,143] Ishikawa et al.[144] have shown a relationship between the use of organochlorine pesticides and the development of Bechet's disease, which is now recognized as a chronic multisystem disease affecting the skin, mucous membranes, eye, joints, CNS, and blood vessels. It is characterized by recurrent ulcers of the mouth, esophagus, and genitalia. It is also associated with iritis. Cases have now been reported of aneurysms with Bechet's disease (Figure 20.26). Pesticide damage identified by Ishikawa is consistent with our studies on the relationship of toxic chemicals in blood vessels.

Our series at the EHC-Dallas involved over 100 people with large vessel involvement. These had either Raynaud's phenomena, transient hemiplegia due to peripheral bilateral spasm, lower extremity claudication, rest pain, or necrosis.

A group of case reports are now summarized to demonstrate pollutant-induced vascular injury.[145]

Case study #1. Rea et al.[145] detailed the case of a 44-year-old female who exhibited large vessel involvement. She was found to be sensitive to 10 foods and 3 synthetic chemicals, all of which appeared to trigger spasm of her femoral arteries. This patient had had pulmonary emboli in the past and a clip placed on the vena cava by this author. She eventually lost her left leg from advanced inflammatory vascular disease (Figure 20.27).

Case study #2. A 26-year-old paperhanger developed a cyanotic, pulseless right lower extremity rapidly progressing to no pulses in either leg. An arteriogram showed normal-sized vessels with areas of severe spasm. Intraarterial papaverine opened the vessels to normal caliber for five minutes but then spasm returned. Oral and intravenous vasodilators did not relieve the spasms, nor did intravenous heparin. Hospitalization in a controlled environment alleviated the cyanosis, with femoral, popliteal, dorsalis pedis and posterior tibial pulses returning to normal. Challenge with food caused no problems, nor did ingested chemicals; however, inhalation challenge with fungicide used in the wallpaper paste that he used daily reproduced the symptoms. This patient changed jobs and continues to live a healthy, vibrant life.

Case study #3. A 43-year-old white female underwent a sudden onset of arterial occlusion of her left leg. Eventually amputation was performed after an attempted bypass was unsuccessful. Microscopic sections of the leg showed that a nonarteriosclerotic fibrosis surrounded by severe inflammation had occluded all arteries. Significant past history revealed two pulmonary emboli, recurrent phlebitis, sinusitis, recurrent gastrointestinal upset, and sensitivity to odors. Six months later her peripheral pulses disappeared, and her remaining leg became cyanotic. An arteriogram showed severe spasm, and the

Table 20.7. Immunological Data of Vascular Chemically Sensitive
Patients and Normals (Mean Values and Differences)

	Patients (70 persons)	Normals (60 persons)	Difference (patients to normal)	Significance (p)
WBC (#/mm³)	7010.00 ± 290.00	7560.00 ± 220.00	No	>0.05
L (#/mm³)	2420.00 ± 90.00	2770.00 ± 91.00	Smaller	<0.005
L(%)	36.20 ± 1.10	37.30 ± 1.10	No	>0.2
T_{11} (#/mm³)	1780.00 ± 75.00	2080.00 ± 74.00	Smaller	<0.005
T_{11} (%)	72.40 ± 1.00	75.20 ± 0.80	Smaller	<0.05
T_4 (#/mm³)	1090.00 ± 48.00	1160.00 ± 43.00	No	>0.1
T_4 (%)	43.70 ± 1.00	42.20 ± 0.70	No	>0.1
T_8 (#/mm³)	560.00 ± 30.00	740.00 ± 38.00	Significantly smaller	<0.0001
T_8 (%)	22.90 ± 0.70	25.80 ± 0.80	Smaller	<0.005
T_4/T_8	2.20 ± 0.15	1.70 ± 0.06	Significantly larger	<0.0001
B (#/mm³)	220.00 ± 19.00	270.00 ± 23.00	No	>0.05
B(%)	9.40 ± 0.90	9.40 ± 0.6	No	—

Source: Rea, W. J., A. R. Johnson, S. Youdim, E. J. Fenyves, and N. Samadi. 1986. T and B lymphocyte parameters measured in chemically sensitive patients and controls. *Clin. Ecol.* 4:11–44. With permission.

patient developed rest pain refractory to antispasmodics. Limb loss appeared imminent. Institutionalization in a relatively particle- and fume-free controlled environment for 5 days resulted in restitution of 4+ dorsalis pedis pulse, normal color, and function of the limb without problems. Challenge with numerous individual foods and chemicals entirely reproduced the above-mentioned symptoms. The patient is now on environmental control 3 years after diagnosis with a symptom-free leg.

Case study #4. A 45-year-old former chocolate factory worker was admitted to the ECU complaining of excruciating leg pain, which rendered him unable to walk. Eleven years prior to the current admission, the patient had experienced an episode of leg pain, chest pain, and shortness of breath that had necessitated hospitalization. Findings indicated heart failure accompanied by pulmonary infiltration and eosinophilia. Steroids were required to bring the situation under control. Eventually, vessel biopsy revealed an inflammation, compatible with periarteritis nodosa or Wegener granuloma. This patient had to retire from his job because of his excruciating leg pain. For 11 years, he had required morphine (1/4 grain every four hours) for pain relief. He had not been able to walk extensively for 5 years and had developed paralysis of his left foot.

Table 20.8. 100 Chemically Sensitive Patients with Vascular Deregulation Following 684 Individual Inhaled Double-Blind Challenges after 4 Days Deadaptation with Total Load Reduced as a Representation of 300 Cases

	Dose (PPM)	No. of challenges	No. reacted with symptoms and signs[a]	Changes			
				BP	P	PFR	Brain function
Ethanol (petroleum derived)	<0.50	100	65	9	4	13	23
Phenol	<0.002	94	63	17	4	10	20
Chlorine	<0.33	55	41	7	2	6	15
Formaldehyde	<0.20	103	69	8	4	8	24
Pesticide (2,4-DNP)	<0.0034	54	49	6	4	9	18
1,1,1-Trichloroethane	ambient	15	12	9	1	5	3
Placebo	saline	263	33[b]	20	3	4	7

a Signs — edema, bruises, petechiae, acne, arterial spasm, cold extremities, and/or cold sensitivity.
b No patient reacted to more than one placebo.

Source: EHC-Dallas. 1990.

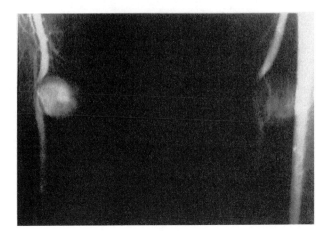

Figure 20.26.   Bilateral superficial femoral artery aneurysms. Behcet's syndrome. (From Bartlett, S. T., W. J. McCarthy, III, A. S. Palmer, W. R. Flinn, J. J. Bergan, and J. S. Yao. 1988. Multiple aneurysms in Behcet's Disease. *Arch. Surg.* 123:1004–1008. With permission.)

On admission to the ECU, this patient exhibited an inability to walk more than a few steps. His left foot was pulseless, the three smaller toes were paralyzed, and the patient could not dorsiflex his feet. Light touch elicited extreme hyperesthesia with excruciating pain. During the first 5 days of environmental control (after removal of all medications), severe withdrawal symptoms occurred. However, on the sixth day peripheral pulses were noted, the toe paralysis disappeared, the color returned, and he was able to walk without problems. No morphine was required after the fourth day. Random challenge of single, chemically less-contaminated foods demonstrated the following sequential reactions:

    0 minutes   — immediate repulsion with nausea at odor
    5 minutes   — itching throat
    35 minutes — (a) general malaise
                  (b) arms and legs aching
                  (c) blueness of hands
                  (d) stomach bloating
    45 minutes — (a) vomiting
                  (b) blueness of hands and feet
    50 minutes — pulse decrease in left foot, unable to walk, toe
                  paralyzed, hyperesthesia

Five foods gave similar reproducible sequential reactions. Chocolate was the worst offender. Challenge with 20 other foods resulted in no reactions.

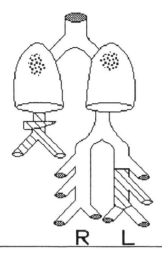

44-year-old white female--arterial spasm and amputation
1. Headache, depression.
2. Stiff neck.
3. Chest pain.
4. Sour eructions.
5. Triggering agents on challenge: beef and chemicals (gas heat and petro chemicals).

Figure 20.27.   44-year-old white female arterial spasm and amputation. Symptoms and signs reproduced with challenge of beef and chemicals (gas heat and petrochemicals) included headache, depression, stiff neck, chest pain, and sour eructations. (From EHC-Dallas. 1979.)

Inhaled odor of cigarette smoke produced symptoms similar to those of the reactive ingested foods. He was treated with avoidance, nutrient supplementation, and antigen injections. This patient did well. His vascular spasm was relieved, and his morphine was discontinued. He was discharged, and 1 year follow-up showed him to be doing well.

Case study #5. Another 36-year-old white female was found to develop peripheral arterial spasm each time she was exposed to certain foods, cigarette smoke, and gas heat. In one episode, she developed a clot in her left anterior tibial artery. The clot was removed, and she did well. She has continued to do well on an avoidance program for 10 years follow-up without medication (Figure 20.28).

Case study #6. This 65-year-old white female was sensitive to 10 foods and 3 chemicals. She developed peripheral arterial spasm of one lower extremity

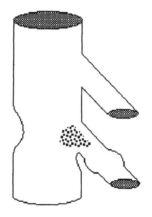

36-year-old white female
1.   Depression.
2.   Sinusitis.

3.   Agents triggering arterial spasm:
 	
      Oral challenge--milk
      Inhaled challenge--cigarettes, gas heat

Figure 20.28.   36-year-old white female. Symptoms and signs reproduced by challenge included depression and sinusitis. Triggers reproducing arterial spasm on oral challenge included milk, on inhaled challenge — cigarettes and gas heat. (From EHC-Dallas. 1979.)

due to corn (Figure 20.29). She repeatedly lost her dorsalis pedis and posterior tibial pulses after corn challenge. She did well on an avoidance program, again illustrating the susceptibiltiy of chemically sensitive patients to environmental triggers.

A study of 12 patients with large vessel vasculitis (nonarteriosclerotic) revealed that this condition may be an isolated vascular entity or, more often, it may be accompanied by additional symptoms related to sensitization of the other smooth muscle systems such as respiratory, gastrointestinal, and genitourinary tracts (Figure 20.30).

Other parts of the vascular system ranging from the heart to precapillary arterioles and the veins were found to be involved in pollutant injury of these 12 patients as was seen in the aforementioned patient. Arrhythmias, spontaneous bruising, and phlebitis, for example, were seen. Foods and chemicals were found to trigger the allergic-type responses these patients experienced (Table 20.9).[145] The following details this study.

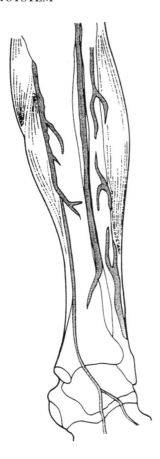

Figure 20.29.   Angiogram of femoral artery showing spasm associated
with sensitivities. This occurred after oral challenge of
corn. (From Rea, W. J. and O. D. Brown. 1987. Cardiovascular disease in response to chemicals and foods. In *Food
Allergy and Intolerance*, eds. J. Brostoff and S. J.
Challacombe. 745. London: Baillière Tindall. With permission.)

From the author's cardiovascular surgery practice, 12 consecutive highly
selected (i.e, nonarteriosclerotic and nonsmoking) patients, aged 22 to 45, with
large-vessel vasculitis were chosen for study. These patients were characterized by loss of peripheral pulses, cyanosis or symptoms in the extremity of
arterial spasm, and mild periorbital and digital edema, usually having a few
acneiform lesions, spontaneous bruising, and/or petechia.

Life-long environmental histories were taken on all patients by Randolph's
method[146] to determine the presence or absence of other symptoms of the
environmental maladaptation syndrome. Specific symptom complexes or diseases (of unknown etiology) involving the smooth muscle or immune system

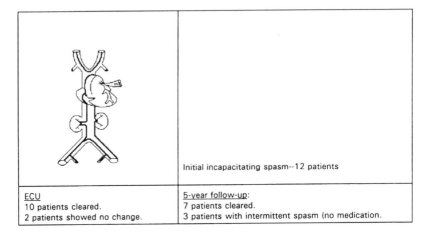

ECU	5-year follow-up:
10 patients cleared.	7 patients cleared.
2 patients showed no change.	3 patients with intermittent spasm (no medication.

Figure 20.30.   Large artery vasculitis occurring after pollutant challenge. (From EHC-Dallas. 1975.)

were sought and recorded. All patients were admitted to the ECU for a minimum of 16 days during the course of testing and treatment.

Pulses, blood pressure, temperature, color of extremities, and tenderness were monitored every 4 hours. Also, signs and symptoms related to the area of the body perfused by the involved artery(ies) were noted. After the patient stabilized in a basal state, free of signs and symptoms (including return of pulses, a heart rate below 80 beats per minute, ability to sleep all night, and loss of hunger), challenge tests with single, chemically less-contaminated source foods were begun.

Double-blind testing of chemical odors was also done. Each patient was given a timed exposure (from 1 to 60 minutes) to the flames of natural gas. Double-blind exposures to the ambient dose fumes of formaldehyde (<0.20 ppm), cigarette smoke, perfume, pine-scented floor cleaner, pesticides (<0.0034 ppm), and ethyl alcohol (<0.50) ppm were performed. Articles that were commonly found in the patient's home or work environment, such as home carpet, foam pillows, and polyester clothes, were also tested. Each exposure was at an ambient concentration, at a distance of 20 inches, with a constant flow of the vapor from a bottle.

The following laboratory tests were performed: complete blood count by the Coulter Model S method; sodium, potassium, chloride, carbon dioxide, blood urea nitrogen, serum protein, calcium, glucose, uric acid, alkaline phosphate, serum glutamic oxaloacetic transaminase, lactic dehydrogenase, and serum calcium by the SMA method; protein electrophoresis by the Helena cellulose acetate method; quantitative immunoglobulins (IgE, IgG, IgA, IgM), complement components 3 and 4, alpha-antitrypsin, and C-reactive protein by the Behring radial immunodiffusion method; total hemolytic serum complements (CH_{100}) and serum complement components by the hemolysis sheep

Table 20.9.   Large Vessel Vasculitis — Triggering Agents

Patient	Foods: oral challenge confirmed by intradermal injection (sensitive/tested)	Chemicals: double-blind inhaled challenge (sensitive/tested)	Systems
1	12/28	13/15	cardiovascular
2	34/48	5/12	vascular
3	78/94	5/13	gastrointestinal vascular cerebral
4	20/36	10/15	respiratory vascular
5	41/57	6/9	vascular cerebral
6	50/66	3/11	vascular
7	25/41	7/15	vascular cerebral
8	10/26	1/10	vascular cerebral
9	42/58	2/10	vascular gastrointestinal
10	3/19	2/10	vascular gastrointestinal
11	0/16	0/8	vascular
12	1/17	0/8	cardiovascular

Source: Rea, W. J., I. R. Bell, and R. E. Smiley. 1975. Environmentally triggered large-vessel vasculitis. In *Allergy: Immunology and Medical Treatment*, eds. F. Johnson and J. T. Spence. 185-198. Chicago: Symposia Specialist. With permission.

cells method (CH_{50}); prothrombin time, partial thromboplastin time, platelets and Lee-White clotting time by the Dade Reagents method; phosphorus by the Hycel method; fibrinogen and fibrinolysins clot lysis by the Biuret quantitative method; fibrin split products by the Burroughs-Wellcome method; C_1-esterase inhibitor by the Behring IEP method; and B and T lymphocytes by the sheep cell rosetting method. All tests were performed upon the patient's entrance into the room and at the beginning and end of testing. C_3 and C_4 were done daily during the period of fasting. Serial IgG, THSC, C_3, and eosinophils were done before and 5 minutes after the onset of reaction, at $1/2$ hour, 2 hours, and at the termination of the reaction.

All patients had many complaints related to smooth-muscle systems other than the symptoms from the involvement of the large blood vessels. Two had respiratory tree involvement, three had gastrointestinal involvement, and one had genitourinary system involvement in addition to the vascular involvement.

Once under environmental control, 10 of the 12 patients cleared their symptoms. This clearing included the subsiding of the inflammation and returning of pulses. All of these patients had their symptoms reproduced, including the reduction of pulses and tenderness over the involved area (from specific challenge tests). These reactions followed challenge exposures to one or more commonly eaten foods and/or chemicals (Table 20.8). Tables 20.10 and 20.11 show findings of the initial laboratory tests. Changes were seen in IgG, eosinophils, WBC, C-reactive protein, total complement and functions, and T lymphocytes. Two patients did not clear and had no adverse reactions. Their symptoms were apparently due to some other problem.

Environmentally triggered large-vessel vasculitis can be an isolated vascular entity or, more often, it is accompanied by additional symptoms related to sensitization of the other smooth-muscle systems such as the respiratory, gastrointestinal, and genitourinary tracts. Frequently, other parts of the vascular system, ranging from the heart to the precapillary arterioles and the veins, are involved. This multiple involvement was apparent in this series where two patients showed arrhythmias, while two others showed spontaneous bruises during the testing. One patient had two episodes of phlebitis. The pattern of triggering and response in all three patients with associated phlebitis, small-vessel involvement and arrhythmias appeared similar in most aspects to previously reported series by the author.[147-149] These data certainly add much weight to the idea of a spectrum of environmentally triggered vascular disease. From this and previous studies, evidence suggests that any one individual can have one or all parts of the vascular tree involved. This study also further substantiates that pollutant injury to vessels occurs.

Large vessel vasculitis may ultimately have more devastating results than other vascular disorders since the blood supply to major organs is affected. Organ ischemia and/or necrosis may result in severe disability or even death.

Carotid Spasm

Rea[150] has seen five chemically sensitive patients with spastic carotid phenomena resulting in transient cerebral vascular accidents. Long-term follow-up has shown them to remain well without medications as long as they avoid their incitants. After 10 years follow-up in the longest and 3 years in the shortest case, no completed strokes have occurred. Of course, the patients have learned how to stop spasm early enough to prevent devastating effects (Table 20.12). The following case study illustrates the devastating effects of pollutant injury on the carotid arteries.

Case study. This patient with large vessel involvement was a 42-year-old surgeon who had served in Vietnam and been sprayed with Agent Orange. He had subsequently developed spells in which arm movements were clonic-like, accompanied by asthma, spontaneous bruising, petechiae, and acneiform lesions. He then began to develop episodes of hemiplegia with diminution of

Table 20.10.   Laboratory Test Results EHC-Dallas

Patient	WBC	Total eosinophil (50 to 200/mm³)	CRP	THSC CH100 (90 to 98%)	C₃ (80 to 120 mg/dL)	C₄ (20 to 40 mg/dL)	Absolute lymphocyte (2000 to 4000/mm³)	T lymphocyte (%)
1	5100	50	Pos.	85	72	36	627	91
2	3900	—	Pos.	94	60	36	545	51
3	8900	156	Neg.	94	88	52	—	—
4	5500	111	Neg.	96	64	59	786	48
5	4900	211	Neg.	80	64	34	—	—
6	7700	67	Neg.	94	74	39	—	—
7	4900	89	Neg.	60	79	52	422	32
8	4400	211	Neg.	96	80	40	—	—
9	6700	154	Pos.	60	82	42	—	—
10	4500	67	Pos.	90	80	40	—	—
11	7700	150	Neg.	90	90	35	1224	53
12	7000	100	+	90	90	30	1491	71

Note: Neg. = negative; Pos. = positive.

Source: Rea, W. J., I. R. Bell, and R. E. Smiley. 1975. Environmentally triggered large-vessel vasculitis. In *Allergy: Immunology and Medical Treatment* ed. F. Johnson and J. T. Spence, 185. Chicago, IL: Symposia Specialist. With permission.

Table 20.11. Immunoglobulin Levels EHC-Dallas

Patient	Age/sex	IgG (800 to 1800 mg/dL)	IgE (10 to 110 IU/mL)	IgA (90 to 450 mg/dL)	IgM (mg/dL)[a]
1	41F	960	25	180	294
2	23F	880	350	75	100
3	59F	900	91	200	370
4	55F	890	75	189	260
5	30F	1250	75	203	291
6	58F	1090	155	306	158
7	37F	600	21	175	90
8	50M	1420	115	273	266
9	44F	1000	70	296	159
10	37M	1240	170	195	296
11	32F	1920	100	246	348
12	50M	1140	35	304	108

[a] F: 70–280 ± 17; m: 60–250 ± 16.

Source: Rea, W. J., I. R. Bell, and R. E. Smiley. 1975. Environmentally triggered large-vessel vasculitis. In *Allergy: Immunology and Medical Treatment*, eds. F. Johnson and J. T. Spence. 185. Chicago: Symposia Specialist. With permission.

Table 20.12.   Transient Hemiplegia

Patient	Side	Number of occurrences	Triggering agents	Long-term without neurological residual (years)
1	R	6	Foods and multiple chemicals	5
2	R	20	Molds, foods, multiple chemicals	3
3	L	3	Foods	10
4	L	2	Chemicals	15
5	R	5	Molds, foods, multiple chemicals	8

Source: EHC-Dallas. 1991.

intensity of his left carotid pulse. A carotid arteriogram revealed a decrease in left carotid and left intracerebral flow due to arterial spasm (Table 20.13; and Figure 20.31).

Double-blind challenges with foods and synthetic chemicals revealed the following sequence of events: immediate right-sided peripheral cyanosis, tenderness in the left neck, decrease of left superficial temporal and carotid pulse, loss of use of the right arm and hand, severe digital edema, and very foggy thinking and memory loss. Triggers included molds, beef, wheat, corn, cane sugar, phenol, formaldehyde, chlorine, pesticide, and car fumes (see Chapter 41).

Case study. Another patient, a 31-year-old laboratory technician, had problems with hemiplegia on the right side. Toxic chemicals, especially the xylene fixative as well as some foods, were found to trigger her symptoms (Figure 20.32). She has done well on an avoidance and an injection program.

Raynaud's Disease

Raynaud's disease refers to any localized peripheral digital vascular spasm or collapse of unknown etiology. It occasionally leads to gangrene.[145] In one study, Rea and Suits[151] were able to identify triggering agents in ten patients (Table 20.14). The culprit agents were found to occur on challenge with 45 of 60 foods and 5 inhaled chemicals. Follow-up over an 8-year period showed clearing of the problem with exacerbations occurring only when massive exposures to the triggering agents occurred. Long-term follow-up of this group has revealed almost a complete resolution of their problem without the use of medication. Only severe exposure has allowed brief recurrences, which were always brought under control without medication and with removal of the

Table 20.13.   Transient Hemaplegia

Patient:	42-year-old white male surgeon

Diagnosis:	1. Loss of function of dominant right hand, weakness of right leg, petechiae, Raynaud's disease of right hand and foot, episodic slurred speech.
	2. Pulmonary symptoms with prior exposure to defoliating agents in Vietnam.

Laboratory:		**Patient**	**Control**
	T lymphocytes	783	$1600 \pm 400/mm^3$
	B lymphocytes	500	$600 \pm 200/mm^3$
	Total hemolytic complement CH_{50}	150	$95 \pm 25\%$

Triggering agents (triggers of symptoms with vascular spasm):
Inhalants: Molds

Foods (oral challenge with intradermal confirmation): Apples, codfish, soybeans, gapes, squash, eggs, potatos, lamb, salmon

Chemicals (double-blind challenge): Chlorine, phenol, natural gas insecticides

Discharge status:	Clear when left unit without medication

Follow-up:	4 year follow-up; still triggered by pesticide exposures.

Source: EHC-Dallas. 1980.

incitants. We have now treated a series of 100 patients with various degrees of Raynaud's who responded similarly.

A case that illustrates the widespread involvement of many pollutants is shown in Table 20.15. This patient presented with gangrene of her fingers on her right hand. After definition and avoidance of triggering agents and an injection program for her inhalant, food, and chemical sensitivities, she has done well without the use of medications.

Hypertension

Hypertension has been noted to have several correctable etiologies such as fixed renal artery stenosis, adrenal tumors, etc. These will not be discussed here, since most textbooks discuss them thoroughly. Rather, emphasis will be placed on known and observed environmental triggers of "essential hypertension".

A.

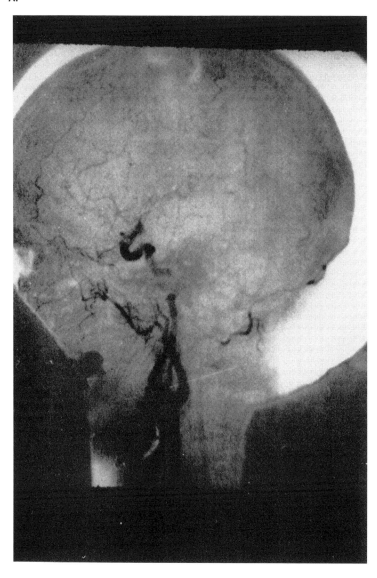

Figure 20.31.   (A) Control — normal carotid. (B) Cerebral arteriograms
revealing decreased carotid and left intracerebral flow due
to arterial spasm. (From Rea, W. J., and O. D. Brown.
1987. Cardiovascular disease in response to chemicals and
foods. In *Food Allergy and Intolerance*, eds. J. Brostoff and
S. J. Challacombe. 746. London: Baillière Tindall. With
permission.)

B.

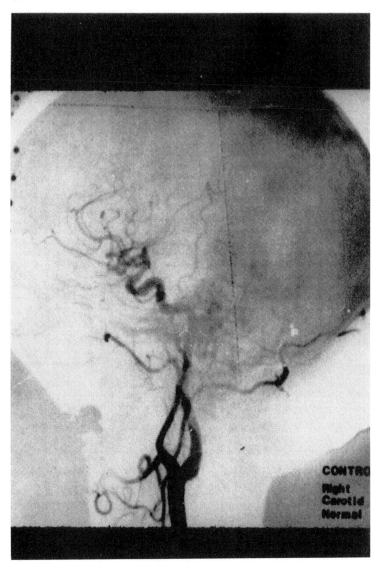

Figure 20.31.   (Continued).

Sodium chloride has long been thought to trigger hypertension. This triggering may be due to sensitivity to sodium and expansion of the extracellular fluid or involve other, as yet undefined, mechanisms.

Calcium has also been related to hypertension. McCarron et al.[152] present findings from their study of the dietary calcium intake of normotensive and hypertensive individuals, and these authors conclude that inadequate calcium

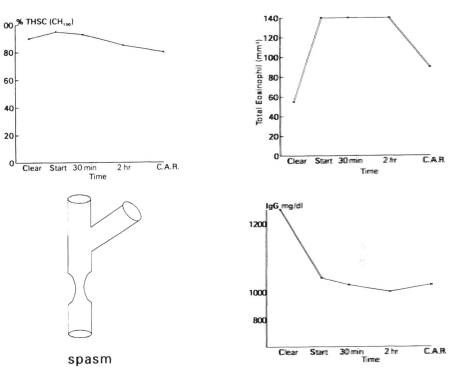

Figure 20.32.   31-year-old white female — salmon challenge. Symptoms and signs reproduced included right neck pain and tenderness, memory loss and confusion, depression and poor judgment, headache, and hemoplegia. Triggers on oral challenge — eggs. Multiple symptoms were confirmed by intradermal challenge test. Triggers on inhaled double-blind challenge were chemicals, gas heat, petrochemicals. All spasms including carotid spasm were reproduced on challenge. (From EHC-Dallas. 1982.)

intake may be a previously unrecognized factor in the development of hypertension. In another study intended to extend understanding of calcium metabolism in essential hypertension by defining extracellular ionized calcium concentrations in patients with untreated essential hypertension, McCarron[153] determined that hypertensive subjects had a small but significant reduction in ionized calcium concentrations.

In addition to calcium deficiency, magnesium deficiency has also been implicated as one of the essential triggers of hypertension. In our series of 200 patients with essential hypertension at the EHC-Dallas, 42% of the patients had intermittent intravenous magnesium challenges that would lower their blood pressure acutely, and 92% were also on oral magnesium supplementation from 500 to 1500 mg per day as long-term treatment.

Table 20.14.   Environmentally Triggered Raynaud's Disease

Patient	Age/sex	Environmental triggers	Long-term without medication and symptoms (years)
1	25/F	Foods, chemicals	4
2	30/M	Foods, chemicals	10
3	35/F	Foods, chemicals	15
4	40/F	Molds, chemicals	5
5	42/F	Molds, food, chemicals	8
6	46/M	Pollens, mold, food, chemicals	6
7	48/F	Foods, chemicals	10
8	41/F	Pollens, mold, foods, chemicals	3
9	35/F	Chemicals	4
10	40/F	Foods, chemicals	2

Source: EHC-Dallas. 1992.

In addition to apparent magnesium deficiency, a series of 53 chemically sensitive patients studied at the EHC-Dallas had excessive high and low intracellular minerals usually an indicator of pollutant injury. This variation of mineral levels compared to the control group was significant to the $p<0.001$ level. Clearly, deranged mineral metabolism is of some significance when studying the etiology of environmentally triggered hypertension (Table 20.16).

The mechanism of morphological changes produced in blood vessels by magnesium deficiency and disordered mineral metabolism is unclear, but the changes almost certainly contribute to the increased arterial resistance. Increased plasma renin activity, blood serotonin level, and urinary aldosterone excretion have been noted in magnesium deficiency. All these factors increase atrial resistance. The effect of magnesium deficiency on blood pressure involves complex interactions. Most experimental models show increased blood pressure with magnesium deficiency. However, some show normal tension and some show low blood pressure. Hypertension with magnesium deficiency has been associated with vitamin D toxicity, eclampsia, hyperadrenalcorticism, and hypoparathyroidism, as shown in the previous section. Pollutant injury to cell membranes shows that the sodium, calcium, and magnesium pumps are disturbed, resulting in inappropriate electrolyte shifts that may be responsible for the hypertension.

It is known that certain foods (i.e., banana, cheese) contain pressor amines, which could cause hypertension. Also, it is well-known that an overload of toxic chemicals (i.e., carbon disulfide) will cause hypertension[154] (see Volume II, Chapter 11[155]). Also, many foods are high in histamine (salmon, strawberries), which will effect blood vessel functions.[156] Other toxic chemicals will cause arterial spasm and trigger the angiotension mechanism. It is also clear that temporary hypertension can be caused by neurogenic stimulation.[157] Here the sympathetic stimulation of the blood vessels occurs. Direct electrical

Table 20.15.   30-Year-Old White Female — Fasted and Challenged in an Environmentally Controlled Area of a Hospital

Diagnosis:	Gangrene of right index finger. Scleroderma Crest syndrome.	

Laboratory:		**Patient**	**Control (N-240; T&B Monoclonal)**
	WBC	4.6	$7.56 \pm 3.45 \times 10^3/mm^3$
	Lymphocytes %	32.0	$37.30 \pm 1.1\%$
	Lymphocytes absolute	1472.0	$2767.00 \pm 91/mm^3$
	$T_{11}\%$	77.0	$75.20 \pm 0.8\%$
	T_{11} absolute	1133.0	$2079.00 \pm 74/mm^3$
	$T_4\%$	59.0	$42.20 \pm 0.7\%$
	T_4 absolute	868.0	$1157.00 \pm 43/mm^3$
	$T_8\%$	13.0	$25.80 \pm 0.8\%$
	T_8 absolute	191.0	$744.00 \pm 38/mm^3$
	T_4/T_8 ratio	4.5	1.7 ± 0.06
	Total B%	12.0	$15.00 \pm 10\%$
	Total B absolute	77.0	$856.00 \pm 655/mm^3$

Triggers of Raynaud's Disease:	Inhalants:	Molds, grasses
	Foods (oral within intradermal confirmation):	Watermelon, walnuts
	Chemicals (double-blind inhaled with intradermal confirmation):	petroleum ethanol (<0.50 ppm) insecticide (<0.0034 ppm) phenol (<0.002 ppm) cigarette smoke (ambient)
Discharge status:	Avoidance with intradermal neutralization for inhalants, foods, chemicals without medication	
Follow-up:	10 year follow-up: (1) no gangrene; (2) no medication; (3) works daily.	

Source: EHC-Dallas. 1979.

stimulation of the sympathetic nerves of the kidneys results in hypertension for days.[158] Prolonged hypertension can be exacerbated by frustration and pain also. In addition, it is well known that increase in aldosterone from adrenal hyperplasia or tumors, increase in cortisol from adrenal hyperplasia or tumors, and increased angiotension from diseased kidneys and narrowing of the renal

Table 20.16. Comparison of the Frequency of Abnormal (2 S.D.) High or Low Red Blood Cell (RBC) Minerals between Multiple Sclerosis Hypertension and Normal Control Group

Groups RBC minerals	M.S. No. abnormal/ total patients	(%)	Hypertension No. abnormal/ total patients	(%)	Normal control No. abnormal/ total controls	(%)	p
Ca	4/12	33.3	8/20	40.0	1/26	3.8	<0.01
Mg	5/12	41.7	6/20	30.0	7/26	26.9	>0.05
K	10/12	83.8	12/19	63.2	1/26	3.8	<0.001
Cu	5/12	41.7	4/20	20.0	2/26	7.7	>0.05
Zn	2/12	16.7	5/20	25.5	1/26	3.8	<0.05
Fe	3/12	25.0	6/20	30.0	1/26	3.8	<0.02
Mn	1/12	8.3	4/20	20.0	1/26	3.8	>0.05
Cr	11/12	91.7	13/20	65.0	1/26	3.8	<0.001
P	3/12	25.0	3/19	15.8	0/26	0.0	>0.05
Se	4/12	33.3	7/20	35.0	3/26	11.5	>0.05
Si	9/12	75.0	9/19	47.4	9/26	34.6	>0.05
S	5/12	41.7	2/19	10.5	6/26	23.1	>0.05
Ba	0/12	0.0	4/19	21.1	1/26	3.8	>0.05

Source: EHC-Dallas. 1990.

artery will cause hypertension.[159] It should be pointed out that some patients with essential hypertension have glomerular nephritis and Finn et al.[160] (see Chapter 22) have shown these to be due to hydrocarbon exposure in mechanics. Also, treatment of essential hypertension emphasizes increase in glomerular filtration and decrease in tubular reabsorption. Many hypertensive agents are vasodilators and act as sympathetic inhibitors or paralyze the renal smooth muscle.

At the EHC-Dallas, we have studied and successfully treated over 200 patients with essential hypertension. One prospective study performed under environmentally controlled conditions of 53 chemically sensitive patients with an age range of 31 to 90 years (mean = 55 years) is now presented.

Of the patients, 52% were outpatients and 48% were hospitalized in the ECU. All patients received meticulous diagnosis with other known causes of hypertension being ruled out. Environmental diagnosis included intradermal testing for biological inhalants, foods, and chemicals; oral challenge for water pollutants, foods, and food contaminants; and inhaled and intradermal challenge for chemical inhalants. The same procedures were applied to inpatients as outpatients, but, in addition, the inpatients were also fasted from 2 to 7 days. All patients prior to treatment were on antihypertensive medication. Evaluation consisted of routine SMA22, CBC, urinalysis, blood hydrocarbons, toxic solvents, intracellular vitamins and minerals, and serum immune parameters. After appropriate diagnosis, all patients were placed on a two-part program. The first part consisted of environmental control, which involved avoidance of offending foods, biological inhalants, and toxic chemicals in air, food, and water as well as injection therapy for biological inhalants, foods, and some chemicals. The second part included nutritional treatment involving appropriate vitamin and mineral supplementation.

The mean change in weight was 4 pounds over the 2 years, which is insignificant as far as blood pressure change goes (Table 20.17).

Laboratory results of this study showed the following: the mean blood pressure before treatment was 177±23/104±13 mmHg. After treatment, it was 121±13/77±8 mmHg. This finding is significant ($p<0.001$) (Figure 20.33). Only 8% of the patients were on medication after treatment up to 2 years follow-up in contrast to 100% of the patients pretreatment.

Pesticide levels were present in both the control and hypertensive groups but showed statistical difference of standard deviation in the means of several subjects.

The blood toxic chemicals were reviewed and compared with the normal blood pressure group. The results showed the average frequency of seven chemicals in the blood of the hypertension group to be much higher than normal blood pressure group of sick patients (patients with MS) ($p<0.01$) and a control group of average people (Table 20.18). The average means of blood toxic chemicals are shown in Figure 20.34. The means of five volatile aromatic hydrocarbons in the hypertension group were higher than the normal blood pressure group ($p<0.05$) (Table 20.19). The means and range of

Table 20.17.   The Means and Standard Deviations of Weight Before
                and After Treatment in 46 Hypertension Patients

Variable name	Number	Mean (lb.)	Sample standard deviation (lb.)	Sample standard deviation error (lb.)
Pretreatment weight	46	176.8	53.8	7.9
Posttreatment weight	46	172.9	52.5	7.7

Source: EHC-Dallas. 1990.

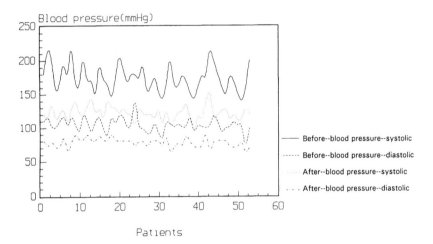

**Figure 20.33.   Comparison of blood pressure in 53 hypertensive patients
before and after environmental treatment. (From EHC-
Dallas. 1990.)**

volatile chlorinated hydrocarbons seen in the 28 patients are shown in Table
20.20. The average frequency of seven chemicals of volatile chlorinated aliphatic
panel in the hypertension group was higher than in the control group ($p<0.05$).
The means of the chlorinated pesticide group were higher than the comparison
group (Table 20.21). The frequency of seven chemicals of volatile aliphatic panel
in the hypertension groups was higher than in the control group ($p<0.01$) (Table
20.22). The means of the toxic chemicals were 105 ppb in the hypertension group
vs. 58 ppb in the normal tensive multiple sclerosis group vs. 80 ppb in the healthy
normal tensive group. The means of volatile aliphatics are shown in Table 20.23.
The frequency of a positive terpene skin test in the hypertension group was
higher than in the control group also ($p<0.01$) (Table 20.24).

The average of blood cholesterol in the hypertensive group was 217 mg/dL
and the normal blood pressure group was 192 mg/dL. The difference of these
two groups was significant ($p<0.05$).

Table 20.18.   Comparison of the Frequency of Blood Toxic Chemicals between Normal Control, Hypertension, and MS Group

Chemicals	Normal control (%)	Hypertension (%)	MS (%)	p
Benzene	38.5	25.0	20.0	>0.05
Toluene	73.1	82.1	66.7	>0.05
Ethylbenzene	3.8	32.1	40.0	<0.01
Xylene	61.5	64.3	26.7	>0.05
Styrene	0.0	21.4	26.7	<0.01
Trimethylbenzene	0.0	3.6	20.0	>0.05
Dichloromethane	38.5	21.4	15.4	>0.05
Chloroform	34.6	28.6	36.4	>0.05
1,1,1- Trichloroethane	69.2	89.3	80.0	>0.05
Trichloroethylene	7.7	39.3	57.1	<0.01
Tetrachloroethylene	80.8	85.7	92.9	>0.05
Dichlorobenzene	0.0	7.1	28.6	>0.05
n-Pentane	42.3	66.7	71.4	>0.05
2,2-Dimethylbutane	30.8	40.0	14.3	>0.05
Cyclopentane	3.8	33.3	28.6	<0.05
2 Methyl pentane	100.0	100.0	100.0	>0.05
3 Methyl pentane	100.0	100.0	100.0	>0.05
n-Hexane	100.0	93.3	100.0	>0.05
n-Heptane	3.8	6.7	0.0	<0.05

Source: EHC-Dallas. 1990.

Results showed significant changes in T & B cell parameters vs. controls (Table 20.25 and Figure 20.35). Abnormal IGs and complements were also seen (Figure 20.36). All patients had sensitivities to biologic inhalants, foods, and chemicals by challenge. This put them in a special group of environmentally triggered hypertension.

Forty-seven patients had double-blind inhaled challenges using the ambient doses of phenol (<0.002 ppm), chlorine (<0.33 ppm), formaldehyde (<0.2 ppm), petroleum derived ethanol (<0.5 ppm), pesticide 2,4-DNP (<0.0034 ppm), and three saline placebos (Figure 20.37). One hundred percent increased their blood pressure with ethanol, formaldehyde, and chlorine challenges, while the blood pressure of 83% increased with phenol and pesticide exposure. Ten percent reacted to placebos ($p<0.001$). Intradermal challenges confirmed these outcomes and, in addition, showed sensitivities to cigarette smoke (83%), orris root (80%), newsprint (59%), perfume (60%), and unleaded diesel fuel (62%). These data strongly suggest a toxic chemical etiology for this environmentally triggered hypertension.

It is clear from this study and that of our other 147 patients that there is a group of chemically sensitive patients who have "essential" hypertension and who have characteristics distinct from a control group. The abnormal differences

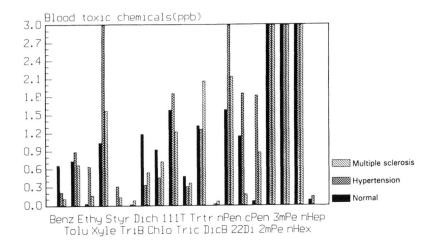

Figure 20.34.   Comparison of the means of blood toxic chemicals be-
                 tween normal control, hypertension, and MS groups. (From
                 EHC-Dallas. 1990.)

Table 20.19.   Comparison of the Means of Volatile Aromatic
                Hydrocarbons between Hypertension Group
                and Normal Blood Pressure Group

Chemicals	Hypertension group	Normal blood pressure group	Difference
Benzene	0.21	0.11	0.10
Toluene	0.99	0.67	0.32
Ethylbenzene	0.64	0.16	0.48
Xylene	5.93	1.57	4.36
Styrene	0.31	0.13	0.18
Trimethylbenzene	0.007	0.073	–0.066

Source: EHC-Dallas. 1990.

are in the volatile aromatic hydrocarbons, aliphatic hydrocarbons, and chlorinated pesticides with resultant differences in T and B cells parameters. There
are multifactorial environmental triggers that relate to hypertension. Toxic
effects of some foods are also known to trigger some cases of hypertension.
These foods include bananas and cheese with their pressor amines (see Volume
II, Chapter 8[161]). In addition, some leukotriene injections have been shown to
ameliorate hypertension in animals. In summary, blood pressure was significantly reduced in these hypertensive chemically sensitive patients both short-
and long-term (24 months), without medication, and using ECU treatment.

Table 20.20.   The Means and Standard Deviations of Volatile Chlorinated Hydrocarbons in 28 Patients with Hypertension

Variable	Mean	Sample standard deviation	Sample standard error
Dichloromethane	0.325	0.80674	0.15246
Chloroform	0.45	0.74759	0.14128
1,1,1-Trichloroethane	1.84643	1.75615	0.33188
Trichloroethylene	0.3	0.44555	0.0842
Tetrachloroethylene	1.25714	1.07512	0.20318
Dichlorobenzene	0.01786	0.06696	0.01265

Source: EHC-Dallas. 1990.

Table 20.21.   Comparison of the Means of Chlorinated Pesticides between Normal Blood Pressure Group and Hypertension Group

Chemicals	Hypertension group	Normal blood pressure group	Difference
α-BHC	0.2	0.007	0.193
β-BHC	0.69	0.36	0.33
DDD	0.22	0	0.22
DDE	4.7	3.2	1.5
DDT	0.06	0.05	0.01
Dieldrin	1.69	0.07	1.62
Endosulfan I	0.009	0.014	−0.005
HCB	0.26	0.36	−0.1
Heptachlor epoxide	0.88	1.08	−0.2
Trans-nonachlor	0.52	0.14	0.38

Source: EHC-Dallas. 1990.

Hypersensitive Vasculitis

The hypersensitivity vasculitides are a group of disorders characterized by small vessel inflammation that have fixed-named pathology and have cellular infiltrates other than the perivascular lymphocytes (Figures 20.38 and 20.39). Manifestations are often mild and self-limited. Although any organ can be involved, the most common is the skin. Palpable purpura with lesions can usually be found on the buttocks, ankles, and legs. Patients who exhibit renal, joint, and gastrointestinal tract involvement are relatively common. Up to 20% show neurologic involvement. Often generalized symptoms such as fever, myalgia, malaise, and anorexia are noted.[162] Causes for some of these symptoms have now been found. Theorell et al.[68] showed occurrences of these vascular pathologies after challenge with molds, cedar, and some foods. At the

Table 20.22.   The Frequency of Volatile Aliphatic
Panel in 15 Patients with Hypertension

Chemicals	No. of positive	Frequency (%)
n-Pentane	10	66.7
2,2-Dimethylbutane	6	40
Cyclopentane	5	33.3
2-Methyl Pentane	15	100
3-Methyl Pentane	15	100
n-Hexane	14	93.3
n-Heptane	1	6.7

Source: EHC-Dallas. 1990.

Table 20.23.   The Means and Standard Deviations of Volatile Aliphatic
Panel in 15 Patients with Hypertension

Volatile aliphatic chemical	Mean	Sample standard deviation	Sample standard deviation error
n-Pentane	5.12	15.34918	3.96314
2,2-Dimethyl butane	1.85333	2.72393	0.70332
Cyclopentane	1.81333	3.5187	0.90852
2-Methyl Pentane	19.34667	18.16189	4.68938
3-Methyl Pentane	45.39333	32.47587	8.38523
n-Hexane	20.07333	14.10099	3.64086
n-Heptane	0.13333	0.5164	0.13333

Source: EHC-Dallas. 1990.

Table 20.24.   Comparison of Terpenes Skin Testing between
Hypertension Group and Normal Blood Pressure Group

Terpenes	Hypertension group (%)	Normal blood pressure group (%)	Difference
Pine	64.1	58.8	5.3
Tree	73.5	50.0	23.5
Grass	82.9	62.5	20.4
Ragweed	67.6	56.3	11.3
Mt. Cedar	79.4	68.8	10.6
Mesquite	68.0	54.5	13.5
Sage	85.0	57.1	27.9

Source: EHC-Dallas. 1990.

Table 20.25. Comparison of the Frequency of Abnormal Immune Parameters between Multiple Sclerosis, Hypertension, and Normal Control Groups

Immune parameters	Multiple sclerosis		Hypertension		Normal (control)	
	No. abnormal/ total patients	(%)	No. abnormal/ total patients	(%)	No. abnormal/ total controls	(%)
WBC	6/16	37.5	7/40	17.5	9/26	34.6
Lym %	0/16	0.0	1/36	2.8	1/26	3.8
Lym C	6/16	37.5	12/37	32.4	6/26	23.1
T_{11} %	5/16	31.3	14/40	35.0	2/26	7.7
T_{11} C	8/16	50.0	15/39	38.5	5/26	19.2
T_4 %	5/16	31.3	12/36	33.3	3/26	11.5
T_4 C	4/16	25.0	9/35	25.7	4/26	15.4
T_8 %	1/16	6.3	5/37	13.5	6/26	23.1
T_8 C	1/16	6.3	6/35	17.1	6/26	23.1
T_4/T_8	2/16	12.5	6/36	16.7	4/26	15.4
Bly %	5/16	31.3	8/38	21.1	4/26	15.4
Bly C	3/16	18.8	5/38	15.8	4/26	15.4

Source: EHC-Dallas. 1990.

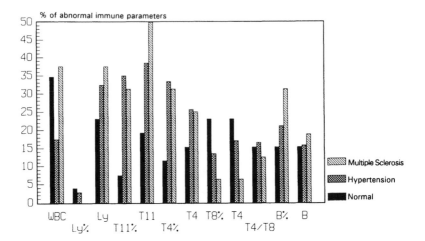

Figure 20.35.   Comparison of the abnormal frequency of immune parameters of normal control, hypertension, and MS groups. (From EHC-Dallas. 1990.)

EHC-Dallas, we have observed such lesions after challenge with phenol, formaldehyde, and beef. The hypersensitivity vasculitides are a diverse group of disorders including serum sickness reactions, Henoch-Schönlein purpura, essential mixed cryoglobunemia, and the connective tissue diseases, particularly rheumatoid and lupus vasculitis.[163] Certain malignancies may also have associated cutaneous vasculitis. Generally, foreign serum proteins can cause serum sickness reactions, but similar reactions may occur after the use of penicillin,[164] sulfonamides,[165] streptomycin,[166] and thiouracils.[166] We have seen a case of Henoch-Schönlein purpura triggered by pollen, dust, molds, foods, and chemicals (Figure 20.40). Clearly, triggering agents have been reported for many cases of arthritis and lupus (see Chapter 23).

Periarteritis Nodosa

Periarteritis Nodosa (PAN) generally follows a prodromal fever, arthralgia, and malaise. It may be manifested as acute gastrointestinal distress, myocardial infarction, neuritis, muscle pain, and/or gangrene of the extremities. It generally presents in the muscular arteries involving all three layers of the arterial wall and adjacent veins. It usually is segmental. Biopsy of skin, subcutaneous nodules, or smooth muscle reveals acute healing vasculitides without giant cells. The infiltration process involves polymorphonuclear leukocytes, eosinophil, and edema followed by fibrinoid necrosis. The areas of fibrinoid necrosis are replaced by fibroblasts, and scar tissue is formed (Figure 20.41).

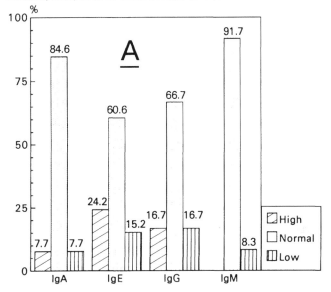

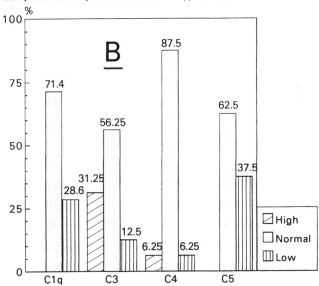

Figure 20.36.   Immune changes in environmentally triggered hypertensive patients. (A) The antibody analysis in 33 cases of hypertension. (B) The complement analysis in 16 cases of hypertension. (From EHC-Dallas. 1990.)

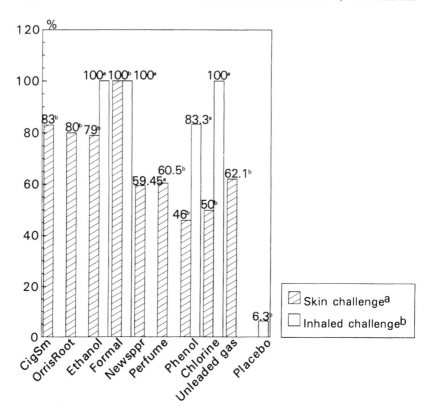

Figure 20.37. Comparison of skin test and inhaled double-blind challenge test in 47 cases of hypertension. Challenge after 4 days of deadaptation in a less-polluted environment with total load decreased. Ethanol, <0.5 ppm; formaldehyde, <0.2 ppm; phenol, <0.002 ppm; chlorine, <0.33 ppm. *Note:* [a]Double-blind challenge; [b]Single-blind challenge. (From EHC-Dallas. 1990.)

Wegener's Granulomatosis

PAN is closely related to Wegener's Granulomatosis, which is characterized by necrotizing granulomas in the respiratory tract and vasculitis of the medium size arteries, veins, arterioles, and venules. Onset of Wegener's Granulomatosis may be acute or chronic. Though pathologically well defined, the etiology of the disease still is obscure. Our recent studies indicate that it is worthwhile looking for incitants in patients diagnosed with this disease.[167]

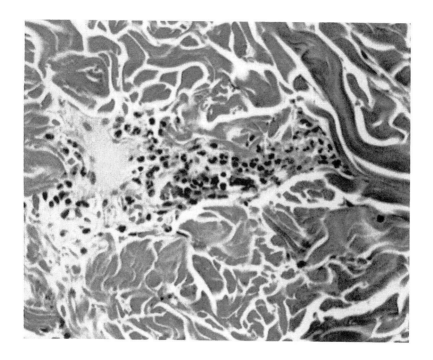

Figure 20.38.   Perivascular inflammation of the deep dermis in a case of
pressure urticaria. Over 20% of the cells are eosinophils.
Note the dilated lymphatic vessel. H&E × 400. (From
Winkelman, R. K. 1987. Food sensitivity and urticaria or
vasculitis. In *Food Allergy and Intolerance*, eds. J. Brostoff
and S. J. Challacombe. 603. London: Ballière Tindall.
With permission.)

Rheumatoid Vasculitis

Diseases such as rheumatoid arthritis have a variety of vascular manifestations and biopsy evidence of vasculitis. Although the etiology of rheumatoid
arthritis is generally not known, we have found evidence of immune changes
in patients with rheumatoid arthritis following food and chemical challenges.
Two controlled series[168,169] that define triggering agents in this illness have
been reported (see Chapter 23).

When rheumatoid patients present, biopsies may be done. Vascular pathology will show either necrotizing type of vasculitis with many polymonuclear
leukocytes or nonnecrotizing type of vasculitis with eosinophils and/or lymphocytes around or in the vessel wall. However, only active lesions will be
positive when biopsy of the petechiae and bruises is done.

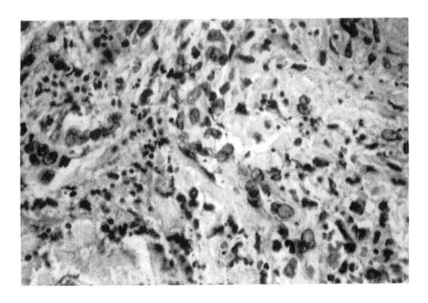

Figure 20.39.   Perivascular and dermal leukocytoclastic vasculitis from skin of a patient with urticarial vasculitis. H&E × 640. (From Winkelman, R. K. 1987. Food sensitivity and urticaria or vasculitis. In *Food Allergy and Intolerance*, eds. J. Brostoff and S. J. Challacombe. 603. London: Ballière Tindall. With permission.)

Challenge tests should be done to define triggering agents. These challenges may be accomplished by various routes including oral, inhalant, and/ or intradermal. Challenge tests must be conducted under controlled and reliable steady-state environmental conditions in order to reduce variability of results. It is important also to take into consideration the acute toxicological tolerance phenomenon by requiring a patient to avoid the suspected incitant for 4 days before challenge. This deadaptation period allows for more accurate and reproducible test results because the patient is in a steady basal state. A comprehensive workup of suspected triggering agents is necessary in order to stop vascular inflammation. The use of an environmental control unit with its reduction of pollutants in air, food, and water allows for the most precise diagnosis and treatment for the environmental aspects of cardiovascular diseases. Since these test conditions are not available in most areas, controlled areas in hospitals and offices may have to be used as a less-than-satisfactory substitute. Clearly, the reduction in total load with subsequent challenge is the key to diagnosis of triggering agents of cardiovascular disease.

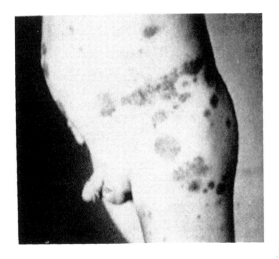

Figure 20.40.   Henoch-Schönlein purpura (anaphylactoid purpura) in a 14-year-old boy, with associated urticarial erythematous lesions with purpura, painful articular swelling and microscopic hematuria. (From Rea, W. J. and O. D. Brown. 1987. Cardiovascular disease in response to chemicals and foods. In *Food Allergy and Intolerance*, eds. J. Brostoff and S. J. Challacombe. 742. London: Ballière Tindall. With permission.)

Eosinophilic Vasculitis

Eosinophilic vasculitis has now been reported in some disease processes, i.e., eosinophilic granulomas and Goodpasture's syndrome. Eosinophils are seen in most cases.[170] These apparently have allergic triggers though they are not well worked out at the present.

Recurrent Phlebitis

In the 1920s, Connors,[171] as cited by Harkavy,[172] noted two patients whose phlebitis was triggered by fish and citrus fruits. Rea's[145] work using the ECU for controlled studies has led to the isolation of numerous triggering agents including foods, chemicals, and inhalants. It has become apparent that, in many chemically sensitive patients, phlebitis is only part of a more generalized and severe vasculitis. Meticulous definition of triggering agents allowed one group of patients to lose their previously intractable phlebitis and to return to work. This group was compared to a control group that did not receive definition of triggering agents and continued to have phlebitis over a 7-year period

A.

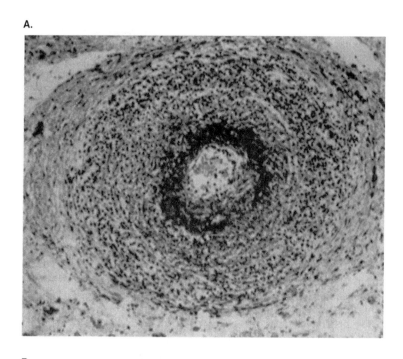

B.

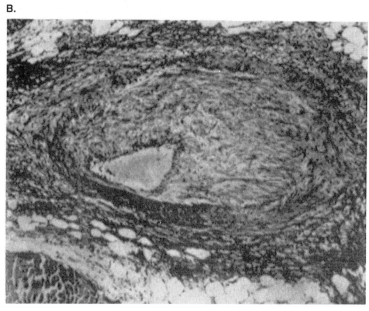

(Table 20.26). Environmentally directed diagnosis and treatment proved to be not only efficacious but also cost effective (Table 20.27).[173]

An impressive number of associated signs and symptoms found in these patients with phlebitis occurred with, or were related to, the environmental maladaptation syndrome over a patient's lifetime. Included were tonsillectomy, recurrent bouts of spontaneous bruising, nasal stuffiness, cold susceptibility, increased sense of smell, adult acne, myalgia, sinusitis, headaches, spastic colon, nonspecific chest pain, bronchitis or bronchopneumonia, overwhelming fatigue, extremity vascular spasm, sore throat, asthma, arrhythmias, cystitis, and depression. It has now been shown in over 100 patients with recurrent phlebitis seen at the EHC-Dallas that the aforementioned patterns of associated symptoms and triggering agents can be found. It is clear that a group of patients who have environmental triggers for their phlebitis does exist.

Cardiac Arrhythmias and Dysfunction

Literature over the past several decades attests to a direct cause and effect relationship between food ingestion and/or chemical inhalation and cardiac arrhythmias, and/or coronary spasm.[174-176] The causal role of coffee and cigarettes in the triggering of atrial arrhythmias is well known. Recent reports show that pollutants can harm the heart. Even fabrics can trigger arrhythmias (Table 20.28).

Seyal[176] showed that the fumes from synthetic fabrics can trigger arrhythmia. In the study of 2000 subjects wearing cotton clothing and 2000 wearing synthetic clothing, he further showed that the mean systolic blood pressure in those wearing synthetics was 135.2 ± 12.1 mmHg and those wearing cotton was 122.7 ± 9.7 mmHg. The differences were significant ($p < 0.05$) (Table 20.29).

Figure 20.41. Polyarteritis nodosa (PAN). (A) Acute inflammatory stage of PAN is characterized by panarteritic and periarteritic inflammation with polymorphonuclear leukocytes, eosinophils and round cells, destruction of vascular tissues and fibrinoid necrosis (dark, amorphous material about lumen). (B) In healing and healed stages of PAN, reparative fibrosis idstorts vascular wall with loss of most of internal lamina (black wavy line) and marked narrowing of original lumen by organized thrombus and reparative fibrosis. (From Rea, W. J. and O. D. Brown. 1987. Cardiovascular disease in response to chemicals and foods. In *Food Allergy and Intolerance*, eds. J. Brostoff and S. J. Challacombe. 743. London: Ballière Tindall. With permission.)

**Table 20.26. Associated Signs and Symptoms of Thrombophlebitis
and Results of Challenge Studies in 10 Patients in an
Environmental Control Unit (EHC-Dallas)**

Associated signs and symptoms reproduced	Offending agents	Phlebitis
Diarrhea, pulse increase (30 beats/min), nasal congestion, bigeminy, multifocal premature ventricular contractions	Beef, chicken, cigarette smoke, shrimp, pork, gas heat, ingested chemicals	Pork, shrimp, inhaled chemicals
Vomiting, pulse increase	Wheat, rice, inhaled chemicals	No
Wheezing, rhinorrhea, red nose, nasal stuffiness, tender muscles, cystitis	Corn, cane sugar, eggs, inhaled chemicals	Corn, eggs, inhaled chemicals
Peripheral pulse from 4 to 1+, tachypnea, shortness of breath, cyanosis, belching	Beef, potatoes, corn, ingested chemicals	Beef, corn, inhaled chemicals
Edema (generalized), tender muscles, colitis, dizzy	Pork, pork fumes, ingested and inhaled chemicals	Inhaled chemicals
Syncope, wheezing, muscle tenderness, hives, paroxysmal atrial tachycardia	Legumes, seafood, cane sugar, wheat, chicken, cigarette smoke, inhaled chemicals	Cigarette smoke, ingested chemicals, inhaled chemicals, seafood
GI bloat, belching, premature ventricular contractions, ventricular tachycardia	Beef, chicken, lettuce, chemicals, inhaled chemicals	Wheat, potatoes
Decrease in pulse left arm only, left neck and arm tenderness, tender over arm veins	Turkey, chicken, peas, beef, cigarette smoke, inhaled chemicals	No

Table 20.26 (Continued). Associated Signs and Symptoms of
Thrombophlebitis and Results of Challenge Studies in 10 Patients
in an Environmental Control Unit (EHC-Dallas)

Associated signs and symptoms reproduced	Offending agents	Phlebitis
Dyspnea, wheezing, eyes watering, hoarse, pulse increase (50 beats/min)	Coffee, peanut butter, cane sugar chemicals	Apple, corn, wheat, inhaled chemicals
Cystitis, diarrhea, skin rash, itching, dyspnea, pulse increase	Corn, wheat, beef, eggs, inhaled chemicals	Chicken, beef, inhaled chemicals

Results of Thrombophlebitis Treated in an Environmental Control Unit

			Exercise			
			Before treatment		After treatment	
	Phlebitis		Times walking around a 10 ft x	Exercycle miles (150 kpm	Walking around a 10 ft x	Exercyle miles (150 kpm
	Cleared	Reproduced	10 ft room	resistance)	10 ft room	resistance)
ECU	10	8	0.5	0	36.6	2.85
Control	0	10 (continued)	2.1	0	3.1	0

Source: Rea, W. J. and O. D. Brown. 1987. Cardiovascular disease in response to chemicals and foods. In *Food Allergy and Intolerance*, eds. J. Brostoff and S. J. Challacombe. 747. London: Baillière Tindall. With permission.

Table 20.27. Cost Effectiveness — Recurrent Phlebitis Divided
into Two Groups

	Number of patients	Total episodes of phlebitis 7 years	Hospitalized 7 years	Medical cost 7 years	Number working
ECU	10	2	0	$2,500.00	10
Control	10	200	114	>$300,000.00	1

Source: EHC-Dallas. 1974–1981.

Table 20.28. Total Score of Individual Symptoms in Two Treatment Groups (Test and Control) at Day 0, 10, 20, and 30 and Relative Difference between the Scores

Symptoms	Test — natural fibers				Control — synthetic fibers				Test			Control		
	Day 0	Day 10	Day 20	Day 30	Day 0	Day 10	Day 20	Day 30	0–10	0–20	0–30	0–10	0–20	0–30
Palpitation	100	41	17	15	100	82	75	75	59	83	85	18	25	25
Extrasystole	100	40	25	18	100	82	75	75	60	75	82	18	25	25
Fatigibility	100	36	19	17	100	100	99	83	64	81	83	0	1	17
Generalized autonomic disturbances	100	60	38	28	100	68	66	60	40	62	72	32	34	40
Precordial oppression	100	73	34	16	100	90	88	81	27	66	84	10	12	19
Sleeplessness	100	35	25	20	100	88	88	75	65	75	80	0	0	13
Stabbing over heart	100	41	38	38	100	100	100	94	59	62	62	0	0	6
Body aches	100	47	25	22	100	100	99	91	53	75	78	0	1	9
Respiratory distress	100	30	30	25	100	80	75	65	70	70	75	5	10	20

Test group — natural fiber

	Total no.	Disappeared	Reduced	No improvement
Palpitations	50	41(82%)	3(6%)	6(12%)
Extrasystole	50	41(82%)	3(6%)	6(12%)
Fatigibility	42	34(80.95%)	2(4.5%)	6(14.35%)

Control group — synthetic fibers

	Total no.	Disappeared	Reduced	No improvement
Palpitations	50	11(22%)	(6%)	36(72%)
Extrasystole	50	11(22%)	3(6%)	36(72%)
Fatigibility	30	4(15.39%)	3(7.69%)	20(76.92%)

Generalized autonomic disturbance	40	24(60%)	10(25%)	6(15%)	44	13(29.54%)	2(20.46%)	22(50%)
Precordial oppression	40	31(77.5%)	5(12.5%)	4(10%)	40	6(15%)	9(7.5%)	31(77.5%)
Sleep disturbance	42	32(76.19%)	4(9.5%)	6(14.23%)	35	4(11.42%)	3(5.71%)	29(82.85%)
Stabbing over heart	35	18(51.41%)	7(20%)	10(28.55%)	32	1(3.12%)	2(−)	31(96.87%)
Body aches	32	22(68.75%)	6(18.75%)	4(12.5%)	38	2(5.2%)	1(2.63%)	35(92.1%)
Respiratory distress	20	12(60%)	6(30%)	2(10%)	17	3(18.75%)	1(6.25%)	12(75%)

Source: Seyal, A. R., M. H. Awan, A. Asghar, and A. Nazir. 1986/87. Psychosomatic cardiovascular disorders: an elusive relationship to the type of clothing worn. *Clin. Ecol.* 4(1):26–30. With permission.

Table 20.29. Systolic Blood Pressure/Quality of Garments — 4000 Patients

	Synthetic garments					Cotton garments			
S. B. P. (mmHG)	No. of individuals	Population frequency (%)	No. having PVCs	PVC frequency (%)	S. B. P. (mmHG)	No. of individuals	Population frequency (%)	No. having PVCs	PVC frequency (%)
90–99	—	—	—	—	90–99	4	(0–2)	—	—
100–109	—	—	—	—	100–109	23	(1–25)	—	—
110–119	182	(9–1)	2	1–09	110–119	862	(43–1)	1	0–116
120–129	376	(19–8)	5	1–329	120–129	847	(42–35)	3	0–345
130–139	1025	(51–25)	16	1–560	130–139	125	(6–25)	2	1–6
140–149	199	(9–95)	6	3–015	140–149	105	(5–25)	2	1–9
150–159	110	(5–5)	7	6–34	150–159	17	(0–85)	1	5–58
160–169	77	(3–85)	6	7–79	160–169	16	(0–8)	—	—
170–179	30	(1–5)	3	10–00	170–179	1	(0–05)	—	—
180–189	1	(0–05)	—	—	180–189				
Total	2000		45	2–25	Total	2000		9	0–45

Source: Seyal, A. R., M. H. Awan, A. Asghar, H. Mumtaz, and A. Nazir. 1986/87. Systolic blood pressure, heart rate and premature ventricular contractions in a population sample: relationship to cotton and synthetic clothing. Clin. Ecol. 4(2):69–74. With permission.

Suspicions as to the clinical importance of chemical air pollution on the cardiovascular system were first aroused when the author of this book observed a 38-year-old physician who had a broad spectrum of symptoms related to smooth muscle sensitization, which included gastrointestinal involvement resulting in bloating, gas, belching, and cramping after each exposure to the fumes from an X-ray developer. He also experienced urinary urgency, chest tightness, and peripheral arterial spasm When he was forced to stay in the area of the X-ray developer, his symptoms progressed from mild uneasiness to shortness of breath and frequent premature ventricular contractions (PVCs). Withdrawal from this environment resulted in cessation of the arrhythmias while on at least 20 separate occasions reexposure resulted in the PVCs.[177]

Suspicions of a more than causal relationship between environmental incitants and the triggering of arrhythmias were further aroused when the author had an opportunity to resuscitate a 69-year-old male truck driver in ventricular fibrillation (Figure 20.42). The patient was hypotensive for 2 days, requiring vasopressors. When he was out of shock, off vasopressors, and satisfactorily recovering, he was fed. Immediately after eating, the patient developed bloating accompanied by PVCs and a drop in blood pressure. After this happened on three different occasions, with each instance appearing life-threatening, an environmental history was obtained. It revealed that during the patient's lifetime every time he ate milk or eggs the aforementioned sequence of events occurred. At age 7 years, the patient had been in a coma for 2 weeks, and upon awakening he had been placed on an elimination diet consisting solely of potatoes. This diet resulted in recovery. Throughout his life, he went on this avoidance diet to clear his symptoms whenever he experienced bloating and irregular heart beats. In addition, during his childhood a similar sequence of events would occur when he was exposed to gasoline fumes. This reaction caused some difficulties during his trucking career, since frequently he would have to stop by the roadside and get away from the truck to allow his symptoms to clear. Avoidance of offending foods and inhaled chemicals subsequently allowed an uneventful recovery.

Further suspicions of a causal relationship between environmental incitants and the onset of illness were aroused when a 32-year-old female with uncontrolled ventricular arrhythmias refractory to drugs was seen at the EHC-Dallas. This patient was clear after 5 days fasting in an environmentally controlled room. On subsequent individual transient exposures (15 seconds to 2 minutes) to ambient chemical fumes such as natural gas, chlorine, or floor cleaner reproduction of arrhythmias occurred immediately.

Since there appeared to be a more than casual relationship between food and chemical susceptibility and cardiac arrhythmias in several patients, a controlled study was initiated.

From this cardiovascular surgeon's practice, 12 consecutive adult patients (ages 12 to 45 years) with unexplained and uncontrolled cardiac arrhythmias, having nonatherosclerotic cardiac disease and having multisymptomatology related to smooth muscle sensitization, were studied. Arteriosclerosis was

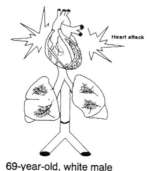

1. 2 MI + ventricular fib.
2. Shortness of breath,
   distention, arrythmia,
   2 pulmonary embolisms
3. Sluggish, confused
4. Potato diet
5. Triggers:
   Oral challenge--milk, eggs
   Inhaled challenge--fuel, oil

69-year-old, white male

Figure 20.42.   69-year-old white male. Symptoms and signs reproduced on challenge included two MI + ventricular fibrillation, shortness of breath, distention, arrhythmia, two pulmonary embolism, sluggishness, confusion. Potato diet and triggers — oral challenge: milk, eggs; inhaled challenge: fuel oil. (From EHC-Dallas. 1976.)

ruled out by the usual methods of electrocardiograms, exercise stress tests, and cardiac catheterization. Competent cardiologists saw the patients initially before instituting environmental control and agreed that the patients had arrhythmias of unknown etiology.

Detailed lifelong ecological histories were taken by Randolph's[178] method in a search for other parts of the environmental maladaptation syndrome. Specific symptom complexes and recurrent inflammations such as sinusitis, laryngitis, hoarseness, bronchitis, cystitis, spastic colon, enteritis, colitis, migraine or vascular headaches, myalgia, arthritis, arthralgia, depression, or phlebitis of unknown etiology were recorded.

To establish cause and effect, an attempt was made to create an environment that was as free as possible from inhaled and ingested contaminants, where triggering agents could be clearly defined. This pollutant-reduced, ECU-type environment was achieved in the usual manner (see Chapter 31).

All patients during the course of diagnosis and treatment were kept in the ECU for at least a period of 16 days. Upon the patients' entrance into the rooms, all medications including anticoagulants, steroids, and anti-arrhythmic drugs were discontinued. Once the patients reached the symptom-free basal state, they could act as their own controls. No medications were given during the stay in the unit with the exception of oral or intravenous bicarbonate of soda and oxygen. Pulses, blood pressure, temperature, bruising, petechiae, and color changes were monitored every 4 hours. After the patients were in a basal state free of signs, including absence of most arrhythmias, normal pulses, and a rate below 80, were able to sleep all night, and had lost their hunger, challenge with a single chemically less-contaminated source food (an individual, pure food whose biological origin was known, i.e., wheat, oats, cane sugar, beet sugar, chicken, duck, turkey, etc.) was begun. The food was selected individually and randomly. In order not to miss delayed reactions, no more than four source foods were tested in one 24-hour period; none was tested closer than 4 hours. As many as 60 individual source foods were tested sequentially during the period in environmental control. Pulses were recorded 5 minutes before and at 5-, 20-, 40-, and 60-minute intervals after meals, and electrocardiograms were taken with the onset of any arrhythmia. The time of onset and duration of reaction after the challenge was noted, and all signs and symptoms were recorded. Testing of the next food was not begun until all previous reactions had subsided.

Because commercial foods are contaminated during their production and processing by synthetic sprays, herbicides, preservatives, artificial colorings and sweeteners, wax and plastic wrappings, and/or other additives, the patients were retested following identification of totally tolerated pure foods. The same foods purchased from the commercial market and cooked on gas stoves in synthetic cookware were used. Individual and accumulative reactions from one to six consecutive meals were observed and recorded.

Testing of odors was done in the following manner: each patient was given a timed exposure of 15 seconds to 60 minutes to the flames of natural gas, odor of cigarette smoke (ambient), chlorine (<0.33), perfume, pine-scented floor wash (ambient), ethyl alcohol (<0.50 ppm), formaldehyde (<0.20 ppm), phenol (<0.0020 ppm), and pesticides (<0.0034 ppm). In addition, the patient was challenged with the fumes of common odor-producing material that he or she breathed daily in the home and work environment, such as carpet, foam pillows, and polyester clothes. Each exposure was at a distance of 30 inches with a constant flow of the vapor. All odors tested were of a double-blind manner in that neither the examiner nor patient knew the content of the container until after the test. Saline was used as a control.

The following laboratory tests were performed: complete blood count by the Coulter Model S method; sodium, potassium, chloride, carbon dioxide, blood urea nitrogen, serum protein, calcium, glucose, uric acid, alkaline phosphate, serum glutamic oxaloacetic transaminase, lactic dehydrogenase, and serum calcium by the SMA method; protein electrophoresis by the Helena Cellulose Acetate method; quantitative immunoglobulins (IgE, IgG, IgA, IgM), C_3, C_4, alpha 1, antitrypsin, and C-reactive protein by the Behring Radial Immuno-Diffusion method; total hemolytic serum complements and serum complements fractions by the Hemolysis Sheep Cells method; prothrombin time, partial thromboplastin time, platelets, and Lee White clotting time by the Dade Reagents method; phosphorus by the Hycel method; fibrinogen and fibrinolysins clot lysis by the Biuret Quantitative method; fibrin split products by the Burroughs-Wellcome method; C_1 esterase inhibitor by the Behring IEP method; and B lymphocytes and T lymphocytes by the Sheep Cell Rosetting method. All tests were performed upon the patient's entrance into the room and at the beginning and end of testing. C_3 and C_4 were done daily during the period of fasting.

There were numerous associated signs and symptoms related to the environmental maladaptation syndrome occurring during the patient's lifetime (Table 20.30). Each patient averaged more than ten distinct, recurrent signs and symptoms.

Distinctive symptomatology related to the smooth-muscle systems was evident in all patients. All had long histories of gastrointestinal, respiratory, and/or genitourinary problems in addition to their cardiac problem; most had at least recent histories of vascular involvement including petechiae, spontaneous bruising, and edema.

Once under environmental control, ten patients cleared their active symptoms, including their ongoing arrhythmias, in 3 to 10 days. Frequently, symptoms were accentuated the first 2 to 3 days. Most patients underwent withdrawal the same as the patients with phlebitis[179] and vasculitis[180] described in previous reports. Each patient initially had hunger followed by complaints of "nervousness", "jitters", and headaches. There were observable signs of agitation, trembling, and depression. Insomnia or disturbed sleep was the rule the first night, which was followed by assorted symptoms such as nausea, diarrhea, wheezing, and/or headaches. Initial accentuation of arrhythmias was common. Backaches, sometimes excruciating, usually signaled the termination of the symptoms. At this time, all significant cardiac arrhythmias and associated signs and symptoms were gone. The patient's feeling of well-being was confirmed by observable signs of animation, walking, or even riding a stationary bicycle without the initiation of arrhythmias. No patient had previously been able to do this.

After testing started, it became quite obvious that extracardiac reactions fell into three categories. The first category consisted of unmistakable signs such as rhinorrhea, nasal stuffiness, hoarseness, cough, wheezes, peripheral blanching, cyanosis, swelling or loss of pulses, bruising, polyuria, fever, increased pulse rate, decreased or elevated blood pressure, petechiae, bruising and phlebitis. The second category consisted of equivocal signs, and the third category had no observed reactions. Two and three were grouped together for statistical purposes and considered as having no observed reactions.

Ten patients clearly had their arrhythmias reproduced on at least three separate occasions. Usually there was a sequential progression of minor signs or symptoms such as rhinorrhea, mild confusion, nose or mouth irritations, nausea, vomiting, or an uneasy feeling preceding the arrhythmias. Many susceptibilities became evident in each patient (Table 20.26). Also, some individual incitants were observed to produce only portions of their symptom complex, while others would produce all the patient's original signs and symptoms. These reactions were further substantiated by the benign asymptomatic course after ingestion of foods or inhalation of food odors that produced no reaction plus the reproducibility of signs by retesting of foods that did cause reactions. Of the reactions occurring within the first 4 hours, 95% started within the first 5 minutes after ingestion, leaving no doubt in the minds of the observing personnel and the patients that there was a direct cause-and-effect relationship. The severe reactions lasted up to 48 hours, with lesser

Table 20.30.   Associated Signs and Symptoms

Patient	Signs and symptoms
1	Shortness of breath, gastric contracture, loss of peripheral pulses, bruising, sinus, GI upset, head foggy, petechiae
2	Chest tight, nose itched and burned, dizzy, hoarseness
3	Leg aches, personality change, drowsy, nausea, vomiting, tingling headache, petechiae
4	GI upset
5	Malaise, mild GI upset
6	Blue hands and feet, nonpitting periorbital and peripheral edema, depression, confusion, headache, dysuria, GI upset, petechiae
7	Headache, depression
8	Sinus, vertigo, urinary frequency and urgency, bruising, cyanosis
9	Nose runs, neck tightness, periorbital edema, petechiae
10	Headache, swelling, hives
11	Headache
12	Headache, shortness of breath

Arrhythmias — spectrum

Sinus tachycardia (above 130/min)	10
Sinus bradycardia (below 45/min)	10
Sinus arrhythmia	11
Atrial fibrillation (P.A.T.)	4
Coronary sinus rhythm	12
1° A.V. block	8
Premature ventricular contraction	8
Ventricular tachycardia	2

Source: EHC-Dallas. 1979.

effects lasting up to 5 days. The moderate reactions lasted 4 to 8 hours, while the mild ones were terminated within a 4-hour period; 100% of the associated signs and symptoms were reproduced.

Testing for ingested chemicals resulted in symptoms and signs after the initial meal in eight patients; two patients took from two to six meals to produce signs, although symptoms were reproduced at least one meal earlier.

All reactions to inhaled chemicals were observed within the first 90 minutes following exposure, and their aftereffects lasted up to 48 hours, although most terminated within a 4-hour period. Nine patients reacted to the fumes of the flame of the gas pilot, reproducing assorted signs and symptoms in all patients and reproducing the cardiac signs in six. The most striking example of gas sensitivity was of a 33-year-old female who had a 15-second exposure to the natural gas pilot. She developed immediate dizziness and staggering. Premature ventricular contractions then started. These were followed by severe leg pain and bruises over both extremities. Finally, bigeminy followed. Recovery took 36 hours.

The results of double-blind exposures to the various inhaled chemicals are shown in Tables 20.31 and 20.32. One can see the widespread involvement of susceptibilities to these chemicals; however, not shown are the results of transient exposures that produced arrhythmias in all patients on at least 2 different occasions.

There was a tendency for the white blood count to be depressed following chemical exposure (Table 20.33). Immunoglobulins, generally, were within normal limits. Alterations in total serum hemolytic complements were seen in eight patients. Frequently, these changes were due to abnormalities in several of the components. C_3s were abnormally lower in nine patients. C_4s were altered in only four patients, all being slightly elevated. Bleeding studies were all normal. C-reactive proteins were positive in seven of eight positive reactors. Total eosinophil counts were generally low on admission and returned to normal with time in the controlled environment; however, two patients had initial elevation. T-cell decrease was seen in seven of seven patients measured, although one was 58%, which was near the control norm. However, the total lymphocyte count was below 1000, probably indicating an abnormality.

With our environment steadily becoming more contaminated by synthetic chemicals, it is important to recognize their influence in inducing disease processes. If one is not aware of the environment's specific effects on the individual, relentless disease occurs without accurate understanding or appropriate intervention. This failure to realize the connection between pollutant exposure and disease onset was graphically demonstrated in one patient who reacted to the 12 different inhaled chemicals used daily in her physician-employer's office. Had she known of the cumulative and specific sensitizing effects of these chemicals on her cardiovascular system, she easily could have prevented years of malaise and $2^1/_2$ years of extremely expensive suffering. Similar long-term effects have been described by other observers.[181-183] The impingement of the chemical environment is lessened markedly by the patient's long-term withdrawal through creating environmentally sound, chemically less-contaminated home environments. Susceptibilities gradually subside, and the patient is able to return to function in the outside world; however, a home oasis must always be preserved.

The ability of food to induce arrhythmias was extensive in this series (Table 20.32). Usually, foods were not perceived as inciting agents, until the patient had been in the relatively fume- and particle-free environment for several days. Apparently, there is an unmasking process derived from the food abstinence that allows the challenge reactions to become acute and definable. Failure to recognize the masking, or adaptation principle, probably explains the present lack of published observations on multiple food incitants. Also, the absence of controlled fume- and particle-free environments adequate enough

Table 20.31. Reaction to Double-Blind Exposure to Fumes of Chemicals (1 to 15-Minute Exposure Ambient Dose) in Environmental Control Unit

Patient	Saline control (3 challenges)	Petroleum alcohol (<0.5 ppm)	Phenol (<0.002 ppm)	Chlorine (<0.33 ppm)	(Mixture) pesticides (<0.134 ppm)	Pine-scented floor wash	Formaldehyde (<0.2 ppm)	Arrhythmia spectrum	
1	−	+	+	+	+	+	+	Sinus tachycardia (above 130/min)	10
2	−	+	+	+	+	+	+	Sinus bradycardia (below 45/min)	10
3	−	−	+	+	+	−	−	Sinus arrhythmia	11
4	−	−	+	+	+	+	+	Atrial fibrillation (PAT)	4
5	−	+	+	+	+	−	−	Coronary sinus rhythm	12
6	−	+	+	+	+	−	−	1° AV block	8
7	−	+	+	−	+	+	+	PVC	8
8	−	+	+	+	+	+	+	Ventricular tachycardia	2
9	−	+	+	+	−	−	−		
10	−	+	+	+	+	+	+		
11	2−, 1+	−	−	−	−	−	−		
12	−	−	−	−	−	−	−		

Source: Rea, W. J. 1978. Environmentally triggered cardiac disease. Ann. Allerg. 40(4):243251. With permission.

Table 20.32.　Incitant Challenges after 4 Days Deadaptation
with Total Load Reduced (ECH-Dallas)

Patient	Reaction to food	Δ heart	Reaction to ingested chemical	(Double-blind) inhaled chemical	Δ heart	(Double-blind) water positive reaction	Δ heart
1	5/30	5	1 meal	12/15	7	1	0
2	14/30	0	1 meal	10/15	4	1	0
3	52/68	5	2 meals	7/15	1	2	1
4	17/83	17	not tested	10/10	10	13	12
5	1/30	1	negative	0/15	0	0	0
6	96/96	10	1 meal	10/15	4	5	0
7	68/68	11	1 meal	15/15	2	13	2
8	69/90	0	1 meal	35/40	2	3	0
9	80/120	10	1 meal	20/35	6	5	1
10	12/28	12	6 meals	2/8	2	0	0
11	0/16	0	0	0/12	0	0	0
12	0/18	0	0	0/12	0	0	0

Source: Rea, W. J. 1978. Environmentally triggered cardiac disease. *Ann. Allerg.* 40(4):243–251. With permission.

to restore the patient to a nonadapted state has precluded adequate study of these complex cases.

Once the patient in our study learned which foods precipitated his or her symptoms, avoidance was undertaken voluntarily and meticulously. Certainly after atherosclerosis is ruled out in patients with arrhythmias of unknown etiology, commonly ingested foods should be suspected and accordingly tested.

Consumption of contaminated water is also a problem in patients with arrhythmias of unknown etiology, and contaminated water does trigger arrhythmias. Certainly, one can speculate as to the causative agents in water. Chlorine is incriminated, since it has been proven in challenge tests to cause sensitivities. Pesticides and numerous other chemical residues not removed at water treatment plants can cause arrhythmias. Apparently, new chemical compounds are now being formed by the interaction of some chemicals in public water supplies, and these could be as yet unidentified inciting agents.

The question arises as to why some spring waters trigger arrhythmias. According to Glaze,[184] naturally occurring phenols result from decayed humus or are acquired by absorption from water running over coal deposits that are present in some spring waters. The tachycardia and sinus arrest seen in one of the patients in the EHC-Dallas that resulted from spring water ingestion also resulted from various dilutions of phenol under the tongue. This case provides circumstantial evidence supporting this concept (see Volume II, Chapter 7[52]).

Emphasis should be placed on the degree of specificity of the chemical triggering in these patients. Initially, it appears that each patient responds nonspecifically to any incitant; yet, as one closely observes each patient, it becomes apparent that they are all generally susceptible to specific derivations of petroleum products and more specifically to one or more of the halogenated hydrocarbons. However, any given individual is more susceptible to certain incitants rather than others within this spectrum. The magnitude of response also appears to be dose related per individual. One individual cannot tolerate a transient exposure while another may require an hour's exposure to produce an adverse reaction. In a given individual, two types of responses occur. The first is a classic allergic type, having memory and specificity triggered by varying small amounts of incitants. The second response is one of an overall nonspecific depressant or irritative effect, which is cumulative and tends to lead to chronicity because it changes overall responses, both to other chemicals and foods, by masking acute responses. It adversely influences food compatibility. This cumulative chemical build-up was seen clinically in these patients, as is evidenced by the fact that long-term withdrawals of multiple incitants tended to decrease overall symptomatology with less food and chemical susceptibility while symptomatology increased as the overall chemical load increased.

As the effects of chemical build-up lessened, it became evident that intra-organ as well as multiple organ sensitization occurred. It appears that myocardial and coronary artery sensitization are only one facet of the whole spectrum of smooth muscle involvement. No patient in this series had isolated heart involvement without symptoms related to other smooth muscle organs; however, in some patients the extracardiac involvement was minimal. Also, most patients had some aspects of peripheral vasculitis, particularly petechiae, periorbital or peripheral nonpitting edema, spontaneous bruising, and/or peripheral vasospasm. The varied types of arrhythmias in this series were further evidence of the effects of chemical build-up on different areas of the heart.

Specificity of intra-organ response was apparent in all patients, since various areas in the heart in a given patient appeared to be more affected than others. One tends to think of the ANS being influenced by environmental incitants, and this certainly was partially true in the patients studied in that some of the sinus brady- and tachycardias could be triggered by environmental stimuli. It was clear that two patients had sensitization of the atrium, or imbalance of the nervous impulses leading to it, and that all cardiac effects were secondary to the patients' stimulation. In themselves, these atrial arrhythmias did not appear to be life-threatening, other than through their indirect effects of lower-organ perfusion due to the resultant brady- or tachycardia. Rapid atrial arrhythmias certainly could be life-threatening by acting as triggering agents for hypoxia which, in the presence of underlying arteriosclerosis in any of the vital organs such as kidney, heart, or brain, could be devastating. Such a reaction appeared to be the case in one patient who, when challenged with an ambient dose of the fumes of natural gas for 1 hour, developed a pulse of 200 with crushing chest pain and S-T depression on EKG. The reaction lasted for 12 hours, while the residual hangover of generalized gastrointestinal upset and cardiac irritability lasted for another 5 days. This reaction necessitated the patient being placed on oxygen during most of this period. Any time he was out of oxygen, tachyarrhythmias occurred. Regardless of whether or not these atrial arrhythmias were life-threatening, their incapacitating effects were evident in the patients. Each felt as though he or she were going to die due to the symptoms of shortness of breath, pounding heart, chest pains, and memory loss or mental confusion. Because of these recurrent symptoms, the patients' capacity to function, even marginally, was impaired. It was extremely important to remove these triggering agents in order to restore the patients to optimum function.

Patients with coronary artery spasm fell into two categories: those with only chest pain and those with EKG changes. Coronary angiograms showed spasm of the large vessels in each case with good distal runoff. Although the angiograms were not done under environmental control during a known food or chemical challenge, it is probable the patient was receiving an inadvertent chemical incitant; the angiographic rooms are a chemically contaminated environment due to their equipment.

Final word awaits angiographic studies under environmentally controlled conditions. Apparently the detrimental effects of the myocardial ischemia in the patients in this study were controlled by the basic collateral myocardial blood supply that each individual is born with and/or the presence of other myocardial disease. Again, the incapacitating effects of the transient ambient chemical incitants were present in one patient (Number 5, Table 20.26) who always required 24 to 65 hours to recover after being challenged. Since she worked around many chemicals to which she was sensitive, her heart was constantly being triggered, and she was ill most of the time.

Ventricular arrhythmias are obviously more life-threatening and are probably the result of a sensitized ventricular myocardium. Patients having these arrhythmias appeared to be in serious difficulty, since they were also refractory or sensitive to all the anti-arrhythmia agents. However, it was clear that, once they were under rigid environmental control, the triggering agents were removed, and their total load was reduced, the heart was able to recover rapidly. Although the patients were not always entirely free of arrhythmias at the end of the fasting, they were clinically asymptomatic, and their arrhythmias continued to improve the longer they were in a controlled environment. Initially, transient exposures still resulted in frightening, but always manageable, situations by withdrawal of incitants and oxygen. Over a period of time, 1 week to 1 year, all arrhythmias disappeared. It was always apparent, though, that massive exposure to the worst inciting agents for a long period of time, 1 or more days, would trigger the still sensitized, but for practical purposes, well heart. Patients in this series are contrasted to Stewart and Hake's[185] patient who survived two episodes of coronary shock only to die on the third because his environment was not controlled. He had underlying arteriosclerotic coronary disease, but received high exposure to fumes of paint remover, which apparently triggered his coronary shock. Unfortunately, in this case, the inciting agent, furniture paint remover, was not determined until after death.

The internal metabolic mechanism disruption is still not clearly defined in these patients. Once the inciting agent gets inside the body, it apparently triggers one or more segments of the basic homeostatic mechanism. In some cases the antigen antibody sequence apparently is activated which, when joined with complement, results in release of mediators that trigger the already primed smooth muscle. Other patients apparently had complement triggered directly via the alternate pathway.

Still other incitants probably directly activated the kinin system[186] resulting in smooth muscle triggering. We would expect that the one patient who had all laboratory studies near normal to have this system involved. Unfortunately, studies on this aspect have not been done as yet in cardiac patients, but the multiplicity of incitants and responses in any one patient suggest involvement of the kinin and prostaglandin system.

Entry of incitants through the fibrinolytic and coagulation mechanism is a possibility but to date has not been demonstrated in these patients. If triggering is involved via these systems, we possibly have not demonstrated changes because they occurred at a local level, and total peripheral blood studies were not altered enough to show a difference.

The presence of positive C-reactive proteins in some patients agrees with Papish's[187] findings in monocellular vasculitis. Unfortunately, due to inaccessibility of the specific area of involved heart muscle, it was impossible to do biopsies to see whether deposition occurred. In the patient with spontaneous bruising as a terminal event of a sequential action after coronary spasm, the serum calcium was depleted along with the positive CRP, which would suggest deposition on the heart and in blood vessels. Previously, deposition of C_3 and C-reactive protein in the vessel wall of those patients with spontaneous bruising has been shown to occur. Hopefully, in the near future, fluorescent studies will be available in our laboratory, enabling us to demonstrate this phenomenon and otherwise completely delineate the terminal event in this clearly defined sequential reaction. Since T-cell decrease occurred, more areas are open to speculation.

In this study, it is obvious from the multitude of incitants and laboratory abnormalities that many facets of the body's homeostatic system are not functioning properly, thus preventing the individual from being able to handle daily environmental insults. This multitude of incitants was immediately evident by the large numbers of foods and individual chemicals that any one patient could not tolerate. Though the basic internal mechanism for derangement is unclear, it is evident that the unmasking processes after chronic adaptation (or maladaptation) is the central issue in these types of patients. New techniques will have to be designed to define properly the basic underlying mechanism derangement.[186]

It is clear from our recent larger series of patients with coronary spasm that most have magnesium deficiency. Recognition of this deficiency has led us to supplement magnesium intravenously for coronary spasm and ventricular arrhythmia. We have had excellent results alleviating symptoms with this treatment.

Several other case studies emphasize the environmentally triggered arrhythmia problem.

Case study #1. A white male had a myocardial infarct at the age of 35 years and a quadruple bypass operation at the age of 37 years. This treatment did not clear his recurrent chest pain and ventricular arrhythmias, and he was refractory to medication. In addition, this patient had recurrent sinusitis, bronchitis, and gastrointestinal upsets. After 5 days in the ECU, he became symptom free, achieved sinus rhythm, and was on no medication. Following the ambient-dose bronchial challenge with formaldehyde (<0.2 ppm), the patient sequentially developed rhinorrhoea, sinus pain, coughing, chest pain, and ventricular ectopic beats (Figure 20.43).

This patient was an undertaker, as was his father. He had grown up in a bedroom above the embalming room. Formaldehyde fumes from this room regularly seeped into his bedroom. Later when he became a mortician, he was exposed to higher levels of formaldehyde. After the discovery of his formaldehyde-induced arrhythmia, he moved his home and office from the mortuary to a home free of formaldehyde. He continued to run his business from his new office, and his symptoms were relieved as a result of his withdrawal from the primary source of his formaldehyde exposures.

This man's ventricular arrhythmia was triggered by formaldehyde, which is extremely difficult to avoid. He remained well without medication during the 2 years following a treatment program that involved environmental control in his home following identification of his triggering agents. His ventricular failure also cleared after a period of time, and he was symptom free without medication for 2 years. He subsequently resumed smoking, which resulted in the reintroduction of formaldehyde, via the smoke, into his system. His heart gradually began to fail, and, eventually, he required a transplant.

Case study #2. A 71-year-old physician developed sudden attacks of passing out accompanied by slurred speech and a fluttering in the chest. Cardiac workup, including coronary angiograms, was negative. Holter monitoring showed multifocal PVCs. The patient was placed in the ECU and fasted for 5 days. His cardiogram returned to normal. Challenge tests were negative for biological inhalants. Food and chemicals triggered his arrhythmias. The patient did extremely well without medication, having no more significant arrhythmias as long as he avoided severe exposures. He lived until the age of 84 years, carrying on an active practice until he died of a metastatic carcinoma (Figure 20.44).

Case study #3. A 51-year-old white male developed rheumatic fever as a child. At the age of 47 years, he had his mitral valve replaced due to mitral insufficiency. He did well for 5 years until he developed supraventricular

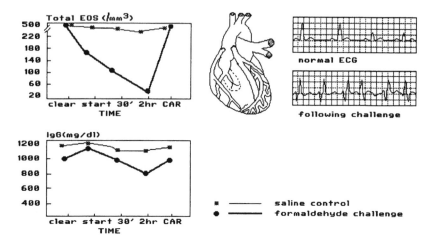

Figure 20.43.   37-year-old white male, double-blind challenge (formaldehyde inhaled, <0.2 ppm) after 4 days deadaptation with
total load decreased in the ECU. Reproduced symptoms
included recurrent PVCs, chest pain, fatigue, sinusitis,
bronchitis, GI upset, and vasculitis. (From EHC-Dallas.
1978.)

arrhythmias that appeared intractable to medical treatment. He also became
sensitive to the available beta-blockers of the time and could not tolerate
them. Therefore, a pacemaker was placed to counteract the severe bradycardia. It was felt that he would need a valve replacement because of the
arrhythmias. Surgery was again scheduled for mitral valve replacement.
However, he was first admitted to the ECU. It was found on history that his
worst episodes of arrhythmias occurred during the high mold season and
when he ate cheese. He was fasted for 5 days in the ECU, and his arrhythmias
disappeared. Challenge with baker's and brewer's yeast and a series of molds
reproduced the arrhythmias. Avoidance of yeast-containing foods and the
taking of mold and yeast injections stopped the arrhythmias and solved the
patient's problem without a new valve replacement (Table 20.34). He has
been well in 15-year follow-up having the same mitral valve that was thought
to cause the arrhythmias.

Case study #4. A 65-year-old white female had recurrent atrial fibrillation
refractory to medical management. She was placed in the ECU, and her
arrhythmias gradually subsided. She was then challenged with biological
inhalants, foods, and toxic chemicals. No inhalants, several foods, and two
toxic chemicals triggered her symptoms. A challenge with soy beans resulted
in changes in her eosinophils and gammaglobulins as well as a reproduction
of her arrhythmias (Figure 20.45).

Oral Food Challenge (Confirmed by intradermal challenge)	PVCs during first 30 min. after challenge
Apple	10
Potatoes	4
Eggs	3
Chicken	5
Beef	203
Rice	15
Peanuts	2
Corn	4

Double-blind, inhaled chemical challenge
(Confirmed by intradermal phenol <0.002 ppm)

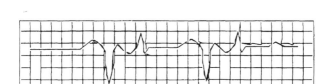

Figure 20.44.   71-year-old, white male surgeon. Premature ventricular arrhythmias reproduced by challenge after 4 days deadaptation in the ECU with the total load decreased. (From EHC-Dallas. 1978.)

Table 20.34.   Case Study — 51-Year-Old White Male

5 year P.T.A. — mitral valve replacement, rheumatic heart
Failure of cardroversion with continuous atrial fibrillation
1 year P.T.A. — onset of incapacitating supraventricular tachycardia
Stopped by Inderal® but always experienced incapacitating weakness and bradycardia
ECU — cheese, yeast → supraventricular tachycardia
Pacemaker — 1976
Only episodes (×5) with cheese and yeast overeating — responds to Inderal®
17-year follow-up — no medication, doing well; original synthetic mitral valve

Source: EHC-Dallas. 1978.

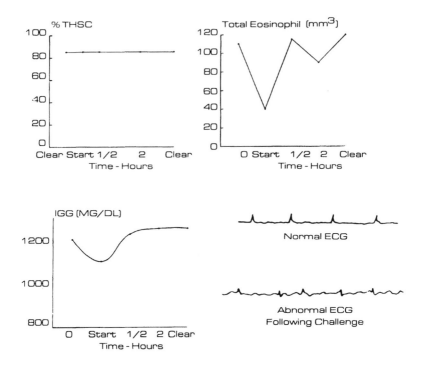

Figure 20.45.   65-year-old white female soybean challenge. Symptoms reproduced: atrial fibrillation and respiratory distress. (From EHC-Dallas. 1978.)

Hertzman et al.[188] reported a case of repeated coronary spasm in a young woman with the toxic chemical-induced eosinophil-myalgia syndrome. This patient developed myocardial damage after a high grade left anterior descending coronary occlusion. Hertzman et al. thought this was due to a toxic metabolite rather than disordered tryptophan metabolism. All medication failed, but the lesion cleared after the administration of magnesium sulfate.

Another case seen at the EHC-Dallas showed spasm triggered by codfish challenge (Figure 20.46). Collectively, these cases emphasize the variety of heart problems that one may discover when looking carefully for environmental triggers.

In summary, it is clear that the triggers of many entities involving the cardiovascular system can now be identified. They can be defined under environmentally controlled conditions and should be defined and dealt with before long-term use of medication is recommended. Once patients have been diagnosed with environmentally induced cardiovascular illness, treatment should follow the usual course. The total body load must be reduced through avoidance of pollutants. Usually, the creation of a home oasis is essential to satisfying this requirement. Also, total body load can be reduced through implementation of nutrition therapy, injection therapy, and heat depuration/physical

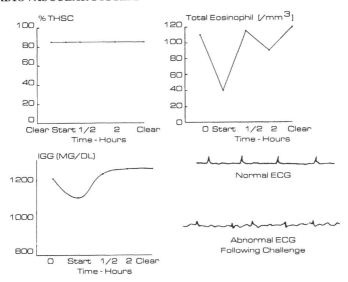

Figure 20.46. Coronary spasm reproduced by codfish challenge. (From Rea, W. J. 1978. Environmentally triggered cardiac disease. *Ann. Allerg.* 40:243–251. With permission.)

therapy. The precise use of these modalities will vary according to individual patient requirements. Volume IV will fully outline treatment options for environmentally induced illnesses and can be consulted for further instruction in devising a treatment plan for patients with cardiovascular system involvement.

REFERENCES

1. Finn, R. 1979. Food allergy. *Lancet* 2:249.
2. Fisher, S. A. 1981. Dermatitis due to the presence of formaldehyde in certain sodium lauryl sulfate (SLS) solutions. *Cutis* 27(4):360–362, 366.
3. Nour-Eldon, R. 1970. Uptake of phenol by vascular and brain tissue. *Microvasc. Res.* 2:224.
4. Rea, W. J. and M. J. Mitchell. 1982. Chemical sensitivity and the environment. *Immun. Allerg. Pract.* 4(5):21–31.
5. Spizer, F. E., D. H. Wegerman, and A. Ramires. 1975. Palpitation rate associated with fluorocarbon exposure in a hospital setting. *N. Engl. J. Med.* 272:624.
6. Taylor, G. S. and W. S. Hern. 1970. Cardiac arrhythmias to aerosol propellants. *JAMA* 219:8.
7. Rea, W. J. 1992. *Chemical Sensitivity. Vol. I. Mechanisms of Chemical Sensitivity.* 47. Boca Raton, FL: Lewis Publishers.
8. Mohrman, D. E. and E. O. Feigl. 1978. Competition between sympathetic vasoconstriction and metabolic vasodilation in the canine coronary circulation. *Circ. Res.* 42:79.
9. Feigl, E. O. 1967. Sympathetic control of coronary circulation. *Circ. Res.* 20:262.

10. Feigl, E. O. 1975. Control of myocardial oxygen tension by sympathetic coronary vasoconstriction in the dog. *Circ. Res.* 37:88.

11. DiSalvo, J., P. E. Parker, J. B. Scott, and F. L. Haddy. 1971. Carotid baroceptor influence on coronary vascular resistance in the anesthetized dog. *Am. J. Physiol.* 221:156.

12. Ely, S. W., D. C. Sawyer, D. L . Anderson, and J. B. Scott. 1981. Carotid sinus reflex vasoconstriction in right coronary circulation of dog and pig. *Am. J. Physiol.* 241:H149.

13. Feigl, E. O. 1987. The paradox of adrengic coronary vasoconstriction. *Perspective* 76(4):737–745.

14. Rea, W. J. and O. D. Brown. 1987. Cardiovascular disease in response to chemicals and foods. In *Food Allergy and Intolerance*, eds. J. Brostoff and S. J. Challacombe. 738. London: Baillière Tindall.

15. Harkavy, J. 1963. *Vascular Allergy and Its Systemic Manifestation.* Washington, D.C.: Butterworths.

16. Pratila, M. G. and V. Pratila. 1978. Anesthetic agents and cardiac electromechanical activity. *Anesthesiology* 49:338.

17. Krishna, G., M. S. Trueblood, and R. R. Paradise. 1975. The mechanism of the positive chronotropic action of diethyl ether on rat atria. *Anesthesiology* 42:312.

18. Merin, R. G., T. Kumazawa, and N. L. Luka. 1976. Myocardial function and metabolism in the conscious dog and during halothane anesthesia. *Anesthesiology* 44:402.

19. Merin, R. G., T. Kumazawa, and N. L. Luka. 1976. Enflurane depresses myocardial function, perfusion, and metabolism in the dog. *Anesthesiology* 45:501.

20. Calverley, R. K., N. T. Smith, C. Prys-Roberts, E. I. Eger, and C. W. Jones. 1978. Cardiovascular effects of enflurane anesthesia during controlled ventilation in man. *Anesth. Analg. (Cleveland)* 57(6):619–628.

21. Eger, E. I., N. T. Smith, R. K. Stoelting, D. J. Cullen, L. B. Kadis, and C. E. Whitcher. 1970. Cardiovascular effects of halothane in man. *Anesthesiology* 32(5):396–409.

22. Merin, R. G. 1969. Myocardial metabolism in the halothane depressed canine heart. *Anesthesiology* 31:20.

23. Merin, R. G. and H. H. Borgstedt. 1971. Myocardial function and metabolism in the methoxyflurane-depressed canine heart. *Anesthesiology* 34:562.

24. Merin, R. G., T. Kumazawa, and N. L. Luka. 1978. Dose dependent depression of cardiac function and metabolism by inhalation anesthetics in chronically instrumented dogs. In *Recent Advances in Studies on Cardiac Structure and Metabolism. Vol. II. Heart Function and Metabolism,* eds. T. Kobayashia, T. Sano, and N. S. Dhalla. 473. Baltimore: University Park.

25. Merin, R. G. 1976. Myocardial metabolic effects of isoflurane, Abstracts of Scientific Papers, Annu. Meet. Am. Soc. Anesthesiologists, 573.

26. Gimeno, A. L., M. F. Gimeno, and J. L. Webb. 1962. Effects of ethanol on cellular membrane potentials and contractility of isolated rat atrium. *Am. J. Physiol.* 203:194.

27. Regan, T. J., G. T. Koroxenidis, C. B. Moschos, H. A. Oldewurtel, P. H. Lehan, and H. K. Hellems. 1966. The acute metabolic and hemodynamic response of the left ventricle to ethanol. *J. Clin. Invest.* 45:270.

28. Wendt, V. E., R. Ajiluni, T. A. Bruce, A. S. Prasad, and R. J. Bing. 1966. Acute effects of alcohol on the human myocardium. *Am. J. Cardiol.* 17:804.

29. Regan, T. J., G. E. Levinson, H. A. Oldewurtel, M. J. Frank, A. B. Weisse, and C. B. Moschos. 1969. Ventricular function in noncardiacs with alcoholic fatty liver: role of ethanol in the production of cardiomyopathy. *J. Clin. Invest.* 48:397.

30. Newman, W. H. and J. F. Valicenti, Jr. 1971. Ventricular function following acute alcohol administration: a strain-gauge analysis of depressed ventricular dynamics. *Am. Heart J.* 81:61.

31. Ahmed, S. S., G. E. Levinson, and T. J. Regan. 1973. Depression of myocardial contractility with low doses of ethanol in normal man. *Circulation* 48:378.

32. Wong, M. 1973. Depression of cardiac performance by ethanol unmasked during autonomic blockage. *Am. Heart J.* 86:508.

33. Regan, T. J. 1971. Ethyl alcohol and the heart. *Circulation* 44:957.

34. Symbas, P. N., D. H. Tyras, R. E. Ware, and B. J. Baldwin. 1972. Alteration of cardiac function by hemodialysis during experimental alcohol intoxication. *Circulation* 45:2.

35. Seeman, P., M. Chau, M. Goldberg, T. Sauks, and L. Sax. 1971. The binding of Ca^{2+} to cell membrane increased by volatile anesthetics (alcohols, acetone, ether) which induce sensitization of nerve or muscle. *Biochim. Biophys. Acta* 225:185.

36. Retig, J. N., M. A. Kirchberger, E. Rubin, and A. M. Katz. 1977. Effects of ethanol on calcium transport by microsomes phosphorylated by cyclic AMP-dependent protein kinase. *Biochem. Pharmacol.* 26:393.

37. Rea, W. J. 1992. *Chemical Sensitivity. Vol. I. Mechanisms of Chemical Sensitivity.* 221. Boca Raton, FL: Lewis Publishers.

38. Balazs, T. 1981. Cardiotoxicity of adrenergic bronchodilator and vasodilating antihypertensive drugs. In *Cardiac Toxicology.* Vol. II., ed. T. Balazs. 63. Boca Raton, FL: CRC Press.

39. Balazs, T. 1981. Cardiotoxicity of adrenergic bronchodilator and vasodilating antihypertensive drugs. In *Cardiac Toxicology.* Vol. II., ed. T. Balazs. 68. Boca Raton, FL: CRC Press.

40. Weiss, L. R., T. Balazs, and S. Krop. 1976. Cardiac lesions induced by hydralazine, minoxidil and isoproterenol in rodents. *Proc. Fed. Am. Soc. Exp. Biol.* 35:534.

41. Lehr, D. 1965. The role of certain electrolytes and hormones in disseminated myocardial necrosis. In *Electrolytes and Cardiovascular Disease*, ed. E. Bajusz. 248. Basel, New York: S. Karger.

42. Lehr, D., M. Krukowski, and R. Colon. 1966. Correlation of myocardial and renal necrosis with tissue electrolyte changes. *JAMA* 197:105.

43. Lehr, D., M. Krukowski, J. Fillisti, and J. Kaplan. 1971. Close correlation between extent of cardiac fibrosis and alterations of myocardial electrolyte content. *J. Am. Geriatrics Soc.* 20:5.

44. Lehr, D., R. Chau, and J. Kaplan. 1972. Prevention of experimental myocardial necrosis by electrolyte solutions. In *Myocardiology*, eds. E. Bajusz and G. Rona. Baltimore: University Park.

45. Lehr, D. and R. Chau. 1973. Changes of the cardiac electrolyte content during development and healing of experimental myocardial infarction. In *Recent Advances in Studies on Cardiac Structure and Metabolism*, ed. N. S. Dhalla. Baltimore: University Park.

46. Lehr, D., R. Chau, and S. Irene. 1975. The possible role of magnesium loss in the pathogenesis of myocardial fibre necrosis. In *Recent Advances in Studies on Cardiac Structure and Metabolism*. Vol. VI, eds. A. Fleckenstein and G. Rona. Baltimore: University Park.

47. Lehr, D. and M. Krukowski. 1963. About the mechanism of myocardial necrosis induced by sodium phosphate and adrenal corticoid over-dosage. *Ann. N.Y. Acad. Sci.* 105:135.

48. Lehr, D. 1981. Studies on the cardiotoxicity of α-and β-adrenergic amines. In *Cardiac Toxicology*. Vol. II, ed. T. Balazs. 78. Boca Raton, FL: CRC Press.

49. Rea, W. J. 1992. *Chemical Sensitivity. Vol. I. Mechanisms of Chemical Sensitivity.* 221. Boca Raton, FL: Lewis Publishers.

50. Weiss, L. R. 1981. The cardiotoxicity of neuroleptic and tricyclic antidepressant drugs. In *Cardiac Toxicology*, Vol. II, ed. T. Balazs. 125. Boca Raton, FL: CRC Press.

51. Heman, E. H. 1981. Cardiotoxicity of antineoplastic drugs. In *Cardiac Toxicology*. Vol. II, ed. T. Balazs. 165–182. Boca Raton, FL: CRC Press.

52. Rea, W. J. 1994. *Chemical Sensitivity. Vol. II: Sources of Total Body Load.* 535. Boca Raton, FL: Lewis Publishers.

53. Wuthrich, B. and L. Fabio. 1981. Acetylsalicylic acid and food intolerance in urticaria, bronchial asthma, and rhinopathy. *Schweiz. Med. Wochenschr.* 1(39):1445–1450.

54. Bunter, R. G., J. H. Carroll, and J. C. Randolph. 1980. Water in the urban environment: real estate lakes. U.S. Dept. of Interior/Geological Survey, 11–19. 49. *Pest. Monit. J.* 14(3):102–107.

55. Iowa Dept. of Air, Water, and Waste Management, Iowa City, Mar. 1986. Little Sioux River synthetic organic compound municipal well sampling survey.

56. Rea, W. J., J. R. Butler, J. L. Laseter, and I. R. DeLeon. 1984. Pesticides and brain-function changes in a controlled environment. *Clin. Ecol.* 2(3):145–150.

57. Gabler, R. et al. 1988. *Is Your Water Safe to Drink?* Mount Vernon: Consumer's Union.

58. National Research Council. 1977. Water hardness and health. In *Drinking Water and Health*. 439–447. Washington, D.C.: National Academy of Sciences.

59. Yevick, P. 1975. Oil pollutants in marine life, Eighth Advanced Seminar, Society of Clinical Ecology, Dallas, TX. Instatape, Tape II.

60. Laseter, J. L., I. R. DeLeon, W. J. Rea, and J. R. Butler. 1983. Chlorinated hydrocarbon pesticides in environmentally sensitive patients. *Clin. Ecol.* 2(1):3–12.

61. Spalding, R. F., G. A. Junk, and J. J. Richard. 1980. Water: pesticides in ground water beneath irrigated farmland in Nebraska. *Pest. Monit. J.* 1(2):70–73.

62. Fourth International Symposium on Magnesium, and American College of Nutrition 26th Annual Meeting, Blacksburg, VA. 1985. 4(3):303–404.

63. Environmental Protection Agency (EPA). 1977. *Is Your Drinking Water Safe?* Washington, D.C.: U.S. EPA Office of Public Affairs.

64. Hare, F. 1905. *The Food Factor in Disease*. 203. London: Longmans.

65. Harkavy, J. 1963. *Vascular Allergy and Its Systemic Manifestations*. Washington, D.C.: Butterworths.

66. Rea, W. J. 1977. Environmentally triggered small vessel vasculitis. *Ann. Allergy* 38:245–251.

67. Rea, W. J., D. W. Peters, R. E. Smiley, R. Edgar, M. Greenberg, and E. Fenyves. 1981. Recurrent environmentally triggered thrombophlebitis: a five-year follow-up. *Ann. Allergy* 47:338–344.

68. Theorell, H., M. Blombock, and C. Kockum. 1976. Demonstration of reactivity to airborne and food antigen in cutaneous vasculitis by variation in fibrino peptide and others, blood coagulation, fibrinolysis, and complement parameters. *Thrombo. Haemo. Sts. (Stattz)* 36:593.

69. Rea, W. J. 1992. *Chemical Sensitivity. Vol. I. Mechanisms of Chemical Sensitivity*. 418. Boca Raton, FL: Lewis Publishers.

70. Zeek, P. M. 1953. Periarteritis nodosa and other forms of necrotizing angiitis. *N. Engl. J. Med.* 248:764.

71. Department of Consumer Affairs. Feb. 1982. The pollutants. In *Clean Up Your Room: A Compendium on Indoor Pollution*. San Francisco: Consumer Affairs.

72. Rea, W. J. 1984. Elimination of oral food challenge reaction by injection of food extracts: a double-blind evaluation. *Arch. Otolaryngol.* 110:248–252.

73. Morin, Y. and P. Daniel. 1967. Quebec beer-drinkers' cardiomyopathy: etiological considerations. *Can. Med. Assoc. J.* 97:926.

74. Kesteloot, H., J. Roelandt, J. Willems, J. H. Claes, and J. V. Joosens. 1968. An enquiry into the role of cobalt in the heart disease of chronic beer-drinkers. *Circulation* 37:854.

75. Sullivan, J., M. Parker, and S. B. Carson. 1968. Tissue cobalt content in beer-drinkers' cardiomyopathy. *J. Lab. Clin. Med.* 71:893.

76. McDermott, P. H., R. L. Delaney, J. D. Eagen, and J. F. Sullivan. 1966. Myocardosis and cardiac failure in men. *JAMA* 198:253.

77. Dingle, J. T., J. C. Heath, M. Webb, and M. Daniel. 1962. The biological action of cobalt and other metals. *Biochem. Biophys. Acta* 5:32.

78. Rea, W. J. 1994. *Chemical Sensitivity. Vol. II. Sources of Total Body Load*. 579. Boca Raton, FL: Lewis Publishers.

79. Nurse, D. D. 1980. Nickel sensitivity induced by skin clips. *Contact Dermatitis* 6(7):497.

80. Norseth, T. 1984. Clinical effects of nickel. In *Nickel in the Human Environment*. 395. Proceedings of a Joint Symposium held at IARC. Lyon, France. March 8–11, 1983, eds. F. W. Sunderman, Jr., A. Aitio, A. Berlin, C. Bishop, E. Buringh, W. Davis, M. Gounar, P. C. Jacquignon, E. Mastromatteo, J. P. Rigaut, C. Rosenfeld, R. Saracci, and A. Sors. London: Oxford University Press.

81. Palme, W. 1975. EKG-Veranderungen bei exogen gasformig intoxikationen, *Z. Gesampte Hyg. Ihre Grenzgeb.* 21:639.

82. Rea, W. J. 1994. *Chemical Sensitivity. Vol. II. Sources of Total Body Load*. 713. Boca Raton, FL: Lewis Publishers.

83. Rea, W. J. 1994. *Chemical Sensitivity. Vol. II. Sources of Total Body Load*. 765. Boca Raton, FL: Lewis Publishers.

84. Magos, L. and J. A. E. Jarvis. 1970. The effects of carbon disulphide exposure on brain catecholamines in rats. *Br. J. Pharmacol.* 39:26.

85. Jarvis, J. A. E. and L. Magos. 1970. Effects of diethyldithiocarbamate and carbon disulphide on brain tyrosine. *J. Pharm. Pharmacol.* 22:936.

86. McKenna, M. J. and V. DiStefano. 1977. Carbon disulfide. II. A proposed mechanism for the action of carbon disulfide on dopamine β-hydroxylase. *J. Pharmacol. Exp. Ther.* 202:253.

87. Tiller, J. R., R. S. F. Schilling, and J. N. Morris. 1968. Occupational toxic factor in mortality from coronary heart disease. *Br. Med. J.* 4:407.

88. Hernberg, S., T. Partanen, C. H. Nordman, and P. Sumari. 1970. Coronary heart disease among workers exposed to carbon disulphide. *Br. J. Ind. Med.* 27:313.

89. Tolonen, M., S. Hernberg, M. Nurminene, and K. Tiitola. 1975. A follow-up study of coronary heart disease in viscose rayon workers exposed to carbon disulphide. *Br. J. Ind. Med.* 32:1.

90. Nurminen, M. 1976. Survival experience of a cohort of carbon disulphide exposed workers from an eight-year prospective follow-up period. *Int. J. Epidemiol.* 5:179.

91. Tolonen, M. S. Hernberg, C. H. Nordman, S. Goto, K. Sugimoto, and T. Baba. 1976. Angina pectoris, electrocardiographic findings and blood pressure in Finnish and Japanese workers exposed to carbon disulfide. *Int. Arch. Occup. Environ. Health* 37:249.

92. Tolonen, M. 1975. Vascular effects of carbon disulfide. A review. *Scand. J. Work Environ. Health* 1:63.

93. Tolonen, M. 1974. Subclinical carbon disulfide poisoning. *Work Environ. Health* 11:154.

94. Sugimoto, K., S. Goto, H. Taniguchi, T. Baba, C. Raitta, M. Tolonen, and S. Hernberg. 1977. Ocular foundus photography of workers exposed to carbon disulfide — a comparative epidemiological study between Japan and Finland. *Int. Arch. Occup. Environ. Health* 39:97.

95. Astrup, P. 1972. Some physiological and pathological effects of moderate carbon monoxide exposure. *Br. Med. J.* 4:447.

96. Price, H. L. and R. D. Dripss. 1965. General anesthetics. In *The Pharmacological Basis of Therapeutics*, 3rd ed., eds. L. S. Goodman and S. Gilman. 71. New York: MacMillan.

97. Teleky, L. 1955. *Gewerbliche Vergitungen*. Berlin: Springer-Verlag.

98. Hoschek, R. 1962. Plotzliche spattodesfalle nach geringfuger trichlorathylen-Einwirkung. *Int. Arch. Gewerbepathol. Gewerbehyg.* 19:319.

99. Kleinfeld, M. and I. R. Tabershaw. 1954. Trichloroethylene toxicity. Report of five fatal case. *Arch. Ind. Hyg. Occup. Med.* 10:134.

100. Schollmeyer, W. 1960. Plotzlicher tod durch trichlorathylen-vergiftung bei einwirkung dieses figtes uber langere zeit. *Arch. Toxicol.* 18:229.

101. Magos, L. 1981. The effects of industrial chemicals on the heart. In *Cardiac Toxicology*. Vol. II, ed. T. Balazs. 208. Boca Raton, FL: CRC Press.

102. Barnes, D. W. and A. E. Munson. 1978. Cadmium-induced suppression of cellular immunity in mice. *Toxicol. Appl. Pharmacol.* 45(1):350.

103. Magos, L. 1981. The effects of industrial chemicals on the heart. In *Cardiac Toxicology*. Vol. II, ed. T. Balazs. 208–209. Boca Raton, FL: CRC Press.

104. Gabliks, J., E. M. Askari, and N. Yolen. 1975. DDT and immunological responses. I. Serum antibodies and anaphylactic shock in guinea pigs. *Arch. Environ. Health* 26(6):305–308.

105. Gabliks, J., T. Al-Tubaidy, and E. Askari. 1975. DDT and immunological responses. III. Reduced anaphylaxis and mast cell population in rats fed DDT. *Arch. Environ. Health* 30(2):81–84.

106. Stokingert, H. E. 1965. Ozone toxicology: a review of research and industrial experience: 1954–1964. *Arch. Environ. Health* 10:719–731.

107. Rea, W. J. 1994. *Chemical Sensitivity. Vol. II. Sources of Total Body Load.* 1011. Boca Raton, FL: Lewis Publishers.

108. Nour-Eldon, R. 1970. Uptake of phenol by vascular and brain tissue. *Microvasc. Res.* 2:224.

109. Rea, W. J. 1992. *Chemical Sensitivity. Vol. I. Mechanisms of Chemical Sensitivity.* 155. Boca Raton, FL: Lewis Publishers.

110. Rea, W. J. and C. W. Suits. 1980. Cardiovascular disease triggered by foods and chemicals. In *Food Allergy: New Perspectives*, ed. J. W. Gerrard. Springfield, IL: Charles C Thomas.

111. Freed, D. J. L., C. H. Buckley, Y. Tsivion, N. Sharon, and D. H. Katz. 1983. Nonallergenic haemolysis in grass pollens and housedust mites. *Allergy* 38:477–486.

112. Gaworski, C. L. and R. P. Sharma. 1978. The effects of heavy metals on (3H) thymidine uptake in lymphocytes. *Toxicol. Appl. Pharmacol.* 46(2):305–313.

113. Romaquera, C. and F. Grimalt. 1980. Sensitization to benzoyl peroxide, retinoic acid, and carbon tetrachloride. *Contact Dermatitis* 6(6):422.

114. Lelbach, W. K. and H. J. Marsteller. 1981. Vinyl chloride associated disease. *Erg. Inn Med. Kinderheilkd.* 47:1–100.

115. Winslow, S. G. 1981. *The Effects of Environmental Chemicals on the Immune System: A Selected Bibliography with Abstracts*, 1–36. Oak Ridge, TN: Toxicology Information Response Center, Oak Ridge National Laboratory.

116. Kinzel, A. 1982. Adenosintriphosphat, die Energie-währung des Lebens. *Biol. Uns. Zeit* 22:48.

117. Chen, B. 1992. Personal communication.

118. von Ardenne, M. 1990. *Oxygen Multistep Therapy: Physiological and Technical Foundations* (trans. by P. Kirby and W. Krüger). New York: Georg Thieme Verlag Stuttgart.

119. Hauss, W. H. 1970. Rolle der Mesenchymzellen in der Pathogenese der Arteriosklerose. *Doc. Angiol.* 2:11.

120. von Ardenne, M. 1990. *Oxygen Multistep Therapy: Physiological and Technical Foundations* (trans. by P. Kirby and W. Krüger). 6–7. New York: Georg Thieme Verlag Stuttgart.

121. von Ardenne, M. 1990. *Oxygen Multistep Therapy: Physiological and Technical Foundations* (trans. by P. Kirby and W. Krüger). 148. New York: Georg Thieme Verlag Stuttgart.

122. von Ardenne, M. 1990. *Oxygen Multistep Therapy: Physiological and Technical Foundations* (trans. by P. Kirby and W. Krüger). 40. New York: Georg Thieme Verlag Stuttgart.

123. Hoelzl-Wallach, D. F. and W. Kreutz. 1976. Discussion of a paper by M. von Ardenne on a colloquium at the Institut für Biophysik und Strahlenbiologie der Universität Freiburg/Br. 14:9.

124. Schmid-Schönbein, H. 1974. Einführung in die Hämorheologie. Fließeigenschaften, Fließbedingungen und Methodik. *Phlebol. u. Proctol.* 2:205.

125. Schmid-Schönbein, H., E. Volger, J. Weiss, and M. Brandhuber. 1975. Effect of O-(ß-hydroxyethyl)-rutosides on the microrheology of human blood under defined flow conditions. *Vasa* 4(3):263–270.

126. Vaupel, P., G. Thews, and P. Wendling. 1976. Kritische Sauerstoff- und Glukoseversorgung maligner Tumoren. *Dtsch. Med. Wochenschr.* 101:1810.

127. Chien, S. 1975. Biophysical behaviour of red cells in suspension. In *The Red Blood Cell*, Vol. II, ed. D. M. Surgenor. New York: Academic Press.

128. Urbaschek, B., H. Fritsch, and J. E. Richter. 1969. Erste Beobachtungen mit dem Rasterelektronenmikroskop in der Frühphase der Endotoxinwirkung an der terminahen. *Strombahn. Klin. Wochenschr.* 47:1166.

129. Schmid-Schönbein, G. W., S. Usami, R. Skalak, and S. Chien. 1980. The interaction of leucocytes and erythrocytes in capillary and postcapillary vessels. *Microvasc. Res.* 19:45.

130. Burton, R. and H. A. Krebs. 1953. The free-energy changes associated with the individual steps of the tricarboxylic acid cycle, glycolysis and alcohol fermentation and with hydrolysis of the pyrophosphate groups of adenosintriphosphate. *Biochem. J.* 54:94.

131. Rapoport, S. M. 1975. *Medizinische Biochemie*, 6th ed. 386. Berlin: Volk und Gesundheit.

132. Hecht, A. 1970. *Einführung in Experimentelle Grundlagen Moderner Herzmuskelpathologie*, 40. Fischer, Jena.

133. Opie, L. H. 1971/72. Substrate utilisation and glycolysis in the heart. *Cardiology* 56:2.

134. von Ardenne, M. 1990. *Oxygen Multistep Therapy: Physiological and Technical Foundations* (trans. by P. Kirby and W. Krüger). 100–109. New York: Georg Thieme Verlag Stuttgart.

135. Hauss, W. H. 1973. Die Rolle des Mesenchyms in der Genese der Arteriosklerose. *Virchows Arch. Abt.* A 359:135.

136. von Ardenne, M. 1990. *Oxygen Multistep Therapy: Physiological and Technical Foundations* (trans. by P. Kirby and W. Krüger). 112. New York: Georg Thieme Verlag Stuttgart.

137. Porter, K. R. and M. A. Bonneville. 1965. *Einführung in die Feinstrutur von Zellen und Geweben*. Berlin: Springer.

138. Rea, W. J. 1994. *Chemical Sensitivity. Vol. II. Sources of Total Body Load*. 837. Boca Raton, FL: Lewis Publishers.

139. Couch, J. R. and J. Wortman. 1984. Anovulatory states as a factor in occurrence of migraine. Paper presented at the Migraine Trust, Fifth International Symposium, London, England.

140. Debled, G. 1989. *L'Andropause: Cause Consequences et Remedes*. Paris: Maloine.

141. Rea, W. J., I. R. Bell, and R. E. Smiley. 1975. Environmentally triggered large vessel vasculitis. In *Allergy: Immunology and Medical Treatment*, eds. F. Johnson and J. T. Spence. 185–198. Chicago: Symposium Specialist.

142. Grant, E. C. 1968. Oral contraceptives, smoking, migraine, and food allergy. *Lancet* 2:581–589.

143. Dickey, J. W., Jr. 1979. Drifting hematomas. *Surg. Gynecol. Obstet.* 148:209.

144. Ishikawa, S., M. Miyata, T. Fukuda, and H. Suyama. 1982. Etiological consideration of Behcet's disease. In *Behcet's Disease: Pathogenetic Mechanism and Clinical Future*, ed. G. Inaba. 25. Tokyo: University of Tokyo Press.

145. Rea, W. J., I. R. Bell, and R. E. Smiley. 1975. Large vessel vasculitis. In *Allergy: Immunology and Medical Treatment*, eds. F. Johnson and J. T. Spence. Chicago: Symposia Specialist.

146. Randolph, T. G. 1962. *Human Ecology and Susceptibility to the Chemical Environment*. Springfield: Charles C Thomas.

147. Rea, W. J. 1976. Environmentally triggered phlebitis. *Ann. Allergy* 37:101.

148. Rea, W. J. 1977. Environmentally triggered small vessel vasculitis. *Ann. Allergy* 38:245.

149. Rea, W. J. 1978. Environmentally triggered cardiac disease. *Ann. Allergy* 40:243.

150. Rea, W. J. 1992. Unpublished data.

151. Rea, W. J. and C. W. Suits. 1980. Cardiovascular disease triggered by foods and chemicals. In *Food Allergy: New Perspectives*, ed. J. W. Gerrard. 99–143. Springfield, IL: Charles C Thomas.

152. McCarron, D. A., C. D. Morris, and C. Cole. 1982. Dietary calcium in human hypertension. *Science* 217:267.

153. McCarron, D. A. 1982. Low serum concentrations of ionized calcium in patients with hypertension. *N. Engl. J. Med.* 397:226.

154. Davidson, M. and M. Feinleib. 1972. Carbon disulfide poisoning: a review. *Am. Heart J.* 83:100.

155. Rea, W. J. 1994. *Chemical Sensitivity. Vol. II. Sources of Total Body Load.* 713. Boca Raton, FL: Lewis Publishers.

156. Jain, V. K. and R. K. Chandra. 1987. Food allergy in adults. In *Handbook of Food Allergies*, ed. J. C. Breneman. 72. New York: Marcel Dekker.

157. Egan, B. M. 1989. Neurogenic mechanisms initiating essential hypertension. *Am. J. Hypertens.* 2(12 Pt 2):357S-362S.

158. Guyton, A. C. 1981. *Textbook of Medical Physiology*. 6th ed. 270. Philadelphia: W. B. Saunders.

159. Jefferies, W. M. 1981. *Safe Uses of Cortisone*. 30. Springfield, IL: Charles C Thomas.

160. Finn, R., A. G. Fennerty, and R. Ahmad. 1980. Hydrocarbon exposure and glomerulonephritis. *Clin. Nephrol.* 14(4):173–175.

161. Rea, W. J. 1994. *Chemical Sensitivity. Vol. II. Sources of Total Body Load.* 579. Boca Raton, FL: Lewis Publishers.

162. Fink, J. N. and A. Fasuci. 1982. Immunologic aspects of cardiovascular disease. *JAMA* 248(20):2716–2721.

163. Katz, P. 1982. Hypersensitivity vasculitis. *AFP* 26(1):171–175.

164. Svedhem, A., K. Alestis, and M. Jertborn. 1980. Phlebitis induced by parenteral treatment with fluxoxacillin and doxacillin: a double-blind study. *Antimicrob. Agents Chemother.* 18:349–352.

165. Reid, J., S. Holt, E. Housley, and D. J. C. Sneddon. 1980. Raynaud's phenomenon induced by sulphasalzine. *Postgrad. Med.* 56:106–107.

166. Cluff, L. E. 1970. Serum sickness and related disorders. In *Harrison's Principles of Internal Medicine*, eds. M. M. Wintrobe, G. W. Thorn, R. D. Adams, I. L. Bennett, Jr., E. Braunwald, K. J. Isselbacher, and R. G. Petersdorf. 374–376. New York: McGraw-Hill.

167. Tumulty, P. A. 1970. Systemic lupus erythematosus. In *Harrison's Principles of Internal Medicine*, eds. M. M. Wintrobe, G. W. Thorn, R. D. Adams, I. L. Bennett, Jr., E. Braunwald, K. J. Isselbacher, and R. G. Petersdorf. 1962–1967. New York: McGraw-Hill.

168. Kroker, G. F., R. M. Stroud, R. Marshall, T. Bullock, F. M. Carroll, M. Greenberg, T. G. Randolph, W. J. Rea, and R. E. Smiley. 1984. Fasting and rheumatoid arthritis: a multi-center study. *Clin. Ecol.* 2(3):137–144.

169. Parish, W. R. 1971. Studies on vasculitis, immunoglobulins, β1C, C-reactive proteins and bacterial antigens in cutaneous vasculitis lesions. *Clin. Allergy* 1:97–110.

170. Rea, W. J. and O. D. Brown. 1987. Cardiovascular disease in response to chemicals and foods. In *Food Allergy and Intolerance*, eds. J. Brostoff and S. J. Challacombe. 744. London: Baillière Tindall.

171. Connors, L. 1920. *Nelson's Loose Leaf Medicine*. Vol. 4. Philadelphia: W. B. Saunders.

172. Harkavy, J. 1963. *Vascular Allergy and Its Systemic Manifestations*. Washington, D.C.: Butterworths.

173. Rea, W. J. 1980. Review of cardiovascular disease in allergy. In *Bi-Annual Review of Allergy*, ed. C. A. Frazier. 282–347. Springfield, IL: Charles C Thomas.

174. Spizer, F. E., D. H. Wegerman, and A. Ramires. 1975. Palpitation rate associated with fluorocarbon exposure in a hospital setting. *N. Engl. J. Med.* 272:624.

175. Taylor, G. S. and W. S. Hern. 1970. Cardiac arrhythmias due to aerosol propellents. *J. Am. Med. Assoc.* 219:8.

176. Seyal, A. R. 1984. Influences of synthetic garments on cardiac rhythm. *Indian Heart J.* 36(1):39–43.

177. Rea, W. J. 1978. Environmentally triggered cardiac disease. *Ann. Allergy* 40(4):243–251.

178. Randolph, T. G. 1962. *Human Ecology and Susceptibility to the Chemical Environment*. Springfield, IL: Charles C Thomas.

179. Rea, W. J. 1976. Environmentally triggered phlebitis. *Ann. Allergy* 37:101.

180. Rea, W. J. 1977. Environmentally triggered large vessel vasculitis. *Ann. Allergy* 38:245–251.

181. Morgan, J. 1977. Personal communication.

182. Randolph, T. G. 1977. Personal communication.

183. Sandberg, D. 1977. Personal communication.

184. Glaze, R. 1976. Personal communication.

185. Stewart, R. D. and C. L. Hake. 1976. Pain remover hazard. *JAMA* 235:288.

186. Bell, I. R. 1985. A kinin model of mediation for food and chemical sensitivity: biobehavioral implications. *Annu. Allergy* 35(4):206–215.

187. Papish, W. E. 1971. Studies on vasculitis, immunoglobulins, BIC, C-reactive proteins and bacterial antigens in cutaneous vasculitis lesions. *Clin. Allergy* 1:92.

188. Hertzman, P. A., G. L. Meddoux, E. M. Sternberg, M. P. Heyes, I. N. Mefford, G. M. Kephart, and G. J. Gleich. 1992. Repeated coronary artery spasm in a young woman with the eosinophilia–myalgia syndrome. *JAMA* 267(21):2932–2934.

189. Schofield, A. T. 1908. A case of egg poisoning. *Lancet* 1:716.

190. Rowe, A. H. 1931. *Food Allergy: In Manifestations, Diagnosis, and Treatment.* Philadelphia: Lea and Febiger.

191. Levi, R. 1979. Is cardiac anaphylaxis a cause of sudden death? January 1979 presentation by the American Heart Association. Anaheim, CA.

192. Balazs, T. 1981. *Cardiac Toxicology.* Vols. I & II. Boca Raton, FL: CRC Press.

193. Fregert, S. 1980. Irritant dermatitis from phenol-formaldehyde resin powder. *Contact Dermatitis* 6(7):493.

194. Suhonen, R. 1980. Contact allergy to dodeayl-di-(amino-ethyl)glyane (Desimex i). *Contact Dermatitis* 6(4):290–291.

21   Gastrointestinal System

INTRODUCTION

Understanding of the environmental influences on the gastrointestinal (GI) tract and various disease processes involving it is rapidly expanding. Environmentally triggered responses of the smooth muscle, the mucosa, and the gut can be local, regional, or systemic. Although multiple symptoms may occur simultaneously, the impact of pollutants on various regions and functions of the GI tract will be discussed separately.

The GI tract has as many nerve cells as the spinal cord and as many endocrine cells as the endocrine system. It also has a large volume of absorptive and secretive cells. This anatomical make-up yields a very complex milieu for receiving environmental triggers and thus allows for complex responses, as are often seen in chemically sensitive individuals.

GASTROINTESTINAL TRACT

The GI tract is quite diverse, stretching from the mouth to the anus. Reactions to pollutant exposure can occur in isolated areas such as the mouth or stomach or in any of the intestinal organs. However, as sensitivity advances, the responses to pollutant excess become much more global with some local, some regional, and a generalized response all occurring within a finite time frame.

Pathophysiology of Pollutant Injury

The pathophysiology of pollutant injury of the GI tract has been studied in a fragmented manner, and the aware physician must fit the clinical situation to the known basic scientific facts. From observing over 20,000 chemically sensitive patients under various degrees of environmental control at the EHC-Dallas, we sense that the gastrointestinal tract is one of the prime conduits for the propagation of chemical sensitivity. The effects of pollutants on smooth muscle, nervous tissue, local mucosa, the microflora, and the neuroendocrine cells will be covered separately in the next section.

Pollutant Effects on the Smooth Muscle

Smooth muscle in different parts of the body has many different characteristics of contraction and, thus, is easily influenced by multiple environmental stimuli. For example, the multi-unit smooth muscle of the large blood vessels contracts mainly in response to nerve impulses, while the large-sized smaller blood vessels, the ureter, the bile ducts, and other glandular ducts have a self-excitatory process controlled by local factors and hormones that trigger rhythmic contractions. Any, or all, of these factors can be influenced by environmental stimuli, and as a result contractile problems similar to those seen in most chemically sensitive patients occur.

Smooth muscle has the ability to maintain a mild steady state, which gives it tone and which may be disturbed in some chemically sensitive individuals. This steady state with mild contractions can be the result of a summation of individual contractile pulses, each initiated by a separate action potential. Mild symptoms of mild spasmotic contractions to symptoms of tetanic contractions produced in either skeletal muscle or by prolonged direct smooth muscle contraction without action potentials are usually caused by local tissue factors or hormones. For instance, prolonged tonic contractions of the gut wall blood vessels, as seen in some chemically sensitive patients, can be caused by secretion of angiotensin, vasopressin, or norepinephrine. In some chemically sensitive individuals, there appears to be an imbalance of secretion of these hormones. This imbalance results in both an abnormal gut wall and vessel tone, which, in turn, produce much bloating and cramping. As a result of this malfunction, this subset of chemically sensitive patients frequently presents with prolonged gut wall contraction or dilation, which can then cause severe constipation or, in some cases, diarrhea.

Nervous signals and smooth muscle membrane action potentials initiate smooth muscle contractions. However, the direct action of the stimulatory factors — local tissue effluents and hormones — on the smooth muscle contractile machinery can also initiate smooth muscle contractions, probably accounting for half of all that occur. Pollutants can alter muscle contractions by changing the environment in which any, or all, of these initiators occur. Therefore, each will be discussed separately.

The myoneural junction in the gut is similar to that of skeletal muscle, except that the muscle forms a syncytium so that, when the nerves fire, the whole tubular unit contracts or relaxes. This firing may have many effects on contraction, one of which is multiple contractions following pollutant stimulation. This state may exist after organophosphate pesticide exposure, when less action potential triggering filaments and a much slower contraction and relaxation occur. Also, 1/500th as much energy is required to sustain the same tension of contraction in smooth muscle as skeletal muscle, thus accounting for the ease with which severe abdominal cramps occur following exposure to small amounts of pollutants, as is seen in chemically sensitive individuals. Membrane and action potentials in the gut are similar to those of skeletal muscle, but quantitative and qualitative differences exist. Pollutant injury due to exposure to organophosphate insecticides and nerve gases, i.e., parathion, malathion, etc., may produce insults to membrane and action potentials in the gut that are similar to those seen in the skeletal muscle. However, the end-organ response will vary, resulting in diverse symptoms such as gas, bloating, cramps, diarrhea, or constipation.

In the gut, the depolarizing process during the action potential of nerve fibers is caused almost entirely by rapid influx of sodium and calcium ions with calcium seemingly being the most important. As shown in Volume I, Chapters 4[2] and 6,[1] ion flux can be severely altered by pollutant overload in the membrane pumps, causing malfunction as seen in the chemically sensitive population. The source of calcium differs in smooth muscle from that of skeletal muscle because the sarcoplasmic reticulum is poorly developed in smooth muscle. Much of the calcium in smooth muscle enters from the extracellular fluid. Because of the time required for the diffusion of calcium, there is a latent period of 200 to 300 milliseconds in the gut, which is 50 times greater than the latent period in the skeletal muscle. To cause relaxation of smooth muscle contractile elements, it is necessary to remove calcium by the calcium pump. If this pump is damaged by pollutants, as happens in many chemically sensitive patients, prolonged contractions in some areas of the gut and relaxation in others occur with resultant cramps, constipation, and bloating. For processing magnesium, a similar mechanism appears to be in place in the chemically sensitive individual. This mechanism prevents absorption and causes cramps. Thus, many chemically sensitive individuals respond well to magnesium supplementation. Local factors that influence gut wall contraction and responses of its blood vessels include local hypoxia, excess carbon dioxides, and increased hydrogen ion concentration, all of which cause vasodilatation. Increased lactic acid, diminished calcium ion concentration, and decreased body temperature also will cause local vasodilitation. Chemically sensitive individuals tend to maintain a low body temperature and be frequently acidotic. Therefore, their gut and blood vessels are often in a state of deregulation, causing muscle dysfunction.

Circulating hormones such as norepinephrine, epinephrine, acetylcholine, angiotensin, vasopressin, oxytocin, serotonin, and histamines may effect smooth muscle contraction. A hormone will cause contraction of the smooth muscle when the muscle cells contain an excitatory receptor and relaxation when they contain an inhibitory receptor. Norepinephrine, vasopression, and angiotensin can cause severe contraction for hours. If a pollutant triggers, or is similar in structure to, norepinephrine or other neurotransmitters, severe excitation or inhibition of the gut results in symptoms of cramps or bloating, as are often seen in chemically sensitive individuals.

Pollutant Effects on the Autonomic Nervous System

As is typical of pollutant damage to other organs, disturbance of the autonomic nervous system (ANS) of the gastrointestinal tract is one of the first events to occur with exposure. In the GI system, the initial response to ingestion or inhalation of chemically contaminated air, food, and water appears in the mouth, nose, pharynx, and esophagus. Dysfunction of the esophageal autonomic nerves accounts for a large portion of the esophageal spasm seen in the chemically sensitive population. Upon entering the blood, pollutants may initially stimulate the muscle, or they may directly stimulate local autonomic nerves, or they may ascend the olfactory nerves to the hypothalamus with impulses passing through the autonomic nerves to the specific gastrointestinal end-organ, finally causing deregulation in the chemically sensitive person.

From the esophagus to the rectum, the intrinsic innervation is affected by pollutants through the enteric plexuses and mucosal blood supply. The plexi are composed of numerous groups of ganglion cells interconnected by networks of fibers that lie between the layers of the muscularis (Auberbach's plexus) and the submucosa (Miesner's plexus). The enteric plexuses contain postganglionic sympathetic and pre- and postganglionic parasympathetic fibers, afferent fibers, and the intrinsic ganglion cells and their processes. Vagal preganglionic fibers form synapses with the ganglion cells whose axons are the postganglionic parasympathetic fibers. The sympathetic preganglionic fibers are already relayed in paravertebral or prevertebral ganglion, so the sympathetic fibers in the plexuses are postganglionic and pass through them to their terminations. The afferent fibers from the esophagus, stomach, and duodenum are carried to the brainstem and cord through vagal and sympathetic nerves supplying these cells in the enteric plexus (Figure 21.1) and, thus, may be frequently triggered by pollutants presenting to the GI mucosa. These plexi are particularly prone to pollutant injury due to their location.

The major portion of neural regulation of visceral functions is carried out through reflex mechanisms. With their central connections in the spinal cord and brainstem, these mechanisms produce a rapid response to pollutant exposure that results in rapid onset of local or regional symptoms. The simplest autonomic reflex consists of an afferent conductor, which is either a visceral or somatic connecting mechanism located mainly in the efferent limb. When local intrinsic gut pollutants trigger this reflex, local contractions or dilitation may occur, causing symptoms in chemically sensitive individuals. Another mechanism of response to pollutant exposure is through the spinal cord up to the medulla oblongata of the brainstem. Here pollutant-stimulated impulses activate various centers such as the vasomotor, vomiting or nausea center, the respiratory center, or the carbohydrate metabolism center. The impulses then proceed to the hypothalamus and down the autonomic nerves to the viscera. Alteration in the function of these reflexes by pollutant overload may result in varied symptoms, including both distal and local reactions in the gut. Raynaud's phenomena, bloating, abdominal pain, abdominal angina, increased appetite,

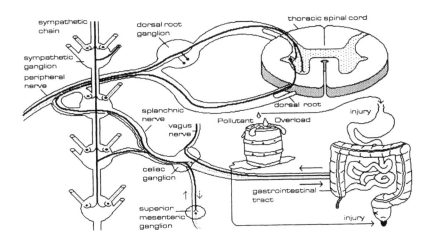

Figure 21.1. Pollutant injury to the innervation of the gastrointestinal tract in the chemically sensitive individual. Pollutants can travel several different routes and cause autonomic deregulation. (1) Pollutants can enter through the olfactory tract and go to the hypothalamus. (2) Pollutants can enter through the lungs and travel to the blood. (3) Pollutants can enter through the gut and go to the spinal cord and/or intestines and travel to the reflex arc or on to the hypothalamus. (4) Pollutants can enter through the skin, travel to the autonomic system, and reach the hypothalamus.

breathlessness, decreased satiety, etc. can then occur in a chemically sensitive individual. Imbalance of hunger and satiety centers in the medulla and hypothalamus or direct stimulation of either by pollutants explains why some chemically sensitive people have ravenous appetites and others become anorexic when they are exposed to toxic chemicals. Deregulation of these centers by pollutant overload suggests why additional peripheral phenomena such as vasospasm are activated in some chemically sensitive patients (see Volume I, Chapter 3[3]). Amphetamines and tobacco smoke are known environmental pollutants that appear to effect the medial hypothalamic area causing autonomic dysfunction (for more information, see Chapter 26, section on the autonomic nervous system) (Figure 21.1).

An intramural network of nonadrenergic and noncholinergic neurons is found in the gut wall containing secretory granules. Several peptides including somatostatin, substance P, VIP, CCK, Met and Leucencephalons, bombesin, calcitonin gene-related peptide, galornin, and neuropeptide Y are found in the gut nerve endings. These substances act as transmitters of signals and regulators of excitability of effector cells.[4] Each type has a characteristic distributional pattern.[5-10] Some have excitatory, and others have inhibitory, patterns. These neuronal populations regulate blood flow and sphincter tone as well as secretory and absorptive functions.[10] Some peptides such as gastrin, CCK, VIP, etc. are present both in the brain and gut. Abnormal triggering by toxic chemicals may well have widespread CNS effects as well as cause long gut problems, as seen in the symptoms of many chemically sensitive patients. Abnormal reactions to these substances have been shown in chemically sensitive patients at the EHC-Dallas who underwent provocation and neutralization injections. Their symptoms can be provoked by injections of minute doses of the individual peptides and eliminated by injections of the neutralizing doses (see Chapter 28 for statistics).

Pollutant Effects on Local Intestinal Mucosa

Pollutant injury may result in inappropriate excess and at times insufficient secretions of the local intestinal mucosa, causing ulcerations or excess mucus production with diarrhea.

Pollutant-induced alteration of the intestinal mucosa may occur in several ways. One possibility is that nutrients needed for formation of appropriate secretions may not diffuse or be transported properly because secretory cells may be damaged. Thus, ATP production may be insufficient to keep the secretions functioning. Also, the endoplasmic reticulum membrane is susceptible to damage. Dysfunction here results in both inadequate protein synthesis and poor xenobiotic detoxication. Further, the golgi apparatus may be damaged preventing mucus secretion. Finally, calcium channels in the membrane may be damaged, either preventing excytosis or limiting the function of the mineral pumps (see Volume I, Chapter 4[2]).

Inappropriate secretion may start in the mouth, where salivary glands do not secrete enough ptyalin (an x amylase). Proper initiation of digestion is then prevented. It has been observed that the chemically sensitive patient usually has a dry mouth with inadequate saliva production. Also, inadequate production of pancreatic juice occurs later in the digestive process.

In contrast to general autonomic pollutant injury, localized injury may also occur upon exposure to a variety of food additives and food breakdown products, as is evidenced by our studies on chemically sensitive patients and the massive use of antacids by the general population.

No part of the gastrointestinal tract mucosa goes unaffected by pollutant injury. Often the GI mucosa loses its ability to ward off pollutants, due either to toxic overload or lowered local resistance as a result of nutrient loss and, thus, less mucus production. This lowered resistance occurs as a result of pollutant destruction of the mucus barrier or loss of immunological tolerance (Figure 21.2). It can also result from the reaction of surface antibodies (e.g., autoantibody) to local cells and cause either an excess or a lack of neurohormones or a loss of membrane integrity with increased permeability. Any of these kinds of damage may occur in chemically sensitive individuals as a result of exposure to various substances such as alcohols, organic chemicals, antiprotease drugs, lectins, trauma, and infections.[11] Once these local resistance

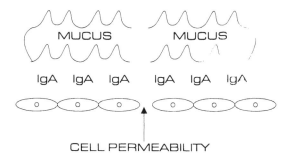

Figure 21.2. Diagram showing changes in the gut barrier resulting from pollutant injury.

factors are altered, a change in permeability occurs. Foods and chemicals can be absorbed in abnormal forms or amounts creating local inflammation (e.g., ulcers) as well as distant pathology (e.g., Raynaud's phenomenon, brain dysfunction). An example of this altered permeability is seen in Hemming's[12] study, where rats were fed beef IgG. Subsequently, IgG was found in the brains of these test animals. It had crossed two membrane barriers that were supposed to keep them out. Apparently, once pollutant damage occurs, nutrients (such as vitamins, amino acids, and minerals) are usually selectively and poorly absorbed resulting in a deficiency or excess of certain nutrients. However, total protein malnutrition may occur in the most severe cases. Also, antibodies may be created against secretory cells, as is seen in the stomach, yielding atrophic gastritis and hypochlorhydria. Antiparietal cell antibodies have been seen in some chemically sensitive patients with breast implants.

Hypochlorhydria that results from either direct pollutant inhibition or autoantibody reaction to the parietal cell is seen in a large portion of the chemically sensitive population. Pollutants may trigger abnormal amounts of vasoactive peptides (including serotonin, vasoactive intestinal peptide, and others), causing problems in the chemically sensitive individual. Bombasin in chili peppers and capsacain in red peppers are well-known triggers of these abnormal responses.

Pollutant Effects on the Intestinal Microflora

Human beings live in a microbial environment and are constantly in contact with a multitude of bacteria and, to a lesser extent, viruses, fungi, and protozoa. Most of these microorganisms are saprophytes, with the soil, air, dust particles, water, and food as their habitat. In general, they do not cause disease in the healthy human. However, when the immune system is compromised by environmental factors, these microroganisms can become pathogens or sensitizers (Table 21.1), as is often seen in a subset of chemically sensitive patients (Figure 21.3).

Table 21.1.   Bacteria Intradermal Testing of 15 Chemically Sensitive Patients with Gastrointestinal Upset

Bacteria intradermal testing			Other antigen testing			
Antigen	Positive	Tested	Positive (%)	Antigen	Positive	Tested	Positive (%)
Alpha Streptocci	2	4	50	Chemical	12	12	100
Beta Streptococci	5	7	71	Inhalants	13	13	100
Enterobacter cloacae	3	3	100	Foods	11	11	100
Streptococcus fecalis	5	6	83				
Brodetella bronch	4	4	100				
Escherichia coli	8	10	80				
Staphylococcus epidermidis	7	8	88				
Corynebacterium xerosis	4	4	100				
Salmonella enteritidis	6	6	100				
Pseudom aerginosa	4	5	80				
Proteus mirabilis	7	7	100				
Proteus vulgaris	7	7	100				
MRV	7	7	100				
BACT antigen mix	2	2	100				
Staphylococcus aureus	4	4	100				
Streptococcus pneumoniae	2	2	100				
Branhamella catarrhalis	2	2	100				
Klebsiella pneumoniae	2	2	100				
Hemophilus influenzae	1	1	100				

Note: Number of males = 1; females = 14. Any range = 22 to 67 years; mean = 46 years.

All microorganisms in a community interact with both the living and the nonliving environment to form an ecological system that has a unique microbial population. In human beings, several ecological niches comprise the normal microbial flora that may be disturbed by pollutant exposure, as is seen in chemically sensitive individuals.

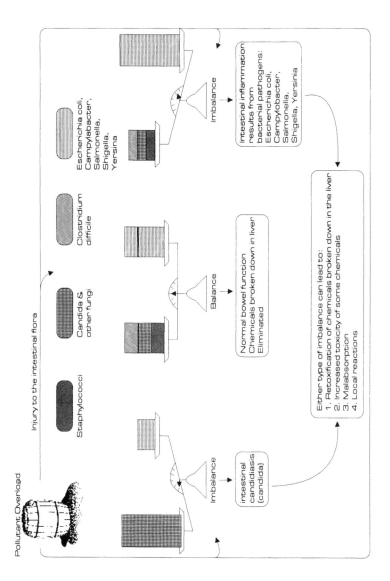

Figure 21.3. Microbial balance.

**Table 21.2.   Aerobic vs. Anaerobic Organisms of Normal Human
Flora Often Imbalanced in the Chemically Sensitive Patient**

Anatomical site	Bacterial concentrations (per mL or gm)	Ratio (anaerobes/aerobes)
Upper airways		
Nasal washings	10^3–10^4	3–5:1
Saliva	10^6–10^8	1:1
Tooth surface	10^{10}–10^{11}	1:1
Gingival crevice	10^{11}–10^{12}	1000:1
Gastrointestinal tract		
Stomach	10^2–10^6	1:1
Proximal small bowel	10^2–10^4	1:1
Ileum	10^4–10^7	1:1
Colon	10^{11}–10^{12}	1000:1
Female genital tract		
Endocervix	10^8–10^9	3–5:1
Vagina	10^8–10^9	3–5:1
Skin		
Surface (per cm^2)	10^4–10^6	100:1
Intrafollicular	10^4–10^8	100:1

Source: Griffiths, B. EHC-Dallas. 1991.

The normal microbial flora comprises a variety of microbial species that are indigenous and transient. The endogenous flora represents those microorganisms that establish permanent resistance on, and in, superficial tissues of the gastrointestinal tract. They can be demonstrated by culture and other techniques to be constantly found in different body areas for a given age, and if disturbed they re-establish themselves (Table 21.2).

The transient flora consists of those microorganisms that are inhabitants (especially of the skin and mucous membranes) for short periods — hours to days. The environment is the source of the transient flora. These microorganisms are of little clinical significance in a host with an established normal flora. However, if the host is compromised, e.g., by the indiscriminate use of antibiotics, as is seen in a subset of chemically sensitive individuals, the resident flora will become suppressed, thus giving the transient flora the opportunity to proliferate and cause disease, either by sensitizing or by causing infections. Many chemically sensitive patients have become sensitive to their own gastrointestinal flora. They will react adversely to individual bacteria such as *Escherichia coli, Proteus vularis*, etc. (Table 21.3).

The importance of maintaining normal flora in the chemically sensitive patient becomes clear when one understands the different beneficial and nonbeneficial relationships of microbes (Tables 21.4, 21.5, 21.6). Beneficial relationships include mutualism, commensalism, immune stimulation, competitive inhibition of nonresident pathogens, and antibiosis. Nonbeneficial

Table 21.3. Microorganisms Found in the Large Intestine

Microorganism	Range of incidence (%)	Microorganism	Range of incidence (%)
Gram negative bacilli (nonspore forming)		Facultative aerobic and anaerobic bacteria	
Bacteroides fragilis (5 subspecies)	100	Gram positive cocci	
Bacteroides melaninogenicus (3 subspecies)	100	*Staphylococcus aureus* (associated with nasal cartilage)	30–50
Bacteroides oralis (2 subspecies)	100	Enterococci (group D Streptococcus)	100
Fusobacterium nucleatum	100	*Streptococcus pyogenes* (groups B, C, F, and G)	0–16
Fusobacterium necrophorum	100		
Gram positive bacilli (with and without spores)		Gram-negative bacilli	
Lactobacilli	20–60	Enterobacteriaceae	100
Clostridium perfringens	25–35		
Clostridium innocuum	5–25	*Escherichia coli*	100
Clostridium ramosum	5–25	Shigella — groups A-D	0–1
Clostridium septicum	5–25	*Salmonella enteritidis* (1400 speicies)	3–7
Clostridium tetani	1–35	*Salmonella typhosa*	0.0001
Eubacterium iimosum	30–70	*Klebsiella* species	40–80
Bifidobacterium bifidum	30–70	Enterobacter species	40–80
Gram positive cocci		*Proteus mirabilia* and other Proteus species	5–55
Peptostreptococcus (aerobic streptococci)	Common	*Pseudomonas aeruginosa*	3–11
Peptococcus (anaerobic staphylococci)	Moderate	*Candida albicans*	15–30

Note: Anaerobic bacteria — 300 times as many anaerobic bacteria as facultative aerobic bacteria (e.g., *E. coli*).

Source: Griffiths, B. EHC-Dallas. 1991.

Table 21.4.   Microorganisms Found in the Oropharynx

Microorganism	Range of incidence (%)
Staphyococcus aureus	35–40
Staphyloccus epidermidis	30–70
Aerobic corynebacteria (diphtheroids)	50–90
Streptococcus pyogenes (group A)	0–9
Streptococcus pneumoniae	0–50
Alpha- or nonhemolytic streptococci	25–99
Neisseria catarrhalis	10–97
Neisseria meningitides	0–15
Hemophilus influenzae	5–20
Hemophilus parainfluenzae	20–35

Source: Griffiths, B. EHC-Dallas. 1991.

Table 21.5.   Microorganisms Found on Skin that May Migrate to the Gastrointestinal Tract of the Chemically Sensitive Individual

Microorganism	Range of incidence (%)
Staphylococcus epidermidis (albus) (coagulase negative)	85–100
Staphylococcus aureus (coagulase positive)	5–25
Streptoccus pyogenes (group A)	0–4
Propionibacterium acnes (anaerobic corynebacteria)	45–100
Aerobic corynebacteria (diphtheroids)	55
Candida albicans	Uncommon
Other candida species, particularly C. parapsilosis	1–15
Gram negative bacteria, e.g., Enterobacteriaceae or Pseudomonas	Uncommon

Source: Griffiths, B. EHC-Dallas. 1991.

relationships include synergistic relationships, production of endotoxins, opportunistic pathogens, and the source of carcinogens. Each will be briefly discussed.

Mutualism is a symbiotic relationship that benefits both the host and parasite. The indigenous flora of the intestines contributes significantly to the overall nutrition of humans. Certain enteric microorganisms synthesize large amounts of the B vitamins and vitamin K, which are absorbed through intestinal mucosa. In response to this synthesis, the intestines provide nutrients, moisture, and optimum growth temperature. Pollutant overload will disturb the indigenous flora of the intestines causing vitamin B deficiency in some chemically sensitive patients (see Volume I, Chapter 6[1]).

**Table 21.6.   Microorganisms Found in the Nose and
Nasopharynx that May Migrate to the Gastrointestinal
Tract in the Chemically Sensitive**

Microorganism	Range of incidence (%)
Staphyococcus aureus	20–85
Staphyloccus epidermidis	90
Aerobic corynebacteria (diphtheroids)	5–80
Streptococcus pneumoniae	0–17
Streptococcus pyogenes (group A)	0.1–5
Alpha-or nonhemolytic streptococci	Uncommon
Neisseria catarrhalis	12
Hemophilus influenzae	12
Gram negative bacteria (Enterobacteriaceae)	Uncommon

Source: Griffiths, B. EHC-Dallas. 1991.

Commensalism is a relationship in which one species benefits, while the other remains unaffected. Pollutant overload may cause an imbalance in this relationship, causing GI malfunction. Immune stimulation that results in a secondary response to certain immuogens will then be promoted. Antibodies are, in general, produced by the host against normal flora. Many of these antibodies cross-react with pathogens, resulting in an amnestic rise in antibody titers that could trigger adverse reactions in the GI tract. Pollutant overload may alter this response, thus allowing an imbalance of microorganisms to occur that then may result in sensitiziation or infection, as seen in a small group of chemically sensitive individuals.

Competitive inhibition of nonresident pathogens may block receptor sites on mucous membranes, thus making them inaccessible to pathogens. This function can be easily disturbed by pollutant overload.

Antibiosis may occur with the production of antagonistic end-products such as hydrogen peroxide, bacteriocidins, and acids that inhibit the establishment of nonresident pathogens. In some chemically sensitive individuals, this function may be disturbed by pollutant exposure.

Synergistic relationships are nonbeneficial. These may mitigate against the host. In this relationship, several microorganisms produce a joint reaction that could not be achieved by a single microorganism. The resulting reaction may enable growth of certain pathogens in tissues in which they would not normally grow, as seen in some chemically sensitive patients.

Production of endotoxins, especially from intestinal flora and resulting in endotoxemia and hypersensitivity to endotoxins, may occur with pollutant overload. These interactions are also nonbeneficial as are opportunistic pathogens that occur when the host defense mechanism is impaired. Then normal flora takes advantage of this opportunity and becomes pathogenic in a variety of manifestations.

Finally, a nonbeneficial relationship between the host and another kind of flora results from exposure to carcinogens that are common in the environment in air, food, or water. Pathogenicity vs. nonpathogenicity depends on the presence or absence of host resistance factors that can be mildly or seriously altered by pollution. Indigenous infections that are caused by the host's normal flora include dental infections, aspiration pneumoniae, intra-abdominal sepsis, nonvenereal infections of the female genital tract, infections of the skin and soft tissues, and sinusitis. These early infections are often seen in a subset of the chemically sensitive population.

Prototype anaerobes are predominant in the normal flora (Table 21.3). The importance of the normal flora in preventing carcinogenesis is not well documented. However, the intestinal flora plays an important role in the chemical transformation of some ingested compounds (see later in this chapter) and is altered in some chemically sensitive patients. Many potential carcinogens become biochemically active only after some modifications. Some intestinal bacteria produce enzymes mediating such modifications. For example, the artificial sweetener Cyclamate® (cyclohexamine sulfate) is said to be modified to the active bladder carcinogen cyclohexamine by bacterial sulfatases.

Studies to classify the microorganisms that are indigenous to human beings are in essence very time-consuming and difficult to do. These limitations coupled with a general lack of interest in, and knowledge about, some microorganisms may have further contributed to the limited investigation of these organisms that has occurred to date. As a result, only sparse documentation on the microorganisms that are indigenous to humans is available, even though fungi, viruses, and protozoa play crucial immunostimulatory roles. Presently, most classifications of these microorganisms are limited to bacteria and, to a lesser extent, yeasts (*Candida* spp.).

The development of microflora in the human occurs in several stages. It is believed that a fetus is sterile in utero. During normal parturition a baby acquires its first microflora of the skin, nose, mouth, and conjunctiva from the mother's vaginal microflora. Immediately after delivery, additional acquisition results from contact with food and attending personnel. Within a few hours after delivery, the oral and nasopharyngeal flora are established. Within 1 to 2 days resident flora of the lower intestinal tract appears.

The sites of resident flora include the skin and adjacent mucous membranes (Table 21.5), the respiratory tract (nose and oropharynx and tonsils) (Table 21.6), digestive tract (mouth and large intestines) (Table 21.7), urinary tract (anterior urethra), and genital system (vagina and external genitalia) (Table 21.8).

Factors governing establishment of flora do not seem to be entirely random. Instead, colonization follows predictable patterns. Anaerobic microorganisms appear within a week postpartum, irrespective of environmental factors. Although a broad and varied make-up of microbes is present in the gut and other organs of the body, and these anaerobes play very crucial roles, immunologically, nutritionally, and metabolically, many laboratories do not report

Table 21.7.   Microorganisms Found in the Mouth
(Saliva and Tooth Surfaces)

Microorganism	Range of incidence (%)
Staphylococcus epidermidis (coagulase negative)	75–100
Staphylococcus aureus (coagulase positive)	Common
Streptococcus mitis and other alpha-hemolytic streptococci	100
Streptococcus salivarius	100
Peptostreptococci	Prominent
Veillonella aicalescens	100
Lactobacilli	95
Actinomyces israeli	Common
Enterobacteriaceae	65
Hemophibus influenzae	25–100
Bacteroides fragilis	Common
Bacteroides-melaninogenicus	Common
Bacteroides oralis	Common
Fusobacterium nucleatum	15–90
Candida albicans	6–50
Treponema dentium and *Borrella refringens*	Common

Source: Griffiths, B. EHC-Dallas. 1991.

anaerobic organisms, unless the physician specifically requests a report of these. The absence of such data results in a skewed view of the GI flora in the chemically sensitive individual. Therefore, to ensure selection of appropriate treatment for the chemically sensitive patient, the physician must order specific, anerobic reporting.

Pollutant injury to the intestinal microflora can have significant effects on the total body metabolism and help regulate the degree of chemical sensitivity that develops. Apart from the expanding recognition of *Candida* as a potential sensitizer and allergen,[13] the bacterial and chemical content of the gut is also being recognized as capable of altering a wide variety of chemical transformations.[14] The resulting alteration of flora then influences the production and absorption of certain nutrients, including vitamins, minerals, and amino acids, and, in turn, the health of the chemically sensitive individual is further compromised.

An example of these pollutant-induced chemical transformations is the decreased transit time seen with antibiotic-induced diarrhea, which reduces intestinal absorption and subsequent metabolism of a wide variety of other xenobiotics. The flora is a very complex interdependent system of several hundred different bacterial and yeast species,[15] which produces about 10^{-8} of

Table 21.8. Microorganisms Found in the Genitourinary Tract

Microorganism	Range of incidence (%)
Kidneys and urinary bladder normally sterile	
Female and male urethra usually sterile	
except for short anterior segment	
Vagina and uterine cervi	
Lactobacilli	50–75
Bacteroides species	60–80
Clostridium species	15–30
Peptostreptococcus	30–40
Bifidobacterium	10
Eubacterium	5
Aerobic corynebacteria (diphtheroids)	47–75
Staphylococcus aureus	5–15
Staphylococcus epidermidis	35–80
Enterococci (group D Streptococcus)	30–80
Streptococcus pyogenes (usually group B)	5–20
Enterobacteriaceae	18–40
Candida albicans	30–50
Trichomonas vaginalis	10–25

Source: Griffiths, B. EHC-Dallas. 1991.

bacteria per gram of feces. According to Kasalow[16] and other investigators,[17] bacterial imbalance will occur with pollutant injury. This imbalance can then cause a spectrum of deregulations. Davis[18] has shown that imbalance of the GI tract may occur, increasing fermentation and alcohol output and thus resulting in both local and systemic reactions in chemically sensitive individuals who are affected.

Most intestinal flora is present distal to the ileocecal valve. Drugs and toxic chemicals that are absorbed high in the gut and then reenter the GI tract through bile logically will have a greater affect on this lower bowel flora.

Basically, the action of the flora on xenobiotics occurs as a reduction reaction in an anaerobic milieu in contrast to oxidative reactions, which require molecular oxygen in the vast majority of other enzymatic pathways. From a chemical standpoint, these reactions of the flora tend to reverse the xenobiotic breakdown effects produced by the liver, such as hydrolyzing of liver conjugates like glucuronides and sulfate esters. Therefore, they may at times release or reactivate toxic chemicals that have already been detoxified by the liver. The imbalance of bacteria and yeasts by toxic pollutants may then produce other toxic pollutants that will exacerbate the chemically sensitive individual. A summary of some of these usually anaerobic chemical transformations that occur in the gut follows.[15]

Not only can some reductions reactivate previously detoxified substances, but they may also activate other toxic substances. Therefore, emphasis is placed on keeping the total ingested toxic load low so that additional toxic substances are not generated or activated in the chemically sensitive individual. For example, azo bond reduction by bacterial flora on compounds such as the early sulfa drugs apparently plays a role in the activation of a chemical to produce the active bacteriostatic agent. A total load of excess azo compounds will exacerbate the chemically sensitve individual; therefore, it is necessary to keep the azo load low. Sulfasalazine is another product that is activated by reduction in this manner. Experiments with germ-free rats show that these rats excrete the original molecule and not the active metabolites that are seen in the normal rat[19] and humans.

The conversion of nitrobenzoic to *p*-amino benzoic acid confirms reduction of nitro groups. This conversion may be the mechanism for the production of methemoglobin caused by nitrobenzene[20] and the mutagenesis of drugs such as metronidazole.[21] The virtual absence of the conversion of nitrobenzene to methemoglobin in germ-free rats lends support to the view that this conversion process is the pathway of reduction of nitro groups by the GI flora.

Nitrobenzene may become toxic by the action of the intestinal microflora and exacerbate chemical sensitivity. Germ-free rats do not make this conversion, nor do rats that have been treated with antibiotics to suppress their bacterial flora.

Chloramphenicol in the intestine can at times cause aplastic anemia due to the nitrol group reduction by the intestinal flora.[22] Lack of reduction may account for the substantially lower incidence of aplastic anemia with parenteral administration of this drug, although some substances may still be excreted by the enterohepatic circulation and subsequent excretion into the bowel. This phenomenon may explain the chemically sensitive patient who can only tolerate oral antibiotics for 1 to 2 days. At the EHC-Dallas, we have observed that many patients who are extremely sensitive to oral antibiotics may tolerate these better when they are administered intravenously.

Metronidazole is used as a trichomonacide and to treat amebiasis and giardiases. However, it is also tumorigenic, mutagenic, cytotoxic, and a radiation sensitizer. Research findings suggest that reduction of the nitro group is the common vehicle producing these effects.[23]

Dehydroxylation, as well as reduction, of catechols is known to occur in the gut. It has potential importance in the metabolism of L-dopa and dopamine. Some of the side effects of L-dopa, specifically, may be produced by dehydroxylation in the gut and may be influenced by the balance of the microflora.

Glycosidases, like glucuronidases (which deconjugate glucuronates formed in the liver), play an important role in the gut flora also. After deconjugation, the lipid solubility of the initial molecule is often restored, and a drug or chemical that had been conjugated and detoxified by the liver can then be reabsorbed, adversely affecting metabolism. This process probably explains why some chemically sensitive patients who eat foods contaminated by toxic chemicals may continue to react until the gastrointestinal tract is finally emptied. This reactivation and reabsorption of toxic chemicals certainly may be one of the reasons patients with large and small bowel dilatation are constantly reacting.

There are many examples of enzymes activating toxic chemicals, which then exacerbate chemical sensitivity. For example, glucosidases act on agents such as cycasin to cause liver tumors in normal rats, but not in germ-free rats. This phenomenon gives good evidence that the flora is essential for release of the toxic metabolite aglycone from cycasin.[24]

When acted on by the gut flora, amygdalin will produce cyanide, which is known to exacerbate chemical sensitivity. Germ-free rats are not affected with the convulsions that will kill normal rats. Amygdalin is the major component of laetrile, and whereas parenteral laetrile will not usually be toxic, cyanide-induced toxicity has been seen when it was given by enema.[25] Many foods such as cassava, almonds, and sorghum contain cyanide and would probably cause similar effects in chemically sensitive individuals.

Lactulose is used in the treatment of portal system encephalopathy. The molecule is hydrolyzed by gut flora and metabolized to acetic and other organic acids. These acids lower the gut pH and cause protonation and reduced absorption of the amines that cause the encephalopathy. Certainly, if the pH is lowered by this or other substances, cerebral effects such as fuzzy thinking, lack of concentration, or memory loss are especially common in a large subset of the chemically sensitive population.

These foregoing effects of the gut flora can play a significant role in the conversion of food products and xenobiotics into potentially toxic metabolites in the chemically sensitive individual. Derangement of bacterial flora may play an important role in the induction of cancer[26, 27] and other diseases such as bacterial overgrowth syndrome and chemical sensitivity. As regards the question of candida toxicity or sensitivity, the competing gut flora has obvious therapeutic significance.

In the 1920s and 30s,[28,29] medical textbooks discussed toxemia due to constipation and excess absorption of toxins from the colon. These books may have been referring to the very entity of unbalanced GI flora and xenobiotic activation. It is clear from the thousands of chemically sensitive patients we have seen at the EHC-Dallas that those who have balanced GI flora with good bowel movements (1 or more times daily) are much healthier. Further, some patients have reported that "cleansing enemas" high into the colon will immediately clear their reactions. Some will take vitamin C enemas with positive effects similar to those effects resulting from administration of intravenous vitamin C. Others' reactions have been observed to clear after diarrhea was reproduced by administration of potassium bicarbonate. We have seen this clearing phenomenon occur at the EHC-Dallas hundreds of times. Randolph[30] has reported and emphasized the importance of clearing the gastrointestinal tract after a pollutant, food, or allergic reaction.

Pollutant Injury to the Neuroendocrine System of the Gut

Chemical messengers play an important role in the control of the physiologic function of the gut, and aberrant function will exacerbate chemical sensitivity. These regulatory chemical messengers (gut hormones and neurotransmitters) are usually biogenic amines or polypeptides of varying chain lengths and are normally present in the nerve terminals of the myoenteric plexus and in the endocrine cells dispersed within the mucosal lining of the gut and glandular derivatives. In the chemically sensitive population, these chemical messengers are particularly prone to pollutant injury as the result of excess exposure to exogenous xenobiotic amines. This exposure can overload the natural detoxification systems, or it can cause direct pollutant injury that results in imbalanced nerve responses. Also, receptor misinterpretation of a xenobiotic as the neurotransmitter or direct triggering of the neurotransmitter by excess pollutant load allows damage to occur. Each endocrine cell type synthesizes, stores, and secretes a specific polypeptide hormone and/or biogenic amine that acts as a chemical messenger in orchestrating the various secretory, motor, and absorptive functions of the gut. Because these cells are complex and multifunctional, each step in this process may be prone to pollutant injury and cause various types of gut dysfunction. Thus, the variability of symptoms seen in many chemically sensitive patients is accounted for.

There are generally two types of endocrine cells: one secretes into the lumen, and the other is buried in the mucosa that is closed and secretes locally. The luminal communications in some cell types have been shown to act as conduits for the release of secretions directly into the lumens. The apical poles of the open cells are generally believed to serve a chemical receptor function. They can monitor subtle pH changes or the composition of the luminal contents and release their secretory products both locally and into the circulation. Therefore, an abnormal bowel flora and balance with excess xenobiotic contact as seen in a subset of chemically sensitive individuals may cause inappropriate triggering with resultant local and systemic effects.[31] The closed type of endocrine cells responds to gut distention or humoral stimuli and, thus, is also prone to pollutant stimuli. In addition to their apical processes, some open cells show elaborate neuronal-like cytoplasmic processes that extend laterally from their mid portions to terminate around neighboring exocrine and endocrine cells, whose functions they effectively modulate.[32] In addition to their endocrine functions, these cells release secretory products that act on adjacent target cells in a paracrine manner (see Chapter 24). Pollutant injury as seen in chemically sensitive individuals may regulate the amount and type of secretion causing disharmony throughout the target cells, which, in turn, can cause gut and peripheral disturbances. At least 17 excretory substances emanating from neuroendocrine cells have already been reported so that pollutant injury to any of these cells may cause mild to gross GI upset. These substances include two types of gastrin, GAWK, somatostatin, secretin GAWK, cholecystokinin, GIP (gastric inhibitory polypeptide), glucagon, P44, neurotensin, 5-HT, substance P, leuenkephalin, histamine, bombasin-like, and pancreatic polypeptides. Although the neuroendocrine cell population in the gut represents different types of cells, these cells share a number of histochemical and ultrastructural characteristics not only among themselves but also with other endocrine cells throughout the body. If pollutant injury occurs, these similarities result in a widespread ripple effect throughout the body and are emphasized in the diffuse symptomatology seen in most chemically sensitive patients. Conversely, pollutant overload outside the gut can trigger a gut response via these cells. This response is often seen in chemically sensitive individuals who inhale fumes, e.g., car exhaust, and immediately develop bloating. Neuroendocrine cells originate from the endodermal cells in the pancreas rather than from the neural crest.

As with all secretory proteins, synthesis of regulatory polypeptides is a highly evolved and intricately controlled intracellular process that is disrupted by multiple pollutant exposure in some chemically sensitive patients. This process involves the nucleus, the granular-endoplasmic reticulum, transport vesicles, golgi apparatus, immature and mature secretory granules, and lysosomes. Chemical processes involved are proteolytic cleavage of endo and carboxy peptides, sulfation, amidation, acetylation, and phosphorylation, all of which may be potentially damaged by xenobiotics as shown in Volume I, Chapter 4.[2] Molecular heterogeneity is a virtual time bomb awaiting triggering or inhibition by an excess xenobiotic load, as seen in some chemically sensitive individuals. Such triggering or inhibition of neuroendocrine cells will cause gut and vascular dysfunction, which is seen in most chemically sensitive patients. Biological potentiating may range from 6 to 8 times in some cells, resulting in an amplification of symptoms once triggering takes place. As seen in some chemically sensitive patients, clinical manifestations of malfunction in this process range from cramps to bloating to constipation to peripheral vascular spasm to brain dysfunction. The pathogenic mechanism for reactive hypoglycemia, which is seen in a large portion of chemically sensitive patients, has now been attributed to the molecular heterogeneity of glucagon. The preferential overproduction of a form of glucagon normally associated with low biogenic potency results in its competition for glucagon receptors in the hepatocyte cell membrane. This triggering results in an inhibition of postprandial glycogenolysis and tends toward reactive hypoglycemia,[33, 34] often seen in the chemically sensitive population. These observations support the need for a controlled diet, which tends to trigger less of the more potent glycogen stimulators in the chemically sensitive individual.

Recently nonautonomous, nonneoplastic endocrine cell hyperplasia of the gut has been described.[35] When hyperplasia occurs, secretions are released locally into the tissues and into the lumen of the gut. An "overspill" phenomenon then occurs, causing inappropriate reactions. This type of response may well acount for the constant and widespread gut dysfunction that occurs in some chemically sensitive individuals. For example, G-cell hyperplasia in rats results in the increased size of the endoplasmic reticulum and the golgi apparatus, with a relative abundance of variants of gastrin. This cellular pattern of hyperplasia is similar to the cellular response to PCBs and phthalates described by Hinton et al.[36] and Price et al.[37] Then, when these responses occur in humans, an additive effect also occurs, causing an exacerbation of chemical sensitivity (see Volume I, Chapter 4[2]).

Gut endocrine cell hyperplasia may occur with prolongation of the half-life of the cells and augmented self-replication and differentiation of uncommitted stem cells. Each mechanism may be triggered by release from the normal restraining influence of an inhibitor of neuroendocrine cell proliferation (postvagotomy-cell hypoplasia), by lack of negative feedback mechanisms (G-cell hyperplasia in hypo or achlorohydric states), or by trophic influence of certain peptides (bombasin-induced G-cell hyperplasia or associated G-cell hyperplasia).[38-44] Most chemically sensitive patients are extremely intolerant of peppers containing capsaicin, which may trigger bombasin, which has been shown to induce G-CCK hyperplasia and to trigger local and widespread symptoms. Also, approximately 50% of chemically sensitive individuals are hypo- or achlorohydric, both of which have been shown to induce G-cell hyperplasia. Hyperplasias have been demonstrated in the gastric antrum after gastric outlet or pyloric obstruction and are believed to be secondary to intragastric retention of food and the chronic stimulation of the antral cells. At the EHC-Dallas, we have seen two chemically sensitive patients who had a large share of their food and chemically triggered symptoms relieved after a pyloroplasty. There is a family tendency in some patients for this type of pyloric hypertrophy.

Hyperplasia of the histamine-containing enterochromaffin-like cells is the most common type of abnormality seen in the stomach. Many environmental substances such as the H_2 blockers (ranitidine, cimetidine, suphotidine, lixtidine), omeprazole (an inhibitor of H^+, K^+- ATPase proton pump in the parietal cells), and cyprofibrate (a hypolipidemic agent with a gastric inhibitory effect) have been shown to trigger ECL cell hyperplasia in rats.[39,43,45] This type of hyperplasia has also been observed in humans as a result of the same agents.[46,47] Many chemically sensitive patients have their symptoms exacerbated by these same agents. Ten percent of all patients who undergo gastroscopy have ECL cell hyperplasia.[48] Once stimulated by pollutants, this type cell can produce carcinoid tumors.

Serotonin secretory enterochromaffin cell (EC) hyperplasia can be seen in small bowel biopsies of untreated people with celiac disease.[49-51] Serotonin increases in the local mucosa, blood, and urine. Serotonin production reverts to normal when the gluten is withdrawn.[52-54] Serotonin neutralization by intradermal injections has often been observed to stop the GI symptoms in some chemically sensitive patients. Perhaps a feedback loop exists that stops excess serotonin production in these cells. Apparently EC numbers increase with chronic gastritis, appendicitis, cholecystitis, and intestinal strictures.[55-57] Many chemically sensitive patients are sensitive to serotonin and have gut inflammation by gastric biopsy that is responsive to removal of certain toxic chemicals and foods. Presumably withdrawal of these chemicals and foods removes the stimulus to these cells and, thus, reverts the hyperplasia. Injections of an appropriate neutralizing dose of histamine and serotonin will cause symptoms to disappear in some chemically sensitive patients (see Chapter 36).

Somatostatin-producing D-cell hyperplasia also occurs and has paracrine as well as endocrine and exocrine function. An inhibitory action of somatostatin on the basal and stimulated gastric secretion occurs by the paracrine modulation of both G-cell and parietal cell functions. The role of D-cell hyperplasia in chemically sensitive individuals is unknown, but patients with this condition should be constantly watched because D-cell hyperplasia may influence chemical sensitivity.

Hyperplasia of other gut endocrine cell types such as CCK-producing I-cell population, GIP-producing K-cells, motlin producing Mo-cells, and the enteroglucagon producing L-cells[58] have been described. They have not, however, been related to chemical sensitivity. Nonetheless, dysfunction in any of these cells could conceivably exacerbate symptoms in the chemically sensitive individual.

Limited study at the EHC-Dallas has shown that intestinal peptides have a significant impact on chemically sensitive individuals. We used intradermal provocation tests for intestinal peptides on 15 patients with recurrent, intermittent gastrointestinal upset (Table 21.9). All 15 showed positive reactions to secretin and vasoactive intestinal peptides, while the other intestinal peptides were found to trigger symptoms less frequently. Bombasin and cholicystkinin were found to trigger not only gastrointestinal symptoms but also laryngeal edema. Our work indicates that the effects of intestinal peptides on chemically sensitive individuals should be further explored.

Clinical Entities of the Gastrointestinal Tract Influenced by Pollutant Injury

Various clinical entities may result from, or be influenced by, pollutant injury. These include dry mouth, bad breath, mouth ulcers, burning of the mouth and tongue, leukoplakia, heartburn, gastritis, gastric and duodenal ulcers, absorption disorders, celiac disease, regional enteritis, diseases of the colon and rectum, irritable bowel syndrome, colitis, and pruritus ani and hemorrhoids. Each will be briefly discussed.

Table 21.9. Peptide Testing in 15 Chemically
Sensitive Patients with Gastrointestinal
Symptoms Reproduced

Antigen	Positive/tested	(%)
Vasoactive intestinal peptide	15/15	100
Secretin	11/11	100
β-Endorphin	10/13	77
Pancreastatin frag. 37–52	10/13	77
Cholecystokinin frag.1–21[a]	7/11	64
Bombesin[a]	9/14	64
Neuropeptide Y	8/13	62
Gastric inhibitory polypeptide	6/10	60
Pancreatic polypeptide	7/12	58
Leucine enkephalin	4/7	57
Peptide YY[a]	6/12	50
Big gastrin I	6/12	50
Motilin	3/7	43
Cholecystokinin frag. 25–33	3/8	38

[a] Laryngeal edema also reproduced.

Source: EHC-Dallas. 1994.

Dry Mouth

Many chemically sensitive patients complain of dry mouth, which may be due to inadequate production of saliva or diminution of secretion by neural-induced stimulation of the salivary ducts. This decrease in saliva initially reduces the potential for proper digestion and often has been observed at the initiation of GI upset in some chemically sensitive patients.

Bad Breath

Unpleasant odors eminating from the mouth frequently occur 1 to 2 hours after reactions in the GI tract of the chemically sensitive individual begin. These reactions may result from exposure to foods or food additives or to contaminants in air or water. Definition and removal of an incitant will usually eliminate the bad breath.

Bad breath may signal chronic illness or dental inflammation. One of the earliest signs of chemical sensitivity is inflammation of the gums with bleeding when the teeth are brushed.

Mouth Ulcers

In some cases of chemical sensitivity, local resistance in the mouth is broken down by a chemical exposure, usually from food but occasionally from inhaled chemicals. In the chemically sensitive individual, mouth ulcers then result. This breakdown has been noted in workers exposed to myrtitanium and chlorine.[59] We have seen many chemically sensitive patients who developed mouth ulcers immediately after eating highly refined, additive-rich, fast foods, yet these patients could eat the same chemically less-contaminated foods without problems. Often this sign serves as a precursor of severe systemic reactions accompanied by regional lymph node swelling, sore throats, sinus flare, and generalized muscle and joint pain. Discomfort may last from 1 to 7 days. Any food to which one has become sensitive can also cause mouth ulcers. Water pollutants also have been seen to trigger mouth ulcers. Ulcers clear rapidly with strict avoidance of incitants. Herpes and other viral and bacterial infections can also cause mouth and lip ulcers in chemically sensitive individuals. Often these ulcers can be cleared and recurrence prevented with fluogen neutralization injections and vitamin C administration as well as elimination of food and chemical triggers that lower local resistance. Occasionally, antiviral substances such as acyclovir help. One thousand to 3000 mg of L-lysine daily have been used effectively to prevent herpes recurrence. Effective use of lysine suggests herpes may be the outcome of deranged local amino acid metabolism. Dental plates and fillings can also cause mouth irritations and ulcers due to localized reaction to their construction materials. If mouth ulcers are recurrent, they usually reflect a disordered immune or enzyme detoxification system. Patients with celiac disease, ulcerative colitis, and regional enteritis frequently have mouth ulcers. The triggering agents of many of these diseases may be foods or toxic chemicals that are absorbed by ingestion or inhalation.

Burning of Mouth and Tongue

Burning of the mouth and tongue has been one of the most difficult problems to treat in environmental medicine. Frequently in the chemically sensitive patient, the triggering agents are multiple foods and chemicals rather than just one or two substances. Burning appears to be the more severe form of pollutant injury, bordering on a regional involvement of the mouth. Often this entity will not clear without total fasting. It may not even clear with time in a clean environment. In some patients, every substance challenged reproduces symptoms. This reaction makes treatment difficult. We have found that, although approximately half of the patients with mouth burning can be cleared, problems remain in the other half.

Leukoplakia

In the chemically sensitive population, leukoplakia is triggered by a host of environmental agents. Certainly cigarette smoke, which is produced from tobacco, a night shade plant, is known to trigger it. Also, we have seen other members of the night shade family cause leukoplakia. In addition to the cigarrette smoke exposure, one patient in our series had a flare-up of leukoplakia when she ate green bell peppers, tomatoes, potatoes, or eggplant. She has become totally free of leukoplakia since she began to avoid religiously all foods in this family and to take neutralizing injections for them.

Heartburn, Esophageal Dysfunction, Reflux, and Dysphagia

Frequently, the chemically sensitive individual becomes dysphagic. Food, and even liquids, will hang up in the upper and mid esophagus. At times, fullness and pain occur with a total inability to swallow. Medication, except nitroglycerin, worsens these symptoms. Fasting, clean and less-polluted environments, and intravenous magnesium will often relieve dysphagia. Specific foods and chemicals are found to reproduce the symptoms upon challenge. Motor disorders of the esophagus often are triggered by environmental pollutants.

Heartburn should be considered a food or chemical reaction until proven otherwise. Gastroesophageal reflux can be triggered by lower esophageal sphincter incompetence with, or without, associated hiatal hernia and may well be the most common etiology. However, we have seen triggering of reflux with loss of sphincter tone by individual exposures to wheat, corn, sugar, coffee, beef, pork, chicken, and other foods in many chemically sensitive patients. Inhaled chemicals such as formaldehyde, chlorine, petroleum alcohol, pesticides, and auto emissions can selectively trigger reflux. Apparently, local sphincter control can be disturbed selectively by specific food and toxic chemicals. Unless stimulated by the toxic chemical or food, the esophagus may show no reflux when evaluated on an esophagogram. In the chemically sensitive patient, esophagitis refractory to standard anti-reflux measures has been brought under control with definition and avoidance of food and chemical triggers.

Gastritis

Acute and chronic inflammatory stomach disease often presents a puzzle in the ill chemically sensitive patient. It is seen frequently with gastroscopy. The etiology is usually unknown, except alcohols and aspirin are well-known triggers. In Poland, Swiatkowski[60] and Swiatkowski et al.[61] have dripped individual foods through the gastroscope and reproduced visible localized swelling and irritation of the gastric mucosa. Studies conducted at the EHC-Dallas using similar methods have confirmed their work in chemically sensitive patients. In addition, the route of administration of nutrients may be important in the pathogenesis of the disease. For example, we have found that some chemically sensitive individuals tolerate nutrients when they are administered through a feeding tube that bypasses the mouth, esophagus, and stomach and is placed into the small intestine. The same foods previously not tolerated by mouth were also tolerated when introduced through the tube. Our experience has been that the difficult case of gastritis will not clear if the triggering agents are not determined and eliminated. Medication treatment is only partially successful.

In addition to the well-known medication-induced gastritis, we have seen gastritis induced by pollens, molds, foods, food additives, water contaminants, and inhaled chemicals such as phenol, pesticides, chlorine, formaldehyde, and petroleum alcohol. Today, it is not enough to put chemically sensitive patients on medication after a diagnosis of gastritis is made. Instead, triggering agents should be sought and eliminated. If triggering agents are identified and eliminated, local effects as well as long-term complications of ill health and overmedication may be eliminated.

Ulcer — Gastric and Duodenal

Even though their physiologies are somewhat different, gastric and duodenal ulcers in the chemically sensitive individual will be discussed together. Therapy for these two types of ulcers is usually identical, with the exception of the use of anticholinergics in the duodenal ulcer. Where there are food intolerances, buffering therapy must be carefully tailored.

A prevalence of peptic ulcers occurs in the spring and fall. Several chemically sensitive patients who appeared to have seasonal exacerbation have been seen at the EHC-Dallas (Table 21.10). One patient we observed had been hospitalized twice a year for ulcers in the spring and fall for 17 years. All forms of ulcer therapy failed until the season was over, and we considered seasonality. We introduced injection therapy for pollens, weeds, trees, and grasses. She seemed to improve until the onset of the spring pollen season when her ulcer pain recurred. An increased dosage of her pollen injection stopped the pain, while cessation of treatment resulted in a recurrence of pain. Reinstitution of the injection therapy eliminated all symptoms. For the first time in 17 years, she was not hospitalized during this season. Her pain recurred with the onset of the fall ragweed season. It was eliminated by increasing her ragweed injection dosage. With maintenance injection therapy, she no longer experiences ulcer exacerbation, and she has not been hospitalized or had an ulcer recurrence in the last 2 years that she has been treated.

Dickey[62] demonstrated peptic ulcers triggered by specific food sensitivity in some individuals. He cited several cases where biopsy of the ulcer revealed many eosinophils. Challenges with milk, wheat, and other foods exacerbated symptoms in his patients. Avoidance of those allowed clearing.

Ingestion of tea, coffee, cola drinks, other caffeine-containing substances, and alcohol increases gastric secretion and sometimes local irritation. Most forms of xanthines will probably increase secretion. Numerous other toxic chemicals may well stimulate gastric secretions or destroy the buffering capacity of the mucosa or the mucosa itself. With so many solvents found in the chemically sensitive patient, it is easy to see why mucolysis could occur. Other chemicals may stimulate hypersecretion resulting in ulcers. Some patients are found to have bacteria that may cause the ulceration. Clearly, the key to

Table 21.10.   Recurrent Seasonal Duodenal Ulcer

Time:	Hospitalization every year for 17 years in the spring and fall
Diagnosis and treatment:	Injection therapy for pollens of weeds, trees, and grasses
Year 1:	Exacerbation of ulcer symptoms in spring and fall relieved by increased in dose of inhalant injections; no hospitalization
Years 2 and 3:	No symptoms year round; no hospitalization

Source: EHC-Dallas. 1988.

treatment is discovering the triggering agents and then eliminating them. Any toxic chemical can be a problem in peptic ulcer and should be considered as part, or all, of the etiology when found to be part of the exposure and in the body.

Disorders of Absorption

Environmentally sensitive patients appear to have a variety of small and large intestinal problems as well as gastric dysfunction. Fifty percent of the chemically sensitive population over 50 have hypochlorhydria. This depletion increases with age, but whether it is due to the aging process or to pollutant injury is unclear. However, many young women who are chemically sensitive due to silicone breast implantation have autoantibodies to parietal cells, suggesting that the pollutant injury caused the hypochlorhydria. Studies by Williams and Weisburger[63] suggest that, among the general population, there is a higher incidence of distal as well as gastric malignancy with the presence of achlorhydria. These studies emphasize the importance of keeping gastric pH in the normal range. Pollutant overload and food sensitivity can trigger hypochlorhydria. Once hypochlorhydria occurs, food will not be broken down properly, and inappropriate digestion and malabsorption is likely to occur. Indications of malabsorption have been found with analysis of amino acids of 200 chemically sensitive individuals. These analyses reveal 80% with amino acid deregulation. Once this cycle of malabsorption and amino acid deregulation starts, long-term nutrient problems may continue, further influencing adversely not only nutrition but also xenobiotic detoxication and detoxification.

In the chemically sensitive patient, bloating is one of the most common symptoms of the malabsorption disorders. It may occur after food or food additive reaction, but often it occurs with inhalation of toxic fumes. Selective malabsorption of various nutrients is a large problem in the chemically sensitive individual. Malnutrition, probably due to malabsorption with some hyper and altered metabolism, is frequently the end-stage of chemical sensitivity in the patients seen in the hospital section of the ECU. Being intolerant of all ingested foods, these patients are severely malnourished, and they require tube feedings or intravenous hyperalimentation or both (see Chapter 37). These patients have experienced severe weight loss and wasting. Their inability to tolerate any foods results in presenting body weights of 60 to 70 pounds. At the EHC-Dallas, we have seen over 200 patients who required this central intravenous regimen, (see Chapter 37).

Some of the pathophysiological basis of the symptoms and signs in the malabsorptive disorders are presented in Table 21.11.

Pollutant-triggered malabsorption syndromes in the chemically sensitive individual fall within several categories: (1) inadequate digestion (usually due to deficiency or inactivation of enzymes from the stomach, pancreas, or duodenum or lack of HCL from the stomach); (2) reduced concentration of intestinal bile salts (due to liver damage, low taurine, cholesterosis or hypomotility state, diabetes, scleroderma, unbalanced bacterial flora in the bowel, precipitation of bile salts); (3) inadequate absorptive surface; (4) lymphatic obstruction; (5) cardiovascular disorders; (6) primary defects in mucosal absorption; (7) biochemical or genetic abnormalities such as disachardase deficiency; and (8) endocrine and metabolic disorders. Too slow or too rapid transit times appear to cause problems. At the EHC-Dallas, chemically sensitive patients with regional enteritis, colitis, and diarrhea disorders were found to have their symptoms triggered by water contaminants, food, food contaminants, inhaled and ingested chemicals, and molds. The following case emphasizes the significance of pollutant exposure upon the GI tract.

> **Case study.** A 20-month-old boy with a life-long history of chronic diarrhea and vomiting was evaluated in the ECU. His mother had reported a loss of amniotic fluid for 1 month before his premature birth at 36 weeks of gestation. All during her pregnancy, the mother had been exposed to the fumes from an oil refinery that was located close to her house. While in utero, the child was also exposed to these fumes as well as chemically contaminated food. He had a history of anemia, recurrent otitis media, sinusitis (diagnosed on X-ray), and hyperactivity. Gastroscopy at 19 days and at 20 months had revealed gastric ulcers and esophageal reflux, and a duodenal biopsy at these times had shown chronic nonspecific inflammation. The child also had atopic eczema, hives, and asthma requiring aerosol treatment. This patient was skin tested and found to react to seven molds, T.O.E., *Candida*, dust, dust mite, dog dander, grain and grain smut, trees, cigarette smoke, weeds, and pine terpenes. He was also sensitive to 24 foods, including cow's milk, eggs, chicken, banana, carrots, rice, and beef.
>
> Laboratory analysis showed a lymphocytosis at 63% with a neutropenia of 30%. IgE was elevated at 257 IU/mL (C = 7 to 100 IU/mL), and serum globulin was depressed at 1.9 mg/dL (C = 2.0 to 3.5 mg/dL).
>
> After being placed in the ECU and started on a monorotational diet, this patient's GI and allergic symptoms decreased significantly. He continued, though, to have intermittent flare-ups on antigen provocation. He was also placed on a program of nutritional supplementation.

Table 21.11. Pathophysiologic Basis for Symptoms and Signs in Malabsorptive Disorders

Symptoms or signs	Pathophysiology
Generalized malnutrition and weight loss	Malabsorption of fat, carbohydrates, and protein → calorie loss
Diarrhea	Impaired absorption or increased intestinal secretion of water and electrolytes; unabsorbed dihydroxy bile acids and fatty acids → lowered absorption of water and electrolytes; excess load of fluid and electrolytes presented to the colon may exceed its absorptive capacity.
Nocturia	Delayed absorption of water; hypokalemia; food and pollutant injury directly to bladder.
Anemia	Impaired absorption of iron, vitamin B_{12}, and folic acid; direct pollutant injury.
Glossitis, cheilosis	Deficiency of iron, vitamin B_{12}, folate, and other vitamins; food, pollutant, and mold injury directly.
Peripheral neuritis	Deficiency of vitamin B_{12}; direct pollutant injury to vitamin E; malabsorption of vitamin E.
Edema	Impaired absorption of amino acids; protein depletion; hypoproteinemia; direct pollutant injury.
Amenorrhea, ↓ libido	Protein depletion and "caloric starvation"; secondary hypopituitarism.
Bone pain	Protein depletion → impaired bone formation → osteoporosis, calcium malabsorption → demineralization of bone → osteomalacia; direct pollutant injury.
Tetany, paresthesias	Calcium malabsorption → hypocalcemia; magnesium malabsorption → hypomagnesemia.
Hemorrhagic phenomena	Vitamin K malabsorption → hypoprothrombinemia.
Weakness	Anemia; electrolyte depletion (hypokalemia); pollutant injury to mitochondria.
Eczema	Food and chemical sensitivity.
Night blindness	Impaired absorption of vitamin A.

Source: Modified from Greenberger, N. J., and K. J. Isselbacher. 1980. Disorders of absorption. In *Harrison's Principles of Internal Medicine*, 9th ed., ed. K. J. Isselbacher, R. D. Adams, E. Braunwald, R. G. Petersdorf, and J. D. Wilson. 1399. New York: McGraw-Hill. With permission.

Follow-up 8 months after discharge found this child to be tolerating foods well with a minimum of symptoms. He was still on his rotary diet of chemically less-contaminated food and taking his injections of food and inhalant antigens. He was growing well and had gained 6½ pounds. His height, which had been at the 5th percentile, had risen to the 80th percentile. His malabsorption syndrome had subsided.

Celiac Disease

Celiac disease is a well-known entity with the environmental trigger of gluten contained in foods, especially wheat. Xenobiotic load has been seen to exacerbate patients with this disease at the EHC-Dallas. Careful observation and treatment revealed that, once the total pollutant load decreased, these patients could occasionally tolerate some wheat products. However, repetition of wheat exacerbated the problem. Even though on a strict gluten-free diet, the chemically sensitive patient has been observed to develop neurological symptoms similar to those present in multiple sclerosis. Therefore, emphasis should be placed on reducing the total pollutant load in these patients.

Regional Enteritis

Alun-Jones and Hunter[64] presented a series of 77 patients whose Crohn's disease was triggered by food and chemicals and was cleared by avoidance of offending substances (Table 21.12). Sixty-four of these patients successfully completed food testing and remained well on an avoidance diet long-term. Randolph[65] and the EHC-Dallas had similar experiences with patients with Crohn's disease. We cannot emphasize enough the importance of strict environmental control, rotary diets, and food and inhalant injection therapy in these patients in order to avoid complications brought on by this disease.

Diseases of the Colon and Rectum

Major changes in bowel habits (constipation, obstipation, or diarrhea) are indicative of diseases or derangements of the colon and rectum. There are two groups of patients who have problems with constipation related to environmental overload. On careful scrutiny, one group appears to have problems from birth on. Most of these patients have a long-colon syndrome which, in addition to poor motility, complicates the problem. Wald[66] has described a series of patients who had subtotal colectomy for the extra long colon. Their constipation was helped in some cases. We have operated on 20 patients who did well with concomitant food and chemical management. Pain was stopped, and about 90% of their constipation cleared. A representative series is presented in Table 21.13 and Figure 21.4.

In a larger series of patients with long colons with constipation (more than 200 subjects) treated at the EHC-Dallas, surgery was unnecessary, and their

Table 21.12A.   Subjective Food Intolerances of 64 Patients
with Crohn's Disease Controlled by Diet

No. of provoking foods	No. of patients	Food	No. intolerant
0	5	Wheat	28
1	11	Dairy products	24
2	6	Brassicas	16
3	10	Maize (corn)	12
4	8	Yeast	11
		Tomatoes	11
5	3	Citrus fruits	10
		Eggs	10
6	5	Tap water	8
		Coffee	8
		Banana	8
7	5	Potatoes	7
		Lamb	7
		Pork	7
8	3	Beef	5
		Rice	5
9	3	Tea	4
>10	5	Fish	3
		Onions	2
		Chicken	1
		Barley	1
		Rye	1
		Turkey	1
		Additives	1
		Alcohol	1
		Chocolate	1
		Shellfish	1
		Turnip	1

symptoms were controlled with ecologic treatment. This group experienced problems at some point in life when pollutant or chemical overload occurred. Chronic constipation in many of these patients seemed to begin gradually. Occasionally, however, some patients developed constipation suddenly as they progressed through life. While some of these patients related onset of constipation to a time of higher stress, the majority remained uncertain as to the factors or exposures that may have influenced onset of their symptoms.

While studying African natives, Burkitt[67] observed that constipation might be due to a lack of fiber in the diet. He noted that these subjects usually had voluminous loose (no diarrhea) bowel movements. However, in the chemically sensitive individual, constipation is not only due to lack of fiber in the diet, but

Table 21.12B. ESR and Orosomucoid Values of 64 Patients with Crohn's Disease Controlled by Diet

Time	ESR (mean ± SD, mm/hour)	Orosomucoid (mean ± SD, %)
Before start of dietary management	34.8 ± 25.6[c] (43)	215.7 ± 86.0[a] (41)
After induction of remission	31.6 ± 24.6[c] (33)	185 ± 77.2[c] (35)
0–6 months	24.8 ± 18.7[b] (52)	156.3 ± 55.7[a] (51)
7–12 months	19.1 ± 13.1[a] (29)	147 ± 62.8[a] (25)
13–18 months	14.4 ± 12.6[a] (15)	135 ± 39.9[a] (15)
19–24 months	15.9 ± 15.8[b] (12)	139 ± 44.4[a] (2)

Notes: Numbers in parentheses refer to number of patients tested; results are of tests done toward the end of the time interval stated.

[a] $p<0.01$.

[b] $p<0.02$.

[c] $p<0.05$ compared with value before start of dietary management (Student's t test).

Source: Data from (1) Alun-Jones, V., R. J. Dickinson, E. Workman, A. J. Wilson, A. H. Freeman, and J. O. Hunter. 1985. Crohn's disease maintenance of remission by diet. *Lancet* 2:177–180. With permission. (2) Alun-Jones, V., and J. O. Hunter. 1987. Irritable bowel syndorme and Crohn's disease. In *Food Allergy and Intolerance*, ed. J. Brostoff and S. J. Challacombe. 555. London: Baillière Tindall. With permission.

Table 21.13. Long Colon Syndrome: Partial Colectomy to Relieve Constipation in Intractable Chemically Sensitive Patients

Age/race/sex	Inhalants	Foods	Toxic chemicals	% Relief of constipation	Diminution of the chemical sensitivity
26/W/F	+	+	+	100	yes
28/W/F	-	+	+	90	yes
37/W/F	-	+	+	60	yes
38/W/F	+	+	+	50	yes
33/W/F	+	+	+	100	yes
36/W/F	+	+	+	100	yes
40/W/F	+	+	+	90	yes
45/W/F	+	+	+	90	yes

Source: EHC-Dallas. 1994.

also it may involve an inability to process animal protein, sensitivities to food and/or inhalants, or infestation of *Candida albicans* or some other type of imbalance in the intestinal flora. It is clear that the more saturated fat included

A.

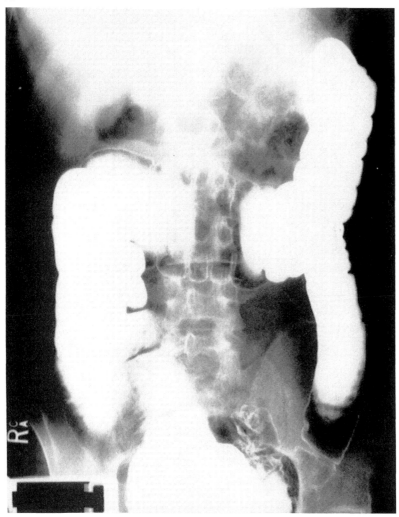

Figure 21.4. The enlarged colon of a 26-year-old white male in the ECU
              — severe constipation intractible to all medical treatment.
              This patient was unable to tolerate any foods. He was treated
              with intravenous hyperalimentation for 3 months. (A) Pre-
              op — long colon. (B) Post-op; 3/4 colon resection produced
              total relief of constipation and food sensitivity. Chemical
              sensitivity lessened. (From EHC-Dallas. 1992.)

in the diet the firmer the stools and the increased tendency toward constipation.
Once constipation occurs, there appears to be an increase in allergic and
hypersensitivity problems as well as other degenerative diseases.

B.

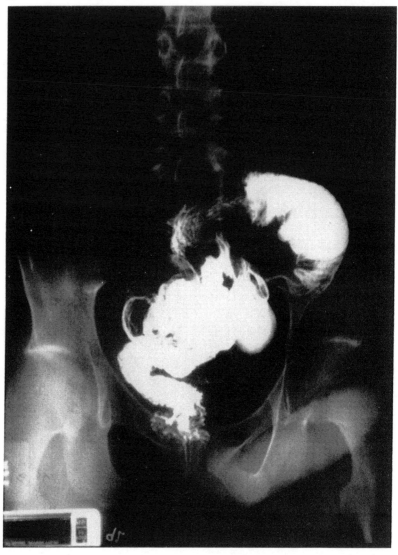

Figure 21.4 (Continued).

Irritable Bowel Syndrome

   Irritable bowel syndrome is particularly influenced by environmental stressors and often can be eliminated by removing them and decreasing the total body pollutant load in the chemically sensitive patient. Difficulties may be triggered not only by ingested substances but also by inhaled ones. Alun-Jones and Hunter[64] found that 70% of their patients with abdominal pain and diarrhea could be successfully managed by diet (Table 21.14). Randolph's[68] and our

Table 21.14. Percentage of IBS Patients Intolerant to Particular Foods

Food	(%)	Food	(%)	Food	(%)
Cereals		**Vegetable**		**Fruit**	
Wheat	60	Onions	22	Citrus	24
Corn	44	Potatoes	20	Apples	12
Oats	34	Cabbage	19	Rhubarb	12
Rye	30	Sprouts	18	Banana	11
Barley	24	Peas	17	Strawberries	8
Rice	15	Carrots	15	Pineapple	8
Dairy Products		Lettuce	15	Pears	8
Milk	44	Leeks	15	Grapes	7
Cheese	39	Broccoli	14	Melon	5
Butter	25	Soya beans	13	Avocado	5
Yogurt	24	Spinach	13	Raspberries	4
Fish		Mushrooms	12	**Miscellaneous**	
White fish	10	Parsnips	12	Coffee	33
Shellfish	10	Tomatoes	11	Eggs	26
Smoked fish	7	Cauliflower	11	Tea	25
Meat		Celery	11	Chocolate	22
Beef	16	Green beans	10	Nuts	22
Pork	14	Cucumber	10	Preservatives	20
Chicken	13	Turnip	10	Yeast	12
Lamb	11	Marrow	8	Sugar–beet	12
Turkey	8	Beets	8	Sugar–cane	12
		Peppers	6	Alcohol	12
				Tap water	10
				Saccharin	9
				Honey	2

Source: Data from (1) Hunter, J. O., E. Workman, and V. Alun Jones. 1985. Dietary studies. In *Topics in Gastroenterology*, Vol. 12, eds. P. R. Gibson and D. P. Jewell, 305–313. Oxford: Blackwell Scientific. With permission. (2) Alun-Jones, V., and J. O. Hunter. 1987. Irritable bowel syndrome and Crohn's disease. In *Food Allergy and Intolerance*, eds. J. Brostoff and S. J. Challacombe. 556. London: Baillière Tindall. With permission.

experience in treating over 10,000 chemically sensitive patients confirm this observation. In our studies, foods were involved in 100% of the irritable bowel cases (Table 21.15). However, we have observed that both biological and chemical inhalants may have a significant impact on irritable bowel syndrome in the majority of chemically sensitive patients who have this entity. Experimental evidence from Finn et al.[69] supports this observation. These researchers studied 58 patients with irritable bowel syndrome by skin prick testing, quantitative total IgE antibodies, and determination of serum antibody reactive with common dietary proteins. They found 48% with positive skin tests and 52% with reactivity to foods as compared to controls.

Table 21.15.   Irritable Bowel Syndrome: 30 Chemically Sensitive Patients, ECU

Environmental triggers found	(%)
Organochlorine pesticides (blood)	20
Intradermal provocation confirmed by inhaled provocation	
Inhalants	
Mold	77
Animal dander	47
Weeds	53
Trees	53
Grasses	53
Terpenes	40
Foods	
Oral challenge confirmed by intradermal challenge	100
Chemicals (formaldehyde, phenol, ethanol, chlorine, pesticides, auto exhaust)	40
Inhaled, double-blind provocation confirmed by intradermal challenge	

Note: Females: 78%; Males: 22%. Mean age: 43 years. Range ages: 6–60 years. Age of onset: <32 years — 60%; >32 years — 40%.

Source: EHC-Dallas. 1986.

Many chemically sensitive patients with constipation find that not only do food and chemicals trigger their problem, but also increased levels of pollens, dust, molds, and toxic fumes that they inhale from the air may trigger their constipation. It is not known whether the inhalation of these substances causes constipation or if swallowing with pollutant entry into the gastrointestinal tract causing local irritation is the route of triggering. However, we have seen several chemically sensitive patients whose colons would go into spasms within 30 seconds of inhaled exposure to toxic chemical fumes such as car exhaust.

Colitis

Cases of chronic nonspecific colitis, Crohn's colitis, and ulcerative colitis triggered by environmental substances have now been seen. In our series of patients with nonspecific colitis, we saw multiple substances trigger the spasm and diarrhea. Though a large percentage of this group of patients was seen to be triggered by specific foods and food and water contaminants, we were impressed by the large number who were triggered by inhaled biological substances and toxic chemicals (Table 21.16). One man in this series noticed that every mold season his injections for molds would not prevent him from developing colitis. Symptoms including cramps and diarrhea would suddenly

Table 21.16.   Nonspecific Colitis: 30 Chemically Sensitive
Patients, ECU

Environmental triggers found	(%)
Chlorinated pesticides (blood)	20
Intradermal provocation confirmed by inhaled provocation:	
Inhalants	
Mold	76
Animal dander	26
Weeds	40
Trees	53
Grasses	53
Terpenes	40
Oral challenge confirmed by intradermal challenge:	76
Foods	
Beef	82
Chicken	72
Corn	72
Egg	70
Wheat	70
Cow's milk	78
Baker's yeast	73
Inhaled challenge, double-blind confirmed by intradermal challenge:	
Chemicals (formaldehyde, phenol, ethanol, insecticide, chlorine)	100
Patients	
Double-blind with 3 placebos	100
Open confirmation (no double-blind, inhaled confirmation)	10
Associated symptoms	
Respiratory	46
Neuro	46

Note: Females: 64%; males: 36%. Mean age: 42 years. Range age: 4 to 71 years.

Source: EHC-Dallas. 1986.

develop. Mold counts were usually high during this time, and adjustment of the dose of his mold injection stopped his colitis.

Whether due to small intestinal or colon spasm, colic in babies has been observed to result from exposure to cow's and goat's milk, synthetic formulas, soy, other foods, and house dust. Also, foods to which the mother is sensitive have been found to cause colic in some breast-feeding babies. If avoidance is possible, it usually results in elimination of symptoms. If elimination through avoidance is not possible, intradermal or sublingual neutralization of the food or breast milk will usually suffice. These observations further substantiate the GI system as a target organ for pollutant activity.

Pruritus Ani and Hemorrhoids

Increased gastrointestinal motility as well as standing upright, causing increased abdominal pressure, seem to contribute to hemorrhoid production in the chemically sensitive individual. Since pollutant injury triggers and deregulates the ANS, pruritis ani and hemorrhoids should be considered entities produced by environmental triggers such as foods, molds, dust, and toxic chemicals. *Candida albicans*, as well as pollutant injury, often triggers pruritus in the chemically sensitive individual. We have seen over 200 chemically sensitive patients whose pruritus and hemorrhoids appeared to be triggered by environmental incitants.

Only the environmental influences of nonmalignant disease have been discussed so far. In this section, the environmental influences upon malignancy will be discussed since some chemically sensitive patients develop malignancy.

Intestinal Malignancy.   Environmental triggers have been implicated in cases of malignancies of the esophagus, stomach, and colon. It is well-known that charcoal cooking in the Far East is related to increased incidence of a cancer of the lower esophagus[70] and stomach. This cancer is probably the result of benzopyrenes released in the cooking process, but it may be due to other enhancers (see Volume II, Chapter 12[71]). High levels of nitrates and nitrites, which form nitrosamine in food and water, are associated with the increased incidence of cancer of the stomach and esophagus in Columbia,[72] Chile,[73] Japan,[74] Iran,[75] China,[76] England,[77] and Hawaii.[77] Williams and Weisburger[63] found that high levels of nitrates in foods coupled with hypochlorhydria was associated with both local gastro and distal malignancies. In China, an increased risk of esophageal cancer has also been associated with ingestion of[78] thermally hot foods and pickled and smoked foods as well as trace mineral deficiencies and diets low in animal products. Further, areas in China with nutrient depleted soil (e.g., selenium) are associated with a high rate of gastric malignancy.[79]

Forty-five mycotoxins are carcinogenic or mitogenic.[80] Mycotoxins are products resulting from the metabolism of molds (see Chapter 17[81]). Aflatoxins come from the molds *Aspergillus flavus* and *parasitians*, which grow in peanuts, corn, cotton seed, and to a lesser extent in walnuts, almonds, and pistachio nuts. These are often found growing in urea foam insulation. Aflatoxin B has been found to induce tumors in ducks, hamsters, mice, rats, and probably liver cancer in humans in Africa.[82] Aflatoxin B is probably the most potent known environmental carcinogen. Hydrazines and hydrazones from mushrooms and the Bracken fern also are carcinogenic.[83] Cycus, which comes from nuts of the palm-like plant, is thought to cause the liver cancer that is common in Guam and Okinawa.[84] Quercetin, kaempferol, and galangin are also mutagens.[85]

Butylated hydroxy toluene (BHT) and butylated hydroxyamsole (BHA), used as food preservatives, induce stomach cancer in rats.[86] Acrylonitrile, a confirmed carcinogen in rats, is transmitted to foods via their packaging and has also been found in sheetrock walls of houses. Polyvinyl chloride (PVC) also used in containers, conduits of foods and water, and wallpaper is a known liver carcinogen.[87] Organochlorine pesticides (e.g., toxaphene, chlordane) cause cancer in mice and other animals. Toxophene is used as a defoliant in the harvesting of cotton. Often, it is still on the cotton at the time of cloth weaving and has been found in public water suppplies (see Volume II, Chapters 7[88] and 12[71]). The National Research Council on Diet, Nutrition, and Cancer[82] concluded that kepone (chlorodecone), toxaphene, hexa-chlorobenzene, and, perhaps, heptachlor with chlordane and lindane present a carcinogenic risk in humans. The latter four often have been found in chemically sensitive patients studied at the EHC-Dallas.

The association of the western refined diet with malignancy appears to be well established now.[67] Low intake of fiber and high intake of nitrites are among the environmental triggers that have been mentioned.[63]

Liver

Since the liver is one of the main detoxifying organs of the body, many environmental triggers may become associated with liver dysfunction in the chemically sensitive patient. Gross dysfunction may be observed only when it is no longer possible to reverse pollutant damage. Therefore, a knowledge of pollutant pathophysiology is necessary to understand the early stages of liver injury (see Volume I, Chapter 4.[2])

Pollutant Injury to Cells and Nutrients[89]

As compared with other liver cells, the hexagonal, epithelial liver cells have multitudinous, and the most diversified, functions. They are the site of the chemical transformations that make body constituents from foodstuffs or their digested breakdown products. They also correlate the three main categories of organic body material so that the liver becomes a great "metabolic pool" of the organism. They are further the central location for xenobiotic detoxification. In the chemically sensitive individual, pollutant overload of this central organ may cause these processes to malfunction, resulting in disordered metabolism of carbohydrates, amino acids, and fats, either alone or collectively. The function of the liver with its capacity to store glycogen, proteins, fats, and vitamins is of the utmost significance for the energy economy of the entire body and can be markedly disturbed by pollutant overload, as seen in the chemically sensitive population (Figure 21.5).

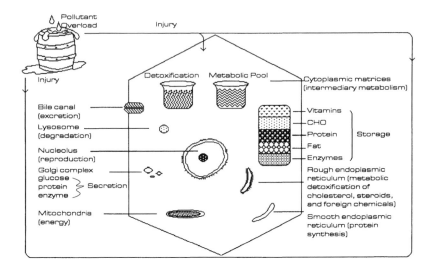

Figure 21.5. Areas of potential pollutant injury to the structure and function of hepatocytes.

The liver stores organic materials not only for its own needs but also to satisfy the needs of distant organs. Imbalance from pollutant overload causes swings in glucose levels and energy for all vital phenomena and results in the varied energy patterns seen in chemically sensitive individuals. The liver cells form many of the serum proteins to provide forces for the oncotic pressure of the plasma or to be used as a transport vehicle for water-insoluble compounds or as coagulating factors or to fulfill enzymatic functions, etc. Once pollutant overload occurs in the areas in the liver responsible for these functions, a vast array of bodily dysfunction may occur often accounting for the varied symptomatology seen in a group of chemically sensitive patients. Some of these dysfunctions may include disordered amino acid metabolism; carbohydrate dysfunction; vitamin, mineral, and enzyme abnormalities; and an inability to detoxify properly (see Volume I, Chapter 6[1]).

The epithelial hepatic cells also protect the chemically sensitive individual from injurious agents by a variety of detoxification processes that yield substances deprived of detrimental properties (detoxified xenobiotics). Pollutant overload may yield a high level of toxic side products or overload the system so that increasing levels of pollutants are in the metabolic pathways, thus disturbing metabolism. This disturbance is frequently seen in chemically sensitive individuals.

The bile, also manufactured by the epithelial cells, contains the characteristic bile pigments, salts of bile acids, cholesterol, and a number of other components. Pollutant overload may alter the conjugation systems so that they may not detoxify chemicals efficiently (see Volume I, Chapter 4[2]). They then yield inappropriate amounts and types of components such as excess cholesterol. Taurine particularly is needed for bile conjugation, and this metabolism may be disturbed by pollutant overload in the chemically sensitive individual (see Volume I, Chapter 6[1]). Peptide conjugation is disturbed by an excess of pollutants, or, if this sulfur-containing amino acid is not available in sufficient quantities, pollutant overload may then proliferate. Methionine is also needed for methyl conjugation reactions and is the most frequently disordered metabolism measured to date in the chemically sensitive individual.

The Kupffer (R-E) cells, besides functioning as endothelial cells like others elsewhere in the organism, represent the quantitatively most important part of the reticulo-endothelial system, which may be disturbed by pollutant overload in the chemically sensitive individual. These cells are involved with the breakdown of hemoglobin to bilirubin. They participate in the formation of gammaglobulin and immunoglobulin bodies, and they act as scavenger cells by removing phagocytosis pigments, bacteria, and other corpuscular or macromolecular elements (Figure 21.6). We have seen reticulo-endothelial dysfunction after pollutant injury from pesticides and other solvents in the chemically sensitive patient. Some develop fragility of the red cells or platelets that results in hemolysis or thrombocytopenia. Hyper- and hypofunction of these cells both in the liver and spleen may result in excess, or deficient, platelets, respectively. Others may develop gammaglobulin deficiency or excess. Sequestering of lymphocytes appears to occur in the reticulo-endothelial system after pollutant injury. Many chemically sensitive patients have been seen to have low T-lymphocytes immediately after pollutant challenge only to have them rapidly return to control levels a few hours after the pollutant is withdrawn. This rapid shift is probably due to sequestering in the reticulo-endothelial system. Transient porphyria is often seen in the chemically sensitive individuals.

Pollutant Injury to Metabolism

Pollutant injury to the metabolism of the chemically sensitive individuals is discussed in Volume I, Chapters 4[2] and 6.[1] However, the specific effects of pollutants on carbohydrates, lipids, proteins, and vitamins and minerals on the gastrointestnal tract will be discussed in the following paragraphs.

Carbohydrates. Carbohydrates, or sugars, are ingested in the form of polysaccharides such as starch or oligosaccharides, e.g., sucrose, the common commercial sugar, or lactose. In the intestine the oligo- and polysaccharides are broken down enzymatically into monosaccharides, the hexoses and pentoses, the simple form of the sugars represented by glucose, fructose, galactose, ribose, and others. Pollutant overload may at times damage the breakdown enzymes forcing nonabsorption or too rapid absorption or absorption of bigger molecular products. The simple sugars enter the liver via the portal vein and are either transferred directly into the general circulation or are retained by the liver where, with the aid of the enzyme hexokinase and adenosine triphosphate (ATP), they are transformed into hexose-6-phosphate. Pollutant injury to the

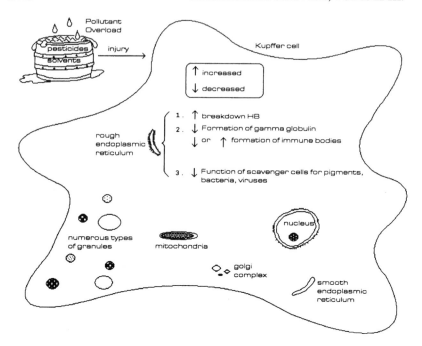

Figure 21.6.   Pollutant injury to the Kupffer cells — reticular-endothelial system.

energy- forming mitochondria may occur in the overloaded chemically sensitive individual either by causing autoantibody reactions or by disturbing their vitamin and mineral metabolism or their membranes, thus disturbing this orderly breakdown by rendering the improper release or decreased quantity of ATP (see Volume I, Chapters 4[2] and 6[1]). Hexose-6-phosphate, in turn, is enzymatically transformed into hexose-phosphate (Figure 21.7), which subsequently loses the phosphate moiety under the influence of phosphorylases. The hexose is then, by enzymatic action, incorporated into the branched chain polymer glycogen. Pollutant overload can not only damage the mitochondria but also the enzymes directly involved in the hexose metabolism. This type of deranged metabolism is seen in many chemically sensitive patients and results in weakness, loss of energy, and irritable bowel. Excess phosphate intake or intake of organophosphate pesticide may imbalance the phosphate pool or render some parts of it toxic to metabolism in the chemically sensitive (see Volume I, Chapter 4,[2] and Volume II, Chapter 13[90]).

Hepatic glycogen is the main form of carbohydrate storage. When glycogenolysis occurs, glycogen is attacked by a debranching enzyme and again esterified with phosphate by a phosphorylase. This series of polymerizing and depolymerizing reactions involves a highly interrelated and complex sequence of enzymatically controlled events that lends itself to deregulation by pollutant injury, as found in the chemically sensitive individual (Figure 21.8). Deficiency or suppression of one of the enzymes involved in this

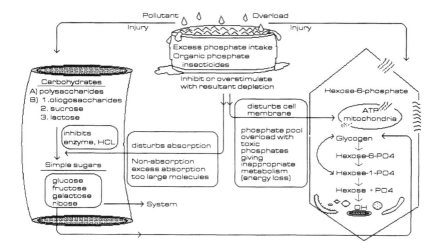

Figure 21.7. Pollutant injury to hexose-6-phosphate metabolism.

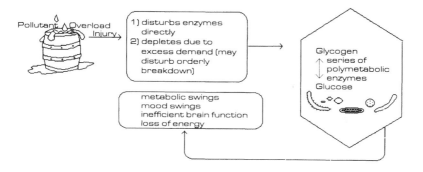

Figure 21.8. Pollutant injury to the enzymes for glycogen metabolism.

reciprocal interaction of glycogenesis and glycogenolysis, such as seen in the congenital enzyme deficiency in von Gierke's disease, may lead to serious or mild metabolic disturbances, the consequences of which are not restricted to the liver alone. Pollutant overload may disturb the system's balance in a mild and chronic manner, which may eventually lead to fixed-named disease, as is frequently seen in the advanced chemically sensitive patient. However, pollutants initially may only mildly alter metabolism resulting in a chronically agitated or fatigued individual.

The quantity of glycogen stored in the liver depends on many factors. High carbohydrate intake increases and a prolonged carbohydrate-deficient diet decreases the glycogen reserves. Either type of intake may be seen in different types of chemically sensitive individuals. Extreme inactivity and exercise have similar effects, respectively. The supply of sugar and the need for it in various organs of the body determine the amount of glycogen in the liver at any given moment. Pollutant injury has been shown to increase the demand for glucose at a minimum of 2 to 3 times; this increased demand, if persistent, strains the liver metabolism over a long-term.[91] Resultant effects in the chemically sensitive individual may be an inability to sustain metabolic equilibrium with food cravings and sudden episodes of weakness. These effects are then followed by inappropriate ingestion of certain types of foods and subsequent reactions to these foods. A vicious cycle of food cravings and malfunctioning metabolism is created.

Not all glucose arriving in the liver from the intestine or released from the breakdown of glycogen enters the bloodstream. Some of the glucose, in the stage of the 6-phosphate ester, is oxidized with the loss of one carbon atom. This process, the hexose monophosphate shunt (HMP) or pentose shunt, yields a five-carbon sugar, a pentose that, besides being broken down for energy purposes over three-carbon compounds, is used for the biosynthesis of nucleotides and nucleic acids. The former consists of a pentose (ribose) or desoxypentose esterified with phosphoric acid and combined with one of two classes of basic substances, the purines and pyrimidines. The nucleotides are polymerized to nucleic acids, which, in combination with certain proteins, become nucleoproteins that are universally present in all cells. Many toxic substances such as benzopyrenes, aflatoxins, cycasins, dimethylnitrosamine, or vinyl chloride can injure the nucleic acids causing benign tumors or cancers in some cases (Figure 21.9) (see Volume I, Chapter 4[2]).

Disturbances of nucleic acids can be observed in some cases of chemical sensitivity. The hexose monophosphate shunt is not specific for the liver, but it is an important feature of the hepatic metabolic functions, since one of the resulting compounds of the shunt is the energy producer adenosine triphosphate (ATP). This ribose- and adenosine-containing compound with high-energy phosphate bonds provides the energy transfer required for many of the biochemical reactions and transformations carried out by the liver. The mitochondria containing these energy substances are easily damaged by pollutant overload, often causing the extreme weakness seen in the chemically sensitive patient. However, the mitochondria are very resilient and return to normal function once the pollutant load is lifted (see Volume I, Chapter 4[2]). This rapid healing phenomenon is apparently one of the reasons the symptoms of the moderately chemically sensitive patient quickly clear once the pollutant load is reduced.

Another part of the glucose in the liver is directly transformed into compounds with three carbon atoms, such as glucaric acid. The three-carbon chain (carbohydrates, fats, and lipids) that results during this transformation is an important interlink in a great number of chemical reactions involving carbohydrate and protein metabolism. The three-carbon chain can also be an end-product used in conjugation reactions for detoxifying chemicals.

Glucaric acid may be elevated in the urine of some chemically sensitive patients. Its terminal product, pyruvic acid, has a major role as precursor for the

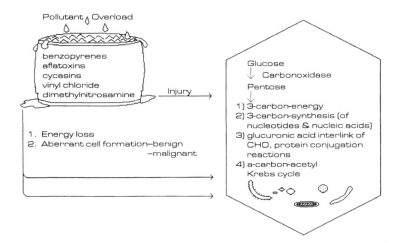

Figure 21.9. Pollutant injury — energy loss, aberrant cell.

two-carbon structures, the acetyl radical and acetate ion, that link carbohydrate and fat metabolism. Together with other short-chain carbon compounds, these two-carbon structures form the "metabolic pool", which is fed by small molecular breakdown products and supplies "building stones" common to all three classes of body constituents — carbohydrates, proteins, and lipids. In the chemically sensitive patient, pollutant injury to these links can cause subtle disturbances in metabolism that result in both over- and underactivity (Figure 21.10). Often food intolerance occurs in the chemically sensitive individual due to this malfunctioning metabolism.

The acetyl and acetate structures never exist in free form but are coupled with other compounds, of which coenzyme A is particularly important (see Volume I, Chapter 4[2]). The latter is a complex composed of the vitamin pantothenic acid, phosphorylated adenosine, and sulfur-containing amine. Bound to the acetyl group, it becomes acetylcoenzyme A, the "active acetate" (more correctly "active acetyl") involved in numerous metabolic reactions, among which are those of the tricarboxylic acid cycle ("Kreb's cycle"). This cycle is the aerobic phase of carbohydrate oxidation and the "final common pathway" of fat and protein metabolism. In the process of detoxification, many pollutants may infringe upon, or travel through, this pathway. Overload of these pathways may cause reactions, and even food intolerance, in chemically sensitive individuals. For example, in chemically sensitive individuals, disturbed acetylation reactions may disorder the metabolism of the Kreb's cycle leading directly to more malfunction, or these reactions may disturb the conjugation acetylation reactions for xenobiotics, resulting in more toxic substances in the blood (see Volume I, Chapter 4[2]). Without appropriate nutrients delivered at appropriate times, the Kreb's cycle will not function as efficiently. Since pollutants can damage the availability of vitamins, minerals, and amino acids in many ways (see Volume I, Chapter 6[1]), this whole cycle is vulnerable to pollutant overload.

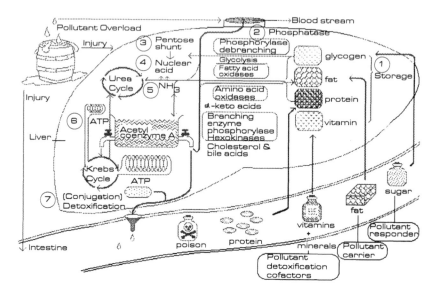

Figure 21.10.   Possible areas of pollutant injury to the metabolic pool of the liver. (1) Storage: fat, CHO, protein, vitamins, minerals; (2) metabolism breakdown, enzymes → imbalance, food reactions; (3) pentose shunt; (4) nucleic acid synthesis (carcinogenesis) + disturbed synthesis; (5) deamination; (6) energy metabolism → disturbed → weakness; (7) detoxification.

In the chemically sensitive, inefficient function of the whole system of metabolic interactions has often resulted in weakness, fatigue, and an inability to function adequately.

Lipids.   Due to the lipophilic character of many toxic chemicals, pollutants can often injure fat metabolism. This xenobiotic lipophilicity allows for high levels of xenobiotics in the neutral fat and phospholipid pool. This xenobiotic overload will often disturb orderly metabolism of the fat compounds, resulting in dysfunction, as is often seen in a subset of chemically sensitive individuals. Pollutant damage to neutral fat and phospholipids is discussed separately because of their diverse functions.

Neutral fat, i.e., glycerol esterified with three long-chain fatty acid molecules, is hydrolyzed by intestinal lipases, but it is not clear whether this hydrolysis, with the split products glycerol and fatty acids going their separate ways (via portal blood or lymphatics, respectively) is complete. Certainly, pollutant injury may disturb this enzyme function or the quantity of enzymes available for complete action. Perhaps only one-third of neutral fat is completely split into glycerol and fatty acids. The water-soluble glycerol is transferred with the portal venous blood to the liver. As a three-carbon compound, it may enter several metabolic pathways. The fatty acids may enter into water-soluble complexes with bile salts and may enter the intestinal wall in such form. This action explains the disturbance of fat digestion in biliary obstruction, the appearance of fatty acids as soap in the feces, etc. The fatty acids may be released in the intestinal wall and may be resynthesized to neutral fat with glycerol. This neutral fat is probably transported via the lymphatics to the peripheral fat depots, as well as to the fat reservoirs of the liver. The neutral fat in the periphery can be transported to the liver if the depots of that organ are exhausted. Being lipophilic, pollutants may be stored in neutral fat. Over a period of time they will gradually fill the available fat carriers and stores throughout the body. In times of stress, when fat is mobilized, it will act as an autotransfuser moving stored toxic chemicals back into the system (see Volume II, Chapter 12,[71] and Volume IV, Chapter 35) (Figure 21.11). At this time, exacerbation of symptoms in the chemically sensitive individual occurs. This phenomenon has been frequently observed in patients being studied under environmentally controlled conditions in the ECU.

In general, fat utilization is much the same in the liver as elsewhere. After hydrolysis, glycerol may be consumed in the manner described above. Fatty acids are broken down by specific oxidases through gradual shortening of the long fatty acid chain, yielding two-carbon metabolites that join the "metabolic pool" and may contribute to the formation of acetylcoenzyme A, thus providing one of the connecting links between fat and carbohydrate metabolism. Fat, especially acting as part of the cell membranes, can be damaged by pollutants creating free radicals by lipid peroxidation, altered membrane permeability, and mineral pump damage (see Volume I, Chapters 4[2] and 6[1]). By a reversal of fatty acid degradation, the liver can synthesize fatty acids and also neutral fat by subsequent esterification with glycerol. Under normal conditions, these anabolic and catabolic processes, as well as the storage and release of liver fat, are in equilibrium. This equilibrium can be displaced in one direction or another in various ways, as, for example, by excess fat consumption, by a deficiency of the so-called lipotropic factors, or by an imbalance between lipogenic factors and lipotropic factors. Included among the lipogenic factors are the sugars, alcohol, and other toxic chemicals. It is important to note that many toxic chemicals are broken down into alcohols (see Volume I, Chapter 4[2]), causing an imbalance of the orderly lipid function and exacerbating the symptoms of chemical sensitivity. In addition, chemically sensitive individuals are known to be intolerant of glycols and other alcohols that apparently cause imbalance in these factors.

Increased metabolic needs manifest during a period of growth, during pregnancy, and/or during severe trauma or chemical exposure when the requirements for lipotropic substances increase. In contrast, the need for lipotropes in some chemically sensitive individuals reduces with lowered metabolic needs, as in a cold climate or subsequent to a strongly restricted total caloric intake (see Volume IV, Chapter 34) (Figure 21.12).

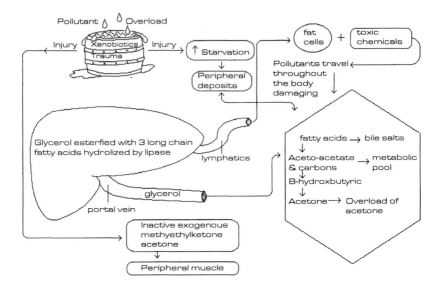

Figure 21.11. Pollutant injury to the metabolism of fatty acids.

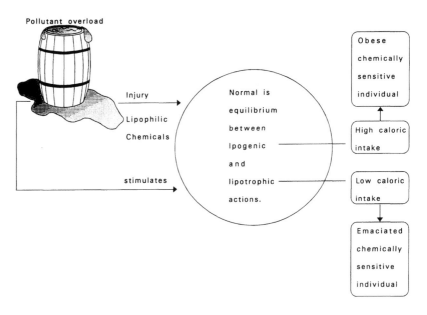

Figure 21.12. Pollutant injury to the lipophilic chemicals.

In the course of the oxidative degradation of the fatty acids by specific oxidases, a four-carbon compound, aceto-acetic acid, and its reduced derivative, β-hydroxybutyric acid, appear, particularly when the degradative process is accelerated under abnormal conditions. β-hydroxybutyric acid may be found in increased amounts in the chemically sensitive individual who is under pollutant stress. These ketone compounds with which the decarboxylated derivative, acetone, is associated, are utilizable in the periphery, especially by the muscular tissue, but not by the liver, which releases them into the blood (see Volume IV, Chapter 34). β-amino isobutyric will pull taurine out of the body through the urine, causing a taurine deficiency. Then xenobiotic conjugation metabolism is diminished and the chemical sensitivity increases (see Volume I, Chapter 5[92]). We suspect that the β-hydroxy butyric acts similarly.

The liver, however, is the only place where ketones are formed as products of certain anabolic and catabolic reactions involved in the oxidation of fatty acids and their synthesis. The most important circumstances under which the ketone bodies rise over their normal low value in the blood are those that force the body to fall back on its energy reserves stored in fat, owing to insufficient carbohydrate reserves in the liver. Starvation and diabetes are typical examples. So, too, is overexposure to toxic substances such as methyl ethyl ketone and acetone, which are exogenous sources and often seen to exacerbate the chemically sensitive individual. Similarly, low sugar and high fat intake favor fat utilization, with resultant ketonemia. Pollutant injury seen in the chemically sensitive population may result in an inability to handle excess ketones. This failure of metabolism may be due to the enzyme damage, zinc, or B_1, B_2, or B_3 deficiency.

Phospholipids, though structurally quite different from the neutral fats and steroidal compounds, are classified as lipids because of some common physical characteristics, especially their solubility, and to some extent because of their physiologic relationships. The synthesis of cholesterol occurs predominantly, though not exclusively, in the liver (Figure 21.13). Its synthesis is the result of the condensation of two-carbon residues involving acetylcoenzyme A. Pollutant overload in the chemically sensitive individual, such as seen with pesticide exposure or pollutant-stimulated excess estrogen, may disturb the synthesis or degradation. Thus, the "metabolic pool" is used in steroid synthesis. Esterification of cholesterol with fatty acids takes place also in the liver, which, in addition, maintains a fairly stable ratio between free and esterified cholesterol. Variation of this ratio and of the absolute amount of cholesterol circulating in the blood has proven to be a sensitive index for functional deficiency of the liver. Over 25% of the chemically sensitive population have disordered cholesterol metabolism, the majority of which return to normal with reduction of total body load. Laseter[93] has shown that arteriosclerotic plaques have the highest content of toxic chemicals in them of any part of the blood vessel. Of course, these chemicals could be attached after passing through the liver, but, in all likelihood, they are there during the formation. Associated with hepatic cholesterol synthesis is the formation of the bile acids, which also contain the steroid nucleus and belong, together with cholesterol, to the main constituents of the bile. Taurine deficiency seen in a high proportion of chemically sensitive

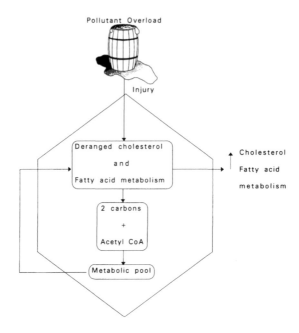

Figure 21.13.   Pollutant injury to the metabolism of cholesterol and fatty acids.

individuals will result in decreased peptide conjugation and, thus, improper excretion of bile salts.

The liver also probably plays a major role in the genesis and turnover of the serum phospholipids. Choline, a lipotropic factor, is a structural component of lecithin. Transmethylation from methyl donors, necessary for the formation of choline and its joining with glycerophosphate and fatty acids to yield lecithin, occurs in the liver. With cephalin, and other phospholipids, the situation is similar. Inadequate methyl compounds for conjugation may occur either due to nutrient deficiency or overutilization of the methylation reactions due to excess pollutant exposure, as seen in many chemically sensitive patients, which then will lead to inappropriate supply of these substances. Chemical sensitivity is aggravated by poor methylation and, at times, may even be caused by it. Poor methylation is the most common deficiency of conjugation measured in the chemically sensitive patient.

Lipotropic substances such as choline favor fat removal from the liver. Probably, choline supports the hepatic oxidation of fatty acids. Another hypothesis is that choline, as a part of the phosphatide lecithin, is essential for the transport of fatty acids from the liver to the blood. The lipotropic action of protein rests mainly, but not entirely, upon those amino acids in the protein molecule that supply methyl groups for the formation of choline, such as methionine and serine. As shown previously, disordered methionine metabolism is frequently seen in the chemically sensitive patient. This disorder can result from the nutrient deficiency or enzyme damage that occurs with toxic overload such as in nitrous oxide overload (see Volume II, Chapter 8[94]) or from direct toxic damage to the amino acid itself or to its vitamin and mineral cofactors. The dependence of lipotropic function on adequate carbohydrate metabolism is seen in various conditions, e.g., in diabetes and starvation when the exhaustion of hepatic glycogen impedes fat removal from the liver. However, insulin deficiency, as in diabetes, exerts a specific effect upon fat oxidation, and, in starvation, fat deposits are transported to the liver. This delicate integration of protein, carbohydrate, and lipid metabolism explains the pathologic features of the liver, as seen after periods of inadequate or unbalanced nutrition present during pollutant overload in chemically sensitive patients.

Proteins are broken down by intestinal enzymes to amino acids that are brought to the liver via portal venous blood. A part of the amino acids serves as constituents for specific cell proteins of the various peripheral tissues. Another portion may be retained in the liver or resynthesized into proteins, which either remain in the liver or proceed into the general circulation as plasma proteins. While all tissues form proteins, production of the plasma albumin, alpha globulin, and fibrinogen seems to be a task reserved for the liver cells. Gammaglobulin appears to be the product of the reticulo-endothelial system. The fate of some amino acids has been successfully investigated. It has been established that amino acids can be attacked by amino acid oxidases, which remove the amino group and are usually B_6 dependent (60% of chemically sensitive patients are B_6 deficient). The resulting deaminated products enter the "metabolic pool," where they are utilized in various ways. For example, the deamination product of glutamic acid is α-ketoglutaric acid, which can be utilized as a "building block" for hexoses or fatty acids. Such an α-keto acid may also be reaminated by B_6 dependent transamination, where amino groups can be transferred from nitrogenous to nonnitrogenous compounds. Supplementation with α-ketoglutaric acid has been an adjunct to aiding rapid detoxification when ammonia scavenging is desired in the chemically sensitive patient. In this manner, pyruvic acid becomes the amino acid alanine. The amino radical removed from amino acids can be carried away as ammonium salts, but, to a greater extent, owing to the omnipresent B_6 dependent transaminating enzyme systems and to the amino group acceptor quality of α-keto acids, they are re-utilized to form a variety of new amino acids. These interactions are only examples of many anabolic and catabolic processes involving amino acids. They are not exclusive liver functions. However, the formation of urea, through the urea cycle, is the main end-product of protein metabolism. This "cycle" ends with production of the amino acid ornithine. The energy for this process derives from adenosine triphosphate, which is normally in the metabolic pool but is often deficient in the chemically sensitive individual due to pollutant injury to the mitochondria. Failure of this exclusively hepatic synthesis of urea, as in liver diseases or pollutant injury, leads to a diminished urea and increased amino acid level in the blood and in the urine. Early stages of derangement of these cycles such as amino acid deregulation are seen in pollutant overload in the chemically sensitive patient and occur frequently (Figure 21.14). Usually there are renal leaks of not only electrolytes but also amino acids in the chemically sensitive patient, as evidenced by ours and Bionostics' (Lisle, Illinois) studies in over 1000 patients (see Volume IV, Chapter 30).

We have seen damage to transamination and deamination that has resulted in severe symptoms after pollutant exposure in the chemically sensitive patient. However, low urea and increased blood amino acids are frequently seen in some patients during and after pollutant exposure resulting in the renal spill. Some chemically sensitive patients develop a strong ammonia body odor and, in fact, have transient elevations of blood ammonia levels following a pollutant exposure and subsequent reaction. This phenomenon has been observed many times in patients in the ECU after they have undergone pollutant challenge or toxic substance mobilization.

Vitamins and Minerals.   Besides storing carbohydrates, fats, and protein, the liver supports the economy of the organism by serving as a depot for vitamins (Figure 21.15). Over 90% of vitamin A is deposited in the liver. The hepatic reserve for other vitamins in the liver is less impressive and to a great extent is shared with other tissues.[95] Pollutant injury can especially deplete the individual of vitamin A and, in some instances, it can deplete the B vitamins (see Volume I, Chapter 6[1]). We have often seen a significant clinical improvement with 50,000 U of parenteral vitamin A. Mineral depletion can also occur with pollutant injury to the liver. For example, zinc may be replaced by cadmium in the liver pool. After an exposure, this replacement compound can damage the glutathione replenishing system and also may be involved in vitamin loss.

Mineral excess can also occur intracellularly after pollutant injury resulting in excess calcium, magnesium, copper manganese, etc. in the chemically sensitive individual. This excess may disturb cellular function, resulting in further metabolic deregulation.

Pollutant Injury to Blood and Water Balance

The liver is a strategic organ, orchestrating intestinal and general circulation, and it harbors, according to its dimensions, a large amount of blood and extracellular fluid. The liver exercises a major influence upon the volume of circulating blood and its constituents. Thus, pollutant injury to this organ may result in generalized vascular dysfunction throughout the body (Figure 21.16). The liver's vascular system serves the proper intrahepatic blood distribution by sphincter actions. The two blood supplies (hepatic artery under high pressure and portal vein under low pressure) are harmonized. The hepatic sinusoids differ from other capillaries in that they have a greater permeability for proteins.[96] Disharmony of this complex vascular system may occur with pollutant

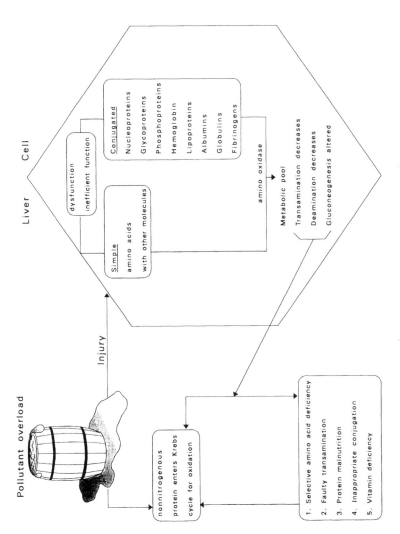

Figure 21.14. Pollutant injury to the metabolism of protein.

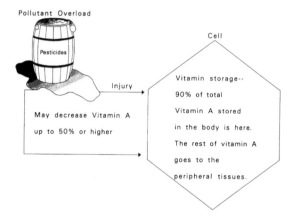

Figure 21.15.   Pollutant injury to the storage of vitamins.

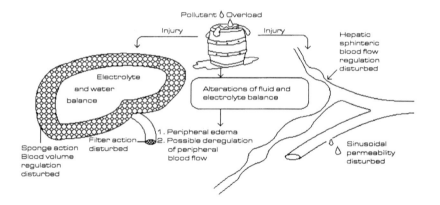

Figure 21.16.   Pollutant injury to the hepatic vascular system.

overload, thereby allowing this most efficient system to cause problems through-out the body that result in stasis and peripheral edema or just vascular deregu-lation.

The liver acts as a sponge or "flood chamber", which can be filled or congested as, for example, right heart failure, or which may be emptied, causing an overload of the pulmonary circulation. The "filter action" of the liver also results from its peculiar anatomic location, since all materials — both nutrient and injurious — absorbed by the intestine are brought to the organ via the portal system. The action of the liver on water and electrolyte balance, which is regulated by the kidneys, lungs, adrenals, and the hypophysis, should not be underestimated, not only because of the large parenchymal mass of the organ but also because all ingested water and salt pass through it before they enter other extracellular departments. Pollutant overload often results in fluc-tuating edema as seen in the chemically sensitive patient. This edema may be partially due to liver dysfunction as well as to endothelial damage in the way the liver handles water and salts.

Pollutant Injury to Liver Detoxifying Processes

The detoxification function of the liver is performed by a variety of chemical reactions, such as oxidation, reduction, degradation, acetylation, acylation, methylation, sulfonation, and conjugation with glucuronic acid of which, especially, the last has been utilized in the functional diagnosis of liver disease. The inactivation of the steroid hormones (androgen, estrogen, progesterone, and corticoids), and thyroid hormones by oxidation and esterification belongs also to this type of liver function (see Volume I, Chapter 4).[2, 89] These functions may be altered in chemically sensitive individuals (see Volume I, Chapter 4[2]).

Sensitivity of the liver to the hepatotoxicity of chemicals is influenced by species, age, sex (more females with high estrogen levels), dieting, and metabolic factors (e.g., high fat, low protein, and deficiencies to specific factors). For example, glutathione deficiency will increase the hepatotoxicity of acetaminophen and bromobenzene and a host of other toxic chemicals such as chlordane. The degree of hepatotoxicity is a function of the rate of active metabolite formation as well as the capacity of the liver to detoxify numerous chemicals.[89]

Early pollutant stimulation by certain chemicals causes metabolic hyperfunction, while other chemicals cause inhibition. Both processes are seen in selected, chemically sensitive patients. Pollutant-induced hyperfunction will be discussed first, followed by a discussion of inhibition (for more details, see Volume I, Chapter 5[97]).

The hepatocyte hypertrophies commonly associated with multiple exposures to numerous chemicals eventually cause many problems, such as the overutilization of nutrient fuels until depletion occurs. The results of pollutant injury to the liver cells due to chemical overload are multiple. For example, the endoplasmic reticulum, the site of microsomal enzymes, enlarges when exposed to such substances as DDT, methylcholanthrene, and phenobarbital (see Volume I, Chapter 4[2]). These chemicals stimulate the endoplasmic reticulum to put out more microsomal enzymes and to convert the toxic lipid soluble substances to less-toxic water-soluble ones. Also, if pollutant stimulation of the endoplasmic reticulum occurs, a metabolite more toxic than the parent compound may be produced (i.e., ethanol to aldehyde and chlordane to heptachlor epoxide). In this case, increased hepatotoxicity occurs, which frequently is seen in the chemically sensitive individual. Further, pollutant stimulation of the liver cell may stimulate the microsomal enzymes that metabolize steroids, and, since there is a feedback loop in the body, more steroids will then be produced because more are removed. As a result, the endocrine system is strained (see Volume I, Chapter 4,[2] and Chapter 24). For example, if microsomal activity increases as a result of toxic exposure (i.e., DDD), the adrenal gland intially speeds up output, but then, as this gland is overtaxed, the adrenal hormones are depleted and adrenal hypofunction ensues. In the chemically sensitive patient, an inability to clear this chlorinated pesticide is frequently seen as is hormone depletion. Finally pollutant stimulation of liver cells increases microsomal acitivity and may also increase drug catabolism, thus rendering normal physiologic doses ineffective.

Porphyria induced by hexachlorobenzene is an example of chemically induced disease. This fungicide induces the mitochondrial enzyme amino levulinic acid synthetase, which is rate limiting in porphyria synthesis. Then porphyrins escape into the plasma, resulting in photosensitization, neurologic and hepatic syndromes, and dark red urine. We have seen transient porphyria in at least six patients with chemical sensitivity, either during a crisis in detoxification or from the initial overexposure. This type of transient porphyria probably is induced by many substances including pesticides and other toxic chemicals. Continued stimulation with hyperfunction will eventually lead to nutrient depletion, resulting in more dysfunction. This multiple etiologic transient porphyria was seen in the following patient.

Case study. For the 7 months previous to her first consultation at the EHC-Dallas, this 28-year-old white female had been experiencing severe epigastric pain, respiratory congestion, eye irritation, dyspeptic symptoms, abdominal pain, swelling, myalgia, fatigue, severe weakness, bruising, forgetfulness, intermittent depression and irritability, insomnia, headaches, and poor concentration. When presenting at the EHC-Dallas, she complained chiefly of severe epigastric pain radiating to the back, nausea, joint pain, and severe weakness. She reported that her illness began in the fall of 1984, shortly after she moved into a new home that had been very tightly constructed. The house had a pressure-treated wood foundation and two wood-burning and oil furnaces. It was still under construction when she moved in. While the house was being built, she frequently visited the construction site because she was intrigued by the building process. Her history revealed she had had tonsillectomy and adenoidectomy at age 9. Her family history was positive for hay fever, eczema, asthma, hives, and hypertension.

Physical examination revealed periorbitol and peripheral edema, yellow color to the skin, epigastric tenderness. Porphyrin screen yielded elevated uroporphyrins at 125 μg/24 hours (C = 0 to 30 μg/24 hours). Laboratory studies showed elevated pophyrins on repeated occasions. In addition, she had depressed RBC at 4.04/mm^3 (C = 4.2 to 5.4/mm^3), depressed HCT at 35.8% (C = 37 to 47%), depressed lymphocytes at 15% (C = 25 to 45%), elevated eosinophils at 18/mm^3 (C = 50 to 300/mm^3), elevated granulocytes at 82% (C = 37 to 75%), elevated total bilirubin at 1.3 mg/dL (C = 0.2 to 1.2 mg/dL), depressed total protein at 5.2 mg/dL (C = 6.0 to 8.5 mg/dL), depressed globulin at 1.8 mg/dL (C = 2.0 to 3.5 mg/dL), and depressed (α-2 globulin at 1.7 mg/dL (C = 2.0 to 3.7 mg/dL). Anti-VCA were at 1:320. Urinalysis yielded large ketones, 1 to 3 WBCs, 1+ bacteria, 2 to 4 epithelial cells, and trace mucus. Twenty-four-hour urine revealed depressed sodium at

27 mEq/24 hours (C = 80 to 180 mEq/24 hours), and depressed magnesium at 5.7 mEq/24 hours (C = 6.0 to 8.5 mEq/24 hours).

The vitamin related enzyme analysis revealed high EGPT at 1.3 index (C<1.25, indicating low B_6), high ETK at 30% (C = 0 to 17%, indicating low B_1), high MMA at 14.3 mg/24 hours (C = 0 to 3 mg/24 hours), indicating low B_{12} and low vitamin D_3 at 23.5 ng/mL (C = 24 to 39 ng/mL).

Her CMI test showed a positive reaction only to *Candida*, with 4 × 4 whealing response. Response to the other seven antigens showed zero whealing, indicating a depressed immune response. Her IgA was normal at 270 mg/dL (C = 85 to 385 mg/dL). IgG was normal at 1175 mg/dL (C = 564 to 1765 mg/dL). IgM was normal at 221 mg/dL (C = 45 to 250 mg/dL).

This patient was placed in the ECU, where she was given nothing by mouth except spring water for 4 days. During this time, she experienced some relief from her abdominal pain. At the time of discharge, she reported that this abdominal pain had subsided on several occasions during her hospitalization when provocative testing was not being performed.

Intradermal antigen challenge showed she had sensitivities to 20 foods, 4 terpenes, 3 hormones, tree mix, weeds mix, grass mix, cotton, fluogen, MRV <7 molds, lake algae, TOE, *Candida*, dust, dust mite, 1 dander, 2 smuts, histamine, and serotonin as well as 5 chemicals, including cigarette smoke, orris root, ethanol, newsprint, and perfume. Double-blind chemical testing in a steel and glass airtight booth resulted in positive reactions to ethanol (<0.50 ppm), phenol (<0.002 ppm), formaldehyde (<0.20 ppm), and chlorine (<0.33 ppm). This patient also had a negative reaction to pesticides (<0.0034 ppm) and saline placebos.

She was treated by avoidance of pollutants in air, food, and water. Her porphyria gradually cleared as her total load decreased, and, except on extreme chemical exposures, she has done well over the last 5 years.

In contrast to metabolic hyperfunction caused by pollutant stimulation liver hypofunction or inhibition of certain enzymes in the liver is the result of exposure to some other toxic chemicals. Also, hypofunction follows continued toxic stimulation, until the nutrient fuels are depleted and then the enzymes become nonfunctional. Both metabolic hyperfunction and liver hypofunction with inhibition of certain enzymes are seen in chemically sensitive patients. A number of chemicals such as cobalt salts, amino glutathimide, and chloramphenicol inhibit the microsomal enzymes, thereby causing delayed metabolism of substances. This inhibition probably results from exposure to many toxic chemicals. Poor metabolism such as sulfoxidase sensitivity is frequently seen in chemically sensitive patients due to pollutant overload (see Volume I, Chapter 4) that may be the result of toxic inhibition.[2]

Toxic chemicals interfere with the excretion of bilirubin by a specific mechanism or as the result of hepatocytic injury. The uptake into hepatocytes is decreased by chemicals competing for binding site Z (transport) or Y (an anion receptor) proteins (e.g., flavospidic acid). Conjugation is inhibited by the antibiotics novciocin and firampicin. Intrahepatic nonobstructive choleostasis is produced by several chemicals, such as estrogen, anabolic steroids, mushroom toxin phalloidin, a-naphthol isothiocynanate (AINT). We have treated six patients with intrahepatic obstructive jaundice who eventually cleared with total load reduction (for more information, see Volume I, Chapter 4[2]).

Neoplasia Caused by Pollutant Injury

The regenerating tissue is usually the first site of malignant change. Most hepatocarcinogens are hepatotoxins and cause degenerative damage. Most hepatocarcinogens are formed from natural substances or synthetic chemicals in the endoplasmic reticulum, the original site of xenobiotic detoxication. Natural carcinogens are aflatoxins from *Aspergillus flavus*, cycosin (an azoglycoside of the tropical cycud plant), and pyirolizidine alkaloids of senecio plants.

Synthetic hepatotoxins are dimethylnitrosamine,[98] thioacetamide,[99] azo dyes[100] used as food colors such as ponceau, and the pesticides aminotriazole[101] and aramite.[102] Others include thorium dioxide[103] in X-ray material and vinyl chloride.[104] Benign adenoma can be introduced by contraceptive steroids.

Many substances can damage hepatocytes. Discussed here are phthalates, benzopyrenes, polychlorinated biphenyls, and heavy metals. These negatively affect the chemically sensitive individual, and, since they are common in the environment, the likelihood of exposure to them with resultant dysfunction is high.

Phthalates. Phthalates can be metabolized by both the mixed function oxidase (MFO) and mitochondrial systems. Phthalates can significantly modify the functioning and inducibility of liver mixed function oxidase (MFO) enzymes.[105] Metabolites of phthalates can alter MFO-induction stimulated by pentobarbital, carbon tetrachloride, or ethanol. They stimulate some MFO enzymes to increased P-450 forms (see Volume II, Chapter 12[71]). They are the stabilizers in plastics and are ingested by eating oil, water, and foods from plastic containers. Phthalates have been found in some chemically sensitive patients at the EHC-Dallas.

Benzopyrenes. Benzopyrenes will also damage mixed function oxidase enzymes. Benzopyrenes are high in charcoal-broiled and smoked foods, vegetable fats and oils, coal-fired heat, and polluted air (especially cigarette and coal smoke) (see Volume II, Chapter 12[71]).

Polychlorinated Biphenyls (PCB). Rats, rabbits, and guinea pigs exposed to PCBs have shown liver damage with fatty infiltrates, central lobe atrophy, and necrosis.[106] PCBs interfere with liver metabolism with their major biochemical effects being the alteration of the mixed function oxidase systems. PCBs contain major contaminants dibenzofuran and dibenzodioxine, both of which are potent mixed function oxidase inducers. PCBs are potent immune suppressants and carcinogens.[107] They can cause lymphoid atrophy in rabbits, chickens, and guinea pigs. A splenic and/or thymic atrophy can occur in monkeys (see Volume II, Chapter 12[71]).

Heavy Metals.   The heavy metals lead, cadmium, and mercury have a high specificity for the sulfhydryl groups. Often these metals inactivate enzymes by binding irreversibly to such groups or by inactivating respiratory electron transfer enzymes. For example, heavy metals can interfere with the glutathione replenishing system, depleting the body of much needed reduced glutathione, which is necessary for detoxication and detoxification. These toxic metals may produce a functional anoxia with lowered ATP production, which may eventually kill the cell where the major detoxification system exists or keep it chronically hypoxic, resulting in the constant weakness experienced by most chemically sensitive patients. Cyanide reacts with the trivalent iron molecules in cytochrome oxidase to inactivate this enzyme (see Volume I, Chapter 4[2]). Carbon monoxide acts as a systemic asphyxiant by binding with hemoglobin and also decreases most detoxifying enzymes to the point of malfunction. Functional hypoxia or anoxia ultimately results in oxidant degradation of cellular membranes. Clearly, chronic exposure to medications along with repeated exposures to other toxic chemicals and molds in air, food, and water could continuously damage the liver until finally total liver failure occurs. Probably of more importance is that, in the majority of chemically sensitive people, chronic exposures to medications and environmental pollutants strain the liver and cause chronic nutrient robbery with resultant total body dysfunctions without overt failure.

Many of the toxic chemicals, such as chloroform, that are isolated in the blood of chemically sensitive patients are detoxified in the liver (see Volume I, Chapter 4[2]). Therefore, a constant strain on the detoxification mechanisms is occurring.

Unfortunately, most peripheral blood enzyme and liver function tests are not sophisticated enough to detect early liver malfunction. However, frequently patients are seen who will have one or two enzymes mildly elevated. This elevation occurs in 40% of chemically sensitive patients. Although evaluation of pollutant injury may be facilitated by information obtained from a liver biopsy, biopsies are not always safe or practical. Therefore, other parameters are better used, including inferences made from a determined exposure and identification of the blood levels of pollutants and the extent of stress these likely put on the liver at any one time. For example, rats fed choline or a diet low in methyl groups can develop liver cancer.[108] This cancer production probably is due to slow methyl conjugation of toxic chemicals, thus allowing them to remain in the body for an extended time, slowly eroding the rat's health. Assuming a similar mechanism exists in humans and is then coupled with the known disordered methionine metabolism found in many chemically sensitive patients, exposure to increased levels of these toxic chemicals should be considered a stress on the liver in spite of normal liver function tests.

Increases in aflatoxins also cause liver cancer in rats.[109] Chemical induction of hepatocarcinoma is governed in part by the level of DNA synthesis in the leukocytes at the time of the carcinogen exposure and also following exposure.[110] It is well-known that vinyl chloride causes hepatocellular sarcoma.[111]

A great variety of clinical manifestations that may belong to some not-well-defined endogenous factors (breakdown products of tissue and blood) or to hormones (thyrotoxicosis, eclampsia) or to drugs and poisons or to bacterial toxins may result from liver injury by toxic substances. Exogenous factors may be divided into four groups as follows: (1) agents such as phosphorus, organic solvents, or plant poisons (muscarine, amanita toxin), absorption of which in significant amounts regularly results in liver injury; (2) agents such as anesthetics, which damage the liver in the presence of other contributing factors (infectious diseases, malnutrition); (3) agents such as sulfonamides, antibiotics, thiouracils, methyltestosterone, and other drugs; (4) agents that primarily impair other organs and affect the liver only secondarily as a result of anoxia, anemia, shock or tissue breakdown. It should be emphasized that not all toxic effects are from the chemicals alone but also may result from malnutrition (see Volume I, Chapter 6[1] for a discussion of the variety of reasons malnutrition might occur and its subsequent impact on the development and severity of chemical sensitivity). The frequent hepatic changes that occur in alcoholics, who are, by definition, chemically sensitive, seem to be primarily the result of malnutrition.[112] Most notably, vitamins B_1, B_2, and B_3 and zinc are deficient in alcoholics with cirrhosis. Similar deficiencies are also present in chemically sensitive individuals who exhibit an inability to handle alcohol.

The liver may respond to toxic agents of any kind with more or less localized degeneration of the liver cell (cloudy swelling, fatty degeneration, "allergic" cholangiolitis). The chemically sensitive patient often presents with a swollen, tender liver, which usually responds to a reduction of total body load. With more severe injuries, necrosis of liver cells develops, presenting the picture of a nonspecific reactive hepatitis and zonal necrosis, mainly centrilobular, which rarely progresses to massive necrosis. Nonspecific reactive hepatitis is the most frequent alteration encountered in biopsy specimens taken from a great variety of cases as well as the chemically sensitive individual. It is characterized by slight, but diffuse, liver cell damage, small focal necrosis with accumulation of segmented leukocytes, mobilization of Kupffer cells, and accumulation of inflammatory cells in the portal triads and sometimes around tertiary cholangioles. All features vary greatly in extent and distribution and are found in gallbladder disease, peptic ulcer, ulcerative colitis and GI carcinoma, in diffuse systemic disorders, in subacute bacterial endocarditis, and in septicemias, pneumonia, and rickettsial and viral diseases. These changes are also prominent and only modified by pigment deposits in malaria. The nonspecific character of the lesion excludes an etiologic diagnosis, and sometimes even the differentiation from a subsiding viral hepatitis is difficult. Nonspecific reactive hepatitis probably represents the anatomic substrate for the abnormalities in hepatic tests frequently found in conditions such as those mentioned. Often biopsies with this aforementioned pathology are misinterpreted as having a viral etiology when, in fact, they are due to chemical overload. The correlation between the functional aberration indicated by the tests and the histologic changes is poor, probably because the morphologic aspect reflects only inadequately the degree of diffuse liver cell damage. However, if one plots liver re-exposure, one can often see definite alteration of the enzymes. This alteration is clearly so on rechallenge with pollutants and also with clearing during rapid detoxification, such as with heat depuration/physical therapy.

In the past, carbon tetrachloride poisoning frequently was observed due to exposure to cleaning fluids. In this poisoning, the liver is enlarged, usually yellow, and the lobular architecture is frequently exaggerated because the central necrosis produces central hyperemia and even hemorrhage. Portal inflammation develops in persons surviving the first week, during which time renal failure with anuria is more often fatal than is hepatic insufficiency. Lesser levels of exposure may cause the nonspecific hepatitis reaction that is seen in the chemically sensitive patient. Survival from severe carbon tetrachloride exposure usually results in chemical sensitivity.

Chloroform poisoning following anesthesia produces a picture similar to carbon tetrachloride poisoning, except that the fatty metamorphosis in the liver is usually far more severe and hemorrhages are more conspicuous. This type of damage is the reason it was discontinued as an anesthetic. Thirty-three percent of chemically sensitive patients have chloroform in their blood in the PPB range. Though these levels may not be enough to give frank chloroform poisoning, they will still put unnecessary demands on the liver cells, straining the detoxification systems.[110] Phosphorous poisoning, also more frequent in previous years, produces similarly fatty liver with a severe necrosis, the localization of which is mainly peripheral. Phosphate levels are frequently elevated in the chemically sensitive patient and in patients who consume large doses of phosphate sodas. These can be very detrimental to the individual's metabolism, causing deregulation of the metabolic pool, although fixed pathology probably will not appear for years.

Clinical Manifestations of Pollutant Injury

Two cases that exemplify pollutant injury to the liver are now presented.

Case study #1. A 47-year-old white female stated that she was exposed to multiple chemicals at work. She entered the ECU with the chief complaint of diarrhea, foot and leg cramps, recurrent gastritis, and intermittent back pain. She had an acute intermittent porphyria.

This patient had had all of her teeth pulled at age 27 years. She had been a two-pack or more per day smoker for 30 years and continued to smoke. Both her father and mother had heart problems. One brother was in good health and two sisters, both of whom had acute intermittent porphyria, were in bad health.

This patient's physical examination revealed an ill-appearing white female with abdominal and back tenderness, mild periorbatal, and pedal edema. Laboratory studies showed elevated WBC at 10,200/mm^3 (C = 4800 to 10,000/mm^3), elevated lymphocytes at 4998/mm^3 (C = 1600 to 4200/mm^3), elevated T$_{11}$ cells at 3898/mm^3 (C = 1260 to 2650/mm^3), elevated T$_4$ percentage at 56% (C = 32 to 54%) and elevated absolute T$_4$ cells at 2799/mm^3 (C = 670 to 1800/mm^3), elevated B lymphocyte percentage at 19% (C = 5 to 18%) and elevated B cells at 949/mm^3 (C = 82 to 477/mm^3). Blood chemistry analysis showed high uric acid at 8.1 mg/dL (C = 1.9 to 6.8 mg/dL), high SGPT at 83 U/L (C = 0 to 53 U/L), high GGTP at 64 U/L (C = 0 to 32 U/L), high triglyceride at 620 mg/dL (C = 0 to 150 mg/dL), and high total cholesterol at 248 mg/dL (C<240 mg/dL). Intradermal skin testing showed her to be sensitive to 7 molds, 28 foods, TOE, *Candida*, dust, dust mite, histamine, serotonin II, grasses, weeds, 7 terpenes, estrone, luteinizing hormone, progesterone, cigarette smoke, orris root, ethanol, formaldehyde, newsprint, perfume, phenol, UNL/diesel exhaust, cotton, fluogen and MRV. Double-blind inhaled challenge in a chemical-free booth showed positive reactions to ethanol, phenol, pesticides, formaldehyde, and chlorine. Placebo challenges were negative. The comparison of inhaled challenge testing and intradermal skin testing is shown in Table 21.17.

This patient was placed in an environmentally controlled room and treated with immunotherapy for foods and inhalants. After treatment, she felt very well. Her blood toxic chemicals decreased as shown in Table 21.18.

Case study #2. This 40-year-old white female was admitted to the ECU with the chief complaint of pain in her chest, dizziness, nausea, and diarrhea. These symptoms appeared to be triggered by working in a small unventilated area that was sprayed weekly with pesticides and that contained many toxic solvents. She continued to work after July 1988. She had lost 103 pounds between this time and 1984 due to an ability to tolerate most foods. This weight loss was progressive, accelerating the last few months before she was seen at the EHC-Dallas. She could only tolerate four vegetables at the time of her admission in November 1988.

Her physical examination revealed emaciation and sores in her mouth, periorbital and peripheral edema, yellow color to the skin, evidence of massive weight loss, and petechiae. She was 5'6" tall and weighed 97 pounds.

She was started on intravenous hyperalimentation consisting of 2000 calories of 50% glucose and 8.5% protein hydrolysate. She gained weight to 125 pounds, but she still could not tolerate food. She had a pathogenic fungi, fusarium, cultured from her mouth. She improved for a couple of weeks after treatment for this, but then she developed nausea and vomiting. This patient was hospitalized in shock with severe acidosis. The following laboratory results were obtained: the urine porphyria test showed hexacarboxy-porphyria was elevated at 24 μg/24 hours (C = 0 to 3 μg/24 hours), pentacarboxyporphyrin was high at 25 μg/24 hours (C = 0 to 5 μg/24 hours), and tetracharboxyporphyrin was elevated at 170 μg/24 hours (C = 0 to 72 to μg/24 hours). Laboratory data revealed low magnesium at 1.4 mg/dL (C = 1.8 to 2.4 mg/dL),

Table 21.17. Comparison of Inhaled Testing and
Intradermal Skin Testing

Chemicals	Double-blind, 15 min. exposure inhaled challenge	Skin challenge confirmation
Ethanol (<0.5 ppm)	+	+
Formaldehyde (<0.2 ppm)	+	+
Phenol (<0.002 ppm)	+	+
Placebo — saline	-	-

Source: EHC-Dallas. 1986.

Table 21.18. Improvement of Blood Toxic Chemicals
after Treatment

Chemicals	Before treatment (9/23/88) (ppm)	After treatment (10/13/88) (ppb)
Toluene	1.5	<0.5
Chloroform	1.6	<1.0
1,1,1-Trichloroethane	1.4	<0.5
n-Pentane	1.7	<1.0
2-Methyl pentane	4.1	<1.0
3-Methyl pentane	7.4	<1.0
n-Hexane	3.0	1.9

Source: EHC-Dallas. 1986.

low cholesterol at 125 mg/dL (C = 150 to 200 mg/dL), low uric acid at 2.1 mg/dL (C = 2.6 to 6.0 mg/dL), high glucose at 143 mg/dL (C = 70 to 115 mg/dL), low creatine at 0.5 mg/dL (C = 0.7 to 1.4 mg/dL), low A/G ratio at 1.4 (C = 1.5 to 2.5), high IgM at 333 mg/dL (C = 45 to 250 mg/dL), low WBC at 2300/mm^3 (C = 4800 to 10,000/mm^3), low RBC at 3.41 × 10^6/mm^3 (C = 4.8 ± 0.6/mm^3). The pathology report of peripheral blood smear and bone marrow showed hypochromic anemia and leukopenia. Hemoglobin electrophoresis showed normal. Her CMI showed positive to three antigens. Her toxic-chemical blood analysis showed increased xylenes at 1.3 ppm, chloroform at 1.6 ppm, 1,1,1, trichloroethane at 1.7 ppm, 2 methyl pentane at 2.3 ppm, 3 methyl pentane at 5.2 ppm, and *n*-hexane at 2.7 ppm.

Intradermal antigen challenge showed she had sensitivities to 52 foods, 7 molds, histamine, serotonin, dust, dust mite, TOE, *Candida*, cotton, weeds, grasses, trees, pine pollen, cedar mix, and primary tree mix.

She was diagnosed with malnutrition, food sensitivity, and porphyria. She was placed in a less-polluted room. She was given intravenous injections of Vitamin C (15 grams per day) and minerals (1 cc of trace elements) and MgSO$_4$ (20 mEq). She also received immunotherapy. She gradually cleared with massive doses of sodium bicarbonate 500 mEq/24 hours. She continued her hyperalimentation and gradually began to eat. Her porphyria cleared completely. After treatment, she felt very well, and her blood cells, including WBC and RBC, increased. Her urine porphyrins became normal.

Many other patients have been seen at the EHC-Dallas with early pollutant injury to the liver. One of the most severe clinical problems is the early diagnosis of pollutant damage to the liver. Unfortunately, liver enzymes are usually elevated later in the disease. However, isolated enzyme elevation should alert the clinician to early liver damage from pollutants. Because microsomal enzymes cannot be measured in the peripheral blood, only liver biopsies give a picture of the microsomal function. This procedure is not usually acceptable in the early pollutant damaged patients with prefixed, permanent malfunction.

Treatment of pollutant overload of the liver can be accomplished by a massive reduction of total body intake of contaminated air, food, and water and replacement of the nutrient fuels.

Gallbladder

Little has been written on pollutant injury to the gallbladder. In our experience at the EHC-Dallas, however, many patients with gallbladder disease have chemical sensitivity and vice versa.

Pollutant Injury to the Autonomic Nervous System (ANS) of the Gallbladder

The sympathetic nerves are derived mainly from the left celiac plexus; therefore, food and chemical sensitivity can have a direct effect by reflex action from the GI tract to the gallbladder or spinal cord as well as the brain. The preganglionic fibers come from the splanchnic nerves. The ganglionic cells concerned in the innervation of the parasympathetic nerves (which come from the gallbladder) of the biliary system are located near, and in, the walls of the bile ducts and in the wall of the gallbladder.

Pollutant Injury to the Gallbladder

Various foods (especially fatty ones) and chemicals may inflame the gallbladder eventually causing cholecystitis. Symptoms may be from mild biliary colic to severe cholecystitis. Many chemically sensitive patients are seen with biliary pain but no stones.

Pancreas

The pancreas has a specialized role in patients with chemical sensitivity, since hypofunction appears to occur often. Mild pancreatic dysfunction may be central in the generation of chemical sensitivity because pollutants may effect pancreatic function, resulting in reduced buffering capacity, decreased digestive enzymes, and occasionally decreased hormones. Since either the endocrine or exocrine function can be selectively involved in pollutant injury, separate patterns of response may occur. Both functions may be involved with more complicated patterns evolving.

Pollutant Injury to the Innervation of the Pancreas[113]

Understanding autonomic innervation to the pancreas is very important in evaluating pollutant injury in chemically sensitive individuals, since most appear to have pancreatic dysfunction. Both the endocrine and exocrine cells are involved with fuel supply and fuel homeostasis, and, when they are deregulated, they cause widespread dysfunction throughout the body.

The pancreas is innervated by the sympathetic and parasympathetic nervous systems. The sympathetic nerves reach the pancreas through the greater and lesser splanchnic trunks, which arise from the fifth to the ninth, and sometimes to the tenth or eleventh, thoracic ganglia. The major sympathetic innervation is through the greater splanchnic nerve. The parasympathetic fibers reach the gland through the vagi. All the nerves in the pancreas, both afferent and efferent, pass through the celiac plexus, and complete excision of the celiac plexus thoroughly denervates the gland. Toxic chemicals such as phenol can injure the plexus, giving results similar to a surgical excision of the plexus. The sympathetic preganglionic fibers terminate in the celiac or superior mesenteric ganglia, whereas the parasympathetic nerves terminate in intrinsic pancreatic ganglia. From the celiac and superior mesenteric ganglia, the nerve fibers proceed along the vessels to the pancreas. The major number accompany the pancreatic duodenal vessels. Some fibers accompany splenic vessels, but most of those that follow the splenic vessels terminate in the spleen. The sympathetic (splanchnic) fibers are distributed only, or at least primarily, to the blood vessels of the pancreas. The parasympathetic fibers (vagi) accompany the vessels as far as the arterioles and then disperse between the pancreatic lobules and around the acini, ultimately finding their endings on individual cells. These nerves serve both the external secreting acini and the islet cells, and the same single fibers may innervate both types of cells. The smooth muscle of the duct is innervated by parasympathetic fibers (Figure 21.17).

Afferent pain fibers from the pancreas are believed to traverse both the sympathetic and vagus pathways, passing through the celiac ganglia, but these fibers are apparently limited to the greater splanchnic nerves. The level of the common areas of pancreatic pain (epigastric, left upper quadrant and back) and the relief of pain by splanchnicectomy are in keeping with this opinion.

Understanding of the influence of nerve stimulation is complicated by the simultaneous action of the nerves on the pancreatic vessels, ducts, and secretory cells, as well as hormonal effects, all of which may be deregulated in the chemically sensitive patient. The studies on innervation of the vascular apparatus

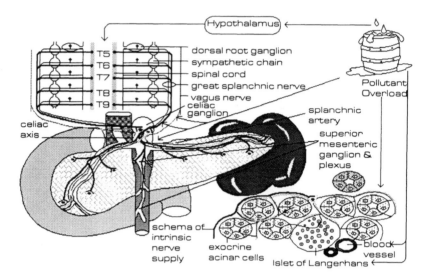

Figure 21.17.   **Pollutant injury to the innervation of the pancreas. Pollutant injury may occur via the hypothalamus, the spinal cord and/or sympathetic chain, the splanchnic and/or vagus nerve, the celiac ganglion, the vascular supply and/or ducts, or the acinar cells may be directly damaged.**

and its surgical interruption may be of some importance for the relief of pain in acute pancreatitis or in understanding early pollutant damage with pancreatitis, suggesting that there is an altering of output with nerve deregulation or severance. The chemically sensitive patient frequently experiences autonomic deregulation, which at times may be analogous to surgical intervention.

Stimulation of the vagus nerves increases the enzyme content of the pancreatic secretions. Most chemically sensitive patients show cholinergic or sympatholytic responses when measured by the iris corder. A slight effect is attributed to sympathetic stimulation with the interference of vascular responses. Both vagotomy and atropine markedly reduce the secretion of enzymes in the pancreatic juice, whereas pilocarpine significantly increases such secretion. Many pollutants give this "vagotomy effect," resulting in a decrease in pancreatic secretion of enzymes. Other pollutants, such as organochlorine pesticides, often act as inhibitors of sympathetic nervous activation, giving sympatholytic effects, while organophosphates usually give cholinergic effects. Stimulation of the ducts should also be considered in exposure studies of this kind, since constriction of the ducts resulting from splanchnic stimulation slows passage of the pancreatic juice, which may lead to inappropriate digestion with gastrointestinal upset and malabsorption, as seen in the chemically sensitive individual. In general, it can be said that left splanchnicectomy (occasionally the right one is necessary) will frequently give relief of pain, and splanchnic or celiac axis anesthetic block frequently results in relief of the severe pain of acute pancreatitis. This sympathectomy effect again emphasizes the loss of sympathetic stimulation of the ducts, which allows for dilation and increases the flow of pancreatic juice. Some pain patterns in chemically sensitive patients reflect a similar situation of stimulation or inhibition. Vagotomy, if effective, acts in two ways. First, it decreases acid secretion in the stomach, thus decreasing secretin formation. Secondly, it directly decreases the stimulation of pancreatic secretion.[114] Some toxic chemicals act on the parasympathetic nerves, producing vagotomy effects.

As in other visceral organs, pollutant injury to the nervous system of the pancreas may occur by absorption of foods and toxins through the intestines with increased liver output after an excess of toxic substances from other pollutant-containing lipid sources has been mobilized. Also, pollutant overload and injury to the nervous system of the pancreas may occur through the nose or blood vessels up to the hypothalamus and then to the autonomic nerves or through the lungs and into the bloodstream and then down to the pancreas and its nerve attachments or to the brain and hypothalamic circuit. All variations have been seen with pollutant injury in chemically sensitive patients (see Chapter 26). Clearly, deregulation of the ANS can occur causing deregulation of the enzyme and endocrine balance. Our studies have shown that many pesticides will cause a cholinergic reaction in 30% of the cases, while more predominantly, 70% of cases have a sympatholytic effect. These effects have about a 55% probability of occurring in chemically sensitive patients with solvents in their blood.[115]

Pollutant Injury — Endocrine Function

At least four types of endocrine cells exist. These are the insulin-producing B-cell, the glucagon-producing A-cell, the somatotrophin-producing D-cell, and the pancreatic polypeptide-producing PP-cell.[116-119] Pollutant overload in any of these cells may exacerbate or occur concomitantly with chemical sensitivity. Therefore, each will be discussed separately.

Insulin-Producing B-Cell. Classic destruction of the islet cells by the drug alloxan is well-known to trigger diabetes. It is probably one of the first recognized chemical inducers of pancreatic dysfunction, after alcohol. Some patients clearly have chemically induced diabetes, while others have a more hazy etiology. However, even the familial types seem to be waiting for the right set of environmental triggers in order to activate their genetic time bomb.[120] We have seen the onset of both adult and childhood diabetes after exposure to a high level of pesticide or natural gas. Often these cases occur without any family history of diabetes.

Most of the endocrine products of the pancreas are intimately involved in the precise control of serum glucose homeostasis. It is well-established that pollutant overload will disturb carbohydrate metabolism, resulting in changes in glucose levels. How this occurs is not totally understood, but it is known that there is a biphasic response to pollutant stimuli with sugars increasing with pollutant exposure two to three times control and then dipping lower than control levels in the second phase before returning to control levels. Since a multitude of factors influence insulin secretion from the B-cells, it is difficult to pinpoint the exact pathways of deregulation. Oral glucose administration induces an enhanced insulin response when compared with equivalent amounts of glucose in vivo. Both neuro and hormonal mechanisms mediate this phenomenon. Pollutant overload certainly is known to disturb the ANS and also hormone regulation. Therefore, one can see how the glucose and carbohydrate regulatory mechanism could be disturbed. Insulin is an anabolic hormone, promoting synthesis of glycogen, protein, and triglyceride in liver, muscle, and fat cells. These effects follow insulin-induced alterations in nutrient transport across cell membranes as well as in the activity of various intracellular enzymes. Pollutant damage occurs frequently at the cellular membrane level thus, again, potentially causing inappropriate transport and activity of the enzymes (Figure 21.18).

Glucagon-Producing A-Cell.   The alpha cells of the islets secrete glucagon, which raises circulatory glucose levels by enhancing hepatic glycogenolysis and gluconeogenesis. These actions oppose those of insulin, thereby minimizing fluctuations in glucose levels despite variability in nutrient intake and energy utilization. Glucagon's actions are especially critical in the maintenance of fasting glucose levels.[121] Glucose inhibits glucagon secretion, but only in the presence of insulin (see Volume IV, Chapter 34). Little is known about the pollutant injury to glycogen secretion and action.

Somatotrophin-Producing D-Cell.   Somatostatin has been localized to the delta cells of the islets and to the peptide-containing nerves, suggesting it has neurotransmitter as well as hormonal function. The latter function is attested by the following facts. First, when given in doses that produce circulatory levels similar to those seen postprandially, somatotrophin causes exogenous or endogenous levels of insulin to decrease. Further, administration of antisomatostatin antiserum enhances the postprandial release of insulin. Insulin is capable of inhibiting somatostatin release. Thus, both insulin and somatostatin may be able to increase their own secretions by inhibiting the release of the other hormone. (For further discussion of pollutant deregulation of somatostatin, see Chapter 24. For further discussion of the immune axis see Chapter 26.)

Pancreatic Polypeptide-Producing PP Cell.   Pancreatic polypeptide is a linear peptide of 26 amino acids with a free NH_2 terminal and a COOH-terminal tyrosine amide.[122] Infusion of PP in humans inhibits exocrine pancreatic secretions and biliary tract motility.[123-126] Its secretion starts immediately after meals. In many chemically sensitive patients, bloating after a protein meal may be the result of inadequate secretion of PP.

1. Excess intake - refined sugar
2. Intestinal malabsorption
3. Liver glucose inadequate or escess release
4. Autonomic nervous system deregulation of islet cells
5. Direct damage to islet cells or their damage to glucogon output and regulation membranes
6. End-organ damage giving inappropriate target
7. Damage to somatostatin output or regulation
8. Feedback to pancreas

Figure 21.18.   Pollutant injury to glucose homeostasis.

The following case study of a 39-year-old female diabetic with a history of extreme lability of diabetic homeostasis with wide swings of blood sugar illustrates pollutant overload causing endocrine pancreatic dysfunction. It further provides an example of how pollutant overload and subsequent good environmental control can make a dramatic difference in a patient's status.

> **Case study.** This patient had had a 95% pancreatectomy for islet cell hyperplasia, and she was allergic to many medications. She also had a history of recurrent pancreatitis and hypersensitivity to many foods, chemicals, and inhalants. In addition, she reacted to the site of insulin injection with inflammation and redness, resulting in subsequent recurrent abscesses.
>
> At one point in her diabetic history, her blood sugar control was showing dangerous swings from hypoglycemia with near-coma to severe hyperglycemia and acidosis, in spite of her extensive knowledge of the disease and the best efforts of both her physician and herself.
>
> On admission to the ECU, her ESR was 57; WBC 14,200, T cells were low at 45% (C = 60 to 80%), B-cells were measured at 45% (C = 20 to 40%), and her glucose was 59. SGOT, LDH, and SGPT were all marginally increased. IgG was decreased at 700 mg/dL (C = 800 to 1800 mg/dL).
>
> She was placed in an environmentally controlled unit, where all chemical exposure was minimized, and she drank only distilled water and ate only chemically less-contaminated food. Her blood sugar went to 20 mg/100mL on fasting after 4 days, and she became comatose. She was revived with intravenous glucose. Her food and chemical sensitivities were determined, and she was placed on a program of avoidance and dietary rotation, along with antigen injection.
>
> At the beginning of her hospitalization, she ate four figs, and her blood sugar soared to 600 mg without insulin. Three weeks later, following treatment with environmental control, 20 figs caused a blood sugar of 90 mg. However, upon mold and chemical challenge testing, her blood sugar fluctuated widely, dipping very low, demonstrating that environmental exposures

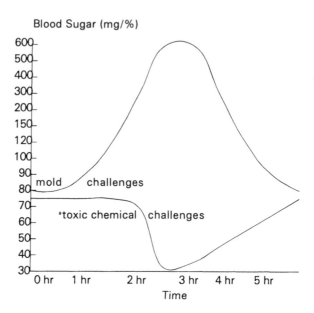

Figure 21.19.   **39-year-old white female with 95% pancreatectomy for islet cell hyperplasia. 15 min. inhaled double-blind challenge after 4 days deadaptation in the ECU. Challenged on separate occasions were [a]formaldehyde (<0.33 ppm), pesticide (2,4-DNP<0.0034 ppm), and phenol (<0.005 ppm). (From EHC-Dallas. 1984.)**

of various types were playing a vital role in her body's homeostasis. Mold exposure increased her blood sugar, while toxic chemical exposure depleted it (Figure 21.19). After hospitalization, she was able to control her diabetes without insulin, a feat that astounded her physicians.

This patient's history illustrates that reactions to foods, chemicals, and inhalants may play a vital role in control of diabetes and that environmental clean-up with appropriate diet and antigen therapy can sufficiently lower the total body load of a diabetic so that the possibility of good control of the disease is greatly enhanced.

Insulin variations may occur by different exposures as well as sensitivity to the insulin itself. There is some evidence that diabetes may be related to chromium deficiency as well as excess refined sugar intake.[127] Clearly immunological dysfunction occurs, and there is a genetic factor in some people.[128] This genetic variation may well be a time bomb waiting for its environmental trigger. Insulin sensitivity does occur, and often it can be reduced by injection neutralization with the dilutions of the problem insulin (see Volume IV, Chapter 36). Also, offending foods must be eliminated to equilibriate blood sugar.

Table 21.19. 45 Chemically Sensitive Diabetics after 4 Days Deadaptation in the ECU with Total Load Decreased

Challenge	Type	(%)
Biological Inhalants	Intradermal Grasses Molds Dust Terpenes	80
Foods	Oral (confirmed by intradermal)	100
Chemicals Formaldehye (<0.20 ppm) Phenol (<0.002 ppm) Pesticide (2,4-DNP) (<0.0034 ppm) Chlorine (<0.33 ppm) Ethanol petroleum (<0.50 ppm) Saline placebos x 3	Intradermal and inhaled double-blind	100

Note: Females: 20; males: 25. Ages: 7 to 70 years. Mean: 40 years.
Source: EHC-Dallas. 1994.

We have treated 45 cases of adult onset diabetes mellitus related, and responsive, to environmental manipulation (Table 21.19). Ages of these patients ranged from 7 to 70 years with a mean of 40 years; 20 were females and 25 males. This high proportion of males is quite unusual for a chemically sensitive population. Eighty percent of the patients were sensitive to biological inhalants; grasses and molds predominated in this group. One hundred percent were sensitive to foods and 100% to inhaled or ingested chemicals. All 45 patients had positive skin tests for phenols, formaldehyde, chlorine, ethanol, newsprint, jet fuel, car exhaust, orris root, and/or men's and women's colognes. Forty patients had inhaled double-blind challenge to phenol (<0.002 ppm); chlorine (<0.33 ppm); formaldehyde (<0.2 ppm); pesticide (2,4-DNP, <0.0034 ppm); petroleum-derived ethanol (<0.50 ppm); and saline placebos. There were a total of 360 challenges with 240 active ingredients and 120 saline placebos. There were 12 placebo reactions and 120 positive chemical reactions. There was no more than one placebo reaction per patient. The significant difference for this group for chemical sensitivity was $p<0.0001$. From this study, it was clear that food and chemical sensitivity existed in these diabetics, and attention to reducing their toal load strongly influenced their long-term clinical course.

Often, insulin use was reduced, or terminated, among those patients studied long-term, and a better control of the diabetes was obtained. It was interesting to note that often challenge with grasses caused blood sugar elevation. In grass season, patients not only had to take their food and inhalant shots, they also had to eliminate from their diet the grass derived foods such as wheat, corn, oats, and rye in order to keep their sugar under control. Also observed was that when a patient was orally challenged with a noncarbohydrate food to which he or she was sensitive, he or she would have an elevated blood sugar. When the same patient was challenged with a carbohydrate to which he or she was not sensitive, his or her blood sugar did not elevate. Often the chemical challenge elevated blood sugar, but sometimes it caused blood sugar depression. We noticed the same vascillation for mold. Clearly, this homeostatic balance between insulin and other hormones can be influenced by environmental manipulation and pollutant exposure.

Pollutant Injury — Exocrine Portion, Enzymes

In the chemically sensitive individual, biochemistry and physiology of the exocrine pancreas may be deranged by pollutant overload, by inappropriate amino acid absorption and synthesis from the duodenum or from enzyme synthesis, and/or by secretion in the pancreas. When it occurs, this deregulation is evident in chemically sensitive individuals. The human exocrine pancreas synthesizes and secretes more protein per gram of tissue than any other organ. Pancreatic juice contains 6 to 12 digestive enzymes in an average daily volume of 2500 mL. The proteolytic enzymes are secreted as the inactive precursors (zymogens) trypsinogen, chymotrypsinogen, proelastase, and procarboxypeptidases A and B. When trypsinogen enters the duodenum, it is activated by enterokinase, an enzyme from duodenal mucosa that cleaves one unique lysine-isoleucine bond of trypsinogen. This limited proteolysis releases an N-terminal hexapeptide and allows the molecule to refold, forming the active site of trypsin. Trypsin can convert trypsinogen to trypsin autocatalytically, but its rate is 2000 times slower than enterokinase. The critical role of enterokinase is proven by the fact that genetic deficiency caused severe protein maldigestion. Abnormal proteins in the stools and blood in indigestion have been seen in the chemically sensitive patient and may account for many of the symptoms of gas and bloating. Frequently this gas and bloating is due to a lack of digestive enzymes because of pollutant inactivation or by the abnormal release of a vasoactive peptide like somatostatin, etc. Enterokinase is released from the microvilli of duodenal mucosa by bile salts and lumenal proteases. Once formed, trypsin activates the other zymogens to produce chymotrypsin, elastase, and oxypeptidases A and B. These five proteases hydrolyze the dietary proteins to di- and tripeptides and amino acids by the time the proteins reach the upper ileum (Figure 21.20).

The reserve capacity of the pancreas is tremendous; 90% of the pancreas must be removed before protein digestion is impaired. However, with pollutant overload, many other functions are disturbed with decreased disturbance of pancreatic function. The zymogens are synthesized in the acinar cells of the

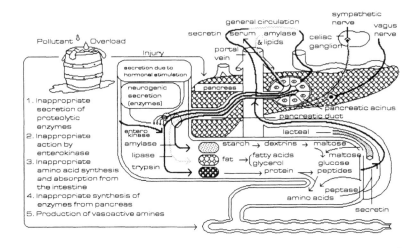

Figure 21.20. Pollutant injury to the exocrine pancreas.

pancreas, where the proteins are assembled on ribosomes attached to the membranes of the rough-surfaced endoplasmic reticulum. As shown in Volume I, Chapter 4,[2] pollutant damage may occur to the endoplasmic reticulum of the chemically sensitive individual.

The enzyme proteins are secreted into the tubules of the endoplasmic reticulum and travel through this tubular system to the golgi apparatus, where the proteins are enclosed in lipoprotein membranes and are condensed into zymogen granules that accumulate at the apex of the cells. Pollutant overload can also cause golgi to dysfunction. The zymogen granules are secreted into the ductal lumen by fusion of their lipoprotein membranes with the membrane of the cell, and their release is initiated by the peptide hormone cholecystokinin (CCK) or by vagal stimulation. The protective mechanism that prevents autodigestion of the pancreas by these proteases is the synthesis of the enzymes as inactive zymogens with investment in lipoprotein membranes within the cell. The synthesis of two trypsin inhibitors, one being present in the acinar cells and the other secreted into pancreatic juice, and the presence of trypsin and other protease inhibitors in the a_1- and a_2-globulin fractions of plasma attempt to give a balance. The trypsin inhibitor in human pancreatic juice acts reversibly, allowing free trypsin to be released in the intestinal lumen. Pollutant injury may occur in the endoplasmic reticulum, golgi apparatus, or any membrane causing inefficient or disordered function. It may be that zymogens are inactivated in chemically sensitive individuals, since many respond to pancreatic enzyme supplementation.

Pancreatic juice contains a ribonuclease and a deoxyribonuclease, but they are not secreted as zymogens. Amylase splits starch to dextrins and maltose. The fat-splitting enzymes are lipase, phospholipase A, and cholesterol esterase. Phospholipase A exists as a zymogen and requires trypsin activation. Mammalian pancreatic enzymes not yet identified in human juice include a lysolecithinase, a vitamin A esterase, and a nonspecific esterase. Conjugated bile salts are important in digestion of lipids. When bile is excluded from the gut, 60% of ingested triglycerides are lost in the stool, whereas only 40% are lost when bile is present but pancreatic juice is diverted. Bile salts inhibit lipase and shift its pH optimum from 9 to 6.5, the pH of jejunal contents. A small peptide, colipase, is secreted in pancreatic juice, binds to lipase, and prevents this inhibition by bile salts. Bile salts function as direct activators of phospholipase A and of cholesterol esterase. Inadequate taurine is seen in many chemically sensitive patients, with a resultant decrease in peptide conjugation of bile salts. This lack of proper salts in the intestine will then prevent orderly pancreatic function with resultant food sensitivity and vitamin A and E malabsorption.

Chemically sensitive individuals often have severe GI upset with bloating. Supplementation with pancreatic enzymes is often necessary in order to stop this process, suggesting that exocrine pancreatic function is disturbed. Pollutant overload may not always stop the production of these enzymes, but it seems to prevent their release or efficient action. Depending on the degree, pollutant damage can cause deregulation of any of these steps of enzyme production and release, thus causing problems in the chemically sensitive patient. Alcohols are one class of pollutants known to cause severe pancreatic damage. The potential for creation of alcohols by pollutant overload is great (see Volume I, Chapter 4[2]).

Cholecystokinin is released by the presence of long chain fatty acids, certain essential amino acids (tryptophan, phenylalanine, valine, methionine), and acid (pH 1 to 2) in the duodenum. This hormone contracts the gallbladder and relaxes the sphincter of Oddi in addition to releasing pancreatic enzymes. Its five C-terminal amino acids are identical to those of gastrin. Thus, by itself it can weakly stimulate gastric acid production. Cholecystokinin usually competes with gastrin for binding to parietal cells, and, therefore, it inhibits gastric acid secretion and motility.

A cephalic phase of pancreatic secretion exists in humans, set off by the sight, smell, and chewing of food and mediated by the vagus nerves. Some chemically sensitive patients with the pancreas as the main target organ have been observed to have their symptoms triggered via this route. Vagotomy reduces pancreatic secretion in response to endogenous stimuli. As discussed in the previous section on innervation, pollutant injury with ANS imbalance may act similarly by stimulating or inhibiting pancreatic secretion or blood supply function. Gastrin and CCK release are lessened, so pancreatic enzyme output decreases. Gastric acid output with release of secretin is reduced, so bicarbonate output falls. However, vagotomy in humans does not result in clinically recognized severe impairment in digestion. Only subtle changes are present.

The decrease in buffering capacity seen in the chemically sensitive patient may well be a function of lower bicarbonate output from chemically stimulated vagotomy effects. Oral supplementation of soda bicarbonate often has been shown to stop chemically induced reactions in these patients.

Gastrin stimulates pancreatic zymogen release and gallbladder contraction because of its structural similarity to CCK. The gastric phase of pancreatic secretion is also mediated by a vagal reflex arc triggered by gastric distention. The chemically sensitive individual bloats during some types of reactions because of the production of either gastric or intestinal dystonia.

Pancreatic polypeptide is found in large endocrine type cells in the islet's periphery. Ingestion of a meal is a potent stimulus for pancreatic polypeptide release. Pancreatic polypeptide does not influence pancreatic islet hormones even when given at pharmacologic doses. However, intravenous pancreatic polypeptide does inhibit pancreatic exocrine secretion when given in doses that raise blood concentrations to those seen after a meal.

Pollutant Injury — Exocrine Portion, Buffering System

Buffering in chemically sensitive individuals is extremely important. Most buffering reactions tend to release acids. Alkali salts are frequently needed in order to buffer adequately an acute reaction in the chemically sensitive individual. The supply of bicarbonate is usually insufficient for the chemically sensitive individual to keep up with an acid load of an acute pollutant reaction.

Water, bicarbonate, and electrolytes are secreted into pancreatic juice by the centroacinar and the ductal cells of the pancreas. The ductal cells can also carry out a chloride/bicarbonate exchange. Water and bicarbonate output is stimulated by secretin, a peptide hormone whose 27 amino acid sequence is closely homologous to glucagon. However, glucagon does not stimulate bicarbonate secretion. Instead, it inhibits enzyme secretion. In addition, secretin stimulates the biliary epithelium to add water and bicarbonate to bile. The human pancreas produces 1500 to 3000 ml juice a day, which is iso-osmotic with plasma and contains the following electrolytes: NA^+, 140 mEq/L; K^+, 6 mEq/L; Ca^{2+}, 1.7 mEq/L; Mg^{2+}, 0.7 mEq/L, and HCO_3, which varies from 27 mEq/L in the resting state to a maximum of 140 mEq/L during maximum secretin stimulation. At times, pollutant stimuli may decrease the output of secretin, which would decrease the availability of bicarbonate. Chemically sensitive people will often exceed their immediate buffering activity after pollutant exposure, requiring sodium bicarbonate, either orally or intravenously, to stop a reaction. This buffering to a more alkaline pH allows more efficient response of the enzyme detoxification systems, since they usually run optimally at an alkaline pH (see Volume I, Chapter 4[2]). Calcium is partly ionized, and it is a partly bound cofactor to enzymes like amylase and deoxyribonuclease.

The pH varies directly with the bicarbonate concentration, ranging from 7.5 to 8.5, and chloride concentration varies inversely with bicarbonate. The bicarbonate output of 120 to 300 mEq per day is sufficient to neutralize gastric acid production, resulting in a pH in the distal duodenum near 7, which is close to the optimal pH for the function of the various pancreatic enzymes. Secretin is released from specialized mucosal cells when the pH of the duodenum falls below 4.5. Its secretion is maximal at pH 3 and is related also to the length of duodenum whose lumen is titrated to pH 3 to 4.5. As shown in Volume I, Chapter 4,[2] the detoxification systems work better at alkaline pHs. Therefore, it is best to keep the chemically sensitive patient more alkaline through administration of trisalts.

Clinical Aspects of Pollutant Injury to the Pancreas

With early pollutant injury, pancreatic dysfunction can be followed in some cases by severe pancreatic pain followed by acute pancreatitis. It is well established that acute pancreatitis can be triggered following exposure to such pollutants as alcohol, thiazide diuretic, furosemide, azathioprine, L-asparaginase, 6-mercaptopurine, estrogens, methyldopa, sulfonamides, tetracyclines, pentamidine, procainamide, valproic acid,[129] or morphine.[130] Other foreign substances may also be candidates for the triggering of pancreatitis, since they may trigger sphincter spasm or may act directly on pancreatic cells. These particularly could be alcohols and solvents. It should be emphasized that many toxic chemicals are broken down into alcohols before leaving the body and may well add to pancreatic dysfunction. The following cases are examples of pollutant triggering of pancreatic dysfunction.

Case study. A 48-year-old white female was admitted to the ECU with complaints of lower abdominal pain, poor concentration, nausea, and vomiting. Her problems had begun in 1958 with an episode of acute pancreatitis, accompanied by hepatitis. These resolved on their own and then reoccurred in 1976. At this time, she also underwent a cholecystectomy. She had three more attacks of jaundice and abdominal pain over several years, each accompanied by elevated amylase. She also had numerous episodes of abdominal pain daily, which incapacitated her. Prior to admission, she was found to be sensitive to chemicals, foods, and inhalants. At that time, she was placed on a 16-food, monorotation diet. She stated that each food seemed to trigger symptoms. For example, most proteins gave her joint and spine pain and poor digestion. Eventually, she was placed on a liberalized rotation diet, with approximately 35 foods neutralized and treated by injection.

Her physical exam was generally normal, except for bimanual examination, which revealed a slightly enlarged uterus, approximately twice its normal size. The pelvic ultrasound showed a 5.5 cm solid mass compatible with a uterine fibroid. She also had tenderness over her mid-abdomen without guarding or rebound.

She was tested for molds, foods, and chemical sensitivity by oral, intradermal, and inhalant challenge. She was found to be sensitive to most of these. For example, oral chicken challenge resulted in muscle cramping, spaciness, and stomach ache. Dust caused a headache and muscle pain.

Abnormal test results yielded depressed alkaline phosphatase at 30 U/L (C = 39 to 117 U/L), elevated SGPT at 40 U/L (C<32 U/L), elevated SGOT at 40 U/L (C<37 u/L), elevated CPK at 714 U/L (C = 4 to 152 U/L), elevated uric acid at 7.2 mg/dL (C = 2.5 to 6.2 mg/dL), elevated EGPT (erythrocyte glutamic pyradine transaminase) at 1.48 index (C<1.25) (indicating low B_6), elevated EGR (erythrocyte glutathione reductase) at 1.48 (C = 0.9 to 1.2 act. coeff.)(indicating low B_2), depressed INMN (1-N-methylnicotinamide) at <1.0 (C = 3.17 mg/24 hours) (indicating low B_3), depressed vitamin C, WBC at 10.1 Ug/10^8 cells (C = 21 to 53 mg/10^8 cells), elevated vitamin D_3 at 44.1 ng/L (C = 19 to 34 ng/L), elevated 5-HIAA quantitative of 32 mg/24 hours (C = 2 to 10 mg/24 hours), depressed 17-ketosteroids at 3.7 mg/24 hours (C = 5 to 15 mg/24 hours), elevated IgM at 260 mg/dL (C = 45 to 250 mg/dL), depressed CH_{100} total complement at 14 units (C>40), depressed C_3 at 77.4 mg/dL (C = 83 to 177 mg/dL), depressed total protein at 4.9 mg/dL (C = 5.9 to 7.8 mg/dL). Her RBC mineral assay revealed low silicone at 0.8 mg/24 hours (C = 1.1 to 1.7 mg/24 hours), strontium at 0.034 mg/24 hours, and serum amylase was elevated at 239 U/L (C = 60 to 222 U/L).

The blood toxic chemicals revealed moderate values of DDT and DDE both at 5.7 ppb (average 4.2 ppb). The total WBC was depressed at 3700/mm^3 (C = 4800 to 10,800/mm^3) and absolute T cells were depressed at 904/mm^3 (C = 1066 to 3197/mm^3). Eosinophil was also depressed at 35/mm^3 (C = 150 to 250/mm^3). The serum tests showed elevated iron at 1.92 ppm (C = 0.5 to 1.5 ppm), and elevated sulfur at 1003 ppm (C = 610 to 950 ppm).

This patient was placed in an environmentally controlled room constructed of porcelain where she consumed nothing by mouth except spring water for 4 days, during which time the cerebral symptoms, such as headache and poor concentration, cleared as did her abdominal pain and vomiting. While in the hospital, this patient experienced pancreatitis after an endoscopic retrograde cholangio-pancreatogram. After this obviously toxic dye-induced reaction reproduced symptoms, her amylase was also found to increase. Also, increases were observed after inhaled chemical (phenol, pesticide, petroleum alcohol) and specific food challenges, when her abdominal pain developed. The specific offending foods and substances were turkey, chicken, and carrots, as well as the ERCP dye and histamine injections. She was also treated with an injection immunotherapy program for these sensitivities and given a rotary diet of tolerated chemically less-contaminated foods (Figure 21.21).

This patient created an oasis at home that was less-polluted. She drank less-contaminated water and ate a rotary diet of less-contaminated foods. She has been asymptomatic for 3 years without the use of medication. She lives a vigorous life and is now the mainstay of her family without medication or recurrence 5 years after treatment in the ECU.

There appear to be no good practical tests to assess total chronic pancreatic dysfunction other than by challenge of the patient with enzymes and then evaluating the clinical response. We stopped doing challenge tests such as secretin when we found that the dose of enzymes and bicarbonate had to be empirically derived. The characteristics of the patient who appears to have exocrine dysfunction are the same as any other chemically sensitive patient. However, these patients are usually universally sensitive to foods and clear with chemical avoidance and food injection therapy. Their major complaints are bloating, gas, and gastrointestinal upset after eating. Trial challenges with

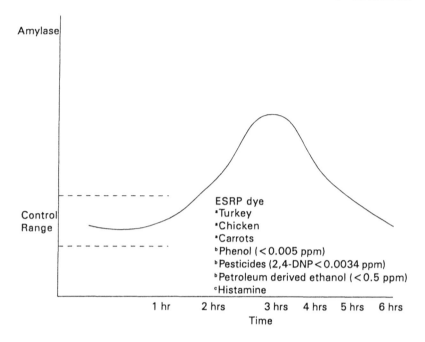

Figure 21.21. **Acute pancreatitis. 48-year-old white female, challenge after 4 days of total load reduction in the ECU. [a]Oral challenge; [b]Inhaled double-blind challenge (15 minute exposure); [c]Intradermal challenge.**

pancreas and other enzymes often relieve the gastrointestinal symptoms, especially the bloating.

Studies on the effects of pollutants on the pancreas have suggested that lack of proper quantities of enzymes can cause problems in chemically sensitive individuals.

Pancreatic duct malfunction and/or stenosis can enhance or adversely alter chemical sensitivity. Stenosis of a sphincter or duct can alter pancreatic dynamics, causing a back-up of exocrine juices and thereby allowing increased absorption in the bloodstream of various mediators and resulting in abnormal vessel sensitivity. Then chemical exposure can trigger the vessel easily, since it is already primed. The following case is a good example of this.

Case study. A 38-year-old woman was investigated in 1981 for recurrent abdominal pain, generalized swelling (especially the left side), and spontaneous bruising. In 1975, she was diagnosed with angioedema secondary to vasculitis.

She also had an extensive history of food and chemical sensitivities, vasomotor instability with vascular spasms, rashes, and difficulty breathing. A skin biopsy in 1978 confirmed chronic perivasculitis. She had previously

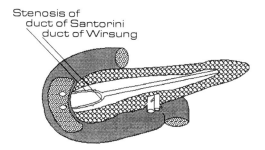

Figure 21.22.   Stenosis of the duct of Santorini and the duct of Wirsung.

been admitted to the hospital on several occasions, and she had found significant improvement in the vasculitis and edema symptoms with dietary modifications and antigen injections.

She was experiencing increasing abdominal pain, distention to certain foods and then to all foods (with analysis) and was noted to have depressed immune function with a reduced total lymphocyte count at 2300/mm³ (3000 to 4000/mm³). T-lymphocytes were 800/mm³ (C = 1400 ± 400/mm³).

CT scan and ultrasound suggested an enlarged pancreatic head, and ERCP showed an unusual bifed pancreas with two ducts, the larger one being stenosed (Figure 21.22).

In 1978, sphincterotomy and sphinctoroplasty were performed at both the sphincters of Santorini and Vater. Pancreatic juice came squirting out when the stenosis was corrected. One week after surgery, her T lymphocytes were 1600/mm³, and most of her symptoms had decreased. One month after surgery, this patient was doing very well, with substantial improvement not only in her abdominal pain and bloating but also in the angioedema and vasculitis symptoms. In addition, her lymphocyte count and distribution returned to normal, with 6300 WBCs and 2785 lymphocytes (44%). T lymphocytes were 1800/mm³. She did notice, however, that exposure to molds (to which she was known to be sensitive) would evoke her pancreatic symptoms. Follow-up at 7 and 15 years revealed that she continues to do well with minimal, if any, gastrointestinal symptoms. This patient, who was previously sensitive to all foods, was now only sensitive to five, and her angioedema and spontaneous bruising had decreased to be insignificant, except on large exposure to toxic chemicals.

This case is an interesting illustration of the potential complicating factors in environmental illness and how lack of pancreatic juices could lead to gastrointestinal and immune dysfunction. There is a good possibility that the chronic, recurrent pancreatic insufficiency played a role in this woman's vasculitis symptoms.

We have also seen patients whose recurring acute pancreatitis was apparently triggered by lectins after they had been primed by previous chemical overexposure. Lectins, which are glycoproteins (see Volume II, Chapter 8[131]), act directly on cell membranes as basic oligopeptides or proteolytic enzymes. Lectins bind to carbohydrate residues on enterocytes, and they lead to epithelial changes and hyperpermeability of the mucosa in experimental situations. This process may facilitate their access to mucosal mast cells. Lectins may also act by bridging the membrane with IgE or by linking either the carbohydrates of the Fc fragments like concanavalin A or the Fab fragments like staphylococcal protein.[132] The following case emphasizes this point.

Case study. A 37-year-old woman had five admissions to hospitals between December 1984 and December 1986 for recurrent vomiting, abdominal pain, weight loss, malnutrition, vertigo, irritable bowel syndrome, and pancreatitis. She was known by previous testing to be very environmentally sensitive to a wide variety of foods and chemicals.

In December 1984, she had consecutive admissions for pancreatitis documented with elevated urinary and serum amylase. An ERCP revealed no evident structural abnormality to account for the recurrent abdominal pain and pancreatitis.

In the hospital, she was placed in a less-polluted environment, rehydrated, and given intravenous nutritional support. Her recurrent pancreatitis had flared after she ingested certain foods containing lectins, so lectin-containing foods such as navy beans, peanuts, wheat, and rice were eliminated from her rotary diet (Table 21.20). This patient was also found to have a low superoxide dismutase at 8.2 (normal = 12.0 to 15.0). WBC was 10,800 with 855 neutrophils and 6% lymphocytes. ESR was 21-hour, and 24-hour urinary amylase was high at 27 U/hour (normal = 0 to 17 U/hour).

With dietary modification, regular antigen therapy, immune stimulations with transfer factor, and environmental cleanup, she improved dramatically for the first time in many years. Whereas she had had chronic abdominal pain, bloating, vomiting, and irritable bowel symptoms for several years, follow-up evaluation showed that these symptoms were greatly reduced, and she had not been vomiting. She gradually increased the foods her rotation diet allowed, tolerating it well. She had fewer sensitivities and was very pleased with her program. At 2 years posttreatment, she was asymptomatic.

Diagnosis of early pollutant injury of the pancreas is important since it may prevent the end-stage disease and then require treatment for chronic pancreatitis and/or diabetes. Therefore, a high index of suspicion is necessary to pinpoint the diagnosis. Certainly withdrawal and challenge testing appear to be the most important diagnostic tool. Measurements of enzyme levels, stool for fats and fiber, and secretin tests have been of little value in our hands.

Treatment of early endocrine or exocrine dysfunction includes dietary manipulation as well as concise environmental control, injection neutralization therapy, nutrient supplementation, and heat depuration/physical therapy (see Volume IV for details).

Table 21.20. White Female, Age 38: ECU after a Minimum of 4 Days Reduction of Total Load in the Deadpated State

Total WBC (/mm³)	Total lymph (/mm³)	T absolute (/mm³)	T (%)	Total (CH$_{100}$) comp (%)	CRP	EOS (/mm³)	C$_3$ (mg/dL)
3300	1176	376	32	80	+	340	60

Wheat challenge amylase (control below 150 per unit)
Control   147 per unit
5 minutes   231 per unit
30 minutes   200 per unit
12 hours   114 per unit
Symptoms
Lower extremity swelling, especially left: 2 years
Recurrent pancreatitis: 1 year
Hospitalization: 1 year
Abdominal surgery: 3X

Note: Food — multiple. Chemical — multiple.

Source: EHC-Dallas. 1986.

REFERENCES

1. Rea, W. J. 1992. *Chemical Sensitivity. Vol. I. Mechanisms of Chemical Sensitivity.* 221. Boca Raton, FL: Lewis Publishers.
2. Rea, W. J. 1992. *Chemical Sensitivity. Vol. I. Mechanisms of Chemical Sensitivity.* 47. Boca Raton, FL: Lewis Publishers.
3. Rea, W. J. 1992. *Chemical Sensitivity. Vol. I. Mechanisms of Chemical Sensitivity.* 17. Boca Raton, FL: Lewis Publishers.
4. Furness, J. B., M. Costa, J. L. Morris, and I. L. Gibbins. 1989. Novel neurotransmitters and the chemical coding of neurons. In *Advances in Physiological Research,* eds. H. Mclennan, J. R. Ledsome, C. H. S. McIntosh, and D. R. Jones. 99. New York: Plenum Press.
5. Schultzberg, M., T. Hokfelt, G. Nilsson, L. Terenius, J. F. Rehfeld, M. Brown, R. Elde, M. Goldstein, and S. Said. 1980. Distribution of peptide- and catecholamine-containing neurons in the gastrointestinal tract of rat and guinea pig: immunohitochemical studies with antisera to substance P, vasoactive intestinal peptide, enkephalins, somatostatin, gastrin/cholecystokinin, neurotensin and dopamine beta-hyroxylase. *Neuroscience* 5:689.
6. Ferri, G. L., P. L. Botti, P. Vezzadini, G. Biliotti, S. R. Bloom, and J. Polak. 1982. Peptide-containing innervation of the human intestinal mucosa: an immunocytochemical study on whole-mount preparations. *Histochemistry* 76:413.
7. Kurian, S. S., G. L. Ferri, J. DeMay, and J. M. Polak. 1983. Immunocytochemistry of serotonin-containing nerves in the human gut. *Histochemistry* 78:523.

8. Costa, M., J. B. Furness, I. J. Smith, B. Davies, and J. Oliver. 1980. An immunohistochemical study of the projections of somatostatin-containing neurons in the guinea pig intestine. *Neuroscience* 5:841.

9. Brodin, E., K. Sjolund, R. Hakanson, and F. Sundler. 1983. Substance P-containing nerve fibers are numerous in human but not in feline intestinal mucosa. *Gastroeterology* 85:557.

10. Sundler, F., R. Hakanson, and S. Leander. 1980. Peptidergic nervous systems in the gut. *Clin. Gastroenterol.* 9(3):517.

11. Rea, W. J. 1992. *Chemical Sensitivity. Vol. I. Mechanisms of Chemical Sensitivity.* Boca Raton, FL: Lewis Publishers.

12. Hemmings, W. A. 1978. The entry into the brain of large molecules derived from dietary protein. *Proc. R. Soc. Lond. B.* 200:175–192.

13. Executive Committee of the American Academy of Allergy and Immunology. 1986. Candidiasis hypersensitivity syndrome. *J. Allergy Clin. Immunol.* 78(2):271–273.

14. Scheline, R. R. 1973. Metabolism of foreign compounds by gastrointestinal microorganisms. *Pharmacol. Rev.* 25(4):451–523.

15. Goldman, P. 1982. Role of the intestinal microflora. In *Metabolic Basis of Detoxification: Metabolism of Functional Groups*, eds. W. B. Jakoby, J. R. Bend, and J. Caldwell. 325. New York: Academic Press.

16. Kasalow, J. 1986. Personal communication.

17. Scheline, R. R. 1973. Metabolism of foreign compounds by gastrointestinal microorganisms. *Pharmacol. Rev.* 25(4):451–523.

18. Davis, S. May 14–17, 1992. Gut fermentation and autobrewery syndrome. Presented at American College of Advancement in Medicine. Dallas, TX.

19. Peppercorn, M. A. and P. Goldman. 1972. The role of intestinal bacteria in the metabolism of salicylazosulfapyridine. *J. Pharmacol. Exp. Ther.* 181:555–562.

20. Reddy, B. G., L. R. Pohl, and G. Krishna. 1976. The requirement of the gut flora in nitrobenzene-induced methemoglobinemia in rats. *Biochem. Pharmacol.* 25:1119–1122.

21. Rosenkranz, H. S. and W. T. Speck. 1975. Mutagenicity of metronidazole: activation by mammalian liver microsomes. *Biochem. Biophys. Res. Commun.* 66:520–525.

22. Glazko, A. J., W. A. Dill, and L. Wolf. 1952. Observations on the metabolic disposition of chloramphenicol (Chloromycetin®) in the rat. *J. Pharmacol. Exp. Ther.* 104:452–458.

23. Goldman, P. 1982. Role of the intestinal microflora. In *Metabolic Basis of Detoxification: Metabolism of Functional Groups*, eds. W. B. Jakoby, J. R. Bend, and J. Caldwell. 330. New York: Academic Press.

24. Spatz, M., D. W. E. Smith, E. G. McDaniel, and G. L. Laqueur. 1967. Role of intestinal microorganisms in determining cycasin toxicity. *Proc. Soc. Exp. Biol. Med.* 124:691–697.

25. Goldman, P. 1982. Role of the intestinal microflora. In *Metabolic Basis of Detoxification: Metabolism of Functional Groups*, eds. W. B. Jakoby, J. R. Bend, and J. Caldwell. 333. New York: Academic Press.

26. Armstrong, B. and R. Doll. 1975. Environmental factors and cancer incidence and mortality in different countries with special reference to dietary practice. *Int. J. Cancer* 15:617–631.

27. Wynder, E. L. and T. Shigematsu. 1967. Environmental factors of cancer of the colon and rectum. *Cancer (Philadelphia)* 20:1520–1561.

28. Duke, W. W. 1925. *Allergy, Asthma, Hay Fever, Urticaria, and Other Manifestations of Reaction.* St. Louis: C. V. Mosby.

29. Vaughan, W. T. 1939. *Practice of Allergy.* St. Louis: C. V. Mosby.

30. Randolph, T. G. 1976. Personal communication.

31. Friddian-Green, R. G., J. Farrel, D. Havlichek, Jr., P. Kothari, and G. Pittenger. 1978. A physiological role for luminal gastrin? *Surgery* 83:663.

32. Larsson, L. I., N. Golterman, L. DeMagistrio, J. F. Rehfeld, and T. W. Schwartz. 1979. Somatostatin cell processes as pathway for paracrine secretion. *Science* 205:1393.

33. Rehfeld, J. F., L. G. Heding, and J. J. Holst. 1973. Increased gut glucagon release as pathogenetic factor in reactive hypolycemia. *Lancet* 1:116.

34. Holst, J. J. and J. F. Rechfeld. 1975. Human circulating gut glucagon: binding to liver cell plasma membranes. In *Gastrointestinal Hormones*, ed. J. C. Thompson. 529. Austin: University of Texas Press.

35. Solcia, E., C. Bordi, W. Creutzfeld, Y. Dayal, A. D. Dayan, S. Falkmer, L. Grimelius, and N. Havu. 1988. Histopathological classification of nonantral gastric endocrine growths in man. *Digestion* 41:185.

36. Hinton, R. H., F. E. Mitchell, A. Mann, D. Chescae, S. C. Price, A. Nunn, P. Grasso, and J. W. Bridges. 1986. Effects of phthalic acid esters on the liver and thyroid. *Environ. Health Perspect.* 70:195–210.

37. Price, S. C., S. Ozalp, R. Weaver, D. Chescoe, J. Mullervy, and R. H. Hinton. 1988. Thyroid hyperactivity caused by hypolipodaemic compounds and polychlorinated biphenyls: the effect of coadministration in the liver and thyroid. *Arch. Toxicol. Suppl.* 12:85–92.

38. Dayal, Y. and H. J. Wolfe. 1984. G-cell hyperplasia in chronic hypercalcemia: an immunocytochemical and morphometric analysis. *Am. J. Pathol.* 116:391.

39. Tielmans, Y., R. Hakanson, F. Sundler, and G. Willems. 1989. Proliferation of enterochromaffin like cells in omeprazole treated hypergastrinemic rats. *Gastroenterology* 96:723.

40. Tielmans, Y., J. Axelson, F. Sundler, G. Willems, and R. Hakanson. 1990. Serum gastrin concentration affects the self-replication rate of the enterochromaffin like cells in the rat stomach. *Gut* 31:274.

41. Dayal, Y., E. F. Vaelkel, A. H. Tashijian, Jr., R. A. Delellis, and H. J. Wolfe. 1977. Antropyloric G-cell hyperplasia in hypercalcemia rabbits bearing the VX$_2$ carcinoma. *Am. J. Pathol.* 89:391.

42. Inokuchi, H., S. Fujimoto, and K. Kawai. 1983. Cellular kinetics of gastrointestinal mucosa with special reference to gut endocrine cells. *Arch. Histol. Jpn.* 46:137.

43. Larsson, H., E. Carlson, H. Mattson, L. Lundell, F. Sundler, G. Sundell, B. Wallmark, T. Watanabe, and R. Hakanson. 1986. Plasma gastrin and gastric enterochromaffin-like cell activation and proliferation: studies with omeprazole and ranitidine in intact and antrectomized rats. *Gastroenterology* 90:391.

44. Ryberg, B., H. Mattson, F. Sundler, R. Hakanson, and E. Carlsson. 1986. Effects of inhibition of gastric acid secretion in rats on plasma gastrin levels and density of enterochromaffin-like cells in the oxyntic mucosa. Supplement, 6th Inst. Symp. Gastrointestinal Hormones. *Can. J. Physiol. Pharmacol.* 110:34.

45. Havu, N. 1986. Enterochromaffin-like cell carcinoids of gastric mucosa in rats after life-long inhibition of gastric secretion. *Digestion* 35(Suppl. 1):42.

46. Lehy, T., M. Mignon, G. Cadiot, L. Eloncer-Blanc, P. Ruszniewski, M. J. Lewin, and S. Bonfils. 1989. Gastric endocrine cell behavior in Zollinger-Ellison patients upon long-term potent antisecretory treatment. *Gastroenterology* 96:1029.

47. Lamberts, R., W. Creuzfeldt, F. Stockman, U. Jacubaschke, S. Maas, and G. Brunner. 1988. Long-term omeprzole treatment in man: effects on gastric endocrine cell populations. *Digestion* 39:126.

48. Bordi, C., T. D'Adda, F. T. Balato, and C. Ferrari. 1988. Carcinoid (ECL cell) tumor of the oxyntic mucosa of the stomach: a hormone dependent neoplasma? *Prog. Surg. Pathol.* 9:177.

49. Challacombe, D. N. and K. Robertson. 1977. Enterochromaffin cells in the duodenal mucosa of children with celiac disease. *Gut* 18:373.

50. Sjolund, K., J. Alumets, N.-O. Berg, R. Hakanson, and F. Sundler. 1979. Duodenal endocrine cells in adult celiac disease. *Gut* 20:547.

51. Sjolund, K., J. Alumets, N.-O. Berg, R. Hakanson, and F. Sundler. 1982. Enteropathy of celiac disease in adults: increased number of enterochromaffin cells in the duodenal mucosa. *Gut* 23:42.

52. Enerback, L., C. Hallert, and K. Norrby. 1983. Raised 5-hydroxytryptamine concentration in enterochromaffin cells in adult celiac disease. *J. Clin. Pathol.* 36:499.

53. Challacombe, D. N., P. D. Dawkins, and P. Baker. 1977. Increased tissue concentrations of 5-hydroxytryptamine in the duodenal mucosa of patients with celiac disease. *Gut* 18:882.

54. Challacombe, D. N., G. A. Brown, S. C. Black, and M. H. Storrie. 1972. Increased excretion of 5-hydroxyindolacetic acid in urine of children with untreated celiac disease. *Arch. Dis. Child.* 47:442.

55. Solcia, E., C. Capella, G. Vassallo, and R. Buffa. 1975. Endocrine cells of the gastric mucosa. *Int. Rev. Cytol.* 42:223.

56. Solcia, E., C. Capella, R. Buffa, L. Usellini, B. Frigerio, and P. Fontana. 1979. Endocrine cells of the gastrointestinal tract and related tumors. In *Pathobiology Annual*, ed. H. L. Iaochim. 163. New York: Raven Press.

57. Lindop, G. B. M. 1983. Enterochromaffin cell hyperplasia and megacolon: report of a case. *Gut* 24:575.

58. Sjolund, K., J. Alumets, N. O. Berg, R. Håkanson, and F. Sundler. 1979. Duodenal endocrine cells in adult celiac disease. *Gut* 20(7):547–552.

59. Rea, W. J. and M. J. Mitchell. 1982. Chemical sensitivity and the environment. *Immunol. Allergy Pract.* 4(5):157–167.

60. Swiatkowski, M. 1982. Przydatność endoskopowej próby prowokacji górnego odcinka przewodu pokarmowego w rozpoznawaniu alergii na pokarmy. Dissertation. Bydgoszcz: Akademia Medyczna.

61. Swiatkowski, M., M. Kurek, J. Kakol, and B. Romański. 1981. Study on the multiorganal manifestations of an acquired form of flour allergy. Xe Congrès d'Interasma (Abstract) Paris. Respiration [Suppl. 1] 42:23.

62. Dickey, L. D. 1976. Gastrointestinal and genitourinary tissue eosinophils and mast cells. In *Clinical Ecology*, ed. L. D. Dickey. 441. Springfield, IL: Charles C Thomas.

63. Williams, G. M. and J. H. Weisburger. 1986. Food and cancer: cause and effect? *Surg. Clin. N. Am.* 66(5):873–879.

64. Alun-Jones, V. and J. O. Hunter. 1987. Irritable bowel syndrome and Crohn's disease. In *Food Allergy and Intolerance*, eds. J. Brostoff and S. J. Challacombe. 555. London: Baillière Tindall.

65. Randolph, T. G. 1980. Personal communication.

66. Wald, A. 1994. Constipation: moving things along at a low cost. In *Difficult Decisions in Digestive Diseases*, 2nd ed., eds. J. S. Barkin and A. I. Rogers. 283. St Louis: C. V. Mosby.

67. Burkitt, D. P., 1980. Fiber in the etiology of colorectal cancer. In *Colorectal Cancer: Prevention Epidemiology and Screening*, eds. D. Schottenfeld, P. Sherlock, and S. J. Winawer. 13–18. New York: Raven Press.

68. Randolph, T. G. 1984. Personal communication.

69. Finn, R., M. A. Smith, G. R. Youngs, D. Chew, P. M. Johnson, and R. M. R. Barnes. 1987. Immunological hypersensitivity to environmental antigens in the irritable bowel syndrome. *Br. J. Clin. Pract.* 41(12):1041–1043.

70. Sugimura, T. 1985. Carcinogenicity of mutagenic heterocyclic amines formed during the cooking process. *Mutat. Res.* 150:33–41.

71. Rea, W. J. 1994. *Chemical Sensitivity. Vol. II. Sources of Total Body Load.* 765. Boca Raton, FL: Lewis Publishers.

72. Cuello, C., P. Correa, W. Haenzell, G. Gordillo, C. Brown, M. Archer, and S. Tannenbaum. 1976. Gastric cancer in Colombia. I. Cancer risk and suspect environmental agents. *J. Natl. Cancer Inst.* 57(5):1015–1020.

73. Armijo, R. and A. H. Coulson. 1975. Epidemiology of stomach cancer in Chile. The role of nitrogen fertilizers. *Int. J. Epidemiol.* 4:301–309.

74. Haenzel, W. F., B. Locke, and M. Segi. 1980. A case-control study of large bowel cancer in Japan. *J. Natl. Cancer Inst.* 64:17–22.

75. Joint Iran-International Agency for Research on Cancer Study Group. Esophageal cancer studies in the Caspian littoral of Iran: results of population studies — a prodrome. 1977. *J. Natl. Cancer Inst.* 59:1127–1138.

76. Coordinating Group for Research on Etiology of Esophageal Cancer in North China. 1975. The epidemiology of esophageal cancer in north China. *Clin. Med. J.* 1:167–183.

77. Palmer, S. and R. A. Mathews. 1986. The role of nonnutritive dietary constituents in carcinogenesis. *Surg. Clin. N. Am.* 66(5):891–915.

78. Soos, K. 1980. The occurrence of carcinogenic polycyclic hydrocarbons in foodstuffs in Hungary. *Arch. Toxicol.* 4(Suppl.):446–448.

79. Levander, O. A., B. Sutherland, V. C. Morris, and J. C. King. 1981. Selenium balance in young men during selenium depletion and repletion. *Am. J. Clin. Nutr.* 34(12):2662–2669.

80. Stoloff, L. 1982. Mycotoxins as potential environmental carcinogens. In *Carcinogens and Mutagens in the Environment. Vol. 1. Food Products*, ed. H. F. Sitch. Boca Raton, FL: CRC Press.

81. Rea, W. J. 1994. *Chemical Sensitivity. Vol. II. Sources of Total Body Load.* 1011. Boca Raton, FL: Lewis Publishers.

82. National Research Council. 1982. *Diet, Nutrition, and Cancer.* Washington, D.C.: National Academy Press.

83. Hirono, I. 1981. Natural carcinogenic products of plant origin. *CRC Crit. Rev. Toxicol.* 8:235–277.

84. Hirono, I., H. Kachi, and T. A. Kato. 1970. A survey of acute toxicity of cycus and mortality rate from cancer in the Miyako Islands, Okinawa. *Acta Pathol. Jpn.* 20:327–337.

85. Brown, J. P. 1980. A review of the genetic effects of naturally occurring flavonoids, anthraquinones and related compounds. *Mutat. Res.* 75:243.

86. Ito, N., A. Hagiwara, M. Shibata, T. Ogiso, and S. Fukushima. 1982. Induction of squamous cell carcinoma in the forestomach of F344 rats treated with butylated hydroxyanisole. *Gann* 73:332–334.

87. Fuchs, G., B. M. Gawell, L. Albanus, and S. Slorach. 1975. Vinyl chloride monomer levels in edible fats. (Swedish, with English summary) *Var Foeda* 17:134–145.

88. Rea, W. J. 1994. *Chemical Sensitivity. Vol. II. Sources of Total Body Load.* 535. Boca Raton, FL: Lewis Publishers.

89. Netter, F. H., ed. 1962. *The CIBA Collection of Medical Illustrations. Vol. III. Digestive System. Part II: Lower Digestive Tract.* 37–39. New York: CIBA Pharmaceutical Company Division of CIBA Corporation.

90. Rea, W. J. 1994. *Chemical Sensitivity. Vol. II. Sources of Total Body Load.* 837. Boca Raton, FL: Lewis Publishers.

91. Olefsky, J. M. 1988. Diabetes mellitus. In *Cecil Textbook of Medicine*, Vol. II, 18th ed., eds. J. B. Wyngaarden and L. H. Smith. 1373–1374. Philadelphia: W. B. Saunders.

92. Rea, W. J. 1992. *Chemical Sensitivity. Vol. I. Mechanisms of Chemical Sensitivity.* 155. Boca Raton, FL: Lewis Publishers.

93. Laseter, J. L. 1988. Personal communication.

94. Rea, W. J. 1994. *Chemical Sensitivity. Vol. II. Sources of Total Body Load.* 579. Boca Raton, FL: Lewis Publishers.

95. Netter, F. H., ed. 1962. *The CIBA Collection of Medical Illustrations. Vol. III. Digestive System. Part II: Lower Digestive Tract.* 36–37. New York: CIBA Pharmaceutical Company Division of CIBA Corporation.

96. Netter, F. H., ed. 1962. *The CIBA Collection of Medical Illustrations. Vol. III. Digestive System. Part II: Lower Digestive Tract.* 35. New York: CIBA Pharmaceutical Company Division of CIBA Corporation.

97. Rea, W. J. 1992. *Chemical Sensitivity. Vol. I. Mechamisms of Chemical Sensitivity.* 1. Boca Raton, FL: Lewis Publishers.

98. Freund, H. A. 1937. Clinical manifestation and studies in parenchymatous hepatitis. *Ann. Int. Med.* 10:1144.

99. Neal, R. A. and J. Halpert. 1982. Toxicology of thiono-sulfur compounds. *Annu. Rev. Pharmacol. Toxicol.* 22:321–339.

100. Yoshimoto, M., M. Yamaguchi, S. Hatano, and T. Watanabe. 1984. Configurational changes in rat liver nuclear chromatin caused by azo dyes. *Food Chem. Toxicol.* 22(5):337–344.

101. Hallewbeck, W. H. and K. M. Cunningham-Burns. 1985. *Pesticides and Human Health.* 13. New York: Springer-Verlag.

102. Popper, H., S. S. Sternberg, B. L. Oser, and M. Oser. 1960. The carcinogenic effects of aramite in rats: a study of hepatic nodules. *Cancer* 13:1035.

103. Looney, W. B. and L. M. Colodzin. 1956. Late follow-up studies after internal deposition or radioactive materials. *JAMA* 160(1):1–3.

104. Pitot, H. C. 1988. Hepatic neoplasia: chemical induction. In *The Liver: Biology and Pathobiology*, 2nd ed., eds. I. M. Arias, W. B. Jakoby, H. Popper, D. Schachter, and D. A. Shafritzs. 1125–1146. New York: Raven Press.

105. Ecobichon, D. J. 1977. Hydrolytic transformation of environmental pollutants. In *Handbook of Physiology*, ed. D. H. K. Lee. 441–454. Baltimore: Williams & Wilkins.

106. Letz, G. 1982. *The Toxicology of PCBs. An Overview with Emphasis on Human Health Effects and Occupational Exposure Health Hazard and Information Service*. State of California: Department of Health Services.

107. Lubet, R. A., B. N. Lemaire, D. Avery, and R. E. Kouri. 1986. Induction of immunotoxicity in mice by polyhalogenated biphenyls. *Arch. Toxicol.* 59:71–77.

108. Shinozuka, H., B. Lombarbi, S. Sell, and R. M. Iammarimo. 1978. Early histological and functional alterations of ethionine liver carcinogenesis in rats fed a choline-deficient diet. *Cancer Res.* 38(4):1092–1098.

109. Degen, G. H. and H. G. Neumann. 1978. The major metabolite aflatoxin B_1 in the rat is a glutathione conjugate. *Chem. Biol. Interact.* 22:239–255.

110. Balazs, T. 1981. Hepatic reactions to chemicals. In *Toxicology: Principles and Practice*. Vol. I, ed. A. L. Reeves. 93–106. New York: John Wiley & Sons.

111. Vainio, H. 1978. Vinyl chloride and vinyl benzene (styrene) — metabolism, mutagenicity, and carcinogenicity. *Chem. Biol. Interact.* 22:117–124.

112. Netter, F. H., ed. 1962. *The CIBA Collection of Medical Illustrations. Vol. III. Digestive System. Part II: Lower Digestive Tract.* 91. New York: CIBA Pharmaceutical Company Division of CIBA Corporation.

113. Netter, F. H., ed. 1962. *The CIBA Collection of Medical Illustrations. Vol. III. Digestive System. Part II: Lower Digestive Tract.* 78–81. New York: CIBA Pharmaceutical Company Division of CIBA Corporation.

114. Netter, F. H., ed. 1962. *The CIBA Collection of Medical Illustrations. Vol. III. Digestive System. Part II: Lower Digestive Tract.* 31. New York: CIBA Pharmaceutical Company Division of CIBA Corporation.

115. Shirakawa, S., W. J. Rea, S. Ishikawa, and A. R. Johnson. 1992. Evaluation of the autonomic nervous system response by pupillographical study in the chemically sensitive patient. *Environ. Med.* 8(4):121–127.

116. Deconinck, J. F., P. R. Potvliege, and W. Gepts. 1971. The ultrastructure of the human pancreatic islets. 1. The islets of adults. *Diabetologia* 7:266.

117. Baum, J., B. E. Simmons, R. H. Unger, and L. L. Madison. 1962. Localization of glucagon in the A-cells in the pancreatic islet by immunofluorescence. *Diabetes* 11:371.

118. Larsson, L. S., F. Sundler, R. Häkansson, H. G. Pollock, and J. R. Kimmel. 1974. Localization of APP, a postulated new hormone to a pancreatic endocrine cell type. *Histochemistry* 42:377.

119. Luft, R., S. Efendic, T. Hökfelt, O. Johansson, and A. Azimura. 1974. Immunohistochemical evidence for the localization of somatostatin-like immunoreactivity in a cell population of the pancreatic islets. *Med. Biol.* 52:428.

120. Goossens, A., P. Heitz, and G. Klöppel. 1991. Pancreatic endocrine cells and their nonneoplastic proliferations. In *Endocrine Pathology of the Gut and Pancreas*, ed. Y. Dayal. 69–71, 76, 80. Boca Raton, FL: CRC Press.

121. Liljenquist, J. E., G. L. Mueller, A. D. Cherrington, U. Keller, J.-L. Chiasson, J. M. Perry, W. W. Lacy, and D. Babinowitz. 1977. Evidence for an important role of glucagon in the regulation of hepatic glucose production in normal man. *J. Clin. Invest.* 59(2):369–374.

122. Kimmel, J. R., H. G. Pollock, R. E. Chance, M. G. Johnson, J. R. Reeve, I. L. Taylor, C. Miller, and J. E. Shively. 1984. Pancreatic polypeptide from rat pancreas. *Endocrinology* 114:1725.

123. Lin, T. M., D. C. Evans, R. E. Chance, and C. F. Spray. 1977. Bovine pancreatic polypeptide: action on gastric and pancreatic secretion in dogs. *Am. J. Physiol.* 232:E311.

124. Taylor, I. L., T. E. Solomon, J. H. Walsh, and M. I. Grossman. 1979. Pancreatic polypeptide metabolism and effect on pancreatic secretion in dogs. *Gastroenterology* 76:524.

125. Greenberg, G. R., R. F. McCloy, T. E. Adrian, V. S. Chadwick, J. H. Baron, and S. R. Bloom. 1978. Inhibition of pancreas and gallbladder by pancreatic polypeptide. *Lancet* 2:1280.

126. Greenberg, G. R., R. F. McCloy, V. S. Chadwick, T. E. Adrian, J. H. Baron, and S. R. Bloom. 1979. Effect of bovine pancreatic polypeptide on basal pancreatic and biliary outputs in man. *Dig. Dis. Sci.* 24(1):11–14.

127. Werbach, M. R., ed. 1988. *Nutritional Influences on Illness: A Sourcebook of Clinical Research.* 175–176. Tarzana, CA: Third Line Press.

128. Miyano, M., K. Nanjo, S. J. Chan, T. Sanke, M. Kondo, and D. F. Steiner. 1988. Use of in vitro DNA amplification to screen family members for an insulin gene mutation. *Diabetes* 37(7):862–866.

129. Steinberg, W. M. 1992. Pancreatitis, In *Cecil Textbook of Medicine*, 19th ed., eds. J. B. Wyngaarden, L. H. Smith, Jr., and J. C. Bennett. 722. Philadephia: W. B. Saunders.

130. Snodgrass, P. J. 1977. Diseases of the pancreas. In *Harrison's Principles of Internal Medicine*, 8th ed., eds. G. W. Thorn, R. D. Adams, E. Braunwald, K. J. Isselbacher, and R. G. Petersdorf. 1636. New York: McGraw-Hill.

131. Rea, W. J. 1994. *Chemical Sensitivity. Vol. II. Sources of Total Body Load.* 579. Boca Raton, FL: Lewis Publishers.

132. Friedman, S. J., A. L. Schroeter, and H. A. Homburger. 1985. IgE antibodies to staphylococcus aureus. *Arch. Dermatol.* 121:869.

22   Genitourinary System

INTRODUCTION

The genitourinary system is quite susceptible to the influence of environmental triggers.[1,2] Many animal and some human studies support this fact.[1,2] Numerous animal studies delineate problems of the genitourinary system that result from chemical exposures. The reproductive parts of both males and females are affected as well as the excretory system. This chapter will discuss the excretory system as well as some parts of the genital system, namely the prostate and seminal vesicles. The rest of the reproductive system, including the testes and sperm, will be covered in Chapters 24 and 28.

PROSTATE AND SEMINAL VESICLES

The prostate is prone to environmental triggers, especially bacteria. However, we have observed that many cases of prostatitis can be triggered by chemical exposure. One should always look for the underlying cause when prostatitis occurs.

Pollutant Effects on the Autonomic Nervous System

No information is available on the pollutant effects of prostatic innervation.

Pollutant Effects on Physiology

The seminal vesicles secrete a mucoid material containing an abundance of fructose and other nutrient substances, prostaglandins, and fibrinogen. The fructose and other substances are of considerable nutrient value for ejaculated sperm until one of them fertilizes the ovum. The prostaglandins aid fertilization by making the cervical mucosa more receptive to sperm and possibly by causing reverse peristalsis in the uterus and fallopian tubes to move the sperm toward the ovaries. Pollutant overload certainly can change the fructose and other nutrient contents of these secretions. As has been shown in Volume I, Chapter 6,[3] it also can alter prostaglandin production, which renders the male less fertile.

The prostate secretes a thin alkaline fluid containing citric acid, calcium, zinc, acid phosphate, a clotting enzyme, and a profibrinolysin. This alkalinity is necessary for sperm to function optimally and to neutralize the vaginal acid milieu. Pollutant injury to this delicately composed secretion will decrease the pH and alter the other components, causing malfunction.

Clinical Manifestations of Pollutant Effects

The prostate reaches stationary size by the age of 20 years and remains there until the age of 40 or 50 years. At that time, it begins to degenerate along with the decreased production of testosterone by the testes, perhaps *because* of the decreased testosterone. Benign prostatic hypertrophy occurs, but this is apparently not caused by testosterone. Cancer of the prostate is extremely common and results in 2 to 3% of all male deaths.

Some information is available on the environmental aspects of clinical entities and prostatic disease. For example, a multipotential carcinogen, 3,2-dimethyl-4-aminobiphenyl (DMAB), was shown to produce prostate carcinoma and atypical hyperplasia.[4] At the EHC-Dallas, we have observed a series of chemically sensitive patients with prostatic dysfunction who were triggered by environmental overload. We have had the opportunity of observing a 45-year-old dentist who had recurrent benign prostatic obstruction, which cleared when he entered the ECU. His symptoms were reproduced after challenge with natural gas fumes. He subsequently had to remove the gas heater from his office in order to stay well. The following is a case report of chronic prostatitis diagnosed and cleared by environmental means.

> **Case study.** A 30-year-old white male was admitted to the Environmental Control Unit (ECU) with the chief complaints of discomfort of the prostate and intermittent burning on urination and lethargy. The patient stated that his prostate felt "full" at all times. He had been treated for gonorrhea approximately 12 years previously, but he had had no recent use of antibiotics. He was previously seen by numerous urologists without significant help for this problem.
>
> His physical exam was normal, except for a tender prostate. He was admitted to the ECU. During the hospital course, this patient was placed in an environmentally controlled room constructed of porcelain. He fasted for $6^1/_2$ days and did not completely clear his symptoms during the fast. However, with a longer time in the ECU, his symptoms did clear.
>
> Detoxification symptoms included food cravings, cold sensitivities, prostate pain, low back pain, and arm, muscle, and joint pain. This patient was treated with magnesium sulfate, commonly used to treat muscle pain, but could not tolerate it. His treatment was arranged for foods, pollens, dusts, and molds by intradermal and avoidance method.

Table 22.1. Chronic Prostatitis in 30-Year-Old White Male Inhaled Double-Blind Challenge after 4 Days Deadaptation with the Total Load Reduced in the ECU

Chemical challenge	Dose (ppm)	Results
Ethanol (petroleum derived)	<0.50	+[a]
Chlorine	<0.33	+
Phenol	<0.002	+
Formaldehyde	<0.20	−
Pesticide (2,4-DNP)	<0.0034	−
Saline	ambient	−
Saline	ambient	−
Saline	ambient	−
Natural gas[b]	open flame	+

[a] Swollen, tender prostate and back pain, burning on urination, and lethargy.

[b] Open challenge.

Source: EHC-Dallas. 1988.

He was challenged with chemically less-contaminated foods and exhibited sensitivity to 24 of them. He also underwent inhaled chemical exposure done in a steel and glass airtight booth under double-blind conditions in the deadapted state with his total load reduced. Chemicals that produced symptoms were natural gas, alcohol, chlorine, and phenol. Chemicals that failed to produce symptoms were formaldehyde and insecticide. Placebos, too, produced no reactions (Table 22.1).

This patient's abnormal tests showed depressed T lymphocyte 1068/mm^3 (1600 to 2000/mm^3) during hospitalization, depressed vitamin B$_1$ at 23 ng/mL (25 to 75 ng/mL), and vitamin B$_6$ at 26 ng/mL (30 to 80 ng/mL). His intradermal skin testing showed he had sensitivity to histamine, serotonin, cigarette smoke, cotton, orris, perfume, six molds, dust, dust mite, TOE, *Candida*, grass, weed, terpenes, trees, and grains. Yeast was a big offender and also reproduced all of his symptoms, including prostatitis.

He was discharged with the diagnosis of prostatitis, general malaise, cephalgia, immunosuppression characterized by T lymphocyte suppression, tinnitus, and inhalant, food, and chemical sensitivities.

After being treated with food and inhalant injections and pollutant avoidance, he felt much better. It is noted that his problem was work-related in that he had a liquor business, and his job necessitated him tasting several wines daily. Even though he spit them out, this minimal contact, acting as a sublingual challenge, was sufficient to trigger his prostatitis. He was discharged on a rotary diversified diet of chemically less-contaminated foods and spring water. Occasional flare-ups of prostatic symptoms, which occured during the first year of outpatient treatment, were successfully managed with administration of oral vitamin C.

Long-term follow-up revealed 15 years without recurrence. This case demonstrates that environmentally controlled surroundings, a rotation diet of chemically less-contaminated foods, and a program of immunotherapy can be of great benefit to chemically sensitive patients with the symptoms of prostatitis.

At the EHC-Dallas, we have seen five cases of chemical sensitivity with recurrent prostatitis. These patients were well worked up (Table 22.2). They were found to have a diverse number of triggering agents that would reproduce their symptoms. These included a spectrum of biological, inhalant, food, and toxic chemical agents. The patients constructed oases and removed the fossil fuel from their houses. Three took injections for inhalants and foods, and all went on a rotary diet of chemically less-contaminated food, avoiding those to which they were sensitive. They also drank less-polluted water. The patients gradually cleared their symptoms without medication and have remained symptom free on long-term follow-up. See Volume II, Chapter 17[5] for a case report due to algae.

Shambaugh[6] and others[7] have suggested that zinc deficiency will trigger prostatic hypertrophy and even some cases of inflammation. It is clear, however, that environmental nutritional influences occur in the prostate and are only now beginning to be appreciated.

KIDNEY AND UPPER UROGENITAL SYSTEM

Since the kidneys control the concentration of the constituents of the body fluids and excrete the end product of bodily metabolism, they are prone to pollutant injury. Glomerular filtration and tubular reabsoprtion parallel each other. Various factors can alter these. The most important are the degree of sympathetic stimulation to the kidneys, the tubular osmotic clearance, the plasma colloid osmotic pressure, the arterial pressure, and the effect of antidiuretic hormone on tubular reabsorption. Each aspect of kidney function will be discussed separately.

Pollutant Effects on the Renal Nervous System

Autonomic innervation of the kidney is through the greater splanchnic and the vagus nerve. These enter the kidney through the renal plexus, which extends from the aortic plexus. Most of the synaptic relays for the sympathetic nervous system come from the celiac ganglion (Figure 22.1).

Sympathetic stimulation has an especially powerful effect on constriction of the afferent arterioles. It greatly decreases glomerular filtration and glomerular pressures. Conversely, a decrease of autonomic stimulation to below normal causes a mild degree of afferent arteriole dilatation, which increases the glomerular filtration rate and increases urine volume. This phenomenon is

Table 22.2. Environmentally Triggered Prostatitis

Age/race/sex	Time of onset to time seen	Biological inhalants	Foods	Toxic chemicals[a]	Results — long-term without medication
43/W/M	9 years — 8 of constant antibiotic use	–	–	+	3 years
45/W/M	5 years	–	–	+	2 years
30/W/M	3 years	+	–	+	6 years
50/W/M	2 years	+[b]	+	+	2 years
55/W/M	5 years	+	+	+	4 years

[a] Natural gas, phenol, formaldehyde, petroleum, alcohol, pesticide, chlorine.
[b] Algae was the prime offender.

Source: EHC-Dallas. 1994.

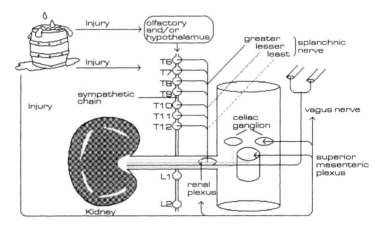

Figure 22.1. Areas of potential pollutant injury to the ANS of the kidney and aorta.

often seen in patients confined to the ECU when their total load has been reduced and they experience diuresis, losing all of their intracellular edema.

Pollutant injury to the ANS usually deregulates the microvasculature of the kidney. As with other organs, deregulation may occur as a result of pollutants entering through the nose and lungs, through the hypothalamus or the gastrointestinal tract, or through other portals of entry. The intricate relationship of the sympathetic nerve supply to the celiac axis suggests one of the reasons that ingested foods and chemicals are so important to the triggering of environmentally triggered hypertension. It also emphasizes why some of the earlier reported cases of hypertension and peripheral allergic vascular phenomena were relieved by sympathectomy.[8] For information on essential hypertension, see Chapter 22.

Pollutant Effects on the Physiology of the Kidney

The kidneys contain about 2.4 million nephrons. If only a few are intially disturbed, damage will likely not be immediately apparent. It is clear in our studies of the chemically sensitive patient at the EHC-Dallas that early pollutant injury can affect any part of the nephron. Often, early pollutant effects seen in the chemically sensitive patient are exhibited in renal leaks of electrolytes such as potassium, magnesium, and calcium, vitamins, and especially amino acids. We have now measured thousands of chemically sensitive patients whose leaks could be measured, and they exhibited the usual symptoms of early deficiency (see Volume I, Chapter 6[3] and Volume IV, Chapter 30). Early pollutant damage can occur to the filtering mechanism of the glomerali and/or to the reabsorption mechanism of the long tubule. The fraction of blood flow through the kidney is 21% of the total blood volume (1200/5600 cc/min), but it may vary from 12 to 30% at any one time. Alteration of this flow by pollutant exposure may alter normal processes in the chemically sensitive individual.

The glomerular membrane has an endothelial layer, a basement membrane, and an epithelial layer. All layers are prone to pollutant injury, perhaps because normal permeability of this membrane is 100 to 1000 times as great as the usual capillary. The tremendous permeability of the glomerular membrane is caused by a special structure. The capillary endothelial cells are perforated by thousands of small holes, while the basement membrane is composed of a mesh work of proteoglycan fibrilluae. The epithelial cells consist of finger-like projections that form slits through which the glomerular filtrate filters.

Experiments have shown that the holes in the endothelial cells are small enough in diameter that they prevent filtration of all particles greater than 16 microns. The mesh work of the basement membrane prevents all filtration of all particles greater than 11 microns, and the slit pores of the epithelial cells prevent filtration of particles of greater than 7 microns. Since plasma proteins are greater than 7 microns, it is possible to prevent filtration of all molecular substances equal to or greater than plasma proteins. One can see how pollutants either combined with plasma proteins or attached to the different proteins around the pores can cause damage and alter filtration. Substances such as cyanides, halogenated hydrocarbons, etc. can either damage the protein or its energy producer or combine with the proteins to form haptens, thus causing malfunction at the kidney or distally. For example, it has been shown that cyanide will be reabsorbed in the kidney and go to the thyroid producing a pseudo-halogen effect, with hypothyroidism resulting (see Volume I, Chapter 4[9]). Also, halogens have been observed to damage the tubules causing magnesium leaks (Figure 22.2).

Pollutant injury can also occur to any area of the kidney per se. The glomerular membrane is impermeable to molecules larger than 4 nm diameter. This corresponds to a molecular weight of approximately 70,000. Injury to the membrane may occur from toxic chemicals, especially pesticides, herbicides, and solvents. As well, certain sensitivity reactions to molds and foods occur. Many toxic chemicals have a lower molecular weight than 70,000. Proximal tubule functions reduce the volume of the glomerular filtrate 70 to 80%. Here the active reabsorption of glucose, sodium, chloride, and potassium, magnesium and iron may be disturbed by pollutant injury. Sulfates, amino acids, and low molecular weight proteins are usually reabsorbed. However, pollutant injury is known to result in amino aciduria[10] as well as electrolyte losses. We have seen losses of magnesium and iron as well as sodium and potassium from pollutant injury in the chemically sensitive patient. Since sodium and water are also reabsorbed in the loop of Henle and develop a concentration gradient in the renal medulla, pollutant damage may disturb the water balance. Some chemicals such as the diuretic furosemide work here. These toxic chemicals cause damage to the water balance and tubular reabsorption and exemplify how the present toxic levels found in the chemically sensitive patient could cause tissue damage.

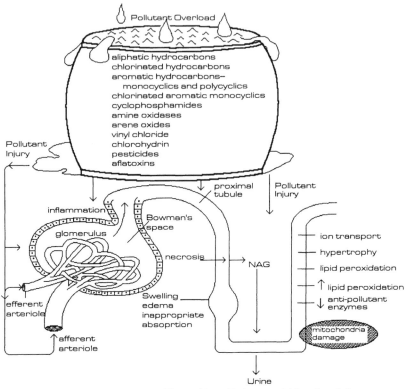

Figure 22.2. Areas of potential pollutant injury to the nephron.

Tubular osmotic pressure. When the osmolar load is great, urine volume increases, as is seen in patients with a high sugar solute load in a diabetes that is out of control. Excess sucrose, manitose, and urea will cause a similar diuresis.

Plasma colloid osmotic pressure. A sudden increase in plasma colloid osmotic pressure instantaneously decreases the rate of fluid volume excretion in the kidneys. The cause of this acceleration is twofold. First, an increase in plasma colloid-osmotic pressure decreases the glomerulo infiltration rate. Second, the plasma colloid osmotic pressure increases tubular reabsorption.

Effect of arterial pressure on rate of fluid volume excretion. If arterial pressure increases from 180 to 200 mmHg, the increase in urine output is sevenfold. When the blood pressure falls from 100 to 60 mmHg, urine output falls to near zero.

Effect of antidiuretic hormone on rate of fluid volume excretion. When excess antidiuretic hormone is secreted by pollutant stimulation of the hypothalmic-posterior pituitary system, as is often seen in the chemically sensitive patient, urinary volume decreases, and intracellular edema increases.

The reason for this decrease in volume is that the antidiuretic hormone causes increased water absorption from the collecting ducts and slightly from the distal tubules. Even though less urine is excreted, it is highly concentrated.

Inappropriate antidiruetic hormone (ADH) syndrome involves long-term secretion of large quantities of ADH causing severe abnormalties in body fluid ionic composition, even though body fluid volume is slightly altered. ADH activates adenyl cyclase in the kidney's epithelial membrane, causing formation of cyclic AMP in the cell cytoplasm, and increased permeability occurs.

Water absorption in the distal loop and collecting duct depends on the antidiuretic hormone from the pituitary. We have shown in Volume I, Chapter 4[9] and Volume III, Chapter 24, how this hormone may be disturbed by pollutant injury (Figure 22.2).

Many substances are secreted in the distal tubule including potassium and hydrogen ions, in addition to most xenobiotics. With the occurrence of pollutant injury due to excessive chemical load, selective potassium, magnesium, iron, phosphate, calcium, sulfate, amino acid, and sodium bicarbonate leaks are seen in chemically sensitive individuals.

If one estimates the minimal amount of pollutants needed to cause damage, one can then see how the present pollutant levels found in the chemically sensitive individual could cause tissue damage. For example, studies show that 1 part per trillion of dioxin may cause tissue damage. Thus, if a chemically sensitive patient were to present with 1 part per billion, he or she would be predictably a candidate for illness, if signs and symptoms were not already present.

Many toxic chemicals have now been shown to cause renal damage. Some will be discussed in the following paragraphs.

The mechanism of glutathione nephrotoxicity was studied in isolated rats. The findings suggest that PCB-induced renal injury may be the result of selective loss of Ca^{2+} from the mitochondria, inhibition of cell respiration, and marked depletion of cellular ATP content, e.g., loss of mitochondrial function.[11] We have seen 30 chemically sensitive patients with isolated leaks of calcium and magnesium presumably due to this type lesion. As shown in Volume I, Chapter 6,[3] a significant proportion of chemically sensitive patients have excess, and/or loss of, calcium due to pollutant damage of the membranes. We have treated hundreds of patients with PCBs in their blood.

2,2,4-Trimethylpentane induces hyaline droplet formation in male[12] rat kidneys. We have treated a segment of patients who have pentanes and hexanes in their blood, and we wonder if early renal damage is occurring similar to that found in the rats. Other chemicals like cyclophosphamide cause nephrotic syndrome in humans.[13-15] Our series of patients with renal damage had not been exposed to cyclophosphamide, though they had been exposed to other toxic chemicals that acted similarly (Table 22.3) . These included the pentanes, hexane, toluene, benzene, and tetrachloroethanes and ethylenes. The effects of cyclophosphamide are decreased by a synthetic prostaglandin E_1. Other toxic chemicals such as chlorambucil induce congenital renal hypoplasia in rats

Table 22.3.   Environmentally Triggered Renal Damage

Age/race/sex	Triggering exposure	Lesion	Triggering agents Biological	Foods	Toxic chemicals
14 years/W/F	Herbicide	Glomerulo	+	+	+
40 years/W/M	Auto exhaust	Nephritis	–	+	+
35 years/W/F	Formaldehyde	Glomerulo	–	+	+
37 years/W/F	Paraquat	Nephritis	+	+	+
33 years/W/F	Organic solvents	Nephritis	+	+	+
30 years/W/F	Anesthetics	Nephritis	+	+	+
10 years/W/M	Unknown	Nephritis	+	+	+
16 years/W/F	Herbicide and pesticide	Nephrosis	+	+	+

Source: EHC-Dallas. 1994.

while renal dysfunction may occur with exposure to nitrogen and malnutrition.[16] The fungicide N-(3,5-dichloropheny)-3-hydroxysuccinamic acid and its intermediate in metabolism, arene oxide, give proximal tubular necrosis in rats.[17] This finding suggests that the lipophilic character of a substance is not a predictor of nephrotoxic potential.[17] It also emphasizes the importance of the history when one studies renal damage in individual patients. One 15-year-old patient who developed nephrosis was exposed to herbicides in her drinking water and her ambient air. The levels in the water were high enough to kill the family dog, which contained lethal levels of the herbicide at autopsy.

In vivo kidney cells were found to be affected by various chemicals (aflatoxin B_1, aroclor 1254, benzidine, benzo[*a*]pyrene, and 20-methylcholanthrene), which induced chromosomal aberrations.[18] Sodium O-phenylphenate produced decreased urinary gamma-glutamyl transpeptidase (gamma GTP), decreased (80%) gamma GTP and alkaline phosphatase in kidney, increased gamma GTP in liver (8 × control), increased G6PD, and significantly decreased glutathione concentrations.[19]

Four cysteine conjugates of chloro- and fluorohexanes (hexachlorobutadiene [HCBC], trifluoroethylene [TFE] and hexafluoropropene [HFP], trichloroethylene [TCE] and tetra- or perchloroethylene [found in 78% of chemically sensitive patients]) had marked effects on the uptake of both organic anion *p*-aminohippuric acid and the cation tetraethylammonium bromide in kidney slices and suggested interference with ion transport while confirming known nephrotoxicity *in vivo*.[20] These studies again support the observation that isolated nutrient leaks through the kidneys occur in many of our chemically sensitive patients.

The kidney of rats exposed to vinyl chloride at 10,000 and 3000 ppm increased in weight at 6 months. Male rats exposed to 30 ppm of hexafluoroisobutylene had decreased body weights and significantly increased kidney weights in 13 weeks.[21] Also, hydrocarbon exposure did not result in linear binding to intratracheal basement membranes of rabbits.[22] These facts indicate that chemically induced kidney damage is not only a function of the concentration of individual exposure but also a function of total body load as well as individual susceptibility to a specific chemical.

Renal xenobiotic metabolism can result in production of free radicals that may covalently bind macromolecules or initiate lipid peroxidation. The mechanisms of renal xenobiotic metabolism may vary in different anatomical regions. The kidney cortex contains a cytochrome P-450 system, while the medulla contains a prostaglandin endoperoxidase. These facts again emphasize why different patients with different exposures will have different areas of damage. Recently, cysteine-conjugated lysate has been implicated in production of reactive intermediates. Metabolic activation may be amplified by accumulation of xenobiotics within renal cells due to tubular concentrating and/or secretory mechanisms that will exacerbate chemical sensitivity. Additionally, renal xenobiotic detoxification can occur by conjugation with glucuronide, sulfate, or glutathione and can be damaged due to this increased load, resulting in an increase in chemical sensitivity.[23] One study found that sensitivity to chloroform correlated with the capacity of the kidney to metabolize CCl_3 to the toxic metabolite phosgene ($COCl_2$).[24] This more toxic breakdown product increased the sensitivity. Selective damage to any of these processes of conjugation and oxidation in chemically sensitive individuals appears to produce a profound increase in their sensitivity.

Clinical Aspects of Pollutant Injury to the Kidney — Upper Urogenital System

Several reports of workers exposed to different toxic chemicals emphasize how excess exposure will trigger disease. For example, chlorophenate-exposed workers in sawmills and workers exposed to aromatic amines had lower hematocrits and microscopic hematuria vs. controls.[25] Methanesulfonate (MMS), styrene-7,8-oxide, styrene, trichloroethylene (TRI), and tetrachloroethylene (PER) caused single strand breaks (SSB) in kidney DNA. In order of potency, the substances able to induce SSB of DNA are MMS, styrene-7,8-oxide, styrene, PER, and TRI. 2-Hexanone and CCl_4 both cause renal damage in rats.[26] Chemically sensitive individuals with these various substances in their blood appear to be vulnerable to damage by them as well as to damage by other toxic substances. The range of effects of toxic chemicals on the kidney may explain why the sensitivities of these individuals increase when their total load increases.

Exposure to nephrotoxic chlorinated organic solvents such as those found in the chemically sensitive patient affects the renal excretion of the enzyme N-acetyl-beta-D-glucosaminidase (N.A.G.). Initially, with exposure, the enzyme increases. This increase is usually caused by necrosis of renal tubular cells and can be used as a marker for nephropathy. Of the 113 workers exposed to trichloroethylene (TRI), trichloroethane, and freon who had urine samples analyzed for N.A.G. activity and trichloroacetic acid concentration, 10% had enhanced N.A.G. similar to diabetics with subclinical nephropathy. Increased N.A.G. activity was observed in previously TRI-exposed persons, which might indicate induction of an autoimmune renal necrosis.[27]

Increase in N.A.G. is also seen in patients exposed to ifosfamide (chemical name: N,3-*bis*[2-chloroethyl]tetrahydro-2H-1,3,2-oxazaphosphorin-2-amine 2-oxide) for cancer chemotherapy.[28] Chlorambucil (chemical name: 4-[*p*-[*bis*(2-chloroethyl)amino]benzenebutanoic acid) causes gestational damage to the metonephric area in rats. This damage occurs on approximately day 11 postexposure.[29]

Finn et al.[30] showed that males (e.g., auto mechanics) constantly exposed to hydrocarbons from car exhaust had a much higher incidence of glomerulonephritis (GMN) and renal failure than other groups (Table 22.4, Figure 22.3). N-butane, isobutane, N-pentane, and isopentane make up 90% of all gasoline hydrocarbon exposures. These substances are found in 85% of chemically sensitive patients. A mixture of 25% of each of these at 4437 ppm for 21 days inhaled by rats produced no renal lesions, but this lack of response may have been due to species differences. There is a strong implication of induction in renal disease in humans, and the wrong experimental conditions or animal probably were selected in this experiment.

Finn et al.[30] performed an occupational survey involving 87 patients with end-stage failure in order to determine their exposure to hydrocarbon-based compounds. They found that 59% of patients with renal failure due to GMN had significant contact with hydrocarbons as compared with 25% of patients with renal failure due to other causes ($p<0.01$). Also, 81% of patients with GMN were male, and of these 64% had significant hydrocarbon exposure as compared with 35% of an age-matched control group of male hospital inpatients with no renal pathology. Finn et al. concluded that occupational exposure to hydrocarbons may account for the predominance of male patients with glomerulonephritis.

According to the survey by Finn et al.,[30] a group of patients with GMN who were on a hemodialysis program had a significantly higher exposure to hydrocarbons as compared with patients in end-stage renal failure due to other causes (non-GMN) and to controls without renal disease. Occupational hydrocarbon exposure was much more common in males but was more frequent in male patients with GMN as compared with age- and sex-matched controls. Occupational exposure was uncommon in females, and none of the 23 females with end-stage renal failure due to non-GMN renal disease in this survey had had a hydrocarbon contact, whereas 2 of the 7 females with GMN had had such a contact. One of these had worked in a paint factory, and the other had worked as a driver.

In a study by Finn et al.,[30] the hydrocarbon exposure of males with GMN was higher than that of males with renal failure due to other causes, but the difference was not statistically significant. This outcome may have been due to some patients in this group having been incorrectly diagnosed as nephrosclerosis with hypertension when their initial pathology may well have been GMN. It is

Table 22.4.   Contact with Hydrocarbons[a]

Group	Total	HC contact	Probability
1. GMN both sexes	37	22 (59%)	
2. Non-GMN both sexes	52	13 (25%)	$p_{1-2} < 0.01$
3. GMN male	30	20 (66%)	
4. Non-GMN male	29	13 (45%)	p_{3-4} NS
5. Control male	53	19 (35%)	$p_{3-5} < 0.01$
6. GMN female	7	2 (29%)	
7. Non-GMN female	23	0-	$p_{6-7} = 0.048$
			(Fisher's exact test)

[a] Contact defined as a minimum of 2 hours a day for 5 days a week for 2 years.

Source: Finn, R., A. G. Fennerty, and R. Ahmad. 1980. Hydrocarbon exposure and glomerulonephritis. *Clin. Nephrol.* 14(4):173–175. With permission.

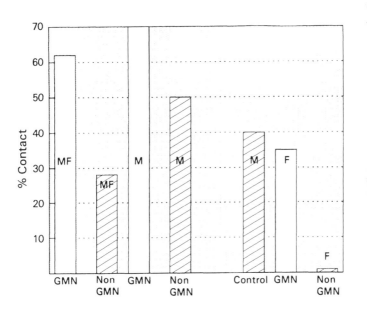

Figure 22.3. Hydrocarbon (HC) exposure in hemodialysis patients with glomerulonephritis (GMN), other forms of renal disease (non-GMN), and male controls without renal disease (M = male; F = female). (From Finn, R., A. G. Fennerty, and R. Ahmad. 1980. Hydrocarbon exposure and glomerulonephritis. *Clin. Nephrol.* 14(4):173–175. With permission.)

of interest that Zimmerman et al.[31] showed a male predominance of 86% of patients with GMN. Following a small survey of patients with Goodpasture's disease, however, Heale et al.[32] suggested that hydrocarbon exposure may account for the male predominance of this disease. A correlation between hydrocarbon exposure and Goodpasture's disease was suspected as long ago as 1963 and has since been confirmed.[32-34]

Three mechanisms to account for hydrocarbon-induced GMN have been suggested. First, by comparison with Goodpasture's disease, chemical damage to alveolar cells by inhaled hydrocarbons may initiate antibody formation to the altered alveolar antigen, which may then share a common structure with that of the glomerular basement membrane, and GMN may, thus, be initiated. Second, via a similar mechanism, renal tubular damage, known to occur following exposure to hydrocarbons,[35] may initiate antibodies to shared glomerular basement antigen. Third, there may be a direct toxic effect of hydrocarbon on the basement membrane similar to that demonstrated on the kidney tubules. Proliferative GMN has been induced experimentally in rats fed on N,N-diacetyl-benzidine.[36]

Evidence for the potential toxic effect of hydrocarbon on other systems is mounting, and exposure may be responsible for conditions as diverse as cardiac arrhythmias, marrow depression, and lymphomas,[37-40] as well as the known association between benzene and leukemia.

Since hydrocarbon exposure is common in our industrial society, such effects probably represent idiosyncratic or sensitivity reactions rather than direct toxic effects. Finn et al.[30] argue that the results of their study suggest that a certain minimum exposure to hydrocarbon is usually required to produce a situation in which sensitization can occur. This minimum exposure probably represents a fairly heavy industrial contact, defined by Finn et al. as a minimum of 2 hours per day for 5 days per week for 2 years of exposure to hydrocarbons. From the therapeutic point of view, it is possible that removal from the hydrocarbon environment may improve or halt the progression of acute and chronic GMN,[41] while, from the prophylactic point of view, routine screening of at risk populations may reduce the incidence of this disease and, thus, lessen the load on expensive hemodialysis and transplant programs.

Females metabolize some toxic chemicals, especially chloroform, faster than males[42] (Table 22.5). Only those males susceptible to chloroform nephrotoxicity appear to have altered renal mixed function oxidase activity. Sex-related differences of the renal cytochrome P-450 occur, with activity higher in males. As well, altered concentrations of ethoxycoumarin o-diethylase occur. Castration of male rats eliminated susceptibility to chloroform nephrotoxicity and reduced renal mixed function oxidase to the levels seen in females.[43] Of patients with chemical sensitivity, 33% have chloroform at levels up to 50 ppb in their blood. This occurrence may be due to either the breakdown of chlorinated hydrocarbons taken in at work and/or home, or it may be generated from the breakdown of the trichloromethane generated from the ingestion of chlorinated water. Gross renal lesions were not seen in these patients, although some of them had abnormally low levels of superoxide dismutase and glutathione peroxidase and increased lipid peroxidases. Enzyme studies were not available in all patients.

Table 22.5. Sex Differences in Kidney Necrosis Following Chloroform Exposure

	No kidney necrosis		Kidney necrosis	
Sex	A	B	A	B
Normal females	80	20		
Castrated males	69	31		
Normal males			31	69
Castrated males treated with testosterone			36	64

Note: A = percent of Bowman's capsules lined with squamous cells. B = percent of Bowman's capsules lined with cuboidal cells.

Source: Eschenbrenner, A. B. and E. Miller. 1945. Sex differences in kidney morphology and chloroform necrosis. *Science* 102(2647):302–303. With permission.

"White spirit solvent" (C_{10}-C_{12}) mixture of branched-chain unsubstituted saturated aliphatic hydrocarbons has produced distal tubular alterations in some patients.[44] Hexachloro-1- and 3-butadiene and its derivatives produce acute necrosis of the pars recta of the proximal renal tubule.[45] These chemicals have been identified in some chemically sensitive patients.

Bromobenzene is nephrotoxic. Early events include decreased glutathione levels and inhibition of mitochondrial respiration. Later, plasma membrane permeability occurs. Induction of the renal mixed function oxidase systems correlates with ultrastructural changes in the proximal tubule.[46] Phenolic bromobenzene exposure leads to renal proximal tubular and renal functional changes.[46,47] Bromobenzene has also been found to increase SGPT activity and to produce severe hepatic necrosis.[48]

Twenty percent of the chemically sensitive patients seen at the EHC-Dallas have bromoforms in their blood. This presence is probably due to the compounds used as fumigants for wheat and wheat flour. Its presence there is emphasized by the results of a recent Texas survey that showed only one source of wheat flour without bromoforms. This source was a brand sold as organic and found mainly in health food stores. However, our patients could also have been exposed to bromoforms via their water supply. Brominated benzene compounds have been found in some drinking waters.

Postnatal methyl mercury exposures caused a precocious onset of ornithine decarboxylase (ODC) responses to trophic agents specifically in the kidney.[49] There were significant renal hypertrophy and increased ODC activity. Live weight was reduced. Methyl mercury, usually associated primarily with specific nervous system damage, also adversely affected biochemical and functional indices of renal organ development.[50] Hexafluoroisobutylene (HRIB) caused renal damage in rats.[28] Hexachlorobutadiene-N-acetylcysteine (HCBD-NAC), adriamycin, and 2-bromoethanimine hydrobromide are renal toxins that showed selective toxicity to the proximal tubules, glomerular epithelial cells, and proximal tubules in humans.[51,52]

1,2-Dibromo-3-chloropropane reduced survival rate and produced renal and hepatic necrosis.[54] Forty-three male printing trade workers exposed to toluene (382 mg/m³ for 6¹/₂ hours) were compared with 43 age-matched male controls. There were no significant changes in renal excretion rates of albumin and beta 2-microglobulin during toluene exposure.[54] No evaluation was made for renal nutritional losses or chemical sensitivity.

Sandberg[55] reported a series of patients with minimum change nephrosis who were triggered by various foods and toxic chemicals. Also, he showed some patients who had membranous glomerulopathy while others had nephritis associated with anaphylactoid purpura due to foods and chemicals (Tables 22.6 to 22.10 and Figures 22.4 to 22.6).

Matsumura and his group reported in 1961, and again in 1971, results of clinical studies of a large number of children with nephrosis who were shown to have sensitivity to foods and in whom the use of limited diets was apparently helpful in achieving prolonged remission of their disease.[55,56] They also reported an association of food sensitivity with postural proteinuria. In 1977, Sandberg's group reported studies of six children with frequently relapsing steroid-dependent minimal change disease (MCD) in whom cow's milk sensitivity was associated with relapse.[57] They subsequently described further studies in the same children and investigation of an additional group of 13 children.[58] Richards and co-workers in 1977 described a girl who had nephrosis as well as a history of asthma, eczema, and urticaria who appeared to have relapsed on ingestion of chicken eggs.[59] Lagrue and his colleagues[60] have reported results of a study of the human basophil degranulation test (HBDT) to investigate food sensitivity in 34 patients with MCD nephrosis. Five food allergens were used (wheat, milk, egg, beef, and pork); 64% of 34 patients had at least one positive test, while these tests were positive only once in a group of 19 blood donors ($p<0.001$). Only nine showed one sensitization; eight reacted to two foods, and three reacted to three foods. All five foods were found to cause reactions in multiple patients; this group included five patients who were reactive to egg, six to milk and beef, and nine to wheat and pork. The HBDT was positive whether patients were in relapse or remission, whether they were on or off corticosteroid therapy, and whether their serum immunoglobulin E was elevated or not. There was little correlation between standard skin tests, radioallergosorbent tests, and HBDT.

Matsumura's group of patients with nephrosis is of importance to our understanding of environmentally induced renal disease since they were not selected for a pattern of frequent relapse. In contrast, the children studied in Miami had had numerous relapses requiring frequent administration of prednisone or had been treated with an immunosuppressive agent such as cyclophosphamide.[56] Matsumura reported a high degree of successful control of the nephrosis in his patients using limited diets. Evidence of food sensitivity was obtained in those patients by individual oral food challenges.

Table 22.6.   Studies of Food Sensitivity in Nephrosis: Laboratory
Values with Long-term Follow-up of Four Children with MCD

Patients	Protein excretion (mg/24 hours)	Albumin TP/albumin (mg/dL)	Cholesterol (mg/dL)	IgG (mg/dL)	BUN/ creatinine (mg/dL)
EN	80	7.2/4.8	—	950	14/0.9
JR	45	6.9/4.3	—	1150	16/1.1
TN	35	7.8/5.2	190	1260	14/1.0
JM[a]	40	7.2/4.9	205	920	11/0.8

Notes: BUN=blood urea nitrogen; TP=total protein.

[a] This patient did not relapse when prednisone was discontinued.

Source: Sandberg, D. 1987. Food sensitivity: the kidney and bladder. In *Food Allergy and Intolerance*, eds. J. Brostoff and S. J. Challacombe. 760. London: Baillière Tindall. With permission.

Table 22.7.   Studies of Food Sensitivity in Nephrosis: Laboratory
Values with Long-term Follow-up 6/19 Patients with MCD[a]

Patients	Protein excretion (mg/24 hours)	TP/albumin (mg/dL)	Cholesterol (mg/dL)	IgG (mg/dL)	BUN/ creatinine (mg/dL)
VL	40	6.2/3.9	201	800	7/0.8
YB	65	7.3/4.2	165	940	9/0.8
TL	195	5.6/3.4	187	695	9/0.4
TW	60	6.2/3.8	230	580	13/0.8
THo[b]	1800	5.4/3.2	330	410	27/1.2
TH	35	6.8/4.5	180	1130	10/0.6

Notes: TP = total protein. BUN = blood urea nitrogen.

[a] These values were obtained from a recent follow-up visit.

[b] These values were representative of the status prior to dialysis and transplantation.

Source: Sandberg, D. 1987. Food sensitivity: the kidney and bladder. In *Food Allergy and Intolerance*, eds. J. Brostoff and S. J. Challacombe. 760. London: Baillière Tindall. With permission.

Relapsing Minimal Change Disease

The six children with frequently relapsing steroid-responsive MCD studied in Miami[57] were shown to be sensitive to cow's milk by intradermal titration skin testing and by *in vivo* alteration of plasma C_3 complement using crossed immunoelectrophoresis. Oral challenge with cow's milk provoked relapse in five patients; Figure 22.7 indicates one patient's response to challenge with cow's milk. The changes in protein excretion following milk

Table 22.8.   Studies of Food Sensitivity in Nephrosis: Follow-up
of Three Patients with Membranous Glomerulopathy

Patients	Protein excretion (mg/24 hours)	TP/albumin (mg/dL)	Cholesterol (mg/dL)	IgG (mg/dL)	BUN/ creatinine (mg/dL)
SS	40	7.4/5.3	175	1150	12/0.8
SL	50	6.8/3.9	195	1250	10/0.8
JW[a]	2410	5.3/2.9	340	520	45/3.7

Notes: TP = total protein. BUN = blood urea nitrogen.

[a] These values were representative of this patient's status shortly before institution of dialysis.

Source: Sandberg, D. 1987. Food sensitivity: the kidney and bladder. In *Food Allergy and Intolerance*, eds. J. Brostoff and S. J. Challacombe. 762. London: Baillière Tindall. With permission.

Table 22.9.   Studies of Food Sensitivity in
Nephrosis: Long-term Follow-up
of Three Children with MCD

Patients	Number of significant relapses	Prednisone therapy[a]	Other problems
EN	2	2	None
JR	1	1	None
TN	0	0	None

[a] Number of relapses since beginning management for food sensitivity, requiring prednisone therapy.

Source: Sandberg, D. 1987. Food sensitivity: the kidney and bladder. In *Food Allergy and Intolerance*, eds. J. Brostoff and S. J. Challacombe. 759. London: Baillière Tindall. With permission.

ingestion were dramatic. Transient changes in serum IgG and IgM were also demonstrated. One girl did not relapse when prednisone was discontinued, although she had required intermittent, frequent prednisone therapy for several years because of relapse whenever prednisone had been previously discontinued. Subsequently, 13 more children with steroid-dependent MCD were studied and reported in 1961,[56] as well as further studies of three of the original group. All 19 had had persistent disease. In nine, treatment with cytotoxic drugs was undertaken because of inadequate control with prednisone or undesirable side-effects of this latter drug. In 17 children, sensitivity to one or more foods was documented. Although these patients were not selected because of a history of allergic disease, seven had asthma, four had eczema, one had

Table 22.10.   Frequency of HLA-B$_{12}$ and Relation with Atopic Features in Children with Steroid Responsive Nephrotic Syndrome

	Group A		Group B		Combined		British blood donors
	(+)	(−)	(+)	(−)	(+)	(−)	(+)
HLA-B$_{12}$ (%)	54	46	44	56	50	50	24
Atopic history (%)	37	3	15	32	29	16	
Total IgE >150 U (%)	48	13	17	32	34	24	

Notes: Group A consisted of 71 children who were consecutive outpatients; group B comprised 72/81 children treated with cyclophosphamide. All were in remission and off treatment. Twenty-seven of group B were included previously in group A, while 45 were a new series not previously evaluated for HLA type and atopy.

Source: Sandberg, D. 1987. Food sensitivity: the kidney and bladder. In *Food Allergy and Intolerance*, eds. J. Brostoff and S. J. Challacombe. 758. London: Baillière Tindall. With permission.

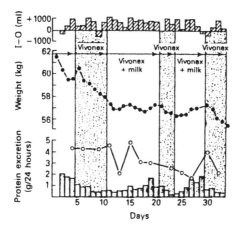

Figure 22.4.   Changes in weight, net water balance and protein excretion during dietary changes in a patient with the nephrotic syndrome associated with membranous glomerulopathy. I–O = 24-hour fluid intake minus urine output; g/24 hours = protein excretion during a 24-hour period. (From Sandberg, D. H., T. F. Mcleod, and J. Strauss. 1980. Renal disease related to hypersensitivity to foods. In *Food Allergy: New Perspectives*, eds. J. W. Gerrard. 144. Springfield: Charles C Thomas. With permission.)

allergic rhinitis, and one had chronic urticaria; 13 of 19 children had positive intradermal skin tests to one or more foods. One child was unresponsive to prednisone; however, he responded well to diet restriction and food extract

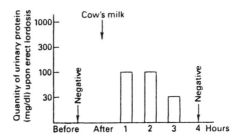

Figure 22.5. Lordotic proteinuria resulting from administration of a single feeding of cow's milk. (From Matsumura, T. 1976. Postural proteinuria. In *Clinical Ecology*, ed. L. Dickey. 233. Springfield: Charles C Thomas. With permission.)

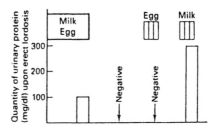

Figure 22.6. Lordotic proteinuria resulting from successive 3-day periods of administration of egg and milk. (From Matsumura, T. 1976. Postural proteinuria. In *Clinical Ecology*, ed. L. Dickey. 233. Springfield: Charles C Thomas. With permission.)

injections with good control of his nephrosis. Although Sandberg's initial report documented provocation of relapse with cow's milk, all but two patients who could not be adequately evaluated were found to be sensitive to multiple foods.[55]

It is clear from these studies that different foods and toxic chemicals influenced changes in the kidneys of these individuals. At times, elimination of the triggering agents reversed or arrested the process. Similar changes also were seen in some patients with positive outcomes at the EHC-Dallas.

Several chemically sensitive patients with pollutant injuries of the kidney have been seen at the EHC-Dallas. These patients' problems ranged from individual electrolyte leaks of magnesium, calcium, potassium, and iron to nephrosis, to glomerulonephritis, to florid renal failure (Table 22.3).

One 10-year-old child presented with minimum change nephrosis refractory to diuretics on 100 mg of hydrocortisone per day with a massive weight gain. He had a severe steroid-induced Cushing's syndrome. He was placed in the ECU. As his total load decreased, his weight spontaneously decreased due

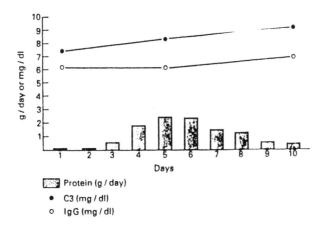

Figure 22.7. Effect of milk ingestion on protein excretion in a child with minimal change nephrotic syndrome. The solid bars depict values for 24-hour protein excretion. (From Sandberg, D. 1987. Food sensitivity: the kidney and bladder. In *Food Allergy and Intolerance*, eds. J. Brostoff and S. J. Challacombe. 759. London: Baillière Tindall. With permission.)

to diuresis. His gammaglobulin and T lymphocytes, which were low, increased to the normal range after a month in the ECU. Challenge tests showed that multiple foods and chemicals triggered his water retention, and a massive weight gain would develop until the food or chemical was eliminated (Figure 22.8).

Patients who have already developed renal failure and need hemodialysis face a new set of problems due to the use of toxic cleaning solutions such as chlorine and formaldehyde as well as emanations from the plastic membranes in the dialysis machines. We have now seen several patients react to these substances, causing an exacerbation in their illness and discomfort. We also have seen several dialysis nurses made ill from the cleaning solutions. A case report of one such nurse follows.

Case study. This 33-year-old white, female nurse worked in a dialysis unit for 5 years without health problems. The dialysis unit was always cleaned with chlorine bleach. Suddenly she developed asthma. She was placed on Ventolin®, aminophyllin, and eventually prednisone. Her asthma, however, continued, and as a side-effect of the prednisone she gained 40 pounds. She became totally incapacitated and was transferred to the EHC-Dallas. She was in such bad shape with a pulmonary flow of 11 cc that she was admitted to the ECU. She was placed on intravenous aminophyllin and magnesium sulfate and fasted for 5 days until her flow came up to 300 cc. She was then fed one food per meal and skin tested for foods and biological inhalants and chemicals. By this time she had developed multiple food, biological inhalant, and chemical sensitivities. Laboratory tests showed blood toxic chloroform,

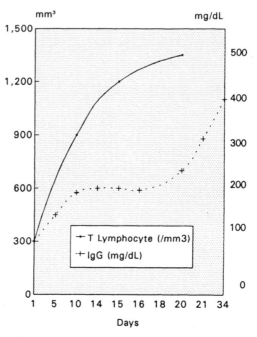

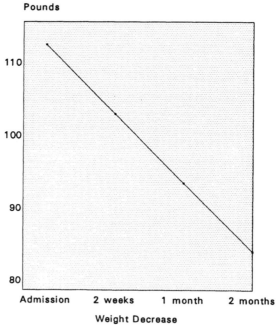

Figure 22.8. Minimum change nephrosis in a 10-year-old white female.
Return of laboratory parameters to normal with time in the
ECU. (From EHC-Dallas. 1980.)

68 ppm (C<0.33 ppm). She was placed in heat depuration/physical therapy and started on nutrient supplementation. The chloroform disappeared from her blood. Her peak flow rose to 450 cc, and she was taken off all prednisone and aminophyllin compounds and returned home well.

BLADDER AND LOWER UROGENITAL SYSTEM

Pollutant injury to the bladder and urethra appears to be more common than originally thought. The presence of neuroendocrine cells at the neck of the bladder allowing for an adverse and immediate set of responses when pollutants enter distally has been reported frequently by our patients.

Pollutant Effects on the Autonomic and Somatic Nervous System

The bladder and urethra function as a single complex unit for storage and expulsion of urine. These functions are controlled by somatic and autonomic nerves with complex central connections allowing voluntary control over micturition. Pollutants may enter through any area including nose, lungs, intestinal tract, or genitourinary system, and they can cause injury that results in problems with urination.

Sensory nerves from the bladder and urethra pass in the pelvic parasympathetic nerves (S_2–S_4) to the spinal cord. From there, fibers ascend in the lateral spinothalamic tracts to micturition centers in the pons and cerebral cortex (Figure 22.9).

However, sacral nerve blockade does not abolish bladder sensation completely because some sensory fibers accompany the sympathetic nerves to the hypogastric plexus. Pollutant injury to the ANS or abnormal stimulation or inhibition of the paraganglion system will cause dysfunction of this mechanism frequently giving a sudden urge to urinate (occasionally with loss of urine) after an acute inhaled toxic exposure. This urgent bladder response is seen frequently in the chemically sensitive patient. Occasionally, a food reaction will give a similar response. Conversely, the opposite response of retention may be triggered.

Detrusor contractions are mediated by parasympathetic cholinergic nerves (S_2–S_4). These travel in the pelvic nerves (nervi erigentes) to the pelvic ganglia from where numerous fibers are distributed to the bladder. Sympathetic nerves (T_{10}–L_2) travel to the bladder via the hypogastric nerves and the pelvic plexus. They are largely vasomotor, but some sympathetic fibers synapse with parasympathetic nerves in the pelvic ganglia and produce inhibition of detrusor contractions. Again, these may malfunction from pollutant injury, causing urinary reduction or the sudden urge to urinate.

The rhabdosphincter is innervated by somatic fibers from S_2 and S_3, which travel with autonomic fibers in the nervi erigentes; the autonomic fibers supply the inner smooth muscle of the urethra. The periurethral muscle sling is also supplied by somatic fibers from S_2 and S_3, but these fibers travel in the

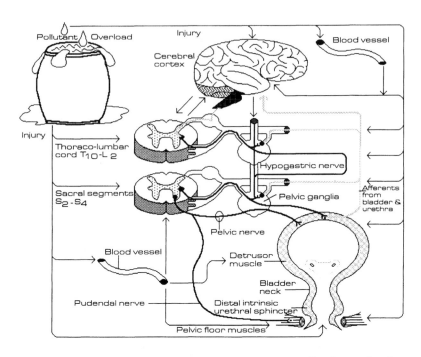

Figure 22.9. Areas of potential pollutant damage affecting urination. (From EHC-Dallas. 1994.)

pudendal nerve. Pollutant injury to these fibers or muscles can result in loss of urine or reactivation depending upon the predominance of imbalance.

Central Connections and the Control of Micturition

During storage of urine in the bladder, the detrusor is inhibited from contracting and its intrinsic properties allow filling with little or no rise in intravesical pressure. Continence is maintained by closure of the bladder neck and by the distal sphincter mechanism. For voiding to occur, contraction of the detrusor muscle must be accompanied by opening of the bladder neck and relaxation of the distal sphincter mechanism (Figure 22.10).

The traditional belief that a simple spinal reflex arc controls micturition is no longer tenable. It is now recognized that a spino-pontine-spinal reflex involving a micturition center in the pons is essential for coordinating detrusor contraction with sphincter relaxation. Influences from higher centers also impinge on the pontine micturition center to bring this reflex under voluntary control.

The anterior portion of the frontal lobe is the cortical area primarily responsible for voluntary control of micturition. During bladder filling, afferent proprioceptive impulses from stretch receptors in the detrusor muscle pass via the posterior roots of S_2–S_4 and the lateral spinothalamic tracts to the frontal

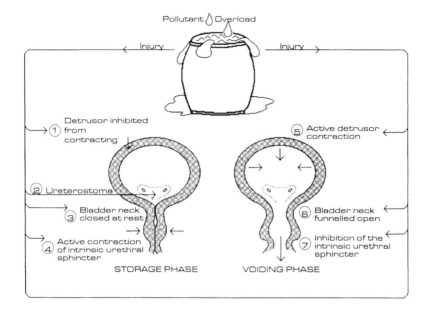

Figure 22.10.   Local areas that pollutants may damage, thereby altering, the voiding cycle. (From EHC-Dallas. 1994.)

cortex. These impulses register the desire to void, and this desire is suppressed by the frontal cortex until a suitable time and place of micturition have been selected. When it is appropriate to pass urine, there is cortical facilitation of the micturition reflex.

During voluntary micturition, relaxation of the urethral sphincter precedes detrusor contraction and produces a marked fall in intra-urethral pressure. There is simultaneous relaxation of the pelvic floor muscles and funneling of the bladder neck. Parasympathetic activity then initiates detrusor contraction and urine flow commences. The presence of urine in the urethra produces reflex facilitation of the detrusor, which helps to sustain its contraction until the bladder is empty.

At the end of micturition, urine flow ceases, the intravesical pressure falls, and the urethral sphincter voluntarily contracts. As the proximal urethra closes, its intrinsic muscle contracts in a retrograde fashion and "milks" any trapped urine back into the bladder. Once these events are complete, inhibition of the micturition reflex is reapplied from higher centers so that the bladder is ready to enter the next filling cycle (Figure 22.11).

Early pollutant injury or stimulation to these areas in the brain along the spinal cord or nerves or in intrinsic muscle may result in dysfunction of urination. Since retention and voiding is complex, the disturbance by pollutants may be complicated. For instance, we have observed that some patients have little problem with their bladder with initial pollutant exposure when the bladder is empty. However, if the bladder is nearly or partially full, they may

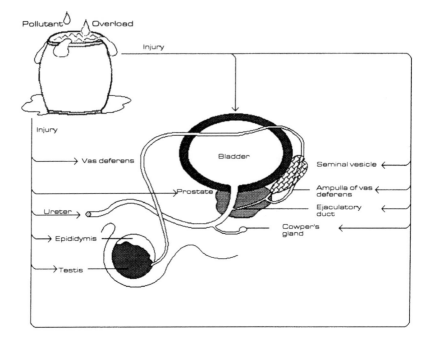

Figure 22.11. Areas of the reproductive system susceptible to pollutant injury. (From EHC-Dallas. 1994.)

have incontinence or an extreme urgency to urinate upon acute pollutant exposure. They will be forced to void immediately, or they cannot control the loss of urine. It is not uncommon to see a severely chemically sensitive individual wearing a diaper because of this phenomenon until his or her problem is solved. Then, as his or her initial load is reduced, these symptoms resolve, and this phenomenon decreases and, finally, ceases. Neuroendocrine cells in the bladder neck area at times can be triggered by central pollutant stimulation in the brain. This anatomical fact appears to account for the sudden urge to urinate seen in many chemically sensitive patients. Conversely, we have observed some patients who find it difficult to empty the bladder. In extreme cases, the chemically sensitive patient has required a catheter for periods of time in order to restore bladder tone. Again, once their total load was decreased and nutritional integrity restored, the process of urination improved.

Pollutant Injury to Mucosa and Physiology

Local pollutant injury to the mucosa may occur allowing for an easy entrance of infection. Noninfectious inflammation may also occur causing chronic irritative symptoms in the chemically sensitive patient.

The blood supply may also be altered by pollutant exposure changing the physiology and giving some bladder dysfunction.

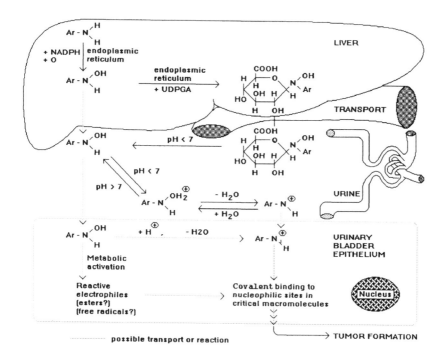

Figure 22.12.   Formation and transport of possible proximate and ultimate carcinogenic metabolites of arylamines for the induction of urinary bladder cancer. Ar = aryl substituent.

Clinical Manifestation of Pollutant Injury

The bladder is frequently a target for environmental incitants (Figure
22.12). The bladder is a target tissue for aromatic amine carcinogens. Bladder
cancer was described in the 1920s in aniline dye workers.[61]

Tamoxifen (chemical name: (Z)-2-[4-(1,2-diphenyl-l-butenyl)phenoxy]-N,N-
dimethyl-2-hydroxy-1,2,3-propanetricarboxylate [1:1]) gives bladder hernia in
rats.[62] Urinary adenosine deaminase binding protein was elevated 5 times over
baseline in patients exposed to cis-diaminedichloroplatinum and ifosfamide.[63]
Women particularly have recurrent cystitis requiring frequent urethral dilatations.

In a series at the EHC-Dallas, 30 patients with recurrent cystitis were
studied. These included 29 females and 1 male. Their ages ranged from 1 to 68
years with a mean of 41 years. These patients were placed in the ECU program.
As their total load decreased, their cystitis cleared, and they had no need for
medication. Intradermal challenge revealed a wide spectrum of sensitivity to
molds, danders, pollens, and terpenes. Oral challenge confirmed by intradermal testing revealed widespread food sensitivity, with corn, yeast, milk, potatoes, eggs, and wheat being the main offenders. Intradermal and inhaled
chemical challenge revealed sensitivities to phenol, formaldehyde, chlorine,
ethanol (petroleum derived), and pesticide (2,4-DNP) (Table 22.11).

Table 22.11. Environmentally Triggered Cystitis: 30 Patients in the ECU

Environmental triggers found	Percentage
Organochlorine pesticides (blood)	33
Intradermal provocation confirmed by inhaled challenge	
Mold	86
Animal dander	23
Smuts	80
Weeds	73
Trees	73
Grasses	70
Terpenes	43
Oral challenge confirmed by intradermal challenge	
Corn	70
Bakers yeast	64
Cow's milk	60
Potatoes	53
Eggs	53
Wheat	46
Chemicals (intradermal and inhaled double-blind challenge)	76[a]
Hormones	26
Associated symptoms: Nervous system — 73%	
Respiratory and cardiovascular — 56%	
Psychological — 43%	

Notes: Females: 97%; males: 3%. Mean age: 41 years; 67% between ages 30 to 49. Range of ages: 1 to 68 years. IgE: 65% normal.

[a] Formaldehyde (<0.20 ppm), phenol (<0.005 ppm), chlorine (<0.33 ppm), pesticide (2,4-DNP) (<0.0034 ppm), ethanol (<0.5 ppm), inhaled double-blind challenge. Some confirmed by intradermal injections.

Source: EHC-Dallas. 1992.

Inhaled double-blind challenge included formaldehyde (<0.2 ppm); chlorine (<0.33 ppm); phenol (<0.0024 ppm); pesticide 2,4-DNP (<0.0034 ppm); petroleum-derived ethanol (<0.50 ppm); perfume; cigarette smoke; newsprint; and diesel, jet, and car fuel.

Laboratory tests showed 35% had elevated IgE; 33% had organochloride pesticides in their blood; and another 33% had low T-lymphocytes. These patients went on an environmental clean-up program and have done well with only an occasional recurrence after inadvertent exposure that cleared without medication. The following case of eosinophilic cystitis emphasizes our findings.

Case study. In October 1982, a 56-year-old female was troubled with severe chronic symptoms of recurrent cystitis; she was found to have significant granular, hemorrhagic trigonitis and cystitis containing inflammatory polyps on cystoscopic exam. Biopsy revealed an eosinophilic infiltrate.

She also had a history of allergic ear, nose, and throat symptoms, with reactions induced by exposure to perfume, soap powder, cigarette smoke, and dust. Her past health also included angina and various gastrointestinal symptoms, including rectal burning.

Laboratory, immune, and general volatile solvents analysis showed nothing unusual. On intradermal skin testing, she reacted to grasses, molds, weeds, terpenes, *Candida*, formaldehyde, cigarette smoke, orris root, and diesel fumes. In addition, she had symptoms evoked by 37 foods, including milk, coffee, pork, onion, corn, yeast, oranges, apples, and chicken. Many of these foods, when tested, reproduced urinary burning and discomfort, which were among her biggest problems.

She started a program of rotation diet, avoiding triggering agents, antigen therapy, and environmental control. Within 2 weeks of starting this program, for the first time in 5 years she was without medication, her bladder symptoms were virtually gone. She suffered one flare-up after indulging in ice cream. Follow-up 2 years later showed that her previously chronic bladder symptoms were now gone, except when she occasionally went off her diet.

Bed Wetting

Enuresis or involuntary emptying of the bladder beyond an age when bladder control should be established has been studied extensively (Table 22.12). Powell et al.[64] reported 82 women with lower urinary tract allergy (Table 22.13). Foods usually caused the problem. Speer,[65] Randolph,[66] Sandberg,[55] Gerrard,[67] Breneman,[68] Pastinszky[69] and Crook[70] have all presented supporting studies relating to food and enuresis. Sanchez-Palacios et al.[71] and Yamada et al.[72] have described food-induced cystitis. Some chemicals were also incriminated in this group. Food colors seem to cause problems in some patients.

Occasionally, formaldehyde can act to sensitize the bladder. Once sensitization occurs, other substances can also trigger the problem.

Case study #1. The following is a case history of a child who as a baby developed severe bladder problems as a baby that resulted in incontinence and recurrent infections. We did not see her until she was 7 years old prior to proposed surgery for a urinary diversion procedure.

This 7-year-old white girl was admitted to ECU with the chief complaints of urinary frequency and increasing wetting at night. She did not know when she dribbled. She also complained of left leg pain and numbness periodically through the day, and her eyes were puffy frequently.

When her mother was 6 months pregnant, her family moved into a new double-wide mobile home, which was proven to have a high-level formaldehyde contamination for $2^1/_2$ years. Since then, her whole family had had unexplained medical problems (see Volume II, Chapter 17,[73] for family case reports).

The patient was born full-term and weighed 7 pounds and 11 ounces. She was diagnosed as normal with a normal hospital stay, but when she was

Table 22.12.   Etiology of Enuresis

Allergy, including bladder spasm
Difficulty in arousal from deep sleep
Nocturnal diuresis after daytime water retention
Genetic factors
Psychological-emotional factors
Nocturnal epilepsy
Urinary tract infection
Increased urine volume secondary to diabetes mellitus or insipidus
Obstructive uropathy
Chronic renal failure with impaired concentrating ability

Source: Sandberg, D. 1987. Food sensitivity: the kidney and bladder. In *Food Allergy and Intolerance*, eds. J. Brostoff and S. J. Challacombe. 765, London: Baillière Tindall. With permission.

Table 22.13.   Foods Causing Lower Urinary Tract Symptoms in 82 Women

Food	% affected
Citrus	60
Tomato	34
Condiments	20
Chocolate	15
Grape	15
Apple	9
Watermelon	7

Source: Sandberg, D. 1987. Food sensitivity: the kidney and bladder. In *Food Allergy and Intolerance*, eds. J. Brostoff and S. J. Challacombe. 765. London: Baillière Tindall. With permission.

brought home she developed a severe vomiting and respiratory problem. At 3 weeks of age, the baby was hospitalized with dehydration, respiratory congestion, and vomiting. She did quite well in the hospital environment. After leaving, though, her symptoms of vomiting, coughing, and congestion continued. When she was 2 years old, her family moved out of that mobile home, and she seemed to be doing better, but at $2^1/_2$ years old she began dribbling between voiding and always before she actually went into the restroom. She developed bladder infections, and she twitched severely in her sleep. She was hospitalized again, and urine tests were run. She had a lower GI and cystoscopy. Nothing was noted to be causing the trouble. The urology staff found what was thought to be neurological problems. A myelogram was run and appeared clear. The neurological consultant said this child must have had damage at birth causing the abnormal reflexes in her legs. A urodynamics test was done, showing 92% efficiency of the bladder. The urologist stated he did not know the cause of her bladder problem. Her physical exam was normal except for swollen eyes and dark circles under her eyes and swollen nose. Her EMG was within normal limits.

The laboratory tests revealed elevated B lymphocytes at 22% (normal = 5 to 18%), and elevated absolute B cells number at 589/mm^3 (normal = 82 to 477/mm^3). Her urine test showed pH at 6, ketone at >80 mg/dL (normal = 0 to 5 mg/dL), and WBC/HPF at 50 to 60 (normal = 0).

Her skin tests showed she had sensitivities to 10 foods, orris root, ragweed, fluogen, MRV, dust, mite, rhizopus, *Candida*, molds, histamine and serotonin. These sensitivities were confirmed on subsequent oral challenge.

Diagnostic chemical challenge was conducted in a steel-and-glass, airtight booth using a double-blind procedure. Chemicals that triggered symptoms including continued bladder pain were pesticides, chlorine, phenol, and formaldehyde. Placebo and ethanol did not produce symptoms (Table 22.14).

She was diagnosed with severe urinary frequency, with recurrent cystitis and incontinence, rhinosinusitis, irritable bowel syndrome, myalgia, toxic brain syndrome, fatigue, and chemical sensitivity.

After a heat depuration/physical therapy detoxification and immunotherapy program, including a formaldehyde-free oasis, she was doing well. One year later, she had gained 10 pounds and gone to school. The bladder problem had completely subsided. Four year follow-up showed no major problems.

Case study #2. This 31-year-old white female exemplifies the multiplicity of triggers to the life-long history that some people have, which can be solved if the triggering agents are defined and removed. This patient was totally incapacitated with recurrent cystitis when she was admitted to the ECU. Her symptoms cleared after a few days. Her worst offenders were natural gas fumes and pesticides, to which she reacted with bladder symptoms up to 48 hours after exposure (see Table 22.15). At two-year follow-up, she was totally well.

PENILE PROSTHESES

Since the early 1970s, inflatable synthetic penile prostheses have been used in the treatment of erectile impotence. There are many different types of synthetic prostheses such as the Scott®, Small-Carrion®, Jonas®, Modified Mentor®, and the Flex-Flat®. Basically these are controlled expansion cylinders that have an outer layer of silicone and usually a middle layer of some other substance such as knitted polypropylene. The indications for insertion of these prostheses are usually impotence due to vascular dysfunction such as results from diabetes mellitus and arteriosclerosis, radical prostatectomy, antihypertensive medication, pelvic irradiation, Peyronie's disease, spinal cord injury, other neuropathy, radical cystectomy, anteroposterior resection of the rectum, hemipelvectomy, penile trauma, priapism, post-kidney transplant,

**Table 22.14.   Inhaled, Double-Blind Challenge:
7-Year-Old White Female after 4 Days
Deadaptation with Total Load Reduction**

Inhaled challenge	Dose (ppm)	Results
Pesticide (2,4-DNP)	<0.0034	+
Formaldehyde	<0.20	+
Chlorine	<0.33	+
Phenol	<0.002	+
Ethanol (petroleum derived)	<0.50	−
Saline	—	—
Saline	—	—
Saline	—	—

Note: Bladder spasm and dribbling reproduced.

Source: EHC-Dallas. 1989.

**Table 22.15.   Intractable Cystitis with Chemical Sensitivity and
Multiple Organ Involvement: White Female, Age 31 Years**

Symptoms:	Foods (multiple):
Recurrent bronchitis — life long	5-second gas exposure
Recurrent myositis — life long	5-second pesticide exposure
Recurrent spastic colon — life long	(same)
Recurrent sinusitis — life long	Chemicals (multiple):
Recurrent chest pain — life long	Writhing in pain — 48 hours
Recurrent bruising — 1 year	Swollen glands
Recurrent cystitis (unmanageable) —	Spontaneous bruising
6 months	Rx as atopics
	No results

Note: IGE (155 IU/mL); IGG (150 mg/dL); IgA (29/mg/dL); CRP (+); THSC (90%);
C$_3$ (52 mg/dL). 10-year follow-up: well, without medication.

Source: EHC-Dallas. 1984.

post-heart transplant, and post-transurethral prostatic resection. There has been a complication rate as high as 22% in 1-year follow-up of patients. The majority of these complications are due to mechanical problems with the prostheses. However, a significant minority of complications have resulted in erosion, adhesions, infections, and inflammation. Since any of these complications can result from sensitivity to synthetics, it is unfortunate that these patients have not been worked up for such sensitivity. In our opinion, patients receiving penile prostheses should be closely monitored following implantation for any signs or symptoms suggesting the development of chemical sensitivity. If such responses do occur, these patients should be immediately worked up for sensitivity.[74-78]

TESTICULAR IMPLANTS

Two hundred children have received testicular implants for cosmetic reasons. Hensle[79] has reported that some of these children developed autoimmune disease.

DIAGNOSIS AND TREATMENT OF POLLUTANT INJURY TO THE GENITOURINARY SYSTEM

Diagnosis and treatment of chemical overload of the urinary system is consistent with pollutant injury to any organ. Diagnosis consists of withdrawal and challenge of incitants. Treatment includes reduction of total body load by heat depuration and physical therapy as well as avoidance of pollutants and substances to which the patient is sensitive such as food, food additives, and biological inhalants. To strengthen the immune system, injection therapy and nutrition supplementation are used. Since extended discussion of each of these aspects will be provided in Volume IV, no further comments will be made here.

REFERENCES

1. Rea, W. J. 1988. Inter-relationships between the environment and premenstrual syndrome. In *Functional Disorders of the Menstrual Cycle*, eds. M. C. Brush and E. M. Goudsmit. 135–157. New York: John Wiley & Sons.
2. Hook, J. B. and W. R. Hewitt. 1986. Toxic responses of the kidney. In *Casarett and Doull's Toxicology: The Basic Science of Poisons*, 3rd ed., eds. C. D. Klaassen, M. O. Amdur, and J. Doull. 310–329. New York: Macmillan.
3. Rea, W. J. 1992. *Chemical Sensitivity. Vol. 1. Mechanisms of Chemical Sensitivity.* 221. Boca Raton, FL: Lewis Publishers.
4. Shirai, T., T. Sakata, S. Fukushima, E. Ikawa, and N. Ito. 1985. Rat prostate as one of the target organs for 3,2'-dimethyl-4-aminobiphenyl-induced carcinogenesis: effects of dietary ethinyl estradiol and methyltestosterone. *Jpn. J. Cancer Res.* 76(9):803–808.
5. Rea, W. J. 1994. *Chemical Sensitivity. Vol. II. Sources of Total Body Load.* 1011. Boca Raton, FL: Lewis Publishers.
6. Shambaugh, G. 1989. Personal communication.
7. Bandlish, U., Br. Prabhakar, and P. L. Wadehra. 1988. Plasma zinc level estimation in enlarged prostate. *Indian J. Pathol. Microbiol.* 31(3):231–234.
8. Coca, A. F. 1949. Antiallergic action of sympathectomy. *Ann. N.Y. Acad. Sci.* 50:807.
9. Rea, W. J. 1992. *Chemical Sensitivity. Vol. 1. Mechanisms of Chemical Sensitivity.* 47. Boca Raton, FL: Lewis Publishers.
10. Davis, J. R. and S. L. Andellman. 1967. Urinary delta-amino-levulinic acid (ALA) levels in lead poisoning. I. A. modified method for the rapid determination of urinary delta-aminolevulinic acid using disposable ion-exchange chromatography columns. *Arch. Environ. Health* 15(1):53–59.
11. Torres, A. M., J. V. Rodriguez, and M. M. Elias. 1987. Urinary concentrating defect in glutathione-depleted rats. *Can. J. Physiol. Pharmacol.* 65(7):1461–1466.

12. Lehman-McKeeman, L. D., and D. Caudill. 1992. Alpha 2u-globulin is the only member of the lipocalin protein superfamily that binds to hyaline droplet inducing agents. *Toxicol. Appl. Pharmacol.* 116(2):170–176.

13. Mohiuddin, J., H. G. Prentice, S. Schey, H. Blacklock, and P. Dandona. 1984. Treatment of cyclophosphamide-induced cystitis with prostaglandin E-2. [letter] *Ann. Intern. Med.* 101(1):142.

14. Garat, J. M., E. Martinez, and F. Aragona. 1985. Open instillation of formalin for cyclophosphamide-induced hemorrhagic cystitis in a child. *Eur. Urol.* 11(3):192–194.

15. Durkee, C. 1980. Bladder cancer following administration of cyclophophamide. *Urology* 16(2):145–148.

16. Kavlock, R. J., B. F. Rehnberg, and E. H. Rogers. 1986. Chlorambucil induced congenital renal hypoplasia: effects on basal renal function in the developing rat. *Toxicology* 40(3):247–258.

17. Rankin, G. O., H. Shih, V. J. Teets, D. W. Nicoll, and P. I. Brown. 1989. Acute nephrotoxicity induced by N-(3,5-dichlorophenyl)-3-hydroxysuccinamic acid in Fischer 344 rats. *Toxicol. Lett.* 48(3):217–223.

18. Al-Sabti, K. 1985. Carcinogenic-mutagenic chemicals induced chromosomal observations in the kidney cells of three cyprinids. *Comp. Biochem. Physiol.* 82C(2):489–493.

19. Nagai, F. and T. Nakao. 1984. Changes in enzyme activities in the urine and tissues of rats fed sodium o-phenylphenate. *Food Chem. Toxicol.* 22(5):361–364.

20. Green, T. and J. Odum. 1985. Structure/activity studies of the nephrotoxic and mutagenic action of cysteine conjugates of chlor- and fluoroalkenes. *Chem. Biol. Interact.* 54(1):15–31.

21. Gad, S. C., G. M. Rusch, R. W. Darr, A. L. Cramp, G. M. Hoffman, and J. C. Peckhan. 1986. Inhalation toxicity of hexafluroisobutylene. *Toxicol. Appl. Pharmacol.* 86(3):327–340.

22. Yamamoto, T. and C. B. Wilson. 1987. Binding of anti-basement membrane after intratracheal gasoline instillation in rabbits. *Am. J. Pathol.* 126:497–505.

23. Pritchard, J. B. and M. O. James. 1982. Metabolism and urinary excretion. In *Metabolic Basis of Detoxication: Metabolism of Functional Groups*, eds. W. B. Jakoby, J. R. Bend, and J. Caldwell. 339–357. New York: Academic Press.

24. Ahmadizadeh, M., C.-H. Kuo, and J. B. Hook. 1981. Nephrotoxicity and hepatotoxicity of chloroform in mice: effect of deuterium substitution. *J. Toxicol. Environ. Health* 8:105.

25. Armeli, G., R. Mattiussi, V. Bareggi, C. Bonfanti, and F. Cazzoli. 1983. Microscopic haematuria and erythrocyte count on sediment of spot urine samples in the surveillance of workers exposed to aromatic amines. *Medicina del Lavoro* 74(3):221–226.

26. Raisbeck, M. F., E. M. Brown, and W. R. Hewitt. 1986. Renal and hepatic interactions between 2-hexanone and carbon tetrachloride in F-344 rats. *Toxicol. Lett.* 31(1):15–21.

27. Brogren, C. H., J. M. Christensen, and K. Rasmussen. 1986. Occupational exposure to chlorinated organic solvents and its effect on the renal excretion of N-acetyl-beta-D-glucosaminidase. *Arch. Toxicol.* (supplement) 9:460–464.

28. Lewis, L. D., L. C. Burton, P. G. Harper, and H. J. Rogers. 1992. Uroepithelial and nephrotubular toxicity in patients receiving ifosfamide/mesna: measurement of urinary N-acetyl-beta-D-glucosaminidase and beta-2-microglobulin. *Eur. J. Cancer* 28A(12):1976–1981.

29. Kavlock, R. J., B. F. Rehnberg, and E. H. Rogers. 1987. Critical prenatal periods for chlorambucil-induced functional alterations of the rat kidney. *Toxicology* 43(1):51–64.

30. Finn, R., A. G. Fennerty, and R. Ahmad. 1980. Hydrocarbon exposure and glomerulonephritis. *Clin. Nephrol.* 14(4):173–175.

31. Zimmerman, S. W., K. A. Groehler, and G. J. Beirne. 1975. Hydrocarbon exposure and chronic glomerulonephritis. *Lancet* 1:199.

32. Heale, W. F., A. M. Matthiesson, and J. F. Niall. 1969. Lung haemorrhage and nephritis (Goodpasture's syndrome). *Med. J. Aust.* 2:355–357.

33. Sprecace, G. A. 1963. Idiopathic pulmonary haemosiderosis. *Am. Rev. Resp. Dis.* 88:330.

34. Seeliger, K. and H. Herland. 1973. The etiology of Goodpasture's syndrome. *Med. Clin.* 68:437–440.

35. Crisp, A. J., A. K. Bhalla, and B. I. Hoffbrand. 1979. Acute tubular necrosis after exposure to diesel oil. *Br. Med. J.* 2(6183):177.

36. Karlin, G. and N. Drommer. 1970. Glonferulonephritis and hydrocarbon exposure. *Arch. Toxicol.* 216:40.

37. Capurro, P. U. 1976. Hydrocarbon exposure and cancer. *Lancet* 2:253.

38. Capurro, P. U. 1978. Solvent exposure and cancer. *Lancet* 1:942.

39. Wyse, D. G. 1973. Deliberate inhalation of volatile hydrocarbons: a review. *Can. Med. Assoc. J.* 108:71.

40. Vianna, J. and A. Polan. 1979. Lymphomas and occupational benzene exposure. *Lancet* 1:1394.

41. Ravnskov, U. and B. Forsberg. 1979. Improvement of glomerulonephritis after discontinuation of solvent exposure. *Lancet* 1:1194.

42. Ahmadizadeh, M., C.-H. Kuo, R. Echt, and B. Hook, Jr. 1984. Effects of polybrominated on arylhydrocarbon hydroxylae activities and chloroform-induced nephrotoxicity and hepatotoxicity in males C57BL/6J and DBA/20 mice. *Toxicology* 31(3/4):343–352.

43. Smith, J. H., K. Maita, S. D. Sleight, and J. B. Hook. 1984. Effect of sex hormone status on chloroform nephrotoxicity and renal mixed function oxidases in mice. *Toxicology* 30(4):305–316.

44. Viau, C., A. Bernard, and R. Lalurverys. 1984. Distal tubular dysfunction in rats chronically exposed to a "white spirit" solvent. *Toxicol. Lett.* 21(1):49–52.

45. Lock, E. A. and J. Ismael. 1985. Effects of the organic acid transport inhibitor probenecid on renal corticol uptake and proximal tubular toxicity of hexachloro-1,3-lentadiene and its conjugates. *Toxicol. Appl. Pharmacol.* 81(1):32–42.

46. Rush, G. F., J. F. Newton, K. Maita, C.-H. Kuo, and J. B. Hook. 1984. Nephrotoxicity of phenolic bromobenzene metabolites in the mouse. *Toxicology* 30(3):259–272.

47. Rush, G. F., C.-H. Kuo, and J. B. Hook. 1984. Nephrotoxicity of bromobenzene in mice. *Toxicol. Lett.* 20(1):23–32.

48. Kuo, C.-H., K. Maita, G. F. Rush, S. Sleight, and J. B. Hook. 1984. Effects of dietary trans-stilbene oxide on hepatic and renal drug-metabolizing enzyme activities and bromobenzene-induced toxicity in male Sprague-Dawley rats. *Toxicol. Lett.* 20(1):13–21.

49. Bartolome, J., A. Grignolo, M. Bartolome, P. Irepanier, L. Lerea, G. Weegelas, W. Whitemore, G. Michalopoulos, R. Kavlock, and T. Slotkin. 1985. Postnatal methylmercury exposure: effects on otogeny of renal and hepatice ornithine decarboxylase responses to trophic stimuli. *Toxicol. Appl. Pharmacol.* 80(1):147–154.

50. Slotkin, T. A., S. Pachman, J. Bartolome, and R. J. Kavlock. 1985. Biochemical and functional alterations in renal and cardiac development resulting from neonatal methylmercury treatment. *Toxicology* 36(2–3):231–241.

51. Ishmael, J. and E. A. Lock. 1986. Nephrotoxicity of hexachlorbutadiene and its glutathione-derived conjugates. *Toxicol. Pathol.* 14(2):258–262.

52. Raguenez-Viotte, G., M. Lahoue, T. Ducastelle, J. P. Morin, and J. P. Fillastre. 1988. CCNU-adriamycin association induces earlier and more severe nephropathy in rats. *Arch. Toxicol.* 61(4):282–291.

53. Hubert, Martin. 1968. *Pesticide Manual.* British Crop Protection Council, Clacks Farm, Boreley, Ombersley, Droitwich, Vorcester, U.K.

54. Nielsen, H. K., L. Krusell, J. Baelum, G. Lindquist, O. Omland, M. Vaeth, S. E. Husted, C. E. Mogensen, and E. Geday. 1985. Renal effects of acute exposure to toluene. A controlled clinical trial. *Acta Med. Scand.* 218(3):317–321.

55. Sandberg, D. 1987. Food sensitivity: the kidney and bladder. In *Food Allergy and Intolerance*, eds. J. Brostoff and S. J. Challacombe. 755–767. London: Baillière Tindall.

56. Matsumura, T. and T. Kuroume. 1961. The role of allergy in the pathogenesis of the nephrotic syndrome. *Jpn. J. Pediatr.* 14:921.

57. Sandberg, D. H., R. M. McIntosh, C. W. Bernstein, R. Carr, and J. Strauss. 1977. Severe steroid-responsive nephrosis associated with hypersensitivity. *Lancet* 1:388.

58. Sandberg, D. H., T. F. McLeod, and J. Strauss. 1980. Renal disease related to hypersensitivity to foods. In *Food Allergy: New Perspectives*, ed. J. W. Gerrard. Springfield: Charles C Thomas.

59. Richards, W., D. Olson, and J. A. Church. 1977. Improvement of idiopathic syndrome following allergy therapy. *Ann. Allergy* 39:332–333.

60. Lagrue, G., J. M. Heslan, D. Belghiti, J. Sainte-Laudy, and J. Laurent. 1986. Basophil sensitization for food allergens in idiopathic nephrotic syndrome. *Nephron* 42:123–127.

61. Pugh, R. C. B. 1985. Lower urinary tract. In *Anderson's Pathology*, Vol. 1, 8th ed., eds. J. M. Kissane and W. A. D. Anderson. 781. St. Louis: C. V. Mosby.

62. Iguchi, T., S. Irisawa, F. D. Uchima, and N. Takasugi. 1988. Permanent chondrification in the pelvis and occurrence of hernias in mice treated neonatally with tamoxifen. *Reprod. Toxicol.* 2(2):127–134.

63. Hacke, M., H. J. Schmoll, J. M. Alt, K. Baumann, and H. Stolte. 1983. Nephrotoxicity of cis-diamminedichloroplatinum with or without ifosfamide in cancer treatment. *Clin. Physiol. Biochem.* 1(1):17–26.

64. Powell, N. B., P. B. Boggs, and J. P. McGovern. 1970. Allergy of the lower urinary tract. *Ann. Allergy* 28:252–255.

65. Speer, F. 1983. *Food Allergy*. 2nd ed. 34. Boston: J. Wright.

66. Randolph, T. G. 1976. Personal communication.

67. Gerrard, J. W. 1980. Nocturnal enuresis. In *Food Allergy: New Perspectives*, ed. J. W. Gerrard. 169. Springfield: Charles C Thomas.

68. Breneman, J. C. 1965. Nocturnal enuresis: a treatment regimen for general use. *Ann. Allergy* 23:185–191.

69. Pastinszky, I. 1959. The allergic diseases of the male genitourinary tract with special reference to allergic urethritis. *Urol. Int.* 9:258–305.

70. Crook, W. G. 1974. Genito-urinary allergy. In *Allergy and Immunology in Childhood*, eds. F. Speer and R. J. Dockhorn. Springfield: Charles C Thomas.

71. Sanchez-Palacios, A., A. Quintero de Juana, J. Martinez-Sagarra, and R. Aparicio Duque. 1984. Eosinophilic food-induced cystitis. *Allergol. Immunopathol.* 12(6):463–469.

72. Yamada, T., H. Taguchi, H. Nisimura, H. Mita, and T. Sida. 1984. Allergic study of interstitial cystitis. (I) A case of interstitial cystitis caused by squid and shrimp hypersensitivity. *Arerugi* 33:264–268.

73. Rea, W. J. 1994. *Chemical Sensitivity. Vol. II. Sources of Total Body Load*. 941. Boca Raton, FL: Lewis Publishers.

74. Merrill, D. C. 1988. Clinical experience with the Mentor inflatable penile prosthesis in 301 patients. *J. Urol.* 140:1424–1427.

75. Kabalin, J. N. and R. Kessler. 1988. Five-year followup of the Scott inflatable penile prosthesis and comparison with semirigid penile prosthesis. *J. Urol.* 140:1428–1430.

76. Mulcahy, J. J. 1988. The Hydroflex self-contained inflatable prosthesis: experience with 100 patients. *J. Urol.* 140:1422–1423.

77. Mulcahy, J. J. 1988. Use of CX cylinders in association with AMS700 inflatable penile prosthesis. *J. Urol.* 140:1420–1421.

78. Peppas, D. S., J. W. Moul, and D. G. McLeod. 1988. Candida albicans corpora abscess following penile prosthesis placement. *J. Urol.* 140:1541–1542.

79. Hensle, T. 1994. Unpublished data. Babies' and Children's Hospital. 3959 Broadway, New York, NY, 10032.

23　　　　　Musculoskeletal System

INTRODUCTION

Environmental influences on disease states of the musculoskeletal system are beginning to be recognized by most physicians. For example, environmental causes of fibromyalgia from solvent exposure or silicon implants have been described. Weather changes are known to influence arthritis. Lupus erythramotosis has been observed to be triggered by chemicals. At the EHC-Dallas, a large number of our patients with chemical sensitivity complain of muscle and fascial aches and pains and of chronic fatigue. These symptoms may continue for years before fixed-named disease becomes evident, or they may remain as a more general chronic pain and fatigue syndrome. In many cases, even where severe fixed-tissue changes are present, damage to the musculoskeletal system may still be quite easily reversed, once the triggering agents are defined and eliminated.

Approximately 40% of the body is skeletal muscle; therefore, pollutant injury to this system may have serious consequences in the chemically sensitive patient in whom musculoskeletal dysfunction is a major factor. Involvement may include chronic fatigue, fibromyalgia, arthralgia, or simple, fleeting aches and pains. Because this system is so large and the effects of pollutants may be broad or localized, a wide range of pollutant-influenced clinical entities occurs in the musculoskeletal system. Each entity will be discussed separately.

POLLUTANT INJURY TO THE AUTONOMIC AND VOLUNTARY NERVOUS SYSTEMS OF THE MUSCULOSKELETAL SYSTEM

Large myelinated nerve fibers innervate the skeletal muscles. These fibers originate in the large motoneurons of the anterior horns of the spinal cord. The nerve ending joins with the muscle fiber at the myoneural junction, approximately at the muscle midpoint, so that the action potential in the fiber travels both directions (Figure 23.1A).

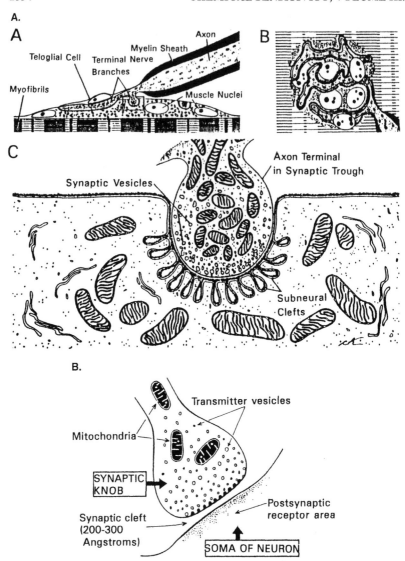

Figure 23.1. (A) Different views of the motor end-plate. A: Longitudinal
section through the end-plate. B: Surface view of the end-
plate. C: Electronmicrographic appearance of the contact
point between one of the axon terminals and the muscle fiber
membrane, representing the rectangular area shown in A.
(B) Physiologic anatomy of the synapse. (From Guyton, A. C.
1981. *Textbook of Medical Physiology*, 6th ed. 138, 565. Philadelphia: W. B. Saunders Co. With permission.)

The nerve fiber branches at its end to form a complex of branching nerve terminals (endplate). These invaginate into the muscle fiber but lie entirely outside the muscle fiber plasma membrane. The entire structure is covered by Schwann cells that insulate the endplate from the surrounding fluids. Pollutant injury may occur in any of these areas, particularly in the membranes of any of the fat soluble areas. Solvents have been seen to damage these areas in some chemically sensitive individuals. The "synaptic gutter" or "trough" is the invagination of the neural membrane into the muscle, and the "synaptic cleft" is the space between the terminal and the fiber membrane. The synaptic cleft is 20 to 30 nm wide and filled with a gelatin-like ground substance through which extracellular fluid diffuses. At the bottom of the gutter are numerous folds of the muscle membrane. These form subneural clefts that greatly increase the surface area where the synaptic transmitter can act (Figure 23.1B). In the axon terminal, many mitochondria supply energy, mainly for the synthesis of acetylcholine, which in turn excites the muscle fiber. If pollutants damage these mitochondria, acetylcholine production decreases, and orderly function will not occur. The acetylcholine is synthesized in the cytoplasm, but it is rapidly absorbed into approximately 300,000 vesicles around the endplate. In the matrix of the subneural clefts are large quantities of cholinesterase that destroy acetylcholine. This particular set of circumstances of anatomy and physiolgy leaves an individual vulnerable to pollutant injury, either by paralyzing or by destroying the production and action of the enzyme or the production and excitation of the stimulator acetylcholine. Arsenic, curare, and organophospate insecticides are known to damage this area in some chemically sensitive patients. When the nerve impulse reaches the myoneural junction, approximately 300 vesicles of acetylcholine are released as the result of a stimulus from the terminals into the synaptic clefts. This release results from movements of calcium rows from the extracellular fluid into the membrane when the vesicles release acetylcholine. Ion flow damage of the membrane as a result of pollutant exposure and/or subsequent hypoxia has been well documented (see Volume I, Chapters 4[1] and 6[2]). The absence of calcium or an excess of magnesium depresses the release of acetylcholine. Often the pollutant-damaged, chemically sensitive patient has low calcium and magnesium with alterations in ion flow.

At the EHC-Dallas, the majority of our chemically sensitive patients have been found to experience physiological changes as a result of excess cholinergic or sympatholytic stimulation due to toxic chemical exposure. The changes due to cholinergic stimulation can be dampened in the chemically sensitive individual by administration of intravenous magnesium sulfate, which relieves muscle spasm (see Volume I, Chapter 6, section on magnesium, for more information[2]). With neurostimulation, acetylcholine is released into the axon terminal and gutter for only one millisecond, which is time enough for the muscle membrane pore to open and the influx of sodium ions to occur. This pore opening is accomplished by the acetylcholine binding to the acetylcholine receptor protein, which undergoes configurational change that allows the pore to open and the endplate potential to increase with the influx of sodium ions into the muscle fiber, which is subsequently triggered. This whole mechanism is prone to pollutant injury, and it is rendered totally nonfunctional by such pollutants as curare or botulinis toxin. The former acts by decreasing the receptivity of the mucous membrane and the latter by decreasing the release of acetylcholine. Many different exogenous compounds including methcholine, carbachol, and nicotine have the same effect on the muscle fiber as acetylcholine. Unlike its effect on acetylcholine, acetyl cholinesterase does not destroy these aforementioned compounds. Therefore, as pollutant overload continues, areas of depolarization also continue to develop, along with other areas of replenishment when acetylcholine is working. This alteration will cause spasm, as is seen in many chemically sensitive patients. Other pollutants may act similarly. Excess stimulation by pollutants or as a byproduct of this whole process may eventually result in muscle fatigue and even flaccid paralysis. Neostigmine, physostigmine, diissopropyl phosphate, and the organophosphate and carbamate pesticides are some of the compounds that stimulate the neuromuscular function by inactivating the cholinesterases. When the cholinesterases are inactivated, the massive increase of acetylcholine can accumulate, causing excessive triggering of the muscle fiber with spasm resulting. This inactivation can even cause laryngeal spasm (see Chapter 18). The fluorophosphates, which are nerve gases, can inactivate cholinesterase for weeks, causing severe problems, even death.

Myesthenia gravis causes paralysis because the neuromuscular junctions become unable to transmit signals from the nerve fibers to the muscle fibers. This inability is thought to be due to the presence of autoantibodies that block the transmission. At the EHC-Dallas, we have treated a few chemically sensitive patients with myesthenia who improved when their chemical triggers were removed. Often physicians mistake the weakness and fatigue observed in many chemically sensitive patients for myesthenia, even though their tensilon test is negative. We emphasize that if the tensilon test is negative, chemical sensitivity, and not just myesthenia, may be present.

Somesthetic messages are either exteroceptive from the skin or proprioceptive from the muscles, tendons, and joints. These proprioceptive messages react to stretch. The nerve fibers bringing proprioceptive messages to the spinal cord are heavily mylenated in contrast to touch fibers, which are medium mylenated, and pain fibers, which are unmylenated. These proprioceptive nerve fibers travel up the spinal cord to the brain. Pollutant stimuli yield a vast array of responses and can even alter limb function. For example, the messages of the leg pour into the nucleate gracilis. The axons of the cells composing the nuclei cross over and ascend through the brainstem as the medial lemniscus ending in the ventero-lateral nucleus of the thalmus. Somesthetic impulses are relayed along axons of the thalmic neurons to the cerebral cortex. The functions of these nerve pathways have usually been narrowly interpreted, or broadly interpreted in chemically sensitive individuals whose health has been altered by pollutant overload. There are usually suggestions that multiple joint aches, muscle aches, and fatigue complaints are psychosomatic in origin, if no other "objective" data can be found. This interpretation is wholly without any scientific proof in the chemically sensitive patient. In fact, our observation of 1000 patients during their stay in the ECU proved a causal relationship. When patients were in the basal state and underwent double-blind inhaled challenge with the active ingredient after 4 days deadaptation with their total load reduced, both signs and symptoms were reproduced. Signs and symptoms, however, were not reproduced during placebo challenge. Apparently, pollutant injury can trigger edema as well as muscle and fascial tension, resulting in triggering of these proprioceptive stretch receptors, which then results in multiple somatic complaints. This cause-and-effect phenomenon can be reproduced under controlled conditions.

POLLUTANT INJURY TO THE METABOLISM OF THE MUSCULOSKELETAL SYSTEM

Pollutant injury to the metabolism of the musculoskeletal system can take many forms and has been seen in the muscle biopsies of some of our chemically sensitive patients at the EHC-Dallas. More commonly, pollutants may damage contractility, membranes, or electrolyte balance. Pollutant damage may occur to the sarcolema, the sarcoplasm, or the sarcoplasmic reticulum. These elements are similar to the cell components (see Volume I, Chapter 4[1]) in the other parts of the body but have specific adaptations for the unique purpose of contraction. For example, the myofibrils are suspended inside the muscle fiber in a matrix of sarcoplasm. This fluid consists of, and contains, large quantities of potassium, magnesium, phosphorus, and protein enzymes. Also present in this fluid are tremendous numbers of mitochondria that lie between, and parallel to, the myofibrils, a placement that is indicative of the great need of contracting myofibrils for large amounts of ATP formed by the mitochondria. The mitochondrial membranes are partitioned and prone to damage by some lipophilic pollutants such as chlorinated pesticides, herbicides, and solvents, which are present in many chemically sensitive individuals. In some cases, the presence of these pollutants often accounts for the tremendous weakness seen in the chemically sensitive patient.

The loss of ATP by mitochondrial damage due to severe pollutant exposure with severe weakness is exemplified by a 34-year-old psychiatrist who could barely walk. She had struggled through medical school barely able to graduate because of muscle weakness. She chose a psychiatric specialty because it required the least physical and mental effort. When she presented to the ECU for treatment, she had been virtually bedridden for 2 years as the result of this muscle weakness. Work-up showed she had seven different chlorinated pesticides in her blood, and she exhibited sensitivity to phenol, formaldehyde, ethanol, chlorine, and pesticide upon inhaled challenge. Muscle biopsy showed 5% of the total normal quantity of mitochondria. This low number of mito-chondria explained her extreme weakness, which was exacerbated by pollutant challenge. Apparently, the mitochondria had been destroyed or paralyzed by the chlorinated pesticides.

Pollutants can cause deranged cell membranes, resulting in sodium, cal-cium, and magnesium imbalance in many chemically sensitive individuals (see Volume I, Chapter 4[1]). In addition to muscle weakness, many chemically sensitive patients have severe muscle cramps and tetany, which are acutely responsive to magnesium or calcium infusions upon intravenous challenge, indicating the presence of the pollutant-derived electrolyte imbalance.

Contractility may also be altered by pollutant injury. Contractility depends on magnesium, ATP, and protein. If these are damaged by pollutants, proper function is impeded.

Many studies have now shown pollutant damage to the musculoskeletal system. For example, changes in body sway (balance) in healthy males exposed to m-xylene were negatively correlated with the intensity of the atmospheric exposure.[3] Fifty-six percent of our chemically sensitive patients have up to 10 ppm of xylene in their blood and have problems with sway and balance. Using computerized posturography, Martinez[4] has shown that many of these chemi-cally sensitive individuals have abnormalities that result in an inability to balance properly. Though sway may be due to muscle imbalance, it also may be due to abnormalities in the CNS, inner ear, or the spinal cord (see Chapter 20). Other abnormalities of the musculoskeletal system have been found after toxic chemi-cal exposure. Wecker et al.[5] showed that human inhalation or ingestion of organophosphate insecticides gives rise to intercostal muscle degeneration.

An impressive body of animal research supports the results of a limited number of human studies that show the effects of toxic chemicals on the musculoskeletal system. For example, trichloroethylene, which is a contami-nant of underground water supplies and is present in the blood of 8% of our chemically sensitive patients, has been associated with skeletal motor changes.[6] Toluene, which is present in the blood of 63% of our patients, has produced altered motor activity in rats.[7] Two pyrethroid insecticides (ismethrine and deltamethrin) have produced dosage-dependent decreases in motor activity. Their effects on the startle response, however, were dissimilar. Dose-depen-dent decreases in amplitude and increases in latency were seen in rats exposed to deltamethin, while cismethrin produced increases in amplitude but no change in latency.[8] Intragastric injection of DDT was found to produce a stimulus-sensitive myoclonus in mice and rats. Unilateral infusions of DDT into rat medullary reticular formation also induced generalized myoclonus, but con-tractions of lesser intensity occurred with injection into cerebellar nuclei, red nucleus, and the inferior olive.[9] Rats exposed to low doses of formaldehyde have led to decreased motor activity as well as neurochemical changes in dopamine.[10] Inhalation of toluene diisocyanate vapors enhances smooth muscle responsiveness in the trachea of guinea pigs.[11]

CLINICAL ENTITIES AFFECTED BY POLLUTANT INJURY

A wide range of clinical entities of the musculoskeletal system are affected by pollutants. Each will be discussed separately. These include myalgia and fatigue, cramps and muscle spasm, fibromyalgia and fibrositis, eosinophilic fasciitis, carpal tunnel syndrome, hyperostosis, arthritis, postpolio syndrome, autoimmune disease, systemic lupus erythematosis, scleroderma, vinyl chloride disease, silica-induced disease, organic solvents and SSc, and implant syndrome. The remainder of this chapter is devoted to separate discussion of the involvement of pollutants in each of these entities.

Myalgia and Chronic Fatigue

We have seen many chemically sensitive patients with myalgia and fatigue that resulted from pollutant exposure. They frequently complain of a flu-like syndrome with muscle aches and pains. The muscles are extremely tender to palpitation. Severe weakness is often seen, but, at times, weakness appears also to be emanating from the CNS rather than the muscles per se. Many of these patients are diagnosed as having chronic fatigue syndrome. In the past, various etiologies, such as Epstein-Barr virus, *Candida albicans*, etc., have been ascribed to these patients, although no real attempts had been made to demonstrate a causal etiology.

Since 5000 chemically sensitive patients have presented at the EHC-Dallas with chronic fatigue and myalgia, the physicians here now use a systematic approach to define etiolgic agents. Using the methods and principles outlined in this book, we are invariably able to identify multiple triggers of biological inhalants, foods, chemicals, viruses, bacteria, parasites, terpenes, and mycotoxins. In addition, nutritional deficiencies are frequently discovered (see Volume I, Chapter 6[2]). Patients with artificial implants, mycotoxin exposure, or chemical exposure are particularly prone to chronic fatigue. A variant of the chronic fatigue and myalgia syndrome, cramps and muscle spasm, will be discussed next.

Muscle Cramps and Spasm

Muscle cramps and spasms appear to be a variant of chronic fatigue syndrome. In the chemically sensitive patient, they are clearly related to electrolyte and pollutant-induced intrinsic muscle dysfunction.

At the EHC-Dallas, we have treated many chemically sensitive patients suffering from a wide range of muscle cramps and spasms that have varied in number, duration, location, and intensity. For example, patients have presented with cramps ranging from mild to severe, disabling tetany. Most of these cramps were accompanied by observable fasciculations and spasms. All patients had other signs of disability related to chemical exposure involving the respiratory, cardiovascular, gastrointestinal, genitourinary, or dermal systems. Two characteristics of the cramps were that they often woke the patient at night and they were relieved occasionally by mild exercise. However, the more severe cramps left the patient in intense pain from minutes to hours. Even after investigation and elimination of triggering agents, residual, constant symptoms recurred until the load had been reduced for months and mineral replacement was complete. Investigation of intracellular deficiency of minerals and then replacement treatment usually eliminated the remaining cramps. Often empirical intravenous challenges were needed in order to assess properly the mineral balance of Ca and Mg, since serum and intracellular blood levels did not indicate the imbalance. The spasm was relieved by intravenous administration of MgCl or $MgSO_4$, usually 10 to 20 mEq, and/or 1 g of calcium gluconate in 250 cc of normal saline, given over a 4-hour period. Of course, sodium and potassium balance had to be maintained at all times. This therapy had to be repeated for several days, until the patients' pool became filled or until a vessel or renal leak was identified.

Over 100 chemically sensitive patients with severe muscle spasm have been studied at the EHC-Dallas. All these patients had similar complaints of cramps, muscle pain, and severe weakness. They were becoming nonfunctional, or their symptoms severely interfered with their ability to work. Of a series of 388 consecutive chemically sensitive patients, 66 had severe muscle cramps (Table 23.1).

The 20 patients who were most severely afflicted had incapacitating tetany. Laboratory studies in these patients showed that 64% had abnormal eosinophils; 58% had elevated IGEs, and 56% of the patients had abnormal complements above or below the 20% standard deviation range. The proportion of these patients with high IGE abnormalities was very high compared to the nonmuscle spasm group of chemically sensitive patients. The cause of so many abnormalities in this group was unclear.

This muscle spasm phenomenon in our patients had two components. First, triggering agents such as pesticides, phenol, chlorine, petroleum alcohol, formaldehyde, and electromagnetic frequencies as well as food were found. Also, disturbances of electrolyte (Na^+, K^+, Ca^{2+}, Mg^{2+}, Li^{2+}) flow through neuromuscular membranes must have occurred. Mineral supplementation was necessary in most of the patients. Some only responded well acutely to intravenous mineral therapy.

A subset of chemically sensitive patients with grossly obvious tetany was studied following a minimum of 4 days of adaptation with their total load

Table 23.1.   Extremity Cramps with Muscle Spasms

Number of consecutive chemically sensitive patients	388
Number with cramps	66 (17%)
Eosinophils/mm^3	64 patients
Number changed more than 20%	19 (30%)
IgE (IU/mL)	58 patients
<10	6 (10%)
10 to 50	25 (43%)
>50	16 (28%)
>200	11 (19%)
Total complement (CH_{100} or CH_{50})	56 patients
Number changed more than 20%	18 (32%)
Chemicals tested[a]	
Formaldehyde (<0.2 ppm)	
Phenol (<0.05 ppm)	
Ethanol (<0.5 ppm)	
Insecticide (2,4-DNP) (<0.0034 ppm)	
Chlorine (<0.33 ppm)	
Saline (placebo dose = 85%)	

[a] Type of substances used for inhaled double-blind challenge after 4 days deadaptation with the total load reduced in the ECU that triggered the cramps

Source: EHC-Dallas. 1986.

decreased (see Chapter 20, Figure 6). These patients were first challenged with single foods orally. Those foods that triggered tetany were isolated. The patients were again challenged for sensitivity to these same foods by the intradermal injection method, and symptom-neutralizing doses that would stop the tetany were identified.

Magnesium deficiency seemed to be the most important in this series of patients. In an environmentally controlled hospital unit, 51 consecutive chemically sensitive patients (12 males and 39 females ranging in age from 13 to 67 years) underwent assessment of magnesium status. The purpose of this study was to evaluate the magnesium status in chemically sensitive patients whose disorders were closely associated with the symptoms of magnesium deficiency. The symptoms included back and neck pain, fine tremors, muscle spasm, anxiety and nervousness, spastic vascular phenomena, ventricular arrhythmia, and fatigue. The results of laboratory tests and intravenous magnesium challenges showed that some patients with chemical sensitivities may have a disturbance of intracellular magnesium. RBC and plasma levels proved to be poor indicators of magnesium status: 2.8% (1/35) positive and 15% (5/33) positive, respectively. Intravenous magnesium challenge appeared to be a more accurate assessment of total body magnesium status (76% positive, 39/51).

Overall clinical improvement with magnesium treatment alone appears to be somewhat low at 45% (23/51), but this treatment is worthy of consideration in those chemically sensitive patients who are magnesium deficient.

Reported symptoms of magnesium deficiency include coarse twitching, hypertonicity, carpal pedal spasm, generalized convulsions, tetany, weakness, confusion, personality changes, nausea, anorexia, lack of coordination, GI upset, alopecia, swollen gums, skin lesions, lesions of small arteries, high blood pressure, strokes, drowsiness, and ventricular arrhythmia. The similarity of symptoms indicated in these seemingly disparate entities led to the thesis that some patients with chemical sensitivity may have a disturbance in their intracellular magnesium. Magnesium assessment was undertaken in a rigorous effort to find relief for this group of patients. The following case study illustrates the magnesium deficiency problem.

> **Case study**. A 55-year-old female treated for chemical sensitivity for 8 years developed low back pain with severe muscle spasms after a fall. Medication did not relieve her pain, which had persisted for 2 years. Her orthopedist thought that she was not absorbing sufficient amounts of magnesium. Challenge with magnesium sulfate at the EHC-Dallas relieved her symptoms and yielded the results summarized in Table 23.2.
>
> Laboratory measurements of immune parameters, vitamin and mineral levels, and the presence of heavy metals yielded the following results: IgE, 25 IU/mL; IgG, 830 mg/dL; IgA, 102 mg/dL; IgM, 109 mg/dL; and eosinophil, 123/mm^3, were found to be within normal limits. Serum levels of calcium, sodium, potassium, copper, zinc, and manganese as well as chromium, phosphorus, selenium, sulfur, and barium were normal. Silicone was found to be slightly low at 1.0 ppm (normal = 1.1 to 1.7 ppm). All heavy metals were found to be below the element detection limit. A chlorinated pesticide screening test[12] showed low levels (<1 ppm) of DDT, heptachlor, and heptachlorepoxide to be present in the blood serum.

At the EHC-Dallas, the 51 consecutive chemically sensitive patients suspected of having symptoms of magnesium deficiency were studied. These patients were deemed too ill to be diagnosed and treated on an outpatient basis by the usual criteria. They were, therefore, admitted to an environmentally controlled hospital unit. A baseline for the purpose of comparing various objective parameters was obtained in the following way: pre- and postchallenge signs and symptoms scores were noted. Forty-seven patients maintained their usual diets, while four fasted. Twenty-four hour food intake was recorded for each patient. Efforts were made to obtain as stable a magnesium intake baseline as was possible. Before challenge, serum and red blood cell magnesium levels were measured by the atomic absorption mass spectrometry method.[13] Twenty-four-hour urine analysis for calcium, magnesium, and creatinine was undertaken. Analysis of urine for magnesium content was also assessed by the atomic absorption mass spectrometry method.[13] Each patient was challenged intravenously with either magnesium chloride or magnesium sulfate (0.2 mEq/kg

Table 23.2.   White Female with Back Pain Relieved by Intravenous MgSO₄ Supplementation

	Patient	Control
10/16/1985		
1 serum Mg	1.6 mEq	1.4 to 2.5 mEq
RBC Mg	34 mEq	39.0 to 49.0 mEq
MgSO₄ IV	20 mEq	
Urine retention: 100%		
Symptoms relieved		
12/12/1985		
Mild symptoms returned		
12/19/85		
MgSO₄IV	20 mEq	
Urine retention: not done		
Symptoms relieved		
2/15/1986		
Mild symptoms returned		
MgSO₄ IV	20 mEq	
Urine retention: not done		
Symptoms relieved		

Source: Rea, W. J., A. R. Johnson, R. E. Smiley, B. Maynara, and O. Dawkins-Brown. 1986. Magnesium deficiency in patients with chemical sensitivity. *Clin. Ecol.* 4(1):17–20. With permission.

of body weight) over a 4-hour period. A second urine collection began at challenge and continued over a 24-hour period. A percentage of magnesium excretion was calculated for each patient in the following manner: the challenged amount of magnesium was added to the amount of baseline excretion and divided into the challenge excretion. This figure was then multiplied by 100 (Figure 23.2).

Magnesium deficiency was defined as less than 80% excretion of the amount of challenged magnesium.[14] Pre- and postmagnesium challenge signs and symptoms were recorded. Pre- and postmagnesium challenge urine excretion values were compared, and pre- and postmagnesium challenge serum and RBC levels were compared. Urine excretion values were also compared to serum RBC values.

Symptoms and Signs Relieved

Forty-eight percent (23/47) of nonfasted patients showed immediate improvement following intravenous magnesium challenge (Table 23.3). Symptoms relieved included back and neck pain, fine tremors, muscle spasm,

$$\% \text{ of excretion } = \frac{\text{Challenge Excretion}}{\text{Baseline excretion plus challenge dose}} \times 100$$

Figure 23.2. Equation for calculation of magnesium deficiency and loss after challenge.

Table 23.3.   Changes in Signs and Symptoms

	Number of changes[a]
Reduced dizziness	1
Improvement in depression	2
Reduced itching	1
Decreased tension	6
Reduced pain	7
Stronger, more energy	6
Decreased anxiety	3
Improvement in sleep	3
No change	17 (14 nonfasted, 3 fasted)
Reaction to magnesium Extreme weakness Felt drugged, unable to move Metallic taste, heavy heart	8[b] (7 nonfasted, 1 fasted)

[a] More than one symptom was relieved in some patients.

[b] Immediate severe reactions in four patients required cessation of magnesium challenge.

Source: Rea, W. J., A. R. Johnson, R. E. Smiley, B. Maynard, and O. Dawkins-Brown. 1986. Magnesium deficiency in patients with chemical sensitivity. *Clin. Ecol.* 4(1):17–20. With permission.

anxiety and nervousness, vascular spasms, ventricular arrhythmia, and fatigue. Adverse reactions to chemical exposures decreased. No fasted patients showed immediate relief.

Excretion of Challenged Magnesium

Less than 40% excretion of challenged magnesium was found in 11% (5/47) of nonfasted patients. Less than 60% excretion was found in 25% (12/47) of nonfasted patients. Excretion of 60 to 80% was found in 42% (20/47) of nonfasted patients. Less than 10% excretion of magnesium was found in 50% (2/4) of fasted patients. Less than 20% excretion was found in 25% (1/4) of fasted patients. Four patients were unable to tolerate the magnesium challenge due to immediate severe reactions, and the administration was stopped (Table 23.4).

Table 23.4. Results of Magnesium Evaluation in 51 Chemically Sensitive Patients

Number of patients	Magnesium (mg) levels	Control
1	Serum 2 S.D. below control	2 ± 0.6 mEq
5/33 (15%)	R.B.C. 2 S.D. below control	44 ± 5 ppm
37/47 (79%)	Mg excretion below 20%	
4/51 (8%)	No Mg excretion level measured[a]	
20/47 (42%)	Mg excretion between 60 and 80%	
12/47 (26%)	Mg excretion between 40 and 60%	
5/47 (11%)	Mg excretion below 40%	

[a] Four patients were unable to tolerate the magnesium challenge due to immediate severe reactions. The challenge was terminated and no excretion measurement was obtained.

Source: Rea, W. J., A. R. Johnson, R. E. Smiley, B. Maynard, and O. Dawkins-Brown. 1986. Magnesium deficiency in patients with chemical sensitivity. *Clin. Ecol.* 4(1):17–20. With permission.

Serum and RBC Magnesium

Serum magnesium was found to be below the normal range of 1.4 to 2.5 mEq in 2.8% (1/35) of patients. Red blood cell magnesium levels were found to be below the normal range of 39 to 49 ppm in 15% (5/33) of patients.

We found a great disparity between the number of patients whose signs and symptoms were alleviated following magnesium supplementation (45%) and the total number of patients calculated to have magnesium deficiency (76%). We propose several reasons for this disparity. First, the patients' symptoms may have been caused by other factors. Second, in some cases, the magnesium deficiency may not have been severe enough in light of biochemical individuality to produce symptoms originally. Third, the patient may not have been treated with a sufficient amount of magnesium to relieve the symptoms. Fourth, lower levels of magnesium retention may not reflect a true magnesium deficiency based on currrent thinking of what normal values of retention are. Finally, the lack of a precise calculated value for dietary intake may be a further complication, although assessment of oral intake would have tended to bias the study toward higher levels of magnesium excretion.

The results of this study suggest that some degree of magnesium deficiency in chemically sensitive people who have muscle cramps may occur and that, for enhancement of their treatment, attention should be paid to their magnesium intake.

A group of 12 additional patients was studied. These patients often had not only tetany and muscle cramps but also frequently severe symptoms of ANS dysfunction including profuse sweating, flushing, tachycardia, and diarrhea, or they would experience sudden drop attacks (Table 23.5).

Table 23.5. Double-Blind Challenges in ECU after 4 Days Reduction of Total Body Load and Deadaptation

Patient	Age	Sex	Original cause	Catatonia tetany	Drop attacks	Autonomic dysfunction (flushing, sweating, tachycardia, bloating, diarrhea)
C.S.	35	F	Chemicals (solvents)	+		+
S.K.	18	F	Chemicals (solvents)	+		+
C.D.	24	F	Gas leak (herbicides)	+		+
S.M.	46	F	Work chemicals	+		
C.B.	31	F	Methyl ethyl ketone	+		+
L.B.	56	F	Unknown	+		
G.L.	41	F	Pesticides			+
J.A.	67	M	Work-type chemical		+	+
F.J.	58	F	Electromagnetic field exp.		+	+
S.S.	52	F	Lifetime chemicals		+	+
L.S.	41	F	Microwave auto accident	+		+
G.D.	32	F	Chemical solvent auto accident	+		+

Source: EHC-Dallas. 1986.

Double-blind challenge showed that symptoms in this small group of patients had their etiology in the following toxic chemicals: phenol (<0.05 ppm), formaldehyde (<0.2 ppm), petroleum alcohol (<0.50 ppm), pesticide (<0.0034 ppm), and chlorine (<0.33 ppm). Also, some symptoms resulted from exposure to electrical magnetic waves. All of these patients developed reproducible signs similar to their original complaints. Most of the patients' original exposures were sudden exposures to toxic chemicals or traumatic situations such as auto accidents.

This group of patients was extremely sensitive, and, during challenge, their symptoms were triggered by an extremely small amount of the substance to which they were sensitive. Some of the patients with tetany also were found to have severe large and small vessel spasm. Their dysfunction was also a result of environmental triggering, and they had lymphocytic infiltrates around the vessel wall when biopsies were done.

From these series, it is clear that a group of patients exists who have muscle cramps, even to the point of tetany, that are triggered by biologic inhalants, food, and chemicals.

Fibromyalgia and Fibrositis

Toxic chemical induction of fibrosis in isolated organs is well documented. Examples include asbestos-induced lung fibrosis[15] and methsergeside-induced retroperitoneal fibrosis.[16] In some individuals, the response is widespread, and a multisystem condition resembling scleroderma or systemic sclerosis (SSc) may be induced by toxic chemicals.

In the chemically sensitive individual who has muscle involvement, the spectrum of symptomatology extends from fibrous tissue aches and pains to symptoms of fibrosis. Whether or not there is a progressive spectrum is unknown.

The cause of primary fibromyalgia is not known. Ödkvist et al.[17] reported a connection between muscle disease and the vestibulo-oculomotor disturbances in patients he treated, many of whom apparently had fibromyalgia. He and others had previously shown many vestibular disturbances in their patients due to toxic volatile organic chemicals. They also had discovered associated muscle problems (see Chapter 18).

Some evidence is accumulating for the environmental triggers of fibromyalgia. Randolph[18] proposed that commonly encountered chemicals, as well as foods, may produce arthritic symptoms in some susceptible individuals. He also observed many cases of fibrositis and fibromyositis associated with these foods and chemicals. Further evidence of chemical involvement has been supplied by Ishikawa et al.[19] who identified the presence of organochlorine and organophosphate pesticides in the serum of patients diagnosed with Behcet's syndrome (see Volume II, Chapter 13[20]). Many patients with myofasciitis appear to be triggered by foods and chemicals.[21] We have performed a prospective study of 30 cases of fibromyositis triggered by double-blind challenges (airborne chemicals and food) (Table 23.6).

Table 23.6.   Thirty Patients with Fibrositis[a]

Incitants	(%)
Intradermal challenge	
Hormones	7
Trees	34
Grasses	27
Weeds	0
Molds	0
Terps	35
Foods[b]	100
Chicken	80
Potatoes	76
Corn	70
Eggs	66
Turkey	63
Chemicals[c]	83
Phenol (<0.02 ppm)	
Formaldehyde (<0.2 ppm)	
Ethanol (<0.5 ppm)	
Chlorine (<0.33 ppm)	
Pesticide (2,4-DNP) (<0.0034 ppm)	
Saline (0.85%)	

[a] Deadapted state with total load reduced for 4 days; studied in the ECU.
[b] Oral challenge confirmed by intradermal challenge.
[c] 15-minute exposure; inhaled challenge double-blind of toxic chemicals confirmed by intradermal skin tests.

Source: EHC-Dallas. 1984.

Stiffness and pain rapidly cleared in the patients in this series once the triggering agents were identified and eliminated. These symptoms only returned when re-exposures occurred.

Over 200 similar patients with fibrositis and fibromyositis have now been studied at the EHC-Dallas with similar results. These patients do extremely well over long-term follow-up of 5 years, remaining free of their symptoms without medications as long as they avoid their triggering agents.

Eosinophilic Fasciitis

Eosinophilic fasciitis (EF) is a disorder characterized by rapidly developing symmetrical inflammation and scleroderma-like sclerosis of the deep fascia, lower subcutis, and dermis. Chiefly affected are the extremities and, in many instances, the face and trunk as well. In contrast to Scleroderma, the fingers, hands, and feet are typically spared. Raynaud's phenomenon, nailfold capillary abnormalities, and internal organ involvement are rarely present. Although the etiology of EF is unknown, some cases, especially in men, appear to have been precipitated by strenuous physical activity such as rapid assumption of physical fitness programs. Other cases of EF recently have been reported due to contamination of a genetically engineered tryptophan used for amino acid supplementation. All age groups can be affected by EF, but the majority of patients are between 30 and 60 years. Clinically, EF symptoms include erythema, edema, and severe induration of the skin and subcutaneous tissues. The overlying skin typically has an "orange-peel" appearance, and exaggerated furrowing over the course of superficial veins is noted in antidependent postures. Arthritis is uncommon, but carpal tunnel syndrome is frequent, and virtually all patients develop joint contractures. Muscle weakness secondary to disuse atrophy and occasionally occurring as an extension of inflammation deep to the fascia is seen.

Laboratory abnormalities include elevation of erythrocyte sedmentation rate, hypergammaglobulinemia, circulating immune complexes, and striking peripheral eosinophilia (typically over 2000/mm^3).[22] Biopsy specimens should include skin to skeletal muscle. Biopsy results reveal edema of the deep fascia and subcutis and infiltration with eosinophils, lymphocytes, and histiocytes in early disease. Later, tissue eosinophils are less conspicuous or are absent, and the predominant feature of biopsied tissue is fibrosis and thickening of the deep fascia, which can extend to the dermis. Etiology of EF is unclear, but, due to the eosinophilia and our experience with fasciitis, we advocate a search for the etiology in environmental triggers.

Carpal Tunnel Syndrome

Carpal tunnel syndrome is seen in many cases of chemical sensitivity. It occurs with various entities like eosinophilic fasciitis, or it may be isolated. Often its occurrence is attributed to a job that requires repetitive use of the hands. Even though repetitive contact may be the trigger, many cases may be caused by vitamin B_6 deficiency. Ellis et al.[23, 24] have studied 22 completed cases of carpal tunnel syndrome. There were 17 bilateral cases for a total of 39 kinds. These patients were treated with 50 to 300 mgm of pyridoxine daily for 12 weeks. All patients, except one, were relieved of pain and paresthesia of their median nerve distribution for a 97.4% cure rate. In a double-blind crossover study, Ellis and Folkers[25] proved that use of 100 to 200 mg of pyridoxine cleared 35 cases of carpal tunnel syndrome. In yet another series, 25 pregnant females had relief of their carpal tunnel syndrome with pyridoxine supplementation.[26] A final study,[25] using EMG and nerve conduction studies along with Preston Punch gauges and agoniometer, revealed that subjects' muscle atrophy past the carpal tunnel could be prevented and reversed with pyridoxine supplementation. Ellis and Folkers also showed by repeated measurements with the geniometer that the tenosynovitis of the carpal tunnel syndrome could be cleared with pyridoxine supplementation.

Hyperostosis

A number of disease states have in common an increase in the mass of bone per unit volume (hyperostosis). Several disease states, including endocrine disorders, can cause hyperostosis. There are known environmental triggers for some patients with hyperostosis. These include radiation, fluoride, phosphorus, beryllium, arsenic, lead, bismuth, and vitamin A excess.[27]

Arthritis

At the EHC-Dallas, we have treated a number of chemically sensitive patients with arthritis that appears to have been triggered by environmental exposures, and although there is no currently available data on the incidence of arthritis resulting from environmental exposures, numerous environmental triggers have been reported to cause some cases. These include infectious agents such as the gonococcal,[28] tuberculosis,[29] pneumococcal[30] and streptococcal;[31] protozoa such as limex amoeba[32] and yeast;[33] and fungus such as *Candida albicans*.[34] Numerous other substances that result in sensitivities to pollens,[35] dust,[33] and foods[36] as well as improper nutrition[37] have also been seen to trigger some cases of arthritis. Estrogenic hormones have been reported to ameliorate some cases of arthritis;[38] therefore, a hormone imbalance may be present in some chemically sensitive arthritic people.

Over 50 years ago, Cook[39] and Talbot[40] first hypothesized a causal relationship between diet and manifestations of arthritis. Many publications have since described associations between food sensitivity and arthritis.[41-55] Randolph[18] proposed that commonly encountered chemicals and foods may produce symptoms of arthritis in some susceptible individuals. A three-center study involved patients who were evaluated under environmentally controlled conditions. They met all of the clinical criteria for classical rheumatoid arthritis, and multiple foods were demonstrated to be triggering agents[56] (Figures 23.3). These patients have had a relatively drug-free course on long-term follow-up. Animal models of dietary-induced arthritis have been reported in recent literature.[57-59] Matsuska et al.[59] reported two cases of acute arthritis associated with isotritinein (retinoic acid) treatment for acne.

Hench and Rosenberg[53] and Rowe[48] cited instances relating rheumatoid arthritis and other rheumatic diseases to occasional cases of food allergy. One provocative observation included a case report wherein a dermatologist documented his own palindromic rheumatism caused by sodium nitrate hypersensitivity.[60] Another report linked black walnut ingestion to clinical exacerbations of Behcet's syndrome and demonstrated cellular hypersensitivity to walnut extract.[21] Ishakawa et al.'s[61] series of patients with Behcet's syndrome and its relationship to organochlorine pesticides emphasizes the potential

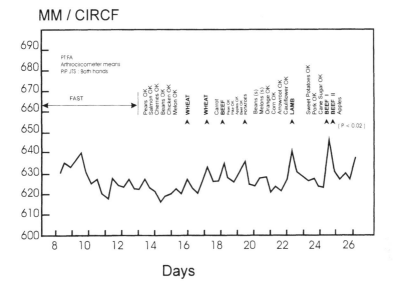

Figure 23.3. 40-year-old white female — arthrocircometer means. PIP joints, both hands. (From EHC-Dallas. 1980.)

environmental effects of xenobiotics upon arthritis. A recent double-blind provocation study used food extracts and other incitants to challenge patients. Joint and muscle reactions were induced in 87.5% of 40 patients.[62] Five of 15 patients with classic rhumatoid arthritis who fasted 7 to 10 days benefitted, whereas only 1 of 10 control patients improved. Fasting patients showed lessened pain, stiffness, and medication requirements. A continuation of a lactovegetarian diet in another series was without consistent benefit,[63] probably because specific food-triggering agents were not identified. Stroud and co-workers[64] at the EHC-Dallas in coordination with Kroker, Randolph, and Carrol working in three separate ECUs confirmed an antirheumatic effect of fasting in patients restricted to an environmentally controlled unit (Tables 23.7 to 23.12; Figures 23.2 to 23.10).

The three-unit study involved patients who met all of the clinical criteria for classical rheumatoid arthritis. These patients were moderately chemically sensitive and had a strong history of failed medical treatment, as well as adverse reactions to the treatment (Table 23.13). Challenge tests demonstrated multiple foods as triggering agents.[56] (Tables 23.13 to 23.16).

A rating scale for evaluating responses was used with OK indicating negative and a range of 1+ indicating minor to 4+ indicating severe (Table 23.8). Pre- and postfasting and pre- and postchallenges were measured by a pain, agility, and swelling index, an arthrocircometer, a dolorimeter, and a joint swelling gauge (Figures 23.11 to 23.13). A weighted food index was developed in order to rate food reactions in an objective manner (Table 23.9). There were 1079 oral food challenges in 35 patients (Table 23.10). The average number of

Table 23.7.   ECU Cooperative Study of RA

Medication	Number of patients	43 Patients (number of adverse reactions)
NSAIAs	42	3
Gold	28	21
Steroids	22	3
D Penicillamine	10	0
Immunosuppressives	3	0

Source: Kroker, G. F., R. M. Stroud, R. Marshall, T. Bullock, F. M. Carroll, M. Greenberg, T. G. Randolph, W. J. Rea, and R. E. Smiley. 1984. Fasting and rheumatoid arthritis: a multicenter study. *Clin. Ecol.* 2(3):137–144. With permission.

Table 23.8.   ECU Cooperative Study of RA Food Challenge Rating Scale

Class	Reaction
OK	Negative
+	Minor
++	Moderate
+++	Major
++++	Severe

Source: Modified from Marshall, R., R. M. Stroud, G. F. Kroker, T. Bullock, F. M. Carroll, M. Greenberg, T. G. Randolph, W. J. Rea, and R. Smiley. 1984. Food challenge effects on fasted rheumatoid arthritis patients: a multicenter study. *Clin. Ecol.* 2(4):181–190. With permission.

foods tested was 30.7/patient. There were a total of 670 foods that elicited no reactions and 409 foods to which the patients reacted. Fruits, vegetables, and meat were the most challenged, followed by cereal grains and fish (Table 23.11). Corn, beef, chicken, fish, turkey, lamb, pork, wheat, spinach, and eggs were found to be the most common foods to produce rheumatoid symptoms (Table 23.12). By the results of this study if one wanted blindly to put an individual with rheumatoid arthritis on a low arthrogenic diet, one would use melon, lima beans, tomato, lettuce, cherries, brocolli, cabbage, oranges, grapes, and peas.[64] This practice is not, however, recommended since, in the long run blind treatment usually results in more new food sensitization and, thus, failure.

This study was significant in that all patients were fasted for a mimimum of 4 days, which allowed them to clear their presenting symptoms and signs.

Table 23.9. ECU Cooperative Study of RA: 35 Patients[a]

	Food analysis
	Derivation
Weighted food index	Derivation
Number OK	Challenges × 1
Number +	Challenges × 2
Number ++	Challenges × 3
Number +++	Challenges × 4
Number ++++	Challenges × 5

[a] Food index $= \dfrac{\text{Total Challenge Points}}{\text{Total Number of Food Challenges}}$

Lowest possible index = 1.0 ("safe food"); highest possible index = 5.0 ("reactive food").

Source: Modified from Marshall, R., R. M. Stroud, G. F. Kroker, T. Bullock, F. M. Carroll, M. Greenberg, T. G. Randolph, W. J. Rea, and R. Smiley. 1984. Food challenge effects on fasted rheumatoid arthritis patients: a multicenter study. *Clin. Ecol.* 2(4):181–190. With permission.

Table 23.10. ECU Cooperative Study of RA

Class	Total reactions
OK	670
+	145
++	150
+++	80
++++	34

Note: 1079 oral food challenges; 35 patients; 2 hospital centers; average number of foods tested per patient = 30.8.

Source: EHC-Dallas. 1984.

They were without medications or physical therapy and exercised at will (Table 23.13). There was a high correlation between patients' clinical improvement and their score for grip strength, dolorimeter, and arthrocircometer (Table 23.14 and Figure 23.14). After the fast, oral challenge testing with the food revealed objective changes after the fast in the dolorimeter, arthrocircometer, and grip strength (Table 23.15).

It was clear from this study that rheumatoid arthritis patients who were moderately severe (Table 23.16) could be improved with fasting in the ECU,

Table 23.11. ECU Cooperative Study of RA: 35 Patients, 2 Hospital Centers[a]

Food type	No. challenges	OK	+	++	+++	++++
Fruits	270	73.7	10.7	8.8	1.8	1.4
Vegetables	334	72.7	11.3	10.1	4.4	1.1
Cereal grains	103	49.5	15.5	14.5	13.5	6.7
Fish	89	48.3	24.7	30.3	13.4	7.8
Meat	165	36.9	25.1	26.0	12.1	4.2
Symptoms elicited (%)						
Musculoskeletal		74				
Neurological		15				
Skin		15				
Gastrointestinal		14				
Neuropsychiatric		11				
ENT		11				
Cardiovascular		6				
Pulmonary		0				
Genitourinary		0				
Arthritic pain		(100)				
Headache		(100)				
Pruritis		(100)				
Nausea, vomiting, abdominal bloating		(100)				
Mental obtundation		(100)				
Laryngitis, blurred vision, tinnitus		(100)				
Palpitations		(100)				

[a] Oral challenge after 4 days deadaptation in ECU with total body load reduced.

Source: EHC-Dallas. 1986.

and their symptoms could be objectively measured after reproduction by oral challenge. Twenty-eight associated symptoms were seen in the 43 patients (Table 23.17).

In a complementary study at the EHC-Dallas, T and B cell parameters were evaluated in a small group of chemically sensitive rheumatoid patients (Table 23.18). There was a statistically significant depression of the supressor cells in the chemically sensitive patients with rheumatoid arthritis vs. the controls. Because this group was small, the results of this study have limited generalization. We do not know if the outcome of a larger study series would reveal supportive findings.

We compared the results of our aforementioned study with a study by Veys et al.[65] and found a statistically significant difference in several parameters in their group of 43 patients and our smaller number (Table 23.19). When T_4/T_8 ratios of both studies were compared, these differences correlated well.

Table 23.12. ECU Cooperative Study of RA: 35 Patients, 2 Hospital Centers, 1077 Food Challenges

Food index (≥10 challenges/food)			
Highest index		Lowest index	
Corn	2.83	Melon	1.17
Beef	2.42	Lima beans	1.20
Chicken	2.28	Tomato	1.21
Fish	2.28	Lettuce	1.23
Turkey	2.24	Cherries	1.23
Lamb	2.24	Broccoli	1.25
Pork	2.05	Cabbage	1.28
Wheat	2.00	Oranges	1.31
Spinach	2.00	Grapes	1.36
Eggs	1.94	Peas	1.38

Source: EHC-Dallas. 1984.

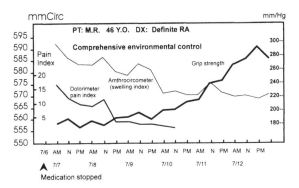

Figure 23.4. 46-year-old white female with rheumatoid arthritis treated with clearing of signs with treatment in the ECU; first 4 days fasting. (From EHC-Dallas. 1986.)

These findings suggest that environmental antigens may play a role in certain immunologic rheumatic diseases. It is generally accepted that the gastrointestinal tract is permeable to macromolecules,[66] that autoimmunity to bovine and food antigens may occur,[67-71] and that nutritional modulation alters experimental autoimmunity.[72-79]

Diet therapy for arthritis has received considerable publicity, and there is objective information about its efficacy beyond the aforementioned studies. Panush et al.[80] undertook a 10-week semi-controlled, double-blind, randomized trial of patients with active rheumatoid arthritis. Twenty-six patients completed the study; eleven were on an experimental diet (a specific, popular diet free of additives, preservatives, fruit, red meat, herbs, and dairy product[s]),

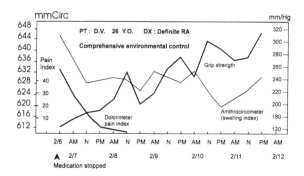

Figure 23.5. 20-year-old white female with rheumatoid arthritis treated
with clearing of signs with treatment in the ECU; first 4 days
fasting. (From EHC-Dallas. 1983.)

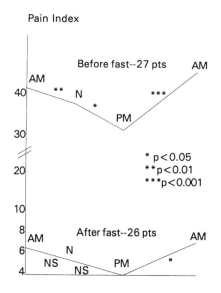

Figure 23.6. Rheumatoid arthritis patients — improvement in the pain
index as measured by the dolorimeter in both hands before
and after fast in the ECU. (From EHC-Dallas. 1983.)

but not a rotary diet, and fifteen were on a "placebo" diet. Of 183 variables
analyzed, no clinically important differences were identified among
rheumatologic, laboratory, immunologic, radiologic, or nutritional findings
between patients on the experimental and placebo diets. Six rheumatoid arthritis patients on the placebo and five on the experimental diet improved by
objective criteria. Improvement averaged 29% for patients on placebo and 32%
for patients on experimental diets. Two patients who improved notably on the
experimental diet elected to remain on it following the study period. They

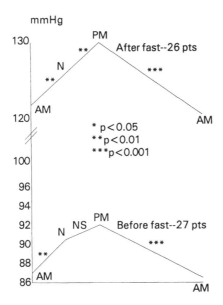

Figure 23.7. **Rheumatoid arthritis patients — improvement in the pain index as measured by the grip strength in both hands before and after fast in the ECU. (From EHC-Dallas. 1983.)**

continued to improve and noted exacerbations of disease when they consumed foods that were not on their study diet. As might be expected, this study failed to provide evidence of objective, overall clinical benefit of a nondetailed and nonspecifically defined diet as followed by a group of patients with longstanding, progressive, active rheumatoid arthritis. However, this data is not inconsistent with the fact that individualized dietary manipulations are beneficial for selected patients with rheumatic disease. In fact, it confirms our observations that blind treatment without individualized dietary and environmental manipulations will result in no statistical change. It is clear from our experience with several hundred patients with rheumatoid and osteo-arthritis at the EHC-Dallas that individualized diagnosis and treatment of the triggering agents is paramount. In our opinion, this is one of the reasons Panush's study[80] failed and the three-center study was shown to be valid.

Individualized diagnosis and therapy were apparent by a wide range of inhalants, foods, and chemicals found in another series of patients with rheumatoid arthritis at the EHC-Dallas. Also, a recent report details a patient with rheumatoid arthritis whose disease remitted following elimination of dairy products, exacerbated during uncontrolled milk and cheese challenges, and again remitted when the foods were withdrawn.[81] A study in our series showed similar findings with different foods in a 28-year-old male (Figure 23.13). At the EHC-Dallas, a second study involving 45 patients showed a similar clearing and response to challenge that the three-center studies showed. This study

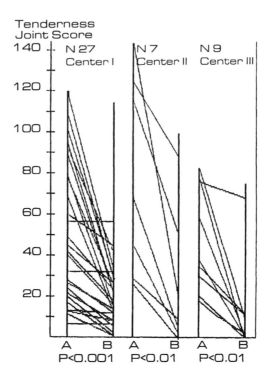

Figure 23.8. Rheumatoid arthritis cooperative ECU study. (A) Before environmental control 52.56 to 36.77. (B) After environmental control 15.87 to 19.62. *p*<0.001. (From EHC-Dallas. 1980.)

performed in the ECU emphasized not only food but also chemical challenge. Thirty percent were males and 70% females in this study. Inhaled double-blind challenges to the ambient doses of toxic chemicals reproduced the arthritis (Table 23.20).

In another study at the EHC-Dallas, many different types of toxic chemicals were shown to trigger rheumatoid arthritis (Table 23.20). These patients were followed long-term (over 2 years), and it was clear that they remained disease free only as long as they avoided their triggering agents. Large, long-term exposures, such as to natural gas and oil heat, easily exacerbated their arthritis.

Other studies have subsequently supported our observations. Sköldstam and Magnusson[82] studied 15 well-nourished patients with rheumatoid arthritis who were randomized to the experimental group. These patients were then fasted for 7 to 10 days. At the end of the fasting period, the Ritchie's index of joint tenderness showed decreases, and the plasma concentration of the acute phase protein orosomucoid was reduced.[83] The most striking effect, however, was the feeling of reduced pain and stiffness reported by almost all fasting patients — an impression that was strengthened by the significant

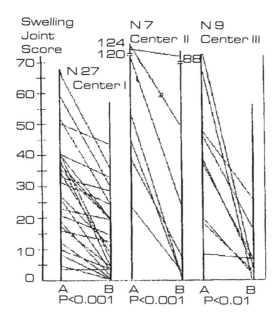

Figure 23.9. Rheumatoid arthritis cooperative ECU study. (A) Before environmental control 37.46 to 26.90. (B) After environmental control 11.79 to 18.20. p<0.001. (From EHC-Dallas. 1980.)

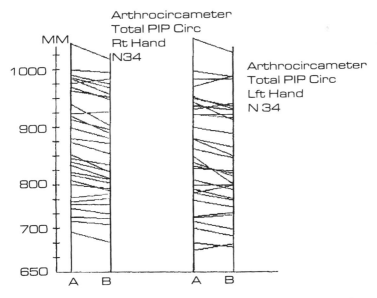

Figure 23.10. Rheumatoid arthritis cooperative ECU study. (A) Before environmental control 865.78 to 81.11. (B) After environmental control 851.50 to 95.25. p<0.001. (From EHC-Dallas. 1980.)

Table 23.13.   ECU Cooperative Study of RA

- Protocol: fasting period
- All medications were decreased except maintenance dose steroids (5.0 to 7.5 mg prednisone) in 3 of 43 patients
- Patients kept NPO except for water, laxatives
- Unrestricted ambulation
- No physical therapy
- Daily sequential measurements (a.m., noon, p.m.) of:
    grip strength
    dolorimeter
    arthrocircameter

Source: Modified from Kroker, G. F., R. M. Stroud, R. Marshall, T. Bullock, F. M. Carroll, M. Greenberg, T. G. Randolph, W. J. Rea, and R. E. Smiley. 1984. Fasting and rheumatoid arthritis: a multicenter study. *Clin. Ecol.* 2(3):137–144. With permission.

Table 23.14.   Arthritis: Changes in Parameters Measured Before and After Fasting in a Controlled Environment

| Measurement | No. | Measurement change during fasting | | |
		Improved	No change	Worse
Grip strength (mmHg)[a]	39	34	1	4
Dolorimeter pain index[a]	39	35	2	2
Swelling index[a]	43	41	1	1
Pain index[a]	43	41	2	0
Sedimentation rate				
westergren (mm/hr)[b]	33	28	1	4
Functional activity index[b]	32	21	6	5
Arthrocircameter[b]				
PIP joint swelling (mm)	35	27	3	5

Correlation Coefficients of Both Hand Measurements on 24 RA Patients at Admission End of Fast, Non- and Reactive Organic Foods, Pre- and Post-Commercial Foods and Discharge[c]

	r	p
Dolorimeter grip strength	−0.97	<0.001
Arthrocircameter grip strength	−0.96	<0.001
Arthrocircameter dolorimeter	0.96	<0.001

[a]   Three hospital centers
[b]   Two hospital centers
[c]   n = 7.

Source: Modified from Kroker, G. F., R. M. Stroud, R. Marshall, T. Bullock, F. M. Carroll, M. Greenberg, T. G. Randolph, W. J. Rea, and R. E. Smiley. 1984. Fasting and rheumatoid arthritis: a multicenter study. *Clin. Ecol.* 2(3):137–144. With permission.

Table 23.15. ECU Cooperative Study of RA

Patient number	Paired means	Before fast (a.m., noon, p.m.)	After fast (a.m., noon, p.m.)	p
40	Grip strength (mmHg)			
	Rt (n = 105)	90.20 ± 54.08	119.31 ± 65.67	<0.001
	Lt (n = 106)	83.21 ± 52.51	115.71 ± 62.45	<0.001
41	Dolorimeter			
	Rt (n = 107)	44.75 ± 39.06	18.78 ± 19.1	<0.001
	Lt (n = 107)	43.49 ± 40.14	19.02 ± 34.93	<0.001
33	Arthrocircameter			
	Rt (n = 33)	866.78 ± 81.14	854.42 ± 86.93	<0.001
	Lt (n = 33)	851.50 ± 98.43	841.51 ± 95.25	<0.001

Source: Kroker, G. F., R. M. Stroud, R. Marshall, T. Bullock, F. M. Carroll, M. Greenberg, T. G. Randolph, W. J. Rea, and R. E. Smiley. 1984. Fasting and rheumatoid arthritis: a multicenter study. *Clin. Ecol.* 2(3):137–144. With permission.

Table 23.16. ECU Cooperative Study of RA Conclusions: 43 Patients, 3 Hospital Centers

Environmental control program: statistically significant ($p<0.001$) improvement in the noted following parameters:
   Swelling index
   Tenderness index
   ESR
   Functional activity score
   Grip strength
   Dolorimeter score
   Arthrocircameter score

Note: Our data suggest that environmental control measures can lessen arthritic activity in certain patients with rheumatoid arthritis.

Source: Modified from Kroker, G. F., R. M. Stroud, R. Marshall, T. Bullock, F. M. Carroll, M. Greenberg, T. G. Randolph, W. J. Rea, and R. E. Smiley. 1984. Fasting and rheumatoid arthritis: a multicenter study. *Clin. Ecol.* 2(3):137–144. With permission.

and concomitant reduction in their daily intake of nonsteroidal antiinflammatory drugs (NSAIDs). Other investigators have presented equivalent results. Trang and colleagues[84] reported subjective improvement in all 13 patients, as did Hafström and colleagues[85] with all of their 14 patients. A decrease was observed in joint inflammatory activity, as assessed with Ritchie's[84,86] and with Lansbury's indices.[84,85,87] The erythrocyte sedimentation rate (ESR) was reduced,[84,85,87] as were various acute phase reacting plasma proteins, orosomucoid,[83,85,87,88] haptoglobin,[84,85,87,88] and complement factor 3.[85,87,88]

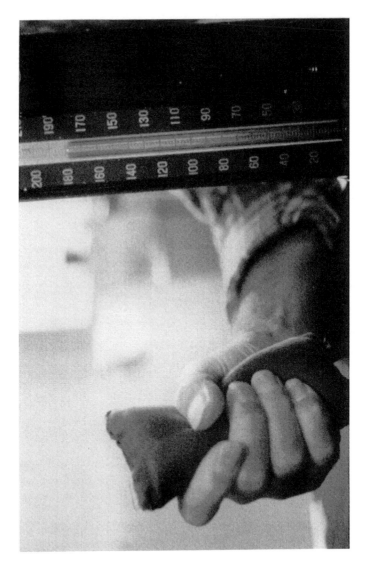

Figure 23.11.　Picture of grip strength measuring device. (From EHC-Dallas. 1980.)

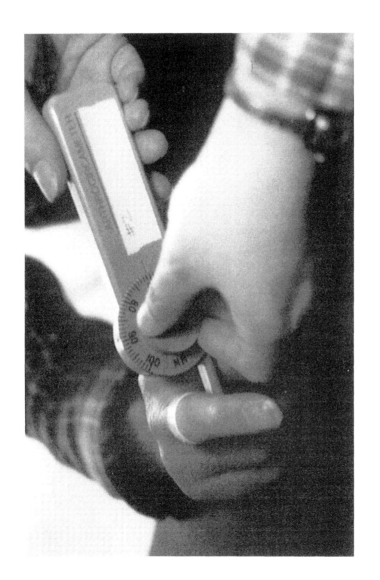

Figure 23.12.   Picture of the arthrocircometer. (From EHC-Dallas. 1980.)

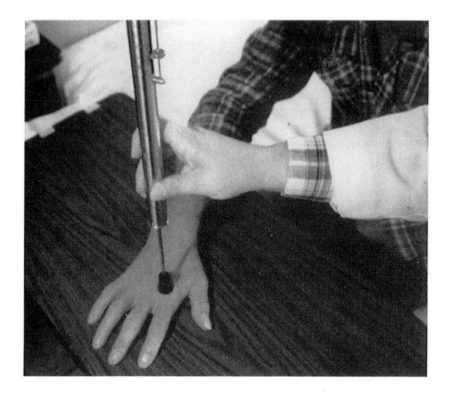

Figure 23.13.   Picture of dolorimeter. (From EHC-Dallas. 1980.)

Sköldstam and Magnusson[82] observed, as we did, that in general the improvement was obtained on the fourth or fifth day of fasting and remained throughout the rest of the fast. Many patients felt a return of pain and stiffness on the day after the fast was concluded, and most of the benefit of the fast was lost within 1 week or so following the patients' return to their usual eating habits. The antirheumatic effects were lost, irrespective of the diet that patients returned to, whether ordinary Western diet,[85] lactovegetarian diet,[83] or strict vegetarian (vegan) diet.[88] Again these investigators proved that nonindividualized diets failed. Similarly, Hamberg et al.[87] reported that 13 patients improved 3 weeks after the conclusion of a 7-day fast, also in comparison to 13 matched patients who had not fasted nor improved. The patients studied by Hamberg et al. ended the fast by taking up a vegan diet, and the special character of this new diet was claimed to preserve the benefit obtained from fasting.

Rheumatoid Ankylosing Spondylitis

Another frequently seen subset of chemically sensitive patients with rheumatoid arthritis is that of rheumatoid ankylosing spondylitis. Ebringer et al.[89] have linked rheumatoid spondylitis to environmental triggers. Further, they

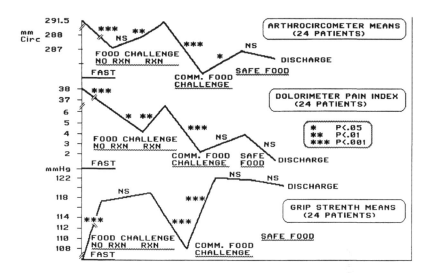

Figure 23.14.   ECU cooperative study—using a comprehensive environmental control program for rheumatoid arthritis. (From
EHC-Dallas. 1980.)

Table 23.17.   ECU Cooperative Study of RA[a]

	No. of patients	Symptoms
Musculoskeletal	31	(Increased myalgias/arthralgias)
Neurological	20	(Headache)
Gastrointestinal	12	(Abdominal pain, nausea)
Neuropsychiatric	10	(Depression, poor concentration)
ENT	5	(Sore throat, earache, etc.)
Complications: electrolyte abnormality	5	(Hyponatremia and/or hypokalemia)

Notes: This study was done in cooperation with the EHC-Dallas, the Chadbourne
Environmental Care Unit of Dr. Murray Carrol and the Henrotin Environmental
Care Unit of Dr. Theron Randolph.

[a] Average duration of fast = 6.2 days in 43 patients; associated symptoms during
fasting of 43 patients.

Source: Modified from Kroker, G. F., R. M. Stroud, R. Marshall, T. Bullock, F. M.
Carroll, M. Greenberg, T. G. Randolph, W. J. Rea, and R. E. Smiley. 1984. Fasting and
rheumatoid arthritis: a multicenter study. Clin. Ecol. 2(3):137–144. With permission.

have suggested that various environmental conditions (e.g., ankylosing
spondylitis, sacroiliitis, symptomatic backache, and gram negative reactive
arthritis) in which an increased frequency of HLA-B27 is observed to be
considered different manifestations of the same syndrome, which they call
"B-27 disease" when it occurs in individuals carrying the marker.[81]

Table 23.18. Immunological Data of RA Patients and Normals (Mean Values and Differences)

	Patients (n = 7)	Normals (n = 60)	Difference (patients to normal)	Significance (p)
WBC (#/mm^3)	7090 ± 810	7560 ± 220	No	>0.2
L (#/mm^3)	2250 ± 280	2770 ± 91	Smaller	<0.05
L (%)	32.1 ± 2.3	37.3 ± 1.1	Smaller	<0.05
T$_{11}$ (#/mm^3)	1600 ± 250	2080 ± 74	Smaller	<0.05
T$_{11}$ (%)	69.6 ± 2.9	75.2 ± 0.8	Smaller	<0.05
T$_4$ (#/mm^3)	1120 ± 210	1160 ± 43	No	>0.4
T$_4$ (%)	47.9 ± 4.0	42.2 ± 0.7	No	>0.05
T$_8$ (#/mm^3)	370 ± 37	740 ± 38	Significantly smaller	<0.0001
T$_8$ (%)	17.1 ± 1.3	25.8 ± 0.8	Significantly smaller	<0.0001
T$_4$/T$_8$	2.96 ± 0.39	1.70 ± 0.06	Significantly larger	<0.001
B (#/mm^3)	170 ± 51	270 ± 23	Smaller	<0.05
B (%)	7.4 ± 1.4	9.4 ± 0.6	No	>0.05

Source: Rea, W. J., A. R. Johnson, S. Youdim, E. J. Fenyves, and N. Samadi. 1986. T and B lymphocyte parameters measured in chemically sensitive patients and controls. *Clin. Ecol.* 4:11–14. With permission.

Table 23.19. Comparison of Immunological Data of RA Patients

	Number of persons	T$_4$%	T$_8$%	T$_4$/T$_8$
EHC-Dallas	7	47.9 ± 4.0	17.1 ± 1.3	2.96 ± 0.39
Veys et al.[a]	35 (very active)	61.0	22.0	3.0
	12 (active)	57.0	23.0	2.2
	5 (clinical remission)	59.0	31.0	1.4

[a] From Veys, E. M., P. Hermans, G. Verbruggen, and H. Mielants. 1983. Immunoregulatory changes in autoimmune disease. *Diag. Immunol.* 1:224–232. With permission.

Source: EHC-Dallas. 1986.

Ebringer et al.[90] have compiled evidence that gram negative bacteria is the trigger of rheumatoid ankylosing spondylitis. The three main bacteria are the gram negative organisms of salmonella, shigella, and yersin types. These have been found in association with reactive arthritis and HLA-B27. Several epidemiological studies have been conducted that clearly establish a temporal relationship between colonic infections and subsequent onset of reactive arthritis in a genetically susceptible host.[91-93] Also, three independent groups[94-96] have published bacteriological studies in which they report increased isolation of fecal *klebsiella* in patients with rheumatoid ankylosing spondylitis during various phases of disease activity (Figure 23.15). Figure 23.16 shows total

Table 23.20.   Rheumatoid Arthritis[a]

No. of patients	Sex		Age	
	M	F	Range	Mean
20	6 (30%)	14 (70%)	20 to 80 yrs.	45 yrs.

	Positive	Negative	p
Chemical	30	70	<0.001
Placebo	2	58	

Notes: Double-blind inhaled challenge with the ambient doses of <0.2 ppm formaldehyde, <0.5 ppm of phenol, <0.0034 ppm of insecticide (2,4-DNP), <0.33 ppm chlorine, <0.5 ppm of petroleum derived ethanol, 3 placebos — saline.

[a] Studied in ECU with 4 days reduction of total load resulting in a deadapted state. Joint swelling, dolorimeter, and grip strength were improved as the arthritis was cleared.

Source: EHC-Dallas. 1986.

serum IGA and C-reactive protein in ankylosing spondylitis patients from the series of Ebringer et al.[90]

It is relevant to note that in Ebringer's studies with his pathogenic model there were no bacterial fragments or antigens at the sites of inflammation or reactive arthritis. The pathogenic agent in the areas of inflammation or arthritis was an antibody evoked by an external antigen (Figures 23.16 and 23.17) but possessing a binding affinity for cross-reacting self antigens and, therefore, acting as a damaging autoantibody that could activate the complement cascade. Circulatory immune complexes are elevated in patients with ankylosing spondylitis, especially during active phases of the disease.

The lymphatic drainage of the colon and rectum occurs through the presacral and paraaortic lymph nodes, which are closely related to the sacro-iliac joints and the lumbar spine. A group from Marseilles[97] has described gross lymph node abnormalities when performing lymphangiograms on patients who are in the early stages of sclerofibrosis. These abnormal nodes produced obstructions to the free flow of lymph. These changes were appearing before the definite bony changes.[97,98]

The findings of the group from Marseilles are significant for patients suffering from environmentally induced ankylosing spondylitis. If these patients are plagued by bacterial and food problems and experiencing abnormalities in drainage via the intestine and colon, then damage to the lymph (immune) system is certain, and the arthritis occurs due to abnormal lymphatic and immune responses.

Ebringer et al.[81] have developed a new approach to prevention and therapy in these patients with ankylosing spondylitis. Initially, Ebringer used the antibiotics co-trimoxazule and phthalyl-sulphathiozole long-term in an attempt

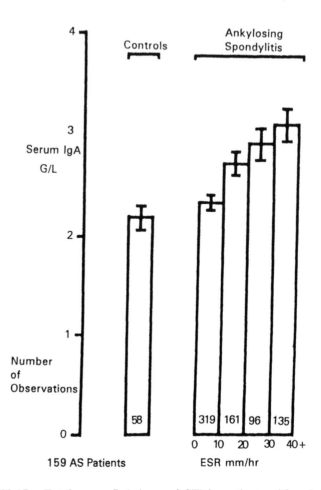

Figure 23.15.   Total serum IgA (mean ± SE) in patients with ankylosing spondylitis and controls. ESR measured only in AS patients. (From Ebringer, A., M. Baines, M. Childerstone, M. Ghuloom, and T. Ptaszynska. 1985. Etiopathogenesis of ankylosing spondylitis and the cross-tolerance hypothesis. In *Advances in Inflammation Research. Vol. 9. The Spondyloarthropathies*, eds. M. Ziff and S. B. Cohen. 116. New York: Raven Press. With permission.)

to reduce the level of acute phase reactants by modulating or changing the gram negative bowel flora. Later, he realized that diets that were low in simple sugars and carbohydrates, but high in proteins, were even more effective (Figure 23.18 to 23.20). He has been highly successful with this approach, treating over 200 patients with dietary manipulation.

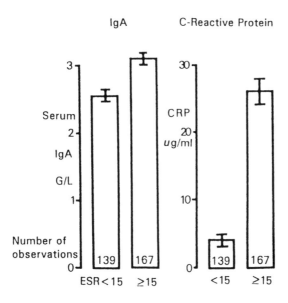

Figure 23.16. Total serum IgA (mean ±Se) and C-reactive protein (mean ±SE) in ankylosing spondylitis patients with elevated and normal ESR. (From Ebringer, A., M. Baines, M. Childerstone, M. Ghuloom, and T. Ptaszynska. 1985. Etiopathogenesis of ankylosing spondylitis and the cross-tolerance hypothesis. In *Advances in Inflammation Research. Vol. 9. The Spondyloarthropathies*, eds. M. Ziff and S. B. Cohen. 117. New York: Raven Press. With permission.)

The following case history demonstrates the variability of triggering agents in the ankylosing spondylitis and the individualized approach to diagnosis and treatment.

Case study. This 22-year-old white female was admitted to the ECU with the chief complaints of spinal arthritis (rheumatoid HLA-B-27), frequent episodes of iritis, tinnitus, poor memory, and muscle and joint pain (for details of iritis, see Chapter 27).

This patient reported that her illness began when she was 14 years old. At this time, she developed iritis and back pain. Shortly after beginning college, she developed neck swelling along with stiffness and headaches and was diagnosed at that time as having viral encephalitis. In 1980, she developed an anaerobic upper respiratory infection, and her blood test revealed toxoplasmosis. During this time, she was also diagnosed with ankylosing spondilitis with positive HLA-B-27. In 1981, she developed menorrhagia, which was corrected with birth control pills, but she eventually developed

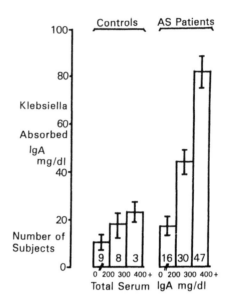

Figure 23.17.   Serum IgA absorption (mean ± SE) with whole *Klebsiella* microorganisms in ankylosing spondylitis patients and healthy controls. (From Ebringer, A., M. Baines, M. Childerstone, M. Ghuloom, and T. Ptaszynska. 1985. Etiopathogenesis of ankylosing spondylitis and the cross-tolerance hypothesis. In *Advances in Inflammation Research. Vol. 9. The Spondyloarthropathies*, eds. M. Ziff and S. B. Cohen. 118. New York: Raven Press. With permission.)

breast tumors that were thought to be linked to the birth control pills. The symptoms snowballed into numerous complaints, and she was unable to gain relief.

   Her family history was positive for hay fever, cancer, eczema, hives, diarrhea and constipation, insect allergies, drug allergies, and arthritis. Her father, age 51, had spinal arthritis and food and chemical sensitivities. Her mother, age 50, suffered from hay fever. The patient had three brothers. One was 28 years old and suffered from spondyloarthritis. Another was 23 years of age and had food sensitivity and hay fever. Her other brother, age 26, was in generally good health.

   After 4 days deadaptation with her total load reduced, this patient was orally and individually challenged with 22 chemically less-contaminated foods; 9 of them caused her symptoms. These positive results were confirmed by individual intradermal skin tests. Her chemical sensitivities were very widespread, and she had inhalant sensitivity to dust, pollens, and molds. Double-blind inhalation challenge with ethanol produced crying, spaciness, headache, inability to talk, bone pain, and blurred vision. All of her arthritis and eye symptoms were reproduced. She also had sensitivity to sulfa, neosynephrine, and antibiotics.

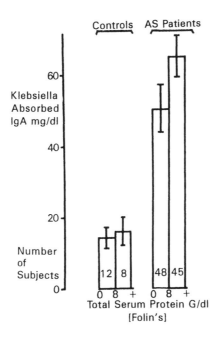

Figure 23.18.   Comparison of serum IgA absorption (mean ± SE) with *Klebsiella* microorganisms in ankylosing spondylitis patients and controls, having high (above 8 g/dL) and low (below 8 g/dL) total serum proteins. (From Ebringer, A., M. Baines, M. Childerstone, M. Ghuloom, and T. Ptaszynska. 1985. Etiopathogenesis of ankylosing spondylitis and the cross-tolerance hypothesis. In *Advances in Inflammation Research. Vol. 9. The Spondyloarthropathies*, eds. M. Ziff and S. B. Cohen. 120. New York: Raven Press. With permission.)

Abnormal test results showed a depressed T helper/suppressor ratio of 0.9 (C = 1.0 to 2.2), elevated T_8 at 43% (C = 20 to 37%), elevated IgM 418 mg/dL (C = 56 to 352 mg/dL), and elevated IgE 400 IU/mL (C = 0 to 180 IU/mL).

She was diagnosed with acute arthritis, rheumatoid-B-27, fibromyositis, acute iritis and uveitis, severe chemical hypersensitivity with atypical syncopal type episodes, immune complex disease (C1q), constipation with redundant colon, atopic disease (inhalant), multiple food hypersensitivities, and psychological factors affecting her physical condition.

This patient was placed in a chemically less-polluted environment and fed nothing by mouth, except spring water for 1 day. During this time, she reported irritability, headaches, and severe hunger pangs. She also reported that the joint pain had improved. Her iritis and arthritis resolved as did the lip chapping. She reported that the headaches and tinnitus exacerbated upon the provocative skin tests. After her admission to the ECU, she became symptom free. She was able to walk, run, and skip for the first time in years. Exercise, which used to incapacitate her, became effortless.

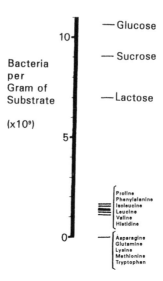

Figure 23.19.   Mean number of *Klebsiella pneumoniae* microorganisms following 24-hour culture using different carbohydrates and amino acids as substrates. (From Ebringer, A., M. Baines, M. Childerstone, M. Ghuloom, and T. Ptaszynska. 1985. Etiopathogenesis of ankylosing spondylitis and the cross-tolerance hypothesis. In *Advances in Inflammation Research. Vol. 9. The Spondyloarthropathies*, eds. M. Ziff and S. B. Cohen. 101–128. New York: Raven Press. With permission.)

In accordance with determined sensitivities, immunotherapy was arranged for 24 foods, trees, grass, weeds, terpenes, cotton, fluogen, MRV, transfer factor, molds, lake algae, *Candida*, dust, dust mite, feathers, smuts, histamine, and serotonin as well as multiple foods. Additional response was achieved with epsom salt as a laxative to clear the gastrointestinal tract. Intravenous infusions of vitamin C and subcutaneous injections of serotonin and histamine were used to clear the symptoms produced by intradermal and oral challenges.

This case revealed that this patient with severe chemical sensitivity and spondylitis had always been triggered by exposure to toxic chemicals. She did well with avoidance of these. She had a reversal of T_4/T_8 ratio, showing altered immoreguality. However, after chemically controlled environmental therapy and immunotherapy, including transfer factor, her symptoms resolved, and her T_4/T_8 ratio returned to normal at 1.4 (C = 1.0 to 2.2). She had a family history of chemical sensitivity, suggesting that her disease process might be associated with genetic factors. It was clear in this patient that many specific environmental triggers set off the genetic time bomb producing her symptoms.

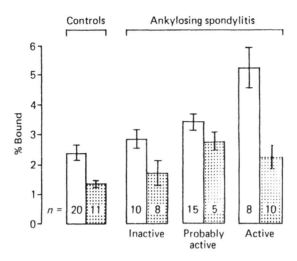

Figure 23.20. Radiobinding assay of human sera from patients with ankylosing spondylitis and healthy controls. Antibodies to *Klebsiella* and *Shigella* (mean ± s.e.) were measured (*n* = number of subjects). Clear bars = *Klebsiella*; dotted bars = *Shigella*. (From Ebringer, A. Aug. 1989. The relationship between *Klebsiella* infection and ankylosing spondylitis. *Baillière's Clin. Rheumatol.* 3(4). With permission.)

Post-Polio Syndrome

Patients with post-polio syndrome were first seen at the EHC-Dallas in the early 1970s. At that time, these patients presented, as many chemically sensitive patients do, with myalgia, arthralgia, and fatigue. Guided by Dickey,[99] who first defined this post-polio syndrome response to environmental incitants in 1971, we appropriately recognized and treated this entity. It was not until later, however, when Halstead's work popularized this "post-polio syndrome" that environmental triggers became commonly accepted as being present in, and constituting the majority of, the etiology of this syndrome.

Other collagen and neuromuscular related clinical entities may be triggered by toxic chemicals. Limited recent information on post-polio patients who have developed chronic fatigue syndrome with weakness and loss of previous ability to function has strongly implicated toxic chemicals and reactions to foods as producers of their symptoms.[100,101] Mandell and Conte[102] subsequently reported a series of patients as did Bailey,[100] who found responses to specific antigen administration.[100] We studied a series of post-polio patients under environmentally controlled conditions and were able to clear and reproduce their symptoms both acutely and long-term (Table 23.21).

**Table 23.21.   Symptoms in Post-Polio Patients
Before and After Treatment[a]**

Patient	Symptoms
Before treatment	
17	lower limb weakness
2	upper limb weakness
9	pain — muscle and bone
After 1 year of treatment	
14	gone
1	recurrent
2	no change

[a]   Age: 31 to 63 yrs. Males = 5; females = 12.

Source: Rea, W. J., A. R. Johnson, E. Fenyves, and J. Butler. 1987. The environmental aspects of the post-polio syndrome. In *Research and Clinical Aspects of the Late Effects of Poliomyelitis: Second Research Symposium on the Late Effects Of Poliomyelitis*. Warm Springs, GA. Sept. 5–7, 1986, eds. L. S. Halstead and D. O. Wiechers. 173–181. White Plains, NY: March of Dimes Birth Defects Foundation. With permission.

The patients either had weakness, pain, or both. Fourteen of these 17 patients cleared their symptoms after an ECU avoidance program. A subsequent series of 15 more patients has shown similar results. The symptom relief has continued for 4 years without the use of medication.

Table 23.22 shows the double-blind challenge of two patients under environmentally controlled conditions. These patients have done extremely well without medications over the past 15 years.

The triggering agents of the aforementioned patients fell into three categories — biological inhalants, foods, and toxic chemicals. Multiple food sensitivity was present in all 17 patients, and 15 patients were sensitive to 7 or more foods. Formaldehyde seemed to be a prime offender in the chemical group. Seven patients were sensitive to chlorine. They were exposed while swimming, which was a part of their routine therapy (Table 23.23).

The associated symptoms seen in the post-polio syndrome patients included those summarized in Table 23.24. Overweight seemed to be one of the major deterrents to complete recovery in some patients with the post-polio syndrome. It was necessary to eliminate excess weight in order to restore proper function in some patients (Table 23.24).

Long-term follow-up over 4 years has further enforced our feeling that many post-polio patients are strongly influenced by environmental triggers.

Table 23.22. Post-Polio Syndrome: Ambient Dose, Inhaled, Double-Blind Toxic Chemical Challenge[a]

Booth	Dose (ppm) Pt. 1[b]	Pt. 2[c]	Pulse change Pt. 1[b]	Pt. 2[c]	Flow meter change Pt. 1[b]	Pt. 2[c]	Symptom change Pt. 1[b]	Pt. 2[c]	EKG Pt. 1[b]	Pt. 2[c]	Reproducible (×3) Pt. 1[b]	Pt. 2[c]
Placebo #1 (spring water)	5 cc	5 cc	0	0	0	0	0	0	0	0	+	+
Placebo #2 (filtered water)	5 cc	5 cc	0	0	0	0	0	0	0	0	+	+
Placebo #3	5 cc	5 cc	0	0	0	0	0	0	0	0	+	+
Petroleum ethanol	<0.5	<0.5	0	0	0	0	0	+	0	0	+	+
Formaldehyde	<0.2	<0.2	>3 SD	>3 SD	>3 SD	0	+	+	0	+	+	+
Phenol	<0.5	<0.5	0	>2 SD	0	0	0	+	+	0	+	+
Chlorine	<0.33	<0.33	0	>2 SD	0	0	0	+	+	0	+	+
Pesticide (2,4-DNP)	<0.0134	<0.0134	>2 SD	>2 SD	0	0	+	+	0	0	+	+

[a] EHC-Dallas. Challenge testing was done in hospital environmental wing, under strictly controlled environmental (ECU) conditions. Patients were deadapted for at least 4 days with reduction of total load.

[b] Patient # 1: 55; WF; symptoms — asthma, weakness.

[c] Patient # 2: 45; WM; symptoms — arrhythmia, weakness, pain

Source: Rea, W. J., A. R. Johnson, E. Fenyves, and J. Butler. 1987. The environmental aspects of the post-polio syndrome. In *Research and Clinical Aspects of the Late Effects of Poliomyelitis: Second Research Symposium on the Late Effects of Poliomyelitis.* Warm Springs, GA. Sept. 5–7, 1986, eds. L. S. Halstead and D. O. Wiechers. 173–181. White Plains, NY: March of Dimes Birth Defects Foundation. With permission.

Table 23.23.   Triggering Agents Defined by Inhaled, Oral, and Intradermal Challenge Under Environmentally Controlled Conditions

Patients	Incitants
15	Inhalants (predominant molds)
17	Chemicals
16	Formaldehyde
14	Cigarette smoke
12	Polio vaccine
9	Terpenes
8	Perfume
8	Petroleum ethanol
7	Chlorine
5	Diesel
4	Phenol
4	Newsprint
17	Foods (sensitive to seven or more, e.g., coffee, tea, cane sugar, wheat, corn, milk, beef, chicken, eggs)

Source: Rea, W. J., A. R. Johnson, E. Fenyves, and J. Butler. 1987. The environmental aspects of the post-polio syndrome. In *Research and Clinical Aspects of the Late Effects of Poliomyelitis: Second Research Symposium on the Late Effects of Poliomyelitis.* Warm Springs, GA. Sept. 5–7, 1986, eds. L. S. Halstead and D. O. Wiechers. 173–181. White Plains, NY: March of Dimes Birth Defects Foundation. With permission.

Table 23.24.   Accessory Symptoms Reproduced by Challenge under Environmentally Controlled Conditions and Alleviated by Long-Term Treatment

Ear, nose, throat	9
Headache, dizziness, recurrent rhinosinusitis	
Gastrointestinal upset	4
Bloating, abdominal cramps	
Weight gain (nonedematous)	3
Edema	100
Respiratory difficulties	
Severe asthma	1
Breathless	4
Shortness of breath	3
Hypertension	2

Source: Rea, W. J., A. R. Johnson, E. Fenyves, and J. Butler. 1987. The environmental aspects of the post-polio syndrome. In *Research and Clinical Aspects of the Late Effects of Poliomyelitis: Second Research Symposium on the Late Effects of Poliomyelitis.* Warm Springs, GA. Sept. 5–7, 1986, eds. L. S. Halstead and D. O. Wiechers. 173–181. White Plains, NY: March of Dimes Birth Defects Foundation. With permission.

Autoimmune Disease

It is clear from our and others' studies that toxic chemicals can induce autoimmunity. A growing body of animal data support the clinical observations we have made of our chemically sensitive patients.

We have found in our clinical work at the EHC-Dallas that dietary manipulation with decreased caloric intake is a viable tool for relieving the symptoms of many patients who have environmentally induced autoimmune disease. Similarly, Fernandes and co-workers[103] offered impressive laboratory evidence for an effect of diet on autoimmune disease. These researchers showed that (NZB X NXW) Fl mice lived up to twice as long when their total food intake was decreased. This dietary effect was accompanied by a marked reduction in anti-DNA levels. Restriction of the caloric intake in these mice was also accompanied by a decrease in the retroviral gp 70-anti-gp 70 immune complex levels.[104] It has also been demonstrated that the autoimmune disease of NZB mice is slowed by deprivation of zinc in the diet.[105] Zinc deficiency in humans and animals has been associated with multiple immmuologic defects[106] but does not seem to prevent autoimmune disease. However, correction of zinc deficiency, along with reduction of the total body pollutant load, has reversed autoimmunity in many of our chemically sensitive patients.

Although we do not entirely agree, based on our studies with environmentally sensitive patients, some investigators have felt that the amount of the two-double-bonded polyunsaturated fatty acid, linoleic acid (LA), in the diet is critical to understanding the mechanism involved in dietary manipulation. The level of intake of this fatty acid has a number of important ramifications above and beyond the fact that it is an essential fatty acid whose absence from the diet can lead to deficiency states. First, linoleic acid is the precursor of arachidonic acid (AA), which is, in turn, the precursor of the prostaglandins (PG) and leukotrienes. Therefore, decreased intake of LA leads to decreased formation of PGs and leukotrienes, and increased intake leads to increased amounts of these important agents in the body.

Both types of change have been seen in chemically sensitive patients and successfully counteracted by use of the rotary diet. The rotary diet used to treat successfully most chemically sensitive patients appears to be an anti-inflammatory diet. The E series prostaglandins and leukotrienes have a proinflammatory effect,[107] which means a rotary diet can minimize the dietary generators of these. The PGEs also have important suppressive effects on both the cellular and humoral immune responses[108] and on the response of a target cell. The fibroblast responds to stimulation by immunologic mediators,[109] so the PGs may be expected to increase or decrease inflammation, immunity, or collagen synthesis, depending on their locus of action and their concentration at that locus. As a corollary, diets that contain either large amounts of LA, leading to the synthesis of increased amounts of PGs, or small amounts of LA, leading to reduced PG synthesis, may be expected to affect inflammation, immunity, and fibrosis. It should not be forgotten, however, that specific food and chemical triggers may also be found in individual patient studies. Many chemically sensitive patients respond to dietary manipulation, which may modulate the mechanism.

Dietary LA could affect autoimmune disease by another mechanism, e.g., an effect on cell membrane function. Cell membranes are dynamic, with membrane lipids passing through either by an ordered or fluid state. The fluidity of membranes and the activity of many membrane enzymes are affected by the ratio of polyunsaturated to saturated fatty acids in the membrane. Presumably related to this phenomenon is the observation, made in both *in vivo* and *in vitro* studies,[110] that the polyunsaturated fatty acid content of the cell membranes of lymphocytes and macrophages has an influence on their functions. Lymphocyte responsiveness is affected by the availability of the essential fatty acids LA and AA,[110] and enrichment of macrophage membranes with saturated fatty acid leads to reduced dendritic activity.[111] They also appear to be affected by toxic chemical exposure due to the chemical lipophilia.

From the previous discussion, it might be expected that the level of intake of LA would have an effect on the course of autoimmune and chronic inflammatory disease, as is seen in some chemically sensitive individuals. There is a growing body of literature demonstrating the importance of the linoleic acid-arachidonic acid-prostaglandin axis in autoimmunity and chronic inflammatory disease. Suppression of adjuvant arthritis in rats by injection of PGE_2[112] and PGE_1[113] has been reported. Subsequently, Zurier and co-workers[114] demonstrated a beneficial effect of PGE_1 on the survival of (NZB X NZW) F_1 mice and on the course of renal disease in these mice. Stackpoole and Martin[115] showed that experimental allergic encephalomyelitis in guinea pigs was almost completely inhibited by oral supplements of essential fatty acid. McLeish et al.[116] observed that injection of AA in mice with an immune complex-induced glomerulonephritis decreased the glomerular damage; and Kunkel et al.[117] found that evening primrose oil, which is very rich in LA, markedly inhibited the development of adjuvant-induced arthritis in rats.

The above reports have established the suppressive effects of PGs or increased intake of polyunsaturated fatty acid on immunologically related musculoskeletal disease. In contrast, a deficient intake of polyunsaturated fat has its own legacy causing a number of illnesses including vascular disease. The rotary diet used to treat the chemically sensitive individual is low in saturated fat and high in unsaturated fat. A number of studies have described benefits of essential fatty acid (EFA)-deficient diets in animal models of immunologically induced inflammatory disease. Adjuvant-induced arthritis was ameliorated in animals deficient in EFA.[118] The inflammation was restored by feeding the rats a small supplement of corn oil. Similar observations have been made with respect to carrageenen-induced inflammation[119] and kaolin-induced pouch granulomas[120] in rats. An EFA-deficient diet, moreover, diminished humoral immune responses to T-cell dependent and T-cell independent antigens in mice. Full restoration of these responses occurred upon switching to a control diet.[121]

Clinical outcomes of treatment of chemically sensitive patients testify to the therapeutic advantage of the rotary diet in reducing inflammatory-inducing fatty acids. This outcome is also documented by various studies in animals. Interesting evidence regarding EFA deficiency and autoimmune disease has been obtained in the NZB/NZW mouse strain. Hurd et al.[122] observed that NZB/NZW mice fed a diet in which all fat was in the form of unsaturated fat had marked prolongation of life, reduced severity of glomerulonephritis, and markedly reduced levels of antinuclear antibodies and anti-DNA antibodies. This improvement appeared to be associated with a deficiency of dietary polyunsaturated fatty acids.

The feeding of "unphysiologic" fatty acids, unphysiologic at least for the population of the western world, has also been associated with improvement of NXB/NZW disease. Prickett et al.[123] recently reported that when NZB/NZW mice were fed a diet rich in eicosapentaenoic acid (EPA), a 5-double-bonded fatty acid present in large quantities in the fish-oil-rich diet of Greenland Eskimos, they experienced a marked prolongation of life and delay in onset of renal abnormalities. These findings were similar to those previously noted by Hurd et al.[122] with the EFA-deficient diet. The reports of Hurd et al.[122] and Prickett et al.[124] present evidence that substitution of polyunsaturated fatty acids in the diet saturated by fatty acid or "unphysiologic" polyunsaturated fatty acids leads to a marked reduction in the severity of the autoimmune disease of the NZB/NZW mouse.

There is, it should be mentioned, a brief report of the successful treatment of the sicca syndrome in humans upon supplementation of the diet with the highly unsaturated evening primrose oil,[125] but the data presented are not detailed enough to evaluate properly. We have successfully treated several cases of sicca syndrome by eliminating the triggering agents (see Chapter 20). It is now being shown that many environmental stimuli are involved in the triggering of collagen diseases.[126]

Systemic Lupus Erythematosis (SLE)

Chemical triggering of lupus-like syndrome has been well established in the literature.[127] Animal studies also show that chemical overload will trigger autoantibody production.[128] Clinical evidence for pollutant involvement in SLE is apparent in the number of chemically sensitive patients with SLE that we have treated at the EHC-Dallas.

There are many therapeutic agents known to be related to SLE or lupus-like syndrome including iodides,[129] heavy metals,[130] dilantin,[131] tridione,[130] isoniazid,[131] pronestyl,[130] hydralazine,[131] methyldopa,[131] antibiotics,[131] sulfonamides,[131] and thiourea derivatives.[114] The disease (or at least an SLE-like syndrome) may be initiated by the use of certain drugs. The known ones are listed in Table 23.25.

The mechanisms of disease induction probably vary depending upon the structure of the drug to which an individual is exposed. Even toxic chemicals

Table 23.25.   Some Drugs Able to Induce
Features of SLE

Related to dose-time administration	More idiosyncratic
Hydralazine	Aminosalicylic acid
Procainamide	D-Penicillamine
Alpha-methyldopa	Griseofulvin
Isoniazid	Penicillin
Chlorpromazine	Ampicillin
Chlorthalidone	Streptomycin
Phenytoin	Sulfonamides
Mephenytoin	Tetracycline
Trimethadione	Methylthiouracil
Primidone	Propythiouracil
Ethosuximide	Phenylbutazone
Carbamazepine	Oxyphenisatin
Phenylethylacetylurea	Practolol
	Tolazamide
	Methysergide
	Reserpine
	Quinidine
	Isoquinazepan
	Guanoxan

Source: Steinberg, A. D. 1985. Systemic lupus erythe-
matosus. In *Text of Medicine*, 17th ed., eds. J. B.
Wyngaarden and L. H. Smith, Jr. 1926. Philadelphia:
W. B. Saunders. With permission.

in foods may induce SLE.[22] The degree to which "ideopathic" SLE is triggered by specific environmental factors is unknown and requires study.

Apparently, many foreign substances can trigger SLE. The following case report is an example of a patient whose symptoms were environmentally induced.

Case study. A 36-year-old white female developed recurrent bouts of vomiting at the age of 5 years. These gave way to migraines at the age of 11 years,
and the latter persisted. At the age of 16 years, she developed polyarthritis,
and a diagnosis of SLE was made. Her disease progressed over the next
several years, with further involvement of the gastrointestinal, genitourinary,
respiratory, and vascular systems. Spontaneous bruising and petechiae occurred together with peripheral edema. She eventually was placed on cortisone and cytotoxic drugs. Antinuclear antibodies and LE preparations were
positive on numerous occasions. She continued to deteriorate. She was placed
in the ECU, and all medications were discontinued. The stiffness and swelling

of her joints gradually disappeared. Her sedimentation rate fell from 63 to 15 mm in 1 week. This sedimentation rate was the lowest it had been for many years. The circumference of her fingers diminished by 1.5 to 2 cm while fasting, reflecting the massive loss of edema. She was able to open and close her hands for the first time in many years. Challenges with 20 different foods precipitated a return of her symptoms; 10 did not. The double-blind inhalation of chemicals such as perfume, phenol, and natural gas also triggered symptoms. She has done well without medications on an avoidance program for several years.

Scleroderma (Systemic Sclerosis)

The EHC-Dallas has a series of chemically sensitive patients whose SSc syndrome was triggered by sensitivity to artificial implants. Breast augmentation with paraffin and silicon can cause SSc-like syndrome (see later discussion of implant syndrome).[132-135]

Numerous environmental triggers have been implicated in SSc onset including vinyl chloride, silica dust, and toxic oil (rapeseed and owaline); organic solvents including aromatic hydrocarbons (e.g., toluene, benzene) and aliphatic chlorinated hydrocarbons (e.g., trichloroethylene, tetrachloroethylene) and aliphatic nonchlorinated hydrocarbons (e.g. naphtha-n-hexane, hexachloroethane, epoxy resins, bleomycin, carbitepa and 5-hydroxitryptophan, diethylpropion hydrachloride, mazindol, and pentazocaine).

Genetic factors such as HLA type, sex, and allotype of drug metabolizing enzymes are also involved in the individual responses to pollutant exposure that could lead to the development of SSc-like illness. Generally, these genetic factors are time bombs awaiting the right set of environmental pollutant exposures in order to trigger SSc. Acetylater, phenotype, hydroxylyser, and sulphoxidizer state partially govern an individual's ability to acetylate amino groups or hydroxylate and oxidase SH groups. Allotype deformation has proven important in influencing the development of drug-induced SLE.

Vinyl Chloride Disease

Vinyl chloride disease is the best worked out for SSc.[132,136] Vinyl chloride is a halogenated derivative of the ethylene family as are trichloroethylene and tetrachloroethylene, both of which also have been implicated in the SSc.[132,137] The HLA system is involved in the susceptiblilty to vinyl chloride disease. Hepatomegaly,[132] Raynaud's phenomena, puffy fingers, skin changes, and derivative bone lesions in the terminal phalanges can occur in people exposed to vinyl chloride.[132,138-144] Lange et al.[145] described skin sclerosis, skin nodules, pseudoclubbing of fingers, acro-osteolysis, Raynaud's, thrombocytopenia, portal fibrosis, impaired hepatic function, and pulmonary fibrosis as a result of vinyl chloride exposure. Exercise, fatigue, muscle and joint pains, nervous system symptoms, and impotence were also described. Forty-seven cases of angiosarcoma were also described.[146,147] Maricq[137] showed vascular instability and eventually capillary abnormalities. We have seen vasculitis in patients exposed to vinyl chloride. Microaneurysms occur in vinyl chloride workers.

The mechanism of action is unknown, but epoxides do form, which result in formation of free radicals. The active epoxide metabolite is a powerful alkylating agent that can bind to cellular macromolecules.[148] The vinyl chloride metabolite might bind to the hapten group causing immunologic changes. Increased immunoglobulins, cryoglobulins and *in vivo* conversion of C_3 and C_4, and low levels of acute antibodies have been formed.[149] HLA-DR5 has been shown in one group.[135,136] Perhaps Al-B8-DR3 haplotype is associated with susceptibility.

Silica Disease

Silica has been known to cause SSc since 1914.[150] This was also shown in gold miners[151,152] as well as some coal miners exposed to silica dust.[132,153,154] The prevalance of SSc in patients with silicosis is estimated to be 110 times that in a normal male population. It appears that, in gold miners, SSc is dose related.[155] It is reported that the South African gold miners have an altered step in tryptophan metabolism.[156] Silica migrating from silicon implants has been found to cause calcium and magnesium deposition in extra-implant areas of the breast, lung, and chest in patients at the EHC-Dallas (Figure 23.21).

Organic Solvents and SSc

Between 1965 and 1981, Walder[157] reported 12 men who were in close contact with solvents before the onset of SSc. Eleven of the 12 were exposed to aromatic solvents. Other isolated case reports have indicated that tricholoroethylene and perchloroethylene cause SSc.[158-160] The Australians reported trilene (a pure form of trichloroethylene) anesthesia to cause SSc. Black and Welsh[132] saw six cases of SSc attributed to anesthetic agents. Eleven other patients with SSc have named environmental agents as the cause of their disorder.

We have been extremely careful with surgery in our chemically sensitive patients, since so many (80%) have levels of trichloroethane, trichloroethylene, and tetrachloroethylene in their blood. We have seen many patients who were made ill by careless choice of anesthesia (see Volume IV, Chapter 41). It is thought that several small, polar, chlorinated hydrocarbons induce subtly different disease. For example, the aromatic hydrocarbons are associated with minimal visceral organ involvement, whereas aliphatic chlorinated derivatives induce systemic disease. In addition, nonchlorinated aliphatic hydrocarbons induce both isolated skin lesions and systemic disease. Epoxy resins induce SSc with muscle weakness, arthralgias, impotence, esophageal and lung involvement. An amino compound, bis (4-amino-3-methyl-cyclohexyl) methane, a clohexylamine, is thought to be the causative agent. Also amino-type

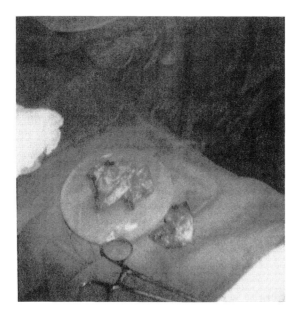

Figure 23.21. Silica migration in breast with calcium deposits. (From EHC-Dallas. 1994.)

local anesthetic,[161] and crumbling urea formaldehyde[162] cause SSc (see Volume I, Chapter 4[1] and Volume IV, Chapter 41).

Bleomycin used for chemotherapy of testicular carcinoma has been shown to cause SSc[132,163-167] associated with abnormalities of tryptophan metabolism, particularly serotonin, as in the carcinoid syndrome. These chemicals have caused SSc.[168-171]

Another group of patients who had environmentally induced solvent disease are the Desert Storm veterans. They were exposed not only to the oil fires and shell and missile explosives but also to oil in washing, phthalates, cooking oil, drinking water, insecticides, skin protectants, kerosene heaters, and many other chemicals. Some of the veterans observed at the EHC-Dallas have developed organic brain dysfunction, arthralgia, fibromyalgia, and gastrointestinal upset.

Diethylpropion hydrochloride and mazindol, both appetite suppressants with sympathomimetic and serotonergic properties, may have a pathogenic role in SSc.[172] Methysergide, which is structurally similar to 5-hydroxy-tryptamine and pentazocine, causes fibrosis.[153]

Implant Syndrome

In addition to many synthetic insertions such as abdominal mesh, vascular grafts, heart valves, false teeth and eyes, and artificial joints, a mamoplasty syndrome from both paraffin and silicon has been associated with SSc. We have seen over 150 patients with chemical sensitivity who have breast implants. One hundred had these removed, which helped reduce their sensitivity markedly (see Chapter 19 and Volume IV, Chapter 41).

A classic model for the induction of chemical sensitivity has now been devised in that the surgical use of artificial implants can induce SSc. Animal and human studies of artificial organs as long as 30 years ago showed tissue reaction causing local fibrosis as well as triggering of the clotting mechanisms. Since the early use of artificial implants was confined to life-threatening procedures where the individual would either die or lose a major organ, this fibrotic or clotting complication was justified. As surgery became more expansive, it was thought that silicone injections and then implants could be used for cosmetic purposes. However, silicone injections of the face and breasts were fraught with many complications including severe inflammations and sloughing. They were, therefore, abandoned. Following these, synthetic meshes of various plastic materials were used in repairs of large hernias. These were moderately successful, and their use spread to repair smaller openings. The following are two case reports of the severe complications that one can develop after mesh implants.

> **Case study #1.** A 45-year-old white female was well until she had a large abdominal hernia repaired with a synthetic mesh graft. When she awoke from surgery, she had severe asthma (which she had never before had) that was intractable to all forms of treatment. She became cortisone dependent, requiring 80 to 100 mg of prednisone per day in order to survive, even though the asthma continued to incapacitate her. She became very odor sensitive, developing all the classic signs and symptoms of chemical sensitivity. She also developed SSc with Raynaud's phenomenon, spontaneous bruising, peripheral and periorbitol edema, and cold sensitivity. This patient underwent challenge testing and was found to be sensitive to toxic chemicals and electromagnetic fields. Removal of the abdominal mesh resulted in immediate cessation of the asthma, which had been present since its insertion 5 years earlier. The chemical sensitivity diminished rapidly, but the electrical sensitivity persisted. The patient lost her need for prednisone and her sensitivity to chemicals. She improved markedly.

> **Case study #2.** This 40-year-old white male college professor already was suffering from severe chemical sensitivity. He developed an inguinal hernia. He told the surgeon not to use synthetics under any circumstances. He awoke from surgery with excruciating pain out of proportion to the surgery and Raynaud's. He immediately developed a swollen, red, inflamed wound, and after many trials of antibiotics the graft had to be removed. He immediately was relieved of the pain and the recurrent infection as well as his Raynaud's, which quickly subsided.

SSc with arthralgia, fibromyalgia, and organic brain syndrome has been seen in the majority of patients with breast implants that we have examined.

Table 23.26. Metal Implant Sensitivity as Determined by Intradermal Challenge and Subsequent Implant Removal

Patient (age/gender/race)	Prostheses	Intradermal test (chromium, nickle, or titanium)	Improved upon removal
22 yrs./W/F	R — clavicular plate	+	yes
70 yrs./W/F	R — femoral plate	+	yes
80 yrs./W/F	Hip prosthesis	+	yes
33 yrs./W/M	Humoral plate	+	yes
40 yrs./W/M	Ulnar plate	+	yes
53 yrs./W/F	R — tibial plate	+	yes
59 yrs./W/M	Female prosthesis	+	yes
40 yrs./W/F	L — tibial plate	+	yes
26 yrs./W/M	L — humoral plate	+	yes
18 yrs./W/F	R — femoral plate	+	yes

Source: EHC-Dallas. 1992.

These symptoms subside with removal of the implants and their capsules, injection therapy for biological inhalants and foods, and heat depuration/physical therapy. (For details of the implant-induced disease, see Chapter 19, section on the breast, and Volume IV, Chapter 41.)

The metal implant syndrome does not appear to cause as severe autoantibody reactions as the silicone does. However, metals can cause local and general inflammation resulting in severe pain and, at times, fibromyalgia (Table 23.26).

DIAGNOSIS AND TREATMENT

Diagnosis of pollutant-related musculoskeletal injury includes challenges (oral, intradermal, or inhaled), blood immune and xenobiotic levels, and various other objective tests. Treatment includes massive avoidance of pollutants in air, food, and water and implementation of a rotary diet, injection therapy for biological inhalants, nutritional replacement, and heat depuration/physical therapy. Because the diagnosis and treatment of disorders of the environmental aspects of musculoskeletal system is no different than those of other systems, no further discussion will be carried out here.

REFERENCES

1. Rea, W. J. 1992. *Chemical Sensitivity, Vol. I. Mechanisms of Chemical Sensitivity*. 47. Boca Raton, FL: Lewis Publishers.
2. Rea, W. J. 1992. *Chemical Sensitivity, Vol. I. Mechanisms of Chemical Sensitivity*. 221. Boca Raton, FL: Lewis Publishers.

3. Savolainen, K., V. Riihim "aki", O. Muona, J. Kekoni, R. Luukkonen, and A. Laine. 1985. Conversely exposure-related effects between atmospheric m-xylene concentrations and human body sense of balance. *Acta Pharmacol. Toxicol. (Copenh)* 57(2):67–71.

4. Martinez, D. M. 1991. Personal communication.

5. Wecker, L., R. E. Mrak, and W. D. Dettbarn. 1986. Evidence of necrosis in human intercostal muscle following inhalation of an organophosphate insecticide. *Fundam. Appl. Toxicol.* 6(1):172–174.

6. Taylor, D. N., K. E. Lagory, D. J. Zaccaro, R. J. Pfohl, and R. D. Laurie. 1985. Effects of trichloroethylene in the exploratory and locomotor activity of rats exposed during development. *Sci. Total Environ.* 47:415–420.

7. Arito, H., N. Tsuruta, K. Nakagaki, and S. Tanaka. 1985. Partial insomnia, hyperactivity and hyperdipsia induced by repeated administration of toluene in rats: their relation to brain monoamine metabolism. *Toxicology* 37(1–2):99–110.

8. Crofton, K. M. and L. W. Reiter. 1984. Effects of two pyrethroid insecticides on motor activity and the acoustic startle response in the rat. *Toxicol. Appl. Pharmacol.* 75(2):318–328.

9. Chung, E. and M. W. Van Woert. 1984. DDT Myoclonus: sites and mechanism of action. *Exp. Neurol.* 85(2):273–282.

10. Boja J. W., J. A. Nielson, E. Foldvary, and E. B. Truitt, Jr. 1985. Acute low-level formaldehyde behavioral and neurochemical toxicity in the rat. *Prog. Neuropsychopharmacol. Biol. Psychiatry* 9(5–6):671–674.

11. McKay, R. T. and S. M. Brooks. 1984. Hyperactive airway smooth muscle responsiveness after inhalation of toluene diioayanate vapors. *Am. Rev. of Resp. Dis.* 129(2):296–300.

12. Laseter, J. L., I. R. DeLeon, W. J. Rea, and J. R. Butler. 1983. Chlorinated hydrocarbon pesticides in environmentally sensitive patients. *Clin. Ecol.* 2:3–12.

13. Tietz, N. W. 1976. *Fundamentals of Clinical Chemistry*. Philadelphia: W. B. Saunders.

14. Jones, J. E., R. Manalo, and E. B. Flink. 1967. Magnesium requirements in adults. *Med. J. Clin. Nutr.* 20:632–635.

15. Department of Labor. 1975. Occupational exposure to asbestos. *Fed. Regist.* 40:47652.

16. Adams, R. D. 1980. Headache. In *Harrison's Principles of Internal Medicine*, eds. K. J. Isselbacher, R. D. Adams, E. Braunwald, R. G. Petersdorf, and J. D. Wilson, 9th ed. 25. New York: McGraw-Hill.

17. Ödkvist, L. M., B. Larsby, R. Tham, and D. Hyden. 1982. Vestibulo-occulomotor disturbances caused by industrial solvents. *Otolaryngol. Head Neck Surg.* 91(5):537–539. Tokyo: University of Tokyo Press.

18. Randolph, T. G. 1976. Ecologically oriented rheumatoid arthritis. In *Clinical Ecology*, ed. L. D. Dickey. 201–212. Springfield, IL: Charles C Thomas.

19. Ishikawa, S., M. Miyato, T. Fukuda, and H. Suyama. 1982. Etiological consideration of Behcet's Disease. In *Behcet's Disease. Pathogenetic Mechanism and Clinical Future*, ed. G. Inaba. 25. Tokyo: University of Tokyo Press.

20. Rea, W. J. 1994. *Chemical Sensitivity. Vol. II. Sources of Total Body Load*. 837. Boca Raton, Fl: Lewis Publishers.

21. Marquardtt, J. L., R. Synderman, and J. J. Oppenheim. 1973. Depression of lymphocyte transformation and exacerbation of Behcet's syndrome by ingestion of English walnuts. *Cell Immunol.* 9:263–272.

22. Wyngaarden, J. B., and L. H. Smith, eds. 1985. *Cecil's Textbook of Medicine,* 17th ed. 1535. Philadelphia: W. B. Saunders.

23. Ellis, J. M, T. Kishi, J. Azuma, and K. Folkers. 1976. Vitamin B_6 deficiency in patients with a clinical syndrome including the carpal tunnel defect. Biochemical and clinical response to therapy with pyridoxine. *Res. Commun. Chem. Pathol. Pharmacol.* 13:743–756.

24. Ellis, J., K. Folkers, M. Levy, S. Shizukuishi, J. Lewandowski, S. Nishii, H. Shubert, and R. Ulrich. 1982. Response of vitamin B_6 deficiency and the carpal tunnel syndrome to pyridoxine. *Proc. Natl. Acad. Sci.* (Med. Sciences) 79(23):7494–7498.

25. Ellis, J. M. and K. Folkers. 1989. March 19–23. Abstract and exhibit presented to Federation of American Societies for Experimental Biology. New Orleans, LA.

26. Ellis, J. 1987. Treatment of carpal tunnel syndrome with vitamin B_6. *South. Med. J.* 80(7):882–884.

27. Krane, S. M. 1977. Hyperostosis, neoplasms, and other disorders of bone and cartilage. In *Harrison's Principles of Internal Medicine*, 8th ed., eds. G. W. Thorn, R. D. Adams, E. Braunwald, K. J. Isselbacher, and R. G. Petersdorf. 2041. New York: McGraw-Hill.

28. Holmes, K. K. 1980. Gonococcal infections. In *Harrison's Principles of Internal Medicine*, 9th ed., eds. K. J. Isselbacher, R. D. Adams, E. Braunwald, R. G. Petersdorf, and J. D. Wilson. 626. New York: McGraw-Hill.

29. Hirschmann, J. V. and B. C. Gilliland. 1980. Osteomyelitis and infections arthritis. In *Harrison's Principles of Internal Medicine*, 9th ed., eds. K. J. Isselbacher, R. D. Adams, E. Braunwald, R. G. Petersdorf, and J. D. Wilson. 1892. New York: McGraw-Hill.

30. Austrian, R. 1980. Pneumococcal infections. In *Harrison's Principles of Internal Medicine*, 9th ed., eds. K. J. Isselbacher, R. D. Adams, E. Braunwald, R. G. Petersdorf, and J. D. Wilson. 605. New York: McGraw-Hill.

31. Bisno, A. L. 1980. Streptococcal infections. In *Harrison's Principles of Internal Medicine*, 9th ed., eds. K. J. Isselbacher, R. D. Adams, E. Braunwald, R. G. Petersdorf, and J. D. Wilson. 616. New York: McGraw-Hill.

32. Harkavy, J. 1963. *Vascular Allergy and Its Systemic Manifestations.* 174. Washington: Butterworths.

33. The Rheumatoid Disease Foundation. Rt. 4, Box 137. Franklin, TN 37064 [letter].

34. Nakamura, Y., T. Masuhara, S. Ito-Kuwa, and S. Aoki. Induction of experimental *Candida* arthritis in rats. *J. Med. Vet. Mycol.* 29(3):179–192.

35. Harkavy, J. 1963. *Vascular Allergy and Its Sytemic Manifestations.* 161, 259. Washington: Butterworths.

36. Panush, R. S. 1990. Food induced ("allergic") arthritis: clinical and serologic studies. *J. Rheumatol.* 17(3):291–294.

37. Sperling, R. I. 1989. Diet therapy in rheumatoid arthritis. *Curr. Opin. Rheumatol.* 1(1):33–38.

38. Chander, C. L. and I. D. Spector. 1991. Oestrogens, joint disease, and cartilage. *Ann. Rheum. Dis.* 50(3):139–140.

39. Cook, R. A. 1918. Hayfever and asthma. *N.Y. Med. J.* 107:577–583.

40. Talbot, F. B. 1917. Role of food idiosyncrasies in practice. *N.Y. Stat. J. Med.* 17:419–425.

41. Turnbull, J. A. 1924. The relationship of anaphylactic disturbances to arthritis. *JAMA* 82:1957–1959.

42. Turnbull, J. A. 1944. Changes in sensitivity to allergenic foods in arthritis. *J. Dig. Dis.* 15:182–190.

43. Pottenger, R. T. 1928. Constitutional factors in arthritis with special reference to incidence and role of allergic disease. *Ann. Intern. Med.* 12:323–333.

44. Zeller, M. 1949. Rheumatoid arthritis-food allergy as a factor. *Ann. Allergy* T:200–239.

45. Kaufman, W. 1953. Food induced allergic musculoskeletal syndromes. *Ann. Allergy* 11:170–184.

46. Zussman, B. M. 1966. Food hypersensitivity simulating rheumatoid arthritis. *South. Med. J.* 59:935–939.

47. Millman, M. 1972. An allergic concept of the etiology of rheumatoid arthritis. *Ann. Allergy* 30:135–141.

48. Rowe, A. H. 1972. Food allergy and the arthropathies. In *Food Allergy: Its Manifestations and Controls*. 435–443. Springfield, IL: Charles C Thomas.

49. Otto, R., W. Reske, I. Leest, V. Puroche, and H. Knoll. 1973. Skin testing with cereal and pulse allergens in progressive chronic polyarthritis. *Deutsche Gesundheitswesen* 28:2001–2003.

50. Service, W. C. 1937. Hydroarthrosis of allergic origin. *Am. J. Surg.* 37:121–123.

51. Lewin, P. and S. J. Taub. 1936. Allergic synovitis due to ingestion of English walnuts. *JAMA* 166:2144.

52. Berger, H. 1939. Intermittent hydroarthrosis with an allergic basis. *JAMA* 112:2402–2405.

53. Hench, P. S. and E. J. Rosenberg. 1941. Palindromic rheumatism. *Proc. Staff Meet. Mayo Clin.* 16:808.

54. Vaughn, W. T. 1943. Palindromic rheumatism among allergic persons. *J. Allergy* 14:256–263.

55. Epstein, S. 1969. Hypersensitivity to sodium nitrate: a major causative factor in a case of palindromic rheumatism. *Ann. Allergy* 27:343–349.

56. Marshall, R., R. M. Stroud, G. F. Kroker, T. Bullock, F. M. Carroll, M. Greenberg, T. G. Randolph, W. J. Rea, and R. E. Smiley. 1984. Food challenge effects on fasted rheumatoid arthritis patients: a multicenter study. *Clin. Ecol.* 2(4):181–190.

57. Cooms, R. R. A. and G. Oldham. 1981. Early rheumatoid-like joint lesions in rabbits drinking cow's milk. *Int. Arch. Allergy Appl. Immunol.* 64:287–292.

58. Mansson, I., R. Norbert, B. Olhagen, and N. E. Bjorklund. 1971. Arthritis in pigs induced by dietary factors. *Clin. Exp. Immunol.* 9:677–693.

59. Matsuska, L. Y., J. Wortsman, and J. J. Pepper. 1984. Acute arthritis during isotretinein treatment for acne. *Arch. Intern. Med.* 144(a):1870–1871.

60. Epstein, S. 1969. Hypersensitivity to sodium nitrate: a major currative factor in a case of palindromic rheumatism. *Ann. Allergy* 27:343–349.

61. Ishakawa, S., M. Miyata, T. Fukuda, and H. Suyama. 1982. Etiological consideration of Behcet's Disease. In *Behcet's Disease: Pathogenetic Mechanism and Clinical Future*, ed. G. Inaba. 25–31. Tokyo: University of Tokyo Press.

62. Mandell, M. and A. Conte. 1980. The role of allergy in arthritis rheumatism, and associated polysymptomatic cerebroviscero-somatic disorders: a double-blind provocation test study. *Ann. Allergy* 44:51.

63. Sköldstam, L., L. Larsson, and F. D. Lindstrom. 1979. Effects of fasting and lactovegetarian diet on rheumatoid arthritis. *Scand. J. Rheumatol.* 8:249–255.

64. Stroud, R. M., G. E. Kroker, W. J. Rea, R. E. Smiley, M. Greenberg, F. M. Carroll, T. M. Bullock, R. T. Marshall, and T. G. Randolph. 1980. Comprehensive environmental control and its effect on rheumatoid arthritis. *Clin. Res.* 28:791A.

65. Veys, E. M., P. Hermans, G. Verbruggen, and H. Mielants. 1983. Immunoregulatory changes in autoimmune disease. *Diagn. Immunol.* 1:224–232.

66. Parke, A. L. and G. R. V. Hughes. 1981. Rheumatoid arthritis and food: a case study. *Br. Med. J.* 282:2027–2029.

67. Walker, W. A. and K. J. Isselbacher. 1974. Uptake and transport of macromolecules by the intestine: possible role in clinical disorders. *Gastroenterology* 67:531–550.

68. Cunningham-Rundles, C., W. E. Brandeis, R. E. Good, and N. K. Day. 1979. Bovine antigens and the formation of circulating immune complexes in selective immunoglobulin A deficiency. *J. Clin. Invest.* 64:272–279.

69. Paganelli, R., R. J. Levinsky, J. Brostoff, and D. G. Wraith. 1979. Immune complexes containing food proteins in normal and atopic subjects after oral challenges and effect of sodium cromoglycate on antigen absorption. *Lancet* 1:1270–1272.

70. Cunningham-Rundles, C., W. E. Brandeis, D. J. Pudifin, N. K. Day, and R. A. Good. 1981. Autoimmunity in selective IgA deficiency: relationship to antibovine protein antibodies, circulating immune complexes and clinical disease. *Clin. Exp. Immunol.* 45:299–304.

71. Paganelli, R., R. J. Levinsky, and D. J. Atherton. 1981. Detection of specific antigen within circulating immune complexes: validation of the assay and its application to food antigen-antibody complexes formed in healthy and food-allergic subjects. *Clin. Exp. Immunol.* 46:44–53.

72. Fernandes, G., E. J. Yunis, D. G. Jose, and R. A. Good. 1973. Dietary influence on antinuclear antibodies and cell-mediated immunity in NZB mice. *Int. Arch Allergy Appl. Immunol.* 44:770–782.

73. Good, R. A., A. West, and G. Fernandes. 1980. Nutritional modulation of immune response. *Fed. Proc.* 39:3048–3104.

74. Chandra, R. K. 1980. Nutritional deficiency, immune responses and infectious illness. *Fed. Proc.* 39:3086–3087.

75. Beach, R. S., M. E. Gershwin, and L. S. Harley. 1981. Nutritional factors and autoimmunity. II. Prolongation of survival in zinc-deprived NZB/W mice. *J. Immunol.* 128:308–313.

76. Good, R. A. 1981. Nutrition and immunity. *J. Clin. Lab. Immun.* 1:3–11.

77. Hurd, E. R., J. M. Johnston, J. R. Okita, P. C. MacDonald, M. Ziff, and J. N. Gilliam. 1981. Prevention of glomerulonephritis and prolonged survival in New Zealand black/New Zealand white F hybrid mice fed an essential fatty acid deficient diet. *J. Clin. Invest.* 67:476–485.

78. Malinow, M. R., E. J. Bandana, and B. Pirofsky. 1981. Systemic-lupus-erythematosus-like syndrome induced by diet in monkeys. *Clin. Res.* 29:626A.

79. Rubin, B. S. and E. Herrington. 1981. Influence of dietary protein on experimental autoimmune disease. *Fed. Proc.* 40:919.

80. Panush, R. S., R. L. Carter, P. Katz, B. Kowsari, S. Longley, and S. Finnie. 1983. Diet therapy for rheumatoid arthritis. *Arthritis Rheum.* 26(4):462–471.

81. Ebringer, A., M. Baines, M. Childerstone, M. Ghuloom, and T. Ptaszynska. 1985. Epiopathogenesis of ankylosing spondylitis and the cross-tolerance hypothesis. In *Advances in Inflammation Research. Vol. 9. The Spondyloarthropathies*, eds. M. Ziff and S. B. Cohen. 101–128. New York: Raven.

82. Sköldstam, L. and K. -E. Magnusson. 1991. Fasting, intestinal permeability, and rheumatoid arthritis. *Rheum. Dis. Clin. N. Am.* 17(2):363–371.

83. Sköldstam, L., L. Larsson, and F. D. Lindström. 1979. Effects of fasting and lactovegetarian diet on rheumatoid arthritis. *Scand. J. Rheumatol.* 8:249.

84. Trang, L. E., O. Lövgren, R. Bendz, and O. Mjös. 1980. The effect of fasting on plasma cyclic adenosine-3',5'-monophosphate in rheumatoid arthritis. *Scand. J. Rheumatol.* 9(4):229–233.

85. Hafström, I., B. Ringertz, H. Gyllenhammar, J. Palmbland, and M. Harms-Ringdahl. 1988. Effects of fasting on disease activity, neutrofil function, fatty acid composition and leucotriene biosynthesis in patients with rheumatoid arthritis. *Arthritis Rheum.* 31(5):585–592.

86. Smith, M. D., R. B. Gibson, and P. M. Brooks. 1985. Abnormal bowel permeability in ankylosing spondylitis and rheumatoid arthritis. *J. Rheumatol.* 12:299.

87. Hamberg, J., O. Lindahl, P. A. Öckerman, and L. Lindwall. 1981. Fasta och hälsokost vid RA — en kontrollerad undersökning. *Swed. J. Biol. Med.* 3:5.

88. Sköldstam, L. 1986. Fasting and vegan diet in rheumatoid arthritis. *Scand. J. Rheumatol.* 15:219.

89. Ebringer, A. 1983. The crosstolerance hypothesis. HLA-B27 and ankylosing spondylitis. *Br. J. Rheumatol.* 22(Suppl. 2):53–66.

90. Ebringer, A. 1989. The relationship between *Klebsiella* infection and ankylosing spondylitis. *Ballière's Clin. Rheumatol.* 3(2):321–338.

91. Friis, J. and A. Svejgaard. 1974. *Salmonella* arthritis and HLA 27. *Lancet* 1:1350.

92. Noer, H. R. 1966. An "experimental" epidemic of Reiter's syndrome. *JAMA* 198:693–698.

93. Paronen, I. 1948. Reiter's disease: a study of 344 cases observed in Finland. *Acta Med. Scand.* 131 (Suppl.):1–114.

94. Hunter, T., G. K. M. Harding, R. E. Kaprove, and M. L. Schroeder. 1981. Faecal carriage of various *Klebsiella* and *Enterobacter* species in patients with active ankylosing spondylitis. *Arthritis Rheum.* 24:106–108.

95. Kuberski, T. T., H. G. Morse, R. G. Rate, and M. D. Bonnell. 1983. Increased recovery of *Klebsiella* from the gastrointestinal tract of Reiter's syndrome and ankylosing spondylitis patients. *Br. J. Rheumatol.* 22 (Suppl. 2):85–90.

96. Warren, R. E., and D. A. Brewerton. 1980. Faecal carriage of *Klebsiella* by patients with ankylosing spondylitis and rheumatoid arthritis. *Ann. Rheum. Dis.* 39:37–44.

97. Delagrange, A. 1968. La lymphographie dans la spondylarthrite ankylosante. M. D. thesis, Marseilles.

98. Fournier, A. M., D. Denizet, and A. Delagrange. 1969. La lymphographie dans la spondylarthrite ankylosance. *J. Radiol. Electrol.* 50:773–784.

99. Dickey, L. 1971. Ecologic illness. *Rocky Mt. Med. J.* 68:23–28.

100. Bailey, A. A. 1985. Post-polio pain: treatment by sublingual immunotherapy. In *Late Effects of Poliomyelitis*, eds. L. S. Halstead and D. O. Wiechers. 181–184. Miami, FL: Symposium Foundation.
101. Rea, W. J., A. R. Johnson, E. Fenyves, and J. Butler. 1987. The environmental aspects of the post-polio syndrome. In *Research and Clinical Aspects of the Late Effects of Poliomyelitis*, eds. L. S. Halstead and D. O. Wiechers. 173–181. White Plains: March of Dimes Birth Defects Foundation.
102. Mandell, M. and A. Conte. 1980. The role of allergy in arthritis, rheumatism and associated polysymptomatic cerebrovisero-somatic disorders: a double-blinded provocation test study. *Ann. Allergy* 44:51.
103. Fernandes, G., E. F. Yunis, D. G. Jose, and R. A. Good. 1973. Dietary influence on antinuclear antibodies and cell-mediated immunity in NZB mice. *Int. Arch. Allergy Appl. Immunol.* 44:770–782.
104. Fernandes, G., S. Izui, N. K. Day, F. J. Dixon, and R. A. Good. 1981. Inhibition of retroviral gp70-anti-gp70 immune complex formation and prolongation of life span by dietary restriction in NZB x NZW F_1 mice. *Fed. Proc.* 40:4204A.
105. Beach, R. S., M. E. Gershman, and L. S. Hurley. 1981. Nutritional factors and autoimmunity: immunopathology of zinc deprivation in New Zealand mice. *J. Immunol.* 126:1999–2006.
106. Ziff, M. 1983. Diet in the treatment of rheumatoid arthritis. *Arthritis Rheum.* 26(4):457–461.
107. Kuehl, F. A. and R. W. Egan. 1980. Prostaglandins, arachidonic acid and inflammation. *Science* 210:978–984.
108. Goodwin, J. S. and D. R. Webb. 1980. Regulation of the immune response by prostaglandins. *Clin. Immunol. Immunopathol.* 15:106–122.
109. Korn, J. H., P. V. Haluska, and E. C. LeRoy. 1980. Mononuclear cell modulation of connective tissue function: suppression of fibroblast growth by stimulation of endogenous prostaglandin production. *J. Clin. Invest.* 65:543–554.
110. Meade, C. J. and J. Mertin. 1978. Fatty acids and immunity. *Adv. Lipid Res.* 16:127–165.
111. Mahoney, E. M., A. L. Hamill, W. A. Scott, and Z. A. Cohn. 1977. Response of endocytosis to altered fatty acyl composition of macrophage phospholipids. *Proc. Natl. Acad. Sci. U.S.A.* 74:4895–4899.
112. Aspinall, R. L. and P. S. Cammarata. 1969. Effect of prostaglandin E_2 on adjuvant arthritis. *Nature* 224:1320–1321.
113. Zurier, R. B. and F. Qualiata. 1971. Effect of prostaglandin E_1 on adjuvant arthritis. *Nature* 234:304–305.
114. Zurier, R. B., D. M. Sayadoff, S. B. Torrey, and N. F. Rothfield. 1977. Prostaglandin E, treatment of NZB/NZW mice. 1. Prolonged survival of female mice. *Arthritis Rheum.* 20:723–728.
115. Stackpoole, A. and J. Martin. 1981. The effect of prostaglandin precursors in in vivo models of cell mediated immunity. Essential fatty acids and prostaglandins. *Prog. Lipid Res.* 20:649–654.
116. McLeish, K. R., A. F. Gohara, L. J. Johnson, and D. L. Sustarsie. 1982. Alteration in immune complex glomerulonephritis by AA. *Prostaglandins* 23:383–389.
117. Kunkel, S. L., H. Ogawa, P. A. Ward, and R. B. Zurier. 1981. Suppression of chronic inflammation by evening primrose oil. Essential fatty acids and prostaglandins. *Prog. Lipid Res.* 20:885–888.

118. Denko, C. W. 1976. Modification of adjuvant inflammation in rats deficient in essential fatty acids. *Agents Actions* 6:636–641.

119. Schellenberg, R. R. and E. Gillespie. 1977. Effects of colchicine vinblastine, griseofulvin and deuterium oxide upon phospholipids by simple and rapid thin layer chromatographic procedure and its application to cultured neuroblastoma cells. *Anal. Biochem.* 80:430–437.

120. Bonta, I. L., M. J. Parnham, and M. J. P. Adolfs. 1977. Reduced exudation and increased tissue proliferation during chronic inflammation in rats deprived of endogenous prostaglandin precursors. *Prostaglandins* 14:295–307.

121. DeWille, J. W., P. J. Fraker, and D. R. Romsos. 1978. Effect of essential fatty acid deficiency and various levels of dietary polyunsaturated fatty acids on humoral immunity in mice. *J. Nutr.* 109:1018–1027.

122. Hurd, E. R., J. M. Johnston, J. R. Okita, P. C. MacDonald, M. Ziff, and J. N. Gilliam. 1981. Prevention of glomerulonephritis and prolonged survival in New Zealand Black/New Zealand White F₁ hybrid mice fed an essential fatty acid-deficient diet. *J. Clin. Invest.* 67:467–485.

123. Prickett, J. D., D. R. Robinson, and A. D. Steinberg. 1983. Effects of dietary enrichment with eicosapentaenoic acid upon autoimmune nephritis in female NZB x NZW Fi mice. *Arthritis Rheum.* 26:133–139.

124. Prickett, J. D., D. R. Robinson, and A. D. Steinberg. 1981. Dietary enrichment with the polyunsaturated fatty acid eicosapentaenoic acid prevents proteinuria and prolongs survival in NZBX NZW F₁ mice. *J. Clin. Invest.* 68:556–559.

125. Horrabin, D. F., A. Campbell, and C. G. McEwen. 1981. Treatment of the sicca syndrome and Sjogren's syndrome with EFA, pyridoxine and vitamin C. Essential fatty acids and prostaglandins. *Prog. Lipid Res.* 20:253–254.

126. Kaiser, W. 1991. Silicon-induced collagenoses. *Dtsch. Med. Wochenschr.* 116(11):433–434. (German)

127. Alarcón-Segovia, D. 1969. Drug-induced lupus syndromes. *Mayo Clin. Proc.* 44:664–681.

128. Blank, M. S., Mendlovic, H. Fricke, E. Mozes, N. Talal, and Y. Shoenfeld. 1990. Sex hormone involvement in the induction of experimental systemic lupus erythematosus by a pathogenic anti-DNA idiotype in naive mice. *J. Rheumatol.* 17:311–317.

129. Tumulty, P. A. 1966. Systemic lupus erythematosus. In *Principles of Internal Medicine*, 4th ed., eds. T. R. Harrison, R. D. Adams, I. L. Bennett, W. H. Resnik, G. W. Thorn, and M. M. Wintrobe. 1370. New York: McGraw-Hill.

130. Mongey, A. -B., and E. V. Hess. 1993. The potential role of environmental agents in systemic lupus erythematosus and associated disorders. In *Dobois' Lupus Erythematosus*, 4th ed., eds. D. J. Wallace and B. H. Hahn. 40–41. Philadelphia: Lea & Febiger.

131. Mannik, M. and B. C. Gilliland. 1980. Systemic lupus erythematosus. In *Principles of Internal Medicine*, 9th ed., eds. K. J. Isselbacher, R. D. Adams, E. Braunwald, R. G. Petersdorf, and J. D. Wilson. 359. New York: McGraw-Hill.

132. Black, C. M. and K. I. Welsh. 1988. Occupationally and environmentally induced sclerodermia-like illness: etiology, pathogenesis, diagnosis, and treatment. *IM* 9(6):136–154.

133. Haustein, J. F. and V. Ziegler. 1985. Environmentally induced systemic sclerosis-like disorders. *Int. J. Dermatol.* 24:147.

134. Sverdrup, B. 1984. Do workers in the manufacturing industry run an increased risk of getting scleroderma? [letter] *Int. J. Dermatol.* 23(9):629.

135. Black, C. M., K. I. Welsh, A. E. Walker, R. M. Bernstein, L. J. Catoggio, A. R. McGregor, and J. K. Lloyd Jones. 1983. Genetic susceptibility to scleroderma-like syndrome induced by vinyl chloride. *Lancet* 1:53–55.

136. Black, C. M., S. Pereira, A. McWhirter, K. Welsh, and R. Caurent. 1986. Genetic susceptibility to scleroderma-like syndrome in symptomatic and asymptomatic workers exposed to vinyl chloride. *J. Rheum.* 13:1059–1062.

137. Maricq, H. R. 1985. Vinyl chloride disease. In *Systemic Sclerosis (Scleroderma)*, eds. C. M. Black and A. R. Myers. 105–113. New York: Gower Medical.

138. Tribukh, S. L., N. P. Tikhomirova, S. V. Levina, and L. A. Kozlov. 1981. Conditions of work and measures of industrial hygeine in the production of, and manufacture from, vinyl chloride plastics, *Gig. Sanit.*, 10:38–44, cited in W. K. Lelbach and H. J. Marsteller. 1981. Vinyl chloride associated disease. *Adv. Int. Med. Pediatr.* 47:1–110.

139. Suciu, I., I. Drejman, and C. Valaczkai. 1963. Investigation of the disease caused by vinyl chloride. *Med. Int.* 15:967–978.

140. Cordier, J. M., M. J. Fievez, and A. Sevrin. 1966. Acro-osteolysis and exposure to vinyl chloride. *Cah. Med. Travail* 4:14–19.

141. Harris, D. K. and W. G. F. Adams. 1967. Acro-osteolysis occurring in men engaged in the polymerization of vinyl chloride. *Brit. Med. J.* 3:712–714.

142. Dinman, B. D., W. A. Cook, W. M. Whitehouse, H. J. Magnuson, A. A. Mich, and T. Ditcheck. 1971. Occupational acroosteolysis: I. An epidemiological study. *Arch. Environ. Health* 22:61–73.

143. Dodson, V. N., B. D. Dinman, W. M. Whitehouse, A. N. M. Nasr, and H. J. Magnuson. 1971. Occupational acroosteolysis. III. A clinical study. *Arch. Environ. Health* 22:83.

144. Markowitz, S. S., J. McDonald, W. Fethiere, M. S. Kerzner. 1972. Occupational acroosteolysis. *Arch. Dermatol.* 106:219.

145. Lange, C. E., S. Juke, G. Stein, and G. Veltman. 1974. Die sogenante Vinylchloridkrankheit-eine berufsbedingte systemsklerose? *Int. Arch. Arbeitsmed.* 32:1–32.

146. Creech, J. L. and M. N. Johnson. 1974. Angiosarcoma of liver in the manufacture of polyvinyl chloride. *J. Occup. Med.* 16:150–151.

147. Suciu I., L. Prodan, E. Ilea, A. Pǎduraru, and L. Pascu. 1975. Clinical manifestations in vinyl chloride poisoning. *Ann. N.Y. Acad. Sci.* 246:53–69.

148. Lelbach, W. K. and H. J. Marsteller. 1981. Vinyl chloride-associated disease. *Erg. Inn. Med. Kinderheilkd* 47:1–110.

149. Ward, A. M., S. Udnoon, J. Watkins, A. E. Walker, and C. S. Darke. 1976. Immunological mechanisms in the pathogenesis of vinyl chloride disease. *Brit. Med. J.* 1(6015):936–938.

150. Bramwell, B. 1914. Diffuse scleroderma: its frequency, its occurrence in stone-masons, its treatment by fibrinolysin, elevations of termperature due to fibrinolysin, elevations of temperature due to fibrinolysin injections. *Edinburgh Med. J.* 12:387–401.

151. Erasmus, L. D. 1957. Scleroderma in gold-miners on the Witwadersrand with particular reference to pulmonary manifestations. *S. Afr. J. Labor Clin. Med.* 3:209–231.

152. Sluis-Cremer, G. K., L. D. Erasmus, S. Hins, S. K. Ludwin, I. Webster, and H. S. Sichel. 1973. Report on incidence of scleroderma (progressive systemic sclerosis) in white South African gold miners. In *National Research Institute for Occupational Disease Report.* South Africa: Pretoria Government Printer.

153. Rodnan, G. P., T. G. Benedek, T. A. Medsger, and R. J. Cammarata. 1967. The association of progressive systemic sclerosis (scleroderma) with Coal Miners' pneumoconiosis and other forms of silicosis. *Ann. Intern. Med.* 66:323–334.

154. Devulder, B., B. Plouvier, J.- Ch. Martin, and L. Lenoir. 1977. L'association sclérodermie-silicose ou syndrome d'Erasmus: aspects cliniques et physio-pathologiques. *Nouv. Presse Med.* 6:2877–2879.

155. Sluis-Cremer, G. K., P. A. Hessel, E. H. Nizdo, A. R. Churchill, and E. A. Zeiss. 1985. Silica, silicosis, and progressive systemic sclerosis. *Brit. J. Ind. Med.* 42(12):838–843.

156. Hankes, L. V., E. DeBruin, C. R. Jansen, L. Vorster, and M. Schmaeler. 1977. Metabolism of ^{14}C-labelled L-tryptophan, L-kynarenine and hydroxy-L-kynurenine in miners with scleroderma. *S. Afr. Med. J.* 51(12):383–390.

157. Walder, B. K. 1983. Do solvents cause scleroderma? *Int. J. Dermatol.* 22:157–158.

158. Sparrow, G. P. 1977. A connective tissue disorder similar to vinyl chloride disease in a patient exposed to perchlorethylene. *Clin. Exp. Dermatol.* 2(1):17–22.

159. Saihan, E. M., J. L. Burton, and K. W. Heaton. 1978. A new syndrome with pigmentation, scleroderma, gynaecomastia, Raynaud's phenomenon and peripheral neuropathy. *Brit. J. Dermatol.* 99:437–440.

160. Reinl, W. 1957. Sklerodermie dureh Trichloroethylene-Einwirkung? *Bull. Hyg.* 32:678.

161. Rose, T., J. Nothjunge, and W. Schlote. 1985. Familial occurrence of dermatomyositis and progressive scleroderma after injection of a local anesthetic for dental treatment. *Eur. J. Pediatr.* 143(3):225–228.

162. Rush, P. J. and A. Chaiton. 1986. Scleroderma, renal failure, and death associated with exposure to urea formaldehyde foam insulation. *J. Rheumatol.* 13(2):475.

163. Keifer, O. 1973. Über die nebenwirkungen der bleomycintherapie auf der haut. *Dermatologica* 146:229–243.

164. Delena, M., A. Guzzon, S. Monfardini, and G. Bonadonna. 1972. Clinical, radiologic, and histopathologic studies on pulmonary toxicity induced by treatment with bleomycin (NSC-125066). *Cancer Chemother. Rep.* 56:343–356.

165. Luna, M. A., W. M. Bedrossian, B. Lichtiger, and P. A. Salem. 1972. Interstitial pneumonitis associated with bleomycin therapy. *Am. J. Clin. Path.* 58:501–510.

166. Cohen, I. S., M. B. Mosher, E. J. O'Keefe, S. N. Klaus, and R. C. De Conti. 1973. Cutaneous toxicity of bleomycin therapy. *Arch. Dermatol.* 107(4):553–555.

167. Finch, W. R., R. B. Buckingham, G. P. Rednan, R. K. Prince, and A. Winkelstein. 1985. Scleroderma induced by bleomycin. In *Systemic Sclerosis (Scleroderma),* eds. C. M. Black and A. R. Myers. 114–121. New York: Gower Medical.

168. Sternberg, E. M., M. H. Van Woert, S. N. Young, I. Magnussen, H. Baker, S. Gauthier, and C. K. Osterland. 1980. Development of a scleroderma-like illness during therapy with L-5-hydroxytriptophan and carbidopa. *N. Engl. J. Med.* 303:782–787.

169. Stachów, A., S. Joblonska, and D. Kencka. 1985. Tryptophan metabolism in scleroderma and eosinophilic fascitis. In *Systemic Sclerosis (Scleroderma)*, eds. C. M. Black and A. R. Myers. 130–134. New York: Gower Medical.

170. Zarafonetis, C. J. D., S. H. Lorber, and S. M. Hanson. 1958. Association of functioning carcinoid syndrome and scleroderma. *Am. J. Med. Sci.* 236:1–14.

171. Fries, J. F., J. A. Lindgren, and J. M. Bull. 1973. Scleroderma-like lesions and the carcinoid syndrome. *Arch. Intern. Med.* 131:550–553.

172. Tomlinson, I. W. and M. I. V. Jayson. 1984. Systemic sclerosis after therapy with appetite suppressants. *J. Rheumatol.* 11(2):254.

24   Endocrine System

INTRODUCTION

There are three limbs of anatomic physiology that may be disrupted or imbalanced by pollutant overload. One limb, the immune system, was discussed in Volume I, Chapter 5.[1] The second limb, the nervous system, is the focus of Chapter 26. The final limb of anatomic physiology that is altered in chemical sensitivity is the endocrine system. The union of the immune, neuro, and endocrine systems occurs at their proximal end in the hypothalamic area with connections to the pituitary gland, pineal gland, thyroid gland, adrenal glands, and ovaries or testicles (see Chapter 26). In addition, these circuits find distal connections with the end organs and the neuroendocrine cells and receptors via the ANS. It is clear that endocrine malfunction as seen in chemically sensitive patients may very strongly affect the neuro-immuno axis. Also, damage to the neuro-immuno axis can affect the endocrine system. This damage can have either adverse or positive effects on the chemically sensitive individual, depending on whether the whole system is well orchestrated or disrupted by pollutant overload. Pollutants have been shown to affect endocrine metabolism in a myriad of ways. Often, for example, pollutant injury is first manifested in dysfunction of ovarian, testicular, adrenal, and thyroid metabolism (Figure 24.1). However, it appears as if the pituitary/hypophysial areas are affected more often, and certainly the severely damaged catabolic states seen in some extremely chemically sensitive patients are the result of a panhypopituitarism or its equivalent.

Multiendocrinopathy is now commonplace in the neglected chemically sensitive individual. At least 10%, and perhaps as many as 20%, of chemically sensitive individuals have polyendocrinopathy. The following case represents this growing problem.

Case study. This 40-year-old white female entered the ECU complaining of bone pain, anxiety attacks, and sensitivities to foods and chemicals. For at least the 10 years previous to this admission, she had experienced intolerance to various foods and chemicals. Initially, she was rigid on her food rotation and environmental control, but, as she improved, she became careless and

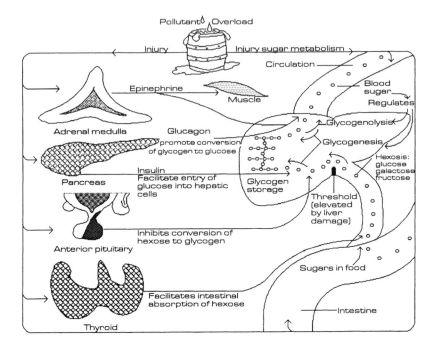

Figure 24.1. Pollutant injury to the endocrine glands may deregulate different aspects of metabolism.

was exposed to many chemicals. She developed premenstrual tension accompanied by irregular periods and antiovarian antibodies. She eventually underwent a bilateral oophectomy, which reduced the antibodies. Five years prior to admission, she developed mild adrenal insufficiency, which occasionally required cortisone supplementation. One year prior to admission to the ECU, she developed antithyroid antibodies. After being placed on thyroid extract, these improved. Three months prior to admission, she developed hypercalcemia and was found to have an abnormally high calcium of 12.5 mg/dL (C = 8.5 to 11.0 mg/dL). This dropped to normal after resection of a parathyroid adenoma and partial thyroidectomy. The patient presented with hypomagnesemia, which responded to intravenous magnesium sulfate supplementation and heat depuration/physical therapy. This woman had retained a large number of toxic chemicals in her blood that appeared to be triggering her problems. Oral, intradermal, and inhaled challenge confirmed her sensitivity to biological inhalants, foods, and chemicals. She gradually improved with institution of environmental control and diet.

This chapter discusses each endocrine gland separately as well as the neuroendocrine cells and the receptor system, all of which may be injured and become imbalanced following pollutant exposure. In the chemically sensitive patient, disturbed physiology of the endocrine system frequently results.

PITUITARY GLAND

Pollutant injury to the pituitary gland has been poorly studied. However, clinical evidence suggests that injury to this organ occurs more often than the clinician has appreciated. The hypothalmic-pituitary axis controls much of the endocrine system. Supression of prolactin and corticosterone has been shown in rats exposed to organochlorine compounds. Endosulfan, an organochlorine pesticide, alters the regulatory mechanism, resulting in abnormal reproduction,[2] and the pesticide DDT and bisphenol A, a breakdown product of plastics, produce an estrogenic effect.[3] These observations set the stage for the kind of endocrine dysfunction that will be discussed in this chapter.

Pollutant Effects on the Nervous System

At least two glands (adrenal medulla and posterior pituitary glands) secrete their hormones in response to direct nerve stimuli. Therefore, if there were a pollutant stimulus to their particular innervation, hormones would be released and symptoms would occur, often to varied degrees in the chemically sensitive individual.

There is much evidence that nerve influences affect the function of the anterior hypophysis, although it probably has no direct innervation. The regulatory neurohumors appear to reach the pituitary from the hypothalamus via the hypophysial portal circulation using neurosecretory substances that act on glandular cells to release anterior pituitary hormones. This idea suggests that the anterior pituitary is governed by a reflex arc comprised of a neural segment ending in the median eminence and a vascular segment formed by the hypophysial portal system. The relationship between the primary capillary plexus of the portal system and the nerve endings suggests that the chemotransmitters passing from nerve to capillary are small peptides in the order of that of the well characterized posterior lobe hormones. Pollutant damage to the nerves, secretory cells, or blood vessels may alter functions, creating generalized body dysfunction as is often seen in chemically sensitive patients. The toxic chemicals can act as false neurotransmitters or overstimulate or suppress the available transmitters or receptors. Evidence for these actions has been found at the EHC-Dallas by using challenges with various neurotransmitters such as serotonin, methacholine, epinephrine, dopamine, etc., which reproduce or ameliorate symptoms in chemically sensitive individuals.

Pollutant Effects on the Physiology

The anterior pituitary acidophil cells produce prolactin (luteotropic hormone, LTH) and somatotropic hormone (growth, STH). The anterior pituitary basophil cells produce thyrotropin (TSH), adrenocorticotropin (ACTH), follicle-stimulating hormone (FSH), luteinizing hormone (LH) or interstitial-cell-stimulating hormone (ICSH), and melanocyte-stimulating hormone (MSH). Stimulation of preoptic and anterior hypothalamic nuclei controls secretion of thyroid stimulatory hormone, while stimulation of the posterior hypothalamus and medial eminence of the infundibulum causes secretion of ACTH. Stimulation of the median eminence of the hypothalamus triggers FSH, and stimulation of the anterior hypothalamus gives release of LH. Other nuclei of the hypothalamus control secretion of growth hormone and prolactin by the anterior pituitary gland. It is evident that pollutant stimulation of these various parts of the hypothalamus could account for many of the various types of responses, such as weakness, fatigue, myalgia, arthralgia, etc., observed in some chemically sensitive individuals. Constant stimulation or suppression of the pituitary causes sustained responses that alter the whole neuro–immune–endocrine axis, causing, exacerbating, or decreasing chemical sensitivity.

Prolactin levels are definitely altered in some patients exposed to toluene (see Volume II, Chapter 12).[4,5] At the EHC-Dallas, we have seen alterations in the other aforementioned pituitary hormones after pollutant injury in chemically sensitive individuals, and we have observed that these hormones often return to normal as the pollutant load and, thus, the chemical sensitivity, decreases.

Secretion of the posterior pituitary involves oxytocin and vasopressin (antidiuretic hormone, ADH) and can occur by direct neurostimulation. ADH is antidiuretic and hypertensive by favoring arteriolar constriction. An increase in the osmotic pressure of the plasma stimulates osmoreceptors in the anterior hypothalamus. From such osmoreceptors, whose membranes may be damaged by pollutants or by other cells stimulated by pollutants, trauma, emotions, or drugs (e.g., tranquilizers, nicotine, morphine), stimuli pass to the cells and fibers of the supra-opticohypophysial nucleus. ADH appears to be formed in the neural cell bodies of the supra-optic nuclei of the hypothalamus and to be passed along the length of axons in combination with a carrier substance. It is released at, or near, the nerve ending in the neural lobe. ADH then enters the bloodstream and goes to the kidney to work on the renal tubules, with subsequent disproportionate readsorption of water. Fifteen percent of the water reabsorbed is under the influence of ADH. Some foods and toxic chemicals can have either a direct or sometimes indirect influence on this hormone. When the chemically sensitive individual fasts in a controlled environment and achieves a significant reduction in his or her total body load, a significant change, with reduction of edema, is often seen. This finding suggests alteration of the ADH effect (see Chapter 34). Also, when a chemically sensitive patient is challenged with specific foods to which he or she is sensitive, swelling will occur. When the patient avoids these foods, however, his or her edema will decrease. This same phenomenon occurs in patients who are sensitive to a specific chemical or who are overloaded with many chemicals.

Pollutant overload may cause deregulation of the hormones that control the activity of the target tissues. This deregulation may occur by alteration of the chemical reactions within the cells, by changes in the permeability of membranes, or by activation of some other specific cellular substances. Hormones achieve their effects by activation of the cyclic AMP of cells, which, in turn, elicits specific cellular functions, and by activation of genes of the cells, which causes the formation of intracellular proteins that initiate specific cellular function. Pollutant exposure can cause injury to both types of effects, causing dysfunction with exacerbation of the chemical sensitivity (see Volume I, Chapter 4[6]). Pollutants may injure the cyclic AMP mechanism in a variety of ways including damaging the hormone receptor site on the cell membrane, the adenyl cyclase enzyme, or the ATP conversion to cyclic 3,5 AMP. The physiological parameters that may be altered are enzyme activation, cell permeability, muscle contraction or relaxation, protein synthesis, and secretions. All of these altered parameters are seen in some chemically sensitive patients.

Pollutant injury of the steroid hormone action on the genes that cause protein synthesis can alter the production of proteins that act as carriers or enzymes. Pollutant injury may occur in the receptor site in the cytoplasm, in the transport process of the hormone to the nucleus, in the alteration of the protein size, in the combination of small protein and hormone in activating specific genes to form messenger RNA, or in the diffusion of messenger RNA back into the cytoplasm where it promotes the translation process of the ribosomes to form new proteins. Many of these alterations are thought to occur in some chemically sensitive patients.

Pollutant injury of the output of other hormones of the pituitary has also been suggested. For example, one sees apparent pituitary ovarian hormone deregulation in some chemically sensitive patients with premenstrual tension syndrome (PMS). Though often the dysfunction is due to pollutant damage to the ovary, correction of the syndrome with administrations of extremely dilute doses of LH may be possible. This correction suggests pituitary deregulation and both have been observed in over 1000 cases of PMS treated at the EHC-Dallas.

Some pesticides have been shown to alter the hypothalamic-pituitary axis control on the endocrine system. Suppression of prolactin and corticosterone due to organochlorine compounds has been shown in rats[7] and appears to occur in some chemically sensitive humans observed at the EHC-Dallas. Endosulphan (an organochlorine pesticide) alters the regulatory mechanism, resulting in abnormal reproduction, and chlordane both reduces pituitary content of the encephalon and produces an estrogenic effect. These changes are often seen in the chemically sensitive patient.

Growth Hormone

Aside from growth hormones (GHs), all major anterior pituitary hormones exert their effects on specific target organs. Therefore, pollutant injury to these growth hormones will be discussed further with the specific target organ. Growth hormone does not function through a specific target organ but instead exers its effects on almost all tissues of the body. Therefore, it will be discussed separately. Growth hormone can have an indirect or direct effect on cartilage and bone growth as well as on other tissues. We have observed, and it has been reported, that certain pesticides and other toxic chemicals can stimulate, and more often suppress, this hormone.[8] Herbicides can cause stunted growth of some young, exposed offspring.[9]

Growth hormone has many generalized effects, which may be altered by pollutant exposure. These alterations include a change in the rate of protein synthesis in all cells of the body, resulting in malnutrition, as seen in some chemically sensitive individuals; a change in the mobilization of fatty acids from adipose tissue, resulting in obesity or hypercholesterolemia; a change in the use of fatty acids for energy, resulting in weakness; and a change in the rate of glucose utilization throughout the body, resulting in blood sugar swings. All of these alterations have been seen at one time or another in the chemically sensitive patient, with resultant decrease in protein and fat stores and excess utilization of carbohydrates. Damage to growth hormone may alter its ability to enhance the transport of amino acids through cell membranes, causing loss of protein synthesis and body wasting, as seen in the severely injured chemically sensitive patient. In addition to pollutant injury to transport defects, injury may also occur to protein synthesis by the ribosomes on which GH has a direct effect. Decreased quantities of RNA will also occur if pollutant injury decreases GH. Growth hormone also decreases catabolism; therefore, pollutant overload may accelerate catabolism, which then negates this protein-sparing effect and results in a debilitated chemically sensitive patient.

Pollutant injury to GH may decrease the release of fatty acids from adipose tissue, also decreasing the protein-sparing effect. Inhibition of GH by pollutant injury may decrease the conversion of fatty acids to acetyl Co-A, with subsequent decrease in energy production as well as decreased acetylation conjugation when GH inhibition is present in the chemically sensitive individual.

Pollutant injury to GH can also alter carbohydrate metabolism, altering, in turn, the cellular metabolism of glucose. There can be an increased utilization of glucose for energy, a prevention of glycogen deposition in cells, and an increased uptake of glucose by the cells, with pollutant injury to GH resulting. This phenomenon results in dramatic blood sugar swings, often seen in a subset of chemically sensitive patients. Dwarfism and gigantion can occur with pollutant injury to GH.

PINEAL GLAND

The pineal gland plays an important role in sexual and reproductive cycles as well as in wake-sleep cycles. It is a vestigial of the third eye, has nerve attachments, and can function as an endocrine organ. Though it is a neuroendocrine organ, it is discussed in the endocrine section because of its unique endocrine effects.

The pineal gland is controlled by the amount of light seen by the eyes each day and, thus, receives direct environmental influences beyond most endocrine glands. A certain amount of darkness will activate the gland. The pineal gland participates actively in regulating the endocrine system and the immune system. It plays a key role in modulating circadian rhythm along with regulating the gonadal steroids, corticosterone, and thyroid hormones.

The pineal gland is one of the circumventricular organs situated in the caudal epithalamus just above the midbrain tegmentum. It has lost all direct connections with the CNS, but it is connected to the ANS. Light passes from the eye to the supraceptic chiasma of the brainstem to the lateral hypothalamus. It proceeds down the sympathetic nerve tracts in the spinal cord to the cervical ganglia, moves along the sympathetics and blood vessels, and terminates directly in the parenchymal cells. Morphologically, the predominant parenchymal cell is the pinealcyte, which has secretory properties of hypothalamic peptides (LH, RH, and somatostatin) and a high concentration of biogenic amines (norepinephrine, serotonin, melatonin, and other indole alkyl amines). The other cell types are astrocytes and are intertwined with pinealcytes. The pineal gland appears to be the only gland that secretes melatonin, which when secreted goes to the anterior pituitary gland to control gonadotropin secretion, which it suppresses. In animals, this suppression happens in the winter. In humans, the cycle is not quite known. However, in some chemically sensitive people, this function has been observed to be much better during the spring and summer months of increased daylight. The more light to which chemically sensitive individuals are exposed during this time, the better balance and function their endocrine system appears to have. The more balanced the endocrine system, the more energy and resistance to toxic chemicals the chemically sensitive individual appears to posess.[10-12]

The formation of melatonin from serotonin through N-acetyl seratonin and subsequent O-methylation takes place in the presence of hydroxyindole-O-methyl transferase (HIOMT). This enzyme differs from catechol-O-methyl transferase (COMT) in that it is found only in or near the pineal gland, and it catalyzes the methylation of hydroxyindoles. COMT, in contrast, is present in most cells and brings about methylation of epinephrine and norepinephrine. The presence of melatonin in peripheral nerves, where HIOMT is not found, suggests that melatonin is a neurohormone released by the pineal gland into the general circulation and subsequently taken up by nerves. Epinephrine, norepinephrine, serotonin, 5-methoxyindole acetic acid and 5-hydroxyindole acetic acid are also found in the pineal gland. Abnormal release of these substances by pollutant stimulation alters both pineal and peripheral functions, exacerbating chemical sensitivity.

Many functions, besides the HIOMT, are characteristic of the pineal gland. For example, blood flow through the pineal is extremely high. This increased blood flow might allow the pineal to be susceptible to more pollutant injury, since blood will transport the toxic chemicals to the gland. HIOMT in the pineal is second only to the kidney on a weight-for-weight basis. An injury may explain the disturbed sleep-wake cycles seen in some chemically sensitive individuals. Also, iodine uptake is second to thyroid uptake. Light causes a profound change in the pineal gland in rats. Melatonin acts as a messenger from light to each end-organ, since the body is dark. When stimulated, it decreases weight and decreases HIOMT, serotonin, RNA synthesis, glycogen, and succinic dehydrogenase activity, and it increases 5-hydroxytryptophan decarboxylase activity. The effects of light on the human pineal gland may be different, but it is evident that lack of natural light is important. Now that humans spend 95% of their time indoors, the many influences of the lack of natural light on the pineal can be observed. For example, Ott[13] reported that lack of light in some cases causes weakness, inability to learn and function, and depression. In contrast, Parry et al.[14] reported that the administration of bright light decreased premenstrual depression. Seasonal affective disorders may be dependent upon light and are seen in some chemically sensitive individuals. The earth's rotation produces a natural biorhythm yearly, 365.25 days and 24-hour cycles. The pineal gland receives the yearly and daily impulses and transmits them to the endocrine system. This gland appears to be the alarm clock for all biorhythms (for more information of the 24-hour cycle, see Volume IV, Chapter 31). Many chemically sensitive individuals experience change in their 24-hour clocks due to work habits or unknown reasons, probably related to pollutant overload, and do not work efficiently due to lack of exposure to natural sunlight. Flying from the west to the east produces much stress on the pineal gland because of the severe change in biorhythms. Lack of nutrients for methyl conjugation due to xenobiotic excess may aid in inappropriate pineal gland function. Faulty methylation of xenobiotics is the primary measurable defect in conjugation seen in chemically sensitive patients.

Melatonin can be collected in the saliva as well as the blood. When metabolized, 75% becomes sulfate, 15% acetylates, and 5% becomes glucuronide. Pollutant injury may alter the conjugation systems, resulting in inappropriate pollutant clearing in some chemically sensitive patients. As a result of faulty clearing, excess melatonin and, thus, altered sleep-wake cycles and immune function may occur.

Many chemically sensitive people have been observed to have severe affective depression that responds well to morning light. This light inhibits their production of melatonin, which has made them depressed, tired, and sleepy. The 509 nm (blue light) shuts off melatonin production. It can be used in a treatment capacity.

Increased melatonin decreases sexual maturity, and anovulatory females have high levels of melatonin at night. Melatonin contains high levels of zinc, and, therefore, pollutant-induced zinc and copper imbalance may trigger melatonin problems. Melatonin may lower cholesterol, especially LDL. Its immunoregulatory effects increase normal killer cell (NK) activity. Melatonin counteracts stress by being an epinephrine antagonist.

NEUROENDOCRINE SYSTEM
(PARAGANGLIA, PARANEURON)

The paraganglia, or paraneurons neuroendocrine system, is comprised of the adrenal medullae and a system of dispersed cell groups that are embryologically co-derived from the neural crest and endocrine cells. The neuroendocrine cells have characteristics shared by neurons and endocrine cells and appear to have a common origin with neurons. These cells are able to produce both substance(s) identical with, or related to, neurotransmitters and protein/polypeptide substance(s) that possess hormonic actions. These cells also possess synaptic vesicle-like and/or neuro-like granules that have a recepto-secretory function, which results in stimuli acting upon the receptor site on the cell membrane. Extra-adrenal neuroendocrine cells are distributed throughout the visceral ANS and are allied closely with sympathochromaffin ganglia.[15-17] These include SIF cells, carotid and aortic body glomeruli, parafollicular cells, adenohypophysial cells, pancreatic islet cells, gastroenteric endocrine cells, parathyroid cells, gastrotory cells, merkel cells, olfactory cells, melanocytes, mast cells, bronchial endocrine cells, hair cells of the inner ear, and Amacrine cells. The variety of cells and functions in this system of neuroendocrine cells constitutes a diffuse network that stretches throughout the body. The diffuseness and unique and varied characteristics of the secretions that are produced upon pollutant stimulation of this system help explain how the chemically sensitive individual might manifest a wide variety of clinical responses including varied symptoms. The importance of understanding this system in relation to chemical sensitivity cannot be overemphasized. Many of the widespread signs and symptoms of some of the chemically sensitive patients can be explained by understanding the pathophysiology of this system. Pollutant triggering of this system may cause appropriate responses at inappropriate times, or it may stimulate the system to be overactive. Pollutant triggering of the neuroendocrine system can also cause system suppression, thus bringing about symptoms that might previously be considered psychosomatic in origin when, in fact, they are the result of pollutant injury.

Paraganglia are richly innervated by autonomic nerve fibers. For example, the parenchymal chief cells synthesize catecholamines from tyrosine[18] and are supported by sustentacular elements analogous to peripheral nerve Schwann cells.[16,19] The sustentacular network can be detected immunocytochemically.[20,21] In extra-adrenal neuroendocrine cells, the sustentacular cells delineate a characteristic cell nest pattern of the chief cells that can be observed both in normal and neoplastic tissues.[17,22,23]

Neuroendocrine Phenotype

Biologically, the paraganglia constitute one of the major functional subsets among the multiple groups of topographically dispersed neuroendocrine

cells. The definition of a "neuroendocrine" phenotype is based upon a conjunction of cytoarchitectural, biosynthetic, and cytofunctional criteria.[24] These
include polyprotein gene expression with an overlapping biosynthetic profile,[25-28] product storage in argyrophilic dense-core secretory granules,[29,30] physiologic activity in monoamine precursor uptake and decarboxylation,[31] and
functional capacities to amplify or generate neural signals either by means of
humoral stimulation[18,32] or by biochemical transduction of microenvironmental
perturbations.[33]

The oligopeptide or monoamine molecules secreted by paraganglion and
other neuroendocrine cells can effect a multiplicity of bioregulatory actions
(Figure 24.2). These actions can be systemic (endocrine function), localized,
adjacent to neuroendocrine cells (paracrine function), or self-regulatory
(autocrine function). Such physiologic variation among neuroendocrine cell
groups provides a sophisticated, multilevel feedback mechanism for modulating microcirculatory dynamics and hormonal responses in target organs.[27,34,35]
For example, paracrine oligopeptide secretions of some neuroendocrine cells
regulate local biosynthetic activity in the gastrointestinal tract,[36] while paracrine
or autocrine interactions of monoamines are thought to underlie the mechanism
of chemoreception in certain extra-adrenal paraganglia.[37] Due to this complexity, pollutant injury to this system may cause deregulation at any level, thereby
causing a chain reaction of dysfunction. This reaction may be only to adjacent
or local cells, or it may go to a distant site, where it produces local, regional,
or generalized symptoms. Depending on the level of release of endocrine
hormones, a chemically sensitive individual might, for example, develop mild
nausea or gut cramps or generalized anxiety. This clinical picture does occur
in some cases of chemical sensitivity.

One role suggested for neuroendocrine cells is that their secretions appear
to influence immune function.[38] Since this function appears to be the case, there
is the potential for additional, widespread effects once pollutant deregulation
occurs. This wide range of effects is often seen in the food- and chemically
sensitive individual who may have mild to severe symptoms due to a pollutant
stimulus. For example, phenol exposure may trigger malaise, headache,
rhinorrhea, and diarrhea in a chemically sensitive patient, while another chemical such as petroleum-derived alcohol might only trigger a localized gut pain.

Often a small, local reaction is triggered only to be followed by the spread
of symptoms to many other organs. For example, chemically sensitive individuals are observed to have hypoglycemic reactions, except that they have
normal blood sugar levels. They become weak and jittery and develop fuzziness
of the head, apparently due to this neuroendocrine effect. In our ECU studies,
we have measured blood sugars in hundreds of patients who had symptoms of
hypoglycemia, only to find that blood sugars were normal. We then injected
different neuro hormones such as serotonin, norepinephrine, epinephrine,
methcholine, somatostatin, and substance P and reproduced the patients' symptoms.

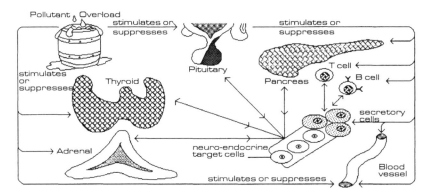

Figure 24.2. Pollutant effect on the neuroendocrine cells in the chemically sensitive individual. *Note*: Pollutants disturb (1) self-regulation; (2) regional regulation; (3) distal regulation; (4) regulation of microcirculation; and (5) the immune system.

Biosynthetic Profile

The biosynthetic products of neuroendocrine cells may be similar or diverse (Table 24.1). They include biologically active monoamines and regulatory oligopeptides as well as a host of common enzymes, structural proteins (such as neurofilaments[39] or intermediate filaments[40]), and membrane markers.[18,41] The cellular products may be stored and later secreted. Many now can be specifically identified in tissue sections with monoclonal antibodies.[39,40,42-45] Very recently, it has become feasible to identify peptide messenger RNAs or neuroendocrine cells in situ, by using labeled nucleic acid probes.[28] Some chemical pollutants resemble these vasoactive substances and cause inappropriate release or direct triggering of target organs. The result of this kind of malfunction is a vicious cycle of too easy triggering with stimulation or inhibition that leads to wave after wave of symptoms such as muscle aches and anxiety in the chemically sensitive patient. Jitteriness, fatigue, fuzzy thinking, and many more symptoms are often seen in a subset of chemically sensitive individuals.

All neuroendocrine cells contain high levels of an enolase enzyme, which is characteristic of neural tissues.[46] Enolase isoenzymes are products of three independent gene loci: alpha, beta, and gamma. The gamma gene is expressed almost exclusively by brain neurons and neuroendocrine cells.[47] This "neuron specific" enolase of gamma subtype can be detected immunocytochemically within the cytoplasmic compartment of all neuroendocrine cells using monoclonal antibodies.[47,48] The other two types may not be as organ specific.

Secretory Granules

The biosynthetic products of neuroendocrine cells are stored within relatively homogeneous cytoplasmic granules of a size range near the limits of

Table 24.1.   Biosynthetic Products of Neuroendocrine Cells

Monoamines	Enzymes
Dopamine[a]	Neuron specific enolase (NSE)[b]
Norepinephrine[a]	Tyrosine hydroxylase
Epinephrine[a]	DOPA decarboxylase
Serotonin	Dopamine beta hydroxylase
Oligopeptides[b]	Phenylethanolamine N-methyltransferase
Adrenocorticotropin (ACTH)	Structural Proteins
Opioid peptides (enkephalins)	Cytochrome B (561)
Calcitonin	Chromagranins A and B[b]
Somatostatin	Intermediate filaments[b]
	Neurofilaments[b]

[a] Secretory products

[b] Immunoreactive substances

conventional light microscopic resolution (0.5 to 2 μM). These granules are highlighted by nonspecific chromaffin, argyrophil, or argentaffin reactions.[19,49,50] Ultrastructurally, the granules are membrane delimited, and their cores are usually electron dense in osmicated tissues.[19,29] This core phenomenon is probably due to the internal concentration of monoamines and associated nucleotides in addition to enkephalins[51] and other oligopeptides. The matrix contains a large proportion of acid soluble proteins categorized as chromaffins.[52] These can be identified immunocytochemically in normal or neoplastic neuroendocrine cells with specific polyclonal or monoclonal antibodies.[30,52-54] Pollutant injury to these membranes may cause deregulation of storage keeping the system in either a depleted or overactive state with the patient being sluggish or overactive or having other symptoms.

Specialization of Neuroendocrine Cells

Biochemical studies, including peptide analyses and molecular genetic studies, indicate that intracellular control of polyprotein gene expression in neuroendocrine cells can be exerted at several levels: transcription, post-transcription, translation, or post-translation.[27] As defined previously, all of the neuroendocrine cells are endowed with the potential to synthesize or store a remarkably broad spectrum of biosynthetic products, some of which resemble neurotransmitter substances of the CNS.[24,28,34,35,55-57]

A consistent range of biosynthetic expression and functional activity is characteristic of neuroendocrine cells: the parenchymal cells produce, store, or secrete catecholamines.[15,17] This presence is associated with the presence of catecholamine-related hydroxylase and decarboxylase biosynthetic enzymes.[18,41] Typically, the storage granules also contain abundant opiate peptides.[43,51,54,57] One subset of chemically sensitive patients has excess catecholamine responses characterized by shakiness, stiffness, tachycardia, etc. and, in fact, some have increased catecholamines in the blood. Upon pollutant challenge, another subset develop reactions similar to opiate effects, becoming sluggish and very sleepy. Abnormal release or failure to clear these substances could account for the increased or prolonged reactions seen in some chemically sensitive patients.

A particular physiologic response of paraganglion parenchymal cells is their exquisite sensitivity to local hypoxia with a consequent rapid release of catecholamines when blood flow or oxygen tension is reduced.[18] Of course, pollutant overload may lead to vasospasm with ensuing local hypoxia that produces the catecholamine effect. Frequently this response is misinterpreted as the aforementioned hypoglycemic response. Two major groups of neuroendocrine cells that have been accessible to direct experimental investigations (intra-adrenal and extra-adrenal paraganglia) exhibit further subspecialization. First, the intra-adrenal paraganglia (adrenal medullae) respond to efferent autonomic stimuli by liberating increased quantities of catecholamines into the venous circulation. In this respect, these parenchymal cells resemble sympathetic neurons and contribute to the mobilization of cardiac, hepatic, and other systemic organ responses to somatic injury or stress as well as to pollutant injury. Indeed, the adrenal medullae are viewed as an integral functional element of the sympathetic nervous system.[18] Also, the intercarotid and aorticopulmonary paraganglia exemplify extra-adrenal paraganglia that transduce chemical changes in the partial pressure of arterial oxygen, carbon dioxide, or pH into efferent nervous signals, which then stimulate ventilation. The exact mechanisms of such chemosensory transduction remain controversial, but they probably involve complementary activities of one or more catecholamines in feedback loops that modulate both internal capillary blood flow and thresholds for afferent neural stimulation.[33,37] One frequently sees chemically sensitive patients who develop breathlessness after pollutant exposure. Pulmonary auscultation and functions are normal in these patients. It appears that there is deregulation of the respiratory center and the paraganglion chemoreceptors in these patients.

Functions of other extra-adrenal paraganglia presumably imitate those of the intercarotid bodies or of the adrenal medullae and peripheral sympathetic neurons, but have not been directly probed. Curiously, many extra-adrenal paraganglia are more conspicuous during fetal life or infancy than in the normal adult.[15,16] (See Chapter 26 for additional information.)

Topography of Neuroendocrine Cell Subsets

The topography of neuroendocrine cell subsets is important in understanding regional symptomatology seen in some chemically sensitive patients.

Paraganglia and other major subsets of the neuroendocrine cells are distinguished empirically, based upon both topography and histogenesis.

The paraganglion cells occur as macroscopically cohesive units in the adrenal medullae and in some major extra-adrenal locations[15] (Table 24.2).

Table 24.2.   Major Families of Extra-Adrenal
Neuroendocrine Cells

Family	Sites
Branchiomeric	Jugulotympanic (middle ear)
	Intercarotid (carotid bodies)
	Intrathyroid
	Laryngeal
	Aortico-pulmonary (superior medastinum)
	Coronary-interatrial (heart base)
Intravagal	Ganglion nodosom
	Nasopharynx
	Angle of mandible
	Inferior vagus nerve distribution
Aortico-sympathetic	Paravertebral
	Intrathoracic
	Retroperitoneal
	Organ of Zuckerkandl
Visceral-autonomic	Gastroduodenal region
	Porta hepatis
	Genital tract
	Urinary bladder
	Cauda equina

During embryonic development, the ancestral stem cells of paraganglia migrate from the ventral neural crest.[58-60] Topographically, they remain associated principally with cervical and mediastinal tissues of ontogenetic gill arch derivation with the paravertebral sympathochromaffin ganglia or with peripheral elements of the ortho- and parasympathetic nervous systems located in the viscera of cerebrospinal axis.[17,61] Anatomically constant groups of extra-adrenal paraganglia thus can be grouped into several broad "families" including the branchiomeric, intravagal, aorticosympathetic, and visceral autonomic.[16,17] Each group will be discussed in detail.

Bronchiomeric Group

The bronchiomeric group includes the jugulotympanic group in the middle ear, the intracarotid bodies, laryngeal group, the aortic-pulmonary bodies in the superior mediastinum, and the coronary-interatrial bodies at the base of the heart. The jugulotympanic group may be affected by pollutant overload locally or distally, both giving a similar response in the middle ear area with resultant pain and, if triggered too long, inflammation. The intracarotid bodies are chemoreceptors and help regulate carbon dioxide and O_2 content to the brain. Pollutant overload can cause deregulation of these bodies with resultant vascular deregulation and deregulation of the respiratory and vasomotor areas of the brain. Laryngeal cell deregulation may result in spasm and spastic type dysphonias. The injury to the aortic-pulmonary bodies may change perception in the baroreceptors, with resultant mild imperception of pressure change. This imperception may account for some environmentally triggered essential hypertension (see Chapter 22). Increased sensitivity may result in symptoms induced by changes in the weather, as seen in some chemically sensitive individuals.

Bronchopulmonary neuroendocrine cells are widely distributed within the bronchopulmonary system, both singly and as tight clusters, which have been referred to as "neuroepithelial bodies".[62] They appear during first trimester differentiation of the fetal lung.[63] Origins of pulmonary neuroendocrine cell groups from both neuroectoderm and gut endoderm have been suggested by immunohistochemical studies as well as patterns of neoplasia.[64-66] Dysfunction may lead to hyperplasia or tumors, both of which have been seen in some chemically sensitive individuals.

Further evidence of neuroendocrine regulation of the bronchopulmonary systems comes from studies of recipients of heart-lung transplants. The recipients of heart-lung transplants have an appropriate level of ventilation during exercise as a result of a disproportionate increase in tidal volume with a reduced respiratory rate.[67] It appears that neurally mediated feedback from the receptors modulates the pattern of ventilation response during exercise. When the exercise response in subjects with denervated hearts has been studied extensively,[68-71] such subjects have been shown to have limitations in maximal heart rate and stroke volume.[72] Abnormality of the pattern of cardiac-rate response in such subjects includes resting tachycardia, a slow heart rate response during mild to moderate exercise and a more rapid response during strenuous exercise, and a prolonged recovery time for the heart after exercise.[69] The greatest heart rate response occurs at a point during exercise,[73] suggesting that circulatory catecholamines rather than autonomic innervation are the predominant influence on the cardiac rate.[69] At the EHC-Dallas, we have seen alterations similar to those just described in some of our chemically sensitive patients.

Bronchopulmonary neuroendocrine cell neoplasms usually present as carcinoid tumors or small cell anaplastic ("oat cell") tumors.[65] Often these patients have flushing and hypertension and other systemic responses. True intrapulmonary paragangliomas are extremely rare.[74] Multicentric "minute paragangliomas", which are incidental autopsy or surgical findings, lack apparent clinical significance[75] and actually may be of pleural origin.[76] However, these minute tumors may explain aberrant function in some people. Perhaps some concentrated effort, coupled with thought and a sophisticated technology, will allow correlation of their presence with some chemically sensitive patients' dysfunction.

Evidence in a subset of pollutant-injured chemically sensitive individuals suggests that responses similar to those seen with bronchopulmonary endocrine tumors such as flushing and hypertension may be partially mediated through the paraganglion ANS.

Coronary and interatrial cells at the base of the heart can account for some of the cardiac and, often especially, the noncardiac pain seen in the chemically sensitive patient. Often this pain is confused with that of myocardial infarction.

Intravagal Neuroendocrine Cells

The intravagal neuroendocrine cells include the ganglion nodosom of the nasopharynx and angle of the mandible and inferior vagus nerve distribution. No one knows the effects of pollutant injury on this area, but many patients do have trouble with their temporal mandibular joint as a result, and sometimes as a trigger, of pollutant overload. Possibly this anatomical fact is a part of the mechanism of triggering. In addition, some chemically sensitive patients present with a feeling of being strangulated in the upper larynx and pharynx. This feeling also occurs after pollutant challenge and could be the result of injury to the ganglion nodes.

Thyroid gland C-cells, characterized by production of calcitonin, are found within the follicles of the thyroid gland. Embryogenetic studies indicate an origin of C-cells from the neuroectoderm or neural crest,[77,78] but some developmental and oncological observations have raised the possibility of an endodermal or a combined endodermal and neural crestic origin.[79,80] Calcitonin effects may be abnormal in the chemically sensitive individual. These effects may be especially apparent in chemically sensitive patients who have muscle spasm and tetany and who have deregulation of their calcium and magnesium balance.

The aortic-autonomic neuroendocrine cells include those of the paravertebral intrathoracic and retroperitoneal areas as well as the organ of Zuckerkandl at the bifurcation of the abdominal aorta. Pollutant injury is probably always secondary to this organ through ANS deregulation, except in those chemically sensitive patients who have aortic bifurcation surgery.

Visceral-autonomic Paraganglion Cells

The visceral-autonomic paraganglion cells include those in the gastroduodenal tract, parahepatic system, genital tract, bladder, and cauda equina.

The gastrointestinal and pancreatic neuroendocrine cells localize within the mucosa of the stomach, the intestine, or the pancreas. In the pancreas, they form both the classical islets and extra-insular collections of hormonally active cells. Subtle differences in the spectrum of regulatory peptide biosynthesis may be related to specific anatomic distribution within the gastrointestinal tract.[64,81] Neuroendocrine cells appear during early differentiation of the gastroenteric pancreatic axis both in normal embryos[78,82] and in teratomas.[64] Other pathobiologic evidence also supports an origin from pluripotential foregut precursors.[81,83] Many pollutant sensitive patients exhibit gastrointestinal upset, which was previously unexplained but now appears to be the result of secretions from these neuroendocrine cells. This system is clearly imbalanced in these patients. Vasoactive intestinal peptide and somatostatin can be released causing this gastrointestinal upset (see Chapter 22).

Genitourinary Paraganglion Cells

Genitourinary tract neuroendocrine cells are associated with intramural neural plexi in the bladder or vagina, probably representing visceral-autonomic paraganglia. Plausibly, these have been related to the sacral-parasympathetic limb of the ANS.[17] In a subset of chemically sensitive individuals, pollutant stimulation of these cells results in a sudden urge to urinate. At the EHC-Dallas, this phenomenon has been demonstrated upon inhaled or injected pollutant challenge and subsequently with injected neurohormones, thus suggesting neuroendocrine deregulation of the genitourinary cells.

Argyrophilic cells, which are immunoreactive for serotonin, somatostatin, and nonspecific enolase, have been demonstrated in the prostatic urethra, prostatic ductules, and prostate acini.[84] These are possibly of endodermal (perhaps hindgut) origin.[85] After an inhaled pollutant exposure, some chemically sensitive patients with recurrent prostatitis report prostate pain and eventually inflammation, which is later followed by bacterial infection (see Chapter 22).

Neuroendocrine cells are found in the normal female genital tract and some ovarian neoplasms;[86] therefore, they probably must be located in the ovaries. Pathways of normal development indicate that some of the latter neoplasms could arise from coelomic epithelium.[86] Intrauterine paragangliomas very rarely have been reported,[87] so the presence of these cells within the uterus is open to question.

The nonasthmatic chemically sensitive patient frequently complains of bladder spasm, lower and upper gastrointestinal spasm, vaginal bleeding, vaginal spasm, peripheral vascular constriction, and a myriad of respiratory signs and symptoms not intimately related to the lung. When neuroendocrine discharge after pollutant exposure occurs, the chemically sensitive patient may have to urinate and defecate suddenly because of spasm in those areas related to these functions. He or she may also experience cold hands and feet with accompanying chest pain and breathlessness. This urgency is compatible with dysfunction of the lower paraganglion system.

Paragangliomas

Paraganglia tumors may give us some insight into chemical release and, thus, the abnormal responses seen in some chemically sensitive individuals.

Solitary or multicentric paragangliomas may arise in relationship to any of the topographic families of paraganglia listed in Table 24.2. Overall, approximately 90% of paragangliomas (paraganglion tumors) arise within the adrenal gland and are alternately classified as "pheochromocytomas" in this location. Knowledge of these and other locations is valuable since they give insight into the extent of the neuroendocrine systems that release their hormones and cause some chemically sensitive patients to react to pollutant exposure.

Extra-adrenal paragangliomas are most common in jugulotympanic, intercarotid, superior mediastinal, and retroperitoneal paravertebral sites, suggesting that these areas are more sensitive to, or prone to, pollutant damage. The majority arise in the head and neck,[17,88-90] where they are allied to anatomic structures of branchial arch ontogeny (branchiomeric paraganglia) or are associated with the vagus nerve. Within the branchiomeric family, intercarotid and jugulotympanic paragangliomas are most common. Laryngeal paragangliomas are relatively uncommon, and intrathyroid paragangliomas are rare.[17,91-93]

Vagal paraganglioma may arise at the level of the ganglion nodosum or in more inferior portions of the vagus nerve. They typically present in the angle of the mandible or near the fossa of Rosenmueller in the posterior pharynx.[94-97]

In the thorax, paragangliomas represented 4% of cases in a large series of neural tumors.[98] Lack et al.[99] reviewed 36 cases of intrathoracic paragangliomas. Mediastinal supra-aortic, aorticopulmonary, or coronary-interatrial paragangliomas may impinge upon the atrial walls or upon the base of the heart and the great vessels. Posterior mediastinal tumors may arise from the paravertebral sympathetic trunk.[30,100,101]

Paragangliomas are probably more common in the retroperitoneum than in the mediastinum.[17,102,103] They occur less commonly in visceral locations supplied by orthosympathetic or cranial-sacral parasympathetic elements: the orbit,[104] duodenum,[105] hepatic ducts,[106] bladder wall,[107] genitourinary tract, or cauda equina.[108] Urinary bladder and cauda equina tumors are probably the most frequent of the latter group. The urinary bladder paragangliomas are often functional and resemble adrenal medullary pheochromocytomas both histologically and cytochemically. Often a positive tissue chromaffin reaction occurs.[109]

Clinical Manifestations of Neuroendocrine Stimulation

In the chemically sensitive individual, the most commonly recognized clinical manifestation of neuroendocrine system stimulation by pollutants is anxiety attacks, or paraganglia surges. In most cases, a symptom complex characterized by headaches, palpitation or tachycardia, and sometimes excessive diaphoresis occurs. These are paroxysmal and often leave the patients with a jittery, shaky feeling with ravenous hunger for a few minutes to several hours. Once the paroxysmal nature of these attacks in the pollutant-injured chemically sensitive patient is recognized, they can be cleared by avoidance of pollutants under environmentally controlled conditions. In addition, alkalinization with soda bicarbonate or treatment with vitamin C may stop the reaction. Once the patient is in a basal state, signs and symptoms can be reproduced by controlled, measured challenges of biological inhalants (pollen, dust, mold), foods, and toxic chemicals, and triggers can be defined. It is a common occurrence to see the chemically sensitive patient become ravenous upon pollutant stimulation.

For example, many patients will experience this phenomenon in areas of high pollution such as airports. They will then eat any food in sight, even if they know it will be harmful to them. Another type of chemically sensitive patient has his or her appetite increasingly triggered in accord with how much he or she eats. For this patient, the end of a normal meal does not signal satiety, but increasing appetite.

Often symptoms of pollutant excess can be terminated by giving just fresh air or magnesium sulfate, ascorbic acid, multiple B vitamins, or trisalts. Other functional manifestations of paraganglia pollutant stimulation may rarely be related to abnormal secretion of neuroendocrine peptides. Reports could include abnormal serum levels of biologically active calcitonin, ACTH vasoactive intestinal peptide, somatostatin, and possibly serotonin. These responses have been affirmed by reports of tumors found in these areas.[44,110-113] Blood levels of serotonin may be increased or decreased.[114] Immunoreactive oligopeptides such as calcitonin,[42] somatostatin,[115] or VIP may be detected in paraganglion tumor cells[116] without evidence of serum elevations or clinical activity. Therefore, at present, it is uncertain whether stimulation of the paraganglion cells by pollutants will give measurable elevations. Clearly, somatostatin effects are seen after pollutant injury, as are the effects of vasoactive intestinal peptides (see Chapter 26).

Laboratory Tests

In the appropriate clinical setting, biochemical tests of serum or urine can be extremely valuable adjuncts for the diagnosis of paragangliomas,[117] but these tests do not yield as clearly defined parameters for hypersecretion in pollutant stimulation. This lack of clear findings in the chemically sensitive individual is probably due both to less output and to local and regional, rather than peripheral, stimulation. Also, levels of hormones may be transient, lasting only a few seconds to minutes. Laboratory tests are not advocated as general screening tools, though elevated catecholamine levels are often registered in the chemically sensitive individual. This elevation occurs in one subset of patients who exhibit brain dysfunction along with anxiety. Provocation tests are much more diagnostic of neuroendocrine abnormalities (Table 24.3).

When symptoms such as anxiety, gas, etc. or signs such as flushing or bloating suggest dysfunction of the neuroendocrine system, follow-up laboratory testing may be clearly indicated. At a minimum, biochemical screening should include a battery of 24-hour urine biochemistries including metanephrines, vanillylmandelic acid (sequential degradation products of norepinephrine or epinephrine), plasma cortisols, and free catecholamines. Results of these tests may vary due to different levels of catechol-O-methyl transferase and monoamine oxidase, which are involved in the sequential catabolic steps, so that reliance on a single test is not advisable.[117] With further study of provocation tests in the chemically sensitive patient, it may become possible to differentiate the various neuroendocrine abnormalities of this population from

Table 24.3.   Neuroendocrine Effects Seen in 100 Chemically Sensitive Patients with Normal Blood Sugars and Provoked by Intradermal Injections Using Diluted Solution

	Dilution	Symptoms and signs
Seratonin	1/5 to 1/125	Swelling and headache
Norepinephrine	1/5 to 1/625	Agitated, jittery, sleepy, depressed
Epinephrine	1/5 to 1/3125	Agitated, jittery, sleepy, depressed
Methacholine	1/5 to 1/125	Agitated, jittery, sleepy, depressed
Substance P	1/5 to 1/125	Myalgia, headache, bloating, cramps
Somatostatin	1/5 to 1/125	Myalgia, headache, bloating, cramps
Vasoactive intestinal peptide	1/5 to 1/3125	Cramps, muscle spasm, headache
Bombasin	1/5 to 1/625	Laryngeal edema, headache, dizziness
Choleocystkinin	1/5 to 1/625	Laryngeal edema
Neuropeptide Y	1/5 to 1/625	Burping, gas
Popamine	1/625 to 1/5	Headache, eyes crossing, chest pain

Source: EHC-Dallas. 1994.

the normal population. Once a patient's basal state is achieved in an ECU, incitants may be determined through pollutant challenge.

Epinephrine may be the predominant catecholamine secreted from the paraganglia. However, stimulation of paraganglia of other extra-adrenal sites usually results in an excess of norepinephrine. Measurement of plasma catecholamines in fasting patients can be the most sensitive and specific test for excess catecholamine production, since fasting decreases catecholamines. Pretreatment with clonidine to suppress physiologic catecholamine elevation in apprehensive or stressed patients increases the specificity of this test.[118] In patients with pheochromocytomas related to multiple endocrine neoplasia, an increase in the fraction of epinephrine may be an early sign of adrenal medullary hyperplasia.[119,120] In a subset of chemically sensitive individuals, intermittently elevated catecholamines such as norepinephrine are often seen. Others such as epinephrine and dopamine are depressed.

Radiography is of little help in hyperplasia, but occasionally it may reveal an increase in the adrenal size. The medullary tissue is normally concentrated in the head and body of the gland, so that expansion of the tail can be an early sign of hyperplasia/neoplasia.[121] Multiple chest films are essential when extra-adrenal paraganglioma of the heart and base or posterior mediastinum is suspected.[98] The following case illustrates the need for further understanding paraganglioma functioning in the chemically sensitive individual.

Case study. This 56-year-old chemically sensitive white female entered the ECU with the chief complaint of episodic hypertension (BP: 190/110 to 220/170 vs. 120/80 mmHg) during quiescent periods. These episodes were accompanied by headache, feelings of anxiety, shortness of breath, jittery feeling, shakiness, and pain over the right adrenal gland. Her past history was significant in that her father had died at age 56 years of an episode of hypertension and back pain. Also, her only daughter was sprayed with Agent Orange and developed mental dysfunction and hirsutism with severe chemical sensitivity.

This patient was highly susceptible to biologic inhalants, requiring biweekly injections of extracts of pollen, dust, molds, weeds, trees, and grasses. In addition, she was sensitive to the additives and preservatives in commercial foods, so she was able to eat only chemically less-contaminated foods. Also, this patient had developed sensitivity to common foods such as beef, pork, wheat, and cane sugar. She required a rotary diversified diet in order to cope with this problem. This patient was very chemically sensitive; she, therefore, required a less-polluted house in which to live and a less-polluted office in which to work. Both of these were located in a less-polluted ranch area of Texas.

For years, this patient did extremely well on this program. Symptoms recurred, however, when she began working with her husband, who was a physician, in his medical practice.

Two years prior to admission to the ECU, she developed episodes of anxiety and eventually hypertension that could not be controlled with medication. When she experienced these episodes, she would have pain over the right adrenal. Endocrine work-up revealed adrenal function to be one and a half to two times normal, suggesting, but not conclusively identifying, adrenal hyperfunction.

CAT scan showed the right adrenal was questionably larger than the left. Due to the family history, the consistency of the patient's recurrent symptoms, with pain over the right adrenal, and the fact that she had been in good environmental control, right adrenalectomy was performed. While the pathology report was not definitive, there was a suggestion of increased chromaffin cells in the adrenal medulla. The weight was still in normal range. Within 48 hours the patient became normotensive with a blood pressure of 110/80 mmHg. This level has remained constant for 5 years without medication. The anxiety attacks and lumbar pain have been absent for 5 years, and the patient has remained well on her avoidance regimen. Provocation tests with epinephrine and norepinephrine reproduced her symptoms.

This case exemplifies the problems of diagnosing and treating imbalances of the neuroendocrine system in the chemically sensitive patient. However, it is clear that persistence in evaluation under environmentally controlled conditions will aid in an improved outcome.

Intra-adrenal paragangliomas of either the sporadic or familial types may be preceded by nodular hyperplasia,[102,122] which is preceded by cellular hyperplasia. This cellular hyperplasia likely was the problem in the aforementioned case. In familial cases, hyperplasia may be accompanied by a shift in the fraction of epinephrine to production.[119,120] Epinephrine biosynthesis depends upon the enzyme phenylethanolamine-N-methyl transferase and reflects a higher level of differentiation.[117] This enzyme may well have been stimulated in this case. Hyperplasia of extra-adrenal paraganglia primarily involves supportive neural elements[123,124] and does not appear to be directly related to subsequent autonomous growth of the parenchymal cells.[23]

Clearly, our and others' observations indicate that pollutant injury with deregulation of the neuroendocrine cells and system plays a significant role in the chemically sensitive patient.

ADRENAL GLANDS

Some physiology of the adrenal glands (Figure 24.3) was described in our discussion of the paraganglia. The physiology is important in understanding adrenal function in the chemically sensitive patient, since so many chemically sensitive patients are thought to be adrenal insufficient, although most really are not. It, therefore, will be given further consideration here. Specifically, the adrenal gland will now be considered in deference to the paraganglia system as a whole.

Pollutant Injury to the Physiology of the Adrenal Glands

The nerve supply of the adrenal glands is through the ANS. The sympathetic preganglionic fibers for these glands are the axons of cells located in the intermediate lateral columns of the lowest two or three thoracic and highest two lumbar segments of the spinal cord. They go down the greater, lesser, and least thoracic and first lumbar splanchnic nerves to the celiac, aorticorenal, and renal ganglia. Most go to the celiac plexus and then to the adrenals. Parts also communicate with the renal and phrenic nerves. Parasympathetic fibers are conveyed to the celiac plexus in the celiac branch at the vagus to the adrenals. The majority of the fibers of the suprarenal plexus enter through or near the hilus through the cortex to the medullary cells. They ramify profusely and terminate around the medullary chromaffin cells. Other fibers invaginate the adrenal vessels. Pollutant injury may occur to the cord, sympathetics, celiac ganglion, the nerves, and the blood vessels, resulting in a myriad of stimulations or suppressions with the resultant dysfunction of the adrenal gland, the rest of endocrine and neuroendocrine systems, the neurological system, and the immune system. These ramifications may reveal why pollutant overload causes many problems in the chemically sensitive individual, resulting in some cases with much diffuse symptomatology.

The adrenal medulla contains columnar cells, which secrete the catecholamines, epinephrine, and norepinephrine. Some islets of these chromaffin cells secrete mostly epinephrine, while others secrete mostly norepinephrine. Preganglion fibers enter the medulla and terminate directly on the parenchymal cells or scattered sympathetic ganglion cells. Again it is evident that deregulation may occur with pollutant injury. This dysfunction may be one of the reasons for the pollution-triggered jitteriness, shakiness, and other symptoms

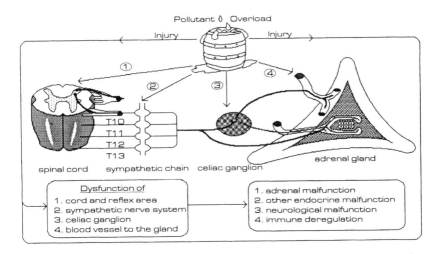

Figure 24.3. Pollutant injury to the adrenal gland.

of epinephrine release often seen in some chemically sensitive patients (see Volume IV, Chapter 30, section on endocrine). Often, chemically sensitive patients present with excess or insufficient levels of norepinephrine, epinephrine, or dopamine, as measured in their urine and blood profiles, indicating injury to their orderly metabolic sequences.

The adrenal cortical cells are divided into three layers. The outer zona glomerulosa produces aldosterone relatively independently (Figure 24.4). Control here is more through the renin angiotensin axis starting at the juxta glomerular apparatus of the kidney. Pollutant injury to this outer cortical layer may result in edema and hypertension. The zona fusciculata and reticularis produce corticosteroids and are under control of ACTH. The adrenal gland receives blood from 30 to 50 small arteries. Obviously, chronic dysfunction of these vessels after direct pollutant damage to any of them or to their ANS attachments or to the pituitary hypophyseal axis may result in the development of subtle adrenal dysfunction. Early pollutant injury may then develop into more deregulation of the endocrine and immune system with a vicious cycle recurring until fixed-named disease occurs. The different types of corticoids are derived from cholesterol; therefore, pollutant injury to cholesterol metabolism might cause inappropriate production.

Pollutant overload to the cortex and cortisol may have widespread effects in the chemically sensitive individual because of known functions of cortisol. Gluconeogenesis may be impaired with the patient not being able to generate the 6- to 15-fold increase that is often needed with pollutant exposure or stress. Transport of amino acids from the extracellular fluid into the liver cells may be altered and not easily converted to glucose. DNA transcription and RNA messenger may be impaired by pollutant injury. Easy mobilization of amino acids from muscles that are under control of the cortex may be impaired,

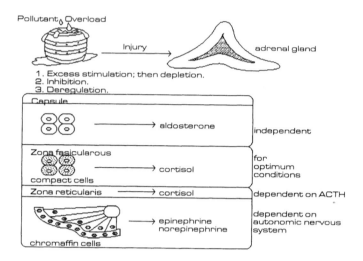

Figure 24.4. Pollutant injury to the adrenal gland.

causing malfunction of adaptation during fasting in the chemically sensitive patient. The rate of glucose utilization, which is usually decreased by cortical cells, may be increased by pollutant injury. NADH oxidation may be accelerated by pollutant stimulation, with rapid and excessive glycolysis causing swings in glucose. This chain of events is frequently seen in some chemically sensitive patients.

Excess cortisone can cause muscle weakness and immune dysfunction due to altering of protein metabolism. Although protein metabolism in the periphery may decrease, it will increase in the liver cells. As observed in 1000 chemically sensitive patients, protein metabolism is disturbed (see Volume IV, Chapter 30). Usually the effects seen are opposite to those produced by cortisol, resulting in a decreased rate of deamination of amino acids by the liver, decreased protein synthesis, decreased formation of plasma proteins by the liver, and the aforementioned decreased conversion of amino acids to glucose.

The mobilization of fatty acids from adipose tissue is influenced by the cortical area and may be altered with pollutant exposure. When a chemically sensitive individual fasts in a controlled environment, a decreased transport of glucose into cells may occur along with an increased mobilization of fats and an increased oxidation of cells. This sequence allows for a more efficient functioning of all systems after the pollutant load has been reduced (see Chapter 34).

Pollutant stimuli can cause an increase in ACTH, which stimulates the cortex and may eventually deplete it of adequate cortisol output. Once tissues are damaged by pollutant exposure, inflammation may occur due to insufficient local amounts of cortisol, which can cause release of lysozymes due to lack of stability of the pollutant-injured lysosomal membranes. Inflammation will result, as seen in the chemically sensitive patient.

Plasma proteins may be lost from the capillary, causing intracellular edema, which is seen in the chemically sensitive individual. Alteration of cortisol output by pollutant overload also will affect the immune system with resultant eosinophilia, as seen in some cases of chemical sensitivity, or an excess of lymphocytopenia may occur, as seen in most others. Lymphocytopenia is due to excess cortisol at the cellular area. This pollutant-derived deregulation of cortisol found in the chemically sensitive patient may result in stimulation or inhibition or atrophy of the lymphoid tissue, which in turn increases or decreases the output of sensitized lymphocytes and antibodies from the lymphoid tissue. Therefore, depending on the type of pollutant stimulation, the chemically sensitive individual may experience increased or decreased immunity to infections. At the EHC-Dallas, we treat more chemically sensitive patients who are resistant to infections. One of the causes of the mild anemia frequently seen in the chemically sensitive patient may result from a decrease in cortisol production.

When ACTH is triggered from the pituitary α- and β-melanocyte stimulating hormone, β-lipotropin and β-endorphin are also secreted. The opiate-like effects of β-endorphins are often seen in a subset of chemically sensitive patients, often accounting for their sluggishness, brain fogginess, and dysfunction. It has been shown in thousands of challenges of foods and chemicals at the EHC-Dallas under environmentally controlled conditions that opiate effects can occur in the chemically sensitive individual. Lipotropin effects by pollutant stimulation are unclear as to status in the chemically sensitive individual. A small subset of chemically sensitive individuals have extreme, adverse reactions to sunlight on their skin. These reactions may be affected by inhibition of melanocyte stimulating hormone.

A whole spectrum of adrenal dysfunction may occur after pollutant injury. At first, this dysfunction may be mild, but, if the stimulation is too long or virulent, gross pathology may be seen. Early diagnosis and treatment is the aim of medical intervention, with the hope of eliminating changes before they become irreversible.

Clinical Picture of Pollutant Injury and Adrenal Dysfunction

The clinical entities seen in adrenal dysfunction due to pollutant overload are either hyper- or hypofunction. Most commonly, however, chemically sensitive patients do not present with either of these, per se. Rather, because they are fragile to most exposures, they demonstrate a generalized instability, suggesting an adrenal regulatory problem. Hyper- and hypofunction are discussed independently.

Hyperadrenalism

The adrenal genital syndrome contains an excess of 17 ketosteroids produced by some kind of biochemical stimulation of the adrenal gland. Masculinization occurs because of the excess hormone. There is an underproduction of cortisol; therefore, ACTH stimulation worsens the adrenal dysfunction.

Some patients have an increase in cortisol through a hyperplasia of the zona fasciculata and zona reticularis. Perhaps environmental pollution may cause Cushing's disease in some patients. Other cases of adrenal virilization have been reported due to pollutant injury, especially from toxic pesticides.[125] Still other patients may have hyperplasia or a small tumor of the zona glomerulosa resulting in aldosteronism.

The cause of increased aldosterone secretion in idiopathic cyclic edema is not clear. Often an excessive secretion of pituitary antidiuretic hormone under the influence of environmental stimuli is present. The exact pathophysiologic mechanisms are complex and aldosterone secretion is not always elevated. When raised, it is often not responsive to stimuli that normally causes its decrease. Food and chemicals have been seen to trigger this edema syndrome as a reaction to challenge. In addition, previously diagnosed idiopathic edema resulting from a variety of environmental causes in patients who completed a stay in the ECU has been cleared.

Adrenal Insufficiency/Hypoadrenalism

There are many stages of adrenal insufficiency seen in the chemically sensitive patient. The most common cause now may be from autoimmune disease, which is usually due to multiple pollutant overload or direct chemical inhibition. There are many mild types of area insufficiency seen in which the whole clinical picture of adrenal insufficiency does not occur in the chemically sensitive patient. The milder form of low adrenal reserve, in which the adrenals are capable of producing enough hormone to maintain an apparently normal state of health in the absence of stress, occurs. This state may be due to chronic malnutrition, chronic excess stimulation of the adrenal with resultant depletion, or sudden, massive, life-threatening stress as seen in massive trauma. With the increased demand on adrenal cortical hormones, mild to moderate symptoms or even circulatory collapse may occur. Usually, this chronic type of chemically sensitive patient first presents with fatigue and/or recurrent infections, an increase in allergic diathesis, a mild autoimmune-like disease, vasculitis, or nausea and vomiting. The ACTH test is then used. As shown in Volume I, Chapter 3,[126] the process of adaptation is important in understanding pollutant injury. Only the adrenal response to adaptation will be discussed here since the clinical features and other ramifications are discussed in Volume I, Chapter 3.[126]

The alarm reaction in pollutant injury is divided into an initial stage of shock and a secondary stage of countershock. Evidence indicates that during the initial stage of pollutant injury, effects characteristic of an acute release of epinephrine are followed by changes characteristic of relative adrenocortical insufficiency (a fall in blood pressure, decreased blood sugar, decreased blood sodium, and increased blood potassium) and lasts from a few minutes to hours. The stage of countershock then occurs, with effects characteristic of increased adrenocorticoid activity (return of blood pressure to normal or above normal; increased blood sugar; restoration of normal or increased level of blood sodium and normal or decreased level of blood potassium; and enlargement of the adrenal cortex). This is the process of adrenal adaptation due to pollutant injury. These effects return to baseline during the first stage of resistance, but reappear during the second stage of exhaustion. In hypophysectomized animals these changes do not occur, indicating that they are mediated by the hypothalamus and pituitary gland through their production of corticotropin releasing factor (CRF) and adrenocorticotrophic hormone (ACTH), respectively. During the recovery phase and as the production of hydrocortisone returns to normal, the production of androgens increases and apparently is related to the healing process, since these hormones that were secreted are known to be anabolic. It is, therefore, evident that adrenal cortical hormones, especially hydrocortisone, play an important role in normal response to pollutant stress.

The critical role of the adrenal cortex in response to pollutant stress in human subjects manifests in a number of ways. A patient with untreated mild adrenal insufficiency or low adrenal reserve may function reasonably well when environmental conditions are optimum, but he or she tends to tire more easily. If strenuous physical exercise is undertaken or a meal skipped, hypoglycemic symptoms as severe as convulsions may develop. Pollutant stress will accentuate this problem. If an infection such as a common cold develops, acute adrenal insufficiency with nausea, vomiting, falling blood pressure, and collapse may occur in the severe cases of chemical sensitivity (rare), while apathy may occur in the milder ones (common). To a point, all of these undesirable developments may be prevented by administration of suitable dosages of hydrocortisone.

It has been demonstrated that normal adrenocortical function is essential for an ability to withstand infections.[127] Several studies have indicated that either too little or too much glucocorticoid can impair resistance to infection, whereas an optimum level of glucocorticoid enhances resistance to infection. However, prolonged use of extra hydrocortisone will damage many systems, resulting in much irreversible damage.

In over 18,000 patients studied at the EHC-Dallas, only two have had overt adrenal insufficiency and responded to proper steroid replacement. However, it is becoming evident that many chemically sensitive patients may have selective adrenal deficiencies. Studies measuring 11-OH and 17 ketosteroids before and then after ACTH stimulation have revealed several patterns of isolated adrenal stepwise alterations (see Volume IV, Chapter 30). Occasionally supplementation (e.g., dehydroepiandrosterone) with one type of steroid synthesis pathway deficiency substances will help the patient recover more rapidly (see Volume IV, Chapter 38; also see Chapter 30 for evaluation of cortical and pituitary function).

PARATHYROID GLAND

The two to six parathyroid glands secrete parathormone, which regulates serum calcium and is stimulated by low calcium and inhibited by high calcium. Thyrocalcitonin secreted by the thyroid neurochemical C-cells appears to be a second hormone that influences calcium by lowering it in the blood. Calcium, which is the opposite of parathormone, increases.

There are four principle actions of the parathormones. First, parathormone inhibits phosphate reabsorption (or enhances phosphate secretion) by the renal tubule. Second, parathormone reabsorbs calcium and phosphate from the bone. Third, parathormone increases calcium absorption from the GI tract. Finally, parathormone enhances calcium reabsorption in the renal tubules. Activity of the hormone parathormone also enhances magnesium excretion. Parathormone can interact with vitamin D, which is altered in 30% (15% excessive, 15% deficient) of our chemically sensitive patients, and pollutant overload may alter absorption of calcium and incorporation of it into bones and cells. Increased phosphate has been observed in the serum of some chemically sensitive individuals. This elevation may be due to a mild inhibition of the parathormone output or excess intake of phosphate. Pollutant injury apparently can occur with excess phosphate intake from soft drinks or organophosphate insecticides. In patients with renal calcium leaks due to pesticide or solvent exposure, a secondary hypoparathyroidism may occur. Pesticide injury to the kidney that results in a magnesium leak that may eventually become a calcium leak is also possible.

An excess production of parathyroid hormone leads to hypercalcemia by increased stimulation of the osteoclastic activity of the bone with a release of calcium and phosphate. Absorption of calcium from the gut and reabsorption of calcium from the renal tubule results in inhibition of the tubular reabsorption of phosphate, causing an excessive loss of phosphate in the urine.

OVARY

In chemically sensitive women, the hormonal system, consisting of three different hierarchies of hormones, has complex responses to pollutant stimuli. Imbalance can occur, resulting in an over- or understimulation of the luteinizing-reducing hormones, the FSH, the LH, and estrogen and progesterone. This imbalance is the most frequent endocrine abnormality observed in the chemically sensitive female.

For all of the sex hormones, cholesterol is the precursor, and thus pollutant injury to the liver with resultant damaged metabolism can cause alteration of hormone output. In all steroid secreting glands, the side chain of cholesterol is degraded to pregnenolone and dehyroepiandosterone. Pregnenolone is transformed into progesterone, which, by degradation of the side chain, becomes testosterone. Due to the high metabolic activity of the ovaries, toxic chemicals are attracted here and have been shown to cause much damage.

Estrogen has been known to have a dampening effect on the cardiovascular system. The late onset of arteriosclerosis in females following menopause is common knowledge. A study by Couch and Wortman[128] supports this observation of the dampening effect. They found a significant occurrence of migraine in pathologically anovulatory females (polycystic ovary, galactorrhea, amenorrhea) vs. pregnant women or those taking birth control pills. It was suggested that this condition might also be a hypothalamic problem. Floroxzymestine, a progesterone compound, also has been shown to help in preventing angioedema.[128] Excess estrogen has been shown to have an adverse effect on vessel walls giving venous inflammation that results in thrombophlebitis and pulmonary emboli.[128] Androgenic hormones are known to influence the vascular tree in preventing some forms of angioedema. Danazole®[129] has been shown to help prevent angioedema in some cases of hereditary angioedema.

Certain toxic chemicals, especially pesticides, have been shown to go to the ovaries in animals and humans (see Volume II, Chapter 13[130]). For example, lindane induced a marked disturbance of the estrous cycle in some animals, prolonging the proestrus phase five to seven times and thereby delaying ovulation.[131] Lindane is used as a delousing medication for children. It has now been shown to be carcinogenic in animals. *In vitro* studies measured effects of the pesticide *o,p'*-DDT and its isomer *p,p'*-DDT on progesterone, showing alterations in production.[132]

The earliest reports of adverse reproductive effects from pollutants began in Europe prior to the 1920s with Hamilton and her counterparts.[133] They described women exposed to very high toxic levels of benzene. These women were unable to bear normal, live children, or they bore children who failed to thrive. Secondary amenorrhea is likely to be a result of toxic substances.[134] Effects of toxic substances on the ovary are germ-cell destruction and receptor modifications. Early onset of menopause has been directly related to cigarette smoking,[135] and it may well be that other toxic chemical exposures may result in the same process. Premature ovarian failure has been experimentally induced using polycyclic hydrocarbons and alkylating agents. A higher risk of cancer of the ovary in cosmetologists and hairdressers may evidence toxic insults;[136] we have seen a high incidence of ovarian difficulties in hairdressers. Data on early-age use of oral contraceptives also often reveals menstrual irregularities[137] as well as breast tumors[138] and venous inflammation.[139]

Impairments induced in the ovaries by sublethal doses of carbaryl and endosulfan have been studied.[2] Exposures produced a reduction in the number of oocytes, a reduction in size and deformity in different stages of oocytes, damage to yolk vesicles in maturing and mature oocytes, an increase in the number of atretic oocytes, development of interfollicular spaces, an increase in the connective tissue of tunica albuginea, and dilation of blood vessels. Effects were dependent on the dose, duration of exposure, and type of pesticide. Doses of endosulfan were found to be more toxic than carbaryl. Apes fed pesticide (arochlor-1254) in their diet developed abnormal estrogen and progesterone levels followed by menstrual irregularities. Dioxin alters the action of estrogen in reproductive organs in a manner that is both age dependent and target organ specific.[140,141] Our group of PMS patients with severe menstrual disturbances and blood levels of toxic organic chemicals failed to show big changes in estrogen and progesterone levels. However, these patients often responded to mini-dose hormonal therapy. Detectable levels of organochlorine pesticides were always present in their blood. Patients did much better as their total load of pesticides and toxic organic hydrocarbons reduced.

During sexual differentiation there are a number of critical periods when the reproductive system is uniquely susceptible to chemically induced changes. At these times, an inappropriate chemical signal can result in irreversible lesions that often result in infertility, whereas similarly exposed young adults are transiently affected. The serious reproductive abnormalities that resulted from human fetal exposure to DES, synthetic hormones, and other drugs provide grim examples of the types of lesions that can be produced by interfering with this process. Furthermore, it is of concern that many of the abnormalities are not expressed during fetal and neonatal life and only became apparent after presenting alternatives in sex differentiation.[142]

In humans, in utero exposure to a hormonally active chemical such as DES, androgen, or progestin results in morphological and pathological alterations or reproductive functions. For example, DES causes cancer, infertility, and severe abnormalities of the cervix, uterus, and Fallopian tubes as well as alterations of reproductive and sex-linked nonreproductive behaviors.[143]

Meyer-Bahlburg et al.[144] and Gray[145] reported that women exposed to DES in utero had less well-established sex-partner relationships, lower sexual desire and enjoyment, decreased sexual excitability, and diminished cortical function. Hines and Shipley[146] found that women exposed to DES in utero showed a more masculine pattern of cerebral lateralization on verbal tests than their sisters. Such sexual differences in specialization of the two hemispheres of the brain for different types of cognitive processing are well documented in humans with men tending toward greater left hemisphere specialization for verbal stimuli than women.[147] Some chemically sensitive patients seen at the EHC-Dallas exhibit characteristics similar to those described by Meyer-Bahlburg et al.[144] and Gray.[145]

Cognitive ability may be different in DES- or other hormonally treated females whose psychopathology includes a tendency toward depression, and anxiety tends to occur.[148,149] An increased incidence occurs in DES-exposed women of immunologic hyperactivity, rheumatic fever, recurrent strep throat, and autoimmune disease.[150,151]

Swaab and Fliers[152] found that the proptic area of the hypothalmus is 2.5 times larger in men than women and contains 2.2 times as many cells. A recent report suggested that homosexual men have female-like INAH-3 brain structures, implying a biological basis for homosexuality.[153]

The presence of excess estrogens in the environment is a cause for concern. In animals, excess estrogen with prenatal or neonatal exposures has dramatic and permanent influences on brain structure and behavior.[154]

Soto et al.[155] tested xenobiotic compounds that were reported to have estrogenic activity[156-158] or were suspected to be estrogens because of their molecular structure. Among them, the mycotoxins zearalenol and zearnlenone have been used as anabolic food additives for cattle and sheep. The phytoestrogen coumestrol and the pesticides DDT (p,p' and o,p'), chlordecone (Kepone®), and 1-hydroxychlordane were all found to be estrogenic (Tables 24.4 and 24.5). Alkyl phenols such as p-nonylphenol (present in modified polystyrene) and pentyl phenols induced estrogen responses.[159] 6-Bromo-2-naphthal and allenoic acid also have estrogen activity.[159]

Endocrine and reproductive changes have been associated with all pesticide types, insecticides (carbaryl, DDT, methoxychlor, aldrin, chlordane, dieldrin, and Kepone®), herbicides (2,4-DNP, 2,4S-T), and fungicides (thiocarbamates such as zineb and maneb).[161] Several authors have claimed that once PCBs have passed through the liver, they become estrogenic.[162,163]

Besides the synthetics with estrogenic activating properties, a number of natural products have steroid-like action. These include some phenolic plant products,[164,165] fusarium-producing resorcylic acid lactones (in stored grains), and isoflavanoids from legumes. These chemicals can alter the sexual differentiation of gonadotropin secretion[164] and the sexual behavior of rodents. The flavonoids are inhibitors of steroid biosynthetics.[166] Of course, the flavonoids are distributed in vegetable products and exhibit many enzymatic activities similar to those of steroids.

Western dietary changes of this century have resulted in higher estrogens. High fat diets result in high intestinal absorption of estrogens, while high fiber diets result in low estrogen levels.[167] Increasing meat intake and reducing vegetable consumption increases estrogens that are more available to steroid-responsive tissues, while concentration of less potent and perhaps estrogen inhibitory isoflavanoid declines. These dietary shifts are associated with increases in the incidence of estrogen-dependent western disease.[168] Estrogen advances menarche in rodents.[169-174] It appears to do the same in humans.

Frem-Titulaer[175] reported that Sáenz de Rodriguez and Pérez-Comas have identified over 1000 children with adrenogenital syndrome from Puerto Rico. This condition appeared to result from ingestion of chicken containing residual toxic chemicals and estrogen compounds that had originally been used to treat the chicken feed. Dramatic suppression with equal potency to estradiol was found.

Pérez-Comas[176] has shown abnormal adrenal and ovarian development in Puerto Rican babies. He has evaluated 1053 patients with abnormal sexual development over a period of 14 years. These patients were from Puerto Rico, the mainland United States, Latin America, and Europe. The most frequent initial diagnoses were premature thelarche,[176] gynecomastia,[177] precocious puberty,[176] and premature pubarche.[176] Other conditions associated with increased estrogen levels were asymmetry of breasts, virginal hyperplasia of breasts, and hirsutism, and five patients were diagnosed with pseudoprecocious puberty associated with hypothyroidism and Down Syndrome.

Table 24.4. Estrogenic Response of MCF7 Cells to Insecticides,
            Phytoestrogens, and Phytohormones

Compound	Concentration[a]	PE[c]	RPE(%)[d]	RPP(%)[e]
Estradiol	30 pM	6.7	100	100
Kepone	30 μM	5.6	81	0.0001
Mirex	30 μM	0.6	—	—
DDT,p,p'	30 μM	5.0	70	0.0001
DDD,o,p'	30 μM	5.8	84	0.0001
1-hydroxychlordene	30 μM[b]	3.1	37	0.0001
Chlordene	30 μM[b]	1.2	4	—
Heptachlor	30 μM[b]	1.5	8	—
Arochlor 1221	30 μM[b]	1.4	7	—
Giberellic acid	30 μM[b]	1.2	4	—
Chlordane	30 μM[b]	1.3	5	—
Zearalenone	3 nM	6.0	88	1
Zearalenol	3 nM	6.3	93	1
Coumestrol	3 μM	6.3	93	0.001

[a] Indicates the lowest concentration needed for maximal cell yield.

[b] Indicates the highest concentration tested in culture.

[c] Proliferative effect is expressed as the ratio between the highest cell yield obtained with the test chemical and the hormone-free control.

[d] Relative proliferative effect, which is calculated as $100 \times$ (PE-1) of the test compound/(PE-1) of E_2. A value of 100 indicates that the compound tested is a full agonist; a value of 0 indicates that the compound lacks estrogenicity at the doses tested, and intermediate values suggest that the xenobiotic is a partial agonist.

[e] Relative proliferative potency is the ratio between the dose of E_2 and that of the xenobiotic needed to produce maximal cell yields $\times$ 100.

Source: Soto, A. M., T-M. Lin, H. Justicia, R. M. Silvia, and C. Sonnenschein. 1992. An "in culture" bioassay to assess the estrogenicity of xenobiotics (E-screen). In *Advances in Modern Environmental Toxicology. Vol. XXI. Chemically-Induced Alterations in Sexual and Functional Development: The Wildlife/Human Connection*, eds. T. Colborn and C. Clement. 302. Princeton, NJ: Princeton Scientific. With permission.

Females were affected by contaminated foods more frequently than males. Total serum estrogen levels were generally increased in 85% of males and 86% of females studied. Prolactin levels were abnormal in 28% of females and 16% of males. FSH levels were increased in 32% of males and 40% of females. LH levels were high in 33% of males and 18% of females.

The years of highest incidence of the adrenogenital syndrome were 1982, 1983, and 1984. Cases have diminished dramatically, with a 50% drop in 1985 when a federal investigation of hormones in meat was being carried out. From

Table 24.5. Response of MCF7 Cells to Natural and Synthetic Estrogens

Compound	Concentration[a]	PE[b]	RPE(%)[c]	RPP[d]
Estradiol	100 pM	4.7	100	100
Diethylstilbestrol	10 pM	5.1	112	1000
1,1-β-Chloromethyl-estradiol	10 pM	5.1	110	1000
Dehydrodoisynolic acid	10 nM	4.8	103	1
Dichloro-doisynolic acid	100 nM	4.6	97	0.1
Allenolic acid	10 nM	4.9	105	1
Hydroxyphenyl-cychlhexanoic acid	10 nM	4.9	105	1

[a] Indicates the lowest concentration needed for maximal cell yield.
[b] Proliferative effect is expressed as the ratio between the highest cell yield obtained with the test chemical and the hormone-free control.
[c] Relative proliferative effect, which is calculated as $100 \times (PE-1)$ of the test compound/(PE-1) of E_2. A value of 100 indicates that the compound tested is a full agonist; a value of 0 indicates that the compound lacks estrogenicity at the doses tested, and intermediate values suggest that the xenobiotic is a partial agonist.
[d] Relative proliferative potency is the ratio between the dose of E_2 and that of the xenobiotic needed to produce maximal cell yields $\times$ 100.

Source: Soto, A. M., T-M. Lin, H. Justicia, R. M. Silvia, and C. Sonnenschein. 1992. An "in culture" bioassay to assess the estrogenicity of xenobiotics (E-screen). In *Advances in Modern Environmental Toxicology. Vol. XXI. Chemically-Induced Alterations in Sexual and Functional Development: The Wildlife/Human Connection*, eds. T. Colborn and C. Clement. 303. Princeton, NJ: Princeton Scientific. With permission.

May 1 to October 31, 1986, 53 new cases were evaluated. Starting in 1986, 45 of them presented with symptomatology of excess hormone.

Overall review of remission data of these patients revealed that a limited diet produced remission in 58% of males and 51% of females. Without diet, remission was observed in 6% of males and 11% of females. The diet was free of poultry, eggs, and meat until estrogen levels became normal. Afterwards, grain-fed poultry and birds were permitted.

Pelvic sonographic abnormalities were observed in 62% of the females affected by hormone-contaminated foods. Ovarian cysts were directly related to increased estrogen levels in 88% of patients with sonographic abnormalities.

Hormonal and clinical remission of Pérez-Comas's patients usually took 3 to 6 months on diet-treated patients. Improvement of ovarian and uterine abnormalities that can be seen by sonogram takes from 6 to 12 months. However, recurrence of this syndrome is frequently due to nonobservance of a diet free of estrogens or hormone-like toxic chemicals. No adequate governmental action has been taken up to the present time.

One case that exemplifies pollutant-triggered, severe ovarian dysfunction is presented.

Case study. This 35-year-old P-4, G-4 white female entered the ECU with the chief complaints of nausea, vomiting, and weight loss for the previous 6 months. Her environmental and past history were significant in that she had grown up in farm country and married a farmer and had lived on a farm all her life. After the birth of each of her children, she noticed a loss of strength and an increase in her premenstrual syndrome. These effects were so severe that she would become nonfunctional several days before her period. She also noticed a lot of irregular and excessive uterine bleeding. She then developed severe nausea, vomiting, and weight loss that first occurred premenstrually but then became constant during the 6 months prior to her admission to the ECU. She became totally food intolerant and lost 30 pounds due to an inability to hold down food. When she was admitted to the ECU, she was emaciated and feeble. Other organ systems were normal, except for areas of excess skin where the weight had been.

Significant laboratory tests revealed chlordane, 5 ppm; heptachlor, 0.3 ppm; DDT, 0.8 ppm; dieldrin, 0.6 ppm; lindane, 3 ppm; hexachlorobenzene, 2 ppm; T lymphocytes, 500/mm^3 (C = 1400 to 2200/mm^3); WBC, 3000/mm^3; B lymphocytes, 4%; and total hemolytic complements, 50% of control.

This patient was placed on intravenous fluid and given nothing by mouth for 1 week. Nausea and vomiting gradually subsided, and she was able to eat an occasional chemically less-contaminated food without vomiting. Treatment with food neutralization injections allowed her to expand her food gradually repertoire, until she could eat a wholesome diet. She was given 15 gms of intravenous, preservative-free vitamin C daily along with multiminerals. Her strength gradually returned, until she could be discharged to her home, where a safe oasis had been built. Over the last 4 years, she found that she had to leave her area of farm county when spraying season started, due to an exacerbation of her symptoms following exposure to the chemicals in the various sprays used. In the fifth year post diagnosis, she was inadvertently exposed to pesticide and developed uterine bleeding that required a hysterectomy. Since then, she has done well.

A broad spectrum of ovarian dysfunction, from failure to conceive to spontaneous abortion to endometriosis to premenstrual syndrome, has been observed in the chemically sensitive female patient. Each condition is discussed in more detail on the following pages.

Premenstrual syndrome is seen frequently in chemically sensitive patients in whom oral contraceptives act as pollutants. Also, PMS is seen in chemically sensitive patients who have been exposed to other external toxic chemicals. Other side effects of oral contraceptives in the chemically sensitive woman are depression, headaches, and anxiety.

Grant[178] has shown that when the hormone balance of a pill is varied and the progesterone strength increased, the side effects (ranging from irregular bleeding to distended veins to irritability to weight gain to arteriole changes to depression to loss of libido) change.

Administration of high progestin and low estrogen causes regular, scanty withdrawal bleeding. Low estrogen and low progesterone cause irregular bleeding, and low progesterone and high estrogen cause vein changes.

Antithrombin III, which prevents clotting, is decreased by estrogen and progesterone administration. It also appears that, after administration, sticky platelets occur.[179]

Vessey[180] demonstrated that women on the pill had a risk of venous thromboembolisms nine times that of women not on the pill. For thrombolic stroke, their increased risk was six times. Increased risk for a variety of other conditions included hemorrhagic stroke, two times; subarachnoid hemorrhage, six times; and myocardial infarction, four times. English and Welsh studies also reported an increased risk of cancer in young women using the birth control pill (Figures 24.5 and 24.6).[181] Grant[182] demonstrated that there is an increased occurrence of gallbladder disease, diabetes, and heart disease in women who take the pill. In this population also there appears to be a change in metabolism resulting in more infections, more food and chemical sensitivity, and more cancer (Table 24.6). Grant's research[183] further discovered that the most common foods triggering PMS headaches were wheat, oranges, eggs, tea, coffee, chocolate, milk, beans, corn, cane sugar, mushrooms, and peas. These findings are similar to our study results at the EHC-Dallas.

Cigarette smoking leads to the distribution of nicotine and its major metabolites, nicotine-1'-N-oxide and cotinine, in the urine, amniotic fluid, breast milk, cord blood, and maternal venous blood of pregnant smokers and their fetuses and neonates. These nicotine products may produce adverse effects on the health of both mother and baby. Additionally, between the 29th and 34th weeks of gestation, a significant increase ($p<0.05$) in the urinary metabolite ratio, cotinine/nicotine-1'-N-oxide, occurs as compared with controls. Studies on female smokers taking the birth contol pill also have shown a significant increase in the cotinine/nicotine-1'-N-oxide ratio ($p<0.02$), as compared with controls. High ratio values (>3) were shown to be due to cotinine excretion. Because this ratio may reflect the relative importance of the α-C and aliphatic N-oxidative pathways, it could have toxicological importance relating to xenobiotics for the mother and fetus.[184,185]

Fetal liver has been shown to be capable of metabolizing (*in vitro*) various tobacco alkaloids via a variety of C- and N-oxidative reactions, which may be of toxicological significance linked not only to tobacco alkaloids but also to other xenobiotics.[185,186]

Sexual Dysfunction

Sexual dysfunction is often seen in pollutant overloaded females. These patients often have little or no sex drive. They fail to be stimulated when

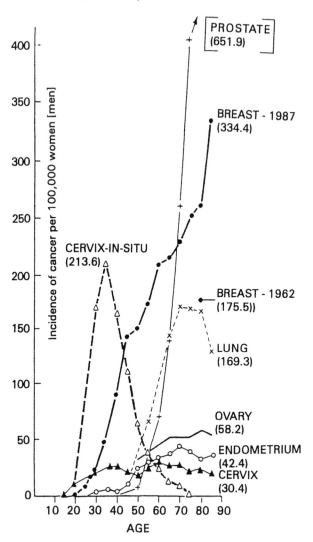

Cancer registrations per 100,000 women (men)

Figure 24.5. Percent increase in cancer registrations per 100,000 men and women between 1965 and 1987. Bracketed figures are the highest rate. (From Grant, E. 1986. *The Bitter Pill: How Safe Is the Perfect Contraceptive*? 119. London: Corgi Books. With permission.)

manual manipulation of the clitoris is attempted. They fail to lubricate or achieve orgasm. In some pollutant-overloaded females, just the opposite, hypersexualism, occurs. Since most sexual function takes place through a series of complex interactions between the neurological and hormonal systems,

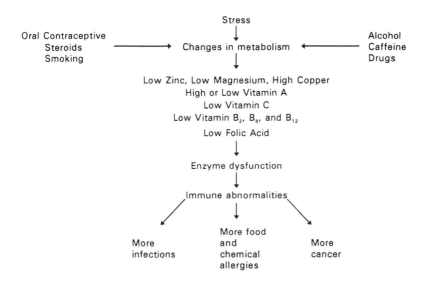

Figure 24.6. Grant's concept of the origin of contraceptive-induced illness. (From Grant, E. 1986. *The Bitter Pill: How Safe Is the Perfect Contraceptive?* 119. London: Corgi Books. With permission.)

Table 24.6. Deaths per 100,000 Women of Those in the Pill-Taking Age Group

	England and Wales (1975)	Oxford/FPA OC users at entry (1968–1979)	R.C.G.P. ever-users (1968–1979)	Walnut Creek all women % (1968–1977)
All deaths	108.5	52.9	87.2	100
Neoplasms	44.3	25.3	30	45
Circulatory	21.9	12.3	29.9	15
Accidents + violence	17.1	4.6	18.2	19
Follow-up			42.0%	86.2%

Source: Grant, E. 1985. *The Bitter Pill.* 110. London: Corgi. With permission.

it is at times difficult to pinpoint one particular pollutant stressor in these patients. However, we have treated enough patients with dysfunction due to specific environmental pollutants to realize that removal of these incitants and a decrease in the patient's total load help to maintain her sexual homeostasis. Many chemically sensitive women report that their mates emanate an odor that probably was acquired at work due to contact with such substances as gasoline, pesticides, and phenols. These odors immediately impede their desire for sex. Once these have been removed by showering or a change in clothes, however, their desire often returns. Before they have sex, some patients require their husbands to be away from work for 2 days to outgas. Other chemically sensitive patients have been observed to have the urge but are so dry vaginally that they need lubrication. Usually saliva works best for lubrication, since artificial and natural oils and jellies can cause reactions. Other women are hyperstimulated by odors. They have an increased sex drive and at times will be somewhat indiscriminate with whom they have sex. The parasympathetic fibers are involved in many of these processes, and stimulation or malfunction has often been observed and measured in the chemically sensitive patient. Although infertility is a problem in some chemically sensitive patients exposed to toxic pesticides, the majority of chemically sensitive patients are able to conceive.

UTERUS AND TUBES

Pollutant injury to the uterus and tubes is common. Premenstrual syndrome, which is partially ovarian and partially uterine, is discussed in this section. It could have just as well been discussed under ovary. The tubes can become inflamed and cause pain. The next section will discuss the environmental aspects of PMS and endometriosis.

Premenstrual Syndrome

Premenstrual syndrome is a complex disorder with a wide range of symptoms that occur regularly before menstruation. Most women find that PMS typically has three symptom patterns. First, symptoms may start 1 to 10 days prior to menstruation and continue until the menstrual period begins; these symptoms may begin as a result of sensitivity to progesterone. Second, symptoms may start at ovulation and disappear in a day or two, reappearing later in the premenstrual phase; this symptom pattern may be due to sensitivity to LH. Finally, PMS symptoms may begin at ovulation, continue steadily into the menstrual period, and disappear only toward the end of menses. This latter type may be due to LH, estrogen, and/or progesterone deregulation. Because symptoms of PMS are so varied and extreme, it can be difficult to identify and treat. However, there are special characteristics that distinguish PMS symptoms from those of other disorders.

The key characteristics in identifying PMS include the following:

1. Time of onset or increased severity of PMS — PMS tends to begin at puberty, after pregnancy, after discontinuing the birth control pill, after amenorrhea, or after tubal ligation.

Table 24.7.   Symptoms of PMS

Feeling "bloated"	Irritability
Feelings of weight increase	Aggression
Breast pain/tenderness	Tension
Skin disorders	Anxiety
Hot flashes	Depression
Headache	Lethargy
Pelvic pain	Insomnia
Change in bowel habits	Change in appetite
Thirst	Crying
Abdominal swelling	Change in sexual desire
Rhinitis	Loss of concentration
Fatigue	Mental confusion
Swelling of hands and feet	Memory loss
	Poor coordination/clumsiness/accidents

2. Painless menstruation — pain during menstruation is NOT related to PMS; however it is, still related to hormone dysfunction.
3. Weight fluctuation (Table 24.7) — temporary monthly gain of 3 to 9 pounds is common.
4. Food cravings — eating binges, especially involving sweet or salty foods; after not eating for 4 to 6 hours, there may be an onset of acute symptoms, panic attacks, or migraine headaches.
5. Lowered tolerance for alcohol — decreased tolerance of alcohol just before onset of the period may be accompanied by alcohol craving.
6. Inability to tolerate birth control pills — women with PMS may experience exacerbation of PMS symptoms when they take oral contraceptives.
7. History of threatened abortion — bleeding in early months of pregnancy followed by successful delivery is common in PMS patients.
8. History of hypertension or toxemia during pregnancy — women who experience a pregnancy complicated by hypertension, toxemia, severe headaches, disturbed vision, or rapidly developing fluid retention, often develop PMS.
9. Increased sex drive — women with PMS often experience an increase in sex drive during the premenstrual phase.

Premenstrual syndrome is not dysmenorrhea, which is pelvic pain during menses. This condition is usually treated with oral contraceptives and prostaglandin inhibitors, which keep the uterus from cramping. PMS is also not endometriosis, which is a condition in which some cells from the lining of the inside of the uterus, instead of passing out of the body during menstrual flow, actually work their way up through the Fallopian tubes into the abdominal cavity where they cause pelvic pain. Finally, PMS is not pelvic inflammatory disease, which is infection of the uterus, fallopian tubes, and ovaries with accompanying pain in the pelvis and lower abdomen, nausea, high fever, and/or rapid pulse.

Premenstrual syndrome is a multifaceted disorder that involves a dysfunction of the patient's immune, enzyme detoxification, and hormonal systems, as well as possibly its nutrient supply. Previous theories have related the problem to a deficiency in progesterone levels or an imbalance in estrogen and progesterone levels, with treatment aimed toward restoring that balance. Laboratory analysis of PMS patients does not always support these theories.

Over 50% of the female patients treated for food and chemical sensitivity in the ECU have PMS symptoms. Laboratory tests for the majority of these patients show normal hormone levels, though this does not mean that they are well balanced at any given time in the cycle. As a result of these findings, we believe that many PMS patients are simply overly sensitive to their own hormones. During the time of their cycle when certain hormone levels are normally high, patients react with physical and emotional symptoms. The reactions are not highly specific to each hormone. The same symptom caused by progesterone in one patient can be related to estrogen or LH in another.

Many other internal and external factors, including nutritional deficiencies, contribute to the disorder in each individual. Therefore, the total load of stress on the body must be considered for effective treatment.

As she interacts with her environment, the chemically sensitive patient with PMS increases her total load as many stressors are continually added. These stressors include such things as outdoor and indoor air pollution, water and food contamination, and emotional and physical stress. Any stress to the immune and enzyme detoxification systems including infection, pregnancy, prior hormone administration, or toxic chemical exposure can upset immune control mechanisms, thereby allowing the inappropriate development of antibodies against any organ in the body.

The EHC-Dallas has treated over 4000 women with some form of PMS. In one consecutive series of 30 patients, the mean age was 34 years with a range of 30 to 39 years in 46% of the patients (Table 24.8); 50% had onset at age 29 years. However, a smaller portion started at or shortly after the onset of menstruation. Usually the mother had a similar history. Another interesting facet of these chemically sensitive patients with PMS was that 90% of them had neurological/psychological symptoms during their premenstrual period, while 66% had chlorinated pesticides in their blood, which may well account for their severe ovarian dysfunction. Seventy percent of these patients were sensitive to trees and 80% to grass. They noted that their PMS was worse during the spring and fall months. Thirty-three percent of the patients were sensitive to terpenes and 86% to toxic chemical exposure such as phenols, formaldehyde, petroleum alcohols, pesticides, and chlorine. It was noteworthy that, when exposed to a series of toxic chemicals by inhalation, the chemically sensitive PMS patient was usually sensitive to just one, rather than a multiple

Table 24.8.   Thirty Chemically Sensitive Patients
with PMS Studied in the ECU after 4 Days
Deadaptation with Reduction of Total Body Load

Challenge studies	Percent positive	Challenge
Trees	70	ID
Grasses	80	ID
Molds	100	ID
Terpenes	33	ID
Foods	100	ID and oral confirmation
Cane sugar	100	
Corn	76	
Wheat	63	
Potato	63	
Tomato	63	
Toxic Chemicals	86	ID
	49	IHDB[a]

Note: Ages: 30–39 years; mean: 34 years. Average age of onset: 29 years. Blood chlorinated pesticides: 100%.

[a] Inhaled double-blind challenge — phenol (<0.005 ppm); chlorine (<0.33 ppm); formaldehyde (<0.2 ppm); pesticide (<0.0034 ppm); saline placebos.

Source: EHC-Dallas. 1986.

series, as seen in the more advanced chemically sensitive patients. One hundred percent of the chemically sensitive PMS patients were sensitive to one or more foods. The main offenders that triggered or exacerbated their syndrome were cane sugar — 100%; corn — 76%; wheat, potatoes, tomatoes, brewer's yeast — 63%; eggs and bananas — 60%; and chicken — 56%. Ninety-five percent of the chemically sensitive PMS patients do well on hormone neutralization, rotary diet with elimination of the worst offenders, and chemical avoidance (see Volume IV, Chapter 36).

The uterus and tubes are particularly susceptible to pollutant injury due to its internal make up of blood vessels and smooth muscle and its direct contact to the external environment. The uterus has a rich supply of immune defense mechanisms, which helps fend off external exposure (Figure 24.7). However, noxious substances can penetrate it and cause severe problems as was shown in the case of the woman with pesticide overexposure.

Vasculitis

Human studies involving three cases of cervical vascular malformation as a cause of antepartum and intrapartum bleeding in DES exposure have been reported.[187] Other abnormal uterine bleeding may be triggered by other

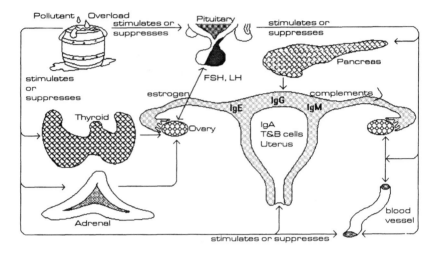

Figure 24.7. Immune and endocrine defense of the uterus against environmental pollutants.

environmental incitants. We have seen numerous cases of uterine pain and bleeding secondary to blood vessel dysfunction from toxic exposure. The uterus may be the sole, or one part of the large, target organ. The following case illustrates the environmentally sensitive patient's problems with uterine bleeding.

> **Case study.** A 36-year-old white female had a severe vasculitis syndrome characterized by spontaneous bruising, peripheral and periorbital edema, and cerebral dysfunction marked by an inability to remember. Also, she had slow mentation. Testing revealed she had multiple food and chemical sensitivities. Biopsy of her spontaneous bruises revealed perivascular lymphocytic infiltrates. This patient developed spontaneous uterine bleeding, which could only be controlled by a hysterectomy. Lesions in the uterus and tubes revealed perivascular lymphocytic infiltrates similar to those seen on the skin bruises. This patient has done well over the past 5 years after moving to a less-polluted area of the United States.

Endometriosis

Endometriosis is characterized by the slow growth of tissue from endometrium in an inappropriate site such as the peritoneum. It is one of the most commonly diagnosed causes of dysmenorrhea and infertility. A fairly common gynecologic disorder, it appears to be on the increase and may be due to increased pollutant exposure. While there were only 20 cases reported before 1921,[188] this disease today affects 5 million women per year in the United States and is more common in industrialized nations and around industrialized centers.[189] In 1992, German researchers reported that women with endometriosis had significantly high levels of polychlorinated biphenyls in their blood.[188] A

Canadian study reported to the Ontario Association of Pathologists in 1985 also showed a high rate of PCBs in monkeys.[188]

Rier et al.[190] showed that the incidence of the reproductive disease endometriosis was determined in a colony of rhesus monkeys chronically exposed O-2,3,7,8-tetrachlorodibenzo-p-dioxin (TCDD or dioxin) for a period of 4 years. Ten years after terminatin of dioxin treatment, the presence of endometriosis was documented by surgical laparoscopy, and the severity of disease was assessed. The incidence of endometriosis was directly correlated with dioxin exposure, and the severity of disease was dependent upon the dose administered (p<0.001). Three of 7 animals exposed to 5 ppt dioxin (43%) and 5 of 7 animals exposed to 25 ppt dioxin (71%) had moderate to severe endometriosis. In contrast, the frequency of disease in the control group was 33%, similar to an overall prevalence of 30% in 304 rhesus monkeys housed at The Harlow Primate Center with no dioxin exposure. This 15-year study indicates that latent female reproductive abnormalities may be associated with dioxin exposure in the rhesus. Therefore, the effects of this toxic chemical may be more diverse than previously reported.

Previous work has described an association of endometriosis in rhesus monkeys following exposure to polychlorobiphenyl (PCB) compounds.[191]. Rier et al.'s[190] results support and extend these findings, since dioxin is used as a reference compound for halogenated aromatic hydrocarbons, including PCBs.[192] Extensive literature has documented the incidence of endometriosis in the rhesus model following exposure to single-energy proton irradiation, mixed-energy proton irradiation, and X-rays.[193-195] In Rier et al's study,[190] endometriosis in dioxin-exposed monkeys was first documented 7 years following the termination of dioxin treatment. Immune system defects are a common probable factor that may contribute to the development of endometriosis in each of these animal models. Indeed, this notion is consistent with human studies, suggesting that immune mechanisms may contribute to the disease process.[196,197] Dioxin has immunosuppressive activities and is a potent inhibitor of T lymphocyte functions.[198-200] In addition, this toxic chemical modulates steroid receptor expression resulting in altered tissue-specific responses to hormones.[141] Chronic immunosuppression in combination with hormonal deregulation may have facilitated the aberrant growth of endometrial tissue within the peritoneum of dioxin-treated animals.[200] There are many theories to explain the pathogenesis of endometriosis, but one of the most important is shown in the schematic representations of Figure 24.8.

There are three major types of endometriotic implants distinguished by their morphological and laparoscopic characteristics: the microscopic or epithelial type plaque, the vesicular and papular type, and the fibrotic nodular type. The microscopic epithelial plaque type of implant is invisible by laparoscopy. At the site of implant, the pelvic peritoneum is replaced by endometriotic epithelium studded with endometriotic stroma. The lesion may be considered as intraperitoneal endometriosis as it is in contact with the peritoneal cavity and secretory products pass directly in the peritoneal cavity.

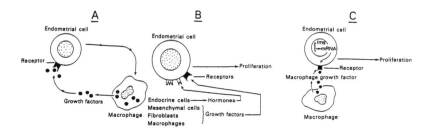

Figure 24.8. Development of endometriosis.

Scanning electron microscopy has shown this type of lesion in 26% of biopsies taken from areas of apparently normal peritoneum in patients with endometriosis.

The vesicular and papular types of implants present at laparoscopy, either as a hemorrhagic or highly vascularized polyp or as a nonpigmented papule. Polyps are composed of an outer layer of endometrial epithelium surrounding a central core of highly vascularized stroma. The contents of polyps may be released into the peritoneal cavity upon rupture. Some papules are lined by active endometrial cells with marked apocrine secretion. These cells cause the cyst to dilate, and they enlarge the mesothelium and submesothelial connective tissues, which then undergo pressure necrosis leading to rupture.

The fibrotic pigmented and nodular type of implant is also known as the "powder burn" lesion and is easily identified as endometriosis. The endometriotic tissue in these implants is usually inactive and embedded in highly pigmented connective tissue.

Comparative studies have demonstrated that endometrium and endometriotic implants differ in their responses to hormonal fluctuations during the menstrual cycle. Concentrations of estrogen and progesterone cytosol receptors are lower in endometriosis than in eutopic endometrium, and no variations occur during the menstrual cycle or hormone therapy. The response to the hormones appears to be related to the site and type of endometriotic lesion. Ovarian endometriosis is less responsive than peritoneal implants, particularly the vesicular type, and is more responsive than fibrotic, nodular implants.

In order to prevent continuous dysmenorrhea in a woman of reproductive age, early diagnoses and treatment of the progression of disease leading to infertility and advanced pathology should be done.

A diagnosis may be entertained when there is dysmenorrhea. Most women expect to experience some pain as a part of their normal monthly period, but women with endometriosis regularly experience 5 days of dysmenorrhea each month, a total of 5 years of pain during their reproductive lifetime. Such women often suffer from dyspareunia and chronic abdominal pain in addition to dysmenorrhea. A large proportion, approximately 65% of patients with endometriosis, experience painful symptoms, whereas 35% do not. In over 60% of endometriosis patients, classic bimanual palpation is negative.

Steroid hormones play a central role in the hormonal management of endometriosis. These patients continue to have debilitating symptoms even after surgical removal of the uterus, adnexa, ovaries, and patches of endometriosis in adjacent areas. These symptoms are wide-ranging, from chronic lower abdominal pain, backache, and urinary problems to dizziness, hot and cold sensations, and an inability to tolerate odors. This group of patients respond to a comprehensive treatment approach consisting of environmental control; avoidance of chemicals; rotation of chemically less-contaminated food; and appropriate injection therapy for environmental, food, and chemical incitants; as well as hormone neutralization treatment (Table 24.9).

Spontaneous Abortion

Studies of various industries have suggested links between occupational exposure to chemicals and reproductive failure. For example, among dentists and dental assistants, there is a significant relationship between total mercury levels in the hair of exposed women and histories of reproductive failure and menstrual cycle disorders.[201] In a review article, Landrigan et al.[202] cited human and animal studies showing the reproductive toxicity effects of ethylene oxide, a biocide used in sterilization of hospital equipment. Nurses handling antineoplastic drugs during the first trimester of pregnancy have higher fetal loss.[203] In the Danish county of Funen, Heidam[204] showed odds ratios for spontaneous abortions were significantly increased among factory workers and painters exposed to such chemicals as nitrous oxide, inorganic mercury, organic solvents, and pesticides. Among women exposed to organic solvents during laboratory work, there was an increased (though not significant) tendency toward miscarriage.[205] In a Swedish study of women working in a rubber plant, unfavorable pregnancy outcomes (threatened abortion, spontaneous abortion, or malformation) were associated with exposure to the tire building process.[206] A study in the plastics industry showed some conflicting results; however, the odds ratio for spontaneous abortions in workers in polyurethane processing factories was increased.[207]

Although there are inherent problems in epidemiological studies such as these cited, laboratory animal research tends to support their conclusions. Solvents such as benzene, which is frequently identified in chemically sensitive individuals, and its derivatives cause spontaneous abortion in rabbits. Some solvents cause a dose-dependent increase in postimplantation loss in rats and mice.[208] One study found pregnant mice exposed to methylmercuric chloride had an increase in fetal toxicity. This study also showed chromosome stickiness and clumping in fetal tissues, leading to reduced mitotic divisions.[209]

Data from industrial accidents (e.g., the leakage of methyl isocyanate in Bhopal, India, in 1984) also show fetal toxicity. Varma[210] reported that a survey of 3270 families 9 months after the accident indicated that 43% of pregnancies did not result in live births. Studies in mice corroborated these findings.[207]

Table 24.9.   Data of 17 Patients with Endometriosis

Associated disease		Skin testing (positive)				High level of toxic chemical in blood	
Disease	No.	Antigen	No.	Antigen	No.	Compound	No.
Fatigue	3	Foods (range: 2 to 43 kinds; mean: 17 kinds)	16	Ethanol	4	Pentane	1
Depression	1	Molds	14	Formaldehye	7	2-Methylpentane	6
Toxic brain syndrome	2	Dust	14	Women's cologne	5	3-Methylpentane	6
Ovarian cyst	1	Mite	14	Men's cologne	4	Cyclopentane	2
Infertility	1	Danders	12	Orris root	8	n-Hexane	4
Cephalgia	3	Candida	11	Newsprint	3	β-BHC	2
Chest wall syndrome	1	T.O.E.	11	Phenol	1	DDE	3
Collagen vascular disease	1	Terpenes	11	Diesel	1	Heptachlor epoxide	1
Autoimmune thyroid	1	Smuts	10	Fluogen	11	Trans-nonachlor	2
Irritable bowel syndrome	2	Tree pollen	14	CMV	1	Trimethylbenzenes	2
Arthralgia	2	Grass pollen	14	EBV	5	Toluene	2
Myalgia	2	Weed pollen	12	Bacteria	1	Ethylbenzene	1
PMS	2	Insecticides	2	Rayon	1	Styrene	1
Engioedema	1	Estrone	10	Polyester	1	Chloroform	1
Allergic rhinosinusitis	7	Leutinizing hormone	6	Wool	2	1,1,1-Trichloroethane	1
Hypothyroidism	1	Progesterone	10	Cotton	7	Trichloroethylene	1
GI upset	1	Natural gas	3	Silk	2	Tetrachloroethylene	1
EMF sensitivity	1	Cigarette smoke	8	MRV	9	HCB	2
Hypoglycemia	1	Conjugated estrogen	1	Algae	4	Oxychlordane	1
Connective tissue disease	1			Perfume	1		
				Linen RV	1		

Inhaled testing		Immune parameters				Immune parameters	
Compound	Positive No.	Parameter	Positive No.	Parameter	No.	Parameter	No.
Chlorine	1	CMI — 3 positive reactions	4	Low total lymphocyte	1	High antimicrosomal AB	2
Phenol	1	CMI — 5 positive reactions	1	Low T_4	1	High antinuclear AB	3
				Low T_8	3		
				High T_{11}	1		
Pesticide	1			High T_{11}	1	High antithyroglobulin	1
Formaldehyde	1			High T_4/T_8	2	High antireticulin AB	1
				Low CH_{50}	1	High thyroglobuline AB	1
				Low C_4	1	Positive hepatitis B surface Ag	1
				IgG — low	1		
				IgG — high	1		
				IgE — high	1		
				High FBV-early antigen IFA	2		
				High EBV-VCA-IgG	4		
				High EBNA IgG	3		
				High EBNA IgM	2		

Table 24.9. (Continued)   Data of 17 Patients with Endometriosis

Blood cell minerals		Pathology	
Compound	No.	Result	No.
Zinc — low	2	Chronic cystic cervicitis, squamous metaplasia, cervix uteri benign	1
Sodium — low	2		
Iron — low	3		
Iron — high	1		
Phosphorus — low	1		
Phosphorus — high	2		
Sulfur — high	1		
Strontium — high	1		
Postassium — high	1		
Calcium — high	1		
Tin — low	1		
Molybdenum — low	1		

Outcome	No.	Treatment	No.
Abnormal iris corder	4	Sauna	1
Abnormal SPECT Scan	1	Rotation diet	9
Surgery for endometriosis	1	Elimination diet	8
Hysterectomy	2	Injection therapy	17
Abnormal equilibrium	1		

Note: Age range = 12 to 59 years. Average age = 35 years.

Source: EHC-Dallas. 1994.

One of the most widely studied prescription drugs in regard to fetotoxicity is DES.[211-213] Research suggests that other relatively common prescription drugs are also fetotoxic. Spontaneous abortions have been reported with the use of oral anticoagulants for patients with cardiac valve prostheses.[214] The use of the anticoagulant warfarin probably is related to early abortions.[215] In a study of renal transplant patients, successful outcome of pregnancy was associated with reduced exposure to warfarin.[216]

In an investigation of fetal exposure to isotretinoin (retinoic acid), Lammer et al.[217] found an unusually high relative risk of major fetal malformations. These authors suggested that the effect is on cephalic neural crest cell activity, since the malformations tend to be craniofacial, cardiac, and thymic in nature. Other studies have shown similar results associated with spontaneous abortion.[218,219]

Other prescription drugs known or suggested to cause spontaneous abortions include imidazole agents, which are used to treat vaginitis;[220] tedral, which is prescribed for upper respiratory infection;[221] and lithium, which is used to control manic-depression.[222] Animal studies suggest that the glucocorticosteroids budesonide and fluocinolone acetonide may be fetotoxic,[223] as may be calcium channel blockers used to prevent premature labor,[224] methyldopa (an antihypertensive),[225] and certain antibiotics.[226] Particularly interesting is a study showing that female mice mated to males that had been treated with methyldopa had a higher incidence of abortion and lower incidences of total implantations, even though mating capacity and fertility were not affected.[222]

Hormones

Progesterone support for *in vitro* fertilization increases the rate of spontaneous abortion.[227] Women who conceive with human gonadotropins have a high rate of spontaneous abortions.[228,229] Animal studies show certain sex steroidal hormones, used clinically for detection of early pregnancy and for supportive therapy of pregnancy, are embryolethal.[230,231]

PCBs

Epidemiologic evidence suggests that PCBs are toxic to reproductive processes.[232] In comparison to a control group, women with missed abortions have shown higher PCB serum levels, with increases in the penta- and hexachlorobiphenyls.[233] Animal studies corroborate these findings.[234]

Pesticides/Herbicides

Workers exposed to pesticides in the grape gardens of Andhra Pradesh show an increase in spontaneous abortions, with cytogenetic studies indicating an increase in chromatid breaks and gaps in chromosomes of peripheral blood.[235]

Results of studies of phenoxy herbicide (Agent Orange) exposure in Vietnam suggests a link between the herbicide and unfavorable outcomes of pregnancy.[236] An increase in chromosome breakage has been observed in exposed males.[237]

Human studies have shown increased incidence of spontaneous abortion related to consumption of alcohol during pregnancy.[238,239] Laboratory studies show nondysjunction chromosomal defects in fertilized eggs of ethanol-exposed female mice[240] and increased rates of spontaneous abortion in pregnant macaques with ethanol administration.[241]

Tetrahydrocannabinol (THC), the principal psychoactive component in marijuana, produced spontaneous abortions when administered to monkeys early in pregnancy;[242] a rapid decrease in chorionic gonadotropin and a decrease in progesterone occurred. When compared to former drug abusers maintained on methadone during pregnancy and to drug-free women, cocaine-using women had a significantly higher rate of spontaneous abortion.[243] At least one study[244] has shown a relationship between caffeine consumption and spontaneous abortion. For more information, see Volume IV, Chapter 30.

Menopause

Early menopause, excessive bleeding during menopause, and exacerbation of peripheral symptoms have been seen in chemically sensitive women. Ovary function will suddenly cease resulting in amenorrhea, dry vaginal mucosa, and early onset of osteoporosis, and arteriosclerosis has been seen in some chemically sensitive individuals. A more common syndrome has been the erratic output of estrogen and progesterone that results in a myriad of symptoms including hot flashes, sweating, anxiety, and an increased appetite. Hot flashes due to hormone irregularity are often confused with those due to chemical and food triggering. We have seen many chemically sensitive women develop hot flashes that were easily eliminated after the offending foods and chemicals were removed.

VAGINA AND VULVA

Though technically not part of the endocrine system, the vagina and vulva can be affected by the various activities of the endocrine system. For completeness, therefore, these organs are discussed here.

Suckling neonatal mice that ingested milk from chlordecone-treated dams (250, 500, or 1000 µg daily) exhibited dose-dependent changes in the vaginal canals similar to those exposed to estradiol (10, 20, or 40 µg). Changes included mucification, keratinization, and desquamation.[245] o,p-Dichloro-diphenyltrichloroethane (o,p'-DDT)-stimulated DNA synthesis and cell division in the uterine luminal epithelium, stroma, and myometrium resulted in uterine hyperplasia.[246]

Environmental incitants can produce observable vaginal irritation, inflammation, and even infection. Spermicidal foams and jellies can trigger vaginal inflammation. Inhaled volatile organic chemicals have been seen to affect the vagina as a target organ, resulting in an increase in inflammation and recurrent infection.[247] Overuse of antibiotics may allow chronic *Candida* infestation to occur.[248] DES administration is the best known cause of vaginal cancer in human offspring.

Food sensitivity frequently triggers a vaginal discharge that is often correlated with nasal discharge. However, other factors can influence problems in the vagina. Severe vaginal pain, for example, can often occur with or without sexual intercourse. Sometimes a patient's vaginal discomfort may be a systemic reaction accompanied by asthma, anxiety, depression, or edema. At other times, vaginal pain with intercourse may result from a woman's sensitivity to her partner's sperm. With such a patient, we have successfully used the partner's sperm in an intradermal injection of an autogenous extract to neutralize the sensitivity, and the woman has then been able to have problem-free intercourse.

Vulvitis also has been seen with severe chemical overload. It can be excrutiatingly painful to the point that patients are unable to eat. Objective evaluation of vulvar burning is difficult, except for occasional mucosal cuts and swelling. Some, though not all, cut and swollen areas are usually tender to touch. As a rule, pain represents a neuropathy, and occasionally rectal manipulation of the muscles of the pelvis will help reduce discomfort. Sometimes these patients also have a burning tongue. Usually these patients are triggered by environmental pollutant and food exposures, and they respond to evironmental manipulation. McKay[249] has described 20 patients with vulvodysnia. She treated them with 10 to 60 mg/daily of amitriptyline with partial success. We have had no success treating environmentally sensitive patients with this drug.

TESTES AND SEMINAL VESICLE

The hypothalamus is responsible for stimulating the anterior pituitary to release gonadotrophin hormones that bring on puberty. Before puberty, the hypothalamus is so extremely sensitive to testosterone that if an individual's testicles are chemically stimulated, releasing even a minute amount of testosterone, inhibition of the entire system occurs. Therefore, pollutant stimulation may imbalance the prepubertal individual, causing a myriad of endocrine, and related, dysfunction. After puberty, the hypothalamus loses this extreme sensitivity to testosterone, which allows the secretory mechanisms of the testes to develop full activity. This process may be altered in some chemically sensitive males who have been seen with sex organ dysfunction.

Some chemically sensitive men begin to exhibit decreasing sexual function in their early 30s and 40s suggesting early climatric. This dysfunction may result in mood swings and depression as well as other physiologic changes in energy, the onset of "pot bellies", and apathy.

The male sexual act may be thwarted in many ways by pollutant injury. Olfactory stimulation by perfumes and toxic chemicals have been reported by many chemically sensitive males to prevent erection and even stop the desire for sex. Others have noted that, even though they can achieve an erection, they often cannot maintain it if sex is occurring in a polluted environment. However, they can maintain an erection for an extended period of time in a less-polluted environment. Apparently, parasympathetic impulses are blocked or dampened by environmental pollution, preventing the erection from being maintained. With pollutant dysfunction, the veins dilate and the artery constricts, with loss of erection.

Other chemically sensitive patients have reported inability to ejaculate, probably indicating pollutant damage to the sympathetic nerves which leave the cord at L-1 and L-2 and passing through the hypogastric plexus. Some cases have been reported to us in which the male was continuously stimulated by certain toxic chemicals or foods that acted as aphrodisiacs and caused an increase in sexual desire and pathologic appetite for indiscriminate sex several times a day.

The understanding of the physiology of testosterone is very important in the chemically sensitive male, because of not only its sexual function but also its anabolic steroid effect. Interstitial cells of Leydig produce testosterone and are influenced by environmental pollutants such as X-ray, heat, and pesticides. In addition to its role in sexual function, testosterone causes hair growth. Loss of testosterone due to pollutant injury will cause patchy hair loss as seen in many chemically sensitive males. Excess testosterone can cause acne. The effect of testosterone on protein formation and muscle development may be minimized in some chemically sensitive patients. As a result, muscle wasting and weakness may develop. Pollutant damage may cause a loss of testosterone, causing the metabolic rate to decrease 5 to 15%, and it may be one of the causes of the cold sensitivity that is seen in most chemically sensitive patients. Low testosterone may adversely effect red blood cells, and it can decrease the readsorption of sodium. Many pollutants have been reported to alter testosterone production.

The effects of toxic chemicals on reproduction are generally more appreciated than their effects on the functions of other organs, although more laboratory data is now becoming available. Studies by Frem-Titulaer et al.[175] discussed in the section on ovaries show the occurrence of precocious puberty in Puerto-Rican males due to contaminants in the chicken feed. Other contaminants show a similar picture. For example, inhaled or ingested methylchloride resulted in uni- or bilateral sperm granulomas in the cauda epididymis, significantly depressed testes weights, significantly lowered testicular spermatid head counts, delayed spermiation, chromatin margination in round spermatids, epithelial vacuolation, luminal exfoliation of spermatogenic cells, and multinucleated giant cells.[249] Additionally, at 7 weeks postexposure, 60 to 70% of spermatogonial stem cells had been killed. Sperm isolated from the vasa deferentia had significantly depressed numbers and an elevated frequency of abnormal sperm head morphology by week 1 postexposure and significantly depressed sperm motility and an elevated frequency of headless tails by 3 weeks postexposure. Methoxychlor, a chlorinated hydrocarbon pesticide, inhibited spermatogenesis, produced degenerative fatty changes in the sustentacular area where the blood-testis barrier could have been affected, resulted in degenerative changes of the spermatogonia and the spermatocytes, and transformed some of these cells into polynucleated and binucleated cells.[251] Myelin figures and degenerate mitochondria were observed in the binucleated cells, while some of the seminiferous tubules were devoid of all cellular elements with the exception of spermatogonia. The epithelium of the ductus epididymis manifested conspicuous cystoplasmic vaculations. The duct was distended with fluid that compressed the lining epithelium. Benomyl, a systemic fungicide, produced age-related decreases in testicular or epididymal weights, decreased epididymal sperm counts, decreased vas deferens sperm concentrations and/or testicular lesions, and increased incidence of diffuse hypospermatocytogenesis.[252] Dibromochloropropane exposure (initial and residual) brought about reduced body and gonadal weight gains and caused hypospermatogenesis or seminiferous tubular atrophy. Acute exposure may produce irreversible injury.[253] DDT [1,1-*bis*(*p*-chlorophenyl)-2,2,2-trichloroethane]-exposed embryos demonstrated significant alterations in both testes and ovaries. The testes consisted of mostly stroma with fewer seminiferous cords, while ovaries contained a larger number of distended nodular cords and differences in the distribution of these cords.[254]

Disorders of development and function of the male reproductive tract have increased in incidence over the past 30 to 50 years.[255] Examples of such disorders include testicular cancer,[256] maldescent (cryptorchidism),[255,257] and urethral abnormalities (hypospadias),[255] and there has been a striking drop in semen volume and sperm counts in normal adult men (Figure 24.9).[258] Since these changes are recent and appear to have occurred in many countries, we presume that they reflect adverse effects of environmental or lifestyle factors on the male rather than, for example, genetic changes in suceptibility.[255] Because testicular cancer, cryptorchidism, and hypospadias are all errors that probably arise during fetal development,[259-261] these abnormalities and reduced sperm counts may have a common etiology.

Treatment of several million pregnant women between 1945 and 1971 with a synthetic estrogen, DES, is now recognized to have led to substantial increases in the incidence of cryptorchidism and hypospadias and decreased semen volume and sperm counts in the sons of these women.[262] DES exposure may also increase the incidence of testicular cancer,[263] and other evidence points to a link between maternal estrogen concentrations and the frequency of testicular cancer[264] and cryptorchidism.[265] The similarity between these effects and the adverse changes in male reproductive development and function over the past 40 to 50 years raises the question of whether the adverse changes are

Testicular descent and development   Hormonal control of the fetal
of the male reproductive tract   testis

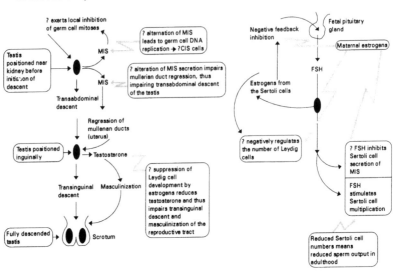

Figure 24.9. Possible mechanisms and physiological pathways via which maternal estrogens could cause impaired development and descent of the testes and other abnormalities of the reproductive tract. Points at which estrogens derived from the mother could exert an adverse effect are indicated by bold lines and arrows. Estrogen administration to pregnant animals or women has been shown to result in all of the indicated abnormalities, but there is no definitive evidence that they occur via the indicated pathways. (From Sharpe, R. M. and N. E. Skakkebaek. 1993. Are oestrogens involved in falling sperm counts and disorders of the male reproductive tract? *Lancet* 341:1392–1395. With permission.)

attributable to altered exposure to estrogens during fetal development. This possibility is not unlikely given the view that humans now live in an environment that can be viewed as a virtual sea of estrogens.[266]

Humans can be exposed to estrogens via inhalation, ingestion, or injection, although few hard facts exist that enable the impact of each of these routes to be evaluated. Variations in diet may have the greatest impact in human absorption of estrogen. It has been argued that the low incidence of breast cancer in Japan and China is related to their high-fiber, low-fat diet,[267] the key argument being that endogenous estrogens (which are implicated in the etiology of breast cancer) that are excreted via the bile are more readily metabolized and reabsorbed from the gut when the gut contains low amounts of fiber. This so-called "enterohepatic" recirculation of estrogens means effectively that a woman on a low-fiber diet would be exposed to more of her endogenous estrogen than a woman on a high-fiber diet. The relative consumption of fats (especially animal fats), proteins, and refined carbohydrates can also affect substantially estrogen excretion and metabolism, and the overall effect of eating a modern western diet is to increase exposure to endogenous estrogens.[267] Whether such a diet in a pregnant woman could expose the male fetus to sufficient estrogen to induce reproductive-tract abnormalities is a matter for speculation. However, there is evidence that links occurrence of cryptorchidism and testicular cancer to endogenous estrogen concentrations in the mother,[264] particularly in first pregnancies when concentrations of bioavailable estrogen can be high.[268] Other studies indicate that breast cancer in the mother is a significant risk factor for testicular cancer in male offspring.[269]

Exogenous Estrogens

Synthetic estrogens pose a hazard to humans because many are manufactured to be orally active and resistant to degradation. DES and other synthetic estrogens were used widely in the livestock industry for 20 to 30 years, and, for at least the first 20 years of their use, it was not recognized that they might pose a risk to humans[270] In 1970, as a result of abnormalities in children born to DES-treated women,[262] procedural changes in the use of DES were introduced to minimize the risk to humans.[271] It remains a matter of conjecture as to how stringently these procedures were adhered to and what the level of human exposure to anabolic estrogens was between the 1950s and 1970s. Orally active anabolic estrogens were banned in Europe in 1981, and many of the anabolic estrogens used now in the livestock industry are not orally active.[272]

The other use of synthetic estrogens that has increased in the past 20 to 40 years is in the oral contraceptive pill (e.g., ethinyl estradiol). There are reports that ethinyl estradiol is detectable in water sources, but there are few data on concentrations in drinking water. However, like DES,[273] ethinyl estradiol does not bind to sex-hormone-binding globulin (SHBG), to which most estrogen in blood is normally bound, which means that it has a very high potency if ingested.

Phytoestrogens

Changes in diet in many countries over the past 40 years may have increased the amount of estrogens in the environment because many plants and fungi contain so-called phytoestrogens,[274,275] and it is well recognized in animals that ingestion of such plants or fungi results in phytoestrogen-induced abnormalities in normal reproductive function.[274] Soya is the richest source of phytoestrogens[275] and its consumption, especially as a substitute for meat protein, has increased enormously in the past 2 or more decades. Attempts have been made to estimate the biological effects of phytoestrogens in humans,[275] and, at face value, the conclusion is that exposure to phytoestrogens alone would probably be insufficient to induce major direct estrogenic effects in most adults. Indeed, phytoestrogens may reduce exposure to endogenous estrogens by stimulating production of SHBG by the liver and thus decreasing the concentration of bioavailable endogenous estrogen.[267]

Estrogens in Milk

It is widely accepted that in developed countries there is too great a consumption of dairy products, a trend that probably started in the 1940s and 1950s. This consumption may lead to changes in estrogen metabolism, as described above, but may have other implications. In modern farming practice most dairy cows are pregnant; however, unlike women, they continue to lactate. Cows' milk therefore contains substantial amounts of estrogens (mainly estrone sulfate).[276] Fortunately, estrogen in pasteurized milk is "lost" during formulation into baby-milk powder,[276] although it is unclear why this occurs or whether the estrogen might appear in other dairy products. It is uncertain to what extent estrone sulfate in cows' milk is absorbed from the gut of an adult or child and whether absorption might be altered by dietary constituents such as fiber.

Estrogenic Chemicals

Many of the chemicals with which we have contaminated our environment in the past 50 years are weakly estrogenic.[266,277] These chemicals are remarkably resistant to biodegradation, are present in our food-chain, and accumulate in our bodies.[276] In various animals, high concentrations of these chemicals have been associated with reproductive abnormalities,[277] which include changes in the semen quality of adult rats exposed neonatally to polychlorinated biphenyls via their mother's milk.[278] Even more disturbing is evidence that a single maternal exposure of rats (on day 15 of pregnancy) to nanogram amounts of the chlorinated hydrocarbon TCDD (dioxin) has no effect on the mother but increases the frequency of cryptorchidism in male offspring and causes dose-dependent reductions in their testicular weight and sperm count in adulthood.[279] The latter effects appear to be a consequence of decreased Sertoli-cell numbers. It is not known whether exposure of humans to these chemicals is sufficient to induce "estrogenic effects" directly or via any of the mechanisms described above, but similar effects are widespread in wild animals.[277]

Changes in human exposure to estrogens are difficult to quantify, especially when the suspected alterations are in the metabolism/bioavailability of endogenous estrogens during pregnancy. The most reasonable (and safest) assumption is that pregnant women (and humans in general) are exposed to more, rather than less, estrogens than was the case 50 years ago. The extent of this increase, its source(s), and its consequences are likely to differ among countries and among individuals if the routes of exposure listed in the table are all valid. Whether increased human exposure to estrogens could account for the increased incidence of abnormalities in male reproductive development and function is unknown, but "weak estrogens" may be more potent in the fetus and neonate than in the adult.[277] Moreover, many (and perhaps all) of the reproductive-tract abnormalities can be brought about by estrogen-induced changes during fetal development, thus providing strong circumstantial support for a role for environmental and dietary estrogens in the development of these abnormalities.[272]

Sertoli Cell Number and Sperm Output

Although there are physiological pathways via which estrogens can impair development of the male reproductive tract, how can impairment of development be linked to falling sperm counts in adult men? The answer, according to studies in animals, is by altering the multiplication of Sertoli cells. Sertoli-cell multiplication occurs during fetal, neonatal, and prepubertal life and is controlled to a large extent by FSH.[280,281] Inhibition of FSH secretion reduces Sertoli-cell multiplication,[280] and in neonates FSH secretion is exquisitely sensitive to inhibiiton by exogenously administered estrogens.[280] At a fixed time in postnatal life, Sertoli-cell multiplication ceases coincident with "maturation" of these cells.[280,282] Importantly, this maturation also coincides with dramatic decreases in the secretion of estrogens and MIS.[280,282] The "fixing" of Sertoli-cell number during prepubertal development has important consequences in adult life because Sertoli cells are responsible for orchestrating and regulating spermatogenesis. Each Sertoli cell can only support a fixed number of germ cells during development into spermatozoa.[283,284] Therefore, the lower the number of Sertoli cells the lower will be the "ceiling" for sperm output. Studies in animals have all shown that alteration of Sertoli-cell number (up[285] or down[284]) in early life determines testicular size and sperm output in adulthood. In such studies the quality of the spermatozoa is not affected, just the quantity. Of particular significance is the fact that estrogen administration to animals in fetal and early neonatal life results in smaller testes and reduced sperm counts in adult life.[280] Moreover, the sons of women exposed to DES during pregnancy show an increased incidence of low sperm counts,[262] consistent with what would be predicted from animal data.[272]

Semen quality data collected systematically from reports published worldwide indicate clearly that sperm density has declined appreciably between 1938 and 1990, although it cannot be concluded whether or not this decline is continuing. Concomitantly, the incidence of some genitourinary abnormalities including testicular cancer and possibly also maldescent and hypospadias have increased. Such remarkable changes in semen quality and the occurrence of genitourinary abnormalities over a relatively short period is more probably due to environmental rather than genetic factors.[272]

Effects of pollutants reported in men were impotency and sterility.[286] Impotency also occurs after chronic exposure to nitrous oxide anesthetics. Reports since the 1970s have linked adverse reproductive effects with anesthetic gases and dibromochloropropane (DBCP).[287] Chemical agents can interfere with the process by which testicular products are formed and, therefore, interfere with functions of a variety of processes throughout the body.[286] Additional complaints associated with the genitourinary system include back pain, secondary to broad ligament swelling, vaginal discharge, vaginitis, urinary frequency, and urgency. According to Hunt,[286] the reproductive system in men and women has an important characteristic. There is cyclic cellular growth and differentiation in each of the tissues, which provides a special susceptibility to toxic effects and may affect the capacity of individuals to develop mature ova and sperm, to provide a suitable environment for the fertilized ovum, and to allow the normal growth and development of the embryo and fetus.

Rosenblum and Rosenblum[288] reported methyltestosterone is effective in preventing episodes of hereditary angioneurotoxic edema (for further information, see Chapter 28).

At the EHC-Dallas, we have identified many men and women with chlorinated pesticides in their blood (see Volume II, Chapter 13,[130] and Chapter 28). Some of these chemicals have appeared to cause congenital abnormalities, especially of the metabolic type. Many have sexual dysfunction relating to their chemical sensitivity. Diagnosis and treatment are similar to that used with any patient with chemical sensitivity. However, the hormone diagnosis and treatment will be discussed in Volume IV, Chapter 38.

THYROID

In the complex patient who has multiple food and chemical sensitivities, endocrine involvement, especially of the thyroid gland, has to be considered. A correlation between Hashimoto's lymphocytic thyroiditis and allergic or autoimmune disease has been identified.[289,290] Evidently, some chemically sensitive patients, especially women, have autoimmune endocrinopathy with associated ovarian, thyroid, and adrenal involvement. Periods of high hormone stress such as puberty, childbirth, and menopause partially account for this endocrinopathy, but evidence for environmental chemical triggers has also been accumulating in recent years. For example, Bahn et al.[291] noted an increased prevalence of thyroid antibodies among workers exposed to polybrominated biphenyls. Gaitan et al.[292] have shown increased incidences of thyroiditis and hypoactive thyroid in people exposed to water drawn from oil shale areas. He has traced this incidence across both the North and South American Continent. Bastomsky has induced thyroid disease in rats treated with polyhalogenated aromatic hydrocarbons and substances such as PCBs and PBBs.[293] Saiffer[294] has reported thyroid dysfunction in many chemically sensitive patients. At the EHC-Dallas, we have also seen several cases.

Agents known to have goitrogenic and/or antithyroid effects on the thyroid of humans and other animal species are widespread (Table 24.10) and have often been found in the chemically sensitive individual.

Table 24.10. Environmental Agents Producing Goitrogenic and/or Antithyroid Effects

| | Goitrogenic/antithyroid effects | | |
| | In vivo | | In vitro |
	Humans	Animals	Systems
Sulfurated organics			
Thiocyanate	+[a]	+	+
Isothiocyanates	NT[b]	+	+
Thioglycosides (goitrin)	+	+	+
Aliphatic disulfides	NT	+	NT
Polyphenols			
Bioflavonoids	NT	+	+
Phenolic and phenolic-carboxylic derivatives			
Resorcinol (1,3-dihydroxybenzene)	+	+	+
Orcinol (5-methylresorcinol)	NT	+	+
2-Methylresorcinol	NT	+	+
4-Chlororesorcinol	NT	+	+
Phloroglucinol	NT	+	+
Pyrogallol	NT	+	+
3,4-Dihydroxybenzoic acid	NT	NT	+
3,5-Dihydroxybenzoic acid	NT	NT	+
3-Chloro-4-hydroxybenzoic acid	NT	NT	+
2,4-Dinitrophenol	NT	+	O[c]
Phthalate esters and phthalic acid derivatives			
Dihydroxybenzoic acids	NT	NT	+
PCBs[d] and PBBs[e]			
PCBs (Aroclor)	NT	+	NT
PBB oxides	+	+	NT
Other organochlorines			
p,p'-DDT[f], p,p'-DDE[g] and dieldrin	NT	+	NT
2,3,7,8-Tetrachlorodibenzo-p-dioxin	NT	+	NT
Polycyclic aromatic hydrocarbons			
3,4-Benzpyrene	Nt	+(?)	NT
3-Methylcholanthrene	NT	+	NT
7,12-Dimethylbenzanthracene	NT	+	NT
Inorganics			
Excess iodine	+	+	+
Lithium	+	+	+

[a] Active.
[b] Nontested.
[c] Inactive.
[d] Polychlorinated biphenyls.
[e] Polybrominated biphenyls.
[f] Dichlorodiphenyltrichloroethane.
[g] Dichlorodiphenyldichloroethylene.

Source: Gaitan, E. 1986. Environmental goitrogens. In The Thyroid Gland — A Practical Clinical Treatise, ed. L. Van Middlesworth. 264. Chicago, IL: Year Book Medical Publishers, Inc. With permission.

Physiology and Pathophysiology

Environmental compounds may cause goiter not only by acting directly on the thyroid gland but also by acting indirectly, altering its regulatory mechanisms and/or the peripheral metabolism and the excretion of thyroid hormones. Price et al.[295] have shown that if rats are challenged with PCBs or other chlorinated compounds, an increase in the response of the cytochrome detoxification mechanism occurs. These compounds remove thyroid from the blood. This increase in metabolism causes thyroid production to increase, and then eventual depletion occurs as the thyroid wears out.

Agents Acting Directly on the Thyroid

Gaitan[296] has reported the various environmental goitrogenic and antithyroid compounds and their sites of action in the thyroid gland. These are shown in Figure 24.10. Goitrogens act in the thyroid gland to interfere with the process of hormonal synthesis, but the mechanism that induces the trophic changes leading to goiter formation is not yet well understood. Antithyroid compounds can be divided into three categories according to the way they act on iodine metabolism in the thyroid.

Class I

The thiocyanates or thiocyanate-like compounds appear primarily to inhibit the active concentrating mechanism of iodine, and their goitrogenic activity can be overcome by iodine administration. These ions have a molecular volume and charge similar to those of iodine. This fact is the reason that they compete with iodine for transport in the follicular cell. Cyanogenic glycosides in a variety of staple foods such as cassava, almonds, apples, walnuts, and sorghum may exert a goitrogenic effect when converted to thiocyanate in the living organism.[297-300] Many chemically sensitive individuals react to these foods upon oral challenge, and some of these foods have been seen to exacerbate thyroiditis in these individuals.

The isothiocyanates act on the thyroid mainly by rapidly converting to thiocyanate. The naturally occurring butyl, allyl, and methyl isothiocyanates do not inhibit the thyroidal peroxidase enzyme. However, isothiocyanates not only use the thiocyanate-metabolic pathway, but also react spontaneously with amino groups, forming disubstituted thiourea derivatives, which produce a thiourea-like antithyroid effect. Additive antithyroidal effects of thiocyanate, isothiocyanate, and the thioglycoside "goitrin" occur with combinations of these naturally occurring goitrogens.

Class II

According to Gaitan,[296] the thiourea or thionamide-like goitrogens interfere with the processes of organification of iodine and coupling of iodotyrosines

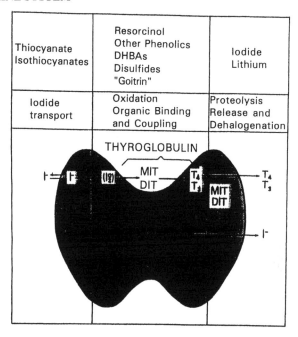

Figure 24.10. Naturally occurring and anthropogenic goitrogens and their site of action in the thyroid gland. DHBAs = dihydrobenzoic acids; I⁻ = iodide; MIT = monoiodotyrosine; DIT = diiodotyrosine; T₄ = thyroxine; T₃ = triiodothyronine. (From Gaitan, E. 1986. Environmental goitrogens. In *The Thyroid Gland — A Practical Clinical Treatise*, ed. L. Van Middlesworth. 265. Chicago: Year Book Medical Publishers. With permission.)

to form the active thyroid hormones, and their action usually cannot be antagonized by iodine. These goitrogens prevent the oxidation of iodine by thyroid peroxidase and impair covalent binding of iodine to thyroglobulin. In small amounts, they inhibit the formation of active thyroid hormones from iodotyrosine precursors. In larger quantities, they impair formation of monoiodotyrosine (MIT) and diiodotyrosine (DIT). These types of compounds do not prevent transport of iodine into the thyroid gland, and pharmacologic doses would be required to suppress iodine uptake. The naturally occurring thiglycoside L-5-vinyl-2-thiooxazolidone, "goitrin", is a main representative in this category. Goitrin is unique in that it does not degrade like other thioglycosides. Goitrin acts *in vitro* on the thyroidal peroxidase (Table 24.11) and I¹²⁵ and organification,[301] and *in vivo* exerts a thionamide-like effect.[298]

Table 24.11. Concentrations of Naturally Occurring Substances Producing I_{50}[a] of Thyroid Peroxidase[b]: Comparison with the Antithyroid Drugs Propylthiouracil and Methimazole

Compound	I_{50} (nmoles/mL)	Potency (propylthiouracil/inhibitor)
6-Propylthiouracil	7.2	1.0
Methimazole	4.2	1.7
L-5-vinyl-2-thiooxazolidone "goitrin"	4.0	1.8
Isothiocyanates		
Butyl	NI^c (1×10^2)	0
Allyl	NI (1×10^2)	0
Methyl	NI (1×10^2)	0
Bioflavonoids		
Catechin	0.72	10.0
Phloretin	0.54	13.3
Hesperitin	3.9	1.8
Hesperidin	61.9	0.1
Phenols		
Resorcinol	0.3	26.7
Phloroglucinol	0.2	37.9
Pyrogallol	3.8	1.9
Orcinol	1.0	7.2
3,4-Dihydroxybenzoic acid	7.0	1.0
3,5-Dihydroxybenzoic acid	2.0	3.6
Phthalates		
Dissobutyl phthalate	NI (1×10^4)	0
Dioctyl phthalate	NI (1×10^4)	0
o-Phthalic acid	NI (1×10^4)	0
m-Phthalic acid	NI (1×10^2)	0

[a] 50% inhibition.
[b] A glucose-glucose oxidase system was used for H_2O_2 generation.
[c] No inhibition.

Source: Gaitan, E. 1986. Environmental goitrogens. In *The Thyroid Gland — A Practical Clinical Treatise*, ed. L. Van Middlesworth. 266. Chicago, IL: Year Book Medical Publishers. With permission.

Naturally occurring bioflavonoids are polyhydroxylated C_6-C_3-C_6 structures. These polyphenols possess intrinsic antithyroid activity. Except for the glycoside hesperidin, its aglycon hesperidin, catechin and phloretin are potent inhibitors of the thyroid peroxidase enzyme. The antithyroid activity appears to be exerted, as in the case of resorcinol, by the presence of two free metapositioned OH-groups in the benzene ring of the benzypyran unit.

According to Gaitan,[296] resorcinol

and its phenolic and phenolic-carboxylic dihydrobenzoic acids (DHBAs) parent compounds,[302,303] and the aliphatic disulfides[297,302] also act on this step of thyroid hormone synthesis. The goitrogenic and antithyroid effects of resorcinol are enhanced when its conjugation with glucuronic acid is affected, which often occurs in the chemically sensitive population. These phenolic compounds apparently can cause or exacerbate chemical sensitivity. Throughout this book, numerous phenolic challenges have been shown to trigger problems in the chemically sensitive.

Class III

Agents in this group interfere with the processes of proteolysis and release of thyroid hormones. The most important representative of this group is iodine.[304-306] An excess intake of iodine, arbitrarily defined as 2 mg or more per day, inhibits the synthesis and release of thyroidal hormones and eventually produces "iodine goiter" and hypothyroidism. Lithium has also been shown to belong to this category.[307-309] Brominated, fluoridated, and chlorinated compounds also can give a pseudohalogen effect and trigger hypothyroidism (see Volume I, Chapter 4[6]). This pseudohalogen effect seems to be relevant in many chemically sensitive patients, who have a tendency toward thyroid dysfunction.

Agents Acting Indirectly on the Thyroid

According to Gaitan,[310] the antithyroid effect of 2,4-dinitrophenol (DNP) is complex. First, it is due in part to an inhibition of the pituitary thyroid-stimulating hormone (TSH) mechanism.[311] In addition, DNP interferes with T_4 binding,[312-314] further decreasing serum T_4 concentration. Polychlorinated biphenyls exert a similar effect.[293,315] Finally, DNP accelerates the disappearance of T_4 from the circulation, exaggerating even more the lowering of serum T_4 concentration.[312] At the EHC-Dallas, we have now challenged over 2000 chemically sensitive patients with the ambient dose of 2,4-DNP (<0.0034 ppm). Symptom reproduction has resulted.

Thyroid hormones are excreted into the intestine in both free and conjugated forms, along with small amounts of their deiodinated metabolites. Glucuronide conjugation occurs mainly in the liver by the action of a UDP-glucuronyltransferase, and sulfate conjugation occurs mainly in the kidney by the action of a sulfate transferase. However, under normal circumstances, little T_4 and T_3 are excreted in conjugated form. As has been shown in many chapters in this book, but those especially on mechanisms, gluconation is often disturbed in the chemically sensitive. This disturbance may account for the occurrence of thyroiditis and hypothyroidism in many chemically sensitive patients.

Potent hepatic microsomal enzyme inducers, PCBs have properties of both the phenobarbital and polycyclic hydrocarbon type of inducer. They greatly enhance the biliary excretion of circulating T_4 as T_4-glucuronide, which then is lost in stools, at least in the rat. This excretion is probably secondary to induction of hepatic microsomal T_4-uridine diphosphate-glucuronyl transferase. Enhanced peripheral metabolism and reduced binding of T_4 to serum proteins in PCB-treated animals result in markedly decreased serum T_4 concentrations, activation of the pituitary-thyrotropin-thyroid axis, and eventually in goiter formation.[293,315] Furthermore, PCB-treated animals exhibit decreased serum T_4, but unchanged T_3, levels. This observation may be explained by increased peripheral deiodination of T_4 to T_3 and increased thyroidal T_3 secretion that results from the state of relative iodine deficiency induced by accelerated metabolism of T_4.

Polybrominated biphenyls (PBBs),[316] dioxin (TCDD),[317,318] and the polycyclic aromatic hydrocarbons (PAH), 3-methylcholanthrene (MCA) and 3,4-benzpyrene (BaP),[319-321] appear to act similarly to PCBs, but there is some indication that PBBs and MCA also interfere with the process of hormonal synthesis in the thyroid gland.

The bioflavonoid phloretin not only inhibits the thyroidal peroxidase (Table 24.11) but it also has been shown to affect the peripheral metabolism of thyroid hormones and phloretin polymers to interact with TSH, preventing its action at the follicular thyroid cell,[322,323] indicating that this class of substances can alter thyroid hormone economy in a complex manner.

General Properties, Distribution, and Epidemiology

Environmental antithyroid and goitrogenic compounds are naturally occurring or anthropogenic. Gaitan[296] has shown that they can be present in foodstuffs, in contaminated water supplies, or in wastewater effluents, or they can be airborne or waste products of industrial processes. Most of these contaminants cause problems in the chemically sensitive individual.

Sulfurated Organics—Thiocyanate (SCN), Isothiocyanates, and Thioglycosides (Goitrin)

Extensive reviews on sources, metabolic pathways, and action of cyanogenic glycosides, thioglycosides, isothiocyanates, and thiocyanates have been published.[296,299,300] The pathways of injury have been discussed in Volume I, Chapter 4.[6]

Thiocyanate and isothiocyanates have been demonstrated as goitrogenic principles in *Cruciferae*. The potent antithyroid compound "goitrin", a thioglycoside, was isolated from yellow turnips and from *Brassica* seeds.

Cyanogenic glucosides (thiocyanate precursors) have also been found in several staple foods (cassava, maize, bamboo shoots, sweet potatoes, lima beans). After their ingestion, these glucosides can be readily converted to SCN by a widespread tissue enzyme. Isothiocyanates, as previously mentioned, not only use the thiocyanate-metabolic pathways, but also react spontaneously with amino groups forming disubstituted thiourea derivatives, which produce a thiourea-like antithyroid effect. Thus, the actual concentration of thiocyanates or isothiocyanates in a given foodstuff may not represent its true goitrogenic potential, nor does the absence of these compounds negate a possible antithyroid effect, because inactive precursors can be converted into goitrogenic agents both in the plant itself or in the animal after its ingestion. Thioglycosides undergo a Lossen arrangement to form isothiocyanate derivatives and in some instances thiocyanate. Therefore, the amount of thiocyanate in the urine is a good indicator of the presence of thioglycosides in food. It has been demonstrated that a mustard oil glucoside, glucobrassicin, yields SCN under the action of thioglucosidase, "myrosinase", an enzyme present in plants. Ingestion of pure progoitrin, a naturally occurring thioglycoside, elicits antithyroid activity in rats and humans in the absence of myrosinase. The antithyroid activity of progoitrin is due to its partial conversion in the animal into the more potent goitrogen, 1,5-vinyl-2-thiooxazolidone or goitrin. This ability of plants and animals readily to convert inactive precursors into goitrogenic agents must be considered when investigating the possible etiologic role of dietary elements in sporadic or endemic goiter. Of course, this is probably one reason that the chemically sensitive individual has trouble tolerating some foods.

Anthropogenically, thiocyanate is found in high concentrations (1 g/L) in wastewater effluents of coal-conversion processes and in body fluids as a metabolite of hydrogen cyanide gas consumed while smoking.[302]

There have been several goiter endemias attributed to the presence of these sulfurated organics in foodstuffs. Two goiter endemias have been ascribed to the presence of these goitrogenic substances in milk.[297] One was in Tasmania, where a seasonal variation in goiter prevalence in school children was noted in spite of adequate iodine intake and in which an isothiocyanate, cheirolin, was suspected as the principal goitrogen. The other occurred in Finland, where goitrin present in cow's milk from the region of endemic goiter was considered the causative factor. Thiocyanate from a cyanogenic glucoside (linamarin) in cassava, a staple food, acting in the presence of extreme iodine deficiency, is thought to be the cause of endemic goiter and cretinism in central Africa.[324] Many chemically sensitive patients have transient thyroid pain and tenderness, which occurs from food and contaminant exposure similar to these reports of periodic goiter. The pain subsides after the pollutant reaction has taken its course over a day or two.

Sporadic goiter and hypothyroidism were also documented in patients on long-continued administration of thiocyanate for treatment of hypertension.[297,325]

Aliphatic Disulfides

The major volatile components of onions and garlic have been identified as small aliphatic disulfides, which have marked antithyroid activity in rats.[297,302] Gaitan[296] has shown that organic disulfides have also been identified as water contaminants in the United States and in water supplying a Colombian district with endemic goiter.[298,326] The most frequently isolated compounds in the United States are dimethyl, diethyl, and diphenyl disulfides, but dimethyl trisulfide, dimethyl sulfoxide, and diphenylene sulfide have also been isolated. Organic sulfide pollutants are also present in high concentration in wastewater effluents of coal-treatment plants.[302]

Polyphenols

Bioflavonoids are C_6-C_3-C_6 aromatic phenolic compounds widely distributed in nature. They are important stable organic constituents of a wide variety of plants. Bioflavonoids in high concentrations are present in polymeric (tannins) and oligomeric (pigments) forms in various staple foods of the Third World (i.e., millet, sorghum, beans, ground nut, etc.).[300,327] This type of polyphenol has been shown to be goitrogenic in rats.[300] Furthermore, according to Gaitan,[296] they are potential immediate precursors of potent phenolic antithyroid monomers. Actually, cyanidin, a naturally occurring compound used as the model subunit of flavonoid-type humic substances (HS), yields the antithyroidmonomers resorcinol, phloroglucinol, and orcinol by reductive degradation.[328]

Decaying organic matter (plants and animals) rich in these phenolic materials becomes the substrate of flavonoid-type HS during the process of fossilization. HS, high molecular-weight complex polymeric compounds, are the principal organic components of soils and waters.[329,330] They are also present in coals, shales, and possibly other carbonaceous sedimentary rocks. Thus, bioflavonoid structures may be the link for phenolic goitrogens in foodstuffs and those present in rocks, soils, and water. Both types trigger chemical sensitivity.

In western Sudan,[331] the prevalence of goiter is higher in remote rural villages, where the staple diet consists of only millet, than in small towns of the same area, where the diet includes a combination of millet, dura, and wheat. Rich in bioflavonoids,[327] millet has been shown experimentally to be goitrogenic.

Phenolic and Phenolic-Carboxylic Derivatives

As stated previously in this chapter, phenols can damage the thyroid in experimental studies and have been found to do so in the chemically sensitive individual. These substances will be discussed in more detail here.

Resorcinol (1,3 Dihydroxybenzene)

The prototype of this group of compounds is antithyroid and goitrogenic both in humans and in experimental animals.[296,300,302] Resorcinol and other parent antithyroid phenolic and phenolic-carboxylic compounds (Tables 24.10 and 11)[296,302] (phloroglucinol, pyrogallol, 5-methylresorcinol [orcinol], 3,4- and 3,5-DHBA and the ortho-[*o*] and meta-[*m*] phthalic acids) are monomeric byproducts of reduction, oxidation, and microbial degradation of HS.[167,328] At the heart of the "humification" process are the chemical and microbial-mediated production and polymerization of phenolic and carboxylic benzene-rings. Up to 70% of flavonoid HS may be made up of these subunits.[329,330] As much as 8% of shale bitumen is constituted by phenols.[302] Phenols are also the major organic pollutants in aqueous effluents from coal conversion processes.[302] Resorcinol and other antithyroid phenolic pollutants comprise as much as 4 g/L in the aqueous effluent from coal liquefaction units.[332] These pollutants may enter community water supplies, constituting a potential environmental goitrogenic factor in humans and other animal species. Coal-conversion waste-waters contain, in addition to phenolics, thiocyanate, and sulfides (S^{2-}),[333,334] antithyroid and goitrogenic properties.

Resorcinol is used industrially in the production of pharmaceuticals, dyes, plasticizers, textiles, resins, and adhesives for wood, plastics, and rubber products (see Volume II, Chapter 12[5]).

Resorcinol has been identified as a water contaminant in the United States.[335] Resorcinol and a substituted resorcinol have also been isolated from water supplies of endemic goiter districts in western Colombia[336] and the coal-rich Appalachian area of eastern Kentucky.[296]

As early as the 1950s, the goitrogenic effect of resorcinol was demonstrated when patients applying resorcinol ointments for the treatment of varicose ulcers developed goiter and hypothyroidism.[337,338] Several observations suggest that resorcinol crosses the human placenta and may cause both goiter and neonatal hypothyroidism.[339]

The presence of halogenated organic compounds with known or potential harmful effects has awakened public health and environmental concerns. These compounds are produced by the chlorination of water supplies, sewage, and power plant cooling waters[340,341] (see Volume II, Chapter 7[342]). Present at μg/L concentrations (parts per billion) in treated domestic sewage and cooling waters, 4-chlororesorcinol and 3-chloro-4-hydroxybenzoic acid possess antithyroid activities. Whether these pollutants exert additive or synergistic antithyroid effects and/or act as "triggers" of autoimmune thyroiditis requires investigation, particularly because more than 60 soluble chloroorganics have been identified in the primary and secondary effluents of typical domestic sewage treatment plants. It is clear that these substances cause many symptoms in the chemically sensitive individual and apparently do influence their thyroid hormone function and balance.

2,4-DNP

Derivatives of DNP are widely used in agriculture and industry. An insecticide, herbicide, and fungicide,[302] DNP is also used in the manufacture of dyes, to preserve timber, and as an indicator; it is also a byproduct of ozonization of parathion.

Administration of 2,4-DNP to human volunteers resulted in rapid and pronounced decline of circulating thyroid hormones.[343,344] The biological significance of this observation and the public health impact of this pollutant on the thyroid are still unknown. We have challenged by inhalation over 2000 patients with 2,4-DNP, showing triggering of a virtual textbook of medicine of symptoms (see Volume IV, Chapters 30 and 31).

Phthalate Esters and Phthalic Acid Derivatives: DHBAs

Phthalates are ubiquitous in their distribution and have been frequently identified as water pollutants.[302] Most commonly, they result from industrial pollution or artificial contamination, including plastic liquid and food containers, but phthalates are also reported to occur naturally in shale, crude oil, petroleum, plants, and as fungal metabolites[345] (see Volume II, Chapter 12[5]).

Although phthalates and phthalic acids do not possess intrinsic goitrogenic activity,[303] they undergo biodegradation by gram-negative bacteria with production of intermediate metabolites, such as DHBA,[302,346] known to possess antithyroid properties.[303] Thus, phthalates, with bacterial intermediates, may become a source of environmental goitrogenic compounds. These abundant pollutants exert deleterious effects on the thyroid of humans and other animal species as evidenced by Price's[295] studies on high doses of phthalates and Chen's[295] with low dose exposure and combination with PCB (Figure 24.11). Electron microscopic study shows extremely enlarged endoplasmic reticulum and vaculization of the mitochondria even in animals who were dosed with low-dose DEHP alone (Figure 24.12).

PCBs and PBBs

PCBs and PBBs are aromatic compounds containing two benzene nuclei with two or more substituent chlorine or bromine atoms (see Volume II, Chapter 12[5]).

Evidence is mounting that dietary PCBs and PBBs have many deleterious effects on health. There is a growing concern and uncertainty about the long-range effects of bioaccumulation and contamination of our ecosystem with these chemicals. The uncertainty extends to the potential harmful effect of these pollutants on the thyroid (see Volume II, Chapter 12[5]). However, we have seen many nonthyroid effects on our chemically sensitive patients and some patients with thyroiditis and thyroid tumors who have PCBs in their blood.

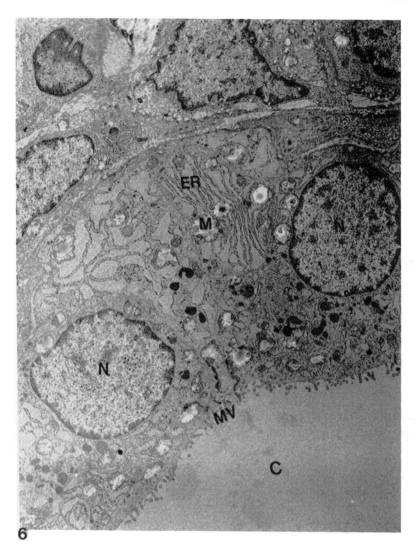

Figure 24.11. Thyroid malfunction in a rat fed DEHP (50 mg/kg) and BHT (25 mg/kg) — ow dose — for 2 weeks. Note enlarged endoplasmic reticulum (ER), mitochondria (M), and disturbed nuclei (N); vaculated microvilli (MV) and colloid (C) are normal. (From B. Chen at Robens Institute Surrey Guildford, England.)

Bahn reported an increased prevalence of primary hypothyroidism (11%) was documented among workers from a plant that manufactured PBBs and PBB oxides.[347] These subjects had elevated titers of antithyroid microsomal antibodies, indicating that hypothyroidism was probably a manifestation of

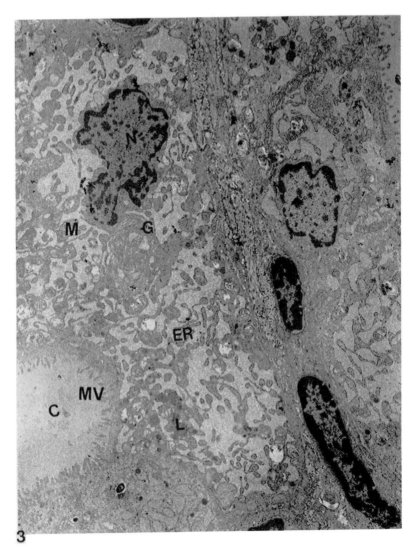

**Figure 24.12.   Rat thyroid changes after being fed DEHP (50 mg/kg) for
2 weeks; endoplasmic reticulum (ER) markedly enlarged;
mitochondria (M) vascularization; Golgi (G) disturbance;
nuclear (N) disruption. Normal: colloid (C) and microvilli
(MV).**

lymphocytic autoimmune thyroiditis, perhaps a PBB-induced pathogenic autoimmune response or exacerbation of underlying subclinical disease. Further,
environmental pollutants operating in genetically predisposed individuals may
trigger the pathogenic mechanisms that lead to goiter formation and autoimmune thyroiditis. The presence of organic goitrogens (resorcinol, substituted

resorcinols, thiocyanates, disulfides) and potential "triggers" of the autoimmune response (PAH and halogenated hydrocarbons) in coal[348-353] and water supplies and endemic areas underlies the need to test this hypothesis.[354-356] The incidence rate of this disorder has steadily increased in the United States during the past 5 decades.[305,357]

Goitrous autoimmune thyroiditis has been observed after administration of carbon tetrachloride and the PAH methylcholanthrene and DMBA to the BUF (Buffalo) inbred strain of rats.[358,359] Similarly, injection of mouse thyroglobulin with bacterial lipopolysaccharide induces autoimmune thyroiditis in "good-responder" mice, whereas "poor-responder" strains develop little pathologic response. "Good" and "poor" responders differ in their H-2 haplotype.[360] A relation between loci in the major histocompatibility complex and susceptibility to autoimmune thyroid disease has also been demonstrated in humans. For instance, the histocompatibility HLA DR[5] antigen is seen with increased frequency in patients with goitrous thyroiditis, whereas atrophic thyroiditis is associated with the HLA-DR3.[358] Thus, environmental pollutants operating in genetically predisposed individuals may also trigger the pathogenic mechanisms that lead to goiter formation and autoimmune thyroiditis. The presence of organic goitrogens (resorcinol, substituted resorcinols, thiocyanates, disulfides) and potential "triggers" of the autoimmune response (PAH and halogenated hydrocarbons) in coal[302,332-334,353,361] and water supplies of endemic areas[296,326,362] underlie the need to test this hypothesis.

Known to cause marked alterations in thyroid gland structure and function of birds,[302] DDT, DDE, and dieldrin induce microsomal-enzyme activity[363] that may affect thyroid hormone metabolism in a way similar to that of the polyhalogenated biphenyls and PAH. The impact of these pollutants on the human thyroid is unknown.

Dioxin (TCDD), one of the most toxic small organic molecules, is a contaminant in the manufacturing process of several pesticides and herbicides, including Agent Orange. Also a potent inducer of hepatic microsomal enzymes, TCDD markedly enhances the metabolism and biliary excretion of T_4-glucuronide.[317,318] Rats treated with TCDD concomitantly develop hypothyroxinemia, increased serum TSH concentrations, and goiter, probably as a result of T_4 loss in the bile.[317] The impact on the thyroid of humans exposed to this agent is unknown, and studies of thyroid function and thyroid hormone metabolism in those cases are necessary.

PAH, 3,4,-BaP, MCA, and 7,12-DMBA

Found repeatedly in food and domestic water supplies, polyaromatic hydrocarbons (PAHs)[343,364,367] (see Volume II, Chapters 7[368] and 8[369]) are present in industrial and municipal waste effluents. They also occur naturally in coal, soils, ground water and surface water and in their sediments and biota. One of the most potent of the carcinogenic PAH compounds, 3,4-BaP, is widely distributed and, as in the case of other PAHs, is not efficiently removed by conventional water treatment processes. Thyroid involvement is best documented and understood.

Genetically predisposed individuals (i.e., HLA DR[5] antigens) appear to have greater susceptibility to organic polycyclic aromatic hydrocarbons (PAH), and microbial water pollutants triggering goiter. Autoimmune thiroiditis does develop after administration of polycyclic aromatic hydrocarbons (PAH) and carbon tetrachloride to the rat.[318,370] Workers exposed to PCBs manifested low thyroxin with increased thyrotropin, and individuals experience repeated exacerbation of thyroiditis with organophosphate exposure. The following is an illustrative case.

> **Case study**. This 58-year-old male electrical worker for 19 years developed a cold thyroid nodule, which was removed. He had a benign thyroid adenoma. He also had fibromyalgia and hypertension. His level of PCB was 10 ppb. These disappeared after the PCB levels became nondetectable following a course of heat depuration treatment. Subsequent challenge showed sensitivity to chlorine and phenol.

A defect in T-suppressor cells in patients with thyroiditis is postulated by Farid[371] and Scherbaum.[372] The EHC-Dallas has seen suppression of the suppressor cells in many of our endocrine chemically sensitive patients.

People with pollutant-triggered thyroid problems will frequently complain of fatigue, transient sore throats, coldness with subnormal body temperatures, and dry skin. They will often have other features of chemical sensitivity including odor sensitivity. They also will have had many viral infections and possibly recurrent *Candida* problems. As discussed in Chapter 19, changes will occur in autoantibodies to the thyroid and the microsomes. In addition, the patient may have an abnormally low thyroid profile.

In summary, many toxic agents in the environment have been shown to cause an imbalance in the endocrine system. Diagnosis and treatment of pollutant-triggered conditions of the endocrine system are consistent with those methods used with other pollutant-related illnesses and are, therefore, not given further discussion here. (See Volume IV, Chapter 38.)

REFERENCES

1. Rea, W. J. 1992. *Chemical Sensitivity. Vol. I. Mechanisms of Chemical Sensitivity*. 155. Boca Raton, FL: Lewis Publishers.
2. Hayes, W. J. 1982. *Pesticides Studied in Man*. 253, 444. Baltimore: Williams & Wilkins.
3. Stone, R. 1994. Environmental estrogens stir debate. *Science* 265:308–310.
4. Anderson, K., O. G. Nilsen, R. Toftgard, P. Eneroth, J. A. Gustafon, N. Battistini, and L. F. Agnati. 1983. Increased amine turnover in several hypothalamic terminal systems and changes in GH, LH, and prolactin secretion in the male rat by low concentrations of toluene. *Neurotoxicology* 4:43–56.

5. Rea, W. J. 1994. *Chemical Sensitivity. Vol. II. Sources of Total Body Load.* 765. Boca Raton, FL: Lewis Publishers.

6. Rea, W. J. 1992. *Chemical Sensitivity. Vol. I. Mechanisms of Chemical Sensitivity.* 47. Boca Raton, FL: Lewis Publishers.

7. Cooper, R. L., R. W. Chadwick, G. L. Rehnberg, J. M. Goldman, K. C. Booth, J. F. Gein, and W. K. McElroy. 1989. Effect of lindane on hormonal control of reproductive function in the female rat. *Toxicol. Appl. Pharmacol.* 99:384–394.

8. Woodard, J. C. and W. S. S. Jee. 1991. Skeletal system. In *Handbook of Toxicologic Pathology,* eds. W. M. Haschek and C. G. Rousseaux. 511–517. San Diego, CA: Academic Press.

9. The damage is strinkingly different from species to species. *Ronment,* Y B7. May 15, 1990.

10. Hayes, D. K., J. E. Pauly, and R. J. Reiter. 1990. *Chronobiology: Its Role in Clinical Medicine, General Biology, and Agriculture. Part B.* New York: John Wiley & Sons.

11. Reiter, R. J. 1982. *The Pineal and Its Hormones: Proceedings of an International Symposium. Jan. 2–9, 1982.* New York: Alan R. Liss.

12. Reiter, R. J. 1982. *The Pineal.* Vol. 7. Montreal: Eden Press.

13. Ott, J. N. 1985. Color and light: their effects on plants, animals and people. *Int. J. Biosoc. Res. Spec. Subj.* 8:1–35.

14. Parry, B. L., N. E. Rosenthal, L. Tamarkin, and T. A. Wehr. 1987. Treatment of a patient with seasonal premenstrual syndrome. *Am. J. Psychiatry* 144:762–766.

15. Couplan, R. E. 1965. *The Natural History of the Chromaffin Cell.* 1–279. London: Longmans, Green & Co.

16. Kjaergaard, J. 1973. *Anatomy of the Carotid Glomus and Carotid Glomus-Like Bodies (Non-Chromaffin Paraganglia). With Electron Microscopy and Comparison of Human Foetal Carotid, Aorticopulmonary, Subclavian, Tempano-jugular, and Vagal Glomera* (translated from the Danish by A. LaCour.). 328. Copenhagen: F. A. D. L.'s Forlag.

17. Gleener, G. G. and P. M. Grimley. 1974. *Tumors of the Extra-Adrenal Paraganglion System (Including Chemoreceptors), Fascicle 9, Atlas of Tumor Pathology (2nd Series).* Washington, D.C.: Armed Forces Institute of Pathology.

18. Carmichael, S. and H. Winkler. 1985. The adrenal chromaffin cell. *Sci. Am.* 253(2):40–49.

19. Grimley, P. M. and G. G. Glenner. 1968. Ultrastructure of the human caroid body. A perspective on the mode of chemoreception. *Circulation* 37:648–665.

20. Ferri, G.-L., L. Probert, D. Cocchia, F. Michetti, P. J. Marangos, and J. M. Polak. 1982. Evidence for the presence of S-100 protein in the glial component of the human enteric nervous system. *Nature* 297:409–410.

21. Lloyd, R. V., M. Blaivas, and B. S. Wilson. 1985. Distribution of chromogranin and S100 protein in normal and abnormal adrenal medullary tissues. *Arch. Pathol. Lab. Med.* 109:633–635.

22. Chaudhry, A. P., J. G. Haar, A. Koul, and P. A. Nickerson. 1979. A nonfunctioning paraganglioma of vagus nerve. *Cancer* 43:1689–1701.

23. Robertson, D. I. and T. P. Cooney. 1980. Malignant carotid body paraganglioma: light and electron microscopic study of the tumor and its metastases. *Cancer* 46:2623–2633.

24. Grimley, P. M. and R. A. DeLellis. 1986. Multisystem neuroendocrine neoplasms. In *Pathology of Incipient Neoplasia*, eds. J. Albores-Savedra and D. Henson. 425. Philadelphia: W. B. Saunders.

25. DeLellis, R. A. and H. J. Wolfe. 1981. The polypeptide hormone-producing neuroendocrine cells and their tumors. *Meth. Achiev. Exp. Pathol.* 10:190–220.

26. Rosenfeld, M. G., S. G. Amara, N. C. Birnberg, J.- J. Mermod, G. H. Murdock, and R. M. Evans. 1983. Calcitonin, prolactin and growth hormone gene expression as model systems for the characterization of neuroendocrine regulation. *Recent Prog. Horm. Res.* 39:305–351.

27. Douglass, J., O. Civelli, and E. Herbert. 1984. Polyprotein gene expression: generation of diversity of neuroendocrine peptides. *Ann. Rev. Biochem.* 53:665–715.

28. Wolfe, H., H. Childers, M. Montminy, R. Goodman, S. Punzak, R. Lechan, R. DeLellis, and A. Tischler. 1985. Use of antisense RNA probes for morphologic detection of peptide-producing cells by in situ hybridization. *Lab. Invest.* 52:77A.

29. Payne, C. M., R. B. Nagle, and V. Borduin. 1984. Methods in laboratory investigation. An ultrastructural cytochemical stain specific for neuroendocrine neoplasma. *Lab. Invest.* 51:350–365.

30. Balsera, E., R. V. Lloyd, S. K. Livingston, M. Lavallee, and H. A. Azar. 1986. Immunohistochemistry and electron microscopy of neuroendocrine neoplasms. *Lab. Invest.* 54:4A.

31. Pearse, A. G. E. 1969. The cytochemistry and ultrastructure of polypeptide hormone-producing cells of the APUD series and the embryologic, physiologic, and pathologic implications of the concept. *J. Histochem. Cytochem.* 17:303–313.

32. Pearse, A. G. E. 1977. The diffuse neuroendocrine system and the APUD concept. Related "endocrine" peptides in brain, intestine, pituitary, placenta, and anuran cutaneous glands. *Med. Biol.* 35:115.

33. Wasserman, K. 1980. Recent advances in carotid body physiology. *Fed. Proc.* 39:2626.

34. Polak, J. M. and S. R. Bloom. 1979. The diffuse neuroendocrine system. Studies of this newly discovered controlling system in health and disease. *J. Histochem. Cytochem.* 27:1398–1400.

35. Snyder, S. H. 1980. Brain peptides as neurotransmitters. *Science* 209:976–983.

36. Larsson, L.-I., N. Goltermann, J. F. Rehfeld, and T. W. Schwartz. 1979. Somatostatin cell processes as pathways for paracrine secretion. *Science* 205:1393–1395.

37. Krammer, E. B. 1978. Carotid body chemoreceptor function: hypothesis based on a new circuit model. *Proc. Natl. Acad. Sci.* 75:2507–2511.

38. Angeletti, R. H. and W. F. Hickey. 1986. A neuroendocrine marker in tissues of the immune system. *Science* 230:89–90.

39. Trojanowski, J. Q. and V. M.-Y. Lee. 1985. Expression of neurofilament antigens by normal and neoplastic human adrenal chromaffin cells. *New Engl. J. Med.* 313:101–104.

40. Sibley, R. K. 1985. The intermediate filament profile of neuro and neuroendocrine neoplasms. *Lab. Invest.* 52:62A.

41. Angeletti, R. H., J. A. Nolan, and S. Zaremba. 1985. Catecholamine storage vesicles: topolgraphy and function. *Trends Biochem. Sci.* 10:240–243.

42. Hassoun, J., G. Monges, P. Giraud, J. F. Henry, C. Charpin, H. Payan, and M. Toga. 1984. Immunohitochemical study of pheochromocytomas. An investigation of methionine-enkephalin, vasoactive intestinal peptide, somatostatin, corticotropin, endorphin, and calcitonin in 16 tumors. *Am. J. Pathol.* 114:56–63.

43. Lloyd, R. V., B. Shapiro, J. C. Sisson, V. Kalff, N. W. Thompson, and W. A. Beierwaltes. 1984. An immunohistochemical study of pheochromocytomas. *Arch. Pathol. Lab. Med.* 108:541–544.

44. Sano, T., H. Saito, H. Inaba, K. Hizawak, S. Saito, A. Yamanoi, Y. Mizunuma, M. Matsumura, M. Yuasa, K. Hiraishi. 1983. Immunoreactive somatostatin and vasoactive intestinal polypeptide in adrenal pheochromocytoma. *Cancer* 52:282–289.

45. Verhofstad, A. A. J., H. W. M. Steinbusch, H. W. J. Joosten, B. Penke, J. Varga, and M. Goldstein. 1983. Immunocytochemical localization of nonadrenaline, adrenaline and serotonin. In *Immunocytochemistry. Practical Applications in Pathology and Biology.* 143–168. Bristol: Wright PSG.

46. Schmechel, D. E. 1985. Subunit of the glycolytic enzyme enolase: nonspecific or neuron specific? *Lab. Invest.* 52:239–242.

47. Lloyd, R. V. and T. F. Warner. 1984. Immunohistochemistry of neuron-specific enolase. In *Advance in Immunochistochemistry*, ed. R. A. DeLellis. 127–140. New York: Masson Publishing.

48. Thomas, P., H. Battifora, and G. Manderino. 1986. Is neuron-specific enolase specific? An immunohistochemical comparison of a monoclonal and a polyclonal antibody against neuron-specific enolase. *Lab. Invest.* 54:63A.

49. Chambers, R. C., M. C. Bowling, and P. M. Grimley. 1968. Glutaraldehyde fixation in routine histopathology. *Arch. Path.* 85:18–30.

50. Smith, D. M. and R. C. Haggitt. 1983. A comparative study of generic stains for carcinoid secretory granules. *Am. J. Surg. Pathol.* 7:61–68.

51. Lewis, R. V., A. S. Stern, S. Kimura, J. Rossier, S. Stein, and S. Udenfriend. 1980. An about 50,000 dalton protein in adrenal medulla: a common precursor of [met] and [leu]-enkephalin. *Science* 200:1450–1461.

52. O'Connor, D. T., D. Burton, and L. J. Deftos. 1983. Chromogranin A: immunohistology reveals its universal occurrence in normal polypeptide hormone producing endocrine glands. *Life Sci.* 33:1657–1663.

53. Wilson, B. S. and R. V. Lloyd. 1984. Detection of chromogranin in neuroendocrine cells with a monoclonal antibody. *Am. J. Pathol.* 115:458–468.

54. Johnson, T. L., R. V. Lloyd, B. Shapiro, J. C. Sisson, and W. H. Beierwaltes. 1985. Cardiac paragangliomas: a clinicopathologic study of four cases. *Lab. Invest.* 52:31A.

55. Tischler, A. S., M. A. Dichter, B. Biales, and L. A. Greene. 1977. Neuroendocrine neoplasms and their cells of origin. *N. Engl. J. Med.* 296:919–925.

56. Roth, J., D. LeRoith, J. Shiloach, J. L. Rosenzweig, M. Lesniak, and J. Havrankova. 1982. The evolutionary origins of hormones, neurotransmitters, and other extracellular chemical messengers. *N. Engl. J. Med.* 306:523–526.

57. DeLellis, R. A., A. S. Tischler, A. K. Lee, M. Blount, and H. J. Wolfe. 1983. Leu-enkephalin-like immunoreactivity in proliferative lesions of the human adrenal medulla and extra-adrenal paraganglia. *Am. J. Surg. Pathol.* 7:29–37.

58. Schimke, R. N. 1980. The neurocristopathy concept: fact or fiction. In *Advances in Neuroblastoma Research. Proceedings of the 2nd Symposium on Advances in Neuroblastoma Research, Philadelphia, 1979,* ed. A. E. Evans. New York: Alan R. Liss.

59. Kissel, P., J. M. Andre, and A. Jacquier. 1981. *The Neurocristopathies.* 262. New York: Masson Publishing.

60. Le Dourain, N. M. 1982. *The Neural Crest.* 259. Cambridge, MA: Cambridge University Press.

61. Llena, J. F. 1983. Paraganlioma in the cerebrospinal axis. *Prog. Neuropathol.* 5:261–276.

62. Stahlman, M. and M. E. Gray. 1984. Ontogeny of neuroendocrine cells in human fetal lung. I. An electron microscopic study. *Lab. Invest.* 51:449–463.

63. Stahlman, M. T., A. G. Kasselberg, D. N. Orth, and M. E. Gray. 1985. Ontogeny of neuroendocrine cells in human fetal lung. II. An immunohistochemical study. *Lab. Invest.* 52:52–60.

64. Bosman, F. T. and J.-W. K. Louwerens. 1981. APUD cells in teratomas. *Am. J. Pathol.* 104:174–180.

65. Gould, V. E., R. I. Linnoila, V. A. Memoli, and W. H. Warren. 1983. Biology of disease. Neuroendocrine components of the bronchopulmonary tract: hyperplasias, dysplasias, and neoplasms. *Lab. Invest.* 5:519–537.

66. Manning, J. T., N. G. Ordonez, H. S. Rosenberg, and W. E. Walker. 1985. Pulmonary endodermal tumor resembling fetal lung. Report of a case with immunohistochemical studies. *Arch. Pathol. Lab. Med.* 109:48–50.

67. Sciurba, F. C., G. R. Owens, M. H. Sanders, B. P. Griffith, R. L. Hardesty, I. L. Paradis, and J. P. Costantino. 1988. Evidence of an altered pattern of breathing during exercise in recipients of heart-lung transplants. *N. Engl. J. Med.* 319:1186–1192.

68. Donald, D. E. and J. T. Shepherd. 1963. Response to exercise in dogs with cardiac denervation. *Am. J. Physiol.* 205:393–400.

69. Savin, W. M., W. L. Haskell, J. S. Schroeder, and E. B. Stinson. 1980. Cardiorespiratory responses of cardiac transplant patients to graded, symptom-limited exercise. *Circulation* 62:55–60.

70. Pope, S. E., E. B. Stinson, G. T. Daughters, 2nd, J. S. Schroeder, N. B. Ingels, Jr., and E. L. Alderman. 1980. Exercise response of the denervated heart in long-term cardiac transplant recipients. *Am. J. Cardiol.* 46:213–218.

71. Campeau, L., L. Pospisil, P. Grondin, I. Dyrda, and G. Lepage. 1970. Cardiac catheterization findings at rest and after exercise in patients following cardiac transplantation. *Am J. Cardiol.* 25:523–528.

72. Clark, D. A., J. S. Schroeder, R. B. Griepp, E. B. Stinson, E. Dong, N. E. Shumway, and D. G. Harrison. 1973. Cardiac transplantation in man: review of first three years' experience. *Am. J. Med.* 54:563–576.

73. Davies, C. T., J. Few, K. G. Foster, and A. J. Sargeant. 1974. Plasma catecholamine concentration during dynamic exercise involving different muscle groups. *Eur. J. Appl. Physiol.* 32:195–206.

74. Carter, D. and J. C. Eggleston. 1979 . *Tumors of the Lower Respiratory Tract, Fascicle 17, Atlas of Tumor Pathology* (2nd Series). Washington, D.C.: Armed Forces Institute of Pathology.

75. Ichinose, H., R. L. Hewitt, and T. Drapanas. 1971. Minute pulmonary chemodectoma. *Cancer* 28:692–700.

76. Costero, I., R. Barroso-Moguel, and A. Martinez-Palomo. 1972. Pleural origin of some of the supposed chemodectoid structures of the lung. *Beitr. Pathol.* 146:351–365.
77. Pearse, A. G. E. and J. M. Polak. 1978. The diffuse neuroendocrine system and the APUD concept. In *Gut Hormones,* ed. S. R. Bloom. 33. Edinburgh: Churchill Livingstone.
78. Andrew, A. 1976. APUD cells, apudomas and the neural crest. *S. Afr. Med. J.* 50:890–898.
79. Jubb, K. V. and K. McEntee. 1959. The relationship of ultimobranchial remnants and derivatives to tumor of the thyroid gland in cattle. *Cornell Veterinarian* 49:41–69.
80. Fernandes, B. J., Y. C. Bedard, and I. Rosen. 1982. Mucus-producing medullary cell carcinoma of the thyroid gland. *J. Clin. Pathol.* 78:536–540.
81. Dayal, Y. 1983. Endocrine cells of the gut and their neoplasms. In *Pathology of the Colon, Small Intestine and Anus,* ed. H. T. Norris. 267–302. New York: Churchill Livingstone.
82. Sidhu, G. S. 1979. The endodermal origin of digestive and respiratory tract APUD cells: histopathologic evidence and a review of the literature. *Am. J. Pathol.* 96:5–20.
83. Cox, W. F. and G. B. Pierce. 1982. The enodermal origin of the endocrine cells of an adenocarcinoma of the rat. *Cancer* 50:1530.
84. Di Sant'Agnese, P. A., K. L. DeMesy Jensen, C. J. Churukian, and M. M. Agarwal. 1985. Human prostatic endocrine-paracrine (APUD) cells. *Arch. Pathol. Lab. Med.* 109:607–612.
85. Di Sant'Agnese, P. A. and K. L. DeMesy Jensen. 1984. Somatostain and/or somatostatin-like immunoreactive endocrine-paracrine cells in the human prostate gland. *Arch. Pathol. Lab. Med.* 108:693–696.
86. Scully, R. E., P. Aguirre, and R. A. DeLellis. 1984. Argyrophilia, scrotonin and peptide hormones in the female genital tract and its tumors. *Int. J. Gynecol. Pathol.* 3:51–70.
87. Young, T. W. and T. V. Thrasher. 1982. Nonchromaffin paraganglioma of the uterus. *Arch. Pathol. Lab. Med.* 106:608–609.
88. Oberman, H. A., F. Holtz, L. A. Sheffer, and J. E. Magielski. 1968. Chemodectomas (nonchromaffin paragangliomas) of the head and neck. *Cancer* 21:838–851.
89. Lack, E. E. 1978. Hyperplasia of vagal and carotid body paraganglia in patients with chronic hypoxemia. *Am. J. Pathol.* 91:497–516.
90. Batsakis, J. D. 1980. Paragangliomas of the head and neck. In *Tumors of the Head and Neck,* ed. J. D. Batsakis. 369–380. Baltimore: Williams & Wilkins.
91. Gallivan, M. V. E., B. Chun, G. Rowden, and E. E. Lack. 1979. Laryngeal paraganglioma. Case report with ultrastructural analysis and literature review. *Am. J. Surg. Pathol.* 3:85–92.
92. Lack, E. E., A. L. Cubilla, and J. M. Woodruff. 1979. Paragangliomas of the head and neck region. *Hum. Pathol.* 10:191–218.
93. Buss, D. H., R. B. Marshall, F. G. Baird, and R. T. Myers. 1980. Paraganglioma of the thyroid gland. *Am. J. Surg. Pathol.* 4:589–593.
94. House, J. M., M. L. Goodman, R. R. Gacek, and G. L. Green. 1972. Chemodectomas of the nasopharynx. *Arch. Otolaryngol.* 96:38–141.

95. Kahn, L. B. 1976. Vagal body tumor (nonchromaffin paraganglioma, chemodectoma, and carotid body-like tumor) with cervial node metastasis and familial association. *Cancer* 38:2367–2377.

96. Tannir, N. M., N. Cortas, and C. Allam. 1983. A functioning catecholamine-secreting vagal body tumor. A case report and review of the literature. *Cancer* 52:932–935.

97. Persson, A. V., J. D. Frusha, P. F. Dial, and E. R. Jewell. 1985. Vagal body tumor: paraganglioma of the head and neck. *Calif. J. Clinicians* 35:232–237.

98. Reed, J. C., K. K. Hallet, and D. S. Felgin. 1978. Neural tumors of the thorax: subject review from the AFIP. *Radiology* 126:9–17.

99. Lack, E. E., R. A. Stillinger, D. B. Colvin, R. M. Groves, and D. G. Burnette. 1979. Aortic-pulmonary paraganglioma. Report of a case with ultrastructural study and review of the literature. *Cancer* 43:269–278.

100. Olson, J. L. and W. R. Salyer. 1978. Mediastinal paragangliomas (aortic body tumor). A report of four cases and a review of the literature. *Cancer* 41:2405–2412.

101. Gallivan, M. V. E., B. Chun, G. Rowden, and E. E. Lack. 1980. Intrathoracic paravertebral malignant paraganglioma. *Arch. Pathol. Lab. Med.* 104:46–51.

102. Melicow, M. M. 1977. One hundred cases of pheochromocytoma (107 tumors) at the Columbia-Presbyterian Medical Center, 1926–1976. A clinicopathological analysis. *Cancer* 40:1987–2004.

103. Lack, E. E., A. L. Cubilla, J. M. Woodruff, and P. H. Lieberman. 1980. Extra-adrenal paragangliomas of the retroperitoneum. A clinicopathologic study of 12 tumors. *Am. J. Surg. Pathol.* 4:109–129.

104. Thacker, W. C. and J. K. Duckworth. 1969. Chemodectoma of the orbit. *Cancer* 23:1233–1238.

105. Perrone, T., R. K. Sibley, and J. Rosai. 1985. Duodenal gangliocytic paraganglioma: an immunohistochemical and ultrastructural study and a hypothesis concerning its origin. *Am. J. Surg. Pathol.* 9:31–41.

106. Miller, T. A., T. R. Weber, and H. D. Appelman. 1972. Paraganglioma of the gallbladder. *Arch. Surg.* 105:637–639.

107. Leestma, J. E. and E. B. Price, Jr. 1971. Paraganglioma of the urinary bladder. *Cancer* 28:1063–1073.

108. Lipper, S. and R. E. Decker. 1984. Paraganglioma of the cauda equina: a histologic, immunohistochemical, and ultrastructural study and review of the literature. *Surg. Neurol.* 22:415–420.

109. Albores-Saavedra, J., M. E. Maldonado, J. Ibarra, and H. A. Rodriguez. 1969. Pheochromocytoma of the urinary bladder. *Cancer* 23:1110–1118.

110. White, M. C. and B. R. Hickson. 1979. Multiple paragangliomata secreting catecholamines and calcitonin with intermittent hypercalcemia. *J. R. Soc. Med.* 72(7):532–535.

111. Apple, D. and K. Kreines. 1982. Cushing's syndrome due to ectopic ACTH production by a nasal paraganglioma. *Am. J. Med. Sci.* 283:32-35.

112. Grizzle, W. E., L. Tolbert, C. S. Pittman, A. L. Siegel, and J. S. Aldrete. 1983. Corticotropin production by tumors of the autonomic nervous system. *Arch. Pathol. Lab. Med.* 108:545–550.

113. Farrior III, J. B., V. J. Hyams, R. H. Benke, and J. B. Farrior. 1980. Carcinoid apudoma arising in a glomus jugulare tumor: review of endocrine activity in glomus jugulare tumors. *Laryngoscope* 90:110–119.

114. Pfeiffer, C. C. 1975. *Mental and Elemental Nutrients.* New Canaan, CT: Keats Publishing.

115. Saito, H., S. Saito, T. Sano, N. Kagawa, K. Hizawa, and K. Tatara. 1982. Immunoreactive somatostatin in catecholamine-producing extra-adrenal paraganglioma. *Cancer* 50:560–565.

116. Tischler, A. S., A. K. Lee, G. Nunnemacher, S. I. Said, R. A. DeLellis, G. M. Morse, and H. J. Wolfe. 1981. Spontaneous neurite outgrowth and vasoactive intestinal peptide-like immunoreactivity of cultures of human paraganglioma cells from the glomus jugulare. *Cell Tissue Res.* 219:543–555.

117. Markel, S. F. and R. M. Johnson. 1975. The clinical features and laboratory diagnosis of functional paragangliomas. *Lab. Med.* 6(10):39–44, 65.

118. Bravo, E. L. and R. W. Gifford, Jr. 1984. Pheochromocytoma: diagnosis, localization and management. *N. Engl. J. Med.* 311:1298–1303.

119. Gagel, R. F., K. E. W. Melvin, A. H. Tashjian, Jr., H. H. Miller, Z. T. Feldman, H. J. Wolfe, R. A. DeLellis, S. Cervi-Skinner, and S. Reichlin. 1975. Natural history of the familial medullary carcinom-phaeochromocytoma syndrome and the identification of pre-neoplastic stages by screening studies: a five-year report. *Trans. Assoc. Am. Phys.* 88:177–191.

120. Hamilton, B. R., L. Landsberg, and R. J. Levine. 1978. Measurement of urinary epinephrine in screening for pheochromocytoma in multiple endocrine neoplasia type II. *Am. J. Med.* 65:1027–1032.

121. Brennan, M. F. 1985. Cancer of the endocrine system. In *Cancer: Principles and Practice of Oncology,* 2nd ed., eds. Vincent T. DeVita, Jr., Samuel Hellman, and Steven A. Rosenberg. Philadelphia: Lippincott.

122. DeLellis, R. A., H. J. Wolfe, R. F. Gagel, Z. T. Friedman, H. H. Miller, D. L. Gang, and S. Reichlin. 1976. Adrenal medullary hyperplasia: a morphometric analysis in patients with familial medullary thyroid carcinoma. *Am. J. Pathol.* 83:177–196.

123. Jago, R., P. Smith, and D. Heath. 1984. Electron microscopy of carotid body hyperplasia. *Arch. Pathol. Lab. Med.* 108:717–722.

124. Fitch, R., P. Smith, and D. Heath. 1985. Nerve axons in carotid body hyperplasia. *Arch. Pathol. Lab. Med.* 109:234–238.

125. Luoma, J. R. May 15, 1990. Scientists are unlocking secrets of dioxins devastating power. The New York Times.

126. Rea, W. J. 1992. *Chemical Sensitivity. Vol. I. Mechanisms of Chemical Sensitivity.* 17. Boca Raton, FL: Lewis Publishers.

127. Jeffries, W. M. 1981. *Safe Uses of Cortisone.* Springfield, IL: Charles C Thomas.

128. Couch, J. R. and J. Wortman. 1984. Anovulatory states as a factor in occurrence of migraine. Paper presented at Migraine Trust, Fifth International Symposium.

129. Gelfand, J. A., R. J. Sherms, D. W. Alling, and M. M. Frank. 1976. Treatment of hereditary angio-edema with Danazol: reversal of clinical and biochemical abnormalities. *N. Engl. J. Med.* 295:1444.

130. Rea, W. J. 1994. *Chemical Sensitivity. Vol. II. Sources of Total Body Load.* 837. Boca Raton, FL: Lewis Publishers.

131. Uphouse, L. 1987. Decreased rodent sexual receptivity after lindane. *Toxicol. Lett.* 39(1):7–14.

132. Thomas, K. B. and T. Colborn. 1992. Organochlorine endocrine disruptors in human tissue. In *Advances in Modern Environmental Toxicology. Vol. XXI. Chemically-Induced Alterations in Sexual and Functional Development. The Wildlife/Human Connection*, eds. T. Colborn and C. Clement. 379. Princeton: Princeton Scientific.

133. Hamilton, A. and H. L. Hardy. 1974. *Industrial Toxicology*, 3rd ed. Acton, England: Publishing Sciences.

134. Mattison, D. R. 1980. Morphology of oocyte and follicle destruction by polycyclic aromatic hydrocarbons in mice. *Toxicol. Appl. Pharmacol.* 53(2):249–259.

135. Everson, R. B., D. P. Sandler, A. J. Wilcox, D. Schreinemachers, D. L. Shore, and C. Weinberg. 1986. Effect of passive exposure to smoking on age at natural menopause. *Br. Med. J. [Clin. Res.]* 293(6550):792.

136. De Coufle, P., J. W. Lloyd, and L. G. Salvin. 1977. Causes of death among construction machinery operators. *J. Occup. Med.* 19(2):123–128.

137. Pérez-Comas, A. 1982. Precocious sexual development in Puerto Rico. *Lancet* 1:1299–1300.

138. Meirik, O. 1986. Oral contraceptives and breast cancer in young women. Some notes on a current controversy. *Acta Obstet. Gynecol. Scand. [Suppl.]* 134:5–7.

139. Muntean, W. 1987. Spontaneous deep vein thrombosis in children and adolescents [letter] *J. Pediatr. Surg.* 22(2):188.

140. DeVito, M. J., T. Thomas, E. Martin, T. H. Umbreit, and M. A. Gallo. 1992. Antiestrogenic action of 2,3,7,8-tetra-chlordibenzo-*p*-dioxin: tissue-specific regulation of estrogen receptor in CDI mice. *Toxicol. Appl. Pharmacol.* 113:284–292.

141. Safe, S., B. Astroff, M. Harris, T. Zacharewski, R. Dickerson, M. Romkes, and L. Biegal. 1991. 2,3,7,8-tetrachlorodibenzo-*p*-dioxin (TCDD) and related compounds as antiestrogens: characterization and mechanisms of action. *Pharmacol. Toxicol.* 69:400–409.

142. Gray, L. E. 1992. Chemical-induced alterations of sexual differentiation: a review of effects in humans and rodents. In *Advances in Modern Environmental Toxicology. Vol. XXI. Chemically-Induced Alterations in Sexual and Functional Development: The Wildlife/Human Connection,* eds. T. Colborn and C. Clement. 203. Princeton, NJ: Princeton Scientific.

143. Gray, L. E. 1992. Chemical-induced alterations of sexual differentiation: a review of effects in humans and rodents. In *Advances in Modern Environmental Toxicology. Vol. XXI. Chemically-Induced Alterations in Sexual and Functional Development: The Wildlife/Human Connection,* eds. T. Colborn and C. Clement. 206. Princeton, NJ: Princeton Scientific.

144. Meyer-Bahlburg, H. F. L., A. A. Ehrhardt, J. F. Feldman, L. R. Rosen, N. P. Veridiano, and I. Zimmerman. 1985. Sexual activity level and sexual functioning in women prenatally exposed to diethylstibestrol. *Psychosomatic Med.* 47(6):497–511.

145. Gray, L. E. 1992. Chemical-induced alterations of sexual differentiation: a review of effects in humans and rodents. In *Advances in Modern Environmental Toxicology. Vol. XXI. Chemically-Induced Alterations in Sexual and Functional Development: The Wildlife/Human Connection,* eds. T. Colborn and C. Clement. 207. Princeton, NJ: Princeton Scientific.

146. Hines, M. and C. Shipley. 1984. Prenatal exposure to diethylstilbestrol (DES) and the development of sexually dimorphic cognitive abilities and cerebral laterilization. *Develop. Psych.* 20(1):81–94.

147. McGlone, J. 1980. Sex differences in human brain asymmetry: a critical survey. *Behav. Brain Sci.* 3:215–263.

148. Hines, M. 1992. Surrounded by estrogens? Considerations for neurobehavioral development in human beings. In *Advances in Modern Environmental Toxicology. Vol. XXI. Chemically-Induced Alterations in Sexual and Functional Development: The Wildlife/Human Connection,* eds. T. Colborn and C. Clement. 269. Princeton, NJ: Princeton Scientific.

149. Hines, M. 1992. Surrounded by estrogens? Considerations for neurobehavioral development in human beings. In *Advances in Modern Environmental Toxicology. Vol. XXI. Chemically-Induced Alterations in Sexual and Functional Development: The Wildlife/Human Connection,* eds. T. Colborn and C. Clement. 274. Princeton, NJ: Princeton Scientific.

150. Blair, P. B. 1992. Immunologic studies of women exposed in utero to diethylstilbestrol. In *Advances in Modern Environmental Toxicology. Vol. XXI. Chemically-Induced Alterations in Sexual and Functional Development: The Wildlife/Human Connection,* eds. T. Colborn and C. Clement. 289. Princeton, NJ: Princeton Scientific.

151. Noller, K. L., P. B. Blair, P. C. O'Brien, and L. J. Mellon. 1988. Increased occurrence of autoimmune disease among women exposed in utero to diethylstilbestrol. *Fertil. Steril.* 49:1080–1082.

152. Swaab, D. F. and E. A. Fliers. 1985. Sexually dimorphic nucleus in the human brain. *Science* 228:1112–1114.

153. LeVay, S. 1991. A difference in hypothalamic structure between heterosexual and homosexual men. *Science* 253:1034–1037.

154. Hines, M. 1992. Surrounded by estrogens? Considerations for neurobehavioral development in human beings. In *Advances in Modern Environmental Toxicology. Vol. XXI. Chemically-Induced Alterations in Sexual and Functional Development: The Wildlife/Human Connection,* eds. T. Colborn and C. Clement. 261. Princeton, NJ: Princeton Scientific.

155. Soto, A. M., T.-M. Lin, H. Justicia, R. M. Silvia, and C. Sonnenschein. 1992. An "in culture" bioassay to assess the estrogenicity of xenobiotics (E-screen). In *Advances in Modern Environmental Toxicology. Vol. XXI. Chemically-Induced Alterations in Sexual and Functional Development: The Wildlife/Human Connection,* eds. T. Colborn and C. Clement. 302. Princeton, NJ: Princeton Scientific.

156. Meyers, C. Y., W. S. Matthews, L. L. Holl, V. M. Kolb, and T. E. Parady. 1977. Carboxylic acid formation from kepone. In *Catalysis in Organic Synthesis,* ed. G. W. Smith. 213–215, 253–255. New York: Academic Press.

157. Palmiter, R. D. and E. R. Mulvihill. 1978. Estrogenic activity of the insecticide kepone on the chicken oviduct. *Science* 201:356–358.

158. Hammond, B., B. S. Katzenollenbogen, N. Kranthammer, and J. McConnell. 1979. Estrogenic activity of the insecticide chlordecone (kepone) and interaction with uterine estrogen receptors. *Proc. Natl. Acad. Sci. U.S.A.* 76:6641–6645.

159. Soto, A. M., T.-M. Lin, H. Justicia, R. M. Silvia, and C. Sonnenschein. 1992. An "in culture" bioassay to assess the estrogenicity of xenobiotics (E-screen). In *Advances in Modern Environmental Toxicology. Vol. XXI. Chemically-Induced Alterations in Sexual and Functional Development: The Wildlife/Human Connection,* eds. T. Colborn and C. Clement. 305. Princeton, NJ: Princeton Scientific.

160. Amdur, M. O., J. Doull, and C. D. Klaassen, eds. 1991. *Casarett and Doull's Toxicology: The Basic Science of Poisons.* 314, 505–506, 565–622. New York: Pergamon Press.

161. Moses, M. 1993. Pesticides. In *Occupational and Environmental Reproductive Hazards: A Guide for Clinicians,* ed. M. Paul. 296. Baltimore: Williams & Williams.

162. Metzler, M. 1985. Role of metabolism in determination of hormonal activity of estrogens: introductory remarks. In *Estrogens in the Environment II — Influences on Development,* ed. J. A. McLachlan. 187–189. New York: Elsevier.

163. Blaich, G., E. Pfaff, and M. Metzler. 1987. Metabolism of diethylstibestrol in hamster hepatocytes. *Biochem. Pharmacol.* 36:3135–3140.

164. Whitten, P. and F. Naftolin. 1991. Dietary plant estrogens: a biologically active background for estrogen action. In *The New Biology of Steroid Hormones,* eds. R. Hochberg and F. Naftolin. 155–167. New York: Raven Press.

165. Whitten, P. L. 1992. Chemical revolution to sexual revolution: historical changes in human reproductive development. In *Advances in Modern Environmental Toxicology. Vol. XXI. Chemically-Induced Alterations in Sexual and Functional Development: The Wildlife/Human Connection,* eds. T. Colborn and C. Clement. 313. Princeton, NJ: Princeton Scientific.

166. Kellis, J. T. and L. E. Vickery. 1984. Inhibition of human estrogen synthetase (aromatase) by flavones. *Science* 225:1032–1034.

167. Adlercreutz, H. 1990. Western diet and western diseases: some hormonal and biochemical mechanisms and associations. *Scand. J. Clin. Lab. Invest.* 50 Suppl. 201:3–23.

168. Trowell, H. C. and D. P. Burkitt. 1983. *Western Diseases: Their Emergence and Prevention.* London: Edward Arnold.

169. Gorski, R. A. 1968. Influence of age on the response to perinatal administration of a low dose of androgen. *Endocrinology* 82:1001–1004.

170. Hines, M., P. Alsum, M. Roy, R. A. Gorski, and R. W. Goy. 1987. Estrogenic contributions to sexual differentiation in the female guinea pig: influences of diethylstilbestrol and tamoxifen on neural, behavioral, and ovarian development. *Horm. Behav.* 21:402–417.

171. Ramirez, V. D. and C. H. Sawyer. 1965. Advancement of puberty in the female rat by estrogen. *Endocrinology* 76:1158–1168.

172. Matsumo, A. and Y. Arain. 1977. Precocious puberty and synaptogenesis in the hypothalamic arcuate nucleus in pregnant mare serum gonadotropin (PMSG) treated immature female rats. *Brain Res.* 129:375–378.

173. Matsumo, A. and Y. Arain. 1980. Sexual dimorphism in "wiring pattern" in the hypothalamic arcuate nucleus and its modifications by neonatal hormonal environment. *Brain Res.* 190:238–242.

174. Lephart, E. D., D. Mathews, J. F. Noble, and S. R. Ojeda. 1989. The vaginal epithelium of immature rats metabolizes androgens through an aromatase-like reaction. Changes through the time of puberty. *Biol. Reprod.* 40:259–267.

175. Frem-Titulaer, L. W., J. F. Cordero, L. Haddock, G. Lebrón, R. Martinez, and J. L. Mills. 1986. Premature theiarche in Puerto Rico: a search for environmental factors. *AJDC* 140:1263–1267.

176. Grant, E. 1985. *The Bitter Pill: How Safe is the "Perfect Contraceptive"?* 38. London: Transworld Publishers.

177. Kimball, A. M., R. Hamadeh, R. A. H. Mahmood, S. Khalfan, A. Muhsin, F. Ghabrial, and H. K. Armenian. 1981. Gynaecomastia among children in Bahrain. *Lancet* 1:671–672.

178. Grant, E. 1985. *The Bitter Pill: How Safe is the "Perfect Contraceptive"?* London: Transworld Publishers.

179. Wright, H. P. 1960. General properties of blood: the formed elements. In *Medical Physiology and Biophysics,* eds. T. C. Ruch and J. F. Fulton. 502–528. Philadelphia: W. B. Saunders.

180. Vessey, M. P. 1980. Female hormones and vascular disease — an epidemiological overview. *Br. J. Fam. Plann.* 6(3):1.

181. Grant, E. 1985. *The Bitter Pill: How Safe is the "Perfect Contraceptive"?* 118. London: Transworld Publishers.

182. Grant, E. 1985. *The Bitter Pill: How Safe is the "Perfect Contraceptive"?* 61–64. London: Transworld Publishers.

183. Grant, E. 1985. *The Bitter Pill: How Safe is the "Perfect Contraceptive"?* 89. London: Transworld Publishers.

184. Hibberd, A. R., V. O'Connor, and J. W. Gorrod. 1978. Detection of nicotine, nicotine-i-N-oxide and cotinine in maternal and foetal body fluids. In *Biological Oxidation of Nitrogen,* ed. J. W. Gorrod. 353–361. New York: Elsevier/North Holland Biomedical.

185. Hibberd, A. R. 1979. Studies on the metabolism and excretion of nicotine and some related compounds. Ph.D. thesis. University of London, U.K.

186. Hibberd, A. R., Y. Abrahams, and J. W. Gorrod. 1980. Metabolism of nicotine, cotinine and nicotine$\Delta^{1'(5')}$ iminium ion by human foetal liver, in vitro. In *Clinical Pharmacy III,* eds. H. Turakka and E. Van der Kleign. 79–88. New York: Elsevier/North Holland Biomedical.

187. Follen, M. M., H. E. Tox, and R. U. Levine. 1985. Cervical vascular malformation as a cause of antepartum and intrapartum bleeding in three diethylstilbestrol-exposed progeny. *Am. J. Obstet. Gynecol.* 153(8):890–891.

188. Witze, A. Feb. 14, 1994. Chemicals suspected in endometriosis: possible role of dioxin, other environmental factors studied. *The Dallas Morning News.*

189. Witze, A. Feb. 14, 1994. Chemicals suspected in endometriosis: possible role of dioxin, other environmental factors studied. *The Dallas Morning News.*

190. Rier, S. E., D. C. Martin, R. E. Bowman, W. P. Dmowski, and J. L. Becker. 1993. Endometriosis in rhesus monkeys (*Macaca mulatta*) following chronic exposure to 2,3,7,8-tetrachlorodibenzo-*p*-dioxin. *Fundam. Appl. Toxicol.* 21:433–441.

191. Campbell, J. S., J. wong, L. Tryphonas, D. L. Arnold, E. Nera, B. Cross, and E. LaBossiere. 1985. Is simian endometriosis an effect of immunotoxicity? Presented at the Ontario Association of Pathologists 48th Annual Meeting, London, Ontario.

192. Poland, A. and J. C. Knutson. 1982. 2,3,7,8-tetrachlorodibenzo-*p*-dioxin and related halogenated aromatic hydrocarbons: examination of the mechanism of toxicity. *Ann. Rev. Pharmacol. Toxicol.* 22:517–554.

193. Wood, D. H., M. G. Yochmowitz, Y. L. Salmon, R. I. Eason, and R. A. Boster. 1983. Proton irradiation and endometriosis. *Aviat. Space Environ. Med.* 54:718–724.

194. Fanton, J. W. and J. G. Golden. 1991. Radiation-induced endometriosis in *Macaca mulatta. Radiat. Res.* 126:132–140.

195. Wood, D. H. 1991. Long term mortality and cancer risk in irradiated rhesus monkeys. *Radiat. Res.* 126:132–140.

196. Dmowski, W. P., D. Braun, and H. Gebel. 1991. The immune system in endometriosis. In *Modern Approaches to Endometriosis*, eds. E. J. Thomas and J. A. Rock. 97–111. Boston: Kluwer Academic.

197. Hill, J. A. 1992. Immunologic factors in endometriosis and endometriosis-associated reproductive failure. *Infertil. Reprod. Med. Clin. N. Am.* 3:583–596.

198. Holsapple, M. P., N. K. Snyder, S. C. Wood, and D. L. Morris. 1991. A review of 2,3,7,8-tetrachlorodibenzo-*p*-dioxin-induced changes in immunocompetence: 1991 update. *Toxicology* 69:219–255.

199. Neubert, R., U. Jacob-Muller, R. Stahlmann, H. Helge, and D. Neubert. 1991. Polyhalogenated dibenzo-*p*-dioxins and dibenzofurans and the immune system. *Arch. Toxicol.* 65:213–219.

200. Tomar, R. S. and N. I. Kerkvliet. 1991. Reduced T-helper cell function in mice exposed to 2,3,7,8-tetrachlorodibenzo-*p*-dioxin (TCDD). *Toxicol. Lett.* 57:55–64.

201. Sikorski, R., T. Juszkiewicz, T. Paszkowski, and T. Szprengier-Juszkiewicz. 1987. Women in dental surgeries: reproductive hazard in occupational exposure to metallic mercury. *Int. Arch. Occup. Environ. Health* 59(6):551–557.

202. Landrigan, P. J., T. J. Meinhardt, J. Gordon, J. A. Lipscomb, J. R. Burg, L. F. Mazzuckelli, T. R. Lewis, and R. A. Lemen. 1984. Ethylene oxide: an overview of toxicologic and epidemiologic research. *Am. J. Ind. Med.* 6(2):103–115.

203. Selevan, S. G., M. L. Lindbohm, R. W. Hornung, and K. Heniminki. 1985. A study of occupational exposure to antineoplastic drugs and fetal loss in nurses. *N. Engl. J. Med.* 313(19):1173–1178.

204. Heidam, L. Z. 1984. Spontaneous abortions among dental assistants, factory workers, painters, and gardening workers: a follow-up study. *J. Epidemiol. Community Health* 38(2):149–155.

205. Axelson, O., B. Johansson, and U. Flodin. 1983. Unidentified risk factor [letter]. *J. Occup. Med.* 25(3):181.

206. Axelson, O., C. Edling, and L. Andersson. 1983. Pregnancy outcome among women in a Swedish rubber plant. *Scand. J. Work Environ. Health* 2(Suppl.):79–83.

207. Lindbohm, M. L., K. Hemminki, P. Kyyrönen. 1985. Spontaneous abortion among women employed in the plastics industry. *Am. J. Ind. Med.* 8(6):579–586.

208. Ungváry, G. and E. Tátrai. 1985. On the embryotoxic effects of benzene and its alkyl derivatives in mice, rats and rabbits. *Arch. Toxicol. [Suppl.]* 8:425–430.

209. Curle, D. C., M. Ray, and T. V. Persaud. 1983. Methylmercury toxicity: in vivo evaluation of teratogenesis and cytogenetic changes. *Anat. Anz.* 153(1):69–82.

210. Varma, D. R. 1987. Epidemiological and experimental studies on the effects of methyl isocyanate on the course of pregnancy. *Environ. Health Perspect.* 72:153–157.

211. Ludmir, J., M. B. Landon, S. G. Gabbe, P. Samuels, and M. T. Mennuti. 1987. Management of the diethylstibestrol-exposed pregnant patient: a prospective study. *Am. J. Obstet. Gynecol.* 157(3):665–669.

212. Menczer, J., M. Dulitzky, G. Ben-Baruch, and M. Modan. 1986. Primary infertility in women exposed to diethylstiboestrol in utero. *Br. J. Obstet. Gynaecol.* 93(5):503–507.

213. Henderson, L. and T. Regan. 1985. Effects of diethylstiboestrol-dipropionate on SCEs, micronuclei, cytotoxicity, aneuploidy and cell proliferation in maternal and foetal mouse cell treated in vivo. *Mutat. Res.* 144(1):27–31.

214. Ben Ismail, M., F. Abid, S. Trablsi, M. Taktak, and M. Fekih. 1986. Cardiac valve prostheses, anticoagulation, and pregnancy. *Br. Heart J.* 55(1):101–105.

215. Lee, P. K., R. Y. Wang, J. S. Chow, K. L. Cheung, V. C. Wong, T. K. Chan. 1986. Combined use of warfarin and adjusted subcutaneous heparin during pregnancy in patients with an artificial heart value. *J. Am. Coll. Cardiol.* 8(1):221–224.

216. O'Donnell, D., H. Sevitz, J. L. Seggie, A. M. Meyers, J. R. Botha, and J. A. Myburgh. 1985. Pregnancy after renal transplantation. *Aust. N.Z. J. Med.* 15(3):320–325.

217. Lammer, E. J., D. T. Chen, R. M. Hoar, N. D. Agnish, P. J. Renke, J. T. Braan, C. J. Curry, P. M. Fernhoff, A. W. Grix, Jr., I. T. Lott, J. M. Richard, and S. C. Sun. 1985. Retinoic acid embryopathy. *N. Engl. J. Med.* 313(14):837–841.

218. Stern, R. S., F. Rosa, and C. Baum. 1984. Isotretinoin and pregnancy. *J. Am. Acad. Dermatol.* 10(5 Pt 1):851–854.

219. Rosa, F. W., A. L. Wilk, and F. O. Kelsey. 1986. Teratogen update: vitamin A congeners. *Teratology* 33:355–364.

220. Rosa, F. W., C. Baum, and M. Shaw. 1987. Pregnancy outcomes after first-trimester vaginitis drug therapy. *Obstet. Gynecol.* 69(5):751–755.

221. Matsuoka, R., E. F. Gilbert, H. Bruyers, Jr., J. M. Optiz. 1985. An aborted human fetus with truncus arteriosus communis — possible teratogenic effect of Tedral. *Heart Vessels* 1(3):176–178.

222. Källen, B. and A. Tandberg. 1983. Lithium and pregnancy. A cohort study on manic depressive women. *Acta Psychiatr. Scand.* 68(2):134–139.

223. Kihlström, I. and C. Lundberg. 1987. Teratogenicity study of the new glucocorticosteroid budesonide in rabbits. *Arzneimittel Forsch.* 37(1):43–46.

224. Olah, M. E. and R. G. Rahwan. 1986. Evaluation of the antiabortifacient and embryotoxic effects of methylenedioxyindene and methylenedioxyindan calcium antagonists. *Gen. Pharmacol.* 17(5):549–552.

225. Kar, R. N., K. Khan, and S. K. Mukherjee. 1984. In vivo mutagenic effect of methyldopa. I. Dominant lethal test in male mice. *Cytobios* 41:151–159.

226. Lynch, P. J. 1984. Abortion in sows after injection of a suspension of penicillin and streptomycin. *Aust. Vet. J.* 61(1):29.

227. Nikkanen, V., P. Katainen, and O. Piiroinen. 1987. Progesterone support of the luteal phase in in vitro fertilization program — a hazard? *Ann. Chir. Gynacol. [Suppl.]* 202:42–44.

228. Bohrer, M. and E. Kemmann. 1987. Risk factors for spontaneous abortion in menotropin-treated women. *Fertil. Steril.* 48(4):571–575.

229. Rock, J. A., A. C. Wentz, K. A. Cole, A. W. Kimball, Jr., H. A. Zacur, S. A. Early, and G. S. Jones. 1985. Fetal malformations following progesterone therapy during pregnancy: a preliminary report. *Fertil. Steril.* 44(1):17–19.

230. Hendrickx, A. G., R. Korte, F. Leuschner, B. W. Neumann, S. Prahalada, A. Poggel, P. E. Binkerd, and P. Günzel. 1987. Embryotoxicity of sex steroidal hormone combinations in nonhuman primates: I. Norethisterone acetate + ethinyl estradiol and progesterone + estradiol benzoate *(Macaca mulatta, Macaca fascicularis,* and *Papio cynocephalus). Teratology* 35(1):119–127.

231. Hendrickx, A. G., R. Korte, F. Leuschner, B. W. Neumann, A. Poggel, P. Binkerd, S. Prahalada, and P. Günzel. 1987. Embryotoxicity of sex steroidal hormones in nonhuman primates: II. Hydroxyprogesterone caproate, estradiol valerate. *Teratology* 35(1):129–136.

232. Rogan, W. J., B. C. Gladen, and A. J. Wilcox. 1985. Potential reproductive and postnatal morbidity from exposure to polychlorinated biphenyls: epidemiologic considerations. *Environ. Health Perspect.* 60:233–239.

233. Bercovici, B., M. Wassermann, S. Cucos, M. Ron, D. Wassermann, and A. Pines. 1983. Serum levels of polychlorinated biphenyls and some organochlorine insecticides in women with recent and former missed abortions. *Environ. Res.* 30(1):169–174.

234. McNulty, W. P. 1985. Toxicity and fetotoxicity of TCDD, TCDF and PCB isomers in rhesus macaques *(Macaca mulatta). Environ. Health Perspect.* 60:77–88.

235. Rita, P., P. P. Reddy, and S. V. Reddy. 1987. Monitoring of workers occupationally exposed to pesticides in grape gardens of Andhra Pradesh. *Environ. Res.* 44(1):1–5.

236. Sterling, T. D. and A. V. Arundel. 1986. Review of recent Vietnamese studies on the carcinogenic and teratogenic effects of phenoxy herbicide exposure. *Int. J. Health Serv.* 16(2):265–278.

237. Kaye, C. I., S. Rao, S. J. Simpson, F. S. Rosenthal, and M. M. Cohen. 1985. Evaluation of chromosomal damage in males exposed to agent orange and their families. *J. Craniofac. Genet. Dev. Biol. [Suppl.]* 1:259–265.

238. Blume, S. B. 1986. Women and alcohol. A review. *JAMA* 256(11):1467–1470.

239. Abel, E. T. 1984. Prenatal effects of alcohol. *Drug Alcohol Depend* 14(1):1–10.

240. Kaufman, M. H. 1983. Ethanol-induced chromosomal abnormalities at conception. *Nature* 302(5905):258–260.

241. Clarren, S. K., D. M. Bowden, and S. J. Astley. 1987. Pregnancy outcomes after weekly oral administration of ethanol during gestation in the pig-tailed macaque *(Macaca nemestrina). Teratology* 35(3):345–354.

242. Asch, R. H. and C. G. Smith. 1986. Effects of delta 9-THC, the principal psychoactive component of marijuana, during pregnancy in the rhesus monkey. *J. Reprod. Med.* 31(12):1071–1081.

243. Chasnoff, I. J., W. J. Burns, S. H. Schnoll, and K. A. Burns. 1985. Cocaine use in pregnancy. *N. Engl. J. Med.* 313(11):666–669.

244. Srisuphan, W. and M. R. Bracken. 1986. Caffeine consumption during pregnancy and association with late spontaneous abortion. *Am. J. Obstet. Gynecol.* 154(1):14–20.

245. Eroschenko, V. P. and F. Osman. 1986. Scanning electron microscopic changes in vaginal epithelium of suckling neonatal mice in response to estradiol or insecticide chlorodecone (kepone) passages in milk. *Toxicology* 38(2):175–185.

246. Rolirson, A. K., W. A. Schmidt, and G. M. Stancel. 1985. Estrogenic activity of DDT: estrogen-receptor for profiles and the responses of individual uterine cell types following o,p-DDT administration. Toxicol. Environ. Health 16(3–4):493–508.

247. EHC-Dallas. 1994. Unpublished data.

248. Graham, E., A. H. Chignell, and S. Eykyn. 1986. Candida endophthalmitis: a complication of prolonged intravenous therapy and antibiotic treatment. J. Infect. 13(2):167–173.

249. McKay, M. 1993. Dysesthetic ("essential") vulvodynia: treatment with amitriptyline. J. Reprod. Med. 38(1):9–13.

250. Working, P. K., J. S. Bus, and T. E. Namm, Jr. 1985. Reproductive effects of inhaled methyl chloride in the male Fischer 344 rat. II. Spermatogonial toxicity and sperm quality. Toxicol. Appl. Pharmacol. 77(1):144–157.

251. Ball, H. S. 1984. Effect of methoxychlor on reproduction systems of the rat. Proc. Soc. Exp. Biol. Med. 176(2):187–196.

252. Carter, S. D., J. F. Hein, G. L. Rehnberg, and J. W. Laskey. 1984. Effect of benomyl on the reproductive development of male rats. J. Toxicol. Environ. Health 13(1):53–68.

253. Sandifer, S. H., R. T. Wilkins, C. B. Loadholt, L. G. Lane, and J. C. Eldridge. 1979. Spermatogenesis in agricultural workers exposed to dibromochloropropane (DBCP). Bull. Environ. Contam. Toxicol. 23(4–5):703–710.

254. Swartz, W. J. 1984. Effects of 1,1-bis-[p-chlorophenyl]-2,2,2-trichloroethane (DDT) on gonadal development in the chick embryo: a histological and histochemical study. Environ. Res. 35(2):333–345.

255. Giwercman, A. and N. E. Skakkeback. 1992. The human testis — an organ at risk? Int. J. Androl. 15:373–375.

256. Osterlind, A. 1986. Diverging trends in incidence and mortality of testicular cancer in Denmark, 1943–1982. Br. J. Cancer 53:501–505.

257. Jackson, M. B. John Radcliffe Hospital Cryptochidism Research Group. 1988. The epidemiology of cryptorchidism. Horm. Res. 30:153–156.

258. Carlsen, E., A. Giwercman, N. Keiding, and N. E. Skakkeback. 1992. Evidence for decreasing quality of semen during past 50 years. Br. Med. J. 305:609–613.

259. Forest, M. G. 1982. Development of the male reproductive tract. In Aspects of Male Infertility, ed. R. de vere White, 1–60. Baltimore: Williams & Wilkins.

260. Hutson, J. M., M. P. L. Williams, M. E. Fallat, and A. Arrah. 1990. Testicular descent: new insights into its hormonal control. In Oxford Reviews of Reproductive Biology, ed. S. R. Milligan. 1–56. Oxford: Oxford University Press.

261. Skakkeback, N. E. 1987. Carcinoma-in-situ and cancer of the testis. Int. J. Androl. 10:1–40.

262. Stillman, R. J. 1982. In utero exposure to diethylstilbestrol: adverse effects on the reproductive tract and reproductive performance in male and female offspring. Am. J. Obstet. Gynecol. 142:905–921.

263. Arai, Y., T. Mori, Y. Suzuki, and H. A. Bern. 1983. Long-term effects of perinatal exposure to sex steroids and diethylstilbestrol on the reproductive system of male mammals. Int. Rev. Cytol. 84:235–268.

264. Prener, A., C.- C. Hsieh, G. Engholm, D. Trichopoulos, and O. M. Jensen. 1992. Birth order and risk of testicular cancer. Cancer Causes Control 3:265–272.

265. Depue, R. H. 1984. Maternal and gestational factors affecting the risk of cryptochidism and inguinal hernia. Int. J. Epidemiol. 13:311–318.

266. Field, B., M. Selub, C.L. Hughes. 1990. Reproductive effects of environmental agents. *Semin. Reprod. Endocrinol.* 8:44–54.

267. Adlercreutz, H. 1990. Diet, breast cancer and sex hormone metabolism. *Ann. N.Y. Acad. Sci.* 595:281–290.

268. Bernstein, L., R. H. Depue, R. K. Ross, H. L. Judd, M. C. Pike, and B. E. Henderson. 1986. Higher maternal levels of free estradiol in first compared to second pregnancy: a study of early gestational differences. *J. Natl. Cancer Inst.* 76:1035–1044.

269. Moss, A. R., D. Osmond, P. Rachetti, F. M. Torti, and V. Gurgin. 1986. Hormonal risk factors in testicular cancer: a case control study. *Am. J. Epidemiol.* 124:39–52.

270. McLachlan, J. A., ed. 1985. *Estrogens in the Environment.* Amsterdam: Elsevier.

271. Lamming, G. E. 1984. Report of the scientific group on anabolic agents in animal production. *Comm. Eur. Commun. Rep. Eur.* 8913:4–25.

272. Sharpe, R. M. and N. E. Skakkeback. 1993. Are oestrogens involved in falling sperm counts and disorders of the male reproductive tract? *Lancet* 341:1392–1395.

273. Sheehan, D. M. and M. Young. 1979. Diethystilbestrol and estradiol binding to serum albumin and pregnancy plasma of rat and human. *Endocrinology* 104:1442–1446.

274. Kaldas, R. S. and C. L. Hughes, Jr. 1989. Reproductive and general metabolic effects of phytoestrogens in mammals. *Reprod. Toxicol.* 3:81–89.

275. Verdeal, K. and D. S. Ryan. 1979. Naturally-occurring oestrogens in plant foodstuffs — a review. *J. Food Protect.* 7:577–583.

276. Hamon, M., I. R. Fleet, and R. B. Heap. 1990. Comparison of oestrone sulphate concentrations in mammary secretions during lactogenesis and lactation in dairy ruminants. *J. Dairy Res.* 57:419–422.

277. Colborn, T. and C. Clement, eds. 1992. In *Advances in Modern Environmental Toxicology. Vol. XXI. Chemically-Induced Alterations in Sexual and Functional Development: The Wildlife/Human Connection,* eds. T. Colborn and C. Clement. Princeton, NJ: Princeton Scientific.

278. Sager, D., D. Girar, and D. Nelson. 1991. Early postnatal exposure to PCBs: spasm function in rats. *Environ. Toxicol. Chem.* 10:737–746.

279. Mably, T. A., D. L. Bjerke, R. W. Moore, A. Gendron-Fitzpatrick, and R. E. Peterson. 1992. In utero and lactational exposure of male rats to 2,3,7,8-tetrachlorodibenzo-*p*-dioxin 3. Effects on spermatogenesis and reproductive capability. *Toxicol. Appl. Pharmacol.* 114:118–126.

280. Sharpe, R. M. 1993. Falling sperm counts in men — is there an endocrine cause? *J. Endocrinol.* 137:357–360.

281. Cortes, D., J. Müller, and N. E. Skakkeback. 1987. Proliferation of Sertoli cells during development of the human testis assessed by sterological methods. *Int. J. Androl.* 10:589–596.

282. Hirobe, S., W.-W. He, M. M. Lee, and P. K. Donahoe. 1992. Müllerian inhibiting substance messenger ribonucleic acid expression in granulosa and Sertoli cells coincides with their mitotic activity. *Endocrinology* 131:854–862.

283. Russell, L. D. and R. N. Peterson. 1984. Determination of the elongate spermatid — Sertoli cell ratio in various mammals. *J. Reprod. Fertil.* 70:635–641.

284. Orth, J. M., G. M. Gunsalus, and A. A. Lamperti. 1988. Evidence from Sertoli cell-depleted rats indicates that spermatid numbers in adults depend on number of Sertoli cells produced during prenatal development. *Endocrinology* 122:787–794.

285. Cooke, S., J. Porcelli, and R. A. Hess. 1992. Induction of increased testis growth and sperm production in adult rats by neonatal administration of the goitrogen prophylthiouracil (PTU): the critical period. *Biol. Reprod.* 46:146–154.

286. Hunt, V. R. 1982. The reproductive system: sensitivity through the life cycle. *Ann. Am. Conf. Ind. Hyg.* 3:53–59.

287. Hunt, V. R. 1979. *Work and the Health of Women.* Boca Raton, FL: CRC Press.

288. Rosenblum, A. H. and P. Rosenblum. 1952. Gastrointestinal allergy in infancy: significance of eosinophiles in the stools. *Pediatrics* 9:311.

289. Eishi, Y. and P. McCullagh. 1988. PVG rats, resistant to experimental allergic thyroiditis, develop high serum levels of thyroidglobulin after sensitization. *Clin. Immunol. Immunopathol.* 49(1):101–106.

290. Mitsunaya, M. 1987. Cytophilic anti-thyroglobulin antibody and antibody-dependent macrophage-mediated cytotoxicity in Hashimoto's thyroiditis. *Acta Med. Okayama* 41(5):205–214.

291. Bahn, A. K., J. L. Mills, P. J. Snyder, P. H. Gann, L. Houten, O. Bialik, L. Hollmann, and R. D. Utiger. 1980. Hypothyroidism in workers exposed to polybrominated biphenyls. *N. Engl. J. Med.* 302:31–33.

292. Gaitan, E., D. P. Island, and G. W. Liddle. 1969. Identification of a naturally occurring goitrogen in water. *Trans. Assoc. Am. Physicians* 132:141–152.

293. Bastomsky, C. H. 1977. Goiters in rats fed polychlorinated biphenyls. *Can. J. Physiol. Pharmacol.* 55:288–292.

294. Saiffer, P. 1988. Personal communication.

295. Price, S. C., S. Ozalp, R. Weaver, D. Chescoe, J. Mullervy, and R. H. Hinton. 1988. Thyroid hyperactivity caused by hypolipodaemic compounds and poly-chlorinated biphenyls: the effect of coadministration in the liver and thyroid. *Arch. Toxicol. [Suppl.]* 12:85–92.

296. Gaitan, E. 1986. Environmental goitrogens. In *The Thyroid Gland — A Practical Clinical Treatise,* ed. L. Van Middlesworth. 263–280. Chicago: Year Book Medical Publishers.

297. Gaitan, E. 1980. Goitrogens in the etiology of endemic goiter. In *Endemic Goiter and Endemic Cretinism,* eds. J. B. Stanbury and B. Hetzel. New York: John Wiley & Sons.

298. Langer, P. and M. A. Greer, eds. 1977. *Antithyroid Substances and Naturally Occurring Goitrogens.* Basel, S. Karger, A. G.

299. Ermans, A. M., N. B. Mbulamoko, F. Delange, R. Ahluwalia, eds. 1980. *Role of Cassava in the Etiology of Endemic Goiter and Cretinism.* IDRC-136e. Ottawa: International Development Research Centre.

300. Van Etten, C. H. 1969. Gotrogens. In *Toxic Constituents of Plant Foodstuffs,* ed. I. E. Liener. New York: Academic Press.

301. Gaitan, E., R. C. Cooksey, D. Matthew, R. Presson. 1983. In vitro measurement of antithyroid compounds and environmental goitrogens. *J. Clin. Endocrinol. Metab.* 56(4):767–773.

302. Gaitan, E. 1992. Adverse effects of environmental pollutant exposure on the thyroid. In *Principles and Practice of Environmental Medicine,* ed. A. B. Tarcher. New York: Plenum Book.

303. Cooksey, R. C., E. Gaitan, R. H. Lindsay, J. Hill, and K. Kelly. 1985. Humic substances: a possible source of environmental goitrogens. *Organ. Geochem.* 8:77–80.

304. Wolff, J. 1969. Iodide goiter and the pharmacologic effects of excess iodide. *Am. J. Med.* 47:101.

305. Gaitan, E. 1975. Iodine deficiency and toxicity. In *Proceedings of the Western Hemisphere Nutrition Congress IV,* eds. Ph. L. White and N. Selvey. Acton, MA: Publishing Sciences Group.

306. Suzuki, H. 1980. Etiology of endemic goiter and iodide excess. In *Endemic Goiter and Endemic Cretinism,* eds. J. B. Stanbury and B. Hetzel. New York: John Wiley & Sons.

307. Lazarus, J. H. and E. H. Bennie. 1972. Effect of lithium on thyroid function in man. *Acta Endocrinol.* 70:166.

308. Emerson, C. H., W. L. Dyson, and R. D. Utiger. 1973. Serum thyrotropin and thyroxine concentrations in patients receiving lithium carbonate. *J. Clin. Endocrinol. Metab.* 36:338.

309. Child, C., G. Nolan, and W. Jubiz. 1976. Changes in serum thyroxine, triiodo-thyronine and thyrotropin induced by lithium in normal subjects and in rats. *Clin. Pharmacol. Ther.* 20:715.

310. Gaitan, E. 1988. Goitrogens. *Bailliere's Clin. Endocrinol. Metab.* 2(3):683–702.

311. Goldberg, R. C., J. Wolff, and R. O. Greep. 1957. Studies on the nature of the thyroid-pituitary interrelationship. *Endocrinology* 60:38.

312. Goldberg, R. C., J. Wolff, and R. O. Greep. 1955. The mechanism of depression of plasma protein bound iodine by 2,4-dinitrophenol. *Endocrinology* 56:560.

313. Wayne, E. J., D. A. Koutras, and W. D. Alexander, eds. 1964. *Clinical Aspects of Iodine Metabolism.* Oxford: Blackwell Scientific Publications.

314. Wolff, J., M. E. Standaert, and J. Rall. 1961. Thyroxine displacement from the serum and depression of serum protein-bound iodine by certain drugs. *J. Clin. Invest.* 40:1373.

315. Bastomsky, C. H., P. V. N. Murphy, and K. Banovac. 1976. Alterations in thyroxine metabolism produced by cutaneous application of microscope immersion oil: effects due to polychlorinated biphenyls. *Endocrinology* 98:1309.

316. Allen-Rowlands, C. F., V. D. Casracane, M. F. Hamilton, and J. Seifter. 1981. Effect of polybrominated biphenyls (PBB) on the pituitary-thyroid axis of the rat (41099). *Proc. Soc. Exp. Biol. Med.* 166:506.

317. Bastomsky, C. H. 1977. Enhanced thyroxine metabolism and high uptake goiters in rats after a single dose of 2,3,7,8-tetrachlorodibenzo-p-dioxin. *Endocrinology* 101:292–296.

318. Potter, C. L., I. G. Sipes, and D. H. Russell. 1983. Hypothyroxinemia and hypothermia in rats in response to 2,3,7,8-tetrachlorodibenzo-p-dioxin administration. *Toxicol. Appl. Pharmacol.* 69:89.

319. Newman, W. C., R. C. Fernandez, R. M. Slayden, R. C. Moon. 1971. Accelerated biliary thyroxine excretion in rats treated with 3-methyl-cholanthrene. *Proc. Soc. Exp. Biol. Med.* 138:899–900.

320. Bastomsky, C. H. and P. D. Papapetrou. 1973. Effect of methylcholanthrene on biliary thyroxine excretion in normal and Gunn rats. *J. Endocrinol.* 56:267.

321. Goldstein, J. A. and A. Taurog. 1968. Enhanced biliary excretion of thyroxine glucuronide in rats pretreated with benzopyrene. *Biochem. Pharmacol.* 17:1049.

322. Melander, A., F. Sundler, and S. H. Ingbar. 1973. Effect of polyphloretin phosphate on the induction of thyroid hormone secretion by various thyroid stimulators. *Endocrinology* 92:1269.

323. Toccafondi, R. S., M. L. Brandi, and A. Melander. 1984. Vasoactive intestinal peptides stimulation of human thyroid cell function. *J. Clin. Endocrinol. Metab.* 58:157.

324. Delange, F. and R. Ahluwalia, eds. 1983. *Cassava Toxicity and Thyroid: Research and Public Health Issues.* IDRC-207e. Ottawa: International Development Research Centre.

325. Roti, E., A. Grundi, and L. E. Braverman. 1983. The placental transport, synthesis and metabolism of hormones and drugs which affect thyroid function. *Endocrine Rev.* 4:11.

326. Gaitan, E. 1973. Water-borne goitrogens and their role in the etiology of endemic goiter. *World Rev. Nutr. Diet* 17:53.

327. Hulse, J. H., ed. 1980. *Polyphenols in Cereals and Legumes.* IDRC-145e. Ottawa: International Development Research Centre.

328. Burges, N. A., H. M. Hurst, B. Walkden. 1964. The phenolic constituents of humic acid and their relationship to the lignin of the plant cover. *Geochem. Cosmo Chim. Acta* 1547:1554.

329. Choudhry, G. G. 1981. Humic substances: I. Structural aspects. *Toxicol. Environ. Chem.* 4:209.

330. Schnitzer, M. and S. Khan. 1972. *Humic Substances in the Environment.* New York: Marcel Dekker.

331. Osman, A. K. and A. A. Fatah. 1981. Factors other than iodine deficiency contributing to the endemicity of goitre in Darfur Province (Sudan). *J. Hum. Nutr.* 35:302.

332. Pitt, W. W., R. L. Jolley, and G. Jones. 1979. Characterization of organics in aqueous effluents of coal-conversion plants. *Environ. Int.* 2:167.

333. Jahnig, C. E. and R. R. Bertrand. 1976. Aqueous effluents from coal-conversion processes. *Chem. Eng. Prog.* 72:51.

334. Klibanov, A. M., T. M. Tu, and K. P. Scott. 1983. Peroxidase-catalyzed removal of phenols from coal-conversion waste waters. *Science* 221:259.

335. Shackelford, W. M. and L. H. Keith. 1976. Frequency of organic compounds identified in water. Athens, GA: Environmental Research Laboratory.

336. Jolley, R. L., E. Gaitan, E. C. Douglas, and L. K. Felker. 1986. Identification of organic pollutants in drinking waters from areas with endemic thyroid disorders and potential pollutants of drinking water sources associated with coal processing areas. *Am. Chem. Soc. Environ. Chem.* 26:59–62.

337. Bull, G. M. and R. Fraser. 1950. Myxoedema from resorcinol ointment applied to leg ulcers. *Lancet* 6:851.

338. Quentin, J. G. and B. M. Hobson. 1961. Varicose ulceration of the legs and myxoedema and goiter following application of resorcinol ointment. *Proc. R. Soc. Med.* 44:164.

339. Walfish, P. G. 1983. Drug and environmentally induced neonatal hypothyroidism. In *Congenital Hypothyroidism,* eds. J. H. Dussault and P. Walker. New York: Marcel Dekker.

340. Jolley, R. L., H. Gorchev, and D. H. Hamilton, Jr., eds. 1978. *Water Chlorination: Environmental Impact and Health Effects,* Vol. 2. Ann Arbor, MI: Ann Arbor Science Publishers.

341. Jolley, R. L., W. W. Pitt, Jr., C. D. Scott, G. Jones, Jr., and J. E. Thompson. 1975. Analysis of soluble organic constituents in natural and process waters by high-pressure liquid chromatography. In *Trace Substance in Environmental Health.* Vol. IX. 247–253. ed. D. D. Hemphill. Columbia, MO: The University of Missouri.

342. Rea, W. J. 1994. *Chemical Sensitivity. Vol. II. Sources of Total Body Load.* 535. Boca Raton, FL: Lewis Publishers.

343. Castor, C. W. and W. H. Beierwaltes. 1955. Depression of serum protein-bound iodine levels in man with dinitrophenol. *J. Clin. Endocrinol. Metab.* 15:862.

344. Nemeth, S. 1958. Short-term decrease of serum protein-bound iodine concentration after administration of 2,4-dinitro-phenol in man. *J. Clin. Endocrinol. Metab.* 18:225.

345. Peakall, D. B. 1975. Phthalate esters: Occurrence and biological effects. *Residue Rev.* 54:1.

346. Keyser, P., B. G. Pujar, R. W. Eaton, and D. W. Ribbons. 1976. Biodegradation of the phthalates and their esters by bacteria. *Environ. Health Perspect.* 18:159–166.

347. Bahn, A. K., J. L. Mills, P. J. Synder, P. H. Gann, L. Houten, O. Bialik, L. Holimann, and R. D. Utiger. 1980. Hypothyroidism in workers exposed to polybrominated biphenyls. *N. Engl. J. Med.* 302:31.

348. Gaitan, E. 1992. Adverse effects of environmental pollutant exposure on the thyroid. In *Principles and Practice of Environmental Medicine,* ed. A. B. Tarcher. 371–387. New York: Plenum Publishing.

349. Pitt, W. W., R. L. Jolley, and G. Jones. 1979. Characterization of organics in aqueous effluents of coal-conversion plants. *Environ. Int.* 2:167.

350. Jahnig, C. E. and R. R. Bertran. 1976. Aqueous effluents from coal-conversion processes. *Chem. Eng. Prog.* 72:51.

351. Klibanov, A. M., T. M. Tu, and K. P. Scott. 1983. Peroxidase-catalyzed removal of phenols from coal-conversion waste waters. *Science* 221:259.

352. Meuzelaar, H. L., W. Windig, A. M. Harper, S. M. Huff, W. H. McClennch, and J. M. Richards. 1984. Pyrolysis mass spectrometry of complex organic materials. *Science* 226:268–274.

353. Morris, S. C., P. D. Moskwitz, W. A. Sevian, S. Silberstein, and L. D. Hamilton. 1979. Coal-conversion technologies: some health and environmental effects. *Science* 206:654–662.

354. Gaitan, E. 1973. Water-borne goitrogens and their role in the etiology of endemic goiter. *World Rev. Nutr. Diet.* 17:53.

355. Gaitan, E. 1983. Endemic goiter in western Colombia. *Ecol. Dis.* 2:295.

356. Gaitan, E. 1993. Antithyroid compounds. In *Thyroid Diseases: Clinical Fundamentals and Therapy,* eds. F. Monaco, M. A. Satta, B. Shapiro, and L. Troncone. 615–625. Boca Raton, FL: CRC Press.

357. Matovinovic, J. and F. L. Trowbridge. 1980. North America. In *Endemic Goiter and Endemic Cretinism*, eds. J. B. Stanbury and B. Hetzel. New York: John Wiley & Sons.

358. Weetman, A. P. and A. M. McGregor. 1984. Autoimmune thyroid disease: developments in our understanding. *Endocrine Rev.* 5:309.

359. Biggazi, P. E. and N. R. Rose. 1975. Spontaneous autoimmune thyroiditis in animals as a model of human disease. *Prog. Allergy* 19:245.

360. Esquivel, P. S., Y. M. Kong, and N. R. Rose. 1978. Evidence for thyroglobulin-reactive T cells in good responder mice. *Cell. Immunol.* 37:14.

361. Meuzelaar, H. L. C., W. Windig, A. M. Harper, S. M. Huff, W. H. McClennen, and J. M. Richards. 1984. Pyrolysis mass spectrometry of complex organic materials. *Science* 226:268.

362. Gaitan, E. 1983. Endemic goiter in western Colombia. *Ecol. Dis.* 2:295.

363. Rogan, W. J., A. Bagniewska, and T. Damstra. 1980. Pollutants in breast milk. *N. Engl. J. Med.* 30:1450.

364. *Safe Drinking Water Committee: Drinking Water and Health,* 1977. ISBN 0–309–02619–9. Washington, D.C.: National Academy of Sciences.

365. Andelman, J. B. and M. J. Sness. 1970. Polynuclear aromatic hydrocarbons in the water environment. *Bull. WHO* 43:479.

366. Pelkonen, O. and D. W. Nebert. 1982. Metabolism of polycyclic aromatic hydrocarbons: etiologic role in carcinogenesis. *Pharmacol. Rev.* 34:189.

367. Lo, M. T. and E. Sandi. 1978. Polycyclic aromatic hydrocarbons (polynuclears) in foods. In *Residue Reviews,* Vol. 69, ed. F. A. Gunther. New York: Springer-Verlag.

368. Rea, W. J. 1994. *Chemical Sensitivity. Vol. II. Sources of Total Body Load.* 539. Boca Raton, FL: Lewis Publishers.

369. Rea, W. J. 1994. *Chemical Sensitivity. Vol. II. Sources of Total Body Load.* 583. Boca Raton, FL: Lewis Publishers.

370. Biggazi, P. E. and N. R. Rose. 1975. Spontaneous autoimmune thyroiditis in animals as a model of human disease. *Prog. Allergy* 19:245.

371. Farid, N. R. 1987. Immunogenetics of autoimmune thyroid disorders. *Endocrinol. Metab. Clin. N. Am.* 16:229–245.

372. Scherbaum, W. A. 1993. Pathogenesis of autoimmune thyroiditis. *Nuklearmediziner* 16:241–249.

25   Integument

INTRODUCTION

The skin comes into contact with many toxic agents. Fortunately, it has a good lipid barrier and is, therefore, not highly permeable. However, some chemicals (particularly if associated with solvents) can affect the skin, either directly or by penetrating it and producing systemic effects. Nerve gases, carbon tetrachloride, pesticides, and other solvents are known to be absorbed through the skin and clearly induce or exacerbate some cases of chemical sensitivity.[1] Organic chemical contaminants in water have recently been shown to be absorbed through the skin[2] (Table 25.1). Many chemically sensitive patients report exacerbation of their illness occurs when they soak in a bathtub filled with unfiltered water.

Skin ailments constitute about 34% of all occupational diseases. These ailments are probably higher for the total number of problems that are environmentally triggered. Because the skin is both a target organ for pollutant injury and a conduit through which some pollutants enter the body and because of the high incidence of skin involvement in patients with chemical sensitivity, this chapter devotes a brief discussion to the physiological involvement of the skin and the mechanisms of action of various chemicals involved in the onset and expression of chemical sensitivity. Also, the commonly seen skin conditions that result from chemical exposure including autoimmune contact dermatitis, acute skin reactions, urticaria, eczema, psoriasis, boils, and skin aging are given specific attention.

POLLUTANT INJURY TO THE PHYSIOLOGY OF THE SKIN

The skin of the chemically sensitive individual is an important interface between the internal organs and environmental pollutants.[3] In some chemically sensitive individuals, however, the skin, consisting of the oily layer, epidermis, dermis, and appendages as well as blood vessels and neuronal components, can

Table 25.1. Estimated Dose and Contribution per Exposure for Skin Absorption vs. Ingestion

Compound	Concentration (mg/L)	Dose (mg/kg)					
		Case 1[a]		Case 2[b]		Case 3[c]	
		Dermal	Oral	Dermal	Oral	Dermal	Oral
Toluene	0.005	0.0002	0.0001	0.0004	0.005	0.002	0.0002
	0.10	0.005	0.003	0.008	0.0095	0.033	0.0045
	0.5	0.02	0.014	0.04	0.048	0.17	0.023
Ethylbenzene	0.005	0.0003	0.0001	0.0004	0.0005	0.002	0.0002
	0.10	0.005	0.003	0.008	0.0095	0.036	0.0045
	0.5	0.03	0.014	0.04	0.048	0.18	0.023
Styrene	0.005	0.0002	0.0001	0.002	0.0005	0.001	0.0002
	0.10	0.003	0.003	0.005	0.0095	0.023	0.0045
	0.5	0.02	0.02	0.02	0.048	0.11	0.023

		Relative contribution (%)					
		Dermal	Oral	Dermal	Oral	Dermal	Oral
Toluene	0.005	67	33	44	56	91	9
	0.10	63	37	46	54	89	11
	0.5	59	41	45	55	89	11

Ethylbenzene	0.005	75	25	44	56	91	9
	0.10	63	37	46	54	89	11
	0.5	68	32	45	55	89	11
Styrene	0.005	67	33	29	71	83	17
	0.10	50	50	35	65	84	16
	0.5	59	41	29	71	83	17

[a] 70 kg adult bathing 15 minutes, 80% immersed (skin absorption); 2 liters water consumed per day (ingestion)

[b] 10.5 kg infant bathed 15 minutes, 75% immersed (skin absorption); 1 liter water consumed per day (ingestion)

[c] 21.9 kg child swimming 1 hour, 90% immersed (skin absorption); 1 liter water consumer per day (ingestion)

Source: Brown, H. S., D. R. Bishop, and C. A. Rowan. 1984. The role of skin absorption as a route of exposure for volatile organic compounds (VOCs) in drinking water. *Am. J. Public Health* 74:379–484. With permission.

be damaged by pollutant injury. Once damage occurs, chemical sensitivity will be exacerbated.

The oily layer of the skin of the chemically sensitive individual can be damaged by contact with solvents in bathing water or by contact with chemicals contained in a variety of common products used in the home or workplace. For example, bleaches, detergents, dishwashing liquids, floor cleaners and waxes, furniture polishes, metal cleaners, oven cleaners, pesticides, shampoos, soaps, toilet cleaners, window cleaners, and scouring pads are household contact irritants. Allergic contact sensitizers are perfumes, cosmetics, preservatives (e.g., parabins), rubber products (mercaptin, thorium), medications (neomycin, benzocaine, mercury), leather (formaldehyde, dichromates), chromes, resins, and nickel. Once the oily layer of the skin has been damaged, superabsorption of these toxic substances may then follow.

Metals known to cause skin reactions include aluminum (Al), antimony (Sb), arsenic (As), beryllium (Be), boron (Bo), cadmium (Cd), chromium (Cr), cobalt (Co), copper (Cu), gold (Au), mercury (Hg), nickel (Ni), palladium (Pd), platinum (Pt), selenium (Se), silica (Si), silver (Ag), tellurium (Te), thallium (Tl), zinc (Zn), and zirconium (Zr).

Toxic chemicals known to produce skin reactions include alcohol, balsam of Peru, benzocaine, butyric acid, cephalosporins, chloramphenicol, chloropromadiethyltoluamide, epoxy resin, gentamycin, lindane, mechlorethamine, menthol, neomycin, nickel, parabins, penicillin, plastic additives, propyline glycol, polysorbate, salicylic acid, sodium sulfate, sulfur dioxide, and streptomycin. Even compounds that are generally considered well-tolerated, such as sodium benzoate, ascorbic acid, and acetic acid, are known to cause skin reactions.

In addition to their ability to damage the oily layer of the skin, toxic chemicals and metals can also damage the epidermis, which is a stratified cellular layer that is constantly evolving and forming stratum corneum, which functions as a protective barrier. Although this barrier is supposed to guard the individual from absorption of harmful substances, it may be ineffective in some chemically sensitive patients.

The dermis, which also may be damaged in some chemically sensitive patients, lies underneath the epidermis and is composed of connective tissue, fibroblasts, collagen, and elastic fibers. It houses the appendages that include the sweat glands, pilosebaceous units, and apocrine glands. In some chemically sensitive patients, all of these often function abnormally. Hence, these individuals are vulnerable to pollutant exposure, and pollutant injury to these appendages can have significant effects. As shown in Chapters 20 and 26 and Volume IV, Chapter 35, the cold nature of the chemically sensitive individual, as manifested by cold skin, is well documented. Their ability to sweat can be delayed or suppressed, and often their skin is dry. This inability to sweat easily may be one of the main reasons that chemicals remain in the body of the chemically sensitive patient, further increasing his or her susceptibility to chemical exposure.

Not only can pollutants damage the oily layer of skin or the epidermis or the dermis separately, but also all of these layers may be simultaneously injured. Often a large subset of chemically sensitive patients is seen with simultaneous damage occurring in various layers of the skin (see Chapter 23).

Body and skin temperatures are controlled by the rate of blood flow and the radiation of heat through small vessels. Although sweating may offer a defense against chemical injury by diluting xenobiotics, it may also increase the state of hydration of the barrier, which will then enhance the absorption of other toxic agents.

Two types of skin are involved in chemical sensitivity. One is the glabrous and the other is the hairy skin. The glabrous skin of the palms and soles has a thick stratum corneum with sweat glands and encapsulated nerve endings, but no hair follicles or sebaceous glands. Many chemically sensitive individuals experience pain when standing on their bare feet, and often the glabrous skin is found to be involved in their discomfort. The hairy type of skin has relatively thin stratum corneum with hair follicles and sebaceous and sweat glands, but no encapsulated nerve endings. It is prone to greater absorption of anything it contacts. The chemically sensitive individuals we have seen frequently experience this kind of excessive absorption.

As a target organ, the skin of the chemically sensitive individual is capable of responding in a variety of pathologic patterns that may involve specific cellular and structural components.[3] The cellular elements of mesenchymal origin are involved in wound healing and play an important role in both the development and maintenance of the fibrous structures. For the chemically sensitive person, these elements also provide a second-line defense against injury by chemical stimuli, physical agents, and microorganisms. Pollutant injury may retard healing in some chemically sensitive inidividuals.

The ground substance in the dermis transfers nutrients and metabolites to and from structures within its surroundings and is strongly influenced by vitamin C, which is depleted in 20% of chemically sensitive individuals. Nutrient deficiency, as seen in chemically sensitive patients, can adversely affect the function in these areas.

The skin has two routes of absorption that can be potentially damaged in the chemically sensitive individual — transfollicular and transepidermal (Figure 25.1). Frequently, the chemical will first pass through the pilosebaceous route and through the epidermis. Chemically sensitive patients with beards are prone to trap pollutants with this facial hair and then absorb them through the skin. Normal skin without a beard offers the best protection against compounds of strong polar nonelectrolytes. It offers poor protection against lipid soluble substances (solvents), low molecular weight compounds, and nonelectrolytes.

Several factors enhance the absorption of chemicals through the skin and are seen to be active in a subset of chemically sensitive patients (Table 25.2). These factors are body site (thin skin, face, genitalia, and folds); integrity of skin damaged by trauma, inflammation, or dehydration; occlusion (wrapping with plastic, wearing dry-cleaned clothes [tetrachloroethylene]); vehicle (urea,

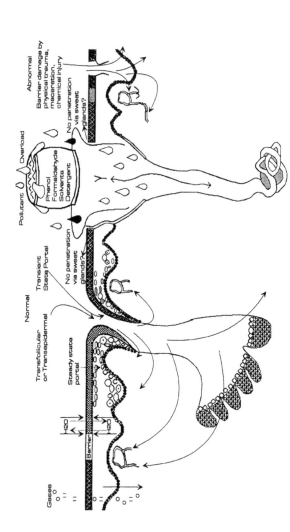

Figure 25.1. Routes of and factors that enhance percutaneous absorption of toxic chemicals and other microbials after pollutant injury in the chemically sensitive individual. Absorption depends on the properties of the skin, the substance(s) to be absorbed, and the environment (vehicle). Absorption is enhanced by the following factors: (1) the breaking barrier, (2) increasing hydration, (3) increasing temperature, (4) the location or site on the body surface, (5) reservoir capabilities, (6) blood flow, and (7) many chemical factors including concentration, molecular size, pH, state of ionization, lipid solubility, and biotransformation systems.

Table 25.2.   Factors Enhancing the Absorption of Chemicals through the Skin

Body site	Thin skin, face, genitalia, and folds
Integrity of skin	Damage by trauma, inflammation, dehydration
Occlusion	Wearing dry-cleaned clothing (tetrachloroethylene contaminated); contact with plastic wrapping
Vehicle	Urea, chloropyrifos, ointment bases
Chemical factors	Solubility, pH concentration
Skin burning chemicals	Alkyl mercury, cement, chromic acid, ethylene oxide, hydrofluoric acid, white phosphorous

chloropyrofos, and ointment bases); and chemical factors (e.g., solubility, pH, and concentration). Also, certain chemicals burn the skin. These include, but are not limited to, alkyl mercury compounds, cement, chromic acid, ethylene oxide, hydrofluoric acid, white phosphorus, and probably many more.

Toxic chemicals are more readily absorbed when in contact with intertriginous areas of the body (groin, inframammary, axillary, perianal) in contrast to the glabrous skin. Toxic chemicals produce varied effects on the skin, e.g., increased melanin has been produced by arsenics,[4] busulfan,[4] tar,[4] and photochemical agents.[4] Storer et al.[5] applied many elements in coal tar that are commonly used in the treatment of skin lesions to the skin of volunteers and then demonstrated the presence of these elements in peripheral blood samples taken from the volunteers (Tables 25.3 and 25.4). The chemically sensitive individuals who have undergone minute exposures of toxic substances experience problems that suggest, in some cases, that superabsorption has occurred. In other cases, however, these problems are probably sensitization reactions.

The skin has a versatile group of defenses against penetration by chemical agents, fluid loss from the body, thermal stress, solar radiation effects, physical trauma, and against infection by microbial agents. In chemically sensitive individuals, these defenses may be inherently weak or defective, making these individuals vulnerable to pollutant exposure. Also, these defenses may be weakened or damaged as a result of an exposure itself. These skin defenses are characterized on the basis of morphological, physiological, and biochemical processes.

Morphologically, the intact stratum corneum provides a significant defense against penetration by chemical agents and against body water loss. It is a physical barrier to invasion by microorganisms, and its surface lipids provide some bacteriostatic protection. The intact skin is infected with difficulty. Pigment components of the melanocytes found among the lowest layers of the epidermis are an important defense against ultraviolet radiation, which causes fragmentation and destruction of the elastic tissue fibers. Ultraviolet and chemical-induced aging of the skin, actinic keratosis, and skin cancer physiologically effect long chain (C_{16-18}) fatty acids derived from sebaceous secretions, which

**Table 25.3.   Polynuclear Aromatic Hydrocarbon Analysis of the
2% Crude Coal Tar in Petrolatum**

Compound no.	Component	Level, PPM (μ/mL)	References
1	Naphthalene	520	
2	Alkylated derivatives (C_1,C_2,C_3) of 1	890	
3	Biphenyl	99	
4	Acenaphthene	260	
5	Alkylated derivatives (C_1,C_2,C_3) of 3 and 4	579	
6	Fluorene	390	
7	Alkylated fluorene	76	
8	Phenanthrene[a]	650	108–112
9	Anthracene[a]	190	108–112
10	Alkylated derivatives (C_1,C_2,C_3) of 8 and 9	288	
11	Fluoranthene[a,b]	400	108–114
12	Pyrene[a]	330	108–112
13	Alkylated derivatives (C_1,C_2,C_3) of 11 and 12	205	
14	Benzo[a]anthracene[a]	560	108–112
15	Chrysene[a,b]	140	108–114
16	Alkylated derivatives (C_1,C_2,C_3) of 14 and 15	228	
17	Benzofluoranthene[a]	420	108–112
18	Benzo[e]pyrene[a,b]	95	108–114
19	Benzo[a]pyrene[a,b]	220	108–114
20	Perylene	40	
21	Benzo(b)thiophene and alkylated derivatives	18	
22	Dibenzothiophene and alkylated derivatives	54	
23	Naphthyl benzo[b]thiophene and alkylated derivatives	44	

[a] Thought to be a carcinogen, cocarcinogen, or tumor initiator.
[b] Positive Ames test result.

Source: Storer, J. S., I. DeLeon, L. E. Millikan, J. L. Laseter, and C. Griffing. 1984.
Human absorption of crude coal tar products. *Arch. Dermatol.* 120:874–877. With
permission.

have bacteriostatic properties on a limited group of organisms (Figure 25.2).
Odd numbered carbon-chain fatty acids such as $C_{9,11,13}$ secreted by the sebaceous glands have fungistatic properties.[3]

Table 25.4. Polynuclear Aromatic Hydrocarbon: Blood Levels
Before and After 2% Crude Coal Tar Application

Compound[b]	Initial/posttreatment level:[a] volunteer no.					References
	1	2	3	4	5	
1	—/4.8	—/—	1.0/—	3.1/4.0	2.2/80.0	
2	—/27.4	0.4/4.9	3.2/4.1	7.1/9.2	5.9/100.0	
3	—/0.6	—/—	—/—	—/—	—/3.1	
4	—/1.9	—/—	—/0.4	—/0.8	—/2.8	
5	—/9.9	—/—	—/—	—/—	—/8.1	
6	—/3.3	—/—	—/—	—/0.04	—/2.6	
8[c]	—/11.0	0.5/0.4	1.4/1.2	1.7/3.1	3.1/6.3	109,111–113, 115–119
9[c]	—/4.7	—/0.08	—/—	—/0.5	—/1.6	109,111–113, 115–119
11[c,d]	—/3.3	—/0.1	—/—	—/0.5	—/1.7	109,111–119
12[c]	—/—	—/—	—/—	—/0.2	—/1.1	109,111–113, 115–119
21	—/13	—/—	—/—	—/13.0	—/217	

[a] Blood levels are in parts per billion (nanograms per milliliter); — indicates no detectable amount.
[b] See Table 25.3 for compound identification.
[c] Thought to be a cocarcinogen or tumor initiator
[d] Indicates a positive Ames test result.

Source: Storer, J. S., I. De Leon, L. E. Millikan, J. L. Laseter, and C. Griffing. 1984. Human absorption of crude coal tar products. Arch. Dermatol. 120:874–877. With permission.

Elastic and collagen fibers provide the skin with physical resiliency and/or a fibrous barrier against trauma. They also provide support for the blood, nerves, and appendages. At times, these fibers are damaged in some chemically sensitive patients, resulting in burning and itching skin.

MECHANISMS OF ACTION OF TOXIC CHEMICALS

The mechanisms of actions of toxic chemicals on the skin of the chemically sensitive are diverse. However, immune and nonimmune responses are involved as shown in Volume I, Chapter 4.[6]

Xenobiotics, including small molecular weight compounds, are capable of any of the four immunologic reactions described in Volume I, Chapter 5.[7] In type I, circulating antibodies produced through mediators of B cells localize in the skin. When an incitant is reintroduced, it causes a release of vasoactive amines, such as histamine from basophils or mast cells, that, in turn, induce a wheal and flare reaction or urticaria. Haptens, such as hexavalent chromin or

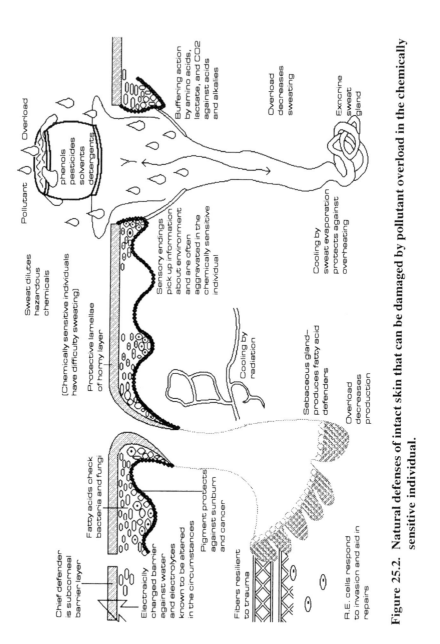

Figure 25.2. Natural defenses of intact skin that can be damaged by pollutant overload in the chemically sensitive individual.

sesquiterpene, form. These are then conjugated with either covalently or noncovalently bound epidermal protein, and this conjugate, in turn, is processed in Langerhans' cells or macrophages that then migrate via the afferent lymphatics into the parycortex of the draining lymph node. There these cells interdigitate with the many T cells exposed to the antigens. The T cells then present the clonal expansion and migrate to the skin. When the skin is reexposed to a hapten, a complete antigen is formed by conjugation and reacts with the T cell. Cytokines are then released. Following the release of cytokines, there is a reaction in the epidermis characterized by an infiltrate of mononuclear cells and microvesicle formation with edema. The response may be followed by an inflammatory reaction in the dermis, as seen in a subset of chemically sensitive individuals. These are the eczematous characteristics of allergic dermatitis.

A primary target cell of toxic chemicals is the T lymphocytes, specifically the T suppressor cells, which are low in a subset of chemically sensitive individuals[8] (see Volume I, Chapter 5,[7] and Volume IV, Chapter 30). These lymphocytes can be damaged and destroyed by free radicals produced by the chemicals. Subsequent production of massive amounts of antibodies may occur, as seen in the rapeseed oil-analine disaster in Spain, or the lymphocytes may just become sensitized with excess antibody production. With either outcome, extreme sensitivity reactions may follow. The mast cells can be triggered by the direct stimulus of a chemical such as DDT. The consequence is lipid peroxidation of the cell membrane and mitochondria, which can result in leakage or inhibition of function with impairment of the electron transport system and functional ATP. After degranulation of the mast cell, inflammatory mediators are released, with subsequent clinical sequelae of vascular spasm and, eventually, inflammation. Mast cells can be destroyed and/or sensitized. This sequence has been observed to be commonplace in the chemically sensitive individual.

As a result of increased total load, increased vascular permeability with subsequent edema, bruising, petechiae, and purpura can occur in a large subset of chemically sensitive individuals. Vascular spasm with blanching or cyanosis can also occur, resulting in Raynaud's phenomena in a large subset of chemically sensitive individuals. This vascular spasm then results in further lipid peroxidation by free radicals with damage to more cell membranes. Peroxidation requires large amounts of oxygen, thereby reducing the availability of oxygen to cells and tissues for metabolism. This deprivation of oxygen results in a state of localized tissue hypoxia, which results in further vascular spasm. This sequence of events explains why many chemically sensitive individuals respond to oxygen supplementation. In addition, this condition of oxygen depletion compromises functional cellular competence and is further stressed by the requirement of supplemental oxygen in the formation of cytochrome P-450 oxidase. Inflammatory vasculitis occurs. Other inflammatory mediators are released via the phospholipid-fatty acid cycle with formation of free radicals, induction of lipid peroxides, peroxidation of cell membranes, and formation of prostaglandins and leukotrienes, leading finally to thrombosis. Some of the mast cell modulators include substances such as cyclic AMP, prostaglandins, phosphodiesterase inhibitor, and calcium channel blockers. Benign mast cell disease can occur, and Meggs[9] has devised a serological indicator (oligoclonal IgG) for diagnosing benign mast cell disease (mastocytosis) vs. malignant mast cell tumors.

Enzyme detoxification systems involved in toxic chemical overload of the skin in chemically sensitive patients include cytochrome P-450 oxidase, super-oxide dismutase, arylhydrocarbon hydrolase, gamma glutamyltranspeptidase, lipoxygenase, epidermal cyclooxygenase, epoxide hydrolase, ornithine decar-boxylase, and uroporphyrinogen decarboxylase (see Volume I, Chapter 4[6]).

Enzyme detoxification nutrients in the skin include glutathione; glutathione peroxidase; vitamins E, C, β-carotene, and A; L-cysteine; methionine; sele-nium; chromium; zinc; copper; manganese; magnesium; and potassium. These may be depleted by direct toxic effects or overutilization of the detoxifying mechanisms or systemic malnutrition, thus causing or propagating the skin disease. The experiences of the EHC-Dallas have encompassed the full spec-trum of skin pathology described in the textbook of medicine. When infection and parasitic organisms have been excluded, most entities in this spectrum can often find their etiology in nutritional imbalance with food and chemical overload.

DERMAL VASCULAR RESPONSES TO POLLUTANT INJURY

The effects of toxic chemicals upon the skin of chemically sensitive individuals are becoming very well known. The early characteristic clinical signs and symptoms of toxic exposure can appear in various forms. A vascular syndrome of acne, petechiae, spontaneous bruising, or purpura as well as cold sensitivity and edema (periorbital, feet, and hands) and peripheral arterial spasm are seen in many patients with chemical sensitivity (Table 25.5).[10] At the EHC-Dallas, where we have seen 200 cases presenting with this vast array of symptoms (see Chapter 20, small vessel vasculitis), we have described this vascular syndrome. Correlating with our observations, 52% of 226 workers exposed to 2,4,5-trichlorophenoxyacetic acid (2–4-5T, a herbicide) had a re-sidual mean duration of chloracne of 26 years.[11] Until evidence to the contrary can be compiled, we at the EHC-Dallas consider adult onset acne to result, in part, from overexposure to toxic chemicals (especially the chlorinated ones). Chemicals proven to cause chloracne are chloronaphthalene,[12] PCBs,[13] poly-chlorinated dibenzofurans,[11] chlorophenols,[11] and chlorobenzene.[11] Also, chloracne has been observed in many soldiers returning from Vietnam who were sprayed with "Agent Orange"[14] (Figure 25.3).

Skin pathologies in humans have recently been associated with a number of different toxic chemicals. Two cases of pemphigus vulgaris were linked to

Table 25.5. Environmentally Triggered
Vascular Response of the Skin Seen in Some
Chemically Sensitive Patients

Acne — usually chloracne
"Spontaneous" bruising
Petechiae
Purpura
Peripheral arterial spasm — Raynaud's phenomena
Peripheral and prescribed edema

Source: EHC-Dallas. 1976.

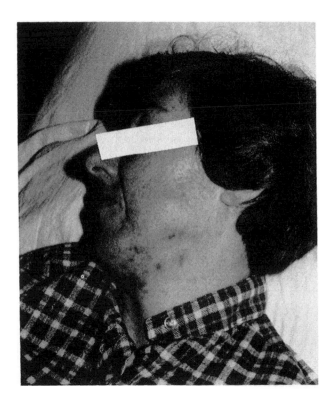

**Figure 25.3. Chloracne in 40-year-old white male exposed to chlorinated
hydrocarbons. (From EHC-Dallas. 1979.)**

known nonoccupational chronic pentachlorophenol (PCP) exposures.[15] There
was a rough correlation between the serum PCP levels and the clinical course
and titer of antibodies. Cole et al.[16] presented a patient who developed chloracne
after exposure to PCP pressure-treated lumber. The authors presumed his

difficulty was through percutaneous absorption of dioxins and furnans, which are known to be contaminants of technical grade PCP.

Some pollutant injury related to the vascular tree manifests as signs and symptoms of the skin. For example, erythemia occurs in some patients exposed to toxic chemicals. Usually, it is nontender and nonirritating. However, occasionally, it burns similar to a sunburn. A diffuse erythema, or flushing, and heat intolerance are seen in a smaller group of chemically sensitive patients. Pruritis, stinging, and burning are often common manifestations of chemical intolerance. One of the cardinal signs of chemical sensitivity is a yellow hue to the skin without jaundice or massive ingestion of β-carotene. The most common offender for the production of the yellow tone is phenol. However, many other chemicals can trigger it. At the EHC-Dallas, the most common offenders for chemical exposure of the skin are formaldehyde, phenol, aromatic hydrocarbons (e.g., pesticides, PCB, PBB), and chlorine.

Systemic vascular manifestations have been demonstrated by many groups including our own.[17,18] Results of large and small end-organ responses were observed in the heart, veins of the legs, and large blood vessels with resultant arrhythmias, phlebitis, and spastic vascular phenomena[19,20] (see Chapter 20). Some important triggering agents were pesticides, phenols, formaldehyde, chlorine, and petroleum alcohol.[21] Skin yellowing accompanied by itching and changes in the immune system is exemplified in the following patient who was challenged with phenol (Figure 25.4).

> **Case study**. Five weeks after exposure to a phenol spill in his workplace, a 26-year-old white male was admitted to the ECU. He had experienced a rapid downhill course characterized by weakness, fatigue, nausea, loss of appetite, mild peripheral and periorbital edema, and a severe yellow color to his skin. His physical exam showed positive findings limited to periorbital edema and 2+ nonpitting edema of the feet and a severe yellow color to his skin with normal color sclera. This patient was fasted for 5 days in the ECU, during which time his color gradually returned to normal, and his edema cleared. Double-blind inhaled challenge of <0.0034 ppm of phenol reproduced the edema and yellow skin color. In addition, his eosinophil count and his complement became abnormal during the challenge. Inhaled challenges with petroleum-derived ethanol (<0.50 ppm) and saline were negative.

Urticaria

Urticaria can be caused by a number of foods and chemicals, including formaldehyde.[22] It can be cleared with fasting diet therapy as was evidenced by Okamoto et al.,[23] who used this treatment with a 28-year-old woman with chronic urticaria that had previously responded only to systemic administration of glucocorticosteroids. This patient's rashes began to decrease on the third therapeutic day and completely disappeared on the eleventh day. Although milder than previous ones, the rashes returned 3 days after her therapy terminated. The overall outcome of fasting therapy with this patient certainly implies

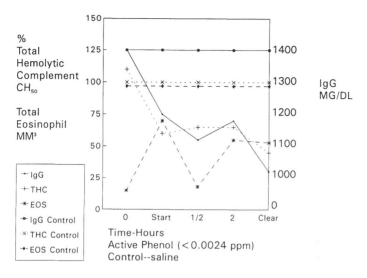

Figure 25.4. 26-year-old white female. Double-blind inhaled challenge after 4 days deadaptation in the ECU with the total load reduced. Symptoms reproduced on challenge included generalized itching; cough; arrhythmia; urgency; hot, sweaty palms with flushing of hands; yellow skin; and edema. (From EHC-Dallas. 1979.)

a causal relationship between intake of certain foods with or without chemical contamination and the onset of urticaria.

The type of eruption described by Okamoto et al.[23] was formerly classified as chronic idiopathic urticaria. At the EHC-Dallas, we have seen both underlying food and chemical sensitivities in chronic idiopathic urticarias. We have treated 100 patients with urticaria resulting from multiple triggering agents. In addition to avoidance, intradermal neutralization injection therapy is usually necessary for full clearing of symptoms. In patients with refractory urticaria, toxic volatile chemicals, such as xylene, are often found in their blood. Rigid avoidance of these chemicals is often essential for recovery to occur.

At the EHC-Dallas, we followed a small, prospective series of 35 patients with urticaria ranging in age from 21 to 90 years, with a mean age of 41 years; 88% were female. Foods played a large role in triggering urticaria as shown in Table 25.6. The average patient was sensitive to 15 foods by oral challenge and 22 by intradermal challenge. Often elimination of these positive foods combined with subcutaneous injection of them markedly decreased the urticara. However, since the patients were all chemically sensitive, reduction of the total load of toxic chemicals was necessary. An abundance of toxic chemicals was found in these patients' blood (Tables 25.7). Both inhaled and intradermal challenge tests confirmed the sensitivity (Tables 25.8 and 25.9). IGEs were elevated in 50% of the patients. Complements were changed in 46% and T cells changed in 52% (Table 25.10).

Table 25.6.   Comparison of Food Sensitivity by Oral and Intradermal
Challenge in 35 Patients with Urticaria

	Oral challenge		Intradermal challenge	
	No. of patients tested	No. of foods to which patient was sensitive	No of patients tested	No. of foods to which patient was sensitive
Total	11	160	35	775
Average	1	15	1	22

Source: EHC-Dallas. 1986.

Table 25. 7.   The Frequency of Toxic Chemicals
in Patients with Urticaria

	No. of patients	No. of positive	Frequency (%)
Pesticides			
Aldrin	16	2	12.5
α-BHC	16	5	31.3
β-BHC	16	9	56.3
DDD	16	0	0.0
DDE	16	15	93.8
Dieldrin	16	10	62.5
Endosulfan I	16	1	6.3
Heptachlor	16	1	6.3
Heptachlor epoxide	16	12	75.0
HCB	16	14	87.5
Transnonachlor	16	5	31.3
Volatile chlorinated hydrocarbons			
Dichloromethane	13	4	30.8
Chloroform	12	6	50.0
1,1,1-Trichloroethylane	8	4	50.0
Trichloroethylene	13	7	53.8
Tetrachloroethylene	13	11	84.6
Dichlorobenzene	10	2	20.0
Volatile Aromatic Hydrocarbons			
Benzene	13	7	53.8
Toluene	13	12	92.3
Ethylbenzene	13	4	30.8
Xylenes	13	9	69.2
Styrene	13	2	15.4
Trimethylbenzene	13	5	38.5

Source: EHC-Dallas. 1986.

Table 25.8. Intradermal and Inhalant Challenge of Chemicals
in 35 Patients with Urticaria After 4 Days Deadaptation
in ECU with Total Load Decreased

Chemicals	Intradermal challenge			Inhalant double-blind challenge		
	No. of tested	No. of positive	(%)	No. of tested	No. of positive	(%)
Molds	35	35	100.0	—	—	—
Cigarette Smoke	29	24	82.8	—	—	—
Orris root	25	20	80.0	—	—	—
Ethanol	25	20	80.0	5	4	80.0
Formaldehdye	20	20	100.0	9	7	77.8
Newsprint	27	18	66.7	—	—	—
Perfume	24	15	62.5	—	—	—
Phenol	16	11	68.8	9	8	88.9
Chlorine	8	4	50.0	8	8	100.0
Pesticides	—	—	—	8	7	87.5
Placebo	—	—	—	11	1	9.1

Note: Ages = 21 to 90 years; 3 females, 4 males.
Source: EHC-Dallas. 1986.

Table 25.9. Terpenes Intradermal Test
in 25 Patients with Urticaria

Terpenes	No. of patients	No. of positive	(%)
Pine	24	17	70.8
Tree	24	15	62.5
Grass	24	20	83.3
Ragweed	25	20	80.0
Mt. Cedar	24	20	83.3
Mesquite	19	14	73.3
Sage	10	9	90.0

Source: EHC-Dallas. 1986.

It was clear from this series that urticarias had multifactorial etiologies. We suggest that when urticaria appears, multiple factors should be sought and defined. The following is a case report of xylene triggering of recurrent intractable urticaria triggered by xylene.

Case study. This 46-year-old white male worked at a plant where he used xylene as an extracting solvent. He had worked there for 7 years without problems. He suddenly developed urticaria that covered his entire body and was refractory to medication. He was admitted to the ECU and fasted for 5

Table 25.10.   The Frequency of Abnormal Immune Parameters
in 36 Patients with Urticaria

Immune parameters	No. of patients tested	No. above normal	No. below normal	No. abnormal	(%)
WBC	35	3	5	8	22.9
Lym (%)	30	2	6	8	26.7
Lym C	33	0	11	11	33.3
T_{11} (%)	36	5	6	11	30.6
T_{11} C	36	4	15	19	52.8
T_4 (%)	20	6	1	7	35.0
T_4 C	20	2	5	7	35.0
T_8 (%)	20	0	3	3	15.0
T_8 C	20	2	2	4	20.0
T_4/T_8	20	2	1	3	15.0
Bly (%)	33	7	5	12	36.4
Bly C	33	6	4	10	30.3

The Frequency of Abnormal Immune Antibodies
in 35 Patients with Urticaria

Immune antibodies	No. of patients tested	No. above normal	No. below normal	No. abnormal	(%)
IgA	17	0	0	0	0.0
IgE	35	10	8	18	51.0
IgG	24	2	0	2	8.3
IgM	18	3	0	3	16.7

The Frequency of Abnormal Complements
in 21 Patients with Urticaria

Complement	No. of patients tested	No. above normal	No. below normal	No. abnormal	(%)
CH_{100}	21	1	4	5	23.8
C_1q	8	0	0	0	0.0
C_2	5	0	2	2	40.0
C_3	15	0	3	3	20.0
C_4	15	4	3	7	46.7
C_5	9	2	0	2	22.2

Source: EHC-Dallas. 1986.

Table 25.11.   46-Year-Old White Male: Inhaled Double-Blind
Challenge after 5 Days of Deadaptation with the
Total Load Reducedin the ECU

Incitant	Dose (ppm)	Reaction
Formaldehyde	<0.20	—
Phenol	<0.0020	—
Ethanol petroleum derived	<0.50	—
Chlorine	<0.33	—
Insecticide 2,4-DNP	<0.0034	—
Xylene	Ambient	Urticaria severe 3 days
Saline 1	—	—
Saline 2	—	—
Saline 3	—	—

Source: EHC-Dallas. 1986.

days. His urticaria cleared. He was then challenged intradermally with pollen, dusts, molds, weeds, trees, and grasses. He was found to be sensitive to some of these at strong dilutions. Both oral and intradermal challenges showed him to be sensitive to some foods. His hives, however, did not return during these tests. He then underwent double-blind inhaled challenge with several toxic chemicals to which he did not react. However, when he was challenged with an ambient dose of xylene, his urticaria was reproduced (Table 25.11). He was kept away from xylene compounds in the workplace and at home for 4 months, after which time he had no recurrence of his urticaria.

ITCHING, STINGING, AND BURNING

Itching seems to be one of the early signs of noncontact chemical sensitivity in approximately half of chemically sensitive patients. Patients describe this as a creepy-crawling feeling as if there were a small insect crawling on their arms and legs. Often scratching exacerbates the problem. Withdrawal from the chemically contaminated environment usually relieves the itching, and reexposure will trigger it again. This sign should not be ignored or suppressed by medication for it can be used as a sentinel for the early diagnosis of a pollutant exposure in the chemically sensitive patient. A sudden stinging may also occur and is often misinterpreted as an insect bite. Usually, a petechiae occurs at this spot.

Burning of the skin is seen in a large subset of chemically sensitive patients. The skin is tender to touch and difficult to clear. However, definition of triggering agents can be done, which will aid in diagnosis.

AUTOIMMUNE CONTACT DERMATITIS

Cosmetics, metals (e.g., nickel), plants, medicaments, fabrics, and many other chemicals have now been shown to produce cutaneous manifestations, e.g., dihydroxydiphenyl methane bisphenol F gives contact dermatitis.[24] Individuals with heavy exposure to these substances are office workers, electroplaters, auto welders, carpenters, and sheet metal workers. Nickel is found in hand tools, bracelets, and steel prostheses, including heart valves. In some individuals, problems associated with contact with these have been found.[25] The most sensitizing epoxy resins are those with a molecular weight of 340. Preservatives such as quaternium 15, IMID 20, parabins, formaldehyde, and glutaraldehyde may be sensitizers found in numerous topical preparations. Machine cutting fluids commonly contain antimicrobials; therefore, most have preservatives such as orthophenylphenol, *p*-chloro-M-xylenol, and formaldehyde-releasing agents.[26] Ethylene diamine dihydrochloride is a stabilizer found in many creams. Glyceryl mono-thio-glycolate (a constituent of acid permanent waves) is a common sensitizer used by beauticians and their clients. Photosensitivity and chronic actinic dermatitis from musk ambrette and after-shave lotion occurs.[27,28] Contact urticaria has resulted from orthophenyl phenate, a preservative used in plaster cast material.[29] Extreme sensitivities of phenolic and thiazide compounds have occurred in metal-working biocides.[30] Rats exposed to methyl mercury chloride have impaired cutaneous sensitivity.[31] This response is similar to that seen in individuals with Minamata disease reported in Japan[32] (see Volume II, Chapter 11[33]). 12–O-tetradecanoyl-13-phorbol acetate (TPA) induces permeability.[34] Contact dermatitis due to maleic hydrazide (MH) in workers handling flue-cured tobacco occurs.[35] Pyrethroids have been assoicated with skin sensory effects characterized by transient itching/tingling sensations.[36]

ACUTE SKIN REACTIONS FROM POLLUTANT EXPOSURE

The most recognized reaction to pollutant exposure is necrolysis. Erythema multiforme (Steven Johnson-type) and severe drug reactions also commonly result from toxic chemical exposures. These well-known entities will not be discussed further here, since they are well described in dermatology texts.

Environmental disasters resulting from chemically contaminated food have revealed well-defined clinical facets, as well as potential mechanisms of some dimensions of the problem of chemical sensitivity. One example of such an event comes from Japan and Taiwan where rice cooking oil was contaminated with polychlorinated biphenyl (a breakdown product of DDT) and polychlorinated dibenzofurans.[37] Those exposed to, and affected by, this contamination presented with acute symptoms—swelling of the upper eyelids, hypersecretion of the meibomian glands, chloracne, and conjunctival pigmentation. Follow-up of these patients a year later revealed that 54% were still ill[38] and that they had clearly developed chemical sensitivity. Two groups of patients evolved. One group had slowly decreasing PCB levels, while the other group remained constant.[39] Similar types of responses, characterized by faster clearing and no clearing, respectively, have been seen in other chemically sensitive patients with skin disorders. Among the chemically sensitive patients seen at the EHC-Dallas, we have observed a group of slow clearers for both PCBs and chlorinated pesticides such as lindane, aldrin, dieldrin, chlordane, heptachlor, benzene hexachloride, and hexachlorobenzene (see Volume I, Chapter 4[6]). Also, some chemically sensitive patients are slow to clear benzene, toluene, xylene, trimethylbenzene, and the clorinated solvents, as is evidenced by studies at the EHC-Dallas (see Volume IV, Chapters 30 and 35). However, we have been able to accelerate their clearance by rigid environmental control measures, vitamin and mineral supplementation, and physical therapy in conjunction with heat depuration and administration of a tolerance moderator such as transfer factor (see Volume I, Chapter 4,[6] and Volume IV, Chapter 39).

Another large incidence of acute skin and systemic problems resulting from toxic chemical overexposure is the Turkish epidemic in which over 3000 people were damaged by eating wheat contaminated with hexachlorobenzene used as a fungicide. These patients initially developed hirsutism, pigmentation, weakness, porphyrinuria, (porphyria cutanea tarda), and bullae. Two years of follow-up showed chronic effects. Neurologic, orthopedic, and dermatological abnormalities were still present. Also, the chemical sensitivity problem persisted. Neurological symptoms included weakness (66%), paresthesias (54%), neuritis (62%), myotonia (49%), and cogwheeling (29%). Orthopedic symptoms were small stature (44%), small hands (64%), painless arthritis (67%), residual scarring of blisters (85 to 90%), pinched fascia (scleroderma-like features) (42%), and enlarged thyroid (32%, of which 59% were females).[40] HCB levels were as high as 2.8 ppm in human milk, averaging 290 ppb. This finding was 140 times the supposedly safe level allowed in cow's milk.

Sixty-four percent of the patients seen in the years 1981 to 1984 at the EHC-Dallas had hexachlorobenzene in their blood. Many of their initial symptoms were similar to the acute symptoms found in the Turks. However, it should be noted that other pesticides, herbicides, and solvents were also found in the blood of most of our patients. These findings might have been the case in the Turkish experience, but concurrent tests were not available at that time to measure as many parameters as we were able to gauge.

A third example of acute toxic chemical contamination of food was the Spanish disaster involving 18,000 people who ingested contaminated rapeseed oil containing denatured aniline.[41] Long-term chemical sensitivity again resulted. Antinuclear antibody titers were positive (1/140 to 1/320) in a high percentage of patients who were affected by this contamination. We have seen similar autoimmune changes in our chemically sensitive patients. Seventeen percent of the chemically sensitive outpatients and 51% of the chemically sensitive inpatients seen at the EHC-Dallas have positive low levels of autoantibodies, suggesting a similar damaging of self-recognition in the immune system. A number of drugs and chemicals are known to produce positive ANA titers and induce systemic lupus erythematosis (SLE)-like syndromes (see

Chapter 23). Long-term follow-up over a 10-year period revealed no progression of autoimmune disease in our patients receiving proper treatment. However, those who did not have definition and removal of the triggering agents continued to worsen.

Other chemicals are known to cause acute autoantibody reactions, as well as other responses. For example, seven patients with SLE who had excerbations of their cutaneous lesions after taking a variety of medications not usually associated with induction of SLE were seen by Pereyo[42] (see Chapter 25). These medications had tartrazine, which is a derivative of analine, and hydrazine derivative (isoniazid, hydralazine, phyenylhydrazine, acetylphenylhydrazine, tartrazine, sulfanilic acid, p-sulfophenylhydrazine, sulfanilamide, aniline).[43,44] Also, saccharin, which is a coal tar derivative, can cause a photosensitization reaction.

Not only are there acute, localized cutaneous manifestations of chemical sensitivity, but also these often reflect an underlying systemic biochemical and immunological pathology. For example, Rozman et al.[45] demonstrated that decreased thyroid hormone levels in hexachlorobenzene induced porphyria in female Sprague-Dawley rats, whereas Phoon et al.[46] showed five patients with liver involvement with erythema multiforme major after exposure to trichloroethylene for 2 to 5 weeks. Another example of systemic manifestation was shown by Doss et al.[47] who found a chronic hepatic disorder induced by long-term industrial exposure to vinyl chloride in 34 workers (see Volume II, Chapter 12,[48] and Chapter 23).

Contact periorbital leukoderma has been seen due to contact with rubber swim goggles. Probably this condition resulted from the breakdown products of neoprene and its glue components.[49] Several patients have been seen at the EHC-Dallas with blisters and contact dermatitis from floating on inner tubes. Isoprene rings are the basis for terpenes, which are discussed in Volume II, Chapter 15.[50] As our study has shown throughout this book, most chemically sensitive patients are sensitive to terpenes upon challenge.

Tumorogenic activity of some chemicals such as 12–O-tetradecanoylphorbol-13-acetate (TPA),[51] phorbol–12-myristate-13 acetate (PMA), benzopyrene, 7,12-dimethylbenzanthracene,[52] and hexachlorobenzene has been observed to trigger hepatocarcinogenicity, as well as purpura, in rats.[53-55] Some of these chemicals may well trigger tumors in humans.

Sixteen patients with known contact allergy to phenol-formaldehyde resins when tested with 3-dihydroxydiphenyl methanes were seen.[56] At least nine experienced acute reactions to 1-dihydroxydiphenyl methanes (HPM) and all reacted to 2,4 (1)-HPM. Three reacted simultaneously to 2,4 (1)-HPM and 4,4 (1)-HPM, and one of these reacted to all 3 HPM. Maibach[57] found no evidence of photoirritation and photosensitization with a glyphosate herbicide in 346 volunteers.

Animal studies have shown DNA polymerase activity in N-hexadecane-induced hyperkeratotic epidermis.[58] Mice were injected with equivalent to human clinical doses of cisplatin, melphalan, and mitoxantrone.[59] Only the melphalan was ulcerogenic when injected undiluted.

ECZEMA

Acute and chronic eczema has been known for ages. The specific immune and nonimmune mechanisms involved in it, however, are now only being defined. At the EHC-Dallas, studies of over 200 chemically sensitive patients with eczema revealed that usually a combination of foods and chemicals triggered this condition. However, the chemical overload seemed to predominate. Figure 25.5 shows a case of severe eczema triggered by inhalants, foods, and chemicals. The most common foods and other substances triggering eczema were wheat, corn, cane sugar, beef, pork, cow's milk, and *Candida albicans*. At the EHC-Dallas, a prospective series of 33 (7 males and 26 females) chemically sensitive patients with eczema, ages 4 to 70 years, revealed the approximate ratio most often seen in overt, unmasked chemical sensitivity. In ten of these patients studied for the presence of chlorinated pesticides, one or more were identified in all, while one or more volatile aromatic hydrocarbons were identified in seven others (Table 25.12). All 33 patients had their chemical sensitivity provoked by intradermal and/or inhaled challenge of ambient doses of toxic chemicals (Table 25.13). In addition, mold provocation appeared to play a significant role in their eczema. All of these patients were triggered by one or more terpenes (Table 25.14). Various immune parameters, including the gammaglobulins, complements, and T and B lymphocytes, were altered in this group of patients (Table 25.15). It is clear from studies by us and others, including Little,[60] Randolph,[61] Monro,[62] Maberly,[63] Zhang,[64] Runow,[65] and Friedrickson,[66] that eczema is frequently present in chemically sensitive individuals. It can be treated with meticulous definition and avoidance of the triggering agents, intradermal injection therapy, nutrient supplementation, and tolerance mediators such as transfer factor. Occasionally, heat depuration and physical therapy are used. These help if the patient's skin can tolerate the heat. Rogers[67] reported a case exemplifying the environmental triggers in atopic dermatitis.

> **Case study.** A 38-year-old white female with 10 years of atopic dermatitis and nasal congestion and 5 years of depression and fatigue was evaluated. She was found to be sensitive to pollen, dust, and mold. Also, she had zinc, chromium, and manganese deficiencies. She was treated with injections for the inhalants. Her nutrient deficiencies were remedied with vitamin supplementation and institution of a macrobiotic diet. Within 5 months, all of her symptoms cleared. She later returned for reevaluation after her palms had turned purple for 2 weeks. Eczema that had cleared with her first course of treatment had returned on her neck. This area of skin also was purple as was the skin in the liver area of her abdomen. She brought a bag to her doctor containing her bed sheets and showed that they were purple where she had lain on them. This patient worked as a hairdresser, and she revealed that the dye of a permanent wave solution she had used on her patrons over the course of the preceding 18 years was purple. Following a second course of treatment that included reduction of total body load and continued implementation of

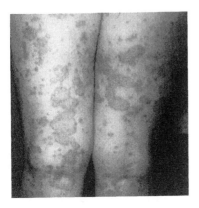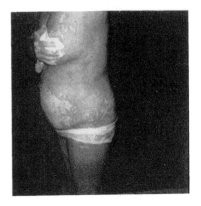

Figure 25.5. Eczema triggered by inhalants, foods, and chemicals. (From EHC-Dallas. 1979.)

a macrobiotic diet, this patient's symptoms diminished. After 2 weeks, she was free of symptoms. She continued the macrobiotic diet and has done well.

This study exemplifies what we have always known. When the body cannot properly metabolize and detoxify chemicals, it stores them. When it is sufficiently unloaded, however, it can begin to depurate old, stored pollutants and dispose of them.

PSORIASIS

Psoriasis is a genetically determined, chronic, epidermal proliferative disease of unpredictable cause. There is an increased prevalence of psoriasis in individuals with HLA antigen BW17, B13, and BW37. Clearly, environmental triggers are able to trip off these genetic time bombs. We have seen yeast triggers as well as food and some toxic chemicals. The basic alteration involves an accelerated cell cycle in an increased number of dividing cells, culminating in rapid epidermal cell accumulation. Cellular turnover is increased sevenfold, and the transit time from the basal layer to the top of the stratum concum is 3 to 4 days rather than the usual 28 days.

Rogers[68] has reported several cases of psoriasis that were triggered by a composite of inhaled molds, yeast, foods, and chemicals. We have seen several causes of psoriasis in our patients at the EHC-Dallas also. This problem seems to result from a combination of the genetic time bomb and environmental triggers.

BOILS

Some chemically sensitive patients develop recurrent boils. Some of the chemically sensitive patients with boils seen at the EHC-Dallas have had the

Table 25.12.   The Frequency of Toxic Chemicals
in Patients with Eczema

	No. of patients	No. of positive	Frequency (%)
Pesticides			
Aldrin	10	0	0.0
α-BHC	10	1	10.0
β-BHC	10	5	50.0
DDD	10	0	0.0
DDE	10	10	100.0
Dieldrin	10	4	40.0
Endosulfan I	10	1	10.0
Heptachlor	10	0	0.0
Heptachlor epoxide	10	8	80.0
HCB	10	8	80.0
trans-Nonachlor	10	5	50.0
Volatile chlorinated hydrocarbons			
Dichloromethane	7	2	28.6
Chloroform	7	4	57.1
1,1,1-Trichloroethylane	7	4	57.1
Trichloroethylene	7	3	42.9
Tetrachloroethylene	7	5	71.4
Dichlorobenzene	7	1	14.3
Volatile aromatic hydrocarbons			
Benzene	7	1	14.3
Toluene	7	6	85.7
Ethylbenzene	7	2	28.6
Xylenes	7	5	71.4
Styrene	7	1	14.3
Trimethylbenzene	7	2	28.6

Source: EHC-Dallas. 1988.

hyper IGE syndrome with an IGE of 2000 IU/mL or above. The triggering agents in these cases and in those cases with the low IGEs are usually multifactorial, being due to biological inhalants, foods, and chemicals. Studies on some of these patients show impaired killing as well as impaired phagocytosis. A select group of these patients will be cleared on an avoidance program with reduction of aspects of pollutants in air, food, and water, along with an injection program for inhalants and foods. A second group responds to transfer factor. The hyper IgE group is the most difficult to treat, and they respond only partially to all of the aforementioned measures.

**Table 25.13.   Intradermal and Inhalant Challenge of Chemicals
in 33 Patients with Eczema after 4 Days Deadaptation
in the ECU with the Total Load Decreased**

| Chemicals | Intradermal test | | | | Double-blind inhalant test | | |
	No. of tested	No. of positive	(%)	Dosage (ppm)	No. of tested	No. of positive	(%)
Molds	33	33	100.0		—	—	—
Cigarette smoke	25	18	72.0		—	—	—
Orris root	21	17	81.0		—	—	—
Ethanol	23	16	69.6	<0.50	1	0	0.0
Formaldehyde	17	17	100.0	<0.20	1	0	0.0
Newsprint	18	12	66.7		—	—	—
Perfume	17	12	70.6		—	—	—
Phenol	15	9	60.0	<0.002	1	1	100.0
Chlorine	9	5	55.6	<0.33	2	0	0.0
Pesticides (2,4-DNP)	—	—	—	<0.0034	2	1	50.0
Placebo	—	—	—		6	1	16.7

Source: EHC-Dallas. 1988.

**Table 25.14.   Terpenes Intradermal Test
in 27 Patients with Eczema**

Terpenes	No. of patients	No. of positive	(%)
Pine	26	15	57.7
Tree	26	21	80.8
Grass	27	19	70.4
Ragweed	26	19	73.1
Mt. Cedar	25	19	76.0
Mesquite	24	17	70.8
Sage	17	15	88.2

Source: EHC-Dallas. 1988.

AGING OF THE SKIN

With the elimination of as many toxic chemicals as possible, the environmentally triggered disease process becomes manageable. Also, maintenance of health and retardation of aging in the chemically sensitive individual may now be possible using ecological principles. It is common to see skin texture improve (and a consequent reduction of the signs of aging) in the environmental unit, where there is a decrease of the patient's chemical and food load.

Table 25.15. The Frequency of Abnormal Immune Parameters
in 23 Patients with Eczema

Immune parameters	No. patients tested	No. above normal	No. below normal	No. abnormal	(%)
WBC	23	1	4	5	22.7
Lym %	17	2	4	6	35.3
Lym C	22	0	6	6	27.3
T_{11} %	23	5	6	11	47.8
T_{11} C	23	2	10	12	55.2
T_4 %	12	2	2	4	33.3
T_4 C	12	1	5	6	50.0
T_8 %	12	0	3	3	25.0
T_8 C	12	0	5	5	41.7
T_4/T_8	12	1	0	1	8.3
Bly %	18	5	3	8	44.4
Bly C	18	5	3	8	44.4
CMI	7	0	2	2	28.6

The Frequency of Abnormal Immune Antibodies
in 29 Patients with Eczema

Immune antibodies	No. patients tested	No. above normal	No. below normal	No. abnormal	(%)
IgA	12	0	4	4	33.3
IgE	29	12	5	17	58.6
IgG	14	0	0	0	0.0
IgM	11	1	0	1	9.1

The Frequency of Abnormal Complements in 17 Patients with Eczema

Complement	No. patients tested	No. above normal	No. below normal	No. abnormal	(%)
CH_{100}	17	0	5	5	29.4
C_1q	4	0	0	0	0.0
C_2	2	1	0	1	50.0
C_3	7	0	1	1	14.3
C_4	7	1	1	2	28.6
C_5	5	1	0	1	20.0

Source: EHC-Dallas. 1988.

Subsequent toxic chemical challenge, in contrast, causes an increase in the aging processes and a resultant alteration in the appearance of the skin. These observations are compatible with those made by Weidruch and Walford[69] who observed the retardation of aging and disease with dietary restrictions in rats. This underfeeding phenomenon of age retardation was also found in studies by McCarter et al.[70] and others[71-76]. McCarter et al.[70] also found that underfeeding allowed peripheral tissue to become sensitive to T_3 in spite of no reduction of minimal oxygen consumption.

These findings are compatible with the recognized role that both genetic and environmental factors play in the aging process. Aging is characterized by a reduced ability to maintain homeostasis after exposure to stressful conditions, and it appears to be accelerated in the chemically sensitive individual whose cells are continuously exposed to exogenous and endogenous stressors like toxic chemicals, ionizing (ultraviolet) and nonionizing radiation, bacteria, virus, etc. The main targets are found among simple cellular components, such as amino acids, nucleotides, and lipids, and high molecular weight structures like cellular membranes and cytoskeletons. Finally, functional damage to cellular districts, e.g., mitochondria, endoplasmic reticulum, plasma membranes, nucleus, etc. are brought about. As a result, cells have developed a variety of defense and repair mechanisms to neutralize these damages and to maintain homeostasis. Unfortunately, however, these may already be damaged in the chemically sensitive individual, thus accelerating the aging process. The most important cellular defense systems appear to be DNA repair mechanisms, antioxidant defense mechanisms (either enzymatic or nonenzymatic), production of heat shock and other stress proteins, and poly (ADP–ribose) polymerase activation (Figure 25.6)[77]

In order to remove or overcome lesions induced in their DNA, cells of prokaryotes and eukaryotes are equipped with certain mechanisms of repair: photoactivities, excision repair, and postreplication repair.[78-91]

Antioxidant defenses are discussed in Volume I, Chapter 4,[6] and will not be elaborated on here. Comments made previously are applicable to the skin.

Heat shock proteins (HSP) are generated by heat and chemical exposure to protect the cell from toxic exposures and may in the long run cause an acceleration of aging.[78,92-96]

Poly (ADP-ribose) synthetic is an abundant and ubiquitous enzyme that cleaves the bond of NAD^+ and the N-glycossolic bond between nicotinamide and ribose rings and then transfers ADP ribosyl.[97-103]

Part of NAD^+ transfers either to chromatin proteins or to another ADP-ribose molecule. This sequencing is described as a cellular defense system that is interconnected and constitutes a network. The proteins and sugars forming this network must be considered together because a single agent can activate multiple pathways. For example, oxidants, besides being counteracted by the enzymatic and nonenzymatic antioxidants, induce HSPs, damage DNA bases (which are excised and substituted by DNA repair enzymes), and cause DNA

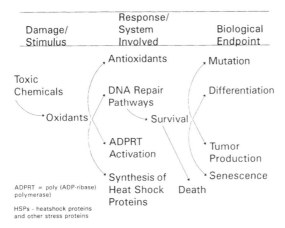

Figure 25.6 The network of cellular defense systems seen in the chemically sensitive individual.

single-strand breaks (which, in turn, activate ADPRT, or ADP-ribosyl transferase). It is important to realize that mechanisms that favor cell survival and the mechanisms that favor cell suicide are part of the same network. The network allows the survival of intact cells and, at the same time, the elimination of severely damaged cells, in order to avoid an excessive accumulation of mutated or transformed cells. The final outcome depends upon the variety of factors described in this book. Excess chemical overload in the chemically sensitive individual appears to accelerate aging, whereas reduction of the total body load appears to decrease it. Senescence has also been shown to suppress tumorigenic activity in many systems.[104-107]

In summary, the skin clearly may be the target organ for numerous toxic chemicals in the chemically sensitive individual. Damage from exposure to these chemicals can result either from direct contact with them or from systemic involvement. Precise definition of triggering agents through challenge testing in a controlled environment with the individual's total body load reduced will aid in the diagnosis and treatment of the individual who has developed chemical sensitivity with skin involvement.

REFERENCES

1. Klaassen, C. D. 1980. Absorption, distribution, and excretion of toxicants. In *Casarett and Doull's Toxicology: The Basic Science of Poisons*, 2nd ed, eds. J. Doull, C. D. Klaassen, and M. O. Amdur. 35–37. New York: Macmillan.

2. Brown, H. S., D. R. Bishop, and C. A. Rowan. 1984. The role of skin absorption as a route of exposure for volatile organic compounds (VOCS) in drinking water. *Am. J. Public Health* 74:379–484.

3. Marks, J. G. 1992. *Contact and Occupational Dermatology*. St. Louis: Mosby Year Book.

4. Fitzpatric, T. B. and H. A. Haynes. 1977. Pigmentation of the skin and disorders of melanin metabolism. In *Harrison's Principles of Internal Medicine*, 8th ed., eds. G. W. Thorn, R. D. Adams, E. Braunwald, K. J. Isselbacher, and R. G. Petersdorf. 278. New York: McGraw-Hill.

5. Storer, J. S., I. Deleon, L. E. Millikan, J. L. Laseter, and C. Griffing. 1984. Human absorption of crude coal tar products. *Arch. Dermatol.* 120:874–877.

6. Rea, W. J. 1982. *Chemical Sensitivity. Vol. I. Mechanisms of Chemical Sensitivity.* 47. Boca Raton, FL: Lewis Publishers.

7. Rea, W. J. 1982. *Chemical Sensitivity. Vol. I. Mechanisms of Chemical Sensitivity.* 155. Boca Raton, FL: Lewis Publishers.

8. Rea, W. J., A. R. Johnson, S. Youdin, E. J. Fenyves, and N. Samadi. 1986. T & B lymphocyte parameters measured in chemically sensitive patients and controls. *Clin. Ecol.* 4(1):11–14.

9. Meggs, W. J., M. Frieri, R. Costello, D. D. Metcalfe, and N. N. Papadopoulos. 1985. Oligoclonal immunoglobulins in mastocytosis. *Ann. Intern. Med.* 103:894.

10. Rea, W. J. and M. J. Mitchell. 1982. Chemical sensitivity and the environment. *Immunol. Allergy Pract.* 4(5):157–167.

11. Moses, M., R. Lilis, K. D. Grow, J. Thornton, A. Fischbein, H. A. Anderson, and I. J. Selikoff. 1984. Health status of workers with past exposure to 2,3,7,8-tetrachlorodibenzo-*p*-dioxin in the manufacture of 2,4,5-trichlorophenoxyacetic acid: comparison of findings with and without chloracne. *Am. J. Ind. Med.* 5(3):161–182.

12. Emmett, E. A. 1986. Toxic responses of the skin. In *Casarett and Doull's Toxicology: The Basic Science of Poisons*, 3rd ed., eds. C. D. Klaassen, M. O. Amdur, and J. Doull. 427. New York: Macmillan.

13. Chang, K. J., K. H. Hsieh, T. P. Lee, and T. C. Tung. 1982. Immunologic evaluation of patients with polychlorinated biphenyl poisoning: determination of phagocyte Fc and complement receptors. *Environ. Res.* 28:329–334.

14. Levy, C. J. 1988. Agent orange exposure and posttraumatic stress disorder. *J. Nerv. Ment. Dis.* 176(4):242–247.

15. Lambert, J., P. Schepens, J. Jianssens, and P. Dockx. 1986. Skin lesions as a sign of subacute pentachlorophenol intoxication. *Acta Derm. Venereol. (Stockh)* 66(2):170–172.

16. Cole, G. W., O. Stone, D. Gates, and D. Culver. 1986. Chloracne from pentachlorophenol preserved wood. *Contact Dermatitis* 15(3):164–168.

17. Rea, W. J., I. R. Bell, and R. E. Smiley. 1975. Environmentally triggered large-vessel vasculitis. In *Allergy: Immunology and Medical Treatment*, eds. F. Johnson and J. T. Spence. Chicago: Symposia Specialist.

18. Rea, W. J. 1977. Environmentally triggered small vessel vasculitis. *Ann. Allergy* 38:245–251.

19. Rea, W. J. 1980. Review of cardiovascular disease in allergy. In *Bi-Annual Review of Allergy*, ed. C. A. Frazier. 282–347. Springfield, IL: Charles C Thomas.

20. Rea, W. J., D. W. Peters, R. E. Smiley, R. Edgar, M. Greenberg, and E. Fenyves. 1981. Recurrent environmentally triggered thrombophlebitis: a five-year following. *Ann. Allergy* 47:337–344.

21. Rea, W. J. and C. W. Suits. 1980. Cardiovascular disease triggered by food and chemicals. In *Food Allergy: New Perspectives*, ed. J. W. Gerrard. Springfield, IL: Charles C Thomas.

22. Andersen, K. E. and H. I. Maibach. 1984. Multiple application delayed onset contact urticaria: possible relation to certain unusual formalin and textile reactions? *Contact Dermatitis* 10(4):227–234.

23. Okamoto, O., I. Murakami, S. Itami, and S. Takayasu. 1992. Fasting diet therapy for chronic urticaria: report of a case. *J. Dermatol.* (H27) 19(7):428–431.

24. Bruze, M. and E. Zimerson. 1985. Contact allergy to dihydroxydiphenyl methanes (bisphenol F). *Derm. Beruf. Umwelt* 33(6):216–220.

25. Rea, W. J. and M. J. Mitchell. 1982. Chemical sensitivity and the environment. *Immunol. Allergy Pract.* 4(5):21–31.

26. Moss, M. 1989. Personal communication.

27. Cirne de Castro, J. L., M. A. Pereira, F. Prates Nunes, and A. Pereira dos Santos. 1985. Musk ambrette and chronic actinic dermatitis. *Contact Dermatitis* 13(5):302–306.

28. Cronin, E. 1984. Photosensitivity to musk ambrette. *Contact Dermatitis* 11(2):88–92.

29. Tuer, W. F., W. D. James, and R. J. Summers. 1986. Contact urticaria to O-phenylphenate. *Ann. Allergy* 56(1):19–21.

30. Andersen, K. E. and K. Hamann. 1984. The sensitizing potential of metalworking fluid biocides (phenolic and thiazole compounds) in the guinea-pig maximization test in relation to patch-test reactivity in eczema patients. *Food Chem. Toxicol.,* 2218:655–660.

31. Wu, M. G., J. R. Ison, J. R. Wecker, and L. M. Lapham. 1985. Cutaneous and auditory function in rats following methyl mercury poisoning. *Toxicol. Appl. Pharmacol.* 79(3):377–388.

32. Tsubaki, T. and K. Irukayama, eds. 1977. *Minimata Disease.* Tokyo: Kedansha.

33. Rea, W. J. 1994. *Chemical Sensitivity. Vol. II. Sources of Total Body Load.* 713. Boca Raton, FL: Lewis Publishers.

34. Nakadate, T., S. Yamamoto, E. Aizu, and R. Kato. 1985. Inhibition of 12–0-tetradecanoylphorbol-13-acetate-induced increase in vascular permeability in mouse skin by lipoxygenase inhibitors. *Jpn. J. Pharmacol.* 38(2):161–168.

35. Herman, N. D., T. W. Hunt, and T. J. Sheets. 1985. Hand harvester exposure to maleic hydrazide (MH) in flue-cured tobacco. *Bull. Environ. Contam. Toxicol.* 34(4):469–475.

36. Cagen, S. Z., L. A. Malley, C. M. Parker, T. H. Gardiner, G. A. Van Gelder, and V. A. Jud. 1984. Pyrethroid-mediated skin sensory stimulation characterized by a new behavioral paradigm. *Toxicol. Appl. Pharmacol.* 76(2):270–279.

37. Fischbein, A., J. N. Rizzo, S. J. Solomon, and M. S. Wolff. 1985. Oculodermatological findings in workers with occupational exposure to polychlorinated biphenyls (PCBs). *Br. J. Ind. Med.* 42(6):426–430.

38. Lü, Y. C. and P. N. Wong. 1984. Dermatological, medical, and laboratory findings of patients in Taiwan and their treatments. *Am. J. Ind. Med.* 5(1–2):81–115.

39. Urabe, H. and M. Asahi. 1985. Past and current dermatological status of yusho patients. *Environ. Health Perspect.* 59:11–15.

40. Cripps, D. J., H. A. Peters, A. Gocmen, and I. Dogramici. 1984. Porphyria turcica due to hexachlorobenzene: a 20 to 30 year follow-up study on 204 patients. *Br. J. Dermatol.* 111(4):413–422.

41. Rush, P. J., M. J. Bell, and A. G. Fam. 1984. Toxic oil syndrome (Spanish oil disease) and chemically induced scleroderma-like conditions [editorial]. *J. Rheumatol.* 11(3):262–264.

42. Pereyo, N. 1979. Tartrazine compounds. *Arch. Dermatol.* 115:508.

43. Pereyo, N. 1982. Tartrazine, a dangerous coloring agent. *Schoch Lett.* 32:78.

44. Pereyo, N. 1986. Hydrazine derivatives and induction of systemic lupus erythematosus. *J. Am. Acad. Dermatol.* 14(3):514–515.

45. Rozman, K., J. R. Gorski, P. Rozman, and A. Parkinson. 1986. Reduced serum thyroid hormone levels in hexachlorobenzene-induced porphyria. *Toxicol. Lett.* 30(1):71–78.

46. Phoon, W. H., M. O. Chan, V. S. Rajan, K. J. Tan, T. Thirumoorthy, and C. L. Goh. 1984. Stevens-Johnson syndrome associated with occupational exposure to trichloroethylene. *Contact Dermatitis* 10(5):270–276.

47. Doss, M., C. E. Lange, and G. Veltman. 1984. Vinyl chloride-induced hepatic coproporphyrinuria with transition to chronic hepatic porphyria. *Klin. Wochenschr.* 62(4):175–178.

48. Rea, W. J. 1994. *Chemical Sensitivity. Vol. II. Sources of Total Body Load.* 765. Boca Raton, FL: Lewis Publishers.

49. Goette, D. K. 1984. Raccoon-like periorbital leukoderma from contact with swim goggles. *Contact Dermatitis* 10(3):129–131.

50. Rea, W. J. 1994. *Chemical Sensitivity. Vol. II. Sources of Total Body Load.* 979. Boca Raton, FL: Lewis Publishers.

51. Skouv, J., B. Christensen, I. Skibshøj, and H. Autrup. 1986. The skin tumor-promoter 12–O-tetradecanoylphorbol-13-acetate induces transcription of the c-fos proto-oncogene in human bladder epithelial cells. *Carcinogenesis* 7(2):331–333.

52. Weiss, H. S., J. F. O'Connell, A. G. Hakaim, and W. T. Jacoby. 1986. Inhibitory effect of toluene on tumor promotion in mouse skin (42240). *Proc. Soc. Exp. Biol. Med.* 181(2):199–204.

53. Lissner, R., W. Branner, K. Bolsen, H. Merk, and G. Goerz. 1985. Influence of N-acetyloysteine on the hexachlorobenzene induced porphyria in rats. *Arznimittelforschung* 35(4):713–715.

54. Wainstok de Calmanovici, R., M. C. Ríos de Molina, M. C. Taira de Yamasato, J. M. Tomio, and L. C. San Martin de Viale. 1984. Mechanism of hexachlorobenzene-induced porphyria in rats. Effect of phenobarbitone pretreatment. *Biochem. J.* 218(3):753–763.

55. Cripps, D. J., H. A. Peters, A. Gocmen, and I. Dogramici. 1984. Porphyria turcica due to hexachlorobenzene: a 20 to 30 year follow-up study on 204 patients. *Br. J. Dermatol.* III (4):413–422.

56. Bruze, M. and E. Zimersoa. 1985. Contact allergy to dihydroxydiphenyl methanes (bisphenol F). *Derm. Beruf. Umwelt* 33(6):216–220.

57. Maibach, H. I. 1986. Irritation, sensitization, photoirritation and photosensitization assays with a glyphosate herbicide. *Contact Dermatitis* 15(3):152–156.

58. Ogura, R., N. Kaneko, and T. Hidaka. 1986. DNA polymerase activity in the N-nexadecane-induced hyperkeratotic epidermis. *Arch. Dermatol. Res.* 278(5):582–585.

59. Dorr, R. T., D. S. Alberts, and M. Soble. 1986. Lack of experimental vesicant activity for the anticancer agents cisplatin, melphalan, and mitoxantrone. *Cancer Chemother. Pharmacol.* 16(2):91–94.

60. Little, C. 1986. Personal communication.

61. Randolph, T. G. 1984. Personal communication.

62. Monro, J. 1989. Personal communication.
63. Maberly, J. 1989. Personal communication.
64. Zhang, H.-Y. 1985. Personal communication.
65. Runow, K. 1992. Personal communication.
66. Friedrickson, P. 1989. Personal communication.
67. Rogers, S. 1993. Personal communication.
68. Rogers, S. 1988. Personal communication.
69. Weidruch, R. and R. L. Walford. 1980. *The Retardation of Aging and Disease by Dietary Restrictions.* 202–226. Springfield, IL: Charles C Thomas.
70. McCarter, R. J., J. T. Herliky, and J. R. McGee. 1989. Metabolic rate and aging: effects of food restriction and thyroid hormone on minimal oxygen consumption in rats. *Aging Clin. Exp. Res.* 1(1):43–47.
71. McCarter, R. J., E. J. Masoro, and B. P. Yu. 1985. Does food restriction retard aging by reducing the metabolic rate? *Am. J. Physiol.* 248:E488-E492.
72. McCarter, R., E. J. Masoro, and B. P. Yu. 1982. Rat muscle structure and metabolism in relation to age and food intake. *Am. J. Physiol.* 11:R89-R93.
73. Culter, R. G. 1986. Superoxide dismutase, longevity and specific metabolic rate. *Gerontology* 29:113–120.
74. Yu, B. P., S. Langaniere, and J. W. Kim. 1988. Influence of life-prolonging food restriction on membrane lipoperoxidation and anti-oxidant status. *Oxy. Radiat. Biol. Med.* 49:1067–1073.
75. Masoro, E. J. 1987. Biology of aging: current state of knowledge. *Arch. Int. Med.* 147:166–169.
76. Yu, B. P., E. J. Masoro, and C. A. McMahan. 1985. Nutritional influences on aging of Fischer 344 rats: I. Physical metabolic and longevity characteristics. *J. Gerontol.* 40:657–670.
77. Franceschi, C. 1989. Cell proliferation, cell death, and aging. *Aging* 1:3–15.
78. Beyreuther, K. T., P. A. Cerutti, B. F. C. Clark, J. M. Delabar, K. Esser, C. Franceschi, T. B. L. Kirkwood, S. I. S. Rattan, J. A. Treton, A. G. Uitterlinden, A. M. Vandenberghe, and J. Vijg. 1988. Molecular biology of aging. *Eurage, Rijswijk* 1–32.
79. Teoule, R. 1988. Radiation-induced DNA damage and its repair. *Int. J. Radiat. Biol.* 51:573–589.
80. Teebor, G. W., R. J. Roostein, and J. Cadet. 1988. The repairability of oxidative free radical mediated damage to DNA: a review. *Int. J. Radiat. Biol.* 54:131–150.
81. Rubin, J. S. 1988. The molecular genetics of the incision step in the DNA excision repair process. *Int. J. Radiat. Biol.* 54:309–365.
82. Cleaver, J. E., C. Borek, K. Milam, and W. F. Morgan. 1987. The role of poly (ADP-ribose) synthesis in toxicity and repair of DNA damage. *Pharm. Ther.* 31:269–293.
83. Hart, R. H. and R. B. Setlow. 1974. Correlation between deoxyribonucleic acid excision repair and lifespan in a number of mammalian species. *Proc. Natl. Acad. Sci. U.S.A.* 71:2169–2173.
84. Licastro, F., C. Franceschi, M. Chiricolo, M. G. Battelli, P. Tabacchi, M. Cenci, M. Barboni, and D. Pallenzona. 1982. DNA repair after gamma radiation and superoxide dismutase activity in lymphocytes from subjects with far advanced age. *Carcinogenesis* 3:45–48.

85. Setlow, R. B. 1982. DNA repair, aging, and cancer. *Natl. Cancer Inst. Monogr.* 60:249–255.

86. Franceschi, C., F. Licastro, M. Chiricolo, M. Zannotti, and M. Masi. 1986. Premature senility in Down's syndrome: a model for and an approach to the molecular genetics of the ageing process. In *Immunoregulation in Ageing*, eds. A. Facchini, J. J. Haaijman, and G. Labò. 77–83. Rijswijk: EURAGE.

87. Vijg, J. and D. L. Knook. 1987. DNA repair in relation to the aging process. *J. Am. Gerontol. Soc.* 35:532–541.

88. Vijg, J. 1987. *DNA Repair and The Aging Process.* 1–149. Rijswijk: TNO Institute for Experimental Gerontology.

89. Kirkwood, T. B. L. 1989. DNA, mutations and aging. *Mut. Res.* 219:1–7.

90. Hayflick, L. 1977. The cellular basis for biological aging. In *Handbook of the Biology of Aging*, eds. C. E. Finch and E. L. Schneider. 159–186. New York: Van Nostrand Reinhold.

91. Ahnstrom, G. 1988. Techniques to measure DNA single-strand breaks in cells: a review. *Int. J. Radiat. Biol.* 54:695–707.

92. Pelham, H. 1985. Activation of heat-shock genes in eukaryotes. *Trends Genet.* 1:31–35.

93. Lindquist, S. 1986. The heat-shock response. *Annu. Rev. Biochem.* 55:1151–1191.

94. Lanks, K. W. 1986. Modulators of the eukaryotic heat shock response. *Exp. Cell. Res.* 165:1–10.

95. Deguchi, Y., S. Negoro, and S. Kishimoto. 1988. Age-related changes of heat shock protein gene transcription in human peripheral blood mononuclear cells. *Biochem. Biophys. Res. Commun.* 157:580–584.

96. Tsuji, Y., S. Ishibashi, and T. Ide. 1986. Induction of heat shock protein in young and senescent human diploid fibroblasts. *Mech. Aging Dev.* 36:155–160.

97. Shall, S. 1988. ADP-ribosylation of proteins: a ubiquitous cellular control mechanism. In *Advances in Translational Modifications of Proteins and Aging*, eds. V. Zappia, P. Galletti, and R. Porta. New York: Plenum Press.

98. Berger, N. A. 1985. Poly (ADP-ribose) in the cellular response to DNA damage. *Radiat. Res.* 101:4–15.

99. Schraufstatter, I. U., D. B. Hinshaw, P. A. Hyslop, R. G. Spragg, and C. G. Cochrane. 1986. Oxidant injury of cells. DNA strand-breaks activate polyadnosine diphosphate-ribose polymerase and lead to depletion of nicotinamide adenine dinucleotide. *J. Clin. Invest.* 77:1312–1320.

100. Carson, D. A., S. Seto, D. B. Wasson, and C. J. Carrera. 1986. DNA strand breaks, NAD metabolism, and programmed cell death. *Exp. Cell Res.* 164:273–281.

101. Gaal, J. C., R. K. Smith, and C. K. Pearson. 1987. Cellular euthanasia mediated by a nuclear enzyme: a central role for nuclear ADP-ribosylation in cellular metabolism. *Trends Biochem. Sci.* 12:129–130.

102. Cossarizza, A., D. Monti, M. Zannotti, and C. Franceschi. 1989 (abstract). Lymphocyte death by oxidative stress and aging: effect of inhibitors of ADP-ribosyl transferase. *J. Cell. Biochem. [Suppl.]* 13C:156.

103. Chapman, M. L., M. R. Zaun, and R. W. Gracy. 1983. Changes in NAD levels in human lymphocytes and fibroblasts during aging and in premature aging syndromes. *Mech. Aging Dev.* 21:157–167.

104. Ames, B. N. and R. L. Saul. 1987. Oxidative DNA damage, cancer and aging. In *Oxygen Radicals and Human Disease*, C. E. Cross (moderator). *Ann. Intern. Med.* 107:536–539.

105. O'Brien, W., B. Stenman, and R. Sager. 1986. Suppression of tumor growth by senescence in virally transformed human fibroblasts. *Proc. Natl. Acad. Sci. USA* 83:8659–8663.

106. Koi, M. and J. C. Barrett. 1986. Loss of tumor-suppressive function during chemically induced neoplastic progression of Syrian hamster embryo cells. *Proc. Natl. Acad. Sci. U.S.A.* 83:5992–5996.

107. Franceschi, C. 1989. Cell proliferation, cell death, and aging. *Aging* 1:3–15.

108. Kipling, M. D. 1976. Soots, tars, and oils as causes of occupational cancer. In *Chemical Carcinogens,* ed. C. E. Searle. 315–323. Washington, D.C.: American Chemical Society.

109. Dipple, A. 1976. Polynuclear aromatic carcinogens. In *Chemical Carcinogens,* ed. C. E. Searle, 245–314. Washington, D.C.: American Chemical Society.

110. Scribner, J. D. 1973. Tumor initiation by apparently non-carcinogenic polycyclic aromatic hydrocarbons. *J. Natl. Cancer Inst.* 50:1717–1719.

111. *Report on Coal Tar.* 1977. 78–107. Washington, D.C.: National Institute for Occupational Safety and Health.

112. van Duuren, B. L. 1976. Tumor-promoting and co-carcinogenic agents in chemical carcinogens. In *Chemical Carcinogens,* ed. C. E. Searle. 24–51. Washington, D.C.: American Chemical Society.

113. McCann, J., E. Choi, E. Yamasaki, and B. Ames. 1975. Detection of carcinogens as mutagens in the *Salmonella*/microsome test: assay of 300 chemicals. *Proc. Natl. Acad. Sci. U.S.A.* 75:5135–5139.

114. Lowe, N. J., J. H. Breeding, and M. S. Wortzman. 1982. New coal tar extract and coal tar shampoos: evaluation by epidermal cell DNA synthesis suppression assay. *Arch Dermatol.* 118:487–489.

115. Andrews, A. W., L. H. Thibault, and W. Lijinsky. 1978. The relationship between carcinogenicity and mutagenicity of some polynuclear hydrocarbons. *Mut. Res.* 51:311–318.

116. Rasmussen, J. E. 1978. The crudeness of coal tar. *Prog. Dermatol.* 12:23–29.

117. Doll, R., R. E. Fisher, E. J. Gammon, W. Gunn, G. O. Hughes, F. H. Tyler, and W. Wilson. 1965. Mortality of gas workers with special reference to cancers of the lung and bladder, chronic bronchitis, and pneumoconiosis. *Br. J. Ind. Med.* 22:1–11.

118. Saperstein, M. D. and L. A. Wheeler. 1979. Mutagenicity of coal tar preparations used in the treatment of psoriasis. *Toxicol. Lett.* 3:325–329.

119. Wheeler, L. A., M. D. Saperstein, and N. J. Lowe. 1981. Mutagenicity of urine for psoriatic patients undergoing treatment with coal tar and ultraviolet light. *J. Invest. Dermatol.* 77:181–185.

26   Nervous System

NERVOUS SYSTEM—INTRODUCTION

The nervous system is integrated with the endocrine system and the immune and nonimmune detoxification systems. For the body to respond properly to pollutant exposure, these systems must be in balance. However, in some cases of chemical sensitivity, the balance between these systems becomes impaired, and the nervous system itself becomes the primary target organ for pollutant injury. Although all chemically sensitive individuals do not experience nervous system involvement, it is a prime target for deposition of many toxic organic chemicals because they are lipophilic. In order to prevent not only brain dysfunction but also end-stage, irreversible brain damage, the clinician must be cognizant of the potential effects of pollutants on the brain. In particular, pesticides and solvents, which are able to disrupt homeostasis, can dramatically affect the neurologically involved chemically sensitive patient.

This chapter emphasizes environmental aspects of both the autonomic and voluntary nervous systems, making specific reference to their influence on anatomy, physiology, and disordered function. In addition, it attempts to demonstrate how to evaluate early pollutant injury in order to establish a method for prevention of environmentally induced illness in the autonomic (ANS), central (CNS), and peripheral nervous systems.

AUTONOMIC NERVOUS SYSTEM

Chemically sensitive individuals who do not first experience a local reaction in response to pollutant exposure appear to experience their first systemic response in the ANS. Frequently, inhibition, triggering, and/or dysfunction of this system are misinterpreted by the clinician as "functional" and nonobjective. This misinterpretation is a gross mistake, for often diagnosis of pollutant injury at this stage is the only time that permanent nerve or other bodily injury can be prevented. When studied under environmentally controlled conditions, signs and symptoms are sufficient to identify objectively the pollutant-triggered responses of the ANS, whether these responses are normal, though prolonged, or dysfunctional in the chemically sensitive patient.

Understanding the anatomy and function of the ANS is necessary to understanding many of the symptoms and signs in chemically sensitive patients, since many of their responses may be due to direct stimulation or inhibition of the ANS. These early symptoms and signs that are produced by pollutants may be prolonged responses of the normal autonomic function, or they may be the result of pathological injury. Prolonged normal responses are much more common and are more easily reversed than pathological injury, but, if no scar tissue has formed, it is possible to reverse pathological injury following removal of the pollutants.

Pollutants can stimulate large portions of the sympathetic nervous system, simultaneously producing a mass discharge of the sympathetic nerves and accounting for the sudden onset of drop attacks seen in some chemically sensitive patients. In contrast, the parasympathetic nervous system has a series of reflexes that are very specific and regional. These particular responses may explain why many chemically sensitive patients experience various and multiple responses such as isolated cardiac, gastrointestinal, pupillary, and genitourinary dysfunctions. In some instances, localized sympathetic responses do occur as a result of pollutant exposure. These responses then result in vasoconstriction of skin vessels, which produces cold hands and feet, which are seen in most chemically sensitive patients.

Although a mass discharge of sympathetic nerves is frequently seen in some chemically sensitive patients following pollutant exposure, a sympatholytic response, as measured by the iris corder, generally occurs (see Chapter 27). This sympatholytic response results in decreased arterial pressure and decreased blood flow to active muscles. This decreased blood flow, in turn, inhibits large muscle exercise, as is often seen in the chemically sensitive patient. Concurrent with the decrease in arterial pressure and blood flow to active muscles are an increase in blood flow to other organs, decreased rates of cellular metabolism, decreased glucose concentration, decreased muscle glycolysis, decreased muscle strength, decreased muscle activity, and decreased blood coagulation. Basically, the sum of these effects is inactivation of the body, reducing the ability of chemically sensitive individuals to respond to both environmental and other stressors.

An example of this type of sympatholytic response is the 55-year-old artist who developed both peripheral voluntary and autonomic nervous system dysfunction, which made it difficult for him to walk. This abnormality was accompanied by difficulty breathing and talking, due to vocal cord nervous dysfunction. His arterial blood pressure would fluctuate between 140/80 and 50/20 mmHg, depending upon pollutant exposure. He would at times pass out when his blood pressure was low and would finally recover with a liter of intravenous saline. This patient was a shop and art teacher. On the job, he had been exposed to various types of solvents, paints, and lacquers. He gradually deteriorated until he became incapacitated.

Challenge testing at the EHC-Dallas revealed positive reactions to molds, pine terpene, tree terpene, grain smut, grass smut, *Candida*, cigarette smoke, chlorine-SL, ethanol, formaldehyde, women's cologne, men's cologne, orris root, news material, and phenol. Two foods, oatmeal and beef, also caused hypotension. He had high magnesium in his blood, low cholesterol, and low CK. His SPECT scan showed locally reduced flow and function in a fairly large area involving the left parietal lobe, a pattern consistent with neurotoxicity, and temporal lobe asymmetry. Iris corder examination showed cholinergic and sympatholytic reaction. Heavy metals in his blood were below reference range. Toxic chemicals in this patient's blood showed high benzene, 2.2 ppb (C<1.0 ppb); 2-methylpentane, 13.8 ppb (C<1.0 ppb); 3-methylpentane, 32.4 ppb (C<1.0 ppb); and n-hexane, 10.7 ppb (C<1.0 ppb). Blood gases showed PO_2-A was low at 74 mmHg (C = 75 to 100 mmHg). Base excess was high at 2.8 mmol/L (C = -2.0 to 2.0 mmol/L). Plasma cortisol, norepinephrine seratonin, and mineral corticoids were all normal. CAT and MRI for tumors were negative.

After exposure, this patient would experience falling blood pressure, which would start slowly at around 130 to 120/80 and decrease to 100/60 to 80/50 to 60 to 40/0 mmHg. At these times, this patient would develop fuzzy thinking and then pass out. This falling blood pressure would become life-threatening in that several times he was found comatose without blood pressure. He was treated with fluornef, constant excess saline (oral and intravenous), leg-wrapping with a gravity-suit, a rotary diet, and a less-polluted environment. This patient is a classic case of chemically induced ANS failure.

The hypothalamus is the highest vegetative center in the brain and is the focal point in triggering the pollutant-driven responses seen in the chemically sensitive patient. It also has functional connections with the forebrain and brainstem as well as the pituitary and pineal glands and the olfactory nerve (Figure 26.1). When the hypothalamus is triggered by pollutants, it may activate any or all of these organs, resulting in varied symptomatology. The hypothalamus is close to, or contains, the supraoptic nucleus, the ventral medial nucleus, the preoptic nucleus, the paraventricular, posterior, dorsamedial, the ventromedial nuclei, the mammillary body, and the interpendulor nucleus. Anterior stimulation of these nuclei produces responses that correspond roughly to parasympathetic responses, while posterior stimulation causes sympathetic responses. Stimulation that produces sympathetic responses results in increased arterial pressure, blood glucose concentration, glycolysis, muscle strength, mental activity, rate of coagulation, and blood flow to active muscles with decreased blood flow to organs that are not needed for rapid activity. Both sympathetic and parasympathetic responses are seen in some pollutant-overloaded, chemically sensitive patients. Frequently, a moderately chemically sensitive individual will get energy bursts characterized by a sympathetic response that may last for a few minutes to hours and then completely give out,

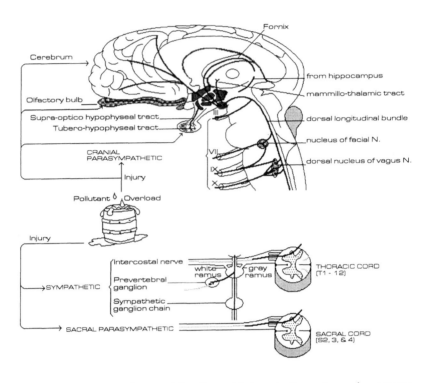

Figure 26.1. Areas of potential pollutant injury to the autonomic nervous system.

resulting in exhaustion that may last for a day or two. This cycle is often reported until the patient is incapacitated if the pollutant(s) are not removed.

The frontal lobe and cornu ammonis usually send messages to one hypothalamic nucleus. Also, afferents from the olfactory bulb go to the supra-optic nuclei and medial forebrain bundle. Stimulation of these anatomical pathways explains why the pollutant-odor chemically sensitive patient has multiple ANS complaints with ultimate triggering of vascular as well as other smooth muscle systems. ANS triggering produces a variety of responses, depending on the nature of the imbalance caused (Table 26.1). For example, generally, if the sympathetic nervous system is triggered, heart rate, blood pressure, sweat, and pupil size increase with bronchial contractility, sputum, saliva, intestinal mobility, urine output, and uterine contractions decreasing. The opposite would be generally true if the parasympathetic nervous system were triggered. The moderate to severely chemically sensitive individual usually cannot sweat or has difficulty in sweating, which apparently helps him retain pollutants and thus increases his sensitivity (see Volume IV, Chapter 35.)

Characteristically, chemically sensitive individuals have a much more fragile ANS. It is more easily triggered, physiologically disrupted, and difficult to keep balanced. Therefore, these patients frequently experience symptoms of autonomic triggering. The posterior lobe of the pituitary gland receives direct tracts from the supra optic nucleus (sympathetic) and the nucleus tuberis (parasympathetic). Therefore, triggering or inhibition of either system results in neuroendocrine dysfunction in most parts of the body. Feedback loops also can keep a cycle of endocrine dysfunction flowing, as is seen in many chemically sensitive patients.

The outflow from the hypothalamus via the mammillary body goes back toward the cerebral cortex to the lower vegetative centers and the hypophysis cerebi. The autonomic outflow, emanating from the hypothalamus and coming from the mammillary body, influences the parasympathetic centers, including cranial nerve VII to the face and cerebral vessels, cranial nerve IX to the pharynx and the region of the carotid sinus, and cranial nerve X, which is the medullary cardiovascular center. Also, pollutant exposure may influence the sacral center. Often, chemically sensitive patients are seen who have apparent triggering and deregulation of the respiratory, vascular, and temperature control centers. These patients will present with breathlessness with normal pulmonary function, severe peripheral vascular deregulation with vasospasm, or with high temperatures with an inability to tolerate heat.

The sympathetic trunks extend from the base at the cranium to the coccyx. The parasympathetic nerves go down the cord in the gray matter, through the white ramus, and either out the peripheral nerve or into the sympathetic ganglion to the viscera. The sympathetic trunk goes from the brain through the ganglion to the sacrum (Figure 26.2).

Some autonomically innervated tissues are supplied with both the parasympathetic and sympathetic nerves, while others are supplied by sympathetic nerves alone, e.g., the adrenal and intracranial vessels, hair follicles, and sweat glands. Probably the peripheral vessels are also included. This complex distribution of autonomic fibers allows the extreme varied responses seen in a group of pollutant-triggered, chemically sensitive individuals. It should be emphasized that the chemically sensitive patient with neurotoxic patterns on SPECT brain scans and with an inability to sweat may well have a pollutant-driven sympathetic response, but possibly could have a cholinolytic pattern triggered extracranially, since the cerebral vessels show a reversible decrease in flow in certain areas of the brain. Vascular sympathetic nerves have both sympathetic and cholinergic fibers that account for these patterns.

Autonomic Reflex Pathways

The reflex mechanisms with their central connections in the spinal cord and brainstem are responsible for the major portion of the neural regulation of visceral functions (Figure 26.3). Therefore, the rapid changes that are often seen in the chemically sensitive patient may result from exposure to noxious pollutant stimuli, which are known to produce a sudden onset of signs and symptoms. Regionalization is usually a function of the parasympathetic system, while generalization is usually a sympathetic response.

Table 26.1 Autonomic Effects on Various Organs of the Body

Organ		Effect of sympathetic stimulation	Effect of parasympathetic stimulation
Eye:	Pupil	Dilated	Constricted
	Ciliary muscle	Slight relaxation	Constricted
Glands:	Nasal	Vasoconstriction and slight secretion	Stimulation of copious (except pancreas) secretion (containing many enzymes for enzyme-secreting glands)
	Lacrimal		
	Parotid		
	Submandibular		
	Gastric		
	Pancreatic		
Sweat glands		Copious sweating (cholinergic)	None
Apocrine glands		Thick, odoriferous secretion	None
Heart:	Muscle	Increased rate	Slowed rate
		Increased force of contraction	Decreased force of contraction (especially of atrium)
	Coronaries	Dilated (β_2); constricted (α)	Dilated
Lungs	Bronchi	Dilated	Constricted
	Blood vessels	Mildly constricted	? Dilated
Gut:	Lumen	Decreased peristalsis and tone	Increased peristalsis and tone
	Sphincter	Increased tone (most times)	Relaxed (most times)
Liver		Glucose released	Slight glycogen synthesis
Gallbladder and bile ducts		Relaxed	Contracted
Kidney		Decreased output and renin secretion	None
Bladder:	Detrusor	Relaxed (slight)	Excited
	Trigone	Excited	Relaxed

		Excretion
Penis	Ejaculation	
Systemic arterioles: Abdominal	Constricted	None
Muscle	Constricted (adrenergic α)	None
	Dilated (adrenergic β$_2$)	
	Dilated (cholinergic)	
Skin	Constricted	None
Blood: Coagulation	Increased	None
Glucose	Increased	None
Basal metabolism	Increased up to 100%	None
Adrenal medullary secretion	Increased	None
Mental activity	Increased	None
Piloerector muscles	Excited	None
Skeletal muscle	Increased glycogenolysis	None
	Increased strength	

Source: Guyton, A. C. 1987. Human Physiology and Mechanisms of Disease. 443. Philadelphia: W. B. Saunders. With permission.

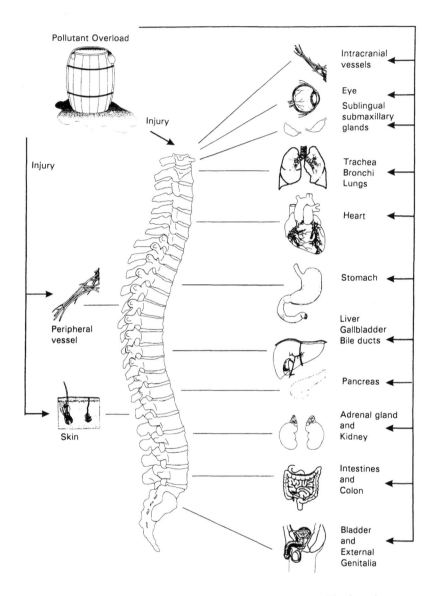

Figure 26.2. Nerve tracts of the ANS and potential areas of dysfunction once pollutant injury has occurred. Specific areas or end-organs may respond while there also may be a generalized imbalance. In some cases, the chemically sensitive individual may have a sequential organ response upon pollutant challenge.

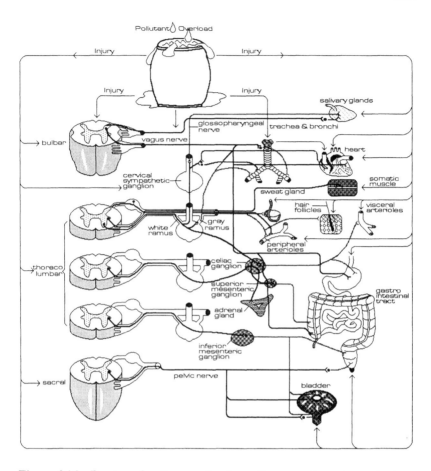

Figure 26.3. Routes of reflex mechanisms that may be triggered after pollutant injury in the chemically sensitive individual. Responses may be isolated to each set of organs, or they may be generalized.

In contrast to the local and regional reflexes, certain higher order reflex and correlation mechanisms are located in the brainstem. These include neuron aggregates in the medulla oblongata, such as the respiratory and vasomotor centers, the center for carbohydrate metabolism, and the major portion of the hypothalamus. All of these areas and their correlated mechanisms may be triggered or inhibited by pollutant stimuli, causing functional deregulation in the chemically sensitive individual.

Impulses that arise in visceral organs are conducted to the higher autonomic centers, mainly through pathways in the spinal cord and brainstem that are made up of short neurons and frequent sympathetic relays.

Cumulatively, these aforementioned facts partially explain the occurrence in some chemically sensitive individuals of food and chemical triggering of the

ANS via the gastrointestinal tract, nose, and lung, the result of which is vasospasm, breathlessness, weakness, shakiness, and a ravenous desire to eat. Usually the weakness and shakiness are clinically misinterpreted as hypoglycemia. However, blood sugar measurements at the EHC-Dallas in hundreds of patients under environmentally controlled conditions revealed normal sugars in all but two patients, suggesting neuroendocrine triggering. Either a portion of, or the whole, neuroendocrine system may also be activated by pollutant exposure, or conversely, neuropeptides released from the end-organ neuroendocrine cells in response to pollutant exposure can activate the rest of the system. The following case report represents a classical ANS response to pollutant exposure.

> **Case study**. This 26-year-old white female entered the ECU with the chief complaints of shortness of breath, cold and blue hands and feet, episodes of bloating, an inability to sweat, migraine headaches, and spells of shaking. She cleared all symptoms after 5 days of fasting. Subsequent testing revealed she was sensitive to 6 molds, 10 foods, and 6 chemicals. Oral challenge with individual foods revealed that cane sugar reproduced her bloating and shaking, while beets reproduced her shortness of breath. Inhaled, double-blind challenge of chlorine <0.33 ppm triggered discoloration and cooling of her hands and feet, her inability to sweat, and bloating. Saline placebos elicited no response. Iris corder measurements of the ANS through the eyes revealed a sympathetic response.

Stimulation of the ANS results in liberation of neurotransmitters, adrenalin (sympathin) from the sympathetic system, and acetylcholine (parasympathin) from the parasympathetic system. Unfortunately, the issue is confused in that some sympathetic ganglia, in having both functions, may liberate acetylcholine. In general, however, stimulation of the sympathetic nerves gives excitation, while stimulation of the parasympathetic nerves results in inhibition of nerve impulses. However, in the gastrointestinal tract, stimulation of the sympathetic nerves gives decreased motility, while stimulation of parasympathetic nerves increases motility and muscle tone. For example, a chemically sensitive patient is exposed to a pollutant in his or her drinking water. The patient immediately develops a sympathetic response of decreased motility of the gastrointestinal tract with severe bloating and distention, while peripherally he or she develops increased heart rate and blood pressure with severe vasospasm of the hands and feet. This type response is frequently seen in a large portion of chemically sensitive individuals. Hours to days later, the patient will finally relieve the stimuli blowout and absorb the gas with relief of the bloating and distention and return to baseline function. These episodes can be incapacitating in the moderately severe chemically sensitive patient.

It should be emphasized that the preganglionic sympathetic nerve fibers contain acetylcholine, and, thus, pollutant exposure in this area results in cholinergic stimulation in contrast to the postganglionic response to pollutants that is usually sympathetic and involves norepinephrine release. The

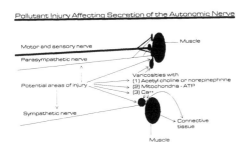

Figure 26.4. Sympathetic ends in connective tissue (fascia) parasympathetic nerve ends in muscle.

post-ganglionic response does not occur in sweat glands, hair, or blood vessels, all of which have a cholinergic response to pollutant stimulation. Therefore, if a chemically sensitive individual were exposed to an organophosphate pesticide, the result might be a confusing set of symptoms as both the sympathetic and parasympathetic nerve impulses could be simultaneously triggered. Also, the varied responses seen in the pesticide-exposed chemically sensitive individual can be attributed to triggering of these dual impulses. In addition to the release of acetylcholine or norepinephrine in response to pollutant exposure, ATP and proper calcium and sodium ion function may be disturbed by exposure. Edema may result with subsequent tissue hypoxia. Further, the mitochondria and ion flow of the ANS may be injured, as seen in some chemically sensitive individuals, causing inappropriate release of these neurotransmitters and subsequent inappropriate ANS response.

The sympathetic fibers usually end in the connective tissue in contrast to the parasympathetic fibers, which end in muscle fibers (Figure 26.4). Because of this anatomical fact, fascial tightness as well as fascial spasms, which frequently result from pollutant exposure, are seen in most chemically sensitive individuals. In fact, following pollutant exposure, a common response in the chemically sensitive individual is fascial spasm with fibromyalgia and fascial pain. Frequently, osteopathic manipulation and massage will release the spasm and the sequestered pollutants with the autonomic responses temporarily improving in the chemically sensitive individual.

Blood vessels, with their sympathetic innervation, contain both adrenergic and cholinergic fibers, while sweat and adrenal glands contain only cholinergic fibers. This anatomic fact would explain the phenomenon of vasospasm with cold hands and feet observed in thousands of chemically sensitive patients. This response would be a sympathetic adrenergic response. Also, the chemically sensitive individual develops an inability to sweat and secrete sufficient quantities of sebum as well as an inability to respond rapidly to injury. Such a simultaneous response to pollutant exposure would be a sympathetic stimulated cholinergic response.

Due to the normal anatomy and physiological function of the ANS, pollutant-induced stimulation or suppression with autonomic deregulation will often produce a diverse and varied response in the chemically sensitive individual. In our iris corder studies of pollutant effects on the ANS of our chemically sensitive patients at the EHC-Dallas, we have demonstrated alteration of this system in excess of 70% of our patients. In addition, clinical peripheral responses such as vasospasm in the hands and legs, distention or bloating of the gut, and bladder spasm suggest that autonomic triggering approaches 100% in the chemically sensitive individual. Recent iris corder studies at the EHC-Dallas have shown 93% triggering in contrast to the earlier 70% due to inclusion of nonspecific pattern changes seen in many individuals. Clinicians who do not thoroughly understand the ramifications of normal ANS anatomy and function are often confused by alterations that result from pollutant exposure. However, an understanding of the anatomic and physiologic construction of the ANS leads to a precise explanation of most of the early complaints (e.g., bloating, gastrointestinal upset, fast heart beat, slow heart beat, urgency of urination) of the chemically sensitive patient before fixed disease occurs. In addition, with better understanding of this anatomy and physiology, the causes and effects of pollutant injury can be measured and reproduced by pollutant challenge under controlled conditions, making possible avenues for prevention.

Understanding of the anatomy and physiology of the ANS leads to the realization that any given chemical may have a sympathetic, cholinergic, cholinolytic, sympatholytic, or combined response, depending on its type and dose as well as total load of other toxic chemicals in an individual at the time of exposure. Organophosphate insecticides are an example of those that usually give a cholinergic or sympatholytic effect in the chemically sensitive individual. (See Chapter 27 for actual measurements.)

Table 26.2 presents a more detailed breakdown of the functions of the ANS. Because of the presence of alpha and beta receptors and the presence of pollutant injury in the ANS, even more complex variations of ANS responses occur, including sympathomimetic, sympatholytic, cholinergic, or cholinolytic responses.

Studies performed on breathing patterns in patients with ANS dysfunction may help us better understand the patterns of response in the chemically sensitive patient.[1,2] McNicholas et al.,[3] for example, investigated the control of breathing in three patients with autonomic deregulation. These patients were free of respiratory and sleep-related symptoms. During wakefulness, their ability to reproduce a breath of a given tidal volume voluntarily did not differ from that of healthy control subjects. However, defects in the metabolic control system were indicated in these patients by lack of a ventilatory response to acute hypoxia. During sleep, the minute volume of ventilation in these patients

Table 26.2. Some Effects of Autonomic System Stimulation from Pollutant Injury Seen in the Chemically Sensitive Individual

	Sympathetic		Parasympathetic	
	α-receptor	β-receptor	Cholinergic	Cholinolytic
Eye				
Radial muscle	Contraction	Relaxation for far vision		
Ciliary muscle	Dilatation		Constriction	Dilatation
Heart		Increased heart rate	Decreased heart rate	
Blood vessels	Constriction	Dilatation	Constriction	Dilatation
Bronchial muscle		Relaxation	Constriction	
Stomach	Decrease in motility	Decrease in motility	Increase in motility	Decrease in motility
Intestine			Dilatation	
Sphincters	Contraction			
Urinary bladder	Contraction	Relaxation		
Detrusor muscle				
Trigone and sphincters				
Sweat glands	Selective stimulation		Decreased activity	
Uterus	Contraction	Relaxation		
Liver	Glycogenolysis			
Muscle		Glycogenolysis	Spasm	Relaxation
Insulin secretion	Inhibition	Stimulation		
Saliva secretion	Inhibition	Stimulation	Inhibition	Stimulation

Source: EHC-Dallas. 1984.

with autonomic dysfunction was not significantly less than that of healthy subjects, but their pattern of breathing was highly irregular, resulting in coefficients of variation of respiratory variables that were two to three times greater than normal ($p<0.05$). This irregularity persisted during slow-wave sleep, when the ventilatory pattern is normally highly regular. Similar problems of variability during sleep are seen in some chemically sensitive patients upon exposure to toxic chemicals. This variability disappears as their toxic load reduces.

Parkinson's patients also used as nonhealthy controls did not have the previously cited irregularities. The combination of findings during wakefulness and sleep in the patients with autonomic dysfunction suggests a defect in the autonomic respiratory rhythm generator of the brainstem. These findings indicate that examination of the breathing pattern during sleep may be useful in the investigation of chemically sensitive patients suspected of having autonomic dysfunction even in the absence of clinical respiratory abnormalities.

We, as well as Monro,[4] have seen a large portion of chemically sensitive patients who have disturbed sleep patterns. Often, these dysautonomic patterns are eliminated in the ECU where the total pollutant load is decreased, and the patterns can then be reproduced by pollutant challenge. Observing patients' loss of sleep disturbance is a valuable clinical tool in following those who have chemical sensitivity. Often, as these patients' chemical sensitivity decreases, their sleep patterns improve.

Emphasis should be placed on reducing the total pollutant load in the chemically sensitive patient in order to prevent ANS failure. As seen in the case report earlier in this chapter, once failure occurs, it appears to be impossible to reverse if fixed-tissue changes have also occurred.

NEUROGENIC VASCULAR RESPONSES TO POLLUTANT STIMULI

The nervous system response to noxious substances entering the body is immediate, with reflex-like rapidity. This pollutant-triggered quick response is the clincial response in the maladapted chemically sensitive individual with overt and florid symptoms and signs in contrast to the adapted normal who does not even perceive the response. This response is apparently mediated through the peripheral and ANS, as evidenced by our studies using the iris corder. When entering through the olfactory nerves, pollutants go directly to, or activate impulses or neurotransmitters to, the brain. In turn, the limbic system, including the amygdaloid and hypophysial pituitary axis, may be activated (Figure 26.1) or this system may be bypassed with stimuli going directly to the anterior or posterior autonomic nuclei. Triggering of the latter impulse pathways appears to be more common than triggering of the limbic system in the general, non-cerebral-involved chemically sensitive individual, while triggering of the limbic system appears to be more frequent in the cerebral-involved chemically sensitive individual. Noxious stimuli occurring anywhere often will affect the pupillary response, due to triggering via the ANS (see Chapter 27). When the noxious stimulus occurs in the periphery, there is a retrograde impulse to the dorsal nerve root ganglia (nodal points) through either the afferent fibers of the peripheral nerves (slow C or rapid delta A) or the gastrointestinal plexus[5] (Figure 26.5). Any of 15 or more neurotransmitters can be released from the neuroendocrine cells of the gut causing varied responses. However, there is a local or regional response rather than a generalized one. For example, the sensory neurotransmitter substance P (a neuropeptide) is released, causing immediate vasodilation and increased permeability of the microcirculation in the area of the nerve, as is often seen in the chemically sensitive patient. Substance P will then activate the non-IgE-mediated release of histamine via the mast cells. In addition, leukotactic factors and leukotrienes are stimulated. Augmentation of macrophage and neutrophil activity also occurs. Somatostatin is released in other cells of the dorsal root, but it also can

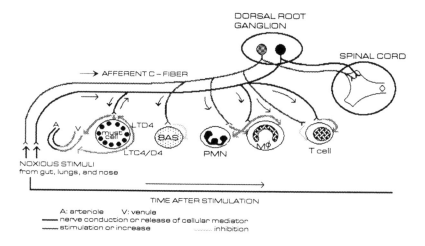

Figure 26.5. Modulation of immunological responses by sensory neuropeptides after pollutant stimuli as seen in some chemically sensitive individuals.

be released from the CNS and the pancreas. Somatostatin apparently acts as an anti-substance P in some of its function, while other functions are complementary. Regardless of function, rapid tachyphylaxis frequently occurs. Recurrent reactions proceed due to a limited amount of production of somatostatin, which suppresses nerve impulses triggered by noxious stimuli. Somatostatin also prevents the recruitment of more basophils, while enhancing the activation of T-lymphocytes and, thus, decreasing the output of lymphokines. This action results in less damage to the vessel wall. Somatostatin inhibits release of histamine from the circulating basophil but not from the mast cells.

Clearly, an increase in, and constant barrage of, noxious substances could continue to trigger substance P, with the resultant vascular and leukocytic components of inflammation occurring and persisting, resulting in the vasculitis usually seen in the chemically sensitive patient. This neurovascular inflammation may explain, in part, both the occurrence of mast cell hyperactivity and the ease with which these cells are triggered by varying and multiple stimuli of ever-decreasing doses.[6] This latter, "hair-trigger" response is frequently seen in the chemically sensitive patient and may partially account for the spreading phenomenon (see Volume I, Chapter 3[7]). Once these events occur, release of leukotrienes, histamines, and possibly other chemical mediators will magnify the responses. These responses, in turn, stimulate substance P creating a vicious cycle of ready triggering that at times may be extremely difficult to terminate. This cyclic pattern of easy triggering is seen in the far-advanced chemically sensitive patient.

Once this cyclic pattern of easy triggering has been set in motion, the immune system may come into play, further propagating the reaction. However, in some instances immune triggering may be the primary event. For example, under environmentally controlled conditions in the ECU at the EHC-Dallas and using 15,000 individual, inhaled double-blind challenges, we studied 2000 chemically sensitive patients who experienced reactions after pollutant challenge. We found that nerve and neuropeptide triggering in these cases was the primary event. Of course, once vascular dysfunction occurs, arterial spasm may follow with resultant local hypoxia. Degeneration may further occur, with the release of free radicals and lysozymes causing increased tissue damage. Other substances that effect vessel wall changes include serotonin and other vasoactive amines, epinephrine, and norepinephrine.

PATHOGENETIC ROLES OF NEUROIMMUNOLOGICAL MEDIATORS (NEUROPEPTIDE TRIGGERING BY POLLUTANT STIMULI)

The multidirectional communication between immune and neuroendocrine systems provides opportunities for coordination of specialized capacities for host defense and enhances expression of abnormalities underlying some neurocontributions to hypersensitivity reactions seen in the chemically sensitive individual.[8-11] Lesions in one system lead to responses in the other systems, often exacerbating responses and contributing to further pathology. For example, injury to the hippocampus alters the numbers and functions of natural killer (NK) and T lymphocytes in the spleen, thymus, and blood.[12-14] Often, suppression of the suppressor T cells, which may cause dysfunction not only of the immune system but also in various parts of the nervous system, is seen in the chemically sensitive patient. Peripheral neuroanatomic circuits that mediate some effects of pollutants on immunity are recognized to be fibers containing specific adrenergic and peptidergic filaments These fibers end in zones of lymphocytes in the thymus, spleen, Peyer's patches, and bone marrow,[15,16] emphasizing both the neuro influences upon the immune system and how pollutant entry through these areas occurs, causing local, regional, and remote triggering, as is often seen in the chemically sensitive individual.

The systemic communications between elements of the immunological and neurological systems are evident both during deficiency states and following the administration of an excess of a mediator from one of these systems, as occurs with pollutant overload in the chemically sensitive individual. The neural effects of pollutant exposure on immune responses may depend on the native state of each system and the direction of the abnormality. For example, antibody response to a foreign substance can be enhanced by epinephrine or by physical stimulation with enough intensity to increase the beta-adrenergic effects on T cells.[17] In contrast, elimination of sympathetic neural influences in athymic mice has been found to permit the development and generation of lymphocytes in thymic remnants.[18] A similar condition may occur in some chemically sensitive individuals, explaining why these patients may have varied immune, neural, and endocrine responses with varied symptom expression, depending on their total body load and the particular part of the nervous system that is triggered or inhibited at any given moment.

Some laboratory observations help explain the extent of nerve involvement with the immune system that will influence function in the chemically sensitive individual. For example, experimentally induced primary afferent denervation of lymph nodes of rats suppressed the response of antibody-producing cells to sheep erythrocytes by over 80%. However, this suppression was reversed by the administration of substance P. In another instance, the synaptic transmission in the superior cervical ganglion of immunized guinea pigs was potentiated specifically by homologous antigen through an HLA receptor-dependent effect of histamine released from mast cells in the ganglion.[19] These experimental outcomes in animals suggest the ways immune deregulation may occur in the chemically sensitive individual. Clinical observation of thoUSAnds of chemically sensitive patients at the EHC-Dallas shows that an interaction between the nervous system and the immune system plays a significant part in their dysfunction. For example, phenol (often found in the chemically sensitive individual) can block the nerve responses to the lymph nodes and the T lymphocytes (often suppressed in the chemically sensitive population). When this blockage occurs, these nodes and lymphocytes become depressed and function poorly. As the chemically sensitive individual's total body pollutant load decreases, however, the nerve then responds again with unblocked impulses. The T_8 cells increase as do their responses. This scenario is often seen at the EHC-Dallas when a chemically sensitive patient improves clinically.

Early immunochemical findings show that distinctive protein antigens are shared by immune and neural cells.

Neurons of the autonomic, primary efferent, and efferent nerves supply the skin, upper airways, lungs, and gastrointestinal tract and contain a range of functionally distinct mediators. The rapid vascular, smooth muscle, and secretory responses to neuropeptides released by noxious stimuli mimic mast cell-dependent, immediate hypersensitivity reactions. At the EHC-Dallas, these responses apparently occur in some chemically sensitive individuals following a chemical exposure. Often they are neurological, mimicking the immune response with a resultant set of similar symptoms.

The neuropeptides (such as vasoactive intestinal peptide [VIP], peptide histidine isoleucine [PHI], and peptide histidine methionine [PHM], substance P [SP] and other tachykinins, calcitonin gene-related peptide [CGRP], and somatostatin [SOM]) that are found in human tissues have been detected in hypersensitivity reactions, and in some chemically sensitive individuals symptoms resulting from deregulation, overproduction, or malfunction of these peptides have been reproduced on pollutant challenge. The most prominent effects of VIP appear to be systemic and pulmonary vasodilation and stimulation of epithelial and glandular secretion,[20] although at the EHC-Dallas, we almost always see abdominal cramps reproduced on challenge. VIP is also an

Table 26.3. Peptide Testing in Patients with Chemical Sensitivity

Formula	Concentrate	Sex	Age (years)	Diagnosis	No.	Positive (%)
$C_{147}H_{238}N_{44}O_{44}$ (vasoactive intestinal peptide)	250 µg/10 mL	F: 17 M: 4	9 to 71[a]	GI: NS: Respiratory CV Endocrine: MS: GU:	21 4 6 8 5 7 1	100
$C_{189}H_{285}N_{55}O_{57}S$ (neuropeptide)	100 µg/10 mL	F: 15 M: 4	9 to 71[b]	GI: NS: Respiratory: CV: Endocrine: MS:	19 6 3 6 3 4	74

[a] Mean = 46 years.
[b] Mean = 42 years.

Source: EHC-Dallas. 1984.

adrenergic-independent bronchodilator of approximately 50-fold greater potency than isoprotenol *in vitro*.[21] The effects of the rest of these substances reflect cholinergic and adrenergic stimulations and may account for some of the varied vascular responses recorded in chemically sensitive individuals. Table 26.3 shows a series of vasoactive intestinal peptides studied in a group of chemically sensitive patients at the EHC-Dallas.

The contributions of neuropeptides to immediate hypersensitivity reactions encompass direct effects not only on target tissue but also on the activation and regulation of tissue cells and mast cells. In some tissues, 85% of subepithelial mast cells are in contact with or within 2 µm of peptidergic nerve fibers containing SP or CGRP (Figure 26.6).[22] This anatomic fact explains the neuroimmune triggering seen in some chemically sensitive individuals.

Neuropeptides influence tissue distribution, proliferation and synthetic responses, and cytotoxic activities of lymphocytes in numerous experimental models.[10] A wide range of lymphocyte functions are inhibited or stimulated by neuroendocrine peptides such as alpha- and beta-endorphins, met- and leu-enkephalins, and adrenocorticotrophic hormone (ACTH). Liberation of these substances by toxic chemical stimulation at times accounts for the low suppressor T-cells seen in some chemically sensitive individuals after pollutant challenge, and, if the endorphins are triggered, the sudden sleepiness seen in some patients with chemical sensitivity also occurs (see Chapter 20).

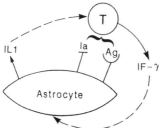

Pathogenetic roles of neuroimmunologic mediators in pollutant stimulation.
Neuroendocrine peptides from immunologic cells:

Mediator	Cell Source	RIA	HPLC	Bioassay	Amino Acid Sequence	Genetic Message
ACTH	Lymphocytes, monocytes	+	+	+	+[a]	+[a]
β-endorphin	Lymphocytes, monocytes	+	+	+	+[a]	+[a]
CRF	Lymphocytes	+	+	+	−	−
Preproenkephalin-derived peptides	CLL lymphocytes, helper T cells	+	−	−	−	+
TSH	Lymphocytes	+	−	−	−	−
Human chorionic gonadotropin	Lymphocytes	+	+	+	−	−
SP	Basophils, mast cells, monocytes, eosinophils	+	+	−	−	−
VIP	Eosinophils, neutrophils	+	+	−	−	−
Bo	Alveolar macrophages	+	+	−	−	−
SOM-like	Basophils, mast cells, monocytes	+	+	−	−	−
VIP-like	Mast cells, basophils	+	+	−	+	+

Note:
[a]These criteria have been met only for lymphocytes.

Neuropeptides associated with diseases of immunity and hypersensitivity often seen in the chemically sensitive individual:

Neuropeptide	Disease	Species	Tissue Source
SP	Arthritis	Rat	Synovium
SP	Asthma	Guinea pig	Pulmonary airways
VIP, Bo	Asbestos-induced pulmonary fibrosis	Rat	Pulmonary parenchyma
CGRP, SOM	Allergic rhinitis	Human	Nasal mucosa
SP	Urticaria	Human	Epidermis
SP	Cutaneous bullous inflammatory diseases	Human	Blister fluid
SP, CGRP	IL-2/LAK cell-induced edema and transudation	Human	Peritoneum, bowel leukocytes

Figure 26.6. Conditioned immunocompetence of neural cells. Ia = immune-associated antigen. Ag = antigen. T = T lymphocytes. If-γ = interferon gamma. (From Goetzl, E. J., S. P. Sreedharan, and W. S. Harkonen. 1988. Pathogenetic roles of neuroimmunologic mediators. *Immunol. Allergy Clin. N. Am.* 8(2):183–200. With permission.)

Murine and human T-lymphocyte functions also affect neuropeptides of the primary afferent, adrenergic, and cholinergic systems such as SP, SOM, VIP, and PHI/M.[10]

The effects of stress *in vivo* and of endogenous neuropeptides *in vitro* on NK and other cytotoxic lymphocyte activities have not been clearly defined.

Macrophages are normal cellular constituents of the nervous system, which increase in number and activity after neural tissue injury so that the myelin is removed phagocytically before axonal regeneration.[23] Unfortunately, many chemically sensitive patients who have artificial implants of substances such as silicon develop antimyelin antibodies, which then produce further dysfunction and deterioration of the myelin before their disease process is arrested. Only recently have macrophages and lymphocytes been recognized for their capacity to regulate neural growth, to inhibit functions, and to induce a state of immune competence in some neural cells. Increased concentrations of the macrophage product interleukin-1 (IL-1) was used after neural injury to stimulate hypothalamic-pituitary functions;[24-26] to stimulate ACTH, LH, TSH, and Somatotropins (GH); and to inhibit prolactin secretion. A case report of toluene overload and altered prolactin function is presented in Volume II, Chapter 12.[27] It demonstrates the concept of neuroendocrine triggering and imbalance in the chemically sensitive individual.

The growth and maturation of oligodendrial cells and astrocytes cells are stimulated by proteins released from antigen and lectin activated T lymphocytes.[28,29] T cells inhibited sympathetic innervation in the spleen of mice, as compared by the higher number of nerve fibers and content of norepinephrine seen in the spleen of athymic nude mice after the introduction of thymocytes.[30] A similar circumstance of stimulation or inhibition may be seen in some chemically sensitive individuals depending on the type of pollutant involved and the nature of the autonomic imbalance.

Many other complex effects of immunological mediators are recorded in neural cells or regulated by neural factors. Included here are the inhibition by alpha melanocyte stimulating hormone of the effects of IL-1 on T lymphocytes and neural cells,[16] the stimulation of neural cell growth by prostaglandin E,[31] and the activation of hypothalamic alpha adrenergic receptors by C5a.[32]

The results from studies of animal models and some human diseases suggest that neuropeptides are involved in disease states of hypersensitivity and altered immunity, as in the chemically sensitive individual. The most comprehensive findings come from animal models of arthritis and asthma. These showed elevated concentrations of neuropeptides in relation to both the development of disease and the prevention or reversal of elements of disease by neuropeptide antagonists. These study results parallel our observations in the chemically sensitive individual that pollutant overload triggers responses from neuropeptide stimulation, and elimination of the pollutants decreases the chemical sensitivity. We have observed the elevation of IL-2 in a large set of chemically sensitive patients. This elevation suggests pollutant triggering and certainly confirms immune derangement of the type just described (see Volume IV, Chapter 30).

Laboratory evidence supports the presence of neuropeptide triggering. For example, repeated application of a mild irritant stimulus to the hind paw of a rat resulted in persistent swelling and tenderness of the opposite hind paw. This swelling was mediated through a neurogenic mechanism attenuated by ablation of either unmyelinated afferent nerves with capsaicin or sympathetic postganglionic efferent nerves by immunosympathectomy.[33] A similar alteration may have occurred in chemically sensitive arthritis patients studied at the EHC-Dallas, since these patients developed immediate swelling upon pollutant challenge (see Chapter 23). The role of neural pathogenesis in the chemically sensitive individual is supported by findings of several forms of immunologically induced arthritis in the general population by a lower nociceptive threshold and a greater intraneuronal content of peptide neuromediators, such as SP.[34] If such a threshold were, in fact, present in the chemically sensitive individual, this lower threshold of triggering would explain the secondary widespread triggering of other systems that is caused by multiple foods and chemicals in environmentally triggered chemically sensitive arthritis and neurological patients after a significant primary pollutant injury. In the same models of immunological arthritis in rats, neonatal destruction of primary afferent neurons reduced and retarded the development of disease, whereas the infusion of SP into lower risk proximal joints increased the severity of joint inflammation and tissue damage.[33] The number of unmyelinated neurons in the saphenous nerves of rats was reduced by doses of gold sodium thiomalate, which alleviates immunologically induced arthritis, but the relative effects of thiomalate on afferent and sympathetic efferent fibers have not been determined.[33] The importance of further defining the involvement of sympathetic neurons is emphasized by the observation that immunologic arthritis was far worse in spontaneously hypertensive rats with high tonic sympathetic activity[33] and by the finding that symptoms and joint function improved when patients with rheumatoid arthritis had sympathetic blockade with guanethidine.[3] The joint abnormalities of chemically sensitive patients at the EHC-Dallas who have arthralgias and arthritis with silicon jaw, breast, and other implants appear to fit this neuroimmune triggering. These patients have toxic patterns on their SPECT brain scans, antimyelin antibodies, low suppressor T lymphocytes, and cholinergic and sympatholytic patterns on their iris corder measurements.

Similar experimental approaches have provided evidence for the roles of SP in rat and guinea pig models of asthma. Pretreatment of animals with specific antagonists of SP significantly reduced the bronchospasm and respiratory mucosal edema evoked by SP that was released through vagal nerve stimulation, chemical irritation, or the application of capsaicin.[35] In addition, chronic treatment with capsaicin depleted the local stores of SP and reduced bronchoconstrictor responses to antigen in sensitized animals.[36] Clearly, chronic stimulation of the nervous and endocrine systems by pollutants, as occurs in the chemically sensitive individual, could cause constant irritation with ongoing release of some neurotransmitters and depletion of others. The result of pollutant-triggered, continuous nervous system stimulation would be many of the symptoms observed in the chemically sensitive patient. In contrast to the lack of knowledge about the concentrations of neuropeptides in the lower airways, the time-courses of appearance of some neuropeptides in nasal lavage fluids have been established in relation to antigen challenge of allergic human subjects.[37] SP, SOM, CGRP, and histamine have been detected in nasal lavage fluids prior to antigen challenge. In allergic patients, but not in normal controls, antigen evoked 3-fold rises in histamine at 15 to 60 minutes, 1.5- to 4.0-fold rises in CGRP at 15 minutes to 24 hours, and more than 2-fold rises in SOM at 6 hours without altering the concentration of SP.[37] Thus, CGRP may make both early and late contributions to allergic nasal congestion, while SOM may regulate events in late hypersensitivity reactions.

Specific lymphokines also can elicit the release of neuropeptides *in vivo*. Infusions of IL-2 in patients with metastatic carcinoma resulted in altered vascular permeability, causing edema and serosal effusions in association with an increase in the intraperitoneal concentration of SP to up to 500 pM. A similar situation exists in some chemically sensitive patients at the EHC-Dallas who have edema triggered by pollutant exposure. In these patients, IL-2s have been noted to be elevated. In contrast, the subsequent intraperitoneal injection of these patients with lymphokine-activated killer cells elicited an increase in the concentration of CGRP in ascitic fluid.[38] In other reports, the high concentrations of neuropeptides found in extracts of lesional tissue or fluid were the sole documentation of a possible pathogenetic pathway. High levels of VIP, together with histamine and serotonin, were found in lung cells of rats exposed to another pollutant, asbestos, for 1 to 6 months in protocols leading to pulmonary fibrosis.[39] SP, but not VIP, was high in the cutaneous fluid of patients with bullous pemphigoid and other vesicular dermatitides, whereas blisters induced on normal skin had no detectable SP.[40] The elicitation of a classical wheal and flare response in human skin by SP, although the flare is histamine-dependent, demonstrated the capacity of neuropeptides alone to mimic immediate hypersensitivity.[10,41] At the EHC-Dallas, we have injected substance P into the skin of some chemically sensitive patients, creating a wheal and flare. Injected substance P also produced some systemic symptoms, which would not abate until somatostatin was injected directly into the substance P-induced wheal.

The possibility that immune and neuroendocrine systems communicate meaningfully in the reactions of host defenses against pollutant injury in the chemically sensitive individual and in some neuroimmunologic diseases in some other populations is supported by findings of anatomic connections and common cellular and molecular pathways (see Chapter 24, section on neuroendocrine). Diverse neuropeptides stimulate and regulate critical elements of immediate hypersensitivity, inflammation, and immunity, with a specificity attributable to distinct subsets of leukocyte receptors. Some lymphokines and monokines evoke primary neuroendocrine responses, stimulate neural cell proliferation, and activate astrocytes to a state of immune competence. In addition, leukocytes generate and secrete neuropeptides and peptides related to neuromediators.

Clearly, deregulation can result from direct toxic chemical stimulation of the neuropeptides, of the nervous system via the lymphocytes, or through the autonomic and peripheral voluntary nervous system due to previously de-scribed routes and through the endocrine glands, as can be seen in the patient with chemical sensitivity. Once triggered, the functions of this communication network explain many of the symptoms and signs that manifest as a result of the various system involvements in the chemically sensitive individual.

Volume I, Chapter 4,[42] and Chapters 18, 24, and 25 further discuss expres-sion of pollutant damage to the lipoxygenase prostaglandin system, enzyme detoxification, and other systems. These chapters explore ways autonomic deregulation may occur and further effect the vasculature with resultant end-organ response affects (see Volume I, Chapter 4,[42] and Chapter 20).

In the body's early response to pollutant stimuli, autonomic dysfunction and peripheral, neural-endocrine immune responses appear to be highly signifi-cant in the expression of the chemical sensitivity. The physician's ability to perceive and act in light of these various responses may prevent fixed end-organ disease.

POLLUTANT INJURY TO THE BLOOD-BRAIN BARRIER

The CNS is thought to be protected from pollutants by the blood-brain barrier. Understanding of this barrier is based upon observations that some substances introduced into the body affect many soft tissue organs (e.g., liver, kidney, muscle), although they are excluded from the brain[43] (Figure 26.7). As a result of pollutant exposure, the blood-brain and peripheral nerve barriers may be altered in the chemically sensitive individual, although some pollutants may penetrate these barriers naturally without alteration of the barrier (e.g., anesthetics).

While the concept of the blood-brain barrier is called into question by most physicians' recognition that some substances are not normally excluded from the brain, e.g., anesthetics, analgesics, and tranquilizers, this concept becomes fur-ther suspect in light of clinical observations at the EHC-Dallas. Here, patients in the deadapted state in the controlled environment revealed subtle brain reactions after the majority of challenge tests, whether the substances tested were foods, biological inhalants, or toxic chemicals and regardless of the final end-organ responses produced by these challenges. If this outcome does not disprove the existence of the blood-brain barrier, it does suggest that, at least in the chemically sensitive individual, the blood-brain barrier is less intact.

Chemical sensitivity appears to be an entity of leaky barriers in the gut, lung, nasal membranes, blood vessel, and nervous system caused by hereditary mucosal defects as well as acquired pollutant injury, thereby accounting for blood membrane changes. Alterations in the blood-brain barrier are one of the reasons that the spreading phenomenon occurs in the chemically sensitive individual (see Volume I, Chapter 3[7]). Increased permeability of the barrier allows for more and more substances to cross and symptoms to occur. It appears, in general, that nonpolar, lipid-soluble compounds usually penetrate the blood-brain barrier more easily, while highly polar compounds tend to be

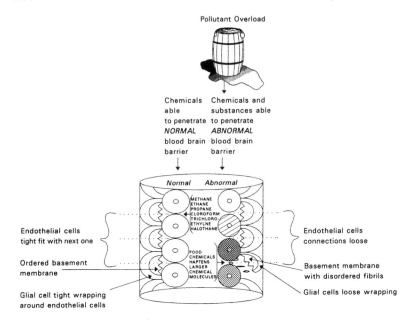

Figure 26.7. Pollutant injury to the blood-brain barrier.

excluded. Many of the toxic chemicals found in the blood of the chemically sensitive patient are lipophilic and, thus, are able to penetrate more rapidly, causing neurological dysfunction. In the immature brain, the "barrier" is generally not as effective, and toxic doses of some compounds, e.g., inorganic lead salts, may accumulate in the CNS of children. In contrast, the peripheral nervous system of an adult develops marked effects from lead because the barrier prevents the chemicals from crossing to the CNS and, hence, effects are seen peripherally. A considerable amount of research and speculation regarding the anatomic features that relate to the functional barriers exists[43] because of efforts to define the blood-brain barrier clearly.

Three concepts are involved in the function of the blood-brain barrier.[43-45] In the first, endothelial cells of blood vessels form tight junctions, have no pinocytotic vessels in their cytoplasm, and have no pores in luminal endothelial membranes. However, small molecules such as methane, ethane, and/or propane may penetrate to the neuron through the junctions and through the cytoplasm of the endothelial cells and glia.[43] These substances do bother chemically sensitive individuals, with even small exposures making them ill. Thus, the ease with which these substances penetrate the blood-brain barrier in the chemically sensitive individual does not necessitate a leaking barrier.

Pinocytotic vesicles may appear in the endothelial cells of the CNS in various pathological conditions that increase the permeability of the blood-brain barrier. These vesicles may then transport some chemicals across the endothelial lining of small blood vessels in the CNS. When blood-brain barrier permeability is increased, some proteins (e.g., horseradish peroxidase or beef IgG or chemicals combined with protein, such as Evans blue dye bound to albumin) may be observed in these intracellular vesicles.[46] Conditions that are known to increase vesicles are postirradiation brain edema,[47] ischemia,[48] and hypertension.[49] The latter two conditions are often observed in the chemically sensitive patient. Thus, permeability increases in some chemically sensitive individuals and the spreading phenomenon occurs, allowing more toxic chemicals to enter the brain and, by this mechanism, increase brain dysfunction.

The function of the blood-brain barrier may also be explained by the presence of glial cells wrapping around capillaries (Figure 26.7). Much of the brain capillary endothelium is invested with astrocytic processes, which compose a mechanical barrier to prevent toxic substances from reaching the neurons in many places. In some brain areas, such as the median eminence of the hypothalamus and the area postrema of the fourth ventricle, however, the capillaries do not have glial wrappings, and, therefore, toxic substances are more readily able to reach the neurons.[43] The majority of chemically sensitive patients show neurovascular instability, which may be due to ANS activation via the hypothalamus or direct injury. This instability may be a precursor of leaks in the barrier due to the continued vessel deregulation. Certainly animal[50] and human[51] studies suggest this idea. The lack of glial wrappings in the medial hypothalamus leaves the parasympathetic part of the hypothalamus slightly more prone to pollutant injury and correlates with our iris corder findings of predominant cholinergic and sympatholytic responses in a large group of chemically sensitive individuals.

A third explanation of the blood-brain barrier involves the extracellular space, which may also act as a barrier. The extracellular basement membrane located between the endothelial cells of the capillary and the glia and neurons is an ordered fibrillar mucoprotein structure and may have unique properties in the brain, as it does in the kidney, allowing it to serve as a "sieve" to transport molecules needed for cell nutrition and to regulate electro-osmotic flow of water while excluding other substances.[43] If pollutant damage does occur, this sieve may alter, allowing other toxicants as well as an improper proportion of nutrients (as seen in some chemically sensitive patients) into the brain and, thus, increase pathological responses (see Volume I, Chapter 6[52]). This alteration of the total barrier may partially explain the spreading phenomenon observed in some cases of chemical sensitivity.

In the peripheral nervous system the blood-neural barrier is variable, and peripheral neuropathy is seen in a large subset of chemically sensitive patients. In both the central and the peripheral nervous system, fenestrated epithelial cells have been found in areas that can be permeated by large molecules such as the aforementioned horseradish peroxidase. In contrast to the greater resistance of some areas of the CNS, the susceptibility of these barrier-free areas to some toxic substances, and even secondary food and biologic inhalant triggering, may be due to differences in ease of penetration of the toxicant through the spaces between the epithelial cells,[53-55] and, again, the absence of, or weakness in, the blood-brain barrier may account for the different symptomatology seen in some chemically sensitive individuals.

POLLUTANT INJURY AFTER PENETRATION OF THE BLOOD-BRAIN BARRIER

After pollutants penetrate brain tissue in the neurologically involved chemically sensitive patient, their effects may still be difficult to determine due to the number of different cell types that potentially may be affected and the individual's total pollutant load. However, SPECT scans of the brain do show disturbances of both flow and function, indicating a toxicity that results in poor brain function in these disturbed areas. These SPECT scans are dependent upon glutathione peroxidase for uptake of the dye for the function phase. Therefore, one can infer a deficiency of this enzyme in these areas of low or no function. Different brain areas usually have different sensitivities to toxicants. These variations reflect the unique biochemistry of the cells as well as the differences in the degree of vascularization of brain areas[43] and the functional demands of pollutant exposure (Figure 26.8). Due to this anatomical variation, pollutants may cause greater damage to some areas than to others, accounting for the varied brain dysfunction seen in some neurologically involved chemically sensitive individuals. However, early signs of pollutant injury in this type of chemically sensitive individual usually are those of short-term memory loss, minor episodes of confusion, inability to concentrate well, and mild clumsiness. An inability to stand on the toes with the eyes closed is often present. Each anatomical variation will be discussed separately.

Unique Biochemistry

The three types of glial cells in the brain differ in their roles in the central nervous system and in their sensitivity to toxic agents. *Astrocytes* are closely associated with neurons in gray matter and are called the nurse cells of neurons because they are thought to be essential in maintaining the stable microenvironment needed for neuronal function. Pollutant damage to these cells creates instability of neuronal function, which is seen in the subset of neurologically involved chemically sensitive individuals. The *oligodendrocyte* in the CNS has a role similar to that of the peripheral Schwann cell and invests neuronal axons with spiral wrappings of myelin. *Microglia* are phagocytic cells with primary functions that resemble peripheral leukocytes, but their response to toxicants has not been widely investigated.[43] Certainly, if these cells are triggered by pollutant stimuli, nutrient deregulation and deficiency may occur.

In the chemically sensitive individual, the variation and degree of vascularization account for some of the variation in sensitivity of different brain areas to hypoxia and toxic chemicals. For example, the globus pallidus has the

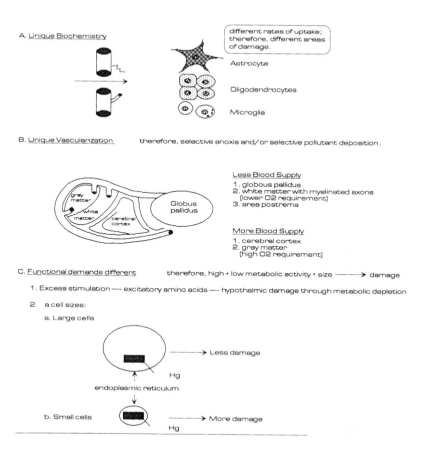

Figure 26.8. Effects of pollutants after penetrating brain tissue.

same density of cell bodies as other neural tissues, but it is more poorly vascularized than the cerebral cortex (hence the name pallidus, meaning pale). White matter is generally less vascularized than gray matter,[56] but the lower oxygen requirement of the myelinated axons, which make up much of the white matter, makes white matter generally less sensitive to pollutants than gray matter with its high cell body-to-neuropil ratio.[43] Triple camera SPECT brain scans on the subset of neurologically involved chemically sensitive patients at the EHC-Dallas usually show gray matter involvement, thus confirming this anatomical principle.

Functional demands on cells may cause different responses to toxicants (e.g., excitatory amino acids may damage hypothalamic neurons by causing excessive stimulation and metabolic exhaustion of the cells).[57] Cell exhaustion appears frequently in the chemically sensitive patient and partially accounts for the onset of various symptoms including weakness, difficulty in arising in the morning, and difficulty in responding to challenging stimuli. Quantitative differences in essential cell components may make one cell type more sensitive than another type, (e.g., small neurons such as granule cells in the cerebellum and visual cortex are preferentially killed when the whole brain is exposed to methyl mercury. This reaction was seen in the Minamata, Japan, disaster (see Volume II, Chapter 11,[58] section on mercury). The amount of cytoplasm and rough endoplasmic reticulum, which binds mercury or any xenobiotic, is less than in larger cells, and thus the small cells may be more likely overwhelmed by their effects.[59]

Studies of the microchemistry of the brain reinforce concepts of its structure as encompassing separate, diverse areas. High concentrations of norepinephrine, serotonin, acetylcholine, and dopamine are found in various pathways of the phylogenetic "old brain," which includes the hypothalamus, reticular formation, basal ganglia, and limbic systems.[43] Pollutants as well as psychogenic drugs have now been shown to alter microchemistry, thereby altering functions.[60,61] Many chemically sensitive individuals who are challenged with these substances experience exaggerated responses. However, administration of the appropriate neutralizing doses of the challenge substance curtails the reactions, suggesting that these individuals undergo an altered sensitivity in these old brain areas.

Studies performed at the EHC-Dallas in controlled environments using intradermal provocation and neutralization techniques as well as oral and inhaled challenge on patients whose total load had been reduced have revealed many persons whose symptoms are reproducible. Intradermal injection of histamine, serotonin, epinephrine, norepinephrine, dopamine, and methylcholine or ingestion or inhalation of some toxic chemicals triggers varied responses. These challenge techniques along with future observations will, perhaps, give us a clinical test for evaluating areas in the brain that are not properly functioning. However, experimental techniques for accurate assessment of substances that stimulate or suppress the transmitters are presently too unsophisticated to allow generalizations regarding the role of brain structures and amine levels in specific functions to be made.

Despite the limitations of present assessment techniques for use in the physician's office, some chemically sensitive patients can reliably and very precisely use their reactions to identify a pollutant to which they have been exposed. They may have, for example, one type reaction after a histamine injection and another type after serotonin injection or when they eat certain foods or are exposed to certain inhaled toxic chemicals. Further, this select number of patients may be able to differentiate whether their reaction is histamine or serotonin induced. At the EHC-Dallas, it has become evident to us that injections of neurotransmitters or injections that cause the release of endogenous neurotransmitters can provoke a neurological response that gives a finite reaction in many chemically sensitive patients.

Certain large cells in the CNS, such as cortical and hippocampal pyramidal cells, cerebellar Purkinje cells, and motor cells in the ventral horn of the spinal cord, have unusually large nuclei, and the DNA is predominantly present as euchromatin, the form of chromatin most closely associated with transcription.[62] These cells often have several nucleoli. All these structural differences point to high metabolic activity in the cells and thus increased susceptibility to anoxic and pollutant damage.[43] The responses seen in the subset of pollutant-triggered, neurologically involved chemically sensitive patients are often those that relate to the functions of these.

Principles of Response after Toxic Exposure

In the chemically sensitive individual, the three principles that govern the responses of the nervous system to pollutants include the following: (1) selective damage to one or more areas or components achieved by selective exposure due to differences in ease of penetration to some cells through barriers, (2) selective anoxia via differences in blood flow and metabolic requirements of some elements, or (3) specific sensitivity resulting from qualitative or quantitative chemical differences in cell components. Identification of the selective nature of the damage from toxic agents is essential in determining the mechanism of action of toxicants, and this nature has further value in analyzing the relation of brain structure to function[43] in various types of neurologically involved chemically sensitive patients.

Reversibility vs. Irreversibility of Cell Damage after Pollutant Exposure

The extent to which irreversible toxicity and sensitivity differ quantitatively and in virulence from the therapeutic actions of drugs is appropriately discussed in numerous reviews and books on the subject of neuropharmacology.[63-65] This aspect will, therefore, not be discussed any further in this text.

The ultimate event in chemically caused irreversible damage to the nervous system of the chemically sensitive individual can be described simply as neuronal death, since the neurons cannot divide and be replaced.[43] The only apparent exception is the cilia of the olfactory nerve endings at the root of the nose, which are continually replaced.[66] These nerves are frequently damaged in the chemically sensitive patient. Though repair of this nervous tissue occurs easily, there frequently appears to be memory in the replacement tissues for certain chemicals, which allows the sensitivity to remain or even to progress. In the chemically sensitive patient in the early stages of illness, a condition in which the cell membranes may be injured, though not severely enough to cause irreversible damage, appears to occur. This series of events is seen frequently posttrauma as well as in the recently induced chemically sensitive individual. Often, normal function may be restored for an individual even after considerable damage from nervous system exposure to toxic substances. Redundancy of function in a population of neurons and plasticity of organization are the methods by which restoration of function is presumed to occur after the death of some neurons.[43] These phenomena allow the process of adaptation to occur (see Volume I, Chapter 3[43]). Long-term adaptation should be guarded against, since it is a process that guarantees that, over time and without incitant removal, total brain dysfunction and end-organ failure will occur. At the EHC-Dallas, we have observed over 10,000 chemically sensitive patients with neurological damage. In these patients, the neurons that were not killed returned to normal function once the total specific pollutant load was decreased and nutritional competence restored.

Conditioning is, in fact, dangerous due to masking of brain damage by the adaptation phenomena.[67] When some neurons die, other cells with the same function may be adequate to maintain normal activity, or failing this, other neurons may assume the needed function. In situations where neither course is possible (for example, in extensive damage to a specialized population of neurons or brain nuclei), some loss of function inevitably results.[43] However, if mild damage occurs, adaptation restores normal function. The continued accommodation reaction of cells depends upon the degree to which metabolism is disturbed by the pollutants as well as their distribution, the intake and maintenance of adequate detoxification and repair nutrients, and the ability to limit additional pollutant exposures.

Some degree of recovery of function usually occurs after nonfatal neurotoxic reactions. When cell death is not involved, the neurotoxic, or sensitivity, reaction lasts only until the toxic agent is removed or metabolized or until cell constituents altered by the toxic exposure have regenerated. These latter exposures account for the continued brain dysfunction seen in some neurologically involved chemically sensitive individuals even though their total load has been reduced and many pollutants removed. In the recently treated neurologically involved chemically sensitive patient, proper distribution of replacement nutrients may be initially difficult to achieve, until all areas of membrane transport and blood-brain barriers have healed. This recovery may require both short- and long-term replacement nutrients. Reversible toxic and sensitivity reactions are often associated with pollutant exposure, but they also occur after therapeutic administration of drugs. Thus, neurotoxicity and sensitivity could be considered to include all undesired nervous system effects of drugs, as well as other chemicals.[43] At the EHC-Dallas, studies of chemically sensitive patients in the ECU who were challenged in the deadapted state after reduction of their total body load reveal that the reactions and responses of these patients to pollutants are finite and clearly defined. In most cases, these reactions usually last from 5 minutes to 4 hours, suggesting a pharmacologic rather than a toxic effect (Figure 26.9).

Generally, a distinction can be made between the pharmacological and toxicological effects that result from exposure to a drug or pollutant, but, at times, differentiation between the two is unclear in the chemically sensitive individual. Characteristically, pharmacological effects are of short duration and completely reversible, while toxicological effects often include irreversible damage. However, it is clear that many reversible toxic changes occurring in neurons or glial cells during therapeutic use of drugs may be closely related to the mechanisms of therapeutic activity. For example, the synaptic clefts

Cell reaction and recovery

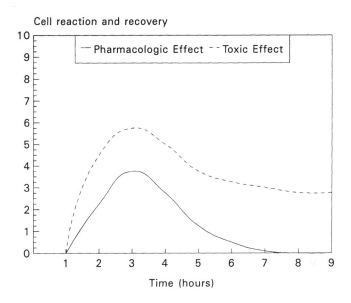

Figure 26.9. Effect patterns of pollutants on the nervous system.

between axons and dendrites of neurons are considered to be especially vulnerable to exogenous chemicals carried by the bloodstream, since the postsynaptic membrane is the site of receptors for chemical transmitters in the nervous system. Many psychoactive drugs are thought to cause psychic changes by altering neuronal transmission, and many neurotransmitters are analogs of psychoactive drugs. Chemical overload creates dysfunction that results in increased vulnerability of the total system with mechanisms and responses similar to those just described (Table 26.4).

This vulnerability seen in some chemically sensitive patients after chemical exposure may involve the body's inability to block normal chemical transmitters going to postsynaptic receptors. The exogenous chemicals can actually act as false transmitters or affect concentrations of transmitters through their influence on synthesis, storage, release, reuptake, or enzymatic inactivation mechanisms. Due to the large number of possible areas of pollutant damage, it is difficult, at times, to pinpoint the exact area of neuronal injury for diagnosis and treatment in the subset of neurologically involved chemically sensitive patients.

If pollutant stimuli continue in the chemically sensitive individual, the pharmacological effects may then become toxic, probably due both to nutrient depletion of the end-organ cells and the structural damage caused directly by the chemical stimuli or the chemical alone. This response may be true for most exogenous chemicals that trigger chemical sensitivity. These effects are proposed to occur in the areas of the CNS in which specific transmitter mechanisms are involved.[43] These include the autonomic areas in the brain, such as the hypothalamus and amygdala, which normally have high concentrations of

Table 26.4.   Ways Pollutant Overload Creates Neural Dysfunction

Blocking the access of normal chemical transmitters to postsynaptic
  receptors
Acting as false transmitters
Affecting concentrations of transmitters through effects on:
  Synthesis
  Storage
  Release
  Reuptake
  Enzymatic inactivation mechanisms

some biogenic amines, particularly serotonin, norepinephrine, dopamine, acetylcholine, and gamma amino butyric acid. Examples of compounds presumed
to act therapeutically by altering neurotransmitters are monoamine oxidase
inhibitors, cholinesterase inhibitors, reserpine, phenothiazines, and L-DOPA.[43]
Organophosphate and carbamate insecticides, known cholinesterase inhibitors,
are also known to exacerbate and cause some chemical sensitivity and are able
to cause pharmacologic as well as neurotoxic effects in these synaptic areas due
to constant exposure as contaminants contained in food and air.

Other drugs, such as general anesthetics, appear to affect neurons generally,
probably through a reversible effect on electrically excitable neuronal membranes.[43] General effects are also seen in halide-generated anesthetics, such as
chloroform or halothane. Some of our patients exhibit this anesthetic effect with
symptoms of sluggishness, sleepiness, weakness, and mental fuzziness. Thirty-
seven percent of our chemically sensitive patients studied in the mid- and late
1980s had detectable chloroform in their blood, which probably induced some
anesthetic effect, accounting for those symptoms previously described. Also
found in the blood of chemically sensitive patients are other toxic chlorinated and
nonchlorinated hydrocarbons that are similar to, or used as, anesthetics. Since
many of these are solvents such as toluene, xylene, 1,1,1-tricholoroethane,
trichloroethylene, and tetrachloroethylene one would expect, and does find,
neurotoxic and anesthetic effects. It is apparent from Butler's studies at the EHC-
Dallas that as these substances leave the blood, mental function is usually
restored.[68] At the EHC-Dallas, we have also observed that even if a patient has
had 20 to 30 years of mental fuzziness from his or her chemical sensitivity,
removal of the toxicants from his or her body will often result in normal or even
supernormal mental function. We have seen some intelligence quotients increase
by as much as 30 points (Table 26.5). This improvement suggests a pharmacological or superpharmacological effect by one or more pollutants (usually multiple) that puts the brain in somewhat of a state of suspended animation or
suppression. Then, as the pollutants are removed, the mental capabilities improve. For a period of time after the removal of pollutants, the patient usually is
extremely supersensitive to exposure to single or multiple toxic chemicals at
ambient doses. Brain function fluctuates with minute pollutant exposure until

Table 26.5. Clearing with Ecological Treatment[a]

WAIS FSIQ scores and differences before and after ecological treatment			Bender scores and differences before and after ecological treatment in the ECU		
Before	After	Difference (+)	Before	After	Difference (−)
106	128	22	18	10	8
123	135	12	24	4	20
129	134	5	11	11	0
104	114	10	15	6	9
116	129	13	8	3	5
106	112	6	15	10	5
130	133	3	15	5	10
117	130	13	24	9	15
112	124	12	14	3	11
106	128	22	20	10	10
115	130	15	15	10	5
108	122	14	15	6	9
106	118	12	16	3	13
138	138	0	20	3	17
115	126	11	15	6	9
105	122	17	9	3	6
118	128	10	10	4	6
102	115	13	24	10	14
108	120	12	11	5	6
128	133	5	14	0	14
112	123	11	12	3	9
123	133	10	15	6	9
117	130	13	24	8	16
109	122	13	8	3	5
122	132	10			
mean = 115	mean = 126.36	mean = 11.36	mean = 15.5	mean = 5.9	mean = 9.6
N = 25		$p<0.001$	N = 24		$p<0.001$

[a] Massive avoidance of pollutants in air, food, and water in the ECU using a rotary diet, intradermal injections, and nutritional supplementation.

Source: Butler, J. R., W. J. Rea, J. L. Laseter, I. R. Deleon, S. G. Wright, and M. J. Milam. 1986. Unpublished data. EHC-Dallas.

overall resistance to pollutants occurs, probably as the individual replaces the nutrients for subcellular and cellular repair and function. During this vulnerable period, very acute brain function alternates with prior dull mental fogginess. Therefore, the patient should be cautioned to avoid new exposures.

Table 26.6.   Chemicals and Symptoms

Chemical	Skin healing	Parasthesia	Tremors	Fatigue
Acrylamide		X	X	
Lead	X	X	X	X
Mercury		X	X	X
OPs		X		X
PBB	X	X		X
PCB	X	X	X	X

Source: EHC-Dallas. 1992.

NEUROLOGICAL EFFECTS OF TOXIC CHEMICALS

The National Institute for Occupational Safety and Health (NIOSH) lists neurotoxic disorders among the top ten work-related diseases. This organization has estimated that 9.8 million people work with organic solvents, many of which are neurotoxic. Table 26.6 presents examples of these.

It is estimated that approximately 20 million workers in the U.S. are exposed to toxic chemicals that can harm the nervous system. Further, another 20 to 40 million family members of these workers are probably exposed to contaminants that are carried home by the workers on their clothing, skin, hair, and body. However, the extent of exposure to toxic chemicals has only begun to be understood. We think these estimates are grossly conservative, since virtually every individual in American society is exposed frequently to some toxic chemicals via their air, food, and water. Toxic chemicals are further present in paints, dyes, adhesives, degreasers, cleaning products, plastics, textiles, and inks (see Volume II, Chapters 7 through 10[69]). Millions of neurotoxic substances are released into the air each year, many of which are the same substances that we have discussed previously (see Volume II, Chapter 9[70]) and are found regularly in the blood of the chemically sensitive. Also, not shown in Table 26.6 are the billions of pounds of toxic pesticides released into the air each year (see Volume II, Chapter 13[71]). The environmental influences of symptom-named functional disturbances and subsequent fixed-named diseases will now be discussed.

Fixed-Named Disease

The final outcome of pollutant injury is cell death that results in fixed-named diseases that often are irreversible. However, if pollutant overload is reduced, many of these named disease processes can be markedly decreased and, at times, arrested. Occasionally, they can be totally reversed. Therefore, the clinician should strive to diagnose and treat environmentally induced illnesses as early in their course as possible.

Known fixed-named diseases that have environmental influences include toxic neuropathy, Parkinson's, Alzheimer's, amyotrophic lateral sclerosis, multiple sclerosis, Huntington's chorea, myesthenia gravis, brain cancer, and other nonmalignant brain dysfunctions. These will now be discussed.

Toxic Neuropathy

In 1856, Delpech[72] described the neurotoxicity of carbon disulfide (CS_2). He identified the symptoms that continue to be observed today (see Volume II, Chapter 11[73]). In contrast to Delpech's observations, Charcot,[74] one of Freud's teachers, contended that these patients' complaints, symptoms, and signs should be more properly ascribed to hysteria (a totally unprovable and still unproven hypothesis). These two views set the stage for the debate that ensued into the early 20th century and may still be occurring in some uninformed groups, usually through ignorance or possibly as a lame defense for polluters. Today, there is strong evidence for the role of environmental triggers in the onset of fixed-named disease. As diagnostic techniques become increasingly precise and experimental evidence continues to be garnered, Charcot's and Freud's ideas have become obsolete. By and large, the argument is no longer focused on whether or not toxic chemicals are able to cause illness. Instead, the debate now centers on identifying the levels of chemicals and individual circumstances necessary to cause harm. While we at the EHC-Dallas continue to reject notions of "safe" threshold levels of exposure, others continue to argue for arbitrary safe levels.

Because there is a long, well-documented history of toxic neuropathy caused by exposure to CS_2, as well as documentation of other mental problems associated with exposure to various chemicals, today's mental health professionals should be checking for exposure to both CS_2 and other toxic inorganic (Pb, HG, Cd, Al) and organic (hydrocarbons, pesticides, etc.) chemicals as well as organic solvents when a patient is encountered who has symptoms such as hallucinations, headaches, lassitude, malaise, indifference, loss of memory, depression, nausea, and psychomotor slowness; decreased vigilance, productivity and intellectual functioning; chronic fatigue; and parathesias. Often these syndromes induced by exposure to organic and inorganic solvents are accompanied by electrophysiological abnormalities (see the section on evaluation in this chapter).

As with many aspects of neurological damage in chemically sensitive patients, generally, a clinician's lack of familiarity with the etiology of CS_2 toxic neuropathy might cause him to miss the presence of such damage, even if the other usual indicators, myocardial ischemia and hypertension, were present.[75,76]

CS_2 is present around oil refineries and sour gas (sulphur containing) fields. It is used extensively in the rubber industry, as it provides elasticity, and it is essential in the manufacturing of viscose rayon and cellophane. Approximately 24,000 workers are estimated to be involved in such manufacturing. (For more information, see Volume II, Chapter 11.[73])

Other neurotoxic compounds such as diethyldithiocarbamate, enhance the neurotoxicity of methylphenyltetrahydropyridine (MPTP) in mice. Savage et al.[77] evaluated the latent neurological effects of organophosphate pesticide poisoning. These authors found no significant differences between poisoned subjects and controls on audiometric tests, ophthalmic tests, electroencephalograms, or on the clinical serum and blood chemistry evaluation. Neuropsychological tests of widely varying abilities, however, revealed apparent differences. Anger[78] found that, of the 588 compounds of known toxicological significance that are most frequently encountered in industry, 25% have documented neurotoxic effects, whereas less than 10% have been linked to cancers and fewer than 5% to cardiovascular toxicity. He further identified the varied neurotoxins emitted in the air to which people can be exposed (see Volume II, Chapter 13[79]).

Parkinson's Disease (PD)

The majority of published data shows that toxic chemicals cause neuropsychological and behavioral problems, but some chemicals also can trigger known fixed-named disease. At the EHC-Dallas, we have 20 years of experience evaluating the effects of chemical exposure on human health. We have found that psychological, behavioral, and learning problems experienced by individuals exposed to toxic compounds are precursors to fixed-named diseases, if these prodromes are allowed to progress unaltered long enough. If these prodromes are not recognized early enough, but covered instead by symptom-relieving medication, the fixed-named disease that does develop will depend upon a person's individual biochemical response to the toxic load. Also, in certain cases, the specific chemicals involved in repeated exposures that eventually lead to the onset of fixed-named disease are important. Examples of toxic exposures that progressed to a fixed-named disease follow.

In 1979, Davis and colleagues[80] described a 23-year-old chemistry graduate who had been manufacturing and intravenously injecting a synthetic heroin. For several months, he produced this drug successfully. Then he devised shortcuts in its synthesis. Following continued injection of this product, he developed marked Parkinsonism.

Langston et al.[81] witnessed similar results when they followed four patients who had a history of heroin abuse and who became symptomatic within a week after starting to use a comparable new synthetic drug. Langston obtained samples of the heroin analog and found the contaminant 1-methyl-4-phenyl-1,2,3,6-tetrahydro-pyridine (MPTP).

Since publication of these reports, both mice[82] and monkeys[83] have been found to develop PD following exposure to MPTP. In the mice study, Gupta found that aged mice responded with PD more than young mice. This finding tends to support Calne's theory that subspecific symptom production is damage manifested with age-related neuron death. The younger mice presumably were able to compensate for the damage because they had more functional neurons

Table 26.7.   Toxic Chemicals Known to Trigger
                    Parkinson's Disease

1-Methyl-4-phenyl-1,2,3,6, tetrahydro-pyridine (MPTP)
Organophosphate pesticides
Petroleum derivatives plus insecticides
Carbon tetrachlorides and carbon disulfide (CS_2)
Cycas circinalis — neurotoxic plant of Guam
Fungicide — hexachlorobenzene

in the substantia nigra than older mice. These findings again emphasize how the adaptive process, if the total toxic load goes unchecked, can result in CNS failure.[84] Cumulatively, the results of these various reports demonstrate the long-term detrimental effects of total load maladaptation.

Recently the list of fixed-named diseases with neurological problems has grown with the hypothesis that Alzheimer's, Parkinsonism, and other moto-neuron diseases could be triggered by environmental toxins (Table 26.7).

Calne et al.[85] postulated that exposure to environmental chemicals could cause subspecific symptom production with damage in specific regions of the CNS. With aging and continuing increase of total body pollutant load, certain individuals would then exhibit the symptoms of damage simultaneously with normal age-related neuronal attrition. Parkinson's, Alzheimer's, and motoneu-ron diseases are all due to selective neuronal degeneration to the substantia nigra, medial basal forebrain, and the motoneurons, respectively. According to Calne's theory, the location of the initial chemical damage in the CNS deter-mines the disease that will manifest in later years. Recently, evidence of just such chemical damage leading to Parkinsonism has come to light.[84]

Calne et al.[86] attempted to prove his theory by using positron emission tomography (PET). This technique enables visualization of dopamine and its metabolites in nigrostriatal nerve endings when using the dopa analog 6-fluorodopa. Calne applied this method to four people who had used MPTP but had not developed symptoms of PD. All of these individuals exhibited the same damage to the nigrostriatal pathway that was apparent in persons with clinical PD. These changes did not, however, occur in normal healthy subjects. These findings provide evidence that MPTP causes subspecific symptom change with brain damage without producing signs of PD.[84] These findings correlate with our observations of some patients at the EHC-Dallas whose diffuse signs of chemical sensitivity occurred and went unrecognized until PD developed.

MPTP is not the only chemical found to cause PD. Davis et al.[87] reported on a 53-year-old crop duster who had worked with organophosphate pesticides for 13 years. He had experienced several episodes of acute organophosphate poisoning. He subsequently began to experience the symptoms of PD, where-upon he quit his job and improved. We have reported similar cases in Volume I, Chapter 6,[52] and Volume II, Chapter 13.[79] Melamed and Lavy[88] reported on a 40-year-old chemist who had been exposed to 20 kg of carbon tetrachloride over a 3-month period of time. He developed progressive akinesia, rigidity, and rest tremor, and was responsive to treatment with Levodopa. Bocchetta and Corsini[89] reported two other incidents of chemical-induced PD. One involved a 41-year-old farmer who became ill after working with pesticides for several years; the other involved a 38-year-old chemist who had worked in a chemical plant making petroleum derivatives and pesticides.[85]

Joubert and Joubert[90] described two patients with organophosphate poisoning both of whom exhibited marked choreiform dyskinesias and one who experienced severe depression and emotional liability.

Many of the symptoms of PD (cerebellar ataxia, resting and intention tremor, cogwheel rigidity, etc.) were widely found in a population of patients exposed to a combination of carbon tetrachloride (CCl_4) and carbon disulfide (CS_2).[91] In this study, the symptoms of neurotoxicity were attributed to the CS_2 rather than CCl_4, but this assessment is only conjecture and is probably due to the combination.

Some types of PD, amyotrophic lateral sclerosis, and Alzheimer's-type dementia had a high incidence among the people of Guam. With the westernization of Guam, though, there has been a dramatic decline in these diseases. Since viral and hereditable factors are absent, environmental factors in the etiology of the disease are implicated. In Guam, the seed of the neurotoxic plant *Cycas circinalis*, which was a traditional source of food and medicine for the natives, is implicated. The results of one study support the hypothesis that *Cycas* exposure plays an important role in the etiology of the Guam disease. In this study, monkeys were fed the *Cycas* amino acid B-N-methylamino-L-alanine, a low potency convulsant that has excitotoxic activity in the mouse brain, which is attenuated by N-methyl-D-aspartate receptor antagonists. Consistent with existing epidemiological and animal data, these animals developed corticomotor neuronal dysfunction, Parkinsonian features, and behavioral abnormalities with chromatolytic and degenerative changes of motorneurons in the cerebral cortex and spinal cord.[92]

A number of recent studies continue to link environmental factors with the onset of PD. For example, Xintaras and Burg[93] and Fleming et al.[94] comment that pesticides and several neurotoxic industrial chemicals have been associated with Parkinsonism and neurological disorders such as peripheral neuropathy, encephalopathy, encephalomyelitis, multiple sclerosis, and amyotrophic lateral sclerosis. Also, Chapman et al.[95] cite a number of studies the results of which indicate a strong relationship between pesticides and Parkinsonism. Further, Peters et al.[96] found Parkinsonism after inadvertent ingestion of seed-corn treated with the fungicide *hexachlorobenzene* (a chemical often found in chemically sensitive patients with PD). The early signs of extrapyramidal disorder include cogwheeling, increased muscle tone, and reduced association movements. Fahn[97] comments on the long association of CS_2 and Parkinsonism.

Case study. In our own studies at the EHC-Dallas, we observed a 49-year-old white male who developed PD or Parkinson-like disease. He had been a salesman for a store dealing in hardware and gardening chemicals for 2 years when he became aware of a sudden and growing tremor and cogwheeling in both his arms and legs. In time, he also noticed paroxysmal vertigo, tachycardia, and shortness of breath. These symptoms usually lessened on Sunday when he was away from work, partially on Monday due to weekend relief, and during vacations; they progressively worsened during the work week with a summit on Friday and Saturday.

His chief complaint, the tremor, was treated and partially relieved by his neurologist with Xanax®, which is a trazolo analog of the 1,4 benzodiazepine class and a symptom-suppressing medication. No effort was made to find the environmental triggering agents. After 15 months of progressive, ongoing deterioration that culminated in his inability to function in the store, he came to the EHC-Dallas for diagnosis and treatment.

At this time, he complained of massive tremor with cogwheeling motion of PD in his right hand and arm, less tremor in his right leg, blurred vision, skin fluorescence that appeared during the night and could be washed away, difficulties in walking and in talking loudly, paroxysmal diurnal and nocturnal dyspnea, coughing up mucus, and paroxysmal tachycardia. He stated that he would very easily become physically exhausted and frequently experience dizziness and feel jittery and depressed. He also had a very short attention span. Three separate neurologists made the diagnosis of PD. Our physical examination of him showed a massive resting tremor, especially in the right arm and hand, equilibrium problems manifested by an inability to stand on his toes with his eyes open or closed and balance, and gait difficulties. Aside from these findings, the examination was not remarkable. His blood pressure was 120/80, and his heart rate was 80 beats per minute.

Blood analysis using Laseter's method showed an extremely high level for hexachlorobenzene (HCB) (16.4 ppb, compared to <0.3 ppb, which was average in the population of the examining laboratory). Beta-benzene hexachloride (2.7 ppb, compared to 0.5 ppb) and DDE (12.8 ppb, compared to 5.6 ppb) were also found to be remarkably high. Besides these three, two chlorinated pesticides—heptachlor epoxide (0.8 ppb, compared to <0.3 ppb) and alpha chlordane (0.9 ppb, compared to <0.3 ppb)—were detected. Also, several volatile chlorinated hydrocarbons were seen in elevated concentrations: dichloromethane (1.7 ppb, compared to <1.0 ppb), chloroform (2.1 ppb, compared to <1.0 ppb), and tetrachloroethylene (1.9 ppb, compared to 1.1 ppb).

This patient underwent the physical therapy/heat chamber depuration program, which included sauna and exercising along with special additions to the diet. He also stayed in the ECU for 6 weeks in order to ensure increased environmental control. His symptoms did not disappear completely, but they did lessen significantly over the course of treatment. This outcome correlates with a reduction in blood levels of chemicals identified in his blood. The chlorinated pesticide HCB was reduced by 85%, beta-BHC by 89%, alpha chlordane by at least 67%, heptachlorepoxide by 62%, and DDE by 54% after 5 weeks of treatment.

Table 26.8. 49-Year-Old White Male Inhaled Double-Blind Challenge after 4 Days Deadaptation with Total Load Decreased under ECU Conditions

Chemical	Dose (ppm)	Parkinson symptoms increased
Phenol	<0.0050	+
Formaldehyde	<0.2	−
Chlorine	<0.33	−
Pesticide (2,4-DNP)	<0.0034	−
Petroleum derived ethanol	<0.50	−
Saline × 3		−

Source: EHC-Dallas. 1986.

Double-blind inhaled challenge under environmentally controlled conditions in the deadapted state with the total load reduced revealed that phenol reproduced his symptoms, while saline placebos did not (Table 26.8). As some of these chemicals (especially HCB and DDE) could be expected to remain at least partially in his system, they were, in our opinion, responsible for the patient's residual symptoms. We, therefore, recommended that this patient undergo a second session of our thermal depuration/physical therapy program. We have now seen at least six cases of PD for which environmental triggers could be found.

Mechanisms for metabolic derangements in PD are unknown. However, reports reveal altered transitional metal metabolism [98] and elevated nigral iron content.[99] Such increased levels of metals might be caused by elevated cysteine concentrations, since this group acts as a metal chelating agent. As discussed in Volume I, Chapter 6,[52] pollutants can disturb cell membranes and cause a selective increase in intracellular minerals in many chemically sensitive patients. Local disturbance of metal metabolism could lead to free radical formation, accumulation, and cellular damage compatible with the suggested pathway for toxicity of MPTP in the basal ganglia.

Steventon et al.[100] have reported that 10% of their patients with PD were poor hydroxylators in contrast to the general population. They found that sulfoxidation of -S-carboxymethyl-L-cysteine, which is accomplished by cysteine dioxygenase, is significantly reduced in PD, where 35.3% of patients produce no S-oxide metabolites as compared with 4.9% of hospital controls and 2.5% of the general population. The endogenous reaction for the hepatic cysteine dioxygenase system is the conversion of cysteine to inorganic sulfate. The supply of this anion is known to be rate-limiting for the formation of sulfate conjugates. Alteration of this pathway has been seen in some chemically sensitive patients with neurological problems at the EHC-Dallas. In agreement with this finding, the patients with PD have reduced capacity for processing para cetamol sulfate when challenged with the parent compound.

These results suggest elevated levels of cysteine and depressed levels of reduced sulfate are available for detoxification. In humans, 95% of the circulating dopamine is sulfate conjugated. Reduced levels of sulfate conjugates of dopamine were found in 27 of the patients with PD studied by Steventon et al., suggesting that this pathway becomes saturated, possibly due to inadequate sulfur supply or excess toxic chemical overload.

Alzheimer's Disease

A chronic deficiency of central cholinergic function has been implicated in Alzheimer's disease[101] and a number of other neurological problems. Pesticides were also found in the brain of some.[94] Recently, Mistry et al.[102] reported on a cholinotoxin that will allow the development of an animal model for neurotoxin-induced Alzheimer's.[77] Aluminum has been found in the brain of some Alzheimer's patients, suggesting at least one environmental trigger and again emphasizing that pollutant injury will result in excess intracellular minerals. In those patients with Alzheimer's and neurological dysfunction and intracellular mineral imbalance studied at the EHC-Dallas, aluminum was one of the most common intracellular minerals found in excess (see Volume I, Chapter 6,[52] section on "minerals"). Salama et al.[103] discussed a case study of a 74-year-old woman with early stage senile dementia who, 2 weeks after exposure to methane, developed severe hypertonia and akinetic mutism. One study[104] suggests neurotoxic damage with nutrient deficiency. One of the hypotheses regarding the etiology of Alzheimer's implicates trace element disturbance.[105] Heafied et al.[104] demonstrated imbalances in eight elements in the Alzheimer's patients as compared with age-matched controls. These imbalances are similar to those seen in our chemically sensitive patients at the EHC-Dallas. Vance et al.[106] studied the concentrations of 17 elements in the hair and nails of 180 patients with Alzheimer's disease and control subjects. He found significant imbalances in Br, Ca, Co, Hg, K, and Zn between the Alzheimer's patients and the control groups. Ward and Mason[107] have also shown trace element changes in patients with Alzhiemer's. They have shown high levels of aluminum and possibly low levels of zinc in the hippocampus. (This disease is probably due not to a single factor but to a combination of many environmental factors.)

Little et al.[108] reported the first long-term, double-blind, placebo-controlled trial of high dose lecithin in senile dementia of the Alzheimer type. Fifty-one subjects were given 20 to 25 g/day of purified soya lecithin (containing 90% phosphatidyl plus lysophosphatidyl choline) for 6 months and followed up for at least another 6 months. Plasma choline levels were monitored throughout the treatment period. There were no differences between the placebo group and the lecithin group, but there was an improvement in a subgroup of relatively poor compliers. These were older and had an intermediate level of plasma choline. Although the effects of lecithin are complex, there may be a "therapeutic window" for the effects of lecithin in Alzheimer's patients, and this window may be more evident in older patients. We have seen a small group of Alzheimer's patients at the EHC-Dallas and have found that only those in the early stages of the disease respond to therapy.

Amyotrophic Lateral Sclerosis (ALS)

Amyotrophic lateral sclerosis (ALS) is the most common of the motoneu-ron diseases. The clinical syndrome typically involves a fatal outcome over 2 to 5 years.

Viruses,[109] metals,[110] endogenous toxins,[111] immune dysfunction,[112] endo-crine abnormalities,[111] impaired DNA repair,[111] altered axonal transport,[113] and trauma[111] have all been etiologically linked with ALS. Tandan and Bradley,[114] however, point out that no convincing research evidence of a causative role for any of these factors has yet been demonstrated. A unifying hypothesis is needed to explain the many diverse factors associated with ALS, and develop-ment of such a hypothesis may well be feasible when the principles of environ-mental medicine are understood.

At the EHC-Dallas, we have seen eight patients who had ALS and in whom environmental triggers were identified. When exposed to certain molds, foods, and toxic chemicals, these patients experienced exacerbated symptoms, while removal of the incitants along with reduction of the total toxic load decreased the signs and symptoms. The following case is a graphic demonstra-tion of food and chemical triggering of ALS.

> **Case study**. This 69-year-old woman presented with the chief complaint of inability to move her arm well. She also had dysphagia and difficulty breath-ing. She had been diagnosed as having ALS by three separate neurologists. Her condition continued to deteriorate in spite of all treatment. Her past history was significant in that for the previous 20 years she had lived next to a large chicken factory farm where there was constant spraying of various pesticides including several organochlorines. After being admitted to the ECU, this patient immediately went into respiratory failure. She had a weight of 70 lbs, a 5'8" frame. Her physical exam revealed she was extremely malnourished, and she had the typical signs of nerve degeneration of ALS. She was placed on a ventilator after intubation. Her admitting immune work-up is summarized in Figure 26.10. In addition to total ventilatory support, this patient was given intravenous hyperalimentation. Initially, she had high lev-els of toxic pesticides in her blood and a suppressed immune system (Figure 26.10). As the pesticides gradually cleared from her blood, her suppressed immune system improved. This patient came off the respirator after the levels of pesticides, which had vigorously increased (apparently due to mobilization from the tissues where they had been previously sequestered) and then dropped to low levels, and the T and B lymphocytes returned to normal. She was discharged in a functional state weighing 95 lbs when all of the pesticides

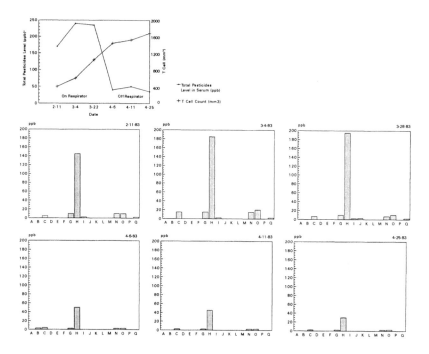

Figure 26.10. Amyotrophic lateral sclerosis correlation of the level of chlorinated pesticides and T cell count in the blood. 69-year-old white female. ªToxicants: β-BHC, DDT, DDe, Dieldrin, HCB, Heptachlor epoxide, Mirex. Bottom bar graphs — key: A = aldrin; B = α-BHC; C = β-BHC; D = gamma-BHC; E = delta-BHC; F = gamma chlordane; G = DDT; H = DDE; I = dieldrin; J = endosulfan I; K = Endosulfan II; L = endrin; M = heptachlor; N = heptachlor epoxide; O = hexachlorobenzene; P = PCB; Q = Mirex. (From EHC-Dallas. 1983.)

were virtually not detectable, except for the DDE, the breakdown product of DDT, which was at 2 ppb (Figure 26.10).

Again sulfoxidation and sulfonation mechanisms seemed to be involved in the detoxification reactions in ALS patients. It should be emphasized that other detoxification mechanisms are also probably involved, but our knowledge is limited.

Multiple Sclerosis

Environmental and genetic factors have been implicated in the etiology and pathogenesis of multiple sclerosis (MS).[115,116] Laborde et al.[117] observed that climate may be involved in the etiology of the disease. Ingalls[118] presented evidence that the time/place clustering of the disease such as was found in a 1983 to 85 outbreak of MS in Key West, FL, was the result of environmental pollution. Murrell et al.[119] review the notion that MS is more prevalent in areas with high sheep populations. Ingalls[120] suggests that seepage of ionic mercury used in dental work may lead to MS in middle age and that cases of unilateral MS may derive from mercury-amalgam fillings in isolated teeth. Also, inhalation of volatile mercury or exhaust fumes of lead additives to gasoline, says Ingalls, may lead to MS. Nanji and Narod[121] report a significant correlation between the prevalence of MS and the consumption of pork fat and other meats. Maas and Hogenhuis[122] believe that MS may be caused by an allergic or other adverse reaction to certain foods, e.g., cocoa products, cola, and coffee.

Still other etiological factors implicated in the onset of MS include an antigen in the swine-flu vaccine,[109,123] a slow-acting virus, and exposure to tuberculosis before the age of 7 years.[124]

Many more recent studies link the onset of MS to exposure to a variety of chemicals. For instance, Blisard et al.[125] present a case study of a man with no previous medical problems who was twice exposed to an insecticide containing chlordane and heptachlor. Neurological symptoms began to develop 6 months to 1 year after exposure. They progressed until his death. The classic findings of multiple sclerosis and peripheral neuropathy were incompatible with the literature on chlordane toxicity but are compatible with chlordane sensitivity. Also, Noseworthy and Rice[126] note that trichlorethylene poisoning mimics MS, while Corsini et al.[127] suggest a significant involvement in grain workers exposed to disulfide-based fumigants.

At the EHC-Dallas, our studies of MS patients suggest food and chemical triggering in some whose disease process had started. We have seen 28 patients of whom 24 were females and 4 were males. They ranged in age from 31 to 70 years, with a mean age of 50 years. There appeared to be a breakdown in the ability of these patients to handle environmental incitants or toxic overloads (Table 26.9). In some patients, these multiple triggers were accompanied by immune abnormalities (Table 26.10).

Gamma globulin changes appeared insignificant because of the number involved. However, there were several abnormalities (Table 26.11). Total complements (CH_{50}) were abnormal in 10 patients, and, in three, C_3 was low (Table 26.12). There was a significant difference ($p<0.05$) between the control group and the MS group for T_{11}. There was a significant difference between 12 immune parameters in the MS group vs. the control group to the $p<0.0001$ level (Table 26.13). Of the intracellular blood cell minerals outside two standard deviations, many were abnormal (Table 26.14).

The striking features between the MS group (PG, 16 patients) and the nonsmoking, nondrinking, nonmedication-taking, control group (CG, 26 healthy individuals) were that, first, the frequency of intracellular potassium that was abnormal was 83.8% (PG) vs. 3.8% (CG) ($p<0.001$). For chromium

Table 26.9.   Review of Multiple Environmental Triggers
in 28 Cases of Multiple Sclerosis

Factors	No. of patients tested	No. of patients positive	(%)
Food sensitivity (oral and intradermal challenge)	23	23	100
Mold sensitivity (intradermal; inhaled)	22	22	100
Chemical sensitivity (intradermal; inhaled	25	24	96
Hormone sensitivity (intradermal)	12	12	100
Terpene sensitivity (intradermal)	18	15	83.3
EB virus (intradermal)	9	9	100
CMV (intradermal)	3	3	100
Measles virus (intradermal)	2	2	100
HBV (intradermal)	2	0	0
Blood pesticides	14	14	100
Blood toxic elements	12	11	91.7
Blood toxic chemicals	15	15	100

Source: EHC-Dallas. 1990.

Table 26.10.   The Frequency of Abnormal Immune Parameters
in 16 Patients with Multiple Sclerosis

Immune parameters	No. of patients tested	No. of patients above normal (2SD)	No. of patients below normal (2SD)	No. of patients abnormal	(%)
WBC	16	0	6	6	37.5
Lym (%)	16	0	0	0	0.0
Lym C	16	0	6	6	37.5
T_{11} (%)	16	1	4	5	31.3
T_{11} C	16	2	6	8	50.0
T_4 (%)	16	4	1	5	31.3
T_4 C	16	1	3	4	25.0
T_8 (%)	16	0	1	1	6.3
T_8 C	16	0	1	1	6.3
T_4T_8	16	2	0	2	12.5
Bly (%)	16	3	2	5	31.3
Bly C	16	1	2	3	18.8
CMI	4	0	0	0	0.0

Source: EHC-Dallas. 1990.

Table 26.11.   The Frequency of Abnormal Immune Antibodies
in 15 Patients with Multiple Sclerosis

Immune antibodies	No. of patients tested	No. of patients above normal (2SD)	No. of patients below normal (2SD)	No. of abnormal	(%)
IgA	13	4	1	5	38.5
IgE	11	1	6	7	63.6
IgG	15	2	2	4	26.7
IgM	13	0	0	0	0.0

Source: EHC-Dallas. 1990.

Table 26.12.   The Frequency of Abnormal and Immune Complex
in 10 Patients with Multiple Sclerosis

Complement	No. of patients tested	No. of patients above normal (2SD)	No. of patients below normal (2SD)	No. of patients abnormal	(%)
CH_{100}	6	0	2	2	33.3
C_1q	10	3	0	3	30.0
C_3	9	0	3	3	33.3
C_4	9	0	0	0	0.0
C_5	8	2	0	2	25.0

Source: EHC-Dallas. 1990.

abnormalities, it was 91.7% (PG) vs. 5.6% (CG) ($p<0.001$); for calcium abnormalities, it was 33.3% (PG) vs. 3.8% (CG) ($p<0.05$). For abnormal zinc, there was 2.0% PG vs. 3.8% (CG) (not significant); copper abnormalities were 41.7% (PG) vs. 7.7% (CG), ($p<0.05$); abnormal silicon was 75% (PG) vs. 34% (CG) ($p<0.01$); abnormal phosphorous was 25% (PG) vs. 75% (CG) ($p<0.05$); abnormal sulfur was 41% (PG) vs. 10% (CG) ($p<0.05$); magnesium was 41% vs. 5% in the CG ($p<0.005$); Na was 9.1% vs. 5% CG (not significant). Chi square showed that there was a significant difference of 14 intracellular blood minerals in the MS group vs. the control group to the $p<0.0003$ level. In contrast, abnormalities for vitamins were on the high side with the exception of B_1, B_2, B_3, and B_6. These low B levels are typical of those usually seen in the chemically sensitive patient (see Volume I, Chapter 6,[52] and Volume II, Chapter 30). All of these abnormalities suggest membrane damage, which is usually triggered by pollutants.

Toxic chlorinated pesticides, volatile aromatic hydrocarbons, volatile chlorinated hydrocarbons, and volatile aliphatic hydrocarbons were found in these patients. The average number of toxic chemicals was 10 per patient with 5

Table 26.13. Chi-Square Test for the Frequency of
Abnormal Immune Parameters Between Normal
Control and Multiple Sclerosis Groups

GRP	Case	Observed	Expected	O-E^2/E
1	1	34.6	30.197	0.642
2	1	37.5	41.903	0.4627
1	2	3.8	1.5915	3.0646
2	2	0	2.2085	2.2085
1	3	23.1	25.3805	0.2049
2	3	37.5	35.2195	0.1477
1	4	7.4	16.2084	4.7869
2	4	31.3	22.4916	3.4496
1	5	19.2	28.9824	3.3018
2	5	50	40.2176	2.3794
1	6	11.5	17.9255	2.3033
2	6	31.3	24.8745	1.6598
1	7	15.4	16.9204	0.1366
2	7	25	23.4796	0.0984
1	8	23.1	12.3133	9.4493
2	8	6.3	17.0867	6.8095
1	9	23.1	12.3133	9.4493
2	9	6.3	17.0867	6.8095
1	10	15.4	11.6851	1.181
2	10	12.5	16.2149	0.8511
1	11	15.4	19.5589	0.8843
2	11	31.3	27.1411	0.6373
1	12	15.4	14.3237	0.0809
2	12	18.8	19.8763	0.0583

Notes: Chi-squared statistics: degrees of freedom (11); Chi-squared (61.0568); $p<0.0001$; Cramer's V (0.3511); contingency coefficient (0.331).

Source: EHC-Dallas. 1990.

pesticides, which contrasts with the nonsensitive individuals we see today who have less than 2 toxic chemicals in their blood (Table 26.15).

Figures 26.11 and 26.12 and Table 26.16 identify the various toxic chemicals found in the blood of our MS patients. Frequencies were also significant (Table 26.17). The means of 11 of 19 chemicals found in the control and MS groups were significantly higher in the MS group to $p<0.0002$ level.

There were significant differences in ethyl benzene, styrene, trichloroethylene, and dichlorobenzene between controls and MS. For ethylbenzene, WC were 3.8% of controls vs. 40% of MS ($p<0.01$). For styrene, no differences in the controls were found, in contrast to 26.7% of the MS group. In the MS group, this finding resulted in a significant difference vs controls ($p<0.001$).

Table 26.14. Intracellular Minerals in 12 Cases of Multiple Sclerosis

Minerals	Below 2 standard deviations			Normal			Above 2 standard deviations		
	No. of patients	No. of patients	(%)	No. of patients	No. of patients	(%)	No. of patients	No. of patients	(%)
Ca	12	0	0	12	8	66.7	12	4	33.3
Mg	12	9	0	12	7	58.3	12	5	41.7
Na	12	1	9.1	11	10	91.1	11	0	0
K	12	1	8.3	12	2	16.7	12	9	75
Cu	12	1	8.3	12	7	58.3	12	4	33.3
Zn	12	2	16.7	12	10	83.3	12	0	0
Fe	12	2	16.7	12	9	75	12	1	8.3
Mn	12	0	0	12	11	91.7	12	1	8.3
Cr	12	11	91.7	12	1	8.3	12	0	0
P	12	1	8.3	12	9	75	12	2	16.7
Se	12	0	0	12	8	66.7	12	4	33.3
Si	12	5	41.7	12	3	25	12	4	33.3
S	12	5	41.7	12	7	58.3	12	0	0
B	12	0	0	12	12	100	12	0	0

Source: EHC-Dallas. 1990.

Table 26.15.   The Average Number of Blood Toxic Chemical Solvents
and Pesticides in 14 Cases of Multiple Sclerosis

	Total no. of chemicals	Total no. of patients tested	Average chemicals per patient	Total no. of control group	Average no. of chemicals in control group	p
Chlorinated pesticides	68	14	5	40	2	p<0.002
Chemical solvents	171	14	10	40	2	p<0.001

Source: EHC-Dallas. 1990.

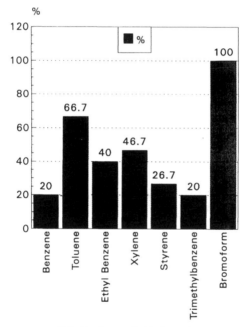

Volatile Aromatic Hydrocarbons

Figure 26.11.   Volatile aromatic hydrocarbons in 15 cases of multiple sclerosis. (From EHC-Dallas. 1990.)

For dichlorobenzene, no differences were found in the control group vs. 28.6% in the MS group (p<0.02) (Figure 26.12). Of the chlorinated pesticides tested in 14 patients, hexachlorobenzene, DDE, and heptachlor epoxide were present in the most (Figure 26.13).

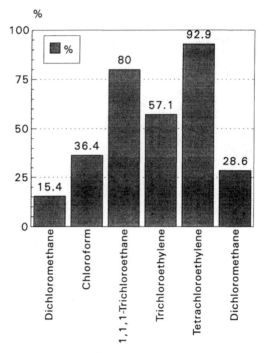

%

Figure 26.12.   Volatile chlorinated hydrocarbons in 14 cases of multiple sclerosis. (From EHC-Dallas. 1990.)

Friedman's two-way anova test showed there was a significant difference for the frequency of total chemicals between MS and normal control groups ($p<0.036$).

Inhaled double-blind challenge in six patients confirmed by provocative skin tests under environmental control showed positive reactions to chemicals with reproduction of the MS symptoms. All of those who underwent inhaled, double-blind challenges reacted to ambient doses of pesticides, phenol, and formaldehyde (Figure 26.14). Seventy-five percent of patients reacted to ethanol, 60% of the patients reacted to chlorine, and 7% reacted to 18 placebo challenges. Intradermal confirmation occurred over 68% of the time.

Further treatment using environmental control showed significant improvement in eight patients, as was evidenced by loss of their signs and symptoms as well as improvement of their T and B lymphocytes (Figure 26.15). The diminution of the chlorinated pesticides in six other patients was evidence of their significant improvement (Figure 26.16). The following case is illustrative of what happened to these patients.

Case study. A 42-year-old white female was bedridden for 3 years as a result of her 7-year bout with MS. She had been refractory to all treatments before

Teble 26.16. Comparison of the Frequency of Blood Toxic Chemicals Between Normal Control and MS Groups

Blood toxic chemicals	Normal control group			MS group			
	No. of tested	No. of positive	(%)	No. of tested	No. of positive	(%)	p
Benzene	26	10	38.5	15	3	20.0	>0.05
Toluene	26	19	73.1	15	10	66.7	>0.05
Ethylbenzene[a]	26	1	3.8	15	6	40.0	<0.01
Xylenes	26	16	61.5	15	7	46.7	>0.05
Styrene[a]	26	0	0.0	15	4	26.7	<0.02
Trimethylbenzenes	26	0	0.0	15	3	20.0	>0.05
Dichloromethane	26	10	38.5	13	2	15.4	>0.05
Chloroform	26	9	34.6	11	4	36.4	>0.05
1,1,1-Trichloroethane	26	18	69.2	10	8	80.0	>0.05
Trichloroethylene[a]	26	2	7.7	14	8	57.1	<0.001
Tetrachloroethylene	26	21	80.8	14	13	92.9	>0.05
Dichlorobenzenes[a]	26	0	0.0	14	4	28.6	<0.02
n -Pentane	26	11	42.3	7	5	71.4	>0.05
2.2-Dimethylbutane	26	8	30.8	7	1	14.3	>0.05
Cyclopentane	26	1	3.8	7	2	28.6	>0.05
2-Methyl pentane	26	26	100.0	7	7	100.0	>0.05
3-Methyl pentane	26	26	100.0	7	7	100.0	>0.05
n-Hexane	26	26	100.0	7	7	100.0	>0.05
n-Heptane	26	1	3.8	7	0	0.0	>0.05

[a] Significant difference.

Source: EHC-Dallas. 1990.

being seen at the EHC-Dallas. Her work-up showed her to be sensitive to the following: biological inhalants — positive for grasses, molds, trees, and weeds; foods — 15 positives: wheat, turkey, chicken, beef, eggs, milk, peanut, green grapes, Brewer's and baker's yeast, coffee, lettuce, potato, soy, corn, and tuna; and toxic chemicals — all terpenes, cigarette smoke, ethanol, formaldehyde, phenol, diesel, newspsrint, and perfume.

Her laboratory results included the following: WBC, 6200; IgG, 1029 mg/dL (C = 564 to 1765 mg/dL); IgM, 145 mg/dL (C = 45 to 250 mg/dL); IgA, 172 mg/dL (C = 85 to 385 mg/dL); total hemolytic complement, 140 (C>70); IgE, 196 IU/mL (C = 10 to 50 IU/mL); total lymphocytes, 2666/mm^3 (C = 1600 to 4200/mm^3); T_{11} lymphocytes, 58% (C = 62 to 86%); absolute T_{11}, 1546/mm^3 (C = 1260 to 3270/mm^3); T_4 helper, 36% (C = 31 to 54%), total count T_4, 960/mm^3 (C = 670 to 2000/mm^3); T_8 suppressor, 22% (C = 17 to 35%); T_8 suppressor absolute count, 587/mm^3 (C = 333 to 1617/mm^3); B lymphocytes, 3% (C = 5 to 18%).

This patient was placed on a rotary diet using chemically less-contaminated foods. She was given injections for inhalants, foods, and chemicals, vitamin and mineral therapy, and good environmental control. She had to remove the gas heat and carpets from her home. Gradually her symptoms cleared.

**Table 26.17.   Comparison of the Means of Blood Toxic Chemicals
Between Normal Control and MS Groups**

Chemicals	Normal control group (ppb)	MS group (ppb)	Difference
Benezene	0.66	0.11	0.55
Toluene	0.73	0.67	0.06
Ethylbenzene	0.02	0.16	–0.14
Xylenes	1.04	1.57	–0.53
Styrene	0.00	0.13	–0.13
Trimethylbenzenes	0.00	0.073	–0.073
Dichloromethane	1.18	0.54	0.64
Chloroform	0.92	0.72	0.20
1,1,1-Trichloroethane	1.58	1.22	0.36
Trichloroethylene	0.47	0.36	0.11
Tetrachlorethylene	1.32	2.06	–0.74
Dichlorobenzenes	0.00	0.057	–0.057
n-Pentane	1.58	2.13	–0.55
2,2-Dimethylbutane	1.14	0.17	0.974
Cyclopentane	0.054	0.87	–0.816
2-Methylpentane	13.36	10.13	3.23
3-Methylpentane	42.87	28.51	14.36
n-Hexane	13.54	9.21	4.33
n-Heptane	0.077	0.00	0.077

Source:EHC-Dallas. 1990.

Over the next few months, she noted that reexposure to these toxic substances caused exacerbations of her MS symptoms; however, she eventually became so well that she took a job teaching and worked on and received a doctorate. She has been totally well for 3 years.

Terpene sensitivity was studied in 17 patients. Up to 68% of the patients were sensitive to mountain cedar (Figure 26.17). All patients were sensitive to at least one terpene.

It is clear from our study that environmental influences play a part in MS, and some patients can be helped and actually put into remission by strict institution of environmental principles.

Huntington's Chorea

Crinnion[84] notes that organophosphate pesticides are a group of compounds that are powerful inhibitors of cholinesterase. In Volume II, Chapter 13[179] we emphasize this fact and stress that some carbamates also cause altered cholinesterase. Their action leads to an accumulation of acetylcholine at the nerve synapses, resulting in overstimulation and then paralysis of neural transmission. The

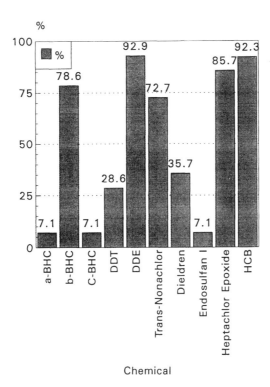

Figure 26.13.   Chlorinated pesticides in 14 cases of multiple sclerosis. (From EHC-Dallas. 1990.)

present effect by organophosphates will supply a cholinergic function deficit that may be associated with such neurologic problems as Alzheimer's,[101] tardive dyskinesia,[128] and Huntington's chorea (HC).[87]

It appears that the genetic defect in HC (like many genetically-related illnesses) is sometimes exacerbated with both toxic chemical and food exposures. We described this "time bomb" effect in Volume I, Chapter 3.[7] We have followed a patient with HC who is definitely affected by toxic exposure.

Case study. In 1984, a 46-year-old white female was admitted to the ECU with the chief complaints of headache, hearing loss, poor equilibrium, severe nausea, and depression.

This patient reported that she had been exposed to chemicals through pesticide spraying one time per month over a period of 6 years. She also reported that when cleaning solvent was used in her workplace she developed a headache, became nauseated, and tended to lose her voice.

This patient had a significant past work history in that she was exposed to toxic chemicals. She had worked as a telephone operator. She had missed a lot of work since 1981, when she developed migraine headaches, and by 1984 she was unable to work because of headaches, allergies, and nausea. At that

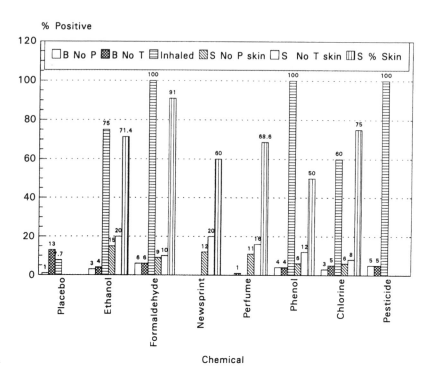

Figure 26.14.   Double-blind inhaled and intradermal challenge of chemicals in six cases of multiple sclerosis. Patient was in the deadapted state with total load reduced. (From EHC-Dallas. 1990.)

time, she lost her hearing, except that she heard voices, which appeared to be hallucinations. She also had ringing in her ears. A neurologist examined her and her family for HC and subsequently made this diagnosis. Several family members were also affected. She was hospitalized and underwent electroconvulsive therapy, but she did not improve. She was allergic to penicillin. She also had had an appendectomy at some time in her past.

The physical exam revealed disequilibrium when she stood on her toes with her eyes closed. Also, mild periorbital edema and obesity were found. Her lumbar spine X-ray showed grossly normal. This patient had an otherwise normal physical exam, except that at times she would show signs of dementia. She was admitted to the ECU with a diagnosis of allergic encephalgia, chemical hypersensitivity, and HC.

This patient was placed in an environmentally controlled room constructed of porcelain fused on steel. She was fed nothing by mouth except spring water for 2 days prior to her admission and 3 days during her stay. She related that her headaches, nausea, loss of hearing, diarrhea, and edema all lessened. The equilibrium problem also improved in that she became able to balance herself with her eyes closed.

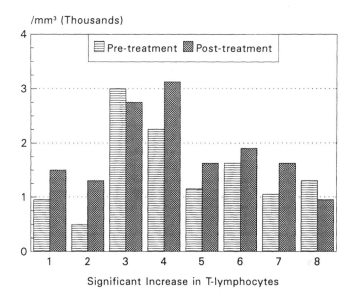

Figure 26.15. Comparison of total T cells in eight cases of multiple sclerosis before and after treatment. (From EHC-Dallas. 1990.)

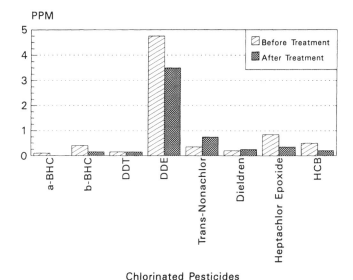

Figure 26.16. Comparison of chlorinated pesticides between before and after treatment in six cases of multiple sclerosis. (From EHC-Dallas. 1990.)

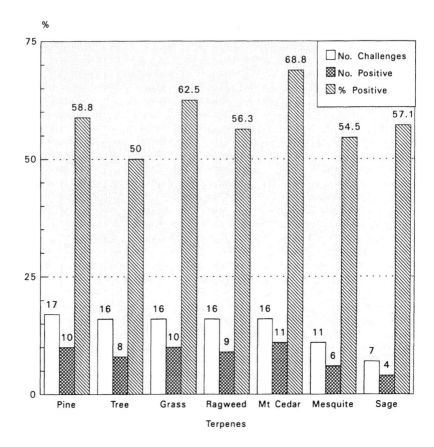

Figure 26.17.   Terpenes sensitivity in 17 cases of multiple sclerosis. (From EHC-Dallas. 1990.)

Laboratory tests showed numerous abnormalities. The RBC mineral analysis showed that her chromium was depressed at nondetectable (C = 0.14 to 0.222 ppm); barium was slightly elevated at 0.099 ppm (C = 0.007 to 0.0062 ppm); aluminum was 0.16 ppm (C = 0.035 ppm). Serum elemental analysis showed a bit lower phosphorus at 105 ppm (C = 107 to 120 ppm) and elevated sulfur at 1013 ppm (C = 610 to 950 ppm).

Her erythrocyte glutathione pyruvate transaminase (EGPT) was elevated at 1.50 index (C<1.25 index), indicating low B_6. The erythrocyte glutathione reductase was elevated at 1.32 act. coeff. (C = 0.90 to 1.20 act. coeff.), indicating low B_2. The vitamin C/WBC was depressed at 17.7 $\mu g/10^8$ cells (C = 21 to 53 $\mu g/10^8$ cells). The whole blood elements test showed depressed calcium at 44.9 $\mu g/mL$ (C = 59 to 61 $\mu g/mL$) and depressed zinc at 5.9 $\mu g/mL$ (C = 7.30 to 7.70 $\mu g/mL$). The 24-hour urine sodium was depressed at 7 mEq/24 hours. (C = 43 to 217 mEq/24 hours). Total blood cholesterol was slightly elevated at 253 mg/dL (C = 120 to 250 mg/dL). Chloride was slightly elevated at 107 mEq/L (C = 95 to 106 mEq/L). Creatine was elevated at 1.5

**Table 26.18.   Huntington's Chorea Inhaled Double-Blind
Challenge After 4 Days Deadaptation
with the Total Load Decreased**

Chemical	Dose (PPM)	Signs and symptoms
Formaldehyde	<0.2	+
Pesticide (2,4-DNP)	<0.0034	+
Phenol	<0.002	+
Ethanol	<0.50	+
Chlorine	<0.33	+
Saline 1	15 cc	0
Saline 2	15 cc	0
Saline 3	15 cc	0

Source: EHC-Dallas. 1990.

mg/dL (C = 0.7 to 1.3 mg/dL). Total protein was depressed at 5.8 g/dL (C = 6.0 to 8.5 g/dL). Globulin was depressed at 1.8 g/dL (C = 2 to 3.5 g/dL). Total complement was depressed at 39.5 CH_{100} units (C>40 U). C_2 was elevated at 18,948 μ/mL (C = 4500 to 10,000 μ/mL). The complement C_3 was depressed at 78.5 mg/dL (C = 83 to 177 mg/dL). C_5 was elevated at 17 mg/dL (C = 7.2 to 15.8 mg/dL). Total WBC count was depressed at 3800/mm³ (C = 4100 to 11,000/mm³). Total lymphocytes were depressed at 1026/mm³ (C = 1200 to 3900/mm³). The anti-EB virus VCA-IgG was at 1/640 titer (C<1/10 titer).

Her inhaled chemical test (Table 26.18) was done in a steel and glass airtight booth by double-blind method after her total load was reduced, and she was maintained in a deadapted state for 1 week. The chemicals that produced symptoms were ethanol (<0.50 ppm), formaldehyde (<0.2 ppm), pesticide (<0.0034 ppm), phenol (<0.002 ppm), and chlorine (<0.33 ppm). Those negative were three saline placebos. The patient was orally challenged with 26 chemically less-contaminated foods. She had sensitivity to five foods.

Intradermal challenge showed she had sensitivity to histamine, serotonin, pollen, five chemicals (including orris root), ethanol, formaldehyde, and phenol. She also had sensitivity to lake algae, five molds, dust, dust mite, three terpenes, two smuts, T.O.E., *Candida*, three animal danders, three trees, three grasses, and thirteen weeds.

During hospitalization, this patient received an immunotherapy program with biological, inhalants, foods, and some chemicals. She also received psychological therapy as well as intravenous and oral vitamin and mineral therapy. As an outpatient, she has lived in a small portable porcelain steel house for the past 4 years. With the exception of mild exacerbation with inadvertent extraneous exposures, she has remained asymptomatic without medication. She has been able to work in a controlled environment helping to care for other disabled patients. She can now even travel to more toxic environments without experiencing health problems. Five years have elapsed since she initially became totally incapacitated with a prognosis of permanent incarceration from her HC.

Myesthenia Gravis

Myesthenia gravis is a disease that may, or may not, be autoimmune. It is often associated with the thymus gland, and biopsies reveal lymphocytes around the muscle fibers and small blood vessels similar to those of the chemically sensitive. The myo-neuro junction is usually the site of involvement. Clinically, there is bulbar and generalized weakness. Antibodies are produced against post-neural-junction acetylcholine receptors. Some cases in adults seem to have been exacerbated by environmental influences. At the EHC-Dallas, we have had the opportunity to study only a small number of patients with myesthenia, and they indeed were sensitive to environmental influences.

Brain Cancer

Epidemiological evidence suggests occupational risks of brain cancer in many industries. Rubber workers, pharmaceutical workers, vinyl chloride workers, chemists, and petrochemical and oil-refinery workers have all shown an excess brain cancer risk based on studies using standardized mortality ratio (SMR) or proportionate mortality ratio (PMR) analysis. Additionally, veterinarians, machinists, lead smelter workers, and aluminum reduction plant workers, all of whom may have increased chemical exposure in the workplace, have an elevated risk of developing brain cancer.[129]

More evidence of the increased risk of brain cancer in specific fields of work comes from a series of prospective studies in the petrochemical industry.[130] These studies emphasize again the need for early evaluation of people exposed to toxicants vs. waiting until end-stage disease develops. Emphasis should be placed on the fact that many of the same chemicals that trigger chemical sensitivity are known carcinogens.

Other Nonmalignant Brain Dysfunctions

Other brain abnormalities have been observed in chemically sensitive patients following toxic exposure. These include arachnoiditis with hydrocephalis, intraventricular hemorrhage, brain damage, and seizures.

Arachnoiditis with Hydrocephalis.   Many nonmalignant, but severely damaging, problems have been reported from toxic chemical exposure, an example of which is a 62-year-old patient who developed severe adhesive spinal arachnoiditis, hydrocephalis, and papilledema following isolthalumate injection myelography. Papilledema and hydrocephalis resolved spontaneously after the removal of the toxic substance.[131]

Intraventricular Hemorrhage.   Hiller et al.[132] concluded that benzyl alcohol used as a preservative in different injectable solutions resulted in toxicity that contributed significantly to both mortality and the occurrence of major intra-ventricular hemorrhage among infants weighing less than 1000 g at birth and that solutions containing benzyl alcohol should never again be used in the care of such infants. Similarly, Benda et al.[133] concluded that, in the treatment of preterm infants, discontinuation of benzyl alcohol solutions, which are linked with increased mortality and incidences of intraventricular hemorrhage, can result in dramatic improvement in outcome for surviving infants in terms of fewer neurologic handicaps.

Brain Damage. McMichael et al.[134] have noted that postnatal blood lead concentration is inversely related to cognitive development in children. They caution, circumspection, however, when making causal inferences from studies of this relation because of the difficulties in defining and controlling confounding effects. However, it is generally well accepted that excess lead in infants and children will cause learning problems and brain dysfunction.

Fernandez et al.[135] reported on a newborn's malformed brain, the cerebral hemispheres of which were completely gone. These investigators believed this deformity was the result of ultra-uterine anoxia that occurred during a critical period of fetal brain development when the baby's mother accidentally inhaled butane gas in her sixth month of pregnancy. Recent reports from the Rio Grande Valley in Texas have shown numerous children born with anencephalus. The cause apparently is multiple environmental agents.

Maizlish et al.[136] discuss their study of 99 pest control workers before and after their work shift, which failed to demonstrate diverse behavioral effects. This group emphasized personal protection equipment and direct supervision. This finding, or its antithesis, contrasts with our findings from observing a series of people who had no protection and were exposed to pesticides. They developed damage that was characterized by behavioral abnormalities.

Choi[137] noted that studies of methyl mercury (MeHg) intoxication have produced clinical evidence suggesting that MeHg poisoning may be lethal. These findings also indicate that lesser degrees of MeHg intoxication may lead to some form of cerebral palsy accompanied by mental retardation.[138-140] In cases of less severe poisoning, an infant may develop psychomotor deficits as the nervous system matures, even though at birth the infant may have appeared completely normal.[139-141]

Abnormal behavioral development in MeHg-intoxicated offspring is well-documented.[142-145] Particularly, significant behavioral abnormalities in the absence of any other overt signs of toxicity have been observed in MeHg treated mice.[142] Choi[137] suggested that a possible morphological basis for these behavioral abnormalities may be the normal distribution of cerebral cortical neurons, which have been observed in the parietal and entorhinal cortices of MeHg-exposed mice.

At the EHC-Dallas, we have observed over 500 children with learning difficulties, attention deficit, and hyperactivity whose symptoms were triggered by food and chemicals. Their symptoms were subsequently relieved by avoidance, intradermal food and biologic inhalant injections, and nutritional therapy.

Seizures.   It is becoming well-established that neurotoxins as well as foods may trigger brainwave activity, which is generalized. At times, however, focal brainwave activity may occur from focal chemical injury, causing seizures. Studies at the EHC-Dallas using multi-channel EEGs and SPECT brain scans have shown a high incidence of increase in electrical activity and neurotoxic patterns in chemically sensitive children with brain dysfunction. We have witnessed seizure activity as a result of both food and chemical triggering (Figure 26.18). Egger et al.[146] and Crayton et al.[147] have also shown seizure-like activity due to food sensitivity.

Primary seizure activity is defined as a syndrome characterized by sudden recurrent changes in consciousness, usually preceded by a sensory warning and often followed by a motor discharge. It is probably true that this definition could be extended to include any recurrent disturbance of consciousness and, in fact, many chemically sensitive patients exhibit recurrent disturbances in consciousness. Also, SPECT brain scans reveal what appears to be a preponderance of temporal lobe seizures in the subset of seizure-prone chemically sensitive patients. Here, even though he may not move, the patient is conscious and able to hear and see.

Obviously not all cases of seizures are due to environmental stimuli, but many may have environmental influences, and some will be triggered by environmental pollutants. The following is an example of an environmentally induced seizure.

Case study. A 33-year-old white male engineer presented at the EHC-Dallas with the chief complaint of recurrent grand mal type seizures. The seizures were characterized by jaw twitching and body jerking, which was usually limited to 4 or 5 times, mostly on the right side. This patient did not lose consciousness. He did have excessive urination and headaches afterward. His previous EEG was abnormal showing diffuse bisynchronous myoclonic activity. His presenting complaints on May 2, 1988, were myoclonic seizures since July 1986, inability to concentrate with mood swings, myalgia and profound fatigue, and "hypoglycemia". His seizures occurred 2 to 4 times a month, lasted for 15 minutes to 3 hours, and occurred more frequently with chemical exposures. He was refractory to medication, and multiple EEGs had confirmed the diagnosis.

His past health history revealed that, in 1982, he had started work for Detroit Edison power utility in the electronics lab with computer systems. In 1983, he transferred to the nuclear power plant as a computer hardware engineer. His work station was located directly above the electronic sensor/control area for the reactor. In 1984, he was exposed to jet engine exhaust (peaking generators) during peak electrical demand. Also in 1984, the parking lot was blacktopped for several weeks, during which time his symptoms were exacerbated, and he was moved to a new pre-fab building still under construction.

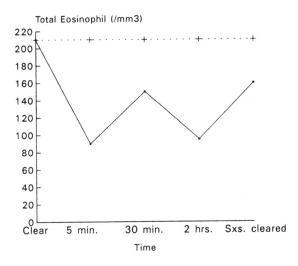

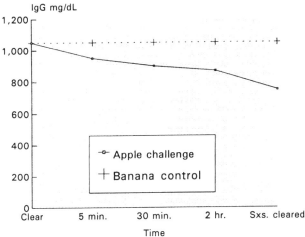

Figure 26.18. Oral food (apple) challenge. 29-year-old white female after at least 4 days deadaptation with total load reduced in the ECU. Symptoms reproduced: grand mal seizures. (From EHC-Dallas. 1988.)

In 1985, he became tired, moody, irritable, and nervous. He had excessive night sweats, poor sleep, weakness, and sudden nausea. He was told he had anxiety, and treatment with tranquilizers, antidepressants, then Mellaril® produced extreme agitation. The first onset of strong jaw trembling and muscle jerking occurred in 1985. These symptoms started in his wrists, then, forearm, and progressed to body jerking. The Mellaril® dose was doubled, and this patient was told he was "creating symptoms". His EEG was abnormal. His neurologist at this time suggested high doses of Dilantin®, Librium®, and

Xanax®. This patient tried these medications and felt worse, so he stopped taking them. He substituted taking vitamins, which seemed to help.

This patient reported a variety of environmental sensitivities at this time, including reactions to perfumes, gasoline, new carpeting, paint fumes, hair spray, cigarette smoke, and many others, all of which caused symptoms. Newspapers induced headache, poor concentration, and a drugged feeling. Car exhaust produced nausea and anxiety. Citrus fruits triggered vomiting and seizures.

This patient's family history included his father, 63 years, with hypoglycemia, hypertension, diabetes, and heart disease. His mother, 62 years, had headaches, mood swings, anemia, cancer, and food sensitivities. His 36-year-old sister had multiple food and chemical sensitivities. His family history was also positive for hayfever, eczema, and arthritis.

At the time of his initial visit, he underwent a physical examination that included evaluation of his neurological system. Normal limits were found, except for an acne-like skin eruption on his back. Laboratory work-up using an intracellular mineral analysis showed low magnesium, potassium, zinc, and selenium. Hematology showed moderate leukopenia (WBC = 3900/cm^3). His erythrocyte superoxide dismutase (ESOD) was decreased to 7.35 units/mgH; Hgb, 12 g/dL (C = 9.4 to 13.4 g/dL). A fat biopsy for chemicals on February 2, 1988, showed polybrominated biphenyls at 603 µg/Kg. (U.S. population mean for polybrominated biphenyls is 15 µg/Kg; lower Michigan mean is 200 µg/Kg; normal is 0.) A second adipose tissue analysis on September 17, 1990, revealed heptachlor epoxide to be 205 ppb; oxychlordane was at 37 ppb; dieldrin was 25 ppb, and DDE was 294 ppb. Blood analysis done on September 6, 1990, showed toluene, xylene, dichloromethane, 1,1,1-trichloroethane, and tetrachloroethylene.

Intradermal testing showed this patient was sensitive to biological inhalants, foods, chemicals, viruses, *Candida*, T.O.E., and neurotransmitters. Neurotransmitter challenge reproduced his symptoms. A challenge injection of 0.01 mL of methacholine 0.0016% (0.01 mL/#3) induced muscle jerking, facial twitching, yawning, difficulty in concentration, and headache.

Good neutralization occurred at 0.20 mL of 0.00064% (0.2 mL/#5), and this dose proved very helpful in reducing the patient's neurological symptoms. The following double-blind inhaled challenge reproduced symptoms including seizures, which were documented on EEG (Figure 26.19).

The patient entered a comprehensive program of detoxification involving exercise, dry heat (specially built all-tile chamber), mineral and electrolyte supplementation, and deep muscle massage. Frequent assessment of vital signs and biochemistry was maintained. For example, PBB was 603 ppb in fat before heat depuration treatment in contrast to 100 ppb in fat after.

This patient underwent a treatment program that included a rotation diet, nutrient supplementation, filtered/bottled water, immunotherapy vaccines, detoxification program, environmental control, and avoidance. The patient was educated in good environmental health practices to give him more control over his symptoms.

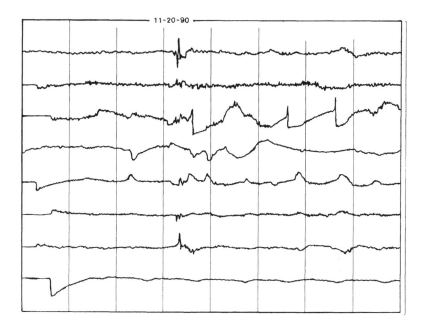

Figure 26.19. Chemical sensitivity — seizure history. 33-year-old male electrical engineer. Ambulatory EEG — chemical challenge. (From EHC-Dallas. 1990.)

Several authors have described the transport metabolism and blood flow in the brain during organophosphate-induced seizures.[148,149] There is a characterization of spontaneous epileptiform discharges induced by organophosphorus anticholinesterases in the in vitro hippocampus (see Volume II, Chapter 13[79]). These type discharges are frequently seen in our chemically sensitive patients.

In reviewing our own previous research of grand mal (22 cases) or petite mal (42 cases) seizures, we found that 60 cases of epilepsy were completely relieved by environmentally directed and manipulated management alone without medication except when massive reexposure occurred. In 41 cases of petite mal or grand mal seizures, patients reported that symptoms were relieved by omission of certain trigger agents such as mold, pollen, dusts, toxic chemicals, and foods from the diet and reduction of their chemical loads. We also demonstrated by double-blind inhaled challenge that these substances would trigger the seizure in individual patients. Food challenge also has been noted to trigger some patients (Figure 26.20).

Early limited single channel EEG studies have served to confuse the issue as to whether the presence of toxic chemical sensitivities may be a cause of environmentally triggered seizures. It appears likely that, unless a patient is challenged with toxic chemicals, as shown in the previous case study, while

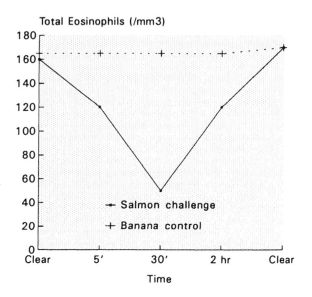

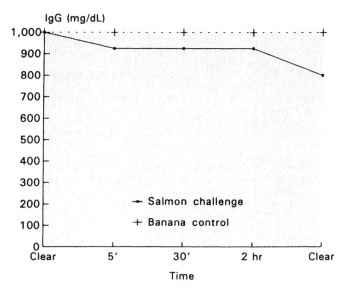

**Figure 26.20.   Oral food (salmon) challenge. 31-year-old white female
challenge after at least 4 days deadaptation with reduction
of total body load in the ECU, EHC-Dallas. Symptoms
reproduced: grand mal seizures. (From EHC-Dallas. 1981.)**

undergoing multichannel EEG testing after deadaptation for at least 4 days
after total load reduction, no definite relationship can be established. However,
there are, in addition to disturbed consciousness, many manifestations associated

with seizures including headache, extreme weakness, migratory paralysis, temporary blindness, visual hallucinations, confusion, partial or total loss of consciousness, defective comprehension, epigastric sensitivities, jerking, local twitching, and more. In addition to seizures, these symptoms have been reproduced by both inhaled and intradermal challenge in this series. Once the patient had an observable seizure, previously established incitants were withdrawn and the patient was not allowed to progress to a full seizure.

In the particular study at the EHC-Dallas, our purpose was to attempt an identification of environmentally triggered seizure patients from other patients. The goal was to see if different criteria parameters other than disturbed consciousness, such as symptom manifestation, were present. The subjects were 46 environmentally triggered patients, 24 of whom reported a history of seizure or were observed in seizure activity and twenty-two other patients with no reported or observed seizures. All patients were assessed along a broad spectrum of physical and psychological dimensions. Those variables that were significantly differentiated in the two groups were seizure and nonseizure patients. The seizure patients showed significantly more depression, headaches, suicide attempts and gestures, memory deficits, impaired concentration, and hormone sensitivity. The nonseizure patients (control group) showed significantly more immune suppression as measured by T lymphocytes, C_3, or vitamin or mineral abnormality. More asthma was seen in this latter group.

Memory deficits and impaired concentration could be considered as representing disturbed consciousness. In this study, these variables were objectively measured over time. Of significance, the standard neuropsychological screening test did not differentiate between the seizure and nonseizure groups, but the Harrel-Butler-profile did (Table 26.19).[150]

While depression is common among environmentally triggered patients, it is also associated with organic brain syndrome and functional psychological disorders. In this study, however, the frequency and severity of depression were differentiating entities between the seizure and nonseizure groups. The more frequent and severe the case of depression, the greater the disturbed psychological profile indicated in seizure patients.

After environmental control, challenge tests with multiple foods and chemicals triggered symptoms in this series of patients who had obvious CNS weakness. The following case exemplifies the type of seizure patient seen in our series at the EHC-Dallas.

Case study. A 61-year-old white male professor of surgery was admitted to the ECU with the chief complaints of serious headache, dizziness, mental confusion, and poor concentration. At the time of his admission to the ECU, he was practicing as a urological surgeon and professor of urology. He was the head of his department, a position he had held for 23 years.

He reported that he had begun to have difficulty in articulation so that he would have to substitute words for those he could not remember. He began to anticipate problems in recall so that he would attempt to marshal alternative

**Table 26.19.   Associated Symptoms and Signs in Seizure-Type
Chemically Sensitive Patients vs. Nonseizure
Chemically Sensitive Patients in the ECU**

	Seizure patients (N = 24)	Nonseizure patients (control) (N = 22)
Showed significantly more	Depression	Immune suppression
	Headaches	Asthma
	Suicide attempts and gestures	Vitamin or mineral abnormality
	Memory deficits	
	Impaired concentration	
	Hormone sensitivity	

Source: Buckley, T. P., J. R. Butler, N. Didriksen, and W. J. Rea. No. 4-8, 1983. Seizure manifestations in ecology patients: neurologic sensitivities. Presented at Society for Clinical Ecology 17th Advanced Seminar in Clinical Ecology and 18th Annual Meeting. The Broadmoor, Colorado Springs, CO. With permission.

words in advance of when he planned to use them in a conversation. He found that reports that had previously taken about 10 minutes to complete were requiring increasingly more time, first 20 to 30 minutes and finally up to an hour. Recall of certain memories, words, conversations, recent events, and the like became less reliable and more erratic. He noticed also minute perceptual-motor problems. Psychomotor speed was slowed a bit because he would have to correct some movement in hand-eye coordination, but once corrected, his speed was as rapid as ever. Perceptual accuracy and discrimination of fine-motor movement would be off momentarily. He was so expert at what he did that no one noticed these subtle differences, except this extremely bright, competent, and conscientious physician himself.

He also began to experience more fatigue than usual, particularly toward the end of the work day and especially after a day in which several hours had been spent in the operating room. At first, he attributed his fatigue to his advancing age, but there was some inconsistency in this instance because there were many times when he seemed to be his old energetic self with concomitant accomplishments to prove it. His wife pointed out another change in his behavior. He had become increasingly argumentative, irritable, and impatient with his family, co-workers, students, and others he encountered daily. This behavior was at considerable variance with his usual calm, friendly, supportive, and accommodating manner. He had always been impatient with himself but not with others.

This patient reported that these symptoms occurred after a prolonged period of time in the operating room, where he was chronically exposed to volatile odors including halothane anesthetics and a cleaning solution that contained glutaraldehyde (a combined formaldehyde and phenol product), which is broken down in the same pathway as formaldehyde. This patient reported that these symptoms had been present for a number of years. He also mentioned that approximately 8 to 9 years previously he had developed seizures for which he had taken Dilantin®. At one point, the Dilantin® was discontinued, and he again collapsed with seizures. He then had to be placed on Tegretol®.

This patient had a history of vasculitis, which was documented on an angiogram. Five years previously he had had lower lip surgery for a precancerous lesion, and he had had a basal cell carcinoma on his right ear removed.

His family history was positive for hay fever, cancer, heart disease, insect allergy, and inhalant allergy.

His chest X-ray and EEG were normal. MRI (magnetic resonance imaging) brain scan showed several areas of old infarction in the cerebellum due to the vasculitis. His physical exam was grossly normal. His neurological exam revealed that the cranial nerves two through twelve were intact. The deep tendon reflexes were 2 to 4+ bilaterally.

This patient was initially placed in the ECU. Except for spring water, he was given no food or drink by mouth for 2 days. He cleared his symptoms while in the ECU. As a part of diagnostic procedures, he was orally challenged on a mono-rotation basis with 13 chemically less-contaminated foods. Only cantaloupe produced his symptoms.

Diagnostic inhaled chemical challenge was conducted in a steel and glass airtight booth using a double-blind procedure. Chemicals that triggered symptoms were formaldehyde (<0.2 ppm), ethanol and Cydex (glutaraldehyde–phenol ambient dose) (<0.5 ppm), pesticide (2,4-DNP) (<0.0034 ppm), chlorine (<0.33 ppm), and phenol (<0.2 ppm). Saline exposure did not produce symptoms (Table 26.20).

This patient had many extraordinary intellectual strengths, but demonstrated significant variability between initial testing and retesting after treatment and within the scatter of tests. His brain function was measured before and after inhaled challenge testing. On exposure to placebos he showed no significant difference in perception or motor functioning, immediate recall, or speed of mental alterations. On exposure to the glutaraldehyde-phenol cleaning solution, however, he demonstrated slowed thinking and memory loss. Visual recall as well as verbal memories were reduced by one half on postchallenge testing. This reaction, which lasted for almost 24 hours, was his worst. On exposure to chlorine, he showed a significant decrease in perceptual organization, nonverbal reasoning, attention to details, alertness, and visual sequencing. These symptoms were similar to those with which this patient first presented and were particularly important findings in light of the fact that he usually scrubbed his hands prior to surgery with a chlorine compound. Exposure to pesticide produced impairment in visual memory, immediate recall, speed of mental operation, and increased latency to response. This patient scored within the normal range on cerebral functioning, including, in general, the maintenance of a normal hand/eye coordination, balance, fine motor movement, and perceptual accuracy and judgment.

His abnormal laboratory tests showed that total lymphocytes were low at 1188/mm^3 (C = 1260 to 2650/mm^3); T$_4$ cells were low at 535/mm^3 (C = 670 to 1800/mm^3). RBC mineral assay revealed elevated magnesium at 61 ppm (C = 39 to 49 ppm), elevated potassium at 89 ppm (C = 60 to 77 ppm), and depressed selenium at 0.1 ppm (C = 0.15 to 0.25 ppm). Aluminum was at 0.76 ppm; lithium was at 0.04 ppm; strontium was 0.16; and zirconium was 0.2 ppm.

Table 26.20. 61-Year-Old White Male with Seizures: Inhaled Double-Blind Challenge After 4 Days Deadaptation With Total Load Reduced in the ECU

Chemical	Dose (PPM)	Signs reproduced
Formaldehyde	<0.2	+
Phenol	<0.002	+
Pesticide (2,4-DNP)	<0.0034	+
Chlorine	<0.33	+
Ethanol (petroleum derived)	<0.5	+
Phenol and gluteraldehyde solution	ambient	4+ lasted 24 hours
Saline 1	ambient	0
Saline 2	ambient	0
Saline 3	ambient	0

Source: EHC-Dallas. 1990.

His lab tests also showed elevated erythrocyte glutathione pyruvate transaminase (EGPT) at 1.5 index (<1.25) (indicating low vitamin B_6), depressed 1-N-methylnicotinamide at 1.4 mg/24 hours (C = 3 to 17) (indicating low vitamin B_3), and depressed vitamin D_3 at 23.5 ng/mL (C = 26 to 41 ng/mL). The blood toxic chemical analysis revealed pentachlorophenol at 10 ppb, DDE at 0.5 ppb, trans-nonachlor at 0.2 ppb, HCB at 0.4 ppb, dichloromethane at 1 ppb, and tetrachloroethylene at 0.7 ppb.

After 2 weeks, this patient was discharged in good condition. His brain function showed marked improvement in all areas of verbal comprehension and memory function. Overall, his IQ, as measured by various tests, improved from high average (110) to very superior classification (>130). He improved significantly and markedly from average (110) to very superior (>130) in the understanding of antecedent-consequent relations, evaluation of past experiences, sound decision-making, organization of knowledge, and utilization of "common sense". He was instructed in detail about how to make his home and work environment ecologically safe. He has remained well for the last year and a half.

Nonfixed Disease (Symptom-Named Responses)

Pathology has been found as a result of exposure to many different chemicals, but with many more there is no specific fixed-tissue damage, just irritative or neurotransmitter triggering, which can be particularly incapacitating in the CNS. This type of triggering is often seen in a subset of chemically sensitive patients. Those agents associated with known damage include the following: agents causing anoxia such as barbiturates, carbon monoxide, cyanide, azide, and nitrogen trichloride; agents damaging myelin such as isonicotinic acid hydrazide (INH, isoniazid), triethyltin, hexachlorophene, lead, thallium, and tellurium; agents causing peripheral axonopathies such as alcohol, acry-lamide, bromophenylacetylurea, carbon disulfide, hexanedione, and organo-phosphorus compounds; agents causing primary damage to perikarya of pe-ripheral neurons such as organomercury compounds, vinca alkaloids, and iminodipropionitrile; agents causing neuromuscular junction of motor nerves such as botulinum toxin, tetrodotoxin, saxitoxin, batrachotoxin, DDT (dichloro-diphenyltrichloroethane), pyrethrins, and lead; and neurotoxicants causing localized CNS lesions such as hepatotoxin, methione sulfoximine, glutamate, gold thioglucose, acetylpyridine, trimethyltin, pyrithiamine, DDT, mercury, and manganese (Table 26.21).

There are many symptoms associated with nervous system dysfunction. They include those presented in Table 26.21. This table, though broad, may not be all-inclusive. Some of the chemicals that may trigger specific responses are also seen in this table. However, these responses only occur with classic cases of individual toxicants. If there are multiple chemicals in an individual's system, a totally different pattern, which is often incapacitating in the chemi-cally sensitive individual, usually emerges.

Headaches

Headaches are discussed in Chapter 18 and, therefore, will not be consid-ered here.

Behavioral Responses

Randolph[151] originally described and discussed "brain fag" as being one of the early symptoms of chemical sensitivity. He described this many years before the chronic fatigue syndrome was reported, but many of the character-istics of chemical sensitivity, such as chronic fatigue, and brain fag are similar. We have observed these patterns in a subset of chemically sensitive patients. He further described and elicited behavior abnormalities after food and chemi-cal challenges.[152]

Studies of organic solvents show important behavioral effects associated with toxicity as well as with sensitivity (Table 26.22). The common miscon-ception that solvent and other toxic exposures do not cause sensitivity and brain dysfunction may well account for much of 20th century humans' need for many kinds of mood-altering drugs and medications, which are usually unnec-essary if the causes of the agitation or depression are found and eliminated. The fact is that solvents do cause mood swings in many cases. They also cause other types of neurological dysfunction. For example, a Swedish study of automobile and industrial spray painters in comparison to a control group found workers had a higher incidence of complaints such as headaches, fatigue, and memory loss problems, as well as impairments in reaction time, manual dexterity, and perceptual speed as measured by psychological testing.[153] The treatment for

Table 26.21.   Reported and Observed Symptoms Associated with Environmental Chemical Exposures

Chemical	Eyes						Skin							
	E1	E2	E3	E4	E5	E6	S1	S2	S3	S4	S5	S6	S7	S8
Acrylamide														
Adiponitrile														
Aniline														
Arsenic														
Arsine														
Benzene														
Bromophenyl-acetylurea														
Carbon disulfide					9									
Carbon monoxide														
Carbon tetrachloride														
Chlordane														
Chlorinated hydroquinolines				9										
Chloroprene														
Cyanide														
DDT														
Dichloroethane														
Dimethyl sulfate														
Dioxin		13												
Ether, diethyl														
Ethylene dichloride					13									
Hexachlorophene														
Hexane														
Hydroquinone														
Kepone														
Lead														
Leptophos						13					66	66	66	66
Manganese														
Mercury														
Methyl bromide														
Methyl chloride			13	49										
Methyl-*n*-butyl ketone														
Methylmercury						13								
Nitrofurans														
Organophosphate esters		13			13	13								
Organophosphates	3		39											
PBB	46		65				3	3	3	3	3	3	3	3

Table 26.21 (Continued). Reported and Observed Symptoms Associated with Environmental Chemical Exposures

Chemical	Eyes						Skin							
	E1	E2	E3	E4	E5	E6	S1	S2	S3	S4	S5	S6	S7	S8
PCB		13					18	18		18	18			
Phenylmercury														
Polychlorinated polycyclics							30	30		30	30		30	
Radiation, ionizing			21											
Sulfur dioxide														
Tetraethyl lead														
Tetraethylthiuram disulfide														
Thallium				13	13									
Toluene														
Trichloroethylene														

Chemical	Associative					Nervous system arousal						
	A1	A2	A3	A4	A5	N1	N2	N3	N4	N5	N6	N7
Acrylamide												
Adiponitrile			13									
Aniline												
Arsenic												
Arsine												
Benzene								13	13			
Bromophenyl-acetylurea												
Carbon disulfide (9)		38	38				38		38	38		
Carbon monoxide			38	38	38							
Carbon tetrachloride			13				13					
Chlordane												
Chlorinated hydroquinolines												
Chloroprene								13	13			
Cyanide											13	
DDT												
Dichloroethane												
Dimethyl sulfate											13	
Dioxin												
Ether, diethyl							13	13	13			
Ethylene dichloride		13					66	66	38	38		
Hexachlorophene												
Hexane							38		38	38		
Hydroquinone												

Table 26.21 (Continued).   Reported and Observed Symptoms Associated with Environmental Chemical Exposures

Chemical	Associative					Nervous system arousal						
	A1	A2	A3	A4	A5	N1	N2	N3	N4	N5	N6	N7
Kepone		13	13				49	49		49		
Lead		13			42							
Leptophos												
Manganese		13										
Mercury		38					13	13	13	13		
Methyl bromide			13			39			39			
Methyl chloride			13	49			65	65				
Methyl-*n*-butyl ketone							18	18				
Methylmercury												
Nitrofurans												
Organophosphate esters			13			30						
Organophosphates	39	39										
PBB												
PCB		18										
Phenylmercury												
Polychlorinated polycyclics							9					
Radiation, ionizing					21							
Sulfur dioxide									13			
Tetraethyl lead												
Tetraethylthiuram disulphide												
Thallium												
Toluene		13	13									
Trichloroethylene												

Chemical	Neuro. funct. physiological				Musculo-skeletal			Gastrointestinal					
	B1	B2	B3	B4	K1	K2	K3	G1	G2	G3	G4	G5	G6
Acrylamide													
Adiponitrile													
Aniline													
Arsenic													
Arsine													
Benzene													
Bromophenyl-acetylurea													
Carbon disulfide	9	38											

Table 26.21 (Continued). Reported and Observed Symptoms Associated with Environmental Chemical Exposures

Chemical	Neuro. funct. physiological				Musculo-skeletal			Gastrointestinal					
	B1	B2	B3	B4	K1	K2	K3	G1	G2	G3	G4	G5	G6
Carbon monoxide													
Carbon tetrachloride													
Chlordane													
Chlorinated hydroquinolines													
Chloroprene													
Cyanide													
DDT													
Dichloroethane			13										
Dimethyl sulfate													
Dioxin													
Ether, diethyl													
Ethylene dichloride	13												
Hexachlorophene													
Hexane													
Hydroquinone													
Kepone													
Lead	66	13	66								66		
Leptophos													
Manganese			13										
Mercury	38	13	38										
Methyl bromide													
Methyl chloride		49	49	49						49			
Methyl-n-butyl ketone													
Methylmercury													
Nitrofurans													
Organophosphate esters		13											
Organophosphates		39	39				39						
PBB	65	65	65		3	3	3	3	3			3	3
PCB (46)	18	18	18		23		23	18	18	23	46		
Phenylmercury													
Polychlorinated polycyclics													
Radiation, ionizing	30					30				30			30
Sulfur dioxide													
Tetraethyl lead													
Tetraethylthiuram disulfide		13											

Table 26.21 (Continued). Reported and Observed Symptoms Associated with Environmental Chemical Exposures

Chemical	Neuro. funct. physiological				Musculo-skeletal			Gastrointestinal					
	B1	B2	B3	B4	K1	K2	K3	G1	G2	G3	G4	G5	G6
Thallium													
Toluene													
Trichloroethylene													

Chemical	Neurological function													
	Sensory						Motor							
	P1	P2	P3	P4	P5	P6	M1	M2	M3	M4	M5	M6	M7	M8
Acrylamide														
Adiponitrile					13		13	13	13					
Aniline														
Arsenic		13						13						
Arsine		5			13			9						
Benzene					13		13							
Bromophenyl-acetylurea								9		9		9		
Carbon disulfide					9	38		38						
Carbon monoxide								13						
Carbon tetrachloride														
Chlordane									13					
Chlorinated hydroquinolines					9			9	9					
Chloroprene														
Cyanide		13												
DDT		13						13						
Dichloroethane									13					
Dimethyl sulfate								13	13					
Dioxin				13										
Ether, diethyl														
Ethylene dichloride									13					
Hexachlorophene									13					
Hexane									13					
Hydroquinone					13									
Kepone							13	13	13					
Lead	66	13			66			13	13	66	38	66	66	66
Leptophos									13					
Manganese				13	13	13			13					
Mercury								38						

Table 26.21 (Continued).   Reported and Observed Symptoms Associated with Environmental Chemical Exposures

Chemical	Sensory						Motor							
	P1	P2	P3	P4	P5	P6	M1	M2	M3	M4	M5	M6	M7	M8
Methyl bromide					13		13							
Methyl chloride						13	13	13				49	49	
Methyl-*n*-butyl ketone				13				13						
Methylmercury	13			13			13	13	13					
Nitrofurans			9				9							
Organophosphate esters	13			13			13	13						
Organophosphates			9	9			9		9		9			39
PBB	65		65	3	65			65				65	65	65
PCB				18	23									
Phenylmercury							13							
Polychlorinated polycylics	30		30	30				30						
Radiation, ionizing														
Sulfur dioxide			13					30						
Tetraethyl lead														
Tetrathylthiuram disulfide				9								9		
Thallium					13									
Toluene	13				13									
Trichloroethylene									13					

Notes: **Eyes** — E1 = eye irritation; E2 = dimness of sight (amblyopia); E3 = double vision (diplopia); E4 = blurred vision; E5 = Eye oscillation (nystagmus); E6 = pupil reactions. **Skin** — S1 = rash; S2 = acne; S3 = sunsensitivity; S4 = darkening or thickening; S5 = discoloration or deformity of nails; S6 = dryness; S7 = increased sweating; S8 = slow or poor healing of cuts. **Associative** — A1 = decreased mental acuity; A2 = impaired memory; A3 = confusion; A4 = disorientation; A5 = slowed functional adolescent development. **Nervous system arousal** — N1 = lethargy; N2 = depression; N3 = nervousness; N4 = irritability; N5 = emotional instability; N6 = hyperactivity; N7 = photophobia. **Physiological responses** — B1 = headaches; B2 = sleeplessness; B3 = sleepiness; B4 = loss of appetite. **Musculoskeletal** — K1 = joint pain; K2 = swelling in joints; K3 = muscular aches and pains. **Gastrointestinal** — G1 = 10 lb or more weight loss; G2 = nausea; G3 = vomiting; G4 = abdominal pain; G5 = abdominal cramps; G6 = diarrhea. **Neurological function: sensory** — P1 = vision impairment; P2 = hearing impairment; P3 = perception changes taste/smell; P4 = burning sensation; P5 = parathesias; P6 = hallucinations. **Motor** — M1 = speech impairment; M2 = muscle weakness; M3 = tremors; M4 = difficulty walking; M5 = seizures; M6 = incoordination/clumsiness; M7 = dizziness; M8 = fatigue.

Source: FASE. 1990. Personal communication.

**Table 26.22.   Psychoneurological Responses with Solvent Profiles
in a Subset of 100 Neurologically Involved
Chemically Sensitive Patients**

Type chemicals	Toxic levels in Blood	Symptoms and signs
Aromatic hydrocarbons	Increased	Cognitive neurobehavior
Aliphatic hydrocarbons	Increased	Functioning decreased
Chlorinated aliphatic hydrocarbons	Increased	Short-term memory
		Coordination decreased
		Motor sequencing decreased
		Somatosensory deficits
		Cognitive dysfunction increased

Note: $p<0.01$ to 0.05.

Source: Baldridge, J. T. and J. R. Butler. 1990. Unpublished data. EHC-Dallas.

these symptoms is removal of the toxicants, not introduction of symptom-suppressing medications, which alone will eventually lead to end-organ failure if the underlying chemical triggers continue.

Like the auto workers and industrial spray painters, shipyard painters have also been shown to be affected by long-term exposure to toxic chemicals. Valciukas et al.[154] found that a group of workers who had had 10 years or more experience involving exposure to ketone, xylene, ethylene, glycol, and mineral spirits scored lower than the general population on tests such as block design and embedded figures. These tests are indicative of impaired CNS function, which is often seen in a subset of chemically sensitive individuals.

Gregersen et al.[155] studied workers who had been exposed to white mineral spirits, toluene, perchloroethylene, and styrene. They found impairments in concentration, attention, and abstract function as well as emotional, mood, and personality changes. These responses are similar to those seen in a subgroup of chemically sensitive patients who were found to have these substances in their blood and who responded adversely after inhaled challenge that followed 4 days of deadaptation.

Hanninen[156] reviewed five studies dealing with long-term occupational exposure to styrene, lead, toluene, and a mixture of organic solvents and observed that in four of the five studies the individuals affected by exposure showed neurotoxic effects involving both cognitive and psychomotor functions. These effects were similar to what we saw in a subset of chemically sensitive individuals. The effects of styrene were limited to perceptual and psychomotor disturbances. The main effects of exposure to solvent mixtures were seen in cognitive functions, while psychomotor retardation was the most pronounced effect in the carbon disulfide group. Butler found similar effects in a group of chemically sensitive patients that he studied at the EHC-Dallas.[157]

Other studies show comparable results. Behari et al.,[158] for instance, described the diagnosis of a case of peripheral neuropathy due to styrene exposure. Also, Mackay and Kelman[159] observed that, after high exposure to styrene vapor, female workers were seen to have a slowing of reaction times. In a within-group comparison of workers exposed to styrene, Muijser et al.[160] found that the most exposed workers had a statistically significant decrease in hearing thresholds at high frequencies than their low-exposure counterparts.

Rosenberg et al.[161] found that toluene abuse causes diffuse pathological and MRI abnormalities, and Aaserud et al.[162] found that workers who experience long-term exposure to CS_2 risk developing toxic encephalopathy that is demonstrable on both neurological and neuropsychological examination. Aaserud found, further, that demonstration of structural changes in the brain caused by CS_2 exposure can be obtained from cerebral CT.

Studies of workers in the paint industry provide further evidence of CNS impairment that may result from solvent exposure. Orbaek et al.[163] found an increase in fatigue, tension, hostility, and memory problems among these workers. Orbaek's study also suggests that exposure to solvents increases the likelihood of CNS disturbance. Fourteen percent of the workers studied by Orbaek et al. had indications of toxic encephalopathy (brain dysfunction) as compared with the unexposed control groups who showed no such signs.

The effects of solvent exposure may be quite long lasting particularly if the secondary as well as primary triggering agents are not removed. Workers who were diagnosed as having disease related to occupational exposure to solvents showed decrements in visual-motor performance and freedom from distractibility when tested 5 or more years after final exposure.[164] Long-term occupational exposure to organic solvents may cause adverse effects to the CNS. One group of 46 patients with toxic encephalopathy (symptoms and test impairment) was studied. At least 5 years after the initial examination the subjects were asked to attend a reexamination that included a structured medical interview and a psychometric investigation. The results indicate that effects on the CNS persist even when exposure has ceased.[165]

Stollery and Flindt's retrospective study[166] of a small group of female workers who were intoxicated by organic solvents suggested that solvent intoxication can cause neuropsychological sequelae lasting for over 8 months. The findings of these studies are contrasted to those of chemically sensitive patients seen at the EHC-Dallas whose brain function usually improved after treatment. The difference in the aforementioned results and ours appears to be that, as much as possible, we eliminated both secondary triggering agents and primary offending chemicals, thus reducing the total environmental pollutant load, while they eliminated the primary offender only. See Figure 26.21 for SPECT scan improvement after environmental control treatment.

"General depression", a functional state of the CNS, commonly results from overexposure to many solvents used in industrial operations as well as exposure to frequently used household chemicals. Often a massive total pollutant load increase will result in general depression. During the detoxification

A.

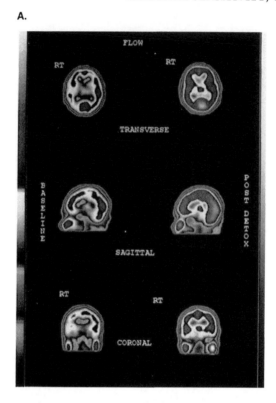

**Figure 26.21.   SPECT brain scan showing improvement of neurotoxic-
ity: (A) neurotoxic, (B) clearing. (From EHC-Dallas. 1992.)**

process, many of the solvents break down into alcohols. Impaired worker
consciousness can be a problem, even though the effect may be completely
reversible on removal of the individual from the toxicant(s). Many fat solvents,
like alcohol, either inhaled or ingested, cause "general depression" characterized initially by drowsiness, difficulty in concentrating, and mood changes.
These symptoms are often seen in a subset of chemically sensitive patients.
Subsequent symptoms include slurred speech, ataxia, and disorientation, followed by loss of consciousness.[43] This depression correlates well with the
patient's mental status after exposure to solvents (toluene, xylenes, 1,1,1-
tricholoroethane, and trichloroethylene produce such symptoms) that we have
found in the blood of a subset of our chemically sensitive patients at the EHC-
Dallas. Clearly, the presence of solvents in their blood usually correlates with
their neurologic signs and symptoms. Brain function test scores do improve
and depression or hyperstimulation lifts as the toxic chemicals are cleared from
the patient's system.

At the EHC-Dallas, Baldridge and Butler[167] correlated psychoneurological
responses with solvent profiles in chemically sensitive patients (Table 26.22).
The solvents that were found in the chemically sensitive patients' blood were

B.

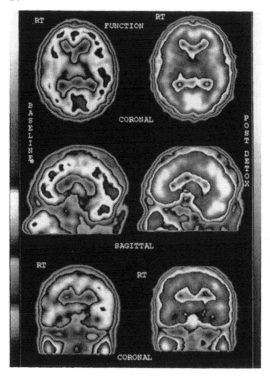

Figure 26.21. (Continued).

aromatic hydrocarbons such as benzene, toluene, xylene, and trimethyl benzene; aliphatics such as pentane, dimethyl butane, cyclopentane, 2 and 3 methyl pentane, hexane, and heptane; and chlorinated aliphatics such as chloroform, 1,1 trichloroethane, trichloroethylene, tetrachloroethylene, and dichloromethane. Results of the study indicated a significant adverse relationship between cognitive neurobehavioral functioning and the number of, and the total parts per billion of, the solvents found in the blood. The symptoms most commonly exhibited included defects in short-term memory, problems with coordination and motor sequencing, somatosensory deficits, and cognitive dysfunction. These correlations were significant to the $p<0.01$ to 0.05 levels. They also correlated intracellular heavy metal and pesticide levels and found no correlation in brain function abnormalities. (See "Emotions" in this chapter.) Other studies on heavy metals at the EHC-Dallas do at times show neurological change. However, other brain function changes were shown in patients exposed to pesticides (see later in this chapter, "Principles Involved in Evaluating Pollutant Caused Nerve and Brain Dysfunction").

Chang[168] demonstrated chronic neurotoxic effects of *n*-hexane on the CNS, including the cerebrum, brainstem, and the spinal cord. We have seen similar problems in a subset of our chemically sensitive patients with not only n-hexane, but also we have noted 2 and 3 methyl pentane and cyclopentane in their blood.[169]

Other studies have also noted nervous system effects of chemicals. Anger et al.,[170] for example, observed that even the low levels of methyl bromide found in fumigation today may produce slight neurotoxic effects. Also, Ferraz et al.[171] determined that chronic exposure to the fungicide maneb may produce symptoms and signs of CNS manganese intoxication.

Obeso et al.[172] identified several toxic chemicals, including bismuth, methyl bromide, cooking oil containing anilines, and tetraethyl lead that produce encephalopathies that, in many cases, resemble the action myoclonus syndrome of posthypoxic encephalopathy. We have seen over 20 patients with myoclonis syndromes and chemical and electrical sensitivity. These conditions have always had multiple triggers by double-blind inhaled challenges, and often they respond to a program of massive avoidance, intradermal neutralization, and nutrient therapy (see Chapter 23). Often the metals need to be neutralized by intradermal injection therapy in these cases, and, when this is accomplished, the patients show dramatic improvement.

McCunney[173] saw three men with relatively unusual manifestations secondary to exposure to trichloroethylene in a degreasing operation in the jewelry industry. Also, toxic encephalopathy, hepatitis, and carpal spasm have been linked to TCE exposure in young, healthy workers. Peters et al.[174] observed three groups of pesticide-exposed workers from three different work facilities who experienced chronic central and peripheral nervous system dysfunction that appeared to be exposure related.

Dager et al.[175] described three cases of idiosyncratic response to occupational solvent exposure, including symptoms characteristic of panic disorder (DSM-III). They discussed the specific treatment and prognostic implications of this panic-like reaction to solvents, and they proposed that sodium lactate infusion be used as an objective test to aid in the diagnosis. This substance provoked the panic attack. We have realized that often these panic attacks are due to a local magnesium deficiency, and we have been able to stop them with intravenous magnesium administration. Working with Finnish rayon workers, Hanninen et al.[176] were clearly able to differentiate between groups on the basis of neuropsychological test measures and clinical manifestations dependent on the degree of exposure to CS_2.

Chloroform and trichloroethylene (TCE), which were seen in the blood of nearly 50% of chemically sensitive individuals in the late 1980s, are two highly volatile neurotoxic chemicals that have been identified as contaminants in tap water. They often are triggering agents when a controlled challenge is performed. The health threat from these toxic chemicals can be from drinking or bathing in water that contains them or from inhaling the air they pollute. They are believed to be carcinogenic and perhaps viral triggers. Chloroform is found in the blood of 37% of our patients, and trichloroethylene is found in the blood of 8%. Also, our studies show considerable levels of other toxic chemicals such as tetrachloroethylene (80%) and trichloroethane (49%). Some studies have confirmed that, at 43°C, shower water releases into the air about 50% of the dissolved chloroform and 80% of the dissolved TCE[177] that it contains. Showers are twice as bad as baths in this regard because the water droplets have a larger surface-to-volume ratio than steaming bath water so that more of the volatiles can vaporize from the shower.

Many environmentally sensitive patients complain of weakness and fatigue after being either in the shower or bath. Both fatigue and extreme weakness have been reproduced consistently in 1500 patients by double-blind inhaled challenge of <0.33 ppm of chlorine in the deadapted state in the controlled environment of the EHC-Dallas over the past 12 years. These patients always complain of being fuzzy-headed and not being able to remember well; apparently, they have developed early anesthetic effects. Patients sensitive to water pollutants have a similar experience (see Volume II, Chapter 7[178] for the challenge studies). Air pollutants are produced from laundry and dishwashers.[177]

Another study in the Los Angeles Basin focused on air pollutants in commuting vehicles.[179] When compared with average outdoor concentrations, the mean "inside auto" level of benzene was found to be 24 times higher than outdoors; lead was 4 times higher, and nickel, manganese, and chromium were each 3 times higher.[179] It should be noted that these levels are still lower than the permissible occupational-exposure limit for an individual pollutant as set by the federal government for an 8-hour period. However, this threshold value is meaningless in individuals who are chemically sensitive. Excess as well as depleted minerals have been seen in many chemically sensitive patients at the EHC-Dallas. (See Volume I, Chapter 6[52].)

The probability that chronic multiple agent "low-level" exposures may bioaccumulate beyond individual safe threshold limits appears not to have been considered until recently. Price et al.[180] and Hinton et al.[181] at the Robens Institute of Occupational Safety and Industrial Hygiene at the University of Surrey have shown additive effects of chlorinated phenols and phthalates in animals. Also, further evidence of such an effect is beginning to accumulate from data gathered on chemically sensitive patients at the EHC-Dallas, where 40% or more have benzene compounds in their blood. Though they may react to each chemical individually upon challenge, patients seem to do worse when exposed to several at one time. At any rate, the exposure to toxic chemicals seems to be pandemic and without general limit at this time. It also appears that exposed workers are most likely to be used as a major part of an early warning system for poisoning and sensitivities from chemicals and/or heavy metals exposure, but, unfortunately, when they have symptoms, these are not heeded until it is too late.

Laboratory animals became restless, ataxic, and dyspneic following subcutaneous or intracerebral injection of a solution of dimethyl sulfoxide (DMSO). Also, these same responses were seen in laboratory animals following implantation of toxins absorbed on talc into various regions of the diencephalon and brainstem. These early changes were followed by depression and immobility. Prior to death, tachypnea and/or convulsions developed. The following case is an example of exposure resulting in advanced brain dysfunction.

Case study. This 45-year-old white female became ill with brain dysfunction, spontaneous bruising, petechiae, and edema. After prolonged exposure to toxic chemicals, her condition worsened. She developed extreme and debilitating deficits in short-term and immediate memory. She became extraordinarily distractible, frequently initiating a task but then forgetting to complete it if she was drawn to a subsidiary task. Her memory of recent events became so feeble that if she subsequently came in contact with her original task, she had no recall of having started the work. For example, she was cleaning out and rearranging the dresser drawers in her bedroom when she was interrupted by the doorbell. She forgot to return to this chore, and when sometime later she entered her bedroom for another purpose, she became frightened at the disarray and thought that an intruder had gone through her things. She contacted her husband immediately, related her alarm to him and thought that perhaps the police should be called. In another instance, she reported burning foods often because she forgot she had started cooking. Also, she forgot she was cooking and turned off the oven. When she was later ready to eat, the food was not cooked. In other episodes, she let the bathtub run over, forgetting she had started the water running. She made the bed without sheets, forgetting to put them on. She went to the grocery store, forgot why she was there, and returned home without buying anything, and she forgot to mail her bills.

This patient's physical examination revealed she suffered from spontaneous bruising, edema, petechiae, and cold extremities. Results of her laboratory tests showed high glutathione peroxidase 8.65 U/NADPH (C = 4.23 to 7.23 U/NADPH) and low superoxide dismutase 4.3 U/mg hemoglobin (C = 12 to 15.4 U/mg hemoglobin). Absolute T cells were 71% (1735/mm^3); T helper (T$_4$) were 49% (total count = 1198/mm^3); suppressor (T$_8$) cells were 20% (total count = 489/mm^3); B lymphocyte were 7% (total count = 171/mm^3). T cell function as measured by cell-mediated immunity skin tests revealed negative reactions to tetanus, diphtheria, streptococcus, tuberculin, trichophyton, candida, and proteus. This patient also proved to be sensitive to a multitude of biological inhalants and foods. After a minimum of 4 days deadaptation with the total load decreased in the ECU, double-blind inhaled challenges with toxic chemicals showed that phenol (<0.002 ppm), formaldehyde (<0.20 ppm), petroleum (ethyl) alcohol (<0.50 ppm), and car exhaust triggered symptoms including spontaneous bruising and brain dysfunction similar to her admitting symptoms (Table 26.23).

This patient was placed in the ECU, and, after several days deadaptation with total load reduction that included fasting, her signs and symptoms gradually cleared and her brain function returned.

EVALUATION OF NERVE AND BRAIN DYSFUNCTION RESULTING FROM POLLUTANT INJURY

Evaluation of brain and nerve dysfunction in order to prevent fixed, irreversible end-organ failure must be performed early after exposure. The exposure(s) must be removed because of the adaptation phenomenon to reduce the total load. The challenge must be performed in the deadapted stated under

Table 26.23. 45-Year-Old White Female
Inhaled Double-Blind Challenge after a
Minimum of 4 Days Deadaptation with the
Total Load Decreased in the ECU

Chemical	Dose (PPM)	Results
Ethanol	<0.50	+
Formaldehyde	<0.20	+
Phenol	<0.002	+
Car exhaust	ambient	+
Pesticide (2,4-DNP)	<0.0034	+
Chlorine	<0.33	−
Saline	ambient	−
Saline	ambient	−
Saline	ambient	−

Note: +: brain dysfunction by objective measurement
signs (petechiae, spontaneous bruising, edema)
−: no reaction.

Source: EHC-Dallas. 1993.

controlled conditions to assure a proper evaluation. Each will be reviewed in the following pages.

Principles Involved in Evaluating Pollutant-Caused Nerve and Brain Dysfunction

Because the adverse reactions to toxic chemicals in the environment involve multiple systems with multiple behavioral effects, it is difficult to determine the "safe" or lowest level of exposure that will not occasion unhealthy behavioral and functional consequences. For the physician, this difficulty is complicated by biochemical individuality of the chemically sensitive patient, the physician's perception of the adaptation phenomenon and redundancy of function, and his or her perception and precise evaluation of the total load at the time of exposure (see Volume I, Chapter 3[7]). Though explained in depth in Volume I, Chapter 3,[7] each of these principles as they apply to specific neurological dysfunction will be discussed separately here.

Biochemical Individuality

Reactions, susceptibility, and responses following chemical exposure are highly individual, so no single effect will necessarily occur in everyone who is exposed, and regardless of the level of exposure, the severity, duration, frequency, or even presence or absence of a given behavioral response, will vary. Also, the extent to which an individual is functioning under all other life stresses (total body stress load), the degree of understanding and acceptance by the examiner of the possibility of environmental, chemically induced behavioral changes, and the ability of the measuring instruments to make fine discriminations in determining "safe" threshold level will affect the evaluation for the chemically sensitive patient. The more sensitive measurement of subtle behavioral change is, the more likely a lower level of "safe" threshold exposures will be identified. (For more information, see Volume I, Chapter 3.[7])

Adaptation

The nervous system possesses considerable redundancy of structure; therefore, adaptation occurs easily and often is the downfall in the chemically sensitive individual, in whom toxicant triggers are not found until late in the disease process. This process of adaptation may mask brain damage until the reserve capacity of the system is exceeded; then permanent, irreversible damage occurs. The physician must never lose sight of this action when evaluating a patient with potential pollutant-induced nervous system malfunction (see Volume I, Chapter 3[7]). The onus is upon the physician to prevent permanent brain damage. Because of this adaptation phenomenon, function is able to return to normal during continuous exposure to toxic substances, if there is enough reserve available for the brain to perform at normal capacity. When clinical symptoms indicate the presence of pollutant damage or when one wants to find them, the patient will need to be evaluated in the nonadapted state with his or her total pollutant load reduced. This evaluation process requires not only that the patient be removed from the prime suspected chemical pollutant but also that he or she be kept from all its relatives and other potential secondary triggers as well (see Volume IV, Chapter 31). In other words, the total body load must be reduced before accurate assessment can be completed. For precise diagnosis, it is crucial that the patient avoid exposure for at least 4 days. Total load reduction is necessary not only because the less the exposure, the less the chance of permanent damage, but also because clearly defined cause and effect relationships can, upon challenge, be demonstrated when the patient is in this state (see Volume I, Chapter 3[7]). At the specific time of pollutant overload, a minimum of 4 days of avoidance is usually adequate to allow the brain to deadapt. As shown in our ECU studies at the EHC-Dallas, however, several additional days are sometimes needed and, in a few cases of severe damage, several months before deadaptation is completed. Evaluation of function before and after total body load is reduced allows for a comparison that will show how much reserve is left and what the potential for recovery is. Often, timed stressed challenge must be done to put enough stress on the individual to bring out the entirety of the defect (Figure 26.22).

Occasionally, when a functional deficit is not detected, the nervous system may sustain damage that can be detected only by cytologic or neurochemical methods. This state is extremely treacherous because permanent damage may

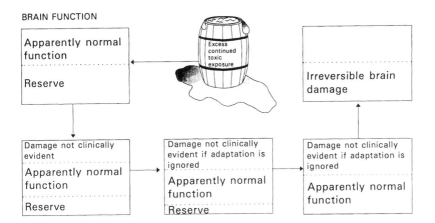

Figure 26.22.   Reduction of brain function with repeated pollutant exposure in the chemically sensitive individual. Adaptation — a way to end-stage, irreversible brain damage.

occur before the individual ever realizes the toxic chemicals to which he or she is exposed are harmful.

The Perception and Precise Evaluation of Total Load
at the Time of Exposure

All efforts should be made to determine subtle disturbance in brain function, since the CNS differs from organs such as the liver in that neurons cannot divide once they have reached a mature state (in contrast to which hepatic cells spend a lifetime regenerating). At the EHC-Dallas, we have been able to show objective changes in our patients' behavior before and after their body's pollutant load has been reduced and the toxic chemicals to which they had been exposed have been withdrawn. Once deadaptation has been achieved, changes often are accentuated and reproduced after an inhaled, oral, or intradermal challenge. In addition to routine pollutant challenge, some patients need to have their responses measured after they are stressed by time or speed constraints in order to bring out hidden, difficult-to-define defects (Figure 26.23).

After challenge, three types of brain responses occur. One is no reaction; the second type is an observable reaction to the pollutant challenge after a reasonable period of time (30 minutes to 3 hours) has elapsed; the third type of response is an observable reaction that occurs only after the patient, following pollutant challenge, is stressed by time constraints (e.g., the same tasks are performed in 5 minutes instead of 30).

After challenge, changes gradually return to normal as the reaction clears. Then another challenge can be performed. Stressed measurements are often necessary because the time for the standard test is insufficient to elicit a

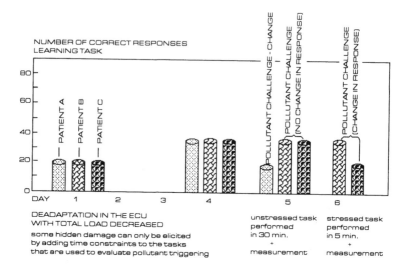

NUMBER OF CORRECT RESPONSES
LEARNING TASK

Figure 26.23.   Latent nervous system damage as measured by timed stressed task performance in addition to pollutant challenge. Some hidden damage can only be elicited by adding time constraints to the tasks that are used to evaluate pollutant triggering. (From EHC-Dallas. 1984.)

response. For example, a focal loss of a small number of granules of cells in the cerebellum from a dose of methyl mercury will not be detected as locomotor changes or even as changes in fine movements. Stressing the individual by forcing both time constraint and energy output, however, will often elicit symptoms and responses of this defect that can be objectively measured. The loss of granule cells in this example represents a reduction in the reserve capacity of the brain, which might well be detectable if additional demands are placed on the system but will not be on normal nonstressed challenge.

It should be emphasized that some individuals' total tolerance for stress is so high that their performance is minimally affected after an ambient pollutant challenge. However, the average individual carrying a similar load could not perform adequately. It is evident that the control groups, which are made up of average, not normal, individuals, may not accurately reflect normal or optimal brain function. Therefore, each individual should act as his or her own control, and measurements should be repeated several times after these patients have achieved their basal deadapted state (see "Challenge Testing" at the end of this chapter).

Evaluation of the Pollutant Exposed Patient

Work-up of the neurologically involved chemically sensitive patient is multifaceted and should be performed in a multistep fashion. Initial history and physical exam is the first step. Laboratory testing is the second. Since the history and physical exam are discussed in depth in Chapter 29, this chapter focuses on the second step of the evaluation process — laboratory work-up. This discussion examines the diagnostic usefulness of CAT and MRI scans, EEGs, nerve biopsies, objective sensory-motor neuropsychological tests, integrative functions, challenge testing, and laboratory and blood analyses.

CAT scans and/or MRI scans can be useful in evaluating the neurologically involved chemically sensitive patient. These should be performed when indicated. Generally, they are minimally helpful in patients with early damage because they show gross anatomical abnormalities that mark advanced development rather than the subtle changes that more frequently occur with initial onset of chemical sensitivity. However, we have seen many patients in the later stages of chemical sensitivity who had abnormal CAT or MRI brain scans, suggesting small areas of infarction. Other patients' scans have shown areas of demyelination and brain atrophy. Some chemically sensitive patients in later stages of their illness have been observed with early cerebral atrophy, which may be of some significance in identification of chemical sensitivity with neurological involvement. However, usually reversibility is too late at this stage. In our series, one young housewife, age 41 years, developed three cerebral infarcts (identified on MRI) after stripping her floor with xylene. Her symptoms were reproduced with double-blind inhaled challenge of xylene. This reaction emphasizes the quickness with which end-stage disease occurs after only inadvertent ambient exposure.

Though early in its use, the triple camera single emission computerized tomography (SPECT) brain scan appears to be efficacious and a much more precise way of defining cerebral neurotoxicity. It can be used at any time during the disease process. This modification of an old modality that can measure flow and function due to its three-dimensional views of small areas of the brain (similar to, but better than, a PET scan) has shown abnormalities in a subset of environmentally sensitive patients with brain dysfunction.

The type of SPECT scan used in diagnosing chemical sensitivity involving neurotoxicity was perfected by Simon and Hickey,[182] working with physicians at the EHC-Dallas. We performed scans on a consecutive series of 400 highly selected chemically sensitive patients who had neurological involvement with symptoms of recent short-term memory loss, episodic confusion, poor concentration, dizziness or imbalance, and an inability to stand on their toes with their eyes closed. These are the patients who will most likely benefit from this diagnostic procedure, and of these 390 neurologically involved, chemically sensitive patients had a neurotoxic pattern and 10 had a chronic fatigue pattern. These patients were compared to 50 patients with other pathologic patterns including depression, schizophrenia, etc. They were also compared to two other groups, one of which was composed of 50 patients with normal patterns and another group of 25 volunteers who were selected because they were totally well and showed normal patterns (Figure 26.24). A clear observable difference was seen in the neurologically involved chemically sensitive patient

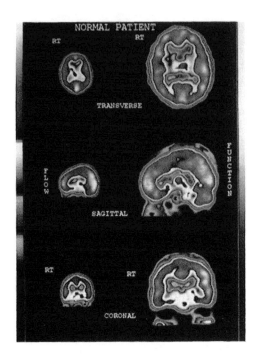

Figure 26.24.   Normal SPECT scan. (From EHC-Dallas. 1991.)

(who showed a discrepancy between flow and function, salt and pepper pattern in the function phase and temporal lobe asymmetry) and controls (who showed uniform uptake and function as well as uniform flow when compared with function).

Similar scan studies by Heuser et al.,[183a] using Xc133 by inhalation, recently confirmed neurotoxicity. In this study, 72 patients (21 males and 51 females) were exposed to pesticides and solvents. These patients were compared with age-matched controls, a group of 36 chronic fatigue patients, and 26 depression patients. Cerebral blood flow was diminished in the young and elderly groups, 45 (7 patients) and 40 (7 patients) mL/min/100g, $p<0.02$, for both groups (Table 26.24). These researchers concluded that patients exposed to chemicals present with diminished cerebral blood flow, worse in the right hemisphere, with random areas of hypoperfusion more prevalent in the dorsal frontal and parietal lobes, and with increased temporal lobe HMPAO uptake. From observations of patients with chronic fatigue and depression that may be the result of a cortical effect due to a vasculitis process, they, as well as we, conclude that these findings are significant.

Clearly, these scan studies and the brain function tests of Butler showed there is a large subgroup of chemically sensitive individuals who have neurotoxic patterns, either by SPECT scans or by their psychological profile. In addition, as shown in Figure 26.25, chemically sensitive patients were found

Table. 26.24. Cerebral Blood Flow in Patients Exposed to
Neurotoxic Chemicals

HMPAO and uptake	Neurotoxics		CFS elderly	Dpr. elderly
	Young	Elderly		
R. orbital frontal			0.59[b]	0.55[b]
R. dorsal frontal	0.63[b]	0.62[a]	0.62[b]	
R. temporal	0.61[b]		0.60[b]	0.64[a]
R. parietal	0.59[b]	0.63[b]		

[a] $p<0.50$.
[b] $p<0.001$.

Source: Heuser, G., I. Mena, C. Thomas, and F. Alamos. 1994. June 5–8.
Presented on the Society of Nuclear Medicine 41st Annual Meeting. Orange
County Convention Center, Orlando, FL. With permission.

to have enhanced neurotoxicity pattern after inhaled double-blind challenges
with formaldehyde (<0.2 ppm), petroleum derived ethanol (<0.50 ppm) or
ethylene glycol, pesticide (2,4-DNP <0.0034 ppm), chlorine (<0.33 ppm),
phenol (<0.002 ppm), and 1,1,1-trichloroethane. In addition, these patients'
SPECT scans worsened on challenge with these various substances such as
toluene. These patterns were similar in the chronically chemically sensitive
patients, the ones with breast implant sensitivity, the soldiers from Desert
Storm, and the patients exposed to the Alaskan oil spill during its clean-up. The
SPECT scan showed patient improvement during environmental control as
well as with treatment in the heat depuration/physical therapy plus environmental control. Figure 26.26 shows the brain scan clearing in a soldier involved
in Operation Desert Storm in Kuwait.

EEGs. The usefulness of EEGs in diagnosing the chemically sensitive patients who have neurological involvement following toxic exposure is currently being evaluated. Krusz,[183b] working with EHC-Dallas physicians, has
shown that multi-channel EEGs seem to be good for locating smaller areas of
damage. They have been useful in defining CNS effects in patients with
advanced chemical sensitivity who also have brain dysfunction. So far, though,
they do not appear to pick up every abnormality, although they can pinpoint
areas of higher electrical activity. The results of multi-channel EEGs and
SPECT scans appear to correlate well. Thus, multichannel EEGs may be a
reliable alternative if SPECT scans are unavailable.

Because they suggest isolated or diffuse areas of dysfunction, multi-channel
EEGs are certainly helpful in diagnosing some patients' illness due to chemical
sensitivity. However, serial, nonmulti-channel EEGs appear to have limited
success in diagnosing some patients as demonstrated by the following study.

Forty workers exposed to significant levels of benzene underwent serial
non-multi-channel electroencephalograms. The findings of these workers' EEGs
were then compared with the findings of control groups of 48 healthy persons,

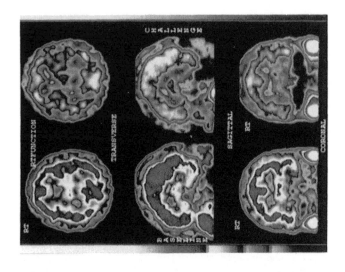

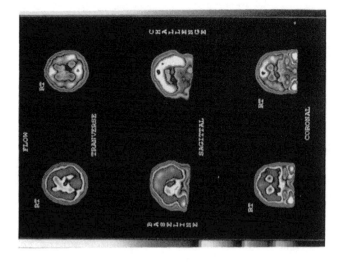

Figure 26.25.   Challenge SPECT brain scan of (L) A and B prechallenge and (R) A and B

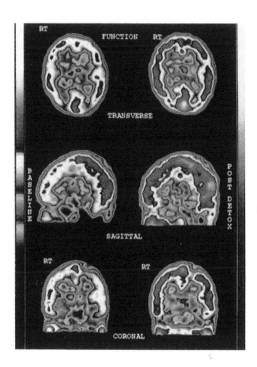

Figure 26.26.   SPECT brain scan improvements after environmental control therapy.

a group of 110 workers exposed to significant levels of toluene and xylene, and a group of 236 workers exposed to vinyl chloride. The individuals exposed to benzene exhibited almost 25% of the abnormal EEGs and 45% of threshold findings. The abnormalities were episodic, diffuse, or a combination of the two. Toluene and vinyl chloride workers did not show any change on this particular test, yet many had clinical brain dysfunction. Hence, the diagnostic limitations of nonmulti-channel EEGs was demonstrated. The rapid onset of deep sleep in 30% of the cases is considered typical for benzene exposure. This particular study identified different EEG features characteristic of the neurotoxic action of said various types of organic solvents and, in so doing, made possible a more efficient diagnostic evaluation of the effects of these chemicals on the CNS.[184]

Studies on chemically sensitive patients at the EHC-Dallas in the late 1980s revealed that 56% had xylene in their blood; 10% had benzene; and 63% had toluene. The majority of these patients had objective CNS dysfunction, including the sleep problems characterized especially by deep as well as troubled sleep. Only 50% had abnormal EEGs, but it should be noted that most did not undergo multi-channel EEGs. All did have brain dysfunction as identified by the Harrel-Butler questionnaire. The blood levels seen in these patients ranged from 0.1 ppb to 200 ppb. We have seen patients who have trouble falling asleep, staying awake, remaining asleep (awaking early from nightmares or night terrors), and waking in the morning. Monro,[185] Randolph,[186] and Little[187] have observed similar patterns.

Nerve Biopsy. To identify chemically sensitive patients with neurological involvement, peripheral nerve and brain biopsy have been performed by some centers, although we have tried to avoid the brain biopsies at the EHC-Dallas. Since damage to either area that is apparent under microscopic examination usually occurs too late to restore proper function, our energy is directed toward identifying and treating patients before permanent changes occur (see textbooks on neuropathology). Volume II, Chapter 12[27] presents a case report of an individual whose neuropathy vanished following completion of an ECU program, even though he had significant changes in his biopsy.

Objective Sensory-Motor Neuropsychological Tests. Objective sensory-motor neuropsychological tests can be useful in diagnosing chemically sensitive patients with neurological involvement. Working with Finnish rayon workers, Hanninen et al.[188] and Tolonen and Hanninen[189] were clearly able to differentiate between groups of workers on the basis of neuropsychological test measures and clinical manifestations dependent on the degree of exposure to CS_2. At the EHC-Dallas, Butler,[190] who has extensive experience in evaluating chemically sensitive patients, has shown this differentiation in patients with pesticides and solvents in their blood. Following examination of more than 4000 individuals, Butler and coworkers have developed criteria that appear unique to those individuals who have chemical sensitivity with neurological involvement. These criteria are useful for studying behavior and emotions and are discussed in these respective sections later in this chapter.

In a two-part study involving the Finnish rayon workers, Hanninen et al.[188] and Tolonen and Hanninen[189] sought to identify a test battery that would be more accurate and less cumbersome than that previously used to diagnose CS_2 intoxication. The result of this effort was a clear delineation of five variables that could accurately demonstrate the brain's deterioration that is causally related to CS_2 poisoning. As defined by Hanninen et al.,[188] these variables include the following: sensorimotor speed, manual dexterity, the tremor of hands and other inabilities to control hand movements, subjects' readiness to respond to emotional stimuli, and originality.

The EHC-Dallas behavioral medicine group has used similar, but perhaps more complete, analyses of brain, neural, and behavioral dysfunction. Butler's group,[191] using psychological and neurological screening techniques developed at the EHC-Dallas, uses a two-part assessment: observation and measurement. This procedure is usually performed before and after environmental control and will be discussed in detail.

Observation.   Initially, a clinical interview is conducted to determine a patient's mental status. During this interview, a general health history from the patient along with a family medical history is secured. Also, the patient describes his or her specific symptoms — their duration, frequency, and intensity. Further questioning and investigation of the possibility of chemical exposure is done.

Occupational history is noted, particularly the patient's job stability and satisfaction. The patient provides evaluation of his or her work performance both before and after he or she became ill. Often, chemically sensitive individuals realize that their performance is deteriorating. However, occasionally, someone else has to point out the decline due to the chemically sensitive individual's loss of mental acuity and perception. Along with occupational history, personal relationships are assessed. Specifically, the stability and longevity of these relationships are evaluated. The patient's relationship to spouse, family, and friends and whether these significant people are supportive or antagonistic or neutral toward the patient and his or her dysfunction are assessed. The patient's marital satisfaction is evaluated.

Descriptive and behavioral observations of the patient that are pertinent to the determination of the mental status are recorded. The patient's physical appearance, such factors as neatness, appropriateness of dress, posture, proportionate height/weight, etc., is noted. Also monitored are the patient's individual mannerisms — excessive or decreased motor activity, wiggling, balance, gait, stance, unusual ticks, etc. Speech, too, is checked for quality, normalcy, articulation, logical progression, etc. A patient's orientation to time, place, person, and his or her attitude (alertness, positiveness, bewilderment, etc.) are further attended to.

Notes of any unusual fluctuations in the affective sphere of the patient are kept. Both mood (euphoria, depression, anger) and affect (behavioral — restrictive, flat, broad, blunted) are monitored. Butler checks his patients' mental productions (spontaneous, blocked, tangential, etc.) and mental state (delusions, obsessions, memory deficits, etc.). Also, opinions of gross perceptual abnormalities (hallucinations, illusions, distortions) are recorded. A patient's level of insight is also considered.

Sherck-O'Connor[192] found that the major contributing factors to total load stress for environmentally sensitive patients at EHC-Dallas were as follows: environmental variables such as chronic high or low dose exposures to chemicals and adverse reactions to airborne inhalants; health factors indicated by chronic recurrent health problems, undue fatigue and probable increase in sympathetic arousal; self-perception stressors including feelings of being slighted, discontentment, and negative thinking; psychological-cognitive variables such as lower intellectual functioning, impaired reasoning, feelings of depression, and loosened emotional controls; interpersonal stressors including strained relationships and perceived lack of support from others; and behavioral stressors such as poor time management, lowered mediation of internal conflicts, and an externalization of blame to others (Table 26.25).

Table. 26.25.   Total Stress Load (TL) for the
Chemically Sensitive Individual[a]

Total stress load inventory mean value analysis			
Group: control (100)	**Inventory**	**Score**	**S.D.**
Self perception (SP)	TLSP	26.31	7.99
Psych-cognitive (PC)	TLPC	19.10	8.20
Behavior (B)	TLB	23.43	8.43
Interpersonal (I)	TLI	22.42	8.92
Health (H)	TLH	21.95	7.43
Nutrition (N)	TLN	28.24	8.00
Environment (E)	TLE	25.81	9.12
External/situational (ES)	TLES	36.53	9.69
Group: chemically sensitive patients (100)	**Inventory**	**Score**	**S.D.**
Self-perception	TLSP	35.53	10.72
Psych-cognitive	TLPC	36.38	11.07
Behavior	TLB	30.08	11.92
Interpersonal	TLI	27.82	12.50
Health	TLH	33.08	7.64
Nutrition	TLN	28.59	9.07
Environment	TLE	36.45	8.57
External/situational	TLES	32.29	9.28

[a]   Major contributing factors include: high or low dose toxic exposure; adverse reaction to airborne inhalants; chronic recurrent health problems, fatigue; increases in sympathetic arousal; self-perception stressor — feeling slighted, discontented, and negative thinking; psychological and cognitive variables — lowered intellectual function, impaired reasoning, depression, loosened emotional controls; interpersonal stressors — strained relationships and lack of support from others; behavioral stressors — poor time management, lowered mediation of internal conflicts, externalization of blame.

Source: Sherck-O'Connor, R. 1986. Total stress load inventory: A validation study. Ph.D. thesis. University of North Texas, Dept. of Behavioral Medicine.

Using the checklist of major symptoms, Butler[190] found that the following complaints occurred 40% of the time in the chemically sensitive patient: overwhelming exhaustion, weariness and easy fatiguability, difficulty in getting started in the morning, difficulty concentrating on work or study, inferior present performance in contrast to prior performance or level of functioning, feelings of not being oneself or not being in control of what is happening to

Table 26.26. Changes in Behavior from Pollutant Damage

Sensory
Motor
Integrative functions
  Symbol formation
  Sensory-motor
  Perceptual-motor
  Emotional stress

oneself, frequent headaches, feelings of losing control of one's destiny, and poor comprehension, though not poor memory. On the physical checklist, headaches, aches and pains, sinus discomfort, clumsiness, easy fatigue, low energy, weakness, restlessness, and mucus and eye problems occurred 40% of the time in the chemically sensitive patients.

Measurement Procedures. The alterations in behavior that can occur as a consequence of toxic damage to the nervous system are arbitrarily divided into several categories based on their functions. This arbitrary classification aids the organization of evaluation in measurements (Table 26.26). These alterations include the following: sensory, motor, and the integrative functions of symbol formation, sensory/motor, perceptual/motor, and emotional states associated with any of the other four categories. Each will be discussed separately. Not all objective observations are considered neuropsychological, but they are included here for completeness.

Sensory Responses. Sensory responses such as startle reflexes, sight, hearing, balance, smell, temperature, touch, feel and pain, vision, smell, and others result from early pollutant stimulation of sensors (Table 26.27). This damage occurs via the autonomic, peripheral, and central nervous system, deregulation through the neurovascular system, or it can occur with increasing toxic damage from central demyelination and peripheral neuropathies. These conditions may fluctuate for years before permanent changes are seen. Recognition of this early stimulation gives physicians a great opportunity for prevention, since functions may be totally reversed with precise definition and elimination of the pollutant triggering agents and with restoration of their nutrient pools.

Table 26.27. Pollutant Damage to Sensory Appatatus

Startle	Smell
Sight	Skin temperature
Hearing	Touch, feel, pain
Balance	

A patient is initially evaluated as described previously. Then he or she is placed in the ECU or a controlled outpatient environment and usually fasted for 3 to 7 days, or placed on a monorotation diet, in order to reduce variables. High intravenous nutrients are given as indicated (see Volume IV, Chapter 37). At this point, the patient can be reevaluated by similar methods. Once the patient is in the basal state, he or she may be challenged with water contaminants, food, food contaminants, or inhaled chemicals (see Volume IV, Chapter 31). Measurements to substantiate reproduced signs can then be repeated after challenge. Though they are always integrated, each evaluation will be discussed separately.

Startle Reflexes. An interesting possibility for evaluation of sensory thresholds in humans as well as other animals is use of the startle response and comparable reflexes that involve the sensory and motor systems.[43] Startle reflexes may be accelerated or slowed in the subgroup of chemically sensitive patients depending on whether the system is stimulated or suppressed.

Vision. Eye changes may be one of the earliest signs of pollutant damage to the nervous system. Total function evaluation including fundoscopy, location of changes in function in the eye, the assessment of visual fields, and the integration of optic nerve function with the sight centers is necessary in some cases of early pollutant damage (see Chapter 27).

In the pollutant damaged individual, dilated veins, spastic arteries and no papilledema are usually seen. Changes in chamber pressure and pupillography may show early signs of pollutant damage. In some patients, chamber pressure will be above 30 mmHg, and, with pollutant exposure, this figure may fluctuate. In some cases, the iris corder demonstrates early damage to the ANS (see Chapter 27). Visual fields may be altered.

Neuronal atrophy in the cortical layers in specific sensory receiving areas can cause loss of sensation such as cortical blindness, even though the remainder of the sensory pathway is intact. These changes can be found early by using the iris corder, and removal of the incitant usually will prevent cortical blindness. Using the iris corder, Ukai, Higashi, and Ishikawa[193] showed visual changes due to exposure to organochlorine and organophosphate pesticides (Figure 26.27). They noted reduced vision, concentric narrowing of the visual fields, with or without central scotoma. They also noted reduction of red sensitivity at the fovea and anomaly of refraction (most of the cases are myopic). The vision of the chemically sensitive patient is difficult to correct with glasses or contact lenses. Rarely is there normal refraction in adult cases. An optic-autonomic peripheral and/or central neuropathy often is present in patients presenting with vision involvement. A severely affected patient has an optic atrophy and/or a pigmentary degeneration of the retina without contraction of the ciliary muscle. Abnormality of the ANS is the most significant initial symptom (see Chapter 27). In our studies at the EHC-Dallas, we have confirmed numerous autonomic changes in chemically sensitive patients with

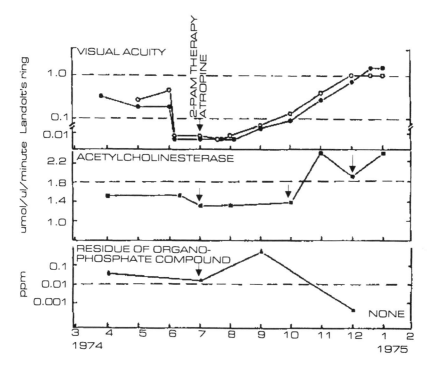

Figure 26.27. Changes in visual acuity, acetylcholinesterase, residue of organophosphate compound before and after 2-PAM therapy with atropine are given. Marked improvement of vision with the elevation of acetylcholinesterase as well as reduction of the residue of organophosphate pesticides are noted. (From Ukai, K., J. T. Higashi, and S. Ishikawa. 1980. Edge-light pupil oxcillation of optic neuritis. *Neuro-ophthalmology.* 1(1):33–43. With permission.)

pesticides, solvents, and other toxic chemicals in their blood (see Chapter 27). The most common abnormalities are cholinergic and sympatholytic followed by cholinolytic and lastly sympathetic. However, it should be emphasized that the iris corder is not efficient in measuring the generalized mass response of sympathetic discharge.

Auditory-Vestibular. Both nerve and bone hearing loss may occur and be measurable by audiometry. Tympanography may show some changes caused by pollutant damage. Ödkvist's revolving chair with recording of ear and eye movement (nystagmus) is helpful in evaluating inner ear function in some chemically sensitive patients.[194] Ödkvist et al.[197] evaluated the inner ear and sometimes parts of the brain with his techniques (see Chapters 18 and 27). Caloric tests may also be of some help.

For evaluation of balance, some physicians such as Martinez are performing posturography, which involves computerized stress balance studies that
can be helpful in generally localizing the areas of pollutant damage either in the
inner ear, the spinal cord, or the brain[196] (see Chapter 18). Posturography is
performed in patients who cannot stand on their toes with their arms out-
stretched and eyes closed. If a patient cannot stand, he usually has neurological
damage that is often related to an increase in pollutant load. Ödkvist[194] has also
suggested that there is a cerebellar inhibition of the vestibular oculomotor
system induced by solvents, and, indeed, this is often the case. We have used
these types of balance tests for patients with toxic damage involving the inner
ear, spinal cord, or CNS.

Examples of sensory change from pollutant damage are lead encephalopathy
and severe poisoning in children from organic mercury compounds, which may
result in marked sensory damage including blindness and hearing loss.[43] Some
of the most sensitive methods of evaluating audio-vestibular damage use
operant conditioning techniques, which require discrimination of light intensity, tone intensity, or pitch. Earlier, less severe changes may be evaluated
using more sensitive techniques that require discrimination of tone, intensity,
and pitch. Examples of the use of these methods can be found in work by
Stebbins and co-workers[197] on damage to the cochlea of monkeys from
dihydrostreptomycin and in the work of Luschei and co-workers[198] on peripheral vision in monkeys intoxicated with methyl mercury. Generally, these
methods are time consuming and difficult to carry out successfully except in
the hands of experienced investigators.

Ödkvist et al.,[194] our group, and several other investigators[199,200] have
shown neurotoxicity to result from exposure to numerous chemicals including
styrene, xylene, toluene, etc. These investigators have used audiometry and
balance studies to evaluate fine changes. Body imbalance and sway seem to be
early signs of pollutant overload in the neurologically involved chemically
sensitive individual.

Segal and Duckert[201] report that when nitrogen mustard (mechlorethamine)
was administered in massive doses (0.6 to 1.5 mg/kg) to cancer patients,
severe, irreversible hearing loss occurred. There have been no reports of
ototoxicity with doses of 0.4 mg/kg or less. These authors described a 36-year-
old man who developed profound sensorineural hearing loss during his first
cycle of chemotherapy for Hodgkin's disease. Cyclophosphamide was substituted for mechlorethamine in subsequent cycles, and his hearing deficit resolved. This case is the first reported one of reversible ototoxicity associated
with currently recommended doses. However, we have seen many cases that
were reversed when the total load was decreased.

Odor Sensitivity.   Odor sensitivity is the prime change in some chemically
sensitive patients. They can smell toxic odors when normal people cannot. The
inhaled challenge, which often will show odor sensitivity, has been described
in Chapter 18, and Volume IV, Chapter 31. In our experience with thousands

of patients in the deadapted state with 3 to 4 days total load reduction who underwent inhaled double-blind challenges, the majority demonstrated objective supersensitivity to such common substances as sulfur fumes, car exhausts, perfumes, and insecticides. These patients could, in fact, at times accurately identify these substances, even when they were exposed to only minute levels. Often odor cannot be detected by the normal individual, but the chemically sensitive individual does react.

Although odor hallucination is extremely rare in chemically sensitive patients, their supersensitivity is often misdiagnosed as precisely that. If odor hallucination is present in the chemically sensitive patient, it usually indicates a space-occupying brain lesion or trauma. Amoore and Venstrom[202] characterized 39 clearly specified odor stimuli by estimating the degree of similarity with each of seven reference standards. These different odor stimuli have been recently extended to include 107 separate chemicals.[203] Whether those and other studies like them[204,205] have any relationship to the chemical problem at this stage is unclear, but highly likely. Decreased sense of smell can also be involved in chemical sensitivity. Here the senses have been damaged until little odor can be perceived.

Skin Temperature. Skin temperature changes can be measured by thermometers and thermography. These methods of assessing toxic damage can be highly informative if performed under rigidly controlled temperatures as well as other controlled environmental conditions.

The chemically sensitive patient usually has markedly decreased skin temperature ranging from the 68°F to low 90s. The physician can initially register this low temperature response by shaking the patient's hand and feeling how cold it is. Skin temperature measurements in our center before and after challenge have shown as much as a 20°F drop under controlled environmental temperature and pollutant conditions. Butler et al.[190] have shown the reduction of body temperature in a large series of chemically sensitive patients at the EHC-Dallas (Table 26.28). Thermograms will often show visually the change in body temperatures at various layers of the skin.

Touch. Sensory neuropathies are commonly accompanied by numbness, tingling, or hypersensitivity to touch.[206] Partial objectivity of these findings can be attained with pressure recording and sensory nerve plotting. Evoked sensory pathway responses can occasionally show change.

Heightened or decreased touch, feel, or pain may be seen in the subset of neurologically damaged chemically sensitive patients. These changes can be ascertained by touching the patient with cotton swabs or various synthetics, using pin prick, using a dolorimeter, or by writing numbers on fingers or hands and asking for interpretations (see Chapter 23). Neurometric responses can be measured objectively by some nerve conduction-like tests. Often these will show a decrease in conduction. Table 26.29 shows a typical case of decreased sensory response in a physician with severe chemical sensitivity. This patient

Table 26.28.   Skin Temperature Changes

300 chemically sensitive patients studied in the ECU
   Temperature range: 68–96°F
   Temperature mean: 84°F

Source: Butler, J. R. and W. J. Rea. 1968. Unpublished data.
EHC-Dallas.

was a practicing obstetrician/gynecologist who gradually developed sensory neuropathy in her hands and feet. She lost the feeling in her hands and had to stop surgery because she could not feel the sutures to tie them. We have measured a series of patients by neurometric analysis and found abnormalities in the nerves of both their hands and feet (Table 26.29). We also found varied sensory responses in various other nerves.

Motor Responses (Gait, Movement).   Tests of motor function are used conjointly to establish the integrity of the motor system since the operant test for sensory function relies on motor movements (see Chapter 18). In our experience, motor involvement is usually late as compared with sensory changes. However, some exposures, such as those from nitrous oxide, may cause an almost pure motor change (see Volume II, Chapter 11[73]). Either demyelinating processes or neuronal damage can produce motor dysfunction. Sloppy gaits and foot drops in some solvent overloaded chemically sensitive patients can be observed. However, these symptoms indicate the patient's disease processes have already developed so far as to disallow prevention of fixed damage. In addition to substances causing mixed-motor and sensory neuropathy, motor fiber demyelination alone is the primary peripheral damage produced by compounds such as isonicotinic hydrazide.[207] Nitrous oxide overload may cause isolated motor damage. The functional pathology seen with damage to motor nerve axons or terminals at the skeletal muscle is weakness or paralysis of the involved muscles. Again, even though motor conduction tests may help diagnosis in some patients, sensory nerve involvement may be in an earlier state of damage with nerve conduction showing nothing.

Clinical observation of the quality and quantity of motor activity including speed of motor activity (slow-accelerated) may be performed to ascertain a degree of nervous system involvement. Monitoring a patient's difficulty starting or stopping motion may also be helpful.

Direct functional tests of nerve function rely on physiological methods of recording conduction velocity or the compound action potential of peripheral nerves.[43] Nerve conduction tests can sometimes reveal a polyneuropathy or other motor changes.

Table 26.30 provides an example of a patient whose peripheral nerves were damaged by solvents. This patient is a 23-year-old white male who was exposed to xylenes and toluene at work. He developed chemical sensitivity and had to go on an environmental control program.

Table 26.29. Neurometer Test

Patient no.	Sex		Age		Side		Location		Current Perception Threshold Grade (H$_2$)
	F	M	Range	Average	Right	Left	Toe	Finger	
10	8	2	32-69 y	48 y	22	12	26	8	1-6.27

Nerve involvement					Condition	
Distal digital branch of the superficial peroneal	Sural	Distal digital branch of the peroneal	Distal digital branch of the median	Distal digital branch of the ulnea	Hyperthesia	Dysfunction
9	7	9	5	4	25	9

Source: FHC-Dallas. 1992.

**Table 26.30.   23-Year-Old White Male — EMG Test — Odor
Sensitivities to Multiple Substances — Fatigue, Numbness of Hands
and Feed, Intractable Nect Pain, and Fasciculations — Solvent
Exposure**

Muscles tested	O–IV	Normal	Moderately large	Large
Lower extremities — motor unit action potential polyphasics				
Adductor group	>0	/		
Quadriceps	−1 to >1	/	/	/
V lateralis				
Intermedius	>0			
Vastus medulis				
Anterior tibialis	−1 to >11		/	/
Peroneal	1 to >11		/	/
External hall, longus	−1 to >11		/	/
Lateral gastroc.	1 to >1		/	/
Med. gastgroc.	1 to >11		/	/
Hamstrings	−1 to >1			
Abductor Hall. Brevis	—		>5	/
Finding suggests polyneuropathy (upper extremities)				
Deltoid	0	/		
Biceps	0	/	/	/
Triceps	1 to >11	/		
Brachioradials	0	/		
Suponitor	0	/		
Extidigitorum	0			
External carpiulnaris	−1 to >11		/	/
External indicius ulnars	−1 to >11		>5	/
Pronator teres	0	/		
Flexor carpiradials	0	/		
Flexor dig. prof.	0	/		
Flexor dig. supf.	−1 to >1	/	/	/
Flexor carp; ulnaris	−1 to >11 +	/	/	/
1st dorsal inter.	−1 to >11	/	/	/
Abductor poll. brevia	1 to 11	/	/	/
Abductor diq quin min.	1 to 11	/		
Upper trapezius	R>L to 0	/		
Levator scapulae	0	/		
Supraspinatous	0	/		
Inferior spinatus	0	/		
Rhomboid	0	/		
Seratus ants	0	/		
Paracervicals C3-C7	0	/		

Note: / = negative.

Source: EHC-Dallas. 1988.

Table 26.31.   Neuropsychological Measures

Bender-Gestalt — Measures perceptural accuracy, judgment, and fine motor movement.

WAIS-R Subscales — Measure auditory and visual recall; higher cognitive functioning, including memory (short and long-term); verbal comprehension, abstract thinking ability, nonverbal reasoning, speed of mental operation, perceptual learning.

Harrel/Butler N/S Comprehensive Neuropsychological Screen — Measures gross brain functioning deficit, including fluid intelligence (rapid shifts), memory functioning (auditory, visual associative), gross motor movement, sensory tactile deficits, construction dyspraxia (ability to reproduce on paper common objects (e.g., cube, triangle, pipe, etc.) on verbal command), abstract reasoning, practical judgment.

Benton Visual Retention Scale — Measures individual ability for visual recall and reproduction; general measure for early deterioration.

Laria-Nebraska — Measures extensive comprehensive test for brain function.

Source: EHC-Dallas. 1986.

Integrative Brain Functions.   There are numerous ways to evaluate integrative brain function. Most of these are by some neuropsychological test (Table 26.31).

Tanquerel des Planche's[208] classic study of factory workers in Paris generated clear results of a "disease" of lead poisoning. He also identified behavioral symptoms or "precursors", as he called them, that heralded the coming syndrome of brain dysfunction from lead poisoning. Since this time, definition of prodromes has become possible. Thus, when evaluating all of the integrative functions today, the physician must remember that one or many triggering agents may be present. Each function, though integrated with the others, will be discussed separately.

Symbol Formation (Cognitive Tests Symbol–Symbol and Symbol–Digit).   These tests that reflect nervous system damage measure performance in the following areas: language and development, verbal comprehension, nonverbal reasoning, trial and error, perceptual learning, visual-motor-spacial three-dimensional complex integration, and integration (integration and incorporation). Language and visual symbols incorporate learning and memory, as the terms are commonly used in behavioral studies. The Wechsler Adult Intelligence Scale-Revised (WAIS-R), with its many important subtests, is used to identify cognitive difficulties. The mental status screen test also helps in this area. Evaluation of learning can take many paths, and each facet may need to be evaluated for proper total work-up.

Learning occurs in all integrated systems, though perhaps to varying degrees. In this sense it could be said that the respiratory center "learns" tolerance to morphine-induced respiratory depression or that various complex visual motor performances such as walking are "learned". Therefore, this type of learning needs to be evaluated. Many common toxic chemicals are now known to alter learning and should not be overlooked during the evaluation of a patient. Some are temporary. For example, our studies on brain function and pesticides show changes in learning, which are improved with the removal of the pesticides. In fact, damage may be permanent in others due to the crosslinking properties of some chemicals such as with formaldehyde, which is known to crosslink proteins in the body, causing irreversible damage.

Kilburn et al.[209] found that formaldehyde impairs memory, equilibrium, and dexterity and that these effects of formaldehyde exposure can last for days after the initial contact. One study showed the neurotoxicity of industrial organic solvents.[210] In this study, organic solvents (particularly styrene, which is used widely in boat building) caused a number of individuals with acute high dose exposure to experience transient narcotic effects on the CNS. These substances certainly impaired learning. These effects are sometimes proportionate to brain dose, which, in turn, is determined by the intensity and duration of exposure, though at other times it may not be related to dose, but to the virulence, of the chemical. The following is an example of impaired brain function with symbol distortion due to styrene.

Case study. A 40-year-old physician experienced tremors and poor coordination in 1983 after purchasing a sailboat and resurfacing the hull with some kind of paint or plastic-containing styrene, the fumes of which made him feel sick. Over a period of time, he found a direct correlation between being around the boat and subsequently feeling ill with symptoms of weakness, fatigue, sleepiness, sluggishness, skin irritation and rash, stiff painful joints, laryngitis, tinnitus, vertigo, and depression. He had been evaluated and tested by neurologists and other specialists, with no conclusive diagnosis other than photosensitivity. When admitted to the ECU, he was steroid-dependent from the medications that he had been taking to control his skin eruptions and joint pains.

During 4 days of reduction of chemical load in the ECU and fasting on spring water, the neurological symptoms of tremor, weakness, tinnitus, sluggishness, sleepiness, and vertigo initially intensified and then subsided, as did his skin and joint lesions. He also required less prednisone.

Initial laboratory analysis showed a depressed calcium and an elevated magnesium (1298 mg, normal = 8.5 on 24-hour urine). WBC was 13,000 with absolute lymphocyte count 1280/mm^3 (8%, which was somewhat decreased from the normal range of 1440 to 4320/mm^3). Total T cells were 742/mm^3 (C<1240/mm^3), a moderate decrease, and total B cells were 269/mm^3 (C<500/mm^3). Vitamin assay revealed deficient B_3, B_6, C, and D_3. Interestingly, this patient was found to have a very low superoxide dismutase at 3.1 units (normal = 12.0 to 15.4) per mg of Hb. Norepinephrine level was 512 units (normal = 125 to 310 units).

Table 26.32. 40-Year-Old White Male Double-Blind Inhaled Challenge After 4 Days of Deadaptation in the ECU with Total Load Reduced 15-Minute Exposure

Chemical	Dosage (PPM)	Results[a]
Phenol	<0.002	+
Formaldehyde	<0.2	+
Pesticide (2,4-DNP)	<0.0034	+
Petroleum-derived ethanol	<0.5	−
Choline	<0.33	−
Saline placebo		−
Saline placebo		−
Saline placebo		−

[a] Objective measurements/signs included Butler's neuropsychological evaluation, blood pressure, pulse, symptom score sheet including an inability to stand up due to vertigo, skin, and joint signs.

Source: EHC-Dallas. 1985.

This patient reacted to a variety of antigens, including molds, dust, dust mite, lake algae, T.O.E., *Candida*, terpenes and nine foods. Inhaled double-blind chemical challenge after 4 days of deadaptation and reduced total body load produced reactions from four chemicals — phenol (<0.002 ppm), form-aldehyde (<0.20 ppm), pesticide (2,4-DNP) (<0.0034 ppm), and ethanol (<0.5 ppm). Three placebos and chlorine challenge were negative. He was found to be very sensitive to cotton, which reproduced symptoms of fatigue, sluggish-ness, sleepiness, parathesiae, myalgia, and burning eyes (Table 26.32).

This patient was treated with environmental clean-up, rotational diet, neutralization shots, and antigen titration for severe sensitivity. Follow-up at 1 month showed that he had improved substantially. He was able to reduce his prednisone dose from 60 mg/day to 20 mg every 4 to to 5 days. He was also able to improve his brain function.

Integrative functions that may or may not have to do with symbol forma-tion but certainly have to do with learning may include many other areas that are often damaged. For example, indecisiveness and poor decision-making are often concomitants of chemical exposure, as is a lessened ability to organize knowledge, evaluation of past experiences, and utilization of "common sense" in the solution of everyday problems. Often, the subset of chemically sensitive patients with cerebral dysfunction have extreme difficulty in making a reason-able decision after pollutant exposure, even though they usually measure far above average in intelligence. These patients know the mechanics of making data-based decisions, but seem immobilized in implementation. Verbal com-prehension and expression are often decreased, and a more restricted range of words, ideas, memories, and conceptualization may also occur. These functions are evaluated by WAIS-R and Harrel-Butler (HB-CNS) comprehensive screening test. The Wechsler Memory scale is used to evaluate some aspects of memory.

Sensory/Motor Integration. Evaluation of sensory/motor integration includes the following:

1. Hand-eye coordination: right hand performance of Santa Ana Dexterity Test for manual dexterity; Bourdon-Wiersa Vigilance Test measures sensory-motor speed (BW speed); Mira Test measures the steadiness of psychomotor performance and indicates the tremor of hands and other inability to control hand movements
2. Sequencing
3. Apraxia and oral construction
4. Latency of response

Unsteadiness of gait may be the result of muscle weakness, although ataxias may also result from damage to visual/motor integration without peripheral neuropathy. The Bender can also be used to help evaluate this status.

Sensory damage alone can affect motor function. Such damage to muscle spindle afferents can be related to ataxia without direct effect or damage to motor fibers. In one case reported in the literature, long-term follow-up (16 years) after a single toxic exposure to trichloroethylene disclosed an MMPI profile and interview that suggested continued depressive symptomatology. Eighteen years after exposure, findings included paresthesia and hypalgesia in the mallor area of the face as well as atrophy of the facial muscles. The deterioration in this case was clearly due to neglected therapy with no attempt at avoidance of the primary or secondary triggering agents or replacement of deficient nutrients and is only mentioned to make the clinician aware that proper evaluations are necessary, and, when findings are positive, appropriate intervention should be initiated.

The spreading phenomenon occurs in other chemically sensitive patients, whereby they then become sensitive to many other chlorinated, and even nonchlorinated, toxic hydrocarbons. Then they develop secondary sensitivities to food and biological inhalants. This spreading reaction is probably due to a limited number of enzymatic detoxification pathways, impaired immune systems, damage to the blood-brain barrier, and many more factors. These patients become overloaded, they are unable to detoxify rapidly enough, and they develop new sensitivities that may become as significant as the primary offender (see Volume I, Chapter 4[42]). Eight percent of our patients have been found to have trichloroethylene in their blood. Related substances such as 1,1,1-trichloroethane (49% of chemically sensitive patients) and tetrachloroethylene (83% of patients) are also detected and found to correlate with neurologic aberrations. At the EHC-Dallas, brain function tests revealed depression of cognitive function, recent memory loss, and loss of ability to remember numbers. Also, vasculitis phenomena were seen in many of these chemically sensitive patients, which explains why such diffuse neurotoxic patterns are found in many chemically sensitive patients.

Many chemical compounds such as inorganic lead salts[211] and organophosphorus compounds[212] such as triorthocresyl-phosphate seem to damage both the sensory and motor systems.

Perceptual/Motor Problems. Perceptual/motor problems may be observed in a slowing of psychomotor speed, decrements in visual-motor coordination, increased clumsiness, decreased perceptual accuracy and judgment, and poorer balance. Sensory perception may be altered also. Learning, memory, and other cognitive functions are included in this group. It appears that recent memory loss, both auditory and visual (e.g., inability to remember numbers), are two of the earliest signs of environmental overload. According to Butler,[190] the Harrel-Butler Neurostatus screen is the best tool for evaluation of these functions.

Butler[190] demonstrated significant changes in IQ and perceptual motor functioning associated with chemical exposure. Using the Harrel-Butler Neurostatus screen in a subset of 50 chemically sensitive patients with clinical brain dysfunction and 100 controls, Butler found that he could identify the chemically sensitive patients 98% of the time with a $p<0.005$ confidence level when he used 19 different evaluations of items such as perceptual motor, gross motor, sensory tactile, and memory function, etc. His scores for the chemically sensitive patients ($\times$ = 35.04 ± 12.43) contrasted to a normal group with a score of 22 ± 8. This same group presented evidence of positive behavioral and personality changes with subsequent treatment in the ECU. Patients gradually cleared with meticulous environmental treatment (Table 26.33).

The personality test used in these studies was the Clinical Analysis Questionnaire (CAQ [641]). Normal ranges on this scale are either <4 or >7. The chemically sensitive patients score similar to patients with cancer or those with severe orthopedic surgical back problems. They are high on the hypochondria scale (8.72 ± 1.54) and anxiety/depression scale (7.04 ± 1.90) and low on the energy/depression scale (7.36 ± 2.13). These findings indicate organic concerns of chemical damage similar to the concerns of the other patients with organic diseases such as ruptured discs or cancer. The median test–retest coefficient across all scales is 0.73 reliability. Validity coefficients for the individual factors, with 0.30 considered important, range from 0.45 to 0.86 for personality traits and from 0.69 to .95 for the clinical factors. These validity coefficients are considered remarkably faithful to their hypothesized structure.

Behavioral Functions of the Neurological System. Evaluation of behavioral functions of the brain in relationship to pollutant injury in the chemically sensitive individual may be difficult and require a deeper understanding of the physiology of the behavior functions of the brain and nervous system. Behavior is a function of the entire nervous system and other metabolic functions. Even the discrete cord reflexes are an element of behavior, and the wakefulness and

**Table 26.33A.   Psychological Checklist Items — Greater than 40%
Chose Three**

Item #	Frequency	Item #	Severity
10	Overwhelming exhaustion or weariness — very easily fatigued	10	Same as left
11	Difficulty getting started in the morning		
14	Difficulty concentating on work or study	14	Same as left
15	Present performance inferior to prior performance or level of functioning	15	Same as left
19	Feelings of "I'm not myself" or "What is happening to me?"		
24	Frequent headaches	24	Same as left
33	Feeling of losing control of one's destiny		
36	Poor comprehension		
37	"This is not me"		
35	Poor memory	35	Same as left

**Table 26.33B.   Physical Checklist Items — Greater than 40%
Chose Three**

Item #	Frequency	Item #	Severity
1	Headaches	1	Same as left
6	Aches and pains	6	Same as left
12	Sinus discomfort		
20	Clumsiness		
31	Easily fatigued	31	Same as left
32	Low energy	32	Same as left
33	Weakness	33	Same as left
34	Restlessness		
42	Mucus		
49	Eye problems		

Table 26.33C.   Mental Status — Neuroscreen — Total Score

50 chemically sensitive patients

Chi square = 35.04

Standard deviation = 12.43

Mean for normal (100) = 22 ± 8

Score of 29 identifies 65% of chemically sensitive patients.

Score of 35 identifies 98% of chemically sensitive patients,

$p<0.005$ — in 19 different items of brain function; e.g., memory, perceptual motor, gross motor, sensory

Table 26.33D. Subject Biographical Data

		(No.)	Mean	SD	(%)
Sex:	Male	12			26.7
	Female	33			73.3
Marital status:	Single	5			11.1
	Married	32			71.1
	Divorced	7			15.6
	Widowed	1			2.2
Age			44.64	11.56	

Source: Butler, J. R. 1993. Personal communication.

sleep cycle is certainly one of the most important of behavioral patterns and is often altered in the neurologically involved chemically sensitive individual. Special types of behavior associated with emotions, subconscious motor and sensory drives, and the intrinsic feelings of punishment and pleasure are performed mainly by subcortical structures located in the basal regions of the brain. Functions of these may be altered in the subset of pollutant-disturbed neurologically involved chemically sensitive patients. Knowledge of the anatomy and physiology of the limbic system will help the clinician evaluate this group. The structures of the limbic system with its relationship to the hypothalamus are presented in Figure 26.28. The key position of the hypothalamus in the limbic system and the other subcortical structure of the limbic system that surrounds it, including the proptic area, the septum, the paraolfactory area, the epithalamus, the anterior nuclei of the thalamus, portions of the basal ganglia, the hippocampus, and the amygdala, are shown in Figure 26.29.

Surrounding the subcortical limbic areas is the limbic cortex composed of a ring of cerebral cortex beginning in the orbitofrontal area on the ventral surface of the frontal lobes, extending upward in front of and over the corpus callosum and downward onto the medial aspect of the cerebral hemisphere to the cingulate gyrus, and, finally, passing behind the corpus callosum and downward out the ventral medial surface of the temporal lobe to the hippocampal gyrus, pyriform area, and uncus. Thus, on the medial and ventral surfaces of each cerebral hemisphere is a ring of paleocortex that surrounds a group of deep structures (intimately associated with overall behavior and with emotions). In turn, this ring of paleocortex functions as a two-way communication and association linkage between the neocortex and the lower limbic structures. It is important to remember that many of the behavioral functions (e.g., somatic excitability) elicited from the hypothalamus and from other limbic structures are mediated through the reticular formation of the brainstem. Therefore, the reticular formation is a very important part of the limbic system, even though anatomically it is considered to be a separate entity.

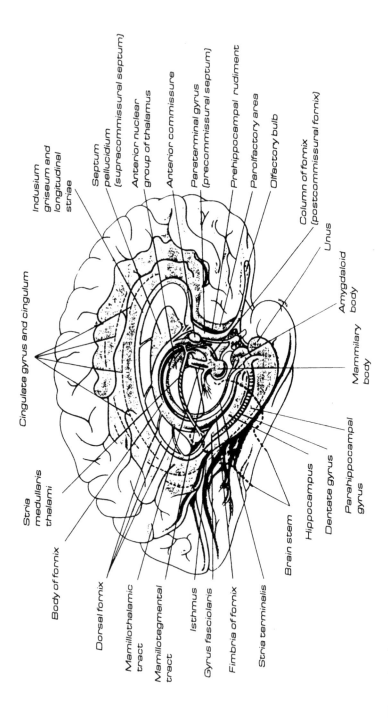

Figure 26.28. Anatomy of the limbic system illustrated by the shaded areas of the figure. (From Guyton, A. C., ed. 1981. *Textbook of Medical Physiology,* 6th ed. 700. Philadelphia: W. B. Saunders. With permission.)

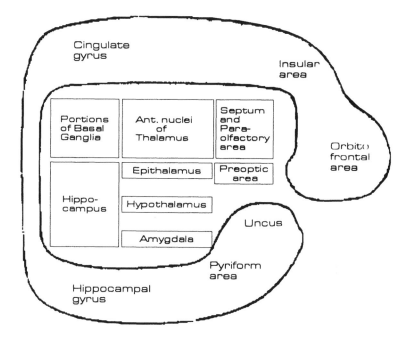

Figure 26.29. The limbic system. (From Guyton, A. C., Ed. 1981. *Textbook of Medical Physiology*, 6th ed. 700. Philadelphia: W. B. Saunders. With permission.)

According to Guyton,[213] the hypothalamus, with the septum and mammillary bodies, is the major output pathway of the limbic system, sending signals in two directions: downward through the brainstem mainly into the reticular formulation of the mesencephalon, pons, and medulla and upward toward many areas of the cerebrum, especially the anterior thalamus and the limbic cortex. These structures control most of the vegetative function of the body as well as many aspects of emotional behavior. Figure 26.30 shows the control centers of the hypothalamus. As noted previously in this chapter, the various vegetative centers are visible by anatomic dissection. The location of the perfornical nucleus and lateral hypothalamical areas, which control not only hunger, blood pressure, and thirst but also many emotional drives, is noteworthy because these functions may be influenced by environmental triggers. Symptoms related to these functions are often triggered in a subset of chemically sensitive patients. Often challenge tests with inhaled chemicals result in ravenous hunger, increased blood pressure, and severe thirst, all of which accompany the induced emotional reaction.

Aside from the vegetative functions of the hypothalamus, experiments have shown that stimulation of, or lesions in, the hypothalamus often have profound effects on emotional behavior, effects which are seen in a small subset of chemically sensitive patients. For example, stimulation of the lateral hypothalamus not only causes thirst and eating but also increases the general

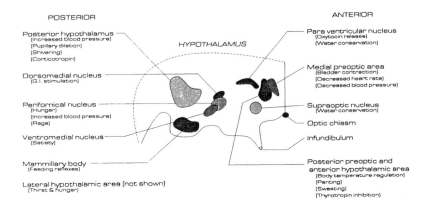

POSTERIOR   ANTERIOR

Posterior hypothalamus
[Increased blood pressure]
[Pupillary dilation]
[Shivering]
[Corticotropin]

HYPOTHALAMUS

Para ventricular nucleus
[Oxytocin release]
[Water conservation]

Dorsomedial nucleus
[G.I. stimulation]

Medial preoptic area
[Bladder contraction]
[Decreased heart rate]
[Decreased blood pressure]

Perifornical nucleus
[Hunger]
[Increased blood pressure]
[Rage]

Supraoptic nucleus
[Water conservation]

Optic chiasm

Ventromedial nucleus
[Satiety]

Infundibulum

Mammillary body
[Feeding reflexes]

Lateral hypothalamic area [not shown]
[Thirst & hunger]

Posterior preoptic and
anterior hypothalamic area
[Body temperature regulation]
[Panting]
[Sweating]
[Thyrotropin inhibition]

**Figure 26.30.   Control centers of the hypothalamus. (From Guyton, A. C.,
                    ed. 1981. *Textbook of Medical Physiology*, 6th ed. 701.
                    Philadelphia: W. B. Saunders. With permission.)**

level of activity of the organism, sometimes leading to overt rage and fighting.
This response has been observed after pollutant challenge in a subset of
chemically sensitive patients. The following case exemplifies this phenomenon.

> **Case study.** A 55-year-old white female presented to the ECU with the chief
> complaints of short-term memory loss, loss of concentration, confusion, and
> episodes of rage. She also exhibited ANS dysfunction with cold hands and
> feet. These symptoms had occurred for the previous 20 years. Blood levels
> and fat biopsies showed an extremely wide variety of chlorinated pesticides
> including hexachlorobenzene, chlordane, heptachlor, heptachlor epoxide, trans-
> nonachlor, dieldren, aldrin, DDT, and DDE. After being placed in the ECU
> and deadapted for 4 days, her total load decreased and her symptoms cleared.
> Inhaled double-blind challenge with a chlorinated pesticide (<0.0034 ppm)
> reproduced both her ANS dysfunction and her rage and memory problems.
> SPECT brain scan revealed a discrepancy in flow and function, "salt and
> pepper" pattern in the rest of the brain, and temporal lobe flow and function
> abnormalities. This patient has improved with good environmental control
> and nutrient replacement. Her episodes of rage have subsided. This case
> certainly emphasizes the effects pollutants can have on the hypothalamus and
> behavior by objective demonstration.

Stimulation of various areas in the brain produces a variety of responses. For instance, stimulation of the most medial portion of the medial hypothalamus immediately adjacent to the third ventricle usually leads to fear and punishment reaction. Interestingly, such responses have been observed in a subset of chemically sensitive patients following food or pollutant challenge, suggesting that the mechanism at work in the general situation is the same one stimulated by specific pollutants. Stimulation of the lateral portion of the medial hypothalamus, the area lying between the medial portion and the third ventricle, can elicit a number of drives and functions such as body temperature, increased sexual drive, thirst, and excessive eating. All of these responses have been elicited after food or pollutant challenge in a subset of chemically sensitive patients, reproducing their admitting complaints. Again, these challenge outcomes suggest that either these areas of the brain have been triggered or autoimmune imbalance has occurred that results in end-organ responses that mimic those that occur when the area is triggered. In the chemically sensitive individual, these triggering agents can only be defined after the total load has decreased, and the patient has been deadapted for at least 4 days. However, it is now generally possible to define which area of the hypothalamus is damaged in some chemically sensitive individuals who have neurological involvement.

In contrast to the active responses produced by stimulation of various brain areas, bilateral lesions in the hypothalamus may cause reduced function in some areas producing dehydration, starvation, and passivity,[214] all of which have been seen in a large subset of neurologically involved chemically sensitive patients. Many neurologically involved chemically sensitive individuals, for example, have short-term memory loss with concomitant temporal lobe involvement on SPECT scans and probably hypothalmic lesions also. Lesions or stimulation in other regions of the limbic system, especially the amygdala, the septal area, and areas of the mesencephalon, often also cause effects similar to those elicited directly from the hypothalamus.

Guyton[214] makes clear that the hypothalamus and limbic systems are particularly concerned with the affective nature of sensory sensitivities of pleasure and pain. These affective qualities are also called reward and punishment, or satisfaction and aversion, by physiologists. Electrical stimulation of certain regions of the brain pleases or satisfies the animal, whereas stimulation of other areas causes terror, pain, fear, defense posture, and all other elements of punishments. Obviously, the two opposite-responding systems greatly affect the behavior of the animal. Observations in the ECU have demonstrated clearing and then initiation of these various responses after pollutant challenge in a subset of the neurologically involved chemically sensitive patients. It should be emphasized that stimulation of a single area of the brain can often elicit opposite responses such as reward or a sense of punishment, depending upon the intensity of the stimulus. Usually weaker stimuli cause the reward impulse, while the stronger stimuli cause the punishment responses. These responses have been observed in some of the chemically sensitive individuals seen at the EHC-Dallas and, in fact, the reward impulse is often seen in the food- or pollutant-addicted, chemically sensitive patient and may well be one of the physiologic reasons for the addiction. In animals, prolonged stimulation of the punishment response can cause them to become very ill and, in some cases, die. The principle centers for pain, punishment, and escape tendencies have been found in the central gray area surrounding the aqueduct of sylvius in the mesencephalon and extending upward into the periventricular structures of the hypothalamus and thalamus. Although the punishment areas in the limbic system are only one-seventh as great as the reward areas, they are often triggered by a larger pollutant stimulus, and, as a result, some of these patients will rapidly worsen and become nonfunctional. In understanding and evaluating the subset of the pollutant-stimulated, neurologically involved, chemically sensitive patients, these actions are especially important. Stimulation of the punishment response can frequently inhibit the reward and pleasure centers completely, illustrating that punishment and fear can take precedence over pleasure and reward. It should be emphasized that often pollutant stimuli will initially trigger a pleasure response in some chemically sensitive patients. These patients will then continue to seek out this pollutant in an addictive manner (as shown in Volume I, Chapter 3[7]) until their adaptation mechanism breaks down and symptoms and signs recur. The patient then becomes fearful and loses memory and concentration and rapidly deteriorates.

Important animal experiments have helped us understand some of the behavior of chemically sensitive individuals. For example, addiction may be produced in animals if pollutant overload occurs. A similar phenomenon has been observed in many chemically sensitive patients, suggesting a lesion in a similar area of the human brain. Often, once food or chemical addiction has occurred in chemically sensitive individuals, they have been observed to actually seek out a pollutant exposure that is causing their problem, ignoring the ill effects because some aspect of their addiction is being satisfied. They seek repetition of exposure to their detriment. This subset of chemically sensitive individuals then has no perception of cause and effect, and this lack of perception results in a downward spiral of constant pollutant exposure until end-organ failure occurs.

According to Guyton,[215] these animal studies have shown that a sensory experience that causes neither reward nor punishment is barely remembered. Electrical recordings have shown that new and novel sensory stimuli always excite the cerebral cortex. But repetition of the stimulus leads to almost complete extinction of the cortical response of the sensory experience, which then does not elicit either a sense of reward or punishment. Thus, the animal becomes habituated to the sensory stimulus and, thereafter, ignores this stimulus. This phenomenon allows for a perfect set up of food and chemical stimuli causing no sensory activation but continuing to do harm until the function of the target organ fails. This process appears to be often occurring in a subset of neurologically involved chemically sensitive patients, with not only end-stage organic brain damage occurring but also end-stage cardiac, respiratory, renal, hepatic, and gastrointestinal failure.

Once food and pollutant stimulation triggers the reward area, damage to the whole system can also occur as a result of repetitious intake of harmful substances. This phenomena has been observed in a subset of chemically sensitive patients.

The reward and punishment centers of the midbrain have much to do with the selection of the information we learn. As a result, injury to any of these areas may distort thinking in the subset of neurologically involved chemically sensitive individuals, allowing for many of the poor choices they make when selecting environments in which to live. They often select toxic environments. Once in these environments, continued exposure produces more adversely altered behavior.

It has been shown that chemicals such as chloropromazine inhibit both the reward and punishment centers thereby greatly decreasing the affective reaction of the animal. It appears that some other xenobiotics may cause a similar response in the chemically sensitive population.

Stimulation of regions of the hypothalamus dorsal to the mammillary bodies greatly excites the reticular activating system and, therefore, causes wakefulness, alertness, and excitement. In addition, the sympathetic nervous system becomes generally excited, increasing arterial pressure, causing pupillary dilatation, and enhancing other activities associated with sympathetic activity. This phenomenon is often temporarily observed after pollutant challenge. The chemically sensitive patient will attempt to maintain this state by taking in more of the harmful pollutant(s) or related ones, such as coffee, colas, or foods, that have been observed to activate this system temporarily. However, repetitious exposure leads to nonactivation and eventually to end-organ failure.

Conversely, stimulation in the septum, in the anterior hypothalamus, or in isolated points of thalamic proteins of the reticular activating system often inhibits the mesencephalic portion of the reticular activating system causing somnolence and sometimes actual sleep. A large subset of chemically sensitive individuals exhibits the sluggishness and sleepiness phenomenon after being exposed to pollutant challenge.

The amygdala seems to be a behavioral awareness area that operates at a semi-conscious level. It also seems to project into the limbic system one's present status in relation both to environment and thought. The amygdala appears to help pattern a person's behavioral response so that it is appropriate for each occasion. We have seen chemically sensitive patients who appear to have an injury in this area after they inhale or ingest certain pollutants. They then just do not respond appropriately to the present situation, e.g., even though they are malnourished, they refuse to take intravenous nutrition or refuse to try other modalities that might improve their situation. One 10-year-old boy became pollutant sensitive and would refuse to eat and would try to kick the wall down. Once his total body pollutant load was reduced in the ECU, he would behave appropriately and eat well. Upon pollutant challenge in the controlled environment, he would revert to this aberrant behavior until the pollutant was withdrawn and the reaction ceased.

According to Guyton,[216] the amygdala is a complex of nuclei located immediately beneath the medial surface of the cerebral cortex in the pole of each temporal lobe, and it has abundant direct connections with the hypothalamus. In a subset of chemically sensitive patients who have brain dysfunction and the usual odor sensitivity, a SPECT scan will show not only a neurotoxic pattern on the cerebral cortex but also hypo- or hyperfunction of this area of the temporal lobe (Figure 26.31). This finding suggests abnormal function of the amygdala, which would correspond with the olfactory stimulation by the pollutant. In lower animals this amygdala complex is concerned primarily with association of olfactory stimuli and with stimuli from other parts of the brain. Indeed, one of the major divisions of the olfactory tract leads directly to a portion of the amygdala called the corticomedial nuclei that lies immediately beneath the cortex in the pyriform area of the temporal lobes. However, in the human the basolateral nuclei has become much more highly developed than the olfactory portion and plays a very important role in many behavioral activities not generally associated with olfactory stimuli. Regardless, the odor sensitive chemically sensitive individual who develops behavior problems after pollutant exposure probably does so at least partially through this pathway. We have observed thousands of neurologically involved chemically sensitive patients in the ECU who were challenged by a single pollutant. They would inhale the pollutant and describe a feeling of an impulse going up the olfactory nerve into, and over, their brain. Then they would lose their concentration, develop short-term memory loss, and other peripheral reactions would follow. Their SPECT scans would show the typical flow-function disparity with a decrease in cerebral blood flow in contrast to function, as well as the "salt and pepper" pattern of neurotoxicity in the rest of the cerebrum. In addition, one temporal lobe would show less flow and function in contrast to the other side. These findings suggest that the amygdala is involved in the pollutant triggering of this type of a subset of neurologically involved chemically sensitive patients.

The amygdala receives impulses from all portions of the limbic cortex, the orbital surfaces of the frontal lobe, the cigulate gyrus, the hippocampal gyrus, and from the neocortex of the temporal, parietal, and occipital lobes, especially from the auditory and visual association areas. Because of these connections, the amygdala is the window through which the limbic system evaluates the place of the person in the world. In turn, the amygdala transmits signals back into these same cortical areas, into the hippocampus, the septum, the thalamus, and especially the hypothalamus. Because of this anatomic set-up, one can see how the impulses going to, and leaving from, the amygdala could easily become imbalanced as a result of pollutant overload. They then could emit temporary pathology, as seen in a subset of chemically sensitive patients once their reaction is triggered.

In general, stimulation of the amygdala not only can cause all the effects of stimulating the hypothalamus, but also can cause tonic movements such as head raising or bending the body, circling movements, clonic rhythmic movements, and different types of movements associated with olfaction and eating

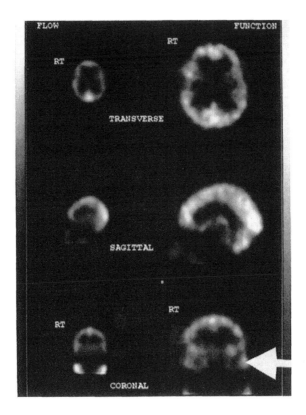

Figure 26.31.   SPECT brain scan showing unequal temporal lobes, which corresponds with short-term memory loss.

such as licking, chewing, and swallowing. Tonic-clonic movement in a subset of chemically sensitive patients has been discussed in the Chapter 23. These movements may partially be due to excess olfactory stimulation of the amygdala by pollutants. A typical case that probably is amygdala stimulation was seen in this series of patients. One in particular illustrates the problem.

Case study. This 41-year-old woman had severe optic nerve degeneration from solvent exposure. She also had peripheral neuritis that caused excruciating pains in her legs. She further developed gastrointestinal upset with malabsorption, which resulted in vitamin, mineral, and amino acid deficiency.

Laboratory results were as follows: blood toxic chemicals — elevated 2-methylpentane, 5.5 ppb (C<0.1 ppb); 3-methylpentane, 21.5 ppb (C<1.0 ppb); n-hexane, 16.4 ppb (C<1.0 ppb); oxychlordane, 0.7 ppb (C<0.3 ppb); trans-nonachlor, 0.3 ppb (C<0.3 ppb); DDE, 11.2 ppb (C<0.3 ppb); DDT, 0.3 ppb (C<0.3 ppb); DDD, 0.4 ppb (C<0.3 ppb); total PCBs, 1.9 ppb (C<0.3 ppb);

benzene, 1.4 ppb (C<1.0 ppb); 1,1,1-trichlorethane, 1.3 ppb (C<1.0 ppb). Vitamins showed low folic acid, 2.3 ng/mL (C = 3.7 to 18.8 ng/mL); high vitamin B_{12}, >2000 pg/mL (C = 200 to 953 pg/mL); high ery. glut. oxa. transaminase (EGOT), 1.48 index (C<1.5 Index); ery. glutathione reductase (EGR), 1.3 act. coeff. (C = 0.9 to 1.20 act. coeff.); ery. transketolase (ETK), 23.3% TPP stim. (C = 0 to 17% TPP stim.), vitamin C/WBC 56 µg/10^8 cells (C = 21 to 53 µg/10^8 cells). Urine elements showed the following: low magnesium, 25 ppm (C = 61 to 174 ppm); sodium, 391 ppm (C = 2316 to 4671 ppm); potassium, 595 ppm (C = 1281 to 3395 ppm); zinc, 0.32 ppm (C = 0.58 to 1.10 ppm); iron, 0.02 ppm (C = 0.059 to 0.120 ppm); chromium, 0.031 ppm (C = 0.064 to 0.132 ppm); molybdenum, 0.036 ppm (C = 0.052 to 0.080 ppm); silicon, 5.0 ppm (C = 14.7 to 31.5 ppm); 0.036 ppm (C = 0.052 to 0.080 ppm); boron, 0.51 ppm (C = 1.91 to 4.41 ppm); phosphorus, 240 ppm (C = 616 to 1092 ppm); strontium, 0.28 ppm (C = 0.147 to 0.259 ppm); arsenic, 0.11 ppm (C = 0.25 to 0.51 ppm); silver, 0.004 ppm (C = 0.0003 to 0.0025 ppm); zirconium, 0.009 ppm (C = 0.0020 to 0.0075 ppm). Head hair mineral analysis showed the following: low chromium, 0.35 ppm (C = 0.78 to 1.00 ppm); iodine, 0.51 ppm (C = 0.63 tp 0.85 ppm); boron, 0.5 ppm (C = 1.0 to 3.0 ppm); lithium, <0.005 ppm (C = 0.018 to 0.095 ppm). Urine amino acid analysis showed the following: low A-amino-N-butyrate, 1.5 µM/24 hours (C = 5.0 to 50.0 µM/24 hours), ammonia, 6027 µM/24 hours (C = 105 to 70.0 µM/24 hours); methionine, 5.3 µM/24 hours (C = 8.0 to 49.5 µM/24 hours); 1-methyl-histidine trace (C = 130 to 930 µM/24 hours); phosphoethanolamine, 13.3 µM/24 hours (C = 20.0 to 95.0 µM/24 hours); tryptophan, 16.4 µM/24 hours (C = 20.0 to 120 µM/24 hours); tyrosine, 25.0 µM/24 hours (C = 40.0 to 220 µM/24 hours).

Multitest CMI showed three positive reactions out of seven antigen challenges with negative control. Total IgE was high — 733 IU/mL (C = 0 to 180 IU/mL). Low WBC, 3300/mm^3 (C = 4800 to 1000/mm^3); low total lymphs, 1023/mm^3 (C = 1600 to 4200/mm^3); low total T_{11}, 798/mm^3 (C = 1260 to 2650/mm^3); T_4, 419/mm^3 (C = 670 to 1800/mm^3); total T_8, 205/mm^3 (C = 333 to 1070/mm^3); total B_4 lymphs, 61/mm^3 (C = 82 to 477/mm^3); high CD_{26}, 54% (C = 0 to 10%). IL_2-R_1, 15% (C = 0 to 10%); low complement CH_{50}, 48% (C = 51 to 150%). SPECT scan impression showed better preservation of flow than function (i.e., uncoupling), a mild pattern of cortical irregularity that correlated with possible mild neurotoxicity, and substantial temporal lobe asymmetry with better preserved flow and function to the right temporal lobe than the left.

Upon exposure to toxic substances such as toluene or 1,1,1-trichloroethane, this patient would develop severe uncontrollable episodes of neck bending and head raising accompanied by tetany of their arms and legs. These episodes would come in waves and would last for 1 to 2 hours until the reaction completed its sequence. Occasionally, the tetany, mental fogginess, and neck bending would last for 2 days. Eventually, with good environmental control, nutrition, neutralizing injections of foods, chemicals, and biological inhalants, the patient improved and did not exhibit these behaviors.

Occasionally stimulation of the amygdala can cause rage, fear, escape, and punishment similar to those responses of hypothalamic stimulation. Finally, excitation of the amygdala can cause sexual activities such as erection, copulatory movement, ejaculation, ovulation, uterine activity, and premature labor. We have treated numerous chemically sensitive males who reported that after the perception of a toxic odor and after a long-term exposure without perception or early adaptation they lost their sexual drive, ability to get an erection, or ability to copulate or ejaculate. They felt that this problem was in their brain and not their genitals. They felt hung-over and not very sharp and experienced difficulty focusing and concentrating. Once they had reduced their total Body Load, their brain seemed to clear and they were immediately able to sustain these functions. We have also treated a few patients who would get a sudden sexual urge and almost uncontrollable erection after certain types of chemical exposure. They would indiscriminately search for relief. Some female chemically sensitive patients also describe an uncontrollable sex drive after pollutant exposure. We have observed these different behaviors after pollutant challenge in the ECU.

The hippocampus originated as part of the olfactory cortex. In animals, it has various roles in determining whether an animal will eat a particular food, whether the smell of a particular object suggests decay or contamination or whether the odor is sexually inviting and in making decisions that are of life and death importance. Thus, very early in the development of the brain, the hippocampus presumably became the critical decision-making neuronal mechanism, determining the degree of importance, and type of importance, of incoming sensory signals. Thus, sensitivity or damage in this area as a result of pollutant exposure could severely alter the health of the chemically sensitive individual. Presumably, as the remainder of the brain developed, the connections from the other sensory areas into the hypothalamus have continued to use this decision-making capability. According to Guyton,[217] the hippocampus is thought to act as an encoding mechanism for translating short-term to long-term memory and also transmitting an additional signal to the long-term memory storage area, directing that storage take place. This area of the brain appears to be damaged in the pollutant overloaded chemically sensitive patient who has short-term memory loss, loss of concentration, and confusion. Almost any type of sensory stimuli causes instantaneous activation of the hippocampus, which, in turn, distributes many signals to the hypothalamus and other parts of the limbic system. Stimulation of the hippocampus can cause almost any one of the aforementioned behavioral patterns. Another feature of the hippocampus is that very weak electrical stimuli can cause local epileptic seizures that persist for many seconds after the stimulation is over, suggesting that the hippocampus can perhaps give off prolonged output signals even under normal functioning conditions. During hippocampal seizures, a person experiences various psychomotor effects including olfactory, visual, auditory, tactile, and other types of hallucinations that cannot be suppressed even though the person has not lost consciousness and knows these hallucinations to be unreal. We have treated at least 30 patients who experience this type seizure after chemical challenge in the controlled environment. Interestingly, most have cleared with long-term environmental therapy. Probably one of the reasons for the hyperexcitability of the hippocampus is that it is composed of a different type of cortex from that elsewhere in the cerebrum, having only three neuronal layers instead of the six found elsewhere. This anatomical finding would allow pollutants easy access. The following case, suggesting hippocampal injury and triggering, was observed at the EHC-Dallas.

> **Case study**. A 65-year-old white female, a physician's wife, presented with the complaint of hearing vintage music from the 1900s at certain times of the day and night. She also had associated symptoms of fatigue. She was placed in the ECU.
>
> This patient fasted for 5 days in the ECU, and her auditory hallucinations cleared as her total load decreased. She then underwent intradermal challenge with biological inhalants. Oral and intradermal challenge for food was done. Her intradermal testing showed she was sensitive to 25 foods, trees, gases, terpenes, 3 chemicals, fluogen, MRV, 5 molds, lake algae, *Candida*, dust, dust mites, 3 danders, and 2 smuts. Inhaled and intradermal challenges were used to test her for chemicals. Inhaled double-blind study showed that pollutants and foods to which she was sensitive would cause her to hear the music that had led her to the EHC-Dallas. Saline placebos and nonreactive foods did not have this effect. The challenge reaction was also accompanied by reproduction of her associated symptom of fatigue. Odor sensitivity was prevalent in this patient. All of her symptoms were reproducible in a double-blind manner and clearly were real. Formaldehyde (<0.020 ppm) would cause a 5-day reaction, with reproduction of all of her symptoms, including the vintage music.

According to Guyton,[218] the most poorly understood portion of the entire limbic system is probably the ring of cerebral cortex called the limbic cortex that surrounds the subcortical limbic structures. This cortex functions as a transitional zone through which signals are transmitted from the remainder of the cortex into the limbic system. Therefore, it is presumed that the limbic cortex functions as a cerebral association area for behavior control. Again, this area could be, and appears to be malfunctioning intermittently in a subset of pollutant-injured, chemically sensitive patients.

In summary, the gustatory and olfactory associations are especially involved in the insular and temporal cortex. In the hippocampal gyri there is a tendency for both complex auditory associations and complex thought associations derived from the general interpretative areas of the posterior temporal lobes. These functions explain why some chemically sensitive individuals have temporal lobe neurotoxicity or other SPECT scan abnormalities as well as behavior problems. There is also reason to believe that sensorimotor associations occur in the cingulate cortex. Also, finally, the orbital frontal cortex presumably aids in the analytical functions of the prefrontal lobes. It certainly is clear that, in a subset of chemically sensitive patients, triggering of this limbic system by pollutants will cause malfunction.

Recently, it has become apparent that some of the synaptic chemical transmitter substances of the brain play especially important roles in behavior and their alteration may trigger abnormal behavior in a subset of chemically sensitive patients with brain dysfunction. Though this field is only beginning to be explored, the chemical transmitter systems that seem to be important are the norepinephrine system, the dopamine system, the serotonin system, the enkephalin-endorphin systems, and the peripheral endocrine system. Although each has been previously discussed, some in the neuroimmune section of this chapter and some in Chapter 24, each will be considered separately here because of their importance in understanding the hormonal triggering that occurs in a large subset of chemically sensitive patients.

The norepinephrine secretory neurons are located in the reticular formation, especially in the locus cerulens, and they send fibers upward through the reticular activating system essentially to all parts of the diencephalon and cerebrum. One area of the hypothalamus that is especially activated by the norepinephrine system is the ventromedial nuclei that, when stimulated, can block many drives such as hunger, thirst, and sex. Therefore, at the same time that the neuroendocrine system activates the cerebral cortex, it blocks other drives that might compete with higher levels of cerebration. It is believed that overstimulation of this system is the cause of the manic phase and underactivity the cause of the depressed phase of manic-depressive psychosis. We have treated a subset of chemically sensitive patients whose depression was relieved by intradermal neutralization injections (very dilute doses — 1/25 to 1/125) of norepinephrine and, conversely, exacerbated by the provoking doses. Once the right dilution was found for these patients, their depression lifted immediately. They would have to take their injections from 1 to 5 times daily until their total pollutant load had been reduced for a period of time, up to 6 months. Then, once their nutrient deficit was corrected, they no longer needed the diluted, neutralizing dose of norepinephrine.

According to Guyton,[219] the dopamine system is located in the substantia nigra and projects to the striate portion of the basal ganglia. This system exerts an important, continuous restraint on the basal ganglion activity. Fibers also project into the area immediately adjacent to the third ventricle and also the lateral hypothalamus. Stimulation of this system increases activity of the lateral hypothalamus, producing enhanced drives for eating and fighting. Conversely, destruction of the dopamine system has almost the same effect as destroying the two lateral hypothalami, causing loss of appetite and other aversive drives.

We know that distraction of this system in the substantia nigra causes Parkinson's Disease. As discussed earlier in this chapter, some cases of Parkinson's Disease have been shown to be triggered by toxic substances such as hexachlorobenzene and cyanus toxins. We also have a subset of brain-involved chemically sensitive patients who respond to intradermal neutralization injections of dopamine. Once the intradermal neutralizing doses are found, their shakiness abates by taking injections (0.05 cc) of the 1/5, 1/25, or 1/125 dilution of the concentrate, one to four times daily. These are needed until their total load is decreased and their nutrient pools replenished.

The serotonin-secreting neurons are located in the medial raphe nuclei of the medulla. Many of these nerve fibers spread downward into the cord where they reduce the input levels of pain signals. They also spread upward into the basal areas of the brain to suppress reticular activating systems as well as other brain activity. This system can promote sleep. The pollutant p-chlorophenylalanine can inhibit serotonin formation, giving prolonged wakefulness. LSD can also antagonize some serotonin functions. We use serotonin neutralizing injections to abate many cerebral-generated symptoms such as mental fogginess, agitation, headaches, and recent memory loss as well as many gut-related symptoms such as cramping and bloating. Neutralizing doses have been used in over 5000 patients. Periodic cessation of these injections has resulted in exacerbation of the chemical sensitivity symptoms. Reinstitution has resulted in ablation of these symptoms.

The encephalin–endorphin system appears to be for suppression of pain. Endorphins are secreted in many areas of the brainstem and thalamus. Excess endorphins clearly make the chemically sensitive patient sluggish and sleepy. Many chemically sensitive patients experience the sensation of being sleepy and are sluggish. When neutralized by injection of a morphine-antagonist neutralization dose, they suddenly wake up. Apparently, a chemical triggering of the encephalin release occurs, and the neutralizing dose stops the production or release of the sleep-producing substance.

The effects of general hormones on behavior are well-known. For example, thyroid hormones increase body metabolism, while deficiency decreases it, making the individual sluggish and slow to respond. Excess thyroid yields hyperactivity. Testosterone and estrogen increase libido. ACTH can cause intense fear. Progesterone produces a calming effect (see Chapter 24 for more information). Most of these hormones are imbalanced in a subset of chemically sensitive individuals, creating a complex situation to diagnose and treat.

Emotional States.   With a little better understanding of the basic anatomy and physiology presented in the previous section, the pathology of emotion in the chemically sensitive individual can also be better understood. It will now be discussed.

The complexity of emotion must be understood first so a better understanding of how to measure and quantify it can be achieved. In our point of view, the cognitive/emotional states encountered in a neurologically involved subset of chemically sensitive patients are reproducible by challenge, are therefore provable, and are not psychosomatic in the general connotation that problems originate from the mind. Health problems that are psychosomatic would make those responses unproven and unprovable. Instead, we view these responses as reflecting complex psycho/biological interactions with certain cognitive/emotional consequences. (See the section in this chapter on the ANS and Chapter 23 for more information.) These emotional cognitive responses in the subset of chemically sensitive patients with emotional problems involve a heightened arousal of the biological amplification systems in the immune, sympathetic, and central and peripheral nervous and neuroendocrine systems. The neuropeptides arise in the gut as well as in the limbic area and hypothalamus and other areas where neurotransmitters are found, such as the pituitary and other endocrine organs as well as the immune axis, making a complex milieu that pollutant exposure could easily upset.

In the chemically sensitive individual, this "hair triggering" of the amplification systems is seen not only in relation to emotions but also in relation to almost any system of the body that is involved. Once these systems are damaged, a few molecules of a substance can trigger any of the amplification systems, resulting in diverse symptomatology in these patients.

It appears that the receptors for the neuropeptides are the key to the biochemistry of emotions in this subset of chemically sensitive individuals. As shown throughout this book, there appears to be a bi- or even multi-directional network of common interaction of immune, neuro, and endocrine systems that are joined together both anatomically and functionally and work as a well-directed symphony. Once a person becomes chemically sensitive, a pollutant stimulus will unbalance this coordination of organs and responses, resulting in severe exacerbation of chemical sensitivity (Figure 26.32). A subset of chemically sensitive patients have more emotional problems than others. The information carriers that integrate the emotional system are the neuropeptides that can be released from the gut, endocrine, neural, or immune cells. Neuropeptides are signaling molecules, and the components that receive the signal are the neuropeptide receptors. Some may be faulty or imbalanced in chemically sensitive individuals. That there are approximately 60 neuropeptides for 60 types of receptor cells tends to make evaluation of both receptor cells and neuropeptides extremely complex. Many neuroendocrine cells exist in the gut and have similar receptors in the brain. It has now been shown that hyperplasia of these gut cells may develop, and additional triggering of, or release of, inordinate amounts of neuropeptides may occur.[220] Biopsy has shown hyperplasia of gut neuroendocrine cells in some of our chemically sensitive patients who were extremely food and chemically sensitive. Foods and toxic chemicals appear to trigger the neuropeptides, resulting in gastrointestinal upset and occasionally pyloric and gastric irritability along with small bowel irritability and exceptionally long colons. If additive, these pollutants cause cerebral symptoms of such as recent memory loss, proprioceptive abnormalities, and emotional lability. Certainly, these pollutants impair judgment.

The molecular unity of receptors appears to be a long polypeptide chain capable of changing forms. When this chain assumes different shapes, either

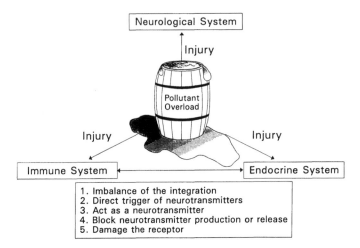

Figure 26.32.   Emotional network in the chemically sensitive individual.

wave-like or particulate, it can gather and retain information. However, some pollutants are able to alter these changing chains and, hence, disrupt this function. It appears that there are identical molecular components for information flow throughout evolution from the protozoa to humans. Therefore, much subhuman data can be used as a scientific baseline for evaluating the subset of chemically sensitive individuals who have emotional problems.

It has been observed that the periaqueductal gray area located around the third ventricle of the brain is filled with opiate receptors of virtually all the studied neuropeptides, making this a control area for pain. It appears that consciousness may be achieved in this area. For example, breathing exercises that relate to the brainstem nuclei may decrease sensitivity in this area. However, in a subset of chemically sensitive individuals, opiate-like effects that create sleepiness, lack of concentration, and dullness occur after toxicant challenge. These reactions are finite, lasting from a few minutes to 24 hours.

These areas of consciousness are full of neuropeptides and neuropeptide receptors, which may become imbalanced in the chemically sensitive individual. As stated previously, the limbic system, which consists of the hypothalamus, the pituitary gland, and the amygdala and other areas, integrates the pain center with the brainstem and the brainstem nuclei. In humans, electrode stimulation of the cortex over the amygdala evokes the whole gamut of emotions. Chemical and electric stimuli have been observed to trigger symptoms in the subset of neurologically involved chemically sensitive patients seen at the EHC-Dallas.

Both the amygdala and the hypothalamus have a 40-fold higher amount of receptors (opiate and others) than any other area in the brain, making them particularly vulnerable to pollutant stimuli. These areas of high numbers of receptors correspond to very specific nuclei and cellular groups identified with such processes as sexual behavior, appetite, and water balance, all of which have been seen to over- or under-respond after pollutant challenge in a subset of chemically sensitive patients with neurological involvement.

The neuropeptides often are analogs of psychoactive drugs and hormones. For example, insulin is not only a hormone from the pancreas but also a neuropeptide made and stored in the brain. Also, insulin receptors are located in the brain. It is now clear that the seat of emotions, the limbic system, is also a focal point for neuroreceptors, and it appears to be triggered often in a subset of chemically sensitive patients.

Neuroreceptors are located in many other areas throughout the body, and these areas may be disturbed by pollutants in some chemically sensitive patients. The nodal points, which are areas of concentration of these receptors, are places where lots of chemical actions take place and, therefore, are prone to pollutant injury. They are located anatomically at places that receive a lot of emotional modulation. For example, one nodal point area is in the dorsal horn of the spinal cord, where information on touch and noxious stimuli is passed to the brain. Other nodal points appear to be areas shown to be part of the neuroendocrine system, as described in Chapter 24. Because of this anatomical function, pollutant injury may imbalance the system, releasing discharges that could be interpreted as disordered emotions.

It appears that the area of entrance for disordered emotions for virtually all senses is a nodal point for neuropeptide receptors. Another example of emotions entering and integrating with neuropeptide receptors is the area in the amygdala that contains angiotensin receptors. When this area is stimulated in an animal, the animal will drink water even when it is full of water. This phenomenon of thirst reactions following pollutant exposure has been observed in chemically sensitive individuals following pollutant challenge in the ECU. Pathologic water drinking follows challenge testing of chemically sensitive patients with a weight gain of as much as 5 to 6 pounds. It has long been known that angiotensin mediates thirst. In individuals with angiotensin, the pollutant-stimulated neuropeptides induce a state of consciousness and then cause alterations in these states. The angiotensin receptors in the brain are identical to those in the kidney. However, the kidney receptors conserve water. Therefore, the same receptor in a different location can conserve or demand water. This function is an example of how a neuropeptide, which perhaps corresponds to a mood state, can integrate brain functions, as shown in a case report earlier in this chapter — see the section on the physiology of the hippocampus. We have also observed that following pollutant challenge, some neurologically involved, chemically sensitive patients develop rage states, some develop hyperactivity, and others develop depression. These responses suggest other neuropeptide triggering.

A further point of neuropeptide triggering is that it influences overall integration of behavior, which seems designed to be consistent with survival, and neuropeptides appear to provide the physiological basis of emotions. There is a striking pattern of neuropeptide receptors in the mood-regulating areas of the brain as well as in their role in mediating communication through the whole body, making neuropeptides the obvious candidate for the biochemical mediation of emotions. According to Pert et al.,[221] each neuropeptide, when occupying receptors at nodal points with the brain and body, may uniquely bias information processing. As a result of triggering of these nodal points by various chemicals, neurologically involved chemically sensitive patients can have varying moods and symptoms, and then, as the pollutant is removed and the reaction subsides, the patient's mood returns to normal. If this process occurs, each neuropeptide may evoke a unique tone that is equivalent to a mood state. An alteration of this unique tone has been observed many times in patients at the EHC-Dallas whose mood was suddenly altered by a chemical, food, or mold challenge. Often, the result was depression or confusion or hyperactivity, or the patient would just be slightly depressed. The endocrine system is integrated into this network. For example, a chemically sensitive premenstrual woman may develop irritability, an increased demand for certain foods, and/or a headache. All of these emotional reactions can now be explained by this network, which has become unbalanced.

As shown in many sections of this chapter and the book as a whole, the immune system must also be integrated into this neurological system. The key property of the immune system is that immune cells often move throughout the body with triggering resulting in emotional activation in a subset of chemically sensitive individuals. The cells of the immune system are in many ways identical to the brain cells in that both have all the receptors. For example, a monocyte comes within "scenting" distance of a neuropeptide, and, because the monocyte has receptors on its cell surface for the neuropeptide, it begins to chemotax toward the chemical. If the monocyte is overstimulated or suppressed by pollutants, emotional instability may occur in a subset of chemically sensitive individuals. The restabilization of the neuroendocrine system, and, thus, emotional stability is emphasized in our studies on transfer factor at the EHC-Dallas. Here not only did the responder patients improve their T cells and their function, but also the depression experienced by over 70% of the subset of neurologically involved chemically sensitive patients dissipated (see Chapter 39).

Neuroreceptors for opiates, PCP, and bombasin are also found on monocytes. These emotion-affecting chemicals actually appear to control the routing and migration of monocytes, which are pivotal in the immune system. They communicate with B and T cells, and they interact in the whole system to fight disease and to distinguish between self and nonself. They may decide which part of the body is a tumor cell to be killed by natural killer cells or which part may be restored. Monocytes not only have neuroreceptors, but they also have neuropeptides. Some subsets of those monocytes make beta endorphins and other opiates. In other words, the monocytes are making the same chemicals that are controlling mood in the brain. They control tissue integrity of the body, and this control results in a brain–body integration, which is disturbed in a large

Table 26.34. **Reproduction of Mood State in Neurologically Involved Chemically Sensitive Patients After 4 Days of Deadaptation with the Total Body Load Reduced Double-Blind Inhaled and Intradermal Challenge (30 Patients)**

	Depression	Hyperactivity
Formaldehyde (<0.20 ppm)[a]	28	2
Phenol (<0.002 ppm)[a]	25	5
Chlorine (<0.33 ppm)[b]	20	10
Pesticide — 2,4-DNP (<0.0034 ppm)	15	15
Ethanol (petroleum-derived <0.50 ppm)[a]	25	5
Saline placebo	2	1
Histamine 1/100 of concentrate[c]	15	15
Serotonin 1/100 of concentrate	15	15
Norepinephrine 1/100 of concentrate[c]	5	25
Epinephrine 1/100 of concentrate[c]	7	23
Meth choline 1/100 of concentrate[c]	15	15
Dopamine 1/100 of concentrate[c]	15	15

[a] Inhaled challenge, confirmed by intradermal challenge.
[b] Inhaled challenge, confirmed by oral challenge.
[c] Intradermal challenge.

Source: EHC-Dallas. 1994.

subset of chemically sensitive individuals. The mind appears to be the coalescence of information flowing through all these parts.

The subset of chemically sensitive individuals with brain dysfunction has this improperly functioning complex circuitry with integration of the neuro–endocrine–immune axis being imbalanced. Some of the excess xenobiotics that are in their bodies may have a similar enough formulation to act as neurotransmitters, and some may trigger neurotransmitters, while others may block neurotransmitter production or damage the neuroreceptors, causing them to function improperly in a hyperactive manner or not at all in their appropriate area. These alterations explain the severe emotional responses seen in some subsets of neurologically involved chemically sensitive patients. As stated previously, histamine, serotonin, norepinephrine, epinephrine, methacholine, and dopamine have not only been found to be altered in some chemically sensitive patients, but injection with these also reproduces their symptoms and then often clears them with the proper neutralizing dose (Table 26.34). It must be emphasized that many chemically sensitive individuals do not have nervous system involvement. Nonetheless, these patients with nervous system dysfunction react with pain and swelling to pollutant exposure. The chemically sensitive patient with environmentally induced arthritis falls into this category. Butler's test reveals they usually have no brain dysfunction or emotional problems, and clinically they do not appear to.

The kindling and sensitization model advocated by Bell et al.[222] integrates the literature of biological psychiatry[223-230] with that of neurotoxicity and sensitivity.[231-236] As has been discussed in Chapter 18, many environmental chemicals gain access to the CNS via the olfactory nerve, which goes to the limbic pathways[237-242] and induces lasting changes in limbic neuronal activity[243] with resultant overall cortical arousal levels.[244,245] Therefore, this initial pollutant stimulus alters a broad spectrum of behavioral and physiological functions, and this alteration is seen in some cases of chemical sensitivity. The lack of a blood-neural barrier in the olfactory system permits a wide range of toxic chemicals (including aromatic hydrocarbon solvents,[241] aluminum,[246] and cadmium[247)] direct access, via the nasal mucosa, to the olfactory bulb. Substances can move to the olfactory tubercle, piriform cortex, periamygdalar and enforhinal cortex, anterior hippocampus, locus coeruleus, raphe nuclei, and diagonal band.[238] According to Bell et al.,[222] kindling is a special type of time-dependent sensitization of olfactory-limbic neurons.[223,229] It involves the ability of a repeated, intermittent stimulus that is initially incapable of eliciting a response eventually to induce large reactions, even to the point of a motor seizure from later applications of the same stimulus. This phenomenon has been observed at the EHC-Dallas and reported throughout this book. The olfactory pathway[248] and limbic structures are especially vulnerable to kindling (sensitization). Time-dependent sensitization (TDS), which is different from kindling, is a phenomenon of amplification of an individual's response to a novel, foreign, and potentially threatening stimulus by the passage of time between the first and later stimuli (typically 2 to 14 days).[229] This TDS phenomenon (differing from kindling by triggering a broader range of behaviors and functions in various physiological systems) as well as kindling certainly fits with our observations of a subset of neurologically involved chemically sensitive patients undergoing challenge in the controlled environment. The marked increase in responses could be enhanced by the aforementioned neuroendocrine amplification system. As in kindling, TDS can develop during repeated, noncontinuous exposures to a given stressor. According to Bell et al.,[222] kindling has a convulsive or subconvulsive endpoint in the limbic structures (a state seen in a small subset of chemically sensitive patients—see the section on amygdala and hippocampus in this chapter), whereas time-dependent sensitization can involve a broader range of behaviors and functions in various physiological systems.[229,249,250]

In our experience at the EHC-Dallas, both types of phenomena have been seen in some of our neurologically involved chemically sensitive patients. In Bell et al.'s model[222] subconvulsive chemical kindling increases in limbic system excitation of the olfactory bulb, amygdala, piriform cortex, and hippocampus and would be the neurobiological mechanism that serves as an amplifier for reactivity to chemical exposures and as an initial common pathway for a range of clinical phenomena including cognitive and affective dysfunctions.[233,236,251] Certainly Butler[190] has demonstrated cognitive and affective disorders in a subset of chemically sensitive patients at the EHC-Dallas who were triggered by pollutants that could fit with her model. Derivative mechanisms would encompass neurophysiological (especially frontal and temporal dysfunctions demonstrated on SPECT brain scans at the EHC-Dallas), autonomic, endocrine, and immune pathways regulated and orchestrated by the limbic systems and connected structures.[252] The result of environmental pollutant exposure is an increase in limbic excitability, which results in plastic changes in excitatory amino acid receptors,[253] changes in dopaminergic pathways and benzodiazepine receptor numbers, and/or failure of GABA-ergic inhibitory function under stress of stimulation.[225,226] Endogenous events may trigger these.[224,226,248] Many environmental pollutants including pesticides,[231,232,242,254-256] flame retardants,[257] formaldehyde, acetone, and ozone[243] can induce or accelerate seizures and/or kindling phenomena in limbic structures.

Studies by Butler[190] at the EHC-Dallas showed that personality/emotional factors often associated with a subset of neurologically involved, long-term chemically sensitive patients include a strong concern for the state of one's health and physical malfunctioning, often to the point of preoccupation or obsession (this behavior alone may lead to the erroneous classification of hysteria or hypochondria by some clinicians). The difference between this subset of chemically sensitive patients and the hysteric is that the chemically sensitive patient can be cleared by pollutant avoidance and provoked by pollutant challenge, whereas the hysteric cannot. Other obvious features of the neurologically involved subset of chemically sensitive patients are often depression, despondency, even despair, and with many patients there will be self-destructive ideation or simply a wishing not to wake up in the morning. Low energy and easy fatiguability, which may be expressed as just feeling tired and worn out, with little zest for life, seldom feeling rested, and with a pervading sense of sadness and gloom, very often occur. These feelings can be explained by observing dysfunction of the neuro–immune–endocrine axis. Irritability, impatience, anger, and dependency along with a tendency to withdraw from others in times of stress are often part of the profile described by Butler.[190] Another general part of the pattern we found was a lowered sense of adequacy and worth, along with a decrease in self-concept and self-esteem. In some neurologically involved chemically sensitive patients, there appears to be a decrement also in coping skills and tolerance of stress, along with lessened ability to reach goals, maintain emotional stability, and to "bounce back". Many chemically sensitive people lose their sense of self-worth as their self-confidence is destroyed. This complex of symptoms may be explained by many different biochemical changes such as inadequate mediator detoxification, particularly lessened methylation, depletion of detoxification enzymes, overstimulation and depletion of neurotransmitters, and many other systems. Decreased methionine metabolism is the most frequent amino acid abnormality seen in the chemically sensitive patient and partially accounts for the excess of norepinephrine seen in a subset of agitated chemically sensitive individuals.

It is clear from our studies at the EHC-Dallas that much emotional dysfunction in a subset of neurologically involved chemically sensitive patients can be triggered by an overload of toxic chemicals, foods, and/or biological inhalants. They can, and have been, reproduced in thousands of double-blind inhaled, intradermal, and oral challenges at the EHC-Dallas. An example is presented in Figure 26.33.

Baldridge et al.,[258] working at the EHC-Dallas and using the Harrell-Butler Neuropsychological Screen Intermediate Psychoneurological Exam, found that the degrees of relationship between neurotoxic chemicals in the blood of chemically sensitive patients and cognitive neurobehavioral functioning were proportional. The number of total levels in parts per billion of the chemicals are adversely related to short-term memory loss, problems with coordination, motor sequencing, somatosensory deficits, and cognitive function. The results of this correlation imply a triggering of neuro–endocrine–immune axis with these resulting symptoms. The toxic chemicals include organochlorine pesticides and aliphatic and aromatic solvents ($p<.01$) (Table 26.35).

Following Butler's[190] advice, physicians at the EHC-Dallas evaluate emotions using several tests. We administer the Clinical Analysis Questionnaire (CAQ) to measure normal and pathological personality factors. Many investigators use the Rorschach test to gauge a subject's readiness to respond appropriately to emotional stimuli in some centers. We have not found this latter test to be efficacious at the EHC-Dallas, even though it is also used to measure a patient's originality and capacity for creative thinking. We monitor changes in life events, which are related to changes in health, with the Social Readjustment Rating Scale (SRRC). Also used are the Total Stress Load Inventory (TSLI) and a checklist of physical and psychological symptoms (see total stress load inventory). These typically show total stress load and various contributing stressors. We use the House–Tree–Person Preventive Test, which measures a person's sense of perspective, proportion, and detail as well as effects of certain mood and anxiety states. Occasionally, the thematic apperception test is used for identification of certain problem themes in their life. Profile of Mood States (POMS) measures transient mood states such as anxiety, tension, depression, anger, vigor, fatigue, bewilderment, and confusion. Occasionally, the MMPI is used for psychological pathology. Since this test was standardized on a psychiatric population data base, which is a different norm group than our patient population, we seldom use it. Health and Wellness Attitude Scale by Diedrickson and Butler[259] is used to assess a patient's positive-negative attitudes toward health or sickness. Pain Questionnaire Scale is used to evaluate the frequency and quality of pain. A stress management inventory scale is used to determine the patient's perception of stressors and how well he or she is coping. The psychological rating scale for the DSM-III is used on all patients to determine if there are any outstanding psychological problems such as suicide ideation (Figures 26.34 and 26.35 and Tables 26.36 and 26.37).

Using the various aforementioned tests and the present questionnaire (Table 26.38), Butler is able to differentiate environmentally triggered patients

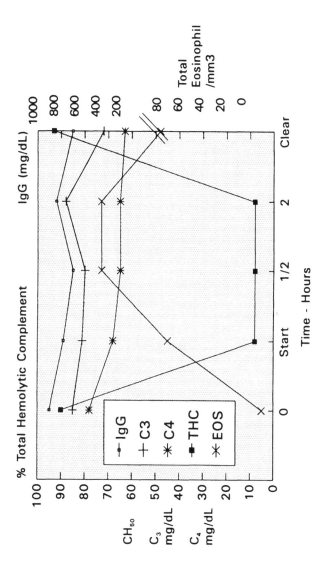

Figure 26.33. 34-year-old white female. Inhaled double-blind challenge with petroleum-derived ethanol (<0.50 ppm) after 4 days deadaptation in the ECU. Active — corn. Control — lettuce. (From EHC-Dallas. 1979.)

Table 26.35.   Person Product Moment Correlation
Coefficients Between the Harrel-Butler Intermediate
Psychoneurological Exam and Toxic Chemical
Blood Levels and Numbers

Chemical variable	Correlation coefficient
Total of all in ppb[c]	0.37[a]
Total of solvents in ppb	0.37[a]
Total in aliphatic hydrocarbons in ppb	0.37[a]
Number of solvents in blood	0.43[b]

[a] $p < 0.05$.

[b] $p < 0.01$.

[c] Parts per billion.

Source: EHC-Dallas. 1992.

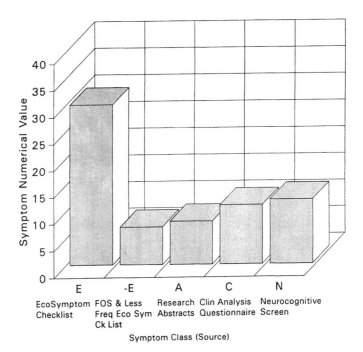

Figure 26.34.   Ecological differential configuration. EDC index: 5.5:1.
Ecological severity index: 1:1. Date of exam: 4/8/93. (From
J. R. Butler, EHC-Dallas. 1994. FOS symptoms from other
sources.)

with neurological reactivity from pure psychiatric patients. Another important
aspect of behavior related to environmental toxins is that of adverse effects on
behavioral development as a result of pre- or postnatal exposure to volatile

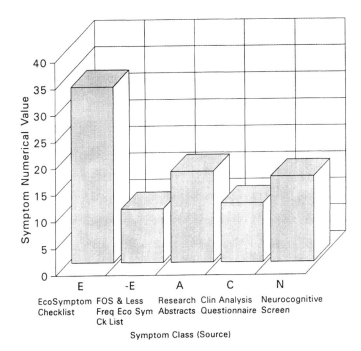

Figure 26.35.   Ecological differential configuration. EDC Index: 3.0:1; Ecological Severity Index: 22:1; Somato Nu Severity Index: 1:0.6. Date of exam: 5/17/93. (From J. R. Butler, EHC-Dallas. 1994.)

chemical and heavy metal toxins. The prenatal exposure may have been at such low levels as to present no difficulty to the mother but still at sufficient intensity to produce harmful effects on her unborn child. These effects were seen in some patients involved in the Minamata Bay mercury disaster. Dowty and Laseter's studies[260] have shown a bioconcentration of toxic organic chemicals in the cord blood of newborns (see Chapter 28). This accumulation of toxic organic chemicals appears to induce behavioral teratology. Such involvement leads to the assessment of the effects of teratogens on human development. Results of such studies show that multiple adverse effects may be seen and include lowered IQ and learning ability, sensory and attentional problems, neuromuscular difficulties, altered activity rates, morphological anomalies, slower growth and smaller size, and other maturational disturbances.

Challenge Testing*.   Inhaled, oral, or intradermal challenge testing can be performed on the chemically sensitive patient with neurological injury. In order to obtain accurate results, the patient must be in the baseline deadapted condition with a reduced total body load. Generally, this state can be achieved

* See Volume IV, Chapter 31, for more detailed information.

Table 26.36.   Classifications for Patients Using Coefficients (Coeff) from Discriminant Analysis

Variable	Questions or items	EXS	PiX	Coeff	Coeff [a]ES	Coeff [a]Pi
Inf		10.00	11.00	0.072	0.72	0.79
Dit Span		7.00	10.00	0.17	1.19	1.70
Dit Sym		7.00	9.00	0.155	1.63	1.40
Verb Mem		86.00	103.00	0.019	1.09	1.95
Del Rec		86.00	105.00	0.013	1.12	1.36
HB-CNS	Total	40.00	26.00	−0.064	−2.58	−1.66
HB-CNS	0.4	1.53	0.38	−0.517	−0.79	−0.19
HB-CNS	0.5	1.51	0.35	−0.671	−1.01	−0.23
HB-CNS	32	1.44	0.41	−0.452	−0.66	−0.19
HB-CNS	0.34	1.58	0.93	−0.307	−0.65	−0.30
				Total scores:	0.2303	4.636

Notes: Inf = information. Dit Span = digit span. Dit Sym = digit symbol. Verb Mem = verbal memory. Del Rec = delayed recall. HB-CNS = Harrell-Butler Comprehensive Neuropsychological Screen. ES = environmentally sensitive patients. X = mean. Pi = psychiatric patients.

[a]   Patients doing no better than 0.1% of the population.

Source: Butler, J. R., E. Lockhart, and W. J. Rea. 1993. Unpublished data. EHC-Dallas.

Table 26.37.   Class Level Discriminant Analysis Between Environmental and Psychiatric Patients

Actual group	No. of cases	Predicted group membership 1	2
Group 1			
Environmental Patients	56	54 96.49%	2 3.51%
Group 2			
Psychiatric Patients	60	1 1.67%	59 98.33%

Source: Butler, J. R., E. Lockhart, and W. J. Rea. 1994. Unpublished data. EHC-Dallas.

in 1 to 7 days in a less-polluted environment such as the ECU. Any of the aforementioned tests can then be performed as a baseline and as many times as desired after challenge. At the EHC-Dallas, over 2000 patients have been evaluated in the ECU using the various brain function parameters before and after challenge. Table 26.39 displays the results of 103 sample chemically sensitive patients who underwent 366 inhaled double-blind challenges and who had brain dysfunction demonstrated objectively. Only the symbol digit modalities and subset were used as objective measures, so more results might have been positive if other measurements had been included. Oral challenges have reproduced brain dysfunction thousands of times. It is clear that some groups, such as the vascular dysfunction patients, usually have brain function changes, while the patients with rheumatoid arthritis do not. The incitants producing the majority of reactions were formaldehyde, phenol, pesticides, chlorine, and foods. Figure 26.36 shows a sample challenge and response.

Intradermal challenges using dilute doses of histamine, serotonin, epinephrine, norepinephrine, methacholine, dopamine, and morphine antagonist have been used in over 5000 chemically sensitive patients with neurological involvement. Many symptoms of brain dysfunction have been reproduced. In addition, intradermal challenges with foods, biological inhalants, and chemicals also have often reproduced the patients' mood states as well as other brain dysfunctions. (See Volume IV, Chapter 36.)

Laboratory and Blood Analyses*.   Blood analyses have been performed on several thousand patients at the EHC-Dallas. Each will be discussed separately.

Solvents.   The EHC-Dallas has performed a series of solvent profiles on 90 patients who exhibited the signs and symptoms of chemical sensitivity with neurological changes including those aforementioned (Table 26.40). Increased levels over the detection limit of 0.3 ppb of many solvents were found including heptane, 15%; *n*-hexane, 66%; 2-methylpentane, 59%; 3-methylpentane, 55%; cyclopentane, 18%; and 2,2-dimethylbutane, 4%. Some substances, such as 2- and 3-methylpentane and hexane, were over 16 times the controls. In addition, lower levels (0.3 to 36 ppb) of trichloroethylene — 8%, tetrachloroethylene — 83%, 1,1,1,trichloroethane — 49%, styrene — 5%, toluene — 41%, benzene — 12%, trimethylbenzene — 7.5%, chloroform — 30% ethylbenzene — 16%, xylene — 38%, and dichloromethane — 15% were also found. Central and peripheral balance studies often correlate with solvent exposure (see Chapter 18). SPECT brain scans correlated with positive balance studies when the patients could not stand on their toes with their eyes closed.

Pesticides.   Levels of organochlorine and organophosphate pesticides at the EHC-Dallas have been shown to correlate with various behavioral abnormalities (see Volume II, Chapters 13,[79] and Volume IV, Chapter 31). Over 1000 chemically sensitive patients studied had organochlorines in their bloods.

Inorganic Chemicals.   Inorganic chemicals such as lead, mercury, and cadmium have been shown in many studies to disturb brain function and cause behavior abnormalities (see Volume II, Chapter 8[261]).

---

* For details, see Volume IV, Chapter 30.

Table 26.38.　Ecological Differential Configuration

Name_____

Date_____

Please mark an X in the appropriate column for *each* of the symptoms or problem areas as they apply to you. Symptoms must occur with sufficient severity and frequency to cause discomfort or to interfere with daily living.

Severe/Frequent	Moderate	Little or None	Symptoms
			1.　I feel tired, weary, or exhausted.
			2.　I have difficulty getting started in the morning.
			3.　I have difficulty concentrating.
			4.　My present performance is inferior to my prior level of functioning.
			5.　I feel "I'm not myself" or "What is happening to me?"
			6.　I have headaches.
			7.　I feel that I am losing control of my destiny.
			8.　My memory is poor — particularly immediate or short-term.
			9.　In general, I comprehend poorly.
			10.　I have a history of physical symptoms of several years.
			11.　I experience a sense that "This is not me."
			12.　I have aches and pains.
			13.　I have sinus discomfort.
			14.　I am clumsy.
			15.　I am easily fatigued.
			16.　I have low energy.
			17.　I experience weakness.
			18.　I am restless.
			19.　Mucus is a problem.
			20.　I have pain in the genital areas and painful urination.
			21.　I have been sickly a good part of my life.
			22.　I lose my voice.
			23.　I have difficulty swallowing.
			24.　I experience deafness.
			25.　I experience blindness.
			26.　I lose consciousness.
			27.　I have seizures.
			28.　I experience paralysis.
			29.　I haved abdominal pain and spells of nausea and vomiting.
			30.　I have shortness of breath.
			31.　I have eye problems.
			32.　My health and bodily dysfunctioning concern me — probably more than average.

Table 26.38 (continued). Ecological Differential Configuration

Severe/Frequent	Moderate	Little or None	Symptoms
			33. I have tended to withdraw socially and perhaps emotionally as well.
			34. When my mind and body are not functioning well, I become frightened/ anxious and depressed — all of which worsen my total ability to function.
			35. When faced with sudden demands, I seem to lack the confidence I once had and become confused, disturbed, depressed, and seem less able to cope.
			36. When my energy is low, my mood and spirits are also low, and I feel sad and despondent.
			37. I seldom awaken full of energy or zest for life but instead feel kind of worn out and blue.
			38. I haved difficulty with speech, cannot express myself well, forget or cannot find the right words.
			39. I seem easily irritated.
			40. I have chest pain and dizziness.
			41. I have phobias.
			42. I haved pain in the back, joints, and extremities.
			43. I cry a lot.
			44. I have anxiety attacks and palpitations.
			45. I am indifferent to, or lack concern for, illness.
			46. I have severe pain, prolonged and constant (no fluctuations).
			47. I am sure that I have a very serious disease of which I have presented a highly detailed account to my physician, but still I am not being treated effectively.
			48. I feel pressured to continue to talk and tell about the symptoms of my disease even through my physician distracts me with questions or inattention.
			49. I have nightmares and panic states.
			50. I have sweating, heart pounding, cold, clammy hands, and dry mouth and am easily startled.
			51. My motivation is lower than before illness.
			52. I have mental slowness — my thought processes seem slow.
			53. I have sleep disturbances.
			54. I have numbness in my hands and feet.

Table 26.38 (continued). Ecological Differential Configuration

Severe/Frequent	Moderate	Little or None	Symptoms
			55. Sensory disturbances (burning sensation, blurred vision, hearing difficulty, tingling, muscle movement, etc.).
			56. Intellectual inefficiency — I do not seem to think or to follow through on thoughts as well as I did before my illness.
			57. Impaired learning — I do not seem to learn as quickly or as effectively as I did before my illness.
			58. Visual construction — I have difficulty constructing what I see with my hands.
			59. Attention — I cannot seem to pay attention as well or as long as I did before my illness.
			60. Concentration — I do not concentrate very well.
			61. Decreased visual-perceptual accuracy — I seem to misjudge distance, bump into things, or misjudge a bit where my hands and feet are going.
			62. Dexterity — I am not as adept, agile, well-coordinated nor as quick with my hands as I was.
			63. Slow response time — going from question to answer or from thought to action takes longer now.
			64. I have odor sensitivity.
			65. Immediate and short-term verbal memory — I cannot seem to remember what I just heard or I forget within a few minutes.
			66. Learning a sequence of instructions and then performing the sequence in proper order is difficult.
			67. My balance is off — I seem unsteady as I stand.
			68. I seem unsteady as I walk, and I do not seem to walk a straight line very well.
			69. Vigilance — can only sustain my attention for a few moments at best.
			70. Sensory-tactile problems — my sense of touch is not very precise.
			71. Working arithmetic problems in my mind seems difficult in terms of speed in accuracy.
			72. I have difficulty in remembering a logical short story or even a conversation.
			73. I have difficulty remembering a picture or design well enough to draw it.

Table 26.38 (continued).   Ecological Differential Configuration

Severe/Frequent	Moderate	Little or None	Symptoms
			74. I have difficulty with abstraction, both thinking and reasoning.
			75. I have a problem with learning a task quickly when it involves looking, listening, remembering, and then performing the task.

Other symptoms or problem areas (describe):

_____

_____

_____

_____

Source: Butler, J. R. 1993. Personal communication.

Table 26.39.   103 Chemically Sensitive Patients with 366 Inhaled Double-Blind Challenges After 4 Days of Deadaptation with Total Load Reduced — Changes on Symbol Digit Modality Over 2 Standard Deviations

Chemical	Dose level (PPM)	Total no. of patients with changes on symbol digit modality subtest[a]
Ethanol-petroleum derived	<0.50	23
Phenol	<0.002	20
Chlorine	<0.33	15
Formaldehyde	<0.20	24
Pesticide (2,4-DNP)	<0.0034	18
1,1,1-Trichloroethane	ambient dose	3
Placebo	Saline — 3 challenges	7[b]

[a] Seven patients reacted to challenge and placebo.

[b] Though challenged with three placebos, no patient reacted to more than one.

Source: EHC-Dallas. 1993.

Immune Tests.   Immune tests in a subset of chemically sensitive patients frequently correlate with brain dysfunction. The most common correlation in our series appeared to be low complements and low T and B lymphocytes. As shown in the previous section on challenge testing, a low complement has often been seen in our patients with nervous dysfunction such as depression.

T lymphocytes are often low in the cerebral reacting chemically sensitive patients. Our studies with immune and brain dysfunction using a massive pollutant avoidance treatment show a correlation and improvement with increase in T lymphocytes to the Bender (75%) and MMRI (80%) in 24 patients (Table 26.41).

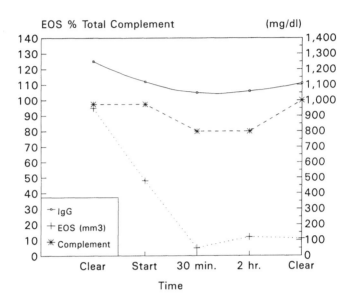

Figure 26.36. 33-year-old white female; inhaled double-blind challenge with chlorine (<0.33 ppm) after 4 days deadaptation in the ECU. Active — corn (Ig, EOS). Control — cabbage (complement). (From EHC-Dallas. 1979–80.)

Table 26.40.   Main Systems Involved in 72 Chemically Sensitive Patients Exposed to Aliphatic Solvents

Systems	No. of patients	(%)
Neurological	68	94
Cardiovascular	10	14
Musculoskeletal	16	22
Respiratory	26	36
Gastrointestinal	21	29
Genitourinary	5	7
Skin	10	14

Source: EHC-Dallas. 1992.

Antipollutant Enzymes.   Glutathione peroxidase, catalase, superoxide dismutase, and lipid peroxides are often abnormal in chemically sensitive patients with brain dysfunction. Usually, the enzymes are depressed and the peroxides are elevated.

Nutrients.   Nutrient deficiency and excess are often seen in chemically sensitive patients with brain dysfunction. Pfeiffer[262] and Wurtman[263] have each

**Table 26.41. Measures of Abnormal
Total T Lymphocyte and Total
Complement in Relation to Measures
of Abnormal Personality and Brain
Cognition — 100 Patients**[a]

% T-lymphocyte mean = 44
Total T-lymphocyte mean = 772/mm³
% B-lymphocyte mean = 26
Total B-lymphocyte mean = 375/mm³
Total complement mean = 115%

[a] Improved immunological parameters were
significantly correlated (<0.01) with improvement on Bender (75%) and MMP) (80%)
scores after avoidanc e therapy.

Source: EHC-Dallas. 1986.

written books to this effect. (See Volume I, Chapter 6,[52] for further discussion
of the experience of the EHC-Dallas.)

TREATMENT OF NERVE AND BRAIN DYSFUNCTION FROM POLLUTANT INJURY

Treatment of pollutant-induced nerve and brain dysfunction involves a
number of modalities. Meticulous avoidance of pollutants in the air, food, and
water and injection therapy for biological inhalants, foods, and selected chemicals should be used if needed. Nutrient supplementation is often necessary.
Heat depuration/physical therapy is used. Immune modulators such as transfer
factor is often helpful. All are discussed thoroughly in the rest of this book (also
see Volume IV).

Mind-body integration appears to be a feasible goal in the treatment of
the environmentally compromised, neurologically involved, chemically sensitive patient, since both psychological and chemical elements may have
devastating effects on the ability of the patient to maintain homeostasis.
Short-term and long-term intervention involving such techniques as positive
guided imagery, cognitive behavioral restructuring, biofeedback, and others
are often used to reduce stress, relieve depression, and initiate positive
cognitive, emotional, and behavioral changes (see Volume IV, Chapter 40).
These modalities appear only to work in the chemically sensitive after the
total pollutant load is decreased.

Results of proper evaluation and treatment have been gratifying. For
example, Rea et al.[191] have shown in 40 chemically sensitive patients proven
by double-blind inhaled challenge after a minimum of 4 days deadaptation with
their total loads decreased in the ECU that brain function measurements before
and after a stay in the ECU showed a statistically significant drop in signs and

symptoms with an improvement in brain function as the chlorinated pesticides decreased (see Volume II, Chapter 13,[79] and Volume IV, Chapter 31).

Other evidence exists that multiple toxic chemicals within a total body load can alter brain function and can be cleared with avoidance. Data from 24, and subsequently 100, ECU patients with pre- and postdetoxification test results on the Wechsler Adult Intelligence Scale-Revised for more cognitive and intelligence parameters and the Bender-Gestalt psychomotor function test has been gathered (Table 26.35).

The results of these tests indicate that significant psychomotor dysfunction accompanies, and is correlated with, environmental overload and that significant recovery of psychomotor and cognitive function occurs following detoxification of environmentally sensitive patients. Also, personality traits and clinical factors are changed significantly in environmentally compromised patients. Detoxification of these patients significantly improves clinical factors and personality traits.

Symptoms, which were recorded by the Physical and Psychological Symptoms Checklist, also improved with environmental treatment.

Personality traits and clinical factors were measured by the Clinical Analysis Questionnaire, and Butler[190] and his group have shown significant improvement.

The personality test used in these studies was the Clinical Analysis Questionnaire (CAQ [641]). The median test–retest coefficient across all scales has a 0.73 reliability. Validity coefficients for the individual factors, with 0.30 considered important, range from 0.45 to 0.86 for personality traits and from 0.69 to 0.95 for the clinical factors. The validity coefficients are considered remarkably faithful to its hypothesized structure.

We have concluded from these results that psychological function correlates significantly with clinical symptoms, blood chemistry, immunological function, and nutritional elements. Stresses from various sources — psychological, chemical, physical, and physiological including endocrinological — increase susceptibility to specific disease processes with symptomatology occurring from the more susceptible physiological or psychological system.

Working at the EHC-Dallas, Butler[190] demonstrated significant changes in IQ and perceptual motor functioning associated with chemical exposure in a group of children. This same group presented evidence of positive behavioral and personality changes with subsequent treatment in the ECU (Table 26.42).

Clearly, patients' inattention and/or limited attention span, maturation scales, memory, school work, and grades improved. Though this improvement was not 100% of the patients, clearly enough positive effects were found to warrant further evaluations. We now have treated over 500 children who were hyperactive and had learning disabilities. The success rate has improved since we now use more environmental control measures and nutrition to increase results. It is quite clear that ECU measures have a significant effect on the hyperactive, learning disabled child.

Using the aforementioned techniques, we have been able to define the learning problems of these patients as affected by their food and chemical

Table 26.42. Study of 13 Patients With Behavioral Problems and
Learning Disabilities

Problem	No. having problems	No. improved		No change
		Mild	Moderate/marked	
Attention span concentration	9	3 (33%)	3 (33%)	1
Motivation	10	1 (10%)	4 (40%)	5
Memory	12	0 (0%)	4 (33%)	8
Interaction with other children	13	2 (15%)	6 (46%)	5
School work and grades	11	3 (27%)	5 (45%)	3

Source: Butler, J. R. and W. J. Rea. 1988. EHC-Dallas.

sensitivities, and we have, thus, been able to solve a formerly unsolvable situation.

Emphasis should be placed on the fact that a spreading phenomenon occurs in these chemically sensitive patients, whereby they then become sensitive to many other chlorinated and even nonchlorinated toxic hydrocarbons as well as food and biological inhalants. These latter act as secondary triggers, which may become as harmful as the original ones. This reaction is probably due to a limited number of enzymatic detoxification pathways, impaired immune systems, and damage to the blood-brain barrier. These patients then become overloaded; they cannot detoxify rapidly enough, and new sensitivities develop (see Volume I, Chapter 4[242]). The clinician must guard against allowing for the adaptation phenomenon to continue to occur, since it masks toxic injury and has the potential for end-stage brain dysfunction to occur. Good and precise environmental control is needed in order to stop this spreading and restore the patient to health.

It is clear from the facts outlined in this chapter that hard evidence exists for the manifestations of a subset of chemically sensitive patients with neurological involvement. The autonomic and voluntary nervous system both may be involved, and good laboratory and clinical facts exist to substantiate this involvement. Much data is now available to demonstrate that pollutants can cause neuro–immune–endocrine dysfunction, and this dysfunction leads to triggering of various parameters, a process that then results in fixed-named disease if the pollutants are not withdrawn and the nutrients are not replaced. Objective evaluation of pollutant injury in the neurologically involved chemically sensitive patient is now possible using the iris corder, neurometer, balance studies, SPECT scans, multichannel EEGs, and the Harrel-Butler neuropsychological screen.

In summary, proper treatment involves massive avoidance of pollutants in air, food, and water. Injection therapy for biological inhalants, foods, and some chemicals is usually necessary. Parenteral supplementation of nutrients is performed. Often heat-depuration/physical therapy and tolerance moderators are used in treatment.

REFERENCES

1. Plum, F. 1970. Neurological integration of behavioral and metabolic control of breathing. In *Breathing: Hering-Breuer Centenary Symposium*, ed. R. Porter. 159–181. London: Churchill.

2. Berger, A. J., R. A. Mitchell, and J. W. Severinghaus. 1977. Regulation of respiration. *N. Engl. J. Med.* 297:138–142, 194–201.

3. McNicholas, W. T., R. Rutherford, R. Grossman, H. Moldofsky, N. Zamel, and E. A. Phillipson. 1983. Abnormal respiratory pattern generation during sleep in patients with autonomic dysfunction. *Am. Rev. Respir. Dis.* 128:429–433.

4. Monro, J. 1989. Personal communication.

5. Payan, D. G., J. D. Levine, and E. J. Goetzl. 1984. Modulation of immunity and hypersensitivity by sensory neuropeptides. *J. Immunol.* 132(4):1601–1604.

6. Foreman, J. and C. Jordan. 1983. Histamine release and vascular changes induced by neuropeptides. *Agents Actions* 13:105.

7. Rea, W. J. 1992. *Chemical Sensitivity. Vol. 1. Mechanisms of Chemical Sensitivity.* 17. Boca Raton, FL: Lewis Publishers.

8. Ader, R., ed. 1981. *Psychoneuroimmunology.* New York: Academic Press.

9. Goetzl, E. J. 1987. Leukocyte receptors for lipid and peptide mediators. *Fed. Proc.* 46(1):190–191.

10. Payan, D. G., J. P. McGillis, and E. J. Goetzl. 1986. Neuroimmunology. In *Advances in Immunology*, eds. F. J. Dixon, K. F. Austen, L. Hood, and J. W. Uhr. 2:199–323. New York: Academic Press.

11. Spector, N. H., ed. 1985. Neuroimmunomodulation (Proceedings of the First International Workshop on Neuroimmunomodulation). Bethesda, Maryland, International. Workshop on Group Neuroimmunomodulation.

12. Brooks, W. H., R. J. Cross, T. L. Roszman, and W. R. Markesbery. 1981. Neuroimmunomodulation: neural anatomical basis for impairment and facilitation. *Ann. Neurol.* 12:56–61.

13. Cross, R. J., W. R. Markesbery, W. H. Brooks, and T. L. Roszman. 1984. Hypothalmic-immune interactions: neuromodulation of natural killer activity by lesioning of the anterior hypothalamus. *Immunology* 51:399–405.

14. Jankovic, B. D. and K. Isakovic. 1973. Neuro-endocrine correlates of immune response. I. Effects of brain lesions on antibody production, arthus reactivity and delayed hypersensitivity in the rat. *Int. Arch. Allergy* 45:360–372.

15. Bulloch, K. and W. Pomerantz. 1984. Autonomic nervous system innervation of thymic-related lymphoid tissue in wildtype and nude mice. *J. Comp. Neurol.* 228:57–68.

16. Cannon, J. G., J. B. Tatro, S. Reichlin, and C. A. Dinarello. 1986. Alpha-melanocyte stimulating hormone inhibits immunostimulatory and inflammatory actions of interleukin-1. *J. Immunol.* 137:2232–2236.

17. Fujiwara, R. and K. Orita. 1987. The enhancement of the immune response by pain stimulation in mice. I. The enhancement effect of PFC production via sympathetic nervous system in vivo and in vitro. *J. Immunol.* 138:3699–3703.

18. Goetzl, E. J., T. Chernov-Rogan, M. P. Cooke, F. Renold, and D. G. Payan. 1985. Endogenous somatostatin-like peptides of rat basophilic leukemia cells. *J. Immunol.* 135:2707–2712.

19. Weinreich, D. and B. J. Undem. 1987. Immunological regulation of synaptic transmission in isolated guinea pig autonomic ganglia. *J. Clin. Invest.* 79:1529–1532.

20. Said, S. I. 1984. Vasoactive intestinal polypeptide (VIP). Current states. *Peptides* 5:143–150.

21. Said, S. I. 1982. Vasoactive peptides in the lung, with special reference to vasoactive intestinal peptide. *Exp. Lung Res.* 3:343–348.

22. Stead, R. H., M. Tomioka, G. Quinonez, G. T. Simon, S. Y. Felten, and J. Bienenstock. 1987. Intestinal mucosal mast cells in normal and nematode-infected rat intestines are in intimate contact with peptidergic nerves. *Proc. Natl. Acad. Sci. U.S.A.* 84:2975–2979.

23. Perry, V. H., M. C. Brown, and S. Gordon. 1987. The macrophage response to central and peripheral nerve injury: A possible role for macrophages in regeneration. *J. Exp. Med.* 165:1218–1223.

24. Bernton, E. W., J. E. Beach, J. W. Holaday, R. C. Smallridge, H. G. Fein. 1987. Release of multiple hormones by a direct action of interleukin-1 on pituitary cells. *Science* 238(4826):519–521.

25. Besedovsky, H., A. del Rey, E. Sorkin, and C. A. Dinarello. 1986. Immunoregulatory feedback between interleukin-1 and glucocorticoid hormones. *Science* 233:652–654.

26. Woloski, B. M., E. M. Smith, W. J. Meyer, III, G. M. Fuller, and J. E. Blalock. 1985. Corticotropin-releasing activity of monokines. *Science* 230(4729):1035–1037.

27. Rea, W. J. 1994. *Chemical Sensitivity. Vol. II. Sources of Total Body Load.* 765. Boca Raton, FL: Lewis Publishers.

28. Benveniste, E. N., J. E. Merrill, S. E. Kaufman, and D. W. Golde. 1985. Purification and characterization of a human T-lymphocyte-derived glial growth-promoting factor. *Proc. Natl. Acad. Sci. U.S.A.* 82:3930–3934.

29. Merrill, J. E., S. Kutsunai, C. Mohlstrom, F. Hofman, J. Groopman, and D. W. Golde. 1984. Proliferation of astroglia and oligodendroglia in response to human T cell-derived factors. *Science* 224(4656):1428–1430.

30. Dafny, N., B. Prieto-Gomez, and C. Reyes-Vasquez. 1985. Does the immune system communicate with the central nervous system? Interferon modifies central nervous activity. *J. Neuroimmunol.* 9:1–12.

31. Benveniste, E. N., S. Kutsunai, and J. E. Merrill. 1986. Immunoregulatory molecules modulate glial cell growth. In *Leukocytes and Host Defense*, eds. J. J. Oppenheim and D. M. Jacobs. 221–226. New York: Alan R. Liss.

32. Williams, C. A., N. Schupf, and T. E. Hugli. 1985. Anaphylatoxin C5a modulation of an alpha-adrenergic receptor system in the rat hypothalamus. *J. Neuroimmunol.* 9:29–40.

33. Levine, J. D., E. J. Goetzl, and A. I. Basbaum. 1987. Contribution of the nervous system to the pathophysiology of rheumatoid arthritis and other polyarthritides. *Rheum. Dis. Clin. N. Am.* 13:369–383.

34. Colpaert, F. C., J. Donnerer, and F. Lembeck. 1983. Effects of capsaicin on inflammation and on substance P content of nervous tissues in rats with adjuvant arthritis. *Life Sci.* 32:1827.

35. Lundberg, J. M., A. Saria, E. Brodin, S. Rosell, and K. Folkers. 1983. A substance P antagonist inhibits vagally induced inflammation and bronchial smooth muscle contraction in the guinea pig. *Proc. Natl. Acad. Sci. U.S.A.* 80(4):1120–1124.

36. Lundberg, J. M. and A. Saria. 1983. Capsaicin-induced desensitization of airway mucosa to cigarette smoke, mechanical and chemical irritants. *Nature* 302:251–253.

37. Walker, K. B., M. H. Serwonska, F. H. Valone, W. S. Harkonen, O. L. Frick, K. H. Seriven, W. D. Ratnoff, J. G. Browning, D. G. Payan, and E. J. Goetzl. 1988. Distinctive patterns of release of neuroendocrine peptides after nasal challenge of allergic subjects with ryegrass antigen. *J. Clin. Immunol.* 8(2):108–113.

38. Schiogolev, S., W. Urba, D. Longo, and E. J. Goetzl. Unpublished data. Contributions of neuropeptides to the altered vascular permeability induced by IL2-LAK cell therapy. *Clin. Res.*

39. Day, R., S. Lemaire, D. Nadeau, I. Keith, and I. Lemaire. 1987. Changes in autacoid and neuropeptide contents of lung cells in asbestos-induced pulmonary fibrosis. *Am. Rev. Respir. Dis.* 136(4):908–915.

40. Wallengren, D., R. Ekman, and H. Moller. 1986. Substance P and vasoactive intestinal peptide in bullous and inflammatory skin disease. *Acta Derm. Venereol. (Stockh)* 66:23–28.

41. Foreman, J. C., C. C. Jordan, and W. Piotrowski. 1982. Interaction of neurotensin with the substance P receptor mediating histamine release from rat mast cells and the flare in human skin. *Br. J. Pharmacol.* 77:531–539.

42. Rea, W. J. 1992. *Chemical Sensitivity. Vol. 1. Mechanisms of Chemical Sensitivity.* 47. Boca Raton, FL: Lewis Publishers.

43. Norton, S. 1986. Toxic responses of the central nervous system. In *Casarett and Doull's Toxicology: The Basic Science of Poisons*, 3rd ed., eds. C. D. Klaassen, M. D. Amdur, and J. Doull. 359–386. New York: Macmillan.

44. Bondareff, W. 1965. The extracellular compartment of the cerebral cortex. *Anat. Rec.* 152:119–127.

45. Kuhlenbeck, H. 1970. *The Central Nervous System of Vertebrates.* Vol. 3. New York: Academic Press.

46. Westergaard, E., B. van Deurs, and H. E. Bondsted. 1977. Increased vesicular transfer of horseradish peroxidase across cerebral endothelim evoked by acute hypertension. *Acta Neuropathol.* 37:141–152.

47. Cervos-Navarro, J. and J. I. Rozas. 1978. The arteriole as a site of metabolic change. *Adv. Neurol.* 20:17–24.

48. Welsh, F. A. and M. J. O'Connor. 1978. Patterns of microcirculatory failure during incomplete cerebral ischemia. *Adv. Neurol.* 20:133–139.

49. Hazama, F., S. Amano, and T. Ozaki. 1978. Pathological changes of cerebral vessel endothelial cells in spontaneously hypertensive, rats with special reference to the role of these cells in the development of hypertensive cerebrovascular lesions. *Adv. Neurol.* 20:359–369.

50. Jacobs, J. M, R. M. MacFarlane, and J. B. Cavanagh. 1976. Vascular leakage in the dorsal root ganglia of the rat, studied with horseradish peroxidase. *J. Neurol. Sci.* 29:95–107.

51. Westergaard, E., B. van Deurs, and H. E. Brondsted. 1977. Increased vascular transfer of horseradish peroxidase across cerebral endothelium evoked by acute hypertension. *Acta Neuropathol. (Berl.)* 37:141–152.

52. Rea, W. J. 1992. *Chemical Sensitivity. Vol. 1. Mechanisms of Chemical Sensitivity.* 221. Boca Raton, FL: Lewis Publishers.

53. Jacobs, J. M., R. M. MacFarlane, and J. B. Cavanagh. 1976. Vascular leakage in the dorsal root ganglia of the rat studied with horseradish peroxidase. *J. Neurol. Sci.* 29:95–107.

54. Jacobs, J. M. 1977. Penetration of systemically injected horseradish peroxidase into ganglia and nerves of the autonomic nervous system. *J. Neurocytol.* 6:607–618.

55. Olney, J. W., Rhee, V., and T. de Gubareff. 1977. Neurotoxic effects of glutamate on mouse area postrema. *Brain Res.* 120:151–157.

56. Friede, F. L. 1966. *Topographic Brain Chemisry.* New York: Academic Press.

57. Olney, J. W. 1971. Glutamate-induced neuronal necrosis in the infant mouse hypothalamus. *J. Neuropathol. Exp. Neurol.* 30:75–90.

58. Rea, W. J. 1994. *Chemical Sensitivity. Vol. II. Sources of Total Body Load.* 735. Boca Raton, FL: Lewis Publishers.

59. Jacobs, J. M., N. Carmichael, and J. B. Cavanagh. 1977. Ultrastructural changes in the nervous system of rabbits poisoned with methyl mercury. *Toxicol. Appl. Pharmacol.* 39:249–261.

60. Rapoport, S. 1976. *Blood-Brain Barrier in Physiology and Medicine.* 129–152. New York: Raven Press.

61. Hanig, J. P. and E. H. Herman. 1991. Toxic responses of the heart and vascular systems. In *Casarett and Doull's Toxicology: The Basic Science of Poisons*, 4th ed., eds. M. O. Amdur, J. Doull, and C. D. Klaassen. 450. New York: Pergamon Press.

62. Arrighi, R. E. 1974. Mammalian chromosomes. In *The Cell Nucleus.* Vol. II, ed. H. Busch. 1–32. New York: Academic Press.

63. National Institute on Drug Abuse. Apr. 1988. Facts about teenagers and drug abuse. NIDA Capsules, C-83–07a. rev.

64. Anthony, D. C. and D. G. Graham. 1991. Toxic responses of the nervous system. In *Casarett and Doull's Toxicology: The Basic Science of Poisons*, 4th ed., eds. M. O. Amdur, J. Doull, and C. D. Klaassen. 434. New York: Pergamon Press.

65. Baselt, R. C., ed. 1982. *Disposition of Toxic Drugs and Chemicals in Man,* 2nd ed. Davis, CA: Biomedical Publications.

66. Monmaney, T. 1987. Are we led by the nose? *Discover* 8:48–56.

67. Ader, R. 1987. *Behavioral Influence on Immunity.* Paper presented at the meeting of the Cabinet of Environmental Medicine, Nashville, TN.

68. Rea, W. J., J. R. Butler, J. L. Laseter, and I. R. DeLeon. 1984. Pesticides and brain function changes in a controlled environment. *Clin. Ecol.* 2(3):145–150.

69. Rea, W. J. 1994. *Chemical Sensitivity. Vol. II. Sources of Total Body Load.* 535–685. Boca Raton, FL: Lewis Publishers.

70. Rea, W. J. 1994. *Chemical Sensitivity. Vol. II. Sources of Total Body Load.* 638. Boca Raton, FL: Lewis Publishers.

71. Rea, W. J. 1994. *Chemical Sensitivity. Vol. II. Sources of Total Body Load.* 101. Boca Raton, FL: Lewis Publishers.

72. Delpech, A. 1856. Note sur les accidents que dévelope, chez les ouvriers en caoutchouc, l'inhalation du sulfure de carbone en vapeur. *Bull. Acad. Imp. Med.* 21:350.

73. Rea, W. J. 1994. *Chemical Sensitivity. Vol. II. Sources of Total Body Load.* 713. Boca Raton, FL: Lewis Publishers.

74. Charcot, J.-M. 1874. De la scherose laterale amyotrophique: Symptomatologie. *Prog. Med.* 29:453.

75. Davidson, M. and M. Feinleib. 1972. Carbon disulfide poisoning: A review. *Am. Heart J.* 83:100.

76. Proctor, N. H. and J. P. Hughes. 1978. *Chemical Hazards of the Workplace.* 149. Philadelphia: J. B. Lippincott.

77. Savage, E. P., T. J. Keefe, L. M. Mounce, R. K. Heaton, J. A. Lewis, and P. J. Burcar. 1988. Chronic neurological sequelae of acute organophosphate pesticide poisoning. *Arch. Environ. Health* 43:38–45.

78. Anger, W. K. 1984. Neurobehavioral testing of chemicals: impact on recommended standards. *Neurobehav. Toxicol. Teratol.* 6:147–153.

79. Rea, W. J. 1994. *Chemical Sensitivity. Vol. II. Sources of Total Body Load.* 837. Boca Raton, FL: Lewis Publishers.

80. Davis, G. C., M. S. Buchsbaum, D. P. van Kammen, and W. E. Bunney. 1979. Chronic parkinsonism secondary to intravenous injection of meperidine analogues. *Psychiatry Res.* 1:249–254.

81. Langston, J. W., P. Ballard, J. W. Tetrud, and I. Irwin. 1983. Chronic Parkinsonism in humans due to a product of meperidine-analog synthesis. *Science* 219:979–980.

82. Gupta, M., B. K. Gupta, R. Thomas, V. Bruemmer, J. R. Sladek, Jr., and D. L. Felton. 1986. Aged mice are more sensitive to 1-methyl-4-phenyl-1,2,3,6-tetrahydropyridine treatment than young adults. *Neurosci. Lett.* 70:326–336.

83. Burns, R. S., C. C. Church, S. P. Markey, M. H. Ebert, D. M. Jacobowitz, and I. J. Kopin. 1983. A primate model of Parkinsonism: selective destruction of dopaminergic neurons in the pars compacta of the substantia nigra by N-methyl-4-phenyl-1,2,3,6-tetrahydropyridine. *Proc. Natl. Acad. Sci. U.S.A.* 80:4546–4550.

84. Crinnion, W. 1988. A brief review of the neurotoxic effects of environmental chemicals. *Townsend Lett. Doctors* 59:256–258.

85. Calne, D. B., A. Eisen, E. McGeer, and P. Spencer. 1986. Alzheimer's disease, Parkinson's disease, and motoneurone disease: abiotrophic interaction between aging and environmental chemicals. *Lancet* 2(8515):1067–1070.

86. Calne, D. B., J. W. Langston, M. R. W. Martin, A. J. Stoessl, T. J. Ruth, M. J. Adam, B. D. Pate, and M. Schulzer. 1985. Positron emission tomography after MPTP: observations relating to the cause of Parkinson's disease. *Lett. Nature* 317:246–248.

87. Davis, K. L., P. A. Berger, L. E. Hollister, and J. D. Barchas. 1978. Cholinergic involvement in mental disorders. *Life Sci.* 22:1865–1871.

88. Melamed, E. and S. Lavy. 1977. Parkinsonism associated with chronic inhalation of carbon tetrachloride. *Lancet* 1(8019):1015.

89. Bochetta, A. and G. U. Corsini. 1986. Parkinson's disease and pesticides. *Lancet* 2:1163.

90. Joubert, J. and P. H. Joubert. 1988. Chorea and psychiatric changes in organophosphate poisoning. A report of two further cases. *S. Afr. Med. J.* 74:32–34.

91. Peters, H. A., R. L. Levine, C. G. Matthews, and L. J. Chapman. 1986. Carbon tetrachloride carbon disulfide (80:20 fumigants) and other pesticides in grain storage workers. *Acta Pharmacol. Toxicol.* 59:(Suppl. 7):535–546.

92. Spencer, P. S., P. B. Nunn, J. Hugon, A. C. Ludolph, S. M. Ross, D. N. Roy, and R. C. Robertson. 1987. Guam amyotrophic lateral sclerosis parkinsonismdementia linked to a plant excitant neurotoxin. *Science* 237(4814):517–522.

93. Xintaras, C. and J. Burg. 1980. Screening and prevention of human neurotoxic outbreaks: Issues and problems. In *Experimental and Clinical Neurotoxicology*, eds. P. Spencer and H. Schaumburg. Baltimore: Williams and Wilkins.

94. Fleming, L., J. B. Mann, J. Bean, T. Briggle, and J. R. Sanchez-Ramos. 1994. Parkinson's disease and brain levels of organochlorine pesticides. *Ann. Neurol.* 36(1):100–103.

95. Chapman, L. J., H. A. Peters, C. G. Matthews, and R. L. Levine. 1987. Parkinsonism and industrial chemicals (letter). *Lancet* 1:332–323.

96. Peters, H. A., A. Gocmen, D. J. Cripps, G. T. Bryan, and I. Dogramaci. 1982. Epidemiology of hexachlorobenzene-induced porphyria in Tuncey: clinical and laboratory follow-up after 25 years. *Arch. Neurol.* 39(2):744–749.

97. Fahn, S. 1977. Secondary parkinsonism. In *Scientific Approaches to Clinical Neurology*, eds. E. S. Goldensohn and S. H. Appel. 732. Philadelphia: Lea and Febiger.

98. Markstein, R. and J.-M. Vigouret. 1989. Is D-1 receptor stimulation important for the anti-parkinson activity of dopamine agonists? In *Early Diagnosis and Preventive Therapy in Parkinson's Disease*, eds. H. Przuntek and P. Riedere. 257–267. New York: Springer-Verlag/Wien.

99. Ben-Shachar, D. and M. B. H. Youdim. 1992. 14. Brain iron and nigrostriatal dopamine neurons in Parkinson's disease. In *Iron and Human Disease*, ed. R. B. Lauffer. 349–363. Boca Raton, FL: CRC Press.

100. Steventon, G. B., M. T. E. Heafield, R. H. Waring, and A. C. Williams. 1989. Xenobiotic metabolism in Parkinson's disease. *Neurology* 39:883–887.

101. Davis, K. L. and J. A. Yesavage. 1979. Brain acetylcholine and disorders of memory. In *Brain Acetylcholine and Neuropsychiatric Disease*, eds. K. L. Davis and P. A. Berger. 205–213. New York: Plenum Press.

102. Mistry, J. S., D. J. Abraham, and I. Hanin. 1986. Neurochemistry of aging. 1. Toxins for an animal model of Alzheimer's disease. *J. Med. Chem.* 29:376–380.

103. Salama, J., R. Gherardi, H. Amiel, J. Poirier, P. Delaporte, and F. Gray. 1986. Postanoxic delayed encephalopathy with leukoencephalopathy and non-hemorrhagic cerebral amyloid angiopathy. *Clin. Neuropathol.* 5(4):153–156.

104. Heafied, M. T., S. Fearn, G. B. Steventon, R. H. Waring, A. C. Williams, and S. G. Sturman. 1990. Plasma cysteine and sulphate levels in patients with motor neurone, Parkinson's and Alzheimer's disease. *Neurosci. Lett.* 110:216–220.

105. Ehmann, W. D., W. R. Markesbery, M. Alanddin, T. Hossain, and E. H. Brubaker. 1986. Brain trace elements in Alzheimer's disease. *Neurotoxicology* 7:197–206.

106. Vance, D. E., W. D. Ehmann, and W. R. Markesbery. 1988. Trace element imbalances in hair and nails of Alzheimer's disease patients. *Neurotoxicology* 9(2):197–208.

107. Ward, N. I. and J. A. Mason. 1986. Neutron activation analysis techniques for identifying elemental status in Alzheimer's disease. Presentation to New Trends in Activation Analysis (7). Conference, Copenhagen.

108. Little, A., R. Levy, P. Chrcaqui-kidd, and D. Hand. 1985. A double-blind, placebo controlled trial of high-dose lecithin in Alzehimer's disease. *J. Neurol. Neurosurg. Psychiatry* 48:742–763.

109. Morris, J. A. 1985. Clinical viral infections and multiple sclerosis. *Lancet* 2:273.

110. Brown, I. A. 1954. Chronic mercurialism: a cause of the clinical syndrome of amyotrophic lateral sclerosis. *AMA Arch. Neurol. Psychiatry* 72:674–681.

111. Jandan, R. and W. E. Bradley. 1985. Amyotrophic lateral sclerosis. Part 2. Etiopathogenesis. *Ann. Neurol.* 18:419–431.

112. Cashman, N. R. and J. P. Antel. 1988. Amyotrophic lateral sclerosis: an immunologic perspective. *Immunol. Allergy Clin. N. Am.* 8(2):331–342.

113. Humphrey, J. H. and M. McClelland. 1944. Cranial nerve palsies with herpes following general anesthesia. *Br. Med. J.* 1:315–318.

114. Tandan, R. and W. G. Bradley. 1985. Amytrophic lateral sclerosis. I. Clinical features, pathology, and ethical issues in management. *Ann. Neurol.* 18:271–280.

115. Ford, H. C. 1987. Multiple sclerosis: A survey of alternative hypotheses concerning altrology, pathogenesis and predisposing factors. *Med. Hypotheses* 24:201–207.

116. McDonald, W. I. 1986. The mystery of the origin of multiple sclerosis. *J. Neurol. Neurosurg. Psychiatry* 49:113–123.

117. Laborde, J. M., W. A. Gando, and M. L. Jcetzen. 1988. Glimate, defused solar radiation and multiple sclerosis. *Soc. Sci. Med.* 27:231–238.

118. Ingalls, T. H. 1986. Trigger for multiple sclerosis. *Lancet* 2:160.

119. Murrell, J. G., P. J. Q'Donoghue, and J. Ellis. 1986. A review of the sheep-multiple sclerosis connection. *Med. Hypotheses* 9:27–39.

120. Ingalls, T. H. 1983. Epidemiology, etiology and prevention of multiple sclerosis: hypothesis and fact. *Am J. Forens. Med. Pathol.* 4:55–61.

121. Nanji, A. A. and G. Narod. 1986. Multiple sclerosis, latitude, and dietary fat: is pork the missing link? *Med. Hypotheses* 20:279–282.

122. Maas, A. G. and L. A. Hogenhuis. 1987. Multiple sclerosis and possible relationships to cocoa: a hypothesis. *Ann. Allergy* 59:76–79.

123. Waisbvren, B. A. 1982. Swine-influenza vaccine (letter). *Ann. Intern. Med.* 97:149.

124. Anderson, E., H. Isager, and K. Hyllested. 1981. Risk factors in multiple sclerosis: tubercular reactivity, age at measles infection, tonsillectomy and appendectomy. *Acta Neurol. Scand.* 63:131–135.

125. Blisard, K. S., M. Kornfeld, P. J. McFreeley, and J. E. Smialek. 1986. The investigation of alleged insecticide toxicity: a case involving chlordane exposure, multiple sclerosis, and peripheral neuropathy. *J. Forens. Sci.* 31:1499–1504.

126. Noseworthy, J. H. and G. P. Rice. 1988. Trichlorethylene poisoning mimicking multiple sclerosis (letter). *Can. J. Neurol. Sci.* 15:87–88.

127. Corsini, G. U., S. Pintus, C. C. Chiueh, J. F. Weiss, I. J. Kpoin. 1985. 1-Methyl-4 phenyl-1,2,3,6-tetrahydropyridine (MPTP) neurotoxicity in mice is enhanced by pretreatment with diethyldith co-carbamate. *Eur. J. Pharmacol.* 119:127–128.

128. Growden, J. H., M. J. Hirsch, R. J. Wurtman, and W. Weiner. 1977. Oral choline administration to patients with tardive dyskinesia. *N. Engl. J. Med.* 297:524–527.

129. Alexander, V., S. S. Leffingwell, J. W. Loyd, R. J. Waxweiler, and R. L. Miller. 1982. Investigation of an apparent increased prevalence of brain tumors in U.S. petrochemical plants. In *Brain Tumors in the Chemical Industry*, eds. I. J. Selikoff and E. C. Hammond. 97–107. New York: The New York Academy of Sciences.

130. Selikoff, I. J. and E. C. Hammond, eds. 1982. *Brain Tumors in the Chemical Industry.* New York: The New York Academy of Sciences.

131. Gupta, S. R., M. H. Naheedy, R. J. O'Hara, and F. A. Rubino. 1985. Hydrocephalus following iophendylate injection myelography with spontaneous resolution: case report and review. *Comput. Radiol.* 9(6):359–364.

132. Hiller, J. L., G. I. Benda, M. Rahatzad, J. R. Allen, D. H. Culver, C. V. Carlson, and J. W. Reynolds. 1986. Benzyl alcohol toxicity: impact on mortality and intraventricular hemorrhage among very low birth weight infants. *Pediatrics* 77:500–506.

133. Benda, G. I., J. L. Hiller, and J. W. Reynolds. 1986. Benzyl alcohol toxicity: impact on neurologic handicaps among surviving very low birth weight infants. *Pediatrics* 77:507–512.

134. McMichael, A. J., P. A. Baghurst, N. R. Wigg, G. V. Vimpani, E. F. Robertson, and R. J. Roberts. 1988. Port Pirie Cohort Study: environmental exposure to lead and children's abilities at the age of four years. *N. Engl. J. Med.* 319(8):468–475.

135. Fernandez, F., A. Perez-Higueras, R. Hernandez, A. Verdu, C. Sanchez, A. Gonzalez, and J. Quero. 1986. Hydranencephaly after maternal butane-gas intoxication during pregnancy. *Dev. Med. Child Neurol.* 28:361–363.

136. Maizlish, N., M. Schenker, C. Weisskopf, J. Seiker, and S. Samuels. 1987. A behavioral evaluation of pest control workers with short-term, low-level exposure to the organophosphate diazinon. *Am. J. Ind. Med.* 12:153–172.

137. Choi, Ben H. 1986. Methylmercury poisoning of the developing nervous system: I. Pattern of neuronal migration in the cerebral cortex. *Neurotoxicology* 7(2):591–600.

138. Snyder, K. D. 1971. Congenital mercury poisoning. *N. Engl. J. Med.* 284:1014–1016.

139. Amin-Zaki, L., S. Ehassani, M. A. Majeed, T. W. Clarkson, R. A. Doherty, M. R. Greenwood, and T. Giovanoli-Jakubczak. 1976. Perinatal methyl mercury poisoning in Iraq. *Am. J. Dis. Child* 130:1070–1076.

140. Harada, M. 1976. Minamata disease. Chronology and medical report. *Bull. Inst. Constitutional Med. Kumamoto Univ.* 25(Suppl.):1–60.

141. Marsh, D. O., G. J. Myers, T. W. Clarkson, L. Amin-Zaki, S. Tikriti, and M. A. Majud. 1980. Fetal methyl mercury poisoning: Clinical and toxicological data on 29 cases. *Ann. Neurol.* 7:348–353.

142. Spyker, J. M., S. B. Sparber, and A. M. Goldberg. 1972. Subtle consequences of methyl mercury exposure: Behavioral deviations in offspring of treated mothers. *Science* 177:621–623.

143. Hughes, J. A. and Z. Annau. 1976. Postnatal behavioral effects in mice after prenatal exposure to methyl mercury. *Pharmacol. Biochem. Behav.* 4:385–391.

144. Su, M. Q. and G. T. Okita. 1976. Behavioral effects on the progeny of mice treated with methyl-mercury. *Toxicol. Appl. Pharmacol.* 38:295–305.

145. Cuomo, V., L. Ambrosi, Z. Annau, R. Cagiano, N. Brunello, and G. Racagni. 1984. Behavioral and neurochemical changes in offspring of rats exposed to methyl mercury during gestation. *Neurobehav. Toxicol. Teratol.* 6:249–254.

146. Egger, J., C. M. Carter, J. Wilson, M. W. Turner, and J. F. Soothill. 1983. Is migraine food allergy? A double-blind controlled trial of oligoantigenic diet treatment. *Lancet* II: 865–869.

147. Crayton, J. W., T. Stone, and G. Stein. 1981. Epilepsy precipitated by food sensitivity: report of a case with double-blind placebo-controlled assessment. *Clin. Electroencephalogr.* 12:192–198.

148. Chambers, J. E. and P. E. Levi. 1992. *Organophosphates: Chemistry, Fate, and Effects.* San Diego: Academic Press.

149. Drewes, L. R. and A. K. Singh. 1985. Transport, metabolism and blood flow in brain during organophosphate induced seizures. *Proc. W. Pharmacol. Soc.* 28:191–195.

150. Harrell, E. H. and J. R. Butler. 1988. Harrell-Butler intermediate neurological examination. Unpublished manuscript.

151. Randolph, T. G. 1945. Fatigue and weakness of allergic origin (allergic toxemia) to be differentiated from "nervous fatigue" or neurasthenia. *Ann. Allergy* 3:418–430.

152. Randolph, T. G. 1974. Demonstrable role of foods and environmental chemicals in mental illness. *Jpn. J. Allergol.* 23:445–459.

153. Elfsson, S., F. Gamberle, T. Hindmarsh, A. Iregren, A. Isaksson, I. Johnson, B. Knave, E. Lydahl, P. Mindus, H. Persson, B. Philipson, M. Steby, G. Struwe, E. Sodenman, A. Wennberg, and L. Widen. 1980. Exposure to organic solvents. A cross-sectional epidemologic investigation on occupationally exposed car and industrial spray painters with special reference to the nervous system. *Scand. J. Work Environ. Health* 6:239–273.

154. Valciukas, J., R. Lilis, R. Singer, L. Glickman, and W. Nicholson. 1985. Neurobehavioral changes among shipyard painters exposed to solvents. *Arch. Environ. Health* 40:47–52.

155. Gregersen, P., B. Angelso, T. Nielsen, B. Norgaard, and C. Uldal. 1984. Neurotoxic effects of organic solvents in exposed workers: an occupational, neuropsychological, and neurological investigation. *Am. J. Ind. Med.* 5(3):201–225.

156. Hanninen, H. 1979. Psychological test methods: sensitivity to long-term chemical exposure at work. *Neurobehav. Toxicol.* 1(Suppl 1):157–161.

157. Butler, J. R. 1985. Personal communication.

158. Behari, M., C. Choudhary, S. Roy, and M. C. Maheshwari. 1986. Styrene-induced peripheral neuropathy. A case report. *Eur. Neurol.* 25:424–427.

159. Mackay, C. J. and G. R. Kelman. 1986. Choice reaction time in workers exposed to styrene vapour. *Hum. Toxicol.* 5(2):85–89.

160. Muijser, H., E. M. Hoogendyk, and J. Hooisma. 1988. The effects of occupational exposure to styrene on high-frequency hearing thresholds. *Toxicology* 49(2–3):331–340.

161. Rosenberg, N. L., B. K. Kleinschmidt-DeMasters, K. A. Davis, J. N. Dreisbach, J. T. Hormes, and C. M. Filley. 1988. Toluene abuse causes diffuse central nervous system white matter changes. *Ann. Neurol.* 23:611–614.

162. Aaserud, O., L. Gjerstad, P. Nakstad, R. Nyberg-Hansen, O. J. Hommeren, B. Ivedt, D. Russell, and K. Rootwelt. 1988. Neurological examination, computerized tomography, cerebral blood flow and neuropsychological examination in workers with long-term exposures to carbon disulfide. *Toxicology* 49:277–282.

163. Orbaek, P., J. Risberg, I. Rosen, B. Haeger-Aronson, S. Hagstadius, N. Hjortsberg, G. Regnell, S. Rehnstrom, K. Svensson, and H. Welinder. 1985. Effects of long-term exposure to solvents in the paint industry. *Scand. J. Work, Environ. Health* 11(Suppl. 2):1–28.

164. Lindstrom, K. 1980. Changes in psychological performances of solvent-poisoned and solvent-exposed workers. *Am. J. Ind. Med.* 1:69–84.

165. Edling, C., K. Ekberg, G. Ahlborg, Jr., R. Alexandersson, L. Barregard, L. Edenvall, L. Nilsson, and B. G. Svensson. 1990. Long-term follow up of workers exposed to solvents. *Br. J. Ind. Med.* 47(2):75–82.

166. Stollery, B. T. and M. L. Flindt. 1988. Memory sequelae of solvent intoxication. *Scand. J. Work Environ. Health* 14:45–48.

167. Baldridge, J. T. and J. R. Butler. 1989. Psychoneurological responses associated with chemicals in serum of environmentally ill patients. Presented to the graduate council of the University of North Texas in partial fulfillment of the requirements for the degree of Master of Arts. Denton, TX.

168. Chang, Y. C. 1987. Neurotoxic effects of *n*-hexane on the human central nervous system: evoked potential abnormalities in n-hexane polyneuropathy. *J. Neurol. Neurosurg. Psychiatry* 50:269–274.

169. Pan, Y., A. R. Johnson, and W. J. Rea. 1987/88. Aliphatic hydrocarbon solvents in chemically sensitive patients. *Clin. Ecol.* 5(3):126–131.

170. Anger, W. K., L. Moody, J. Burg, W. S. Brightwell, B. J. Taylor, J. M. Russo, N. Dickerson, J. V. Setzer, B. L. Johnson, and K. Hicks. 1986. Neurobehavioral evaluation of soil and structural fumigators using methyl bromide and sulfuryl fluoride. *Neurotoxicology* 7(3):137–156.

171. Ferraz, H. B., P. H. Bertolucci, J. S. Pereira, J. G. Lima, and L. A. Andrade. 1988. Chronic exposure to the fungicide maneb may produce symptoms and signs of CNS manganese intoxication. *Neurology* 38:550–553.

172. Obeso, J. A., C. Viteri, J. M. Martinez-Lage, and C. D. Marsden. 1986. Toxic myoclonus. *Adv. Neurol.* 43:225–230.

173. McCunney, R. J. 1988. Diverse manifestations of trichlorethylene. *Br. J. Ind. Med.* 45:122–126.

174. Peters, H. A., R. L. Levine, C. G. Matthews, and L. J. Chapman. 1988. Extrapyramidal and other neurologic manifestations associated with carbon disulfide fumigant exposure. *Arch. Neurol.* 45:537–540.

175. Dager, S. R., J. P. Holland, D. S. Cowley, and D. L. Dunner. 1987. Panic disorder precipitated by exposure to organic solvents in the workplace. *Am. J. Psychiatry* 144:1056–1058.

176. Hanninen, H., M. Nurminen, M. Tolonen, and T. Marelin. 1978. Psychological tests as indicators of excessive exposure to carbon disulfide. *Scand. J. Psychol.* 19:163–174.

177. Raloff, J. 1986. Toxic showers and baths. *Science News* 130:190.
178. Rea, W. J. 1994. *Chemical Sensitivity. Vol. II. Sources of Total Body Load.* 535. Boca Raton, FL: Lewis Publishers.
179. Raloff, J. 1986. How polluting is commuting? *Sci. News* 130:190.
180. Price, S. C., S. Ozalp, R. Weaver, D. Chescoe, J. Mullervy, and R. H. Hinton. 1988. Thyroid hyperactivity caused by hypolipodaemic compounds and polychlorinated biphenyls: the effect of coadministration in the liver and thyroid. *Arch. Toxicol. Suppl.* 12:85–92.
181. Hinton, R. H., F. E. Mitchell, A. Mann, D. Chescoe, S. C. Price, A. Nunn, P. Grasso, and J. W. Bridges. 1986. Effects of phthalic acid esters on the liver and thyroid. *Environ. Health Perspect.* 70:195–210.
182. Simon, T. R. and D. C. Hickey. 1992. Personal communication.
183a. Heuser, T., I. Mena, C. Thomas, and F. Alamos. 1994. June 5–8. Presented at the Society of Nuclear Medicine 41st Annual Meeting. Orange County Convention Center, Orlando, FL.
183b. Krusz, J. 1994. Personal communication.
184. Kellerov, A. V. 1985. Electroencephalographic findings in workers exposed to benzene. *J. Hyg. Epidemiol. Microbiol. Immunol.* 29(4):337–346.
185. Monro, J. 1990. Personal communication.
186. Randolph, T. G. 1990. Personal communication.
187. Little, C. H. 1990. Personal communication.
188. Hanninen, H., M. Nurminen, M. Tolonen, and T. Martelin. 1978. Psychological tests as indicators of excessive exposure to carbon disulfide. *Scand. J. Psychol.* 19:163–174.
189. Tolonen, M. and H. Hanninen. 1978. Psychological tests specific to individual carbon disulfide exposure. *Scand. J. Psychol.* 19:241–245.
190. Butler, J. R. 1990. Personal communication.
191. Rea, W. J., J. R. Butler, J. L. Laseter, and I. R. DeLeon. 1984. Pesticides and brain-function changes in a controlled environment. *Clin. Ecol.* 2(3):145–150.
192. Sherck-O'Connor, R. 1986. Total stress load inventory: a validation study. Ph.D. thesis, University of North Texas, Department of Behavioral Medicine, Denton, TX.
193. Ukai, K., J. T. Higashi, and S. Ishikawa. 1980. Edge-light pupil oxcillation of optic neuritis. *Neuro-ophthalmology* 1(1):33–43.
194. Ödkvist, L. M., B. Larsky, R. Tham, H. Ahlfedt, B. E. Anderson, and S. R. C. Liegren. 1982. Vestibulo oculomotor disturbances in humans exposed to styrene. *Acta Otolaryngol.* 94:487–493.
195. Ödkvist, L. M., L. M. Bergholtz, B. Larsby, R. Tham, B. Eriksson, and C. Edling. 1985. Solvent-induced central nervous system disturbances appearing in hearing and vestibulo-oculomotor tests. *Clin. Ecol.* 3(3):149–153.
196. Rea, W. J. and D. M. Martinez. 1995. Use of computerized balance testing in chemically sensitive patients. In press.
197. Stebbins, W. C., W. W. Clark, R. D. Pearson, and N. G. Weiland. 1973. Noise- and durg-induced hearing loss in monkeys. *Adv. Otorhinolaryngol.* 2:42–63.
198. Luschei, E., N. K. Mottet, and C. M. Shaw. 1977. Chronic methyl mercury exposure in the monkey (*Macaca mulatta*): behavioral tests of peripheral vision, signs of neurotoxicity, and blood concentration in relation to dose and time. *Arch. Environ. Health* 32:126–131.

199. Larsby, B., L. M. Ödkvist, D. Hyden, and S. R. C. Liedgren. 1976. Disturbances of the vestibular system by toxic agents. *Acta Phys. Scand. Suppl.* 440:108–114.

200. Axelson, O., M. Hane, and C. Hogstedt. 1976. Neuropsychiatric ill health in workers exposed to solvents — a case-control study. *Lakartidningen* 73:322–329.

201. Segal, G. M. and L. G. Duckert. 1986. Reversible mechlorethamine-associated hearing loss in a patient with Hodgkin's disease. *Cancer* 57:1089–1091.

202. Amoore, J. E. and D. Venstrom. 1966. Sensory analysis of odour qualities in terms of the stereochemical theory. *J. Food Sci.* 31:118–128.

203. Amoore, J. E. and D. Venstrom. 1967. Correlations between stereochemical and organoleptic analysis of odorous compounds. In *Olfaction and Taste*, ed. T. Hayasi. New York: Pergamon Press.

204. Hirsh, A. 1990. Personal communication.

205. Harper, R., E. C. B. Smith, D. G. Land, and N. M. Griffiths. 1968. A glossary of odour stimuli and their qualities. *P. & E. O. R.* 22–37.

206. Vazuka, F. A. 1962. *Essentials of the Neurological Examination.* 31. Philadelphia, PA: Smith Kline & French Laboratories.

207. Pitts, F. W. 1977. Tuberculosis: Prevention and therapy. In *Current Concepts of Infection Diseases*, eds. E. W. Hook, G. L. Mandell, J. M. Gwalthey, Jr., and M. A. Sande. 181–194. New York: John Wiley & Sons.

208. Tanquerel des Planches and Louis Jean Charles Marie. 1848. *Lead Diseases: A Treatise from the French of L. Tanquerel Des Planches, with Notes and Additions on the Use of Lead Pipe and Its Substitutes.* Lowell, MA: Daniel Bixby and Co.

209. Kilburn, K. H., R. Warshaw, and J. C. Thornton. 1987. Formaldehyde impairs memory, equilibrium, and dexterity in histology technicians: effects which persist for days after exposure. *Arch. Environ. Health* 42(2):117–120.

210. Lilis, R., W. V. Lorimer, S. Diamond, and I. J. Selikoff. 1978. Neurotoxicity of styrene in production and polymerization workers. *Environ. Res.* 15:133–138.

211. Proctor, N. H. and J. P. Hughes. 1978. *Chemical Hazards of Workplace.* 307. Philadelphia, PA: J. B. Lippincott Co.

212. Hayes, W. J. 1982. *Pesticides Studied in Man.* 284–435. Baltimore: Williams & Wilkins.

213. Guyton, A. C. 1981. *Textbook of Medical Physiology.* 700. Philadelphia: W.B. Saunders.

214. Guyton, A. C. 1981. *Textbook of Medical Physiology.* 56. Philadelphia: W.B. Saunders.

215. Guyton, A. C. 1981. *Textbook of Medical Physiology.* 704. Philadelphia: W.B. Saunders.

216. Guyton, A. C. 1981. *Textbook of Medical Physiology.* 705. Philadelphia: W.B. Saunders.

217. Guyton, A. C. 1981. *Textbook of Medical Physiology.* 699. Philadelphia: W.B. Saunders.

218. Guyton, A. C. 1981. *Textbook of Medical Physiology.* 706. Philadelphia: W.B. Saunders.

219. Guyton, A. C. 1981. *Textbook of Medical Physiology.* 707. Philadelphia: W.B. Saunders.

220. Dyal, Y. 1991. Neuroendocrine cells of the gastrointestinal tract: introduction and historical perspectives. In *Endocrine Pathology of the Gut and Pancreas*, ed. Y. Dayal. 13. Boca Raton, FL: CRC Press.

221. Pert, C. B., M. R. Ruff, R. J. Weber, and M. Herkenham. 1985. Neuropeptides and their receptors: a psychosomatic network. *J. Immunol.* 35:2.

222. Bell, I. R., C. S. Miller, and G. E. Schwartz. 1992. An olfactory-limbic model of multiple chemical sensitivity syndrome: possible relationships to kindling and affective spectrum disorders. *Biol. Psychiatry* 32:218–242.

223. Post, R. M. 1980. Minireview. Intermittent versus continuous stimulation: effect of time interval on the development of sensitization or tolerance. *Life Sci.* 26:1275–1282.

224. Post, R. M., D. R. Rubinow, and J. C. Ballenger. 1984. Conditioning, sensitization, and kindling: implications for the course of affective illness. In *Neurobiology of Mood Disorders*, eds. R. Post and J. Ballenger. 432–466. Baltimore: Williams & Wilkins.

225. Adamec, R. E. and C. Stark-Adamec. 1983. Limbic kindling and animal behavior — implications for human psychopathology associated with complex partial seizures. *Biol. Psychiatry* 18:269–293.

226. Adamec, R. E. 1990. Does kindling model anything clinically relevant? *Biol. Psychiatry* 27:249–279.

227. Adamec, R. E. 1991. Individual differences in temporal lobe sensory processing of threatening stimuli in the cat. *Physiol. Behav.* 49:455–464.

228. Adamec, R. E. and C. Stark-Adamec. 1986. Limbic hyperfunction, limbic epilepsy, and interictal behavior. In *The Limbic System. Functional Organization and Clinical Disorders*, eds. B. K. Doane and K. E. Livingston. 129–145. New York: Raven Press.

229. Antelman, S. M. 1988. Time-dependent sensitization as the cornerstone for a new approach to pharmacotherapy: drugs as foreign/stressful stimuli. *Drug Dev. Res.* 14:1–30.

230. Hudson, J. I. and H. G. Pope. 1990. Affective spectrum disorder: does antidepressant response identify a family of disorders with a common pathophysiology? *Am. J. Psychiatry* 147:552–564.

231. Joy, R. M. 1982. Mode of action of lindane, dieldrin, and related insecticides in the central nervous system. *Neurobehav. Toxicol. Teratol.* 4:813–823.

232. Gilbert, M. E. 1992. Neurotoxicants and limbic kindling. In *Vulnerable Brain and Environmental Risks. Vol. 1. Malnutrition and Hazard Assessment*, eds. R. L. Isaacson and K. S. Jensen. New York: Plenum.

233. Ashford, N. A. and C. S. Miller. 1991. *Chemical Exposures. Low Levels and High Stakes*. New York: Van Nostrand Reinhold.

234. Bell, I. R. 1975. A kinin model of medication for food and chemical sensitivities: biobehavioral implications. *Ann. Allergy* 35:206–215.

235. Bell, I. R. 1982. *Clinical Ecology. A New Medical Approach to Environmental Illness*. Bolinas, CA: Common Knowledge Press.

236. Bell, I. R. 1992. Neuropsychiatric and biopsychosocial mechanisms in multiple chemical sensitivity: an olfactory-limbic system model. In National Research Council Report: *Multiple Chemical Sensitivities. Addendum to Biologic Markers in Immunotoxicology*. 89–108. Washington, D.C.: National Academy Press.

237. Cain, D. P. 1974. The role of the olfactory bulb in limbic mechanisms. *Psychol. Bull.* 81:654–671.

238. Shipley, M. T. 1985. Transport of molecules from nose to brain: transneuronal anterograde and retrograde labeling in the rat olfactory system by wheat germ agglutin-horseradish peroxidase applied to the nasal epithelium. *Brain Res. Bull.* 15:129–142.

239. Cheng, Y. S., Y. Yamada, H. C. Yeh, and D. Swift. 1988. Diffusional deposition of ultrafine aerosols in a human nasal cast. *J. Aerosol. Sci.* 19:741–751.

240. Ryan, C. M., L. A. Morrow, and M. Hodgson. 1988. Cacosmia and neurobehavioral dysfunction associated with occupational exposure to mixtures of organic solvents. *Am. J. Psychiatry* 145:1442–1445.

241. Ghantous, H., L. Dencker, J. Gabrielsson, B. R. G. Danielsson, and K. Bergman. 1990. Accumulation and turnover of metabolites of toluene and xylene in nasal mucosa and olfactory bulb in the mouse. *Pharmacol. Toxicol.* 66:87–92.

242. Gilbert, M. E. and C. M. Mack. 1989. Enhanced susceptibility to kindling by chlodimeform may be mediated by a local anesthetic action. *Psychopharmacol.* 99:163–167.

243. Bokina, A. I., N. D. Eksler, A. D. Semenenko, and R. V. Merkuryeva. 1976. Investigation of the mechanism of action of atmospheric pollutants on the central nervous system and comparative evaluation of methods of study. *Environ. Health Perspect.* 13:37–42.

244. Lorig, T. S. and G. E. Schwartz. 1988. Brain and odor. I. Alteration of human EEg by odor administration. *Psychobiology* 16:281–284.

245. Lorig, T. S., G. E. Schwartz, K. B. Herman, and R. D. Lane. 1988. Brain and odor. II. EEG activity during nose and mouth breathing. *Psychobiology* 16:285–287.

246. Perl, D. P. and P. F. Good. 1987. Uptake of aluminum into central nervous system along nasal-olfactory pathways. *Lancet* 1:1028.

247. Hastings, L. and J. E. Evans. 1991. Olfactory primary neurons as a route of entry for toxic agents into the CNS. *Neurotoxicology* 12:707–714.

248. Sato, M., R. J. Racine, and D. C. McIntyre. 1990. Kindling: basic mechanisms and clinical validity. *Electroencephalogr. Clin. Neurophysiol.* 76:459–472.

249. Antelman, S. M., J. E. Cunnick, D. T. Lysle, A. R. Caggiula, S. Knopf, D. J. Kocan, B. S. Rabin, and D. J. Edwards. 1990. Immobilization 12 days (but not one hour earlier) enhanced 2-deoxy-d-glucose-induced immunosuppression: evidence for stressor-induced time-dependent sensitization of the immune system. *Prog. Neuropsychopharmacol. Biol. Psychiatry* 14:579–590.

250. Kalivas, P. W. and C. D. Barnes, eds. 1988. *Sensitization in the Nervous System.* Caldwell, NJ: Telford Press.

251. Dager, S. R., J. P. Holland, D. S. Cowley, and D. L. Dunner. 1987. Panic disorder precipitated by exposure to organic solvents in the work place. *Am. J. Psychiatry* 144:1056–1058.

252. Mesulam, M. M. 1985. *Principles of Behavioral Neurology.* Philadelphia: F. A. Davis.

253. Huang, Y. Y., A. Colino, D. K. Selig, and R. C. Malenka. 1992. The influence of prior synaptic activity on the induction of long-term potentiation. *Science* 255:730–733.

254. Joy, R. M., L. G. Stark, and T. E. Albertson. 1983. Proconvulsant actions of lindane: effects on afterdischarge thresholds and durations during amygdalid kindling in rats. *Neurotoxicology* 4:211–220.

255. Gilbert, M. E. 1992. Proconvulsant activity of endosulfan in amygdala kindling. *Neurotoxicol. Teratol.* 14:143–149.

256. Gilbert, M. E. 1992. A characterization of chemical kindling with the pesticide endosulfan. *Neurotoxicol. Teratol.* 14:151–158.

257. Tilson, H. A., B. Veronesi, R. L. McLamb, and H. B. Mathews. 1990. Acute exposure to tris(2-chloroethyl)phosphate produces hippocampal neuronal loss and impairs learning in rats. *Toxicol. Appl. Pharmacol.* 106:254–269.

258. Baldridge, J. T. 1989. Psychoneurological responses associated with chemicals in serum of environmentally ill patients. Presented to the graduate council of the University of North Texas in partial fulfillment of the requirements for the degree of Master of Arts, Denton, TX.

259. Diedrickson, N. and J. R. Butler. 1990. Personal communication.

260. Dowty, B. J. and J. L. Laseter. 1976. The transplacental migration and accumulation in blood of volatile organic constituents. *Pediatr. Res.* 10:696–701.

261. Rea, W. J. 1994. *Chemical Sensitivity. Vol. II. Sources of Total Body Load.* 579. Boca Raton, FL: Lewis.

262. Pfeiffer, C. C. 1975. *Mental and Elemental Nutrients: A Physician's Guide to Nutrition and Health Care.* New Canaan, CT: Keats Publishing.

263. Wurtman, R. J. 1979. *Disorders of Eating: Nutrients in Treatment of Brain Diseases.* New York: Raven Press.

27   Pollutant Injury to the Eye

INTRODUCTION

Many chemically sensitive patients have difficulty seeing. These patients experience vision problems such as blurring, tunnel vision, and light flashes. They are also bothered by floaters when they are in areas with high pollutant levels such as underground parking garages, conference halls, public buildings, etc. In addition, some chemically sensitive individuals suffer from direct pollutant damage that results in named eye diseases such as glaucoma, iritis, optic neuritis, cataracts, and others. Attention to the powerful effects of pollutant exposure on chemically sensitive individuals not only will aid treatment of the various fixed-named eye diseases that may result in select cases but also will often altogether eliminate their development.

POLLUTANT INJURY TO THE NERVOUS SYSTEM OF THE EYE

The eye has two significant types of innervation. The first is via the cranial nerves, and the second is through the autonomic nerves. Both systems may be significantly affected by pollutant exposure with subsequent injury resulting.

Cranial Nerves

Several cranial nerves may be involved in pollutant injury to the eye. The most important, of course, is the optic nerve, since it is the nerve of vision. All will be discussed in the following pages.

Second Cranial Nerve

The optic nerve is, of course, the key nerve in pollutant injury of the eye with damage ranging from diminished vision to total blindness, as seen in some chemically sensitive patients. This damage can occur at any place from the optic disc along the nerve to the visual center in the brain.

Third Cranial Nerve

The oculo-motor nerve innervates all eye muscles except those innervating the superior oblique muscle, the lateral external rectus muscle, the ciliary muscle, and the iris sphincter. Pollutant injury can occur to this nerve causing eye movement dysfunction.

Fourth Cranial Nerve

The trochlear nerve innervates the superior oblique muscle, which primarily rotates the eye outward and downward. Injury to this nerve will cause eye movement dysfunction.

Fifth Cranial Nerve

The trigeminal nerve (first and second branch) is responsible for sensation to the upper eyelid (first) and the lower lid (second) and for corneal sensation (second). Pollutant injury will derange this function.

Sixth Cranial Nerve

The abducens innervates the lateral external rectus muscle, which rotates the eyeball outward. Dysfunction resulting from pollutant damage to this area may occur.

Autonomic Nerves

Pollutant damage to the actions of the autonomic nerves of the eye appears to mirror much pollutant damage of the ANS throughout the body of the chemically sensitive individual. Once the ANS is triggered or inhibited by pollutants, changes in the eye responses will frequently follow. By our studies with the iris corder, 70% of systemic responses are reflected by changes in the ANS of the eye. These changes occur because the autonomic fibers both enter and leave the visual pathways near the chiasma (Figure 27.1) with the majority of fibers making connections in the supraoptic nuclei of the hypothalamus. The supraoptic nucleus then makes elaborate connections and is in a position to relay visual and retinal impulses into the ANS. The nucleus then connects with the neurohypophysis, the paraventricular nuclei, the mammillary bodies, and the nuclei. Efferent centrifugal fibers may also arise from the supraoptic nucleus (see Chapter 26). The retina hypothalamic pathways are involved with the photoneuric endocrine reflexes and circadian cycles and often malfunction in chemically sensitive individuals.

In general terms, the posterior part of the hypothalamus seems essential for pressure functions associated with sympathetic activity, while the anterior part of the hypothalamus seems to be devoted to parasympathetic activity.

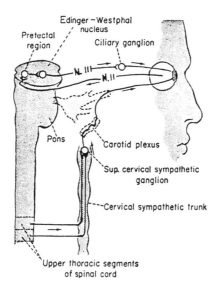

Figure 27.1. **Autonomic fibers leaving and entering visual pathways near the chiasma with the majority of fibers making connection in the super optic nucleus of the hypothalamus. (From Guyton, A. C. 1981. *Textbook of Medical Physiology*, 6th ed. 761. Philadelphia: W. B. Saunders. With permission.)**

Pollution-induced stimulation of the posterior part of the hypothalamus induces vasoconstriction, heat production, increased metabolism, and pupillary dilatation.

Of neuro-ophthalmologic interest is the influence of the hypothalamus on the intrinsic muscles to the eye, the position of the lids, the secretion of tears, the vasomotor system, and the intraocular pressure. Many chemically sensitive patients find a diminution or blurring of vision immediately upon entering a contaminated area such as a hotel or garage or polluted room, presumably due to triggering of these systems in the body.

An increase in intraocular pressure is independent of systemic blood pressure and could account for this diminution of vision. A fall in the intraocular pressure associated with sympathetic activities including local elevation of blood pressure, vasoconstriction, and pupillary dilatation results from stimulation near the anterior column of the fornix. Changes such as these may be seen in chemically sensitive individuals upon pollutant exposure.

It has been suggested that a drop in intraocular pressure is related to decreased production of aqueous fluid, which is a consequence of decreased blood flow in the ciliary body that is probably due to local vasospasm. Stimulation of the cervical sympathetic nerve produces a similar effect and is more likely due to rapid decreases in pressure as seen in the treated chemically sensitive individual with glaucoma.

The autonomic elements in the brainstem are connected with the third, seventh, ninth, tenth, and the eleventh cranial nerves, which are concerned with pupillary, lacrimal, salivary, cardiovascular, respiratory, alimentary, and other visceral activities. These connections allow impulse communication with the eye and, thus, are a mirror for peripheral pollutant injury in many chemically sensitive patients.

The postganglionic pupillary fibers emerge from the first and second thoracic roots and the synapse in the supracervical ganglion. Care has to be taken during cervical sympathectomy that the upper part of this ganglion is not injured, or a lid drop may occur. These fibers are a long pathway that has importance diagnostically. For example, pancoast lung tumors impinging on the cervical plexus of the ANS can produce lid drop, and surgical removal of tumors can be hazardous, resulting in permanent lid drop. Often pollutant injury will cause a similar lid drop, which has been seen in some chemically sensitive patients seen at the EHC-Dallas. In the chemically sensitive individual, this drop suggests autonomic ganglion or nerve damage from the pollutant.

In the eye, the peripheral ANS is divided into preganglionic and postganglionic parts separated from each other by synapses. The preganglionic parts are cholinergic. For example, their response could be initiated by acetylcholine triggered by pollutant stimulation. The postganglionic pathways can be either cholinergic (parasympathetic) or adrenergic. For example, their response could be initiated by adrenalin (sympathetic)-triggered pollutant induction. The former appears to be triggered more often than the latter in the chemically sensitive individual.

In the parasympathetic system of the eye, preganglionic fibers run with cranial nerves III, VII, IX, and X. The synapses take place in a peripherally situated ganglion for the ciliary muscle and sphincter of the pupil. The preganglionic fibers originate in the Edinger-Westphal nucleus in the III nuclear region. They run with the third cranial nerve to the ciliary ganglion where they undergo synapsis. The postganglionic fibers run with the short ciliary nerves to the eyeball. The parasympathetic fibers for the lacrimal gland originate in the nucleus of the IX cranial nerve and synapse in the sphenopalatine ganglion (Figure 27.2). The synapses of the nerve fibers for the dilatation of the pupils take place in the superior cervical ganglion.

Certain chemicals and pollutants are known to stimulate the parasympathetic nervous system of the eye (Figures 27.3 and 27.4). These contaminants include pilocarpine and mecholyl, eserine, ubretid, and organophosphate insecticides. They cause constriction of the pupil, which is frequently seen in the chemically sensitive patient. In fact, one third of the chemically sensitive patients presenting at the EHC-Dallas have miosis. Known suppressors of the parasympathetic nervous system of the eye include atropine, tropicamide, and cyclogyl and botulinus toxin. Botulinus toxin will block motor nerves as well, since the effects of the toxin are not entirely parasympathetic. Exposure to botulinus toxin will allow the pupil to be dilated by suppression of the sphincter muscle (Figure 27.4).

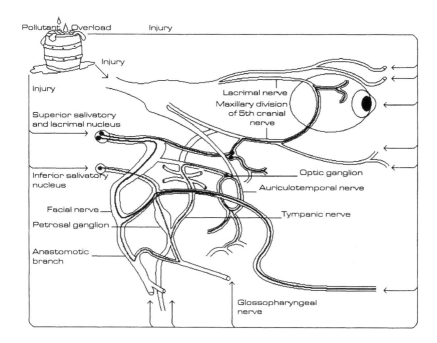

Figure 27.2. Potential areas of pollutant injury of the ocular nerves and ocular autonomic nerves.

Certain pollutants are known to stimulate the sympathetic nervous system. These include epinephrine, phenylephrine, norepinephrine, tyramine, and cocaine (Figure 27.5).

Other pollutants, including guanethidines, reserpine, *d*-methyl-dopa, and bethanidine, are known to suppress the sympathetic nervous system (Figure 27.5). A large portion of chemically sensitive individuals (25%) have measureable sympatholytic effects.

It is clear that many toxic chemicals, including organochlorine insecticides and aliphatic and aromatic solvents, can cause imbalance in the ANS. Depending upon dosage, timing, and combination with other toxic chemicals, these toxic chemicals may stimulate or suppress either part of the ANS. It should be emphasized that the responses of the ANS may be physiologic, pharmacologic, or toxic. A physician's awareness of these types of responses in the chemically sensitive patient allows for early diagnosis and treatment of pollutant injury before it becomes permanent.

At the EHC-Dallas, we measured blood organochlorines and solvents in a series of chemically sensitive patients.[1] When studying them with the iris corder, described later, we found that these patients had predominantly cholinergic and sympatholytic changes in their ANS functions. Some patients had cholinolytic effects, and very few had sympathetic stimulation (Tables 27.1 to 27.3).

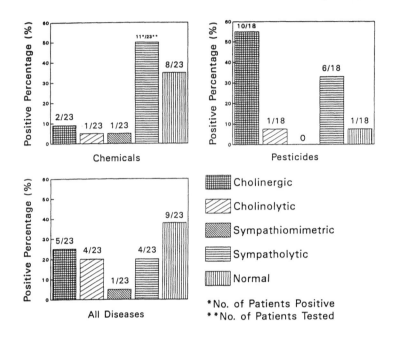

Figure 27.3. Autonomic changes as measured by the iris corder in patients with chlorinated pesticides and/or toxic volatile organic chemicals in their blood. (From S. Shiragawa. EHC-
Dallas. 1988.)

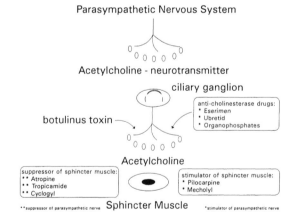

Figure 27.4. Pollutant chemicals known to alter the parasympathetic
nervous system.

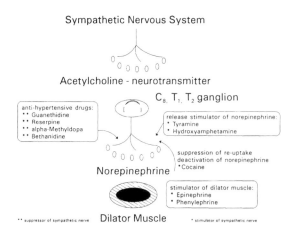

Figure 27.5. Pollutant chemicals known to alter the autonomic nervous system.

NUTRIENT DEFICIENCY IN THE EYE

Nutrient deficiency in the chemically sensitive patient can occur in many ways and result in much pathology (see Volume I, Chapter 6[2]). Poor quality food with poor intake, malabsorption, competitive absorption, inappropriate transport, inappropriate membrane transfer into the cell, hyper- or altered metabolism, and/or inappropriate elimination are some of the ways that the eye of the chemically sensitive individual becomes nutrient depleted. In addition, direct toxic injury may occur to the nutrients themselves, causing depletion to occur.

When considering nutrient depletion in the eye, one must realize that most conditions due to deficiency involve inadequacy of more than one nutrient. However, here each potential deficiency is discussed separately.

Vitamin A Deficiency

Deficiency of vitamin A usually first manifests as night blindness. More severe deficiency leads to xerophthalmia, keratomalacia, and hyperakoratosis of the skin. Because vision is dependent on adequate amounts of retinal visual purple rhodopsin, the chief changes in night vision occur when light intensity is low. Rhodospin is a compound of a protein combined with retinal aldehyde of vitamin A (see Volume I, Chapter 6[2]). When little or no light falls upon the retina, an increasing amount of visual purple is produced. Since the sensitivity of the retina to light is proportional to the amount of visual purple present in the retina, there is an increased sensitivity during exposure to low illumination (dark adaptation). In a patient with a vitamin A deficiency, less visual purple

Table 27.1. Comparison of the Chemically Sensitive Groups and Healthy Volunteer Controls with Pupillography

Patient types	A1	CR	T2	VC	T5	VD	Number	Mean age
Chemical solvents	34.2 ± 7.6	0.47 ± 0.07	211.6 ± 30.1	43.4 ± 8.6[a]	2438.7 ± 891.7[a]	10.6 ± 2.9	23	41.5
Chlorinated pesticide	28.6 ± 9.9[b]	0.45 ± 0.11	215.7 ± 53.3	37.8 ± 10.1[c]	2667.6 ± 1239.3[d]	8.4 ± 2.3[c]	20	45.4
Healthy volunteers control	36.9 ± 6.0	0.48 ± 0.07	205.4 ± 20.6	49.6 ± 8.5	1881.8 ± 555.5	12.1 ± 3.2	18	29.7

[a] $p < 0.03$ measured in the controlled environment (mean ± SD).
[b] $p < 0.006$ measured in the controlled environment (mean ± SD).
[c] $p < 0.001$ measured in the controlled environment (mean ± SD).
[d] $p < 0.02$ measured in the controlled environment (mean ± SD).

Source: Shirakawa, S., W. J. Rea, S. Ishikawa, and A. R. Johnson. 1992. Evaluation of the autonomic nervous system response by pupillographical study in the chemically sensitive patient. Environ. Med. 8(4):121–127. With permission.

**Table 27.2. Frequency of Autonomic Nerve
Disturbance by Pupillographical
Result in Each Group**

	No. of patients	No. of patients with AND
Chemical solvents	23	15 (65%)
Chlorinated pesticides	20	18 (90%)
Total	43	33 (77%)

Note: AND = autonomic nerve disturbances.

Source: Shirakawa, S., W. J. Rea, S. Ishikawa, and A. R. Johnson. 1992. Evaluation of the autonomic nervous system response by pupillographical study in the chemically sensitive patient. *Environ. Med.* 8(4):121–127. With permission.

**Table 27.3. Distribution of Autonomic Nerve Disturbance
in Each Group**

Autonomic nerve disturbance	Chemical solvents	Chlorinated pesticides	Total
Cholinergic	2 (13.3%)	7 (38.9%)	9 (27.3%)
Cholinolytic	1 (6.7%)	1 (5.6%)	2 (6.1%)
Symphathiomimetic	1 (6.7%)	0 (0%)	1 (3.0%)
Symphatholytic	11 (73.3%)	10 (55.5%)	21 (63.6%)
	15	18	33

Source: Shirakawa, S., W. J. Rea, S. Ishikawa, and A. R. Johnson. 1992. Evaluation of the autonomic nervous system response by pupillographical study in the chemically sensitive patient. *Environ. Med.* 8(4):121–127. With permission.

can be regenerated in low illuminations so that the extent to which the eye can adapt in the dark is lessened. The first symptoms of this deficiency usually appear at dark.

A second important result of vitamin A deficiency is metaplasia of the epithelial tissue. The normal epithelia of the conjunctiva and cornea, as well as air passages and skin, becomes replaced with a stratified squamous keratinizing epithelium. The conjunctiva and cornea become keratinized (xerophthalmia), and later the cornea becomes softened and infected and finally perforated. The first symptoms of this damage usually appear at dusk. The patient complains of a gritty feeling in his or her eyes, and this feeling is followed by lacrimation, photophobia, and conjunctivitis. Bitot's spots of pearly or mucoid areas occur in the conjunctiva most frequently on the equator just lateral to the cornea (see Volume I, Chapter 6, sections on vitamins for physiology[3]). Fifteen percent of the chemically sensitive patients seen at the EHC-Dallas have been found to be deficient in vitamin A, suggesting vitamin A deficiency may be an early sign of chemical sensitivity in some individuals.

Vitamin B$_1$ Deficiency

Deficiency of vitamin B$_1$ is responsible at least in part for Beriberi, Wernieke's encephalopathy, Korsakoff's syndrome, polyneuritis of pregnancy, and alcoholic neuritis.

Cranial and sympathetic nerves to the eye may be involved in vitamin B$_1$ deficiency. The nerves undergo Wallarian degeneration. The muscles supplied by these nerves undergo atrophy with cloudy swelling and fatty degeneration. The most common lesion is retrobulbur neuritis, but ophthalmoplegia is sometimes seen.

In patients with Wernieke's syndrome, diplopia and photophobia occur as a result of vitamin B$_1$ deficiency. Nystagmus and ophthalmoplegia is common. Failing vision and ptosis also may occur, and diplopia is occasionally found.

With thiamine deficiency, which has been found in at least 20% of the chemically sensitive population, thiamine pyrophosphate may be insufficient to serve as a transient carrier in the detoxication of aldehydes. A build-up of aldehydes in some areas of the eye may then occur, thus causing cross-linking of proteins, which may result in altered vision in some chemically sensitive patients.

Vitamin B$_2$ (Riboflavin) Deficiency

Riboflavin deficiency, found in more than 20% of chemically sensitive individuals, can result in eye symptoms similar to those in pellagra, but in this case these symptoms will be refractory to niacin administration. The lesions of the eye begin with circumcorneal injection. The limbal vessels proliferate, anastomosis occurs, and loop arcades form. This process extends until the cornea is invaded with capillary twigs and loops.

The first symptoms of ocular damage due to vitamin B$_2$ deficiency are itching and burning. Lacrimation and photophobia with blepharospasm and complaints of eye strain and blurred vision may also occur. Examination of the eyes of chemically sensitive patients deficient in vitamin B$_2$ reveals reduced visual acuity, suggesting the presence of retrobulbar neuritis. Iritis is relatively rare in these patients. Cataracts appear in experimental deficiencies of B$_2$ in animals.[4]

Riboflavin is needed to form the coenzyme FAD, which is needed in oxidative phosphorylation reactions and flavin dehydrogenases. It is an indirect antioxidant working with glutathione reductase and superoxide dismutase to minimize free radical damage generated from light entering the eye. It is also a cofactor in monoxygenase reactions and aldehyde dehydrogenase. Therefore, riboflavin deficiency may result in inappropriate aldehyde breakdown in the eye with subsequent excess aldehyde that results in cross-linking of protein and visual impairment.

Vitamin B₃ (Niacin) Deficiency

If severe enough, vitamin B_3 deficiency, seen in at least 20% of chemically sensitive patients, results in pellagra, which is rarely seen in the western world today. Visual disturbances resulting from B_3 deficiency, however, are common. Frequently, they take the form of failing vision with retrobulbar neuritis, possibly leading to optic atrophy. Diplopia and cataract may also occur. Photophobia, lacrimation, and blepharitis have been seen.

Niacin is an integral part of NADP and NADPH, which are coenzymes for many metabolic and detoxication reactions. They are essential for the microsomal enzymes, especially cytochrome P-450, to function. Therefore, niacin deficiency may cause inappropriate detoxications of xenobiotics in the chemically sensitive individual, resulting in continuous circulation of toxic chemicals. In turn, various eye pathologies may be seen. Niacin is also important in the beta-oxidation of fatty acids, which are vital in supplying energy to maintain the integrity of the eye cells.

Vitamin B₅ (Patothenic Acid) Deficiency

Pantothenic acid is bound to coenzyme A and is a transient carrier of acetyl groups. It is important in acetylation of specific aromatic amines as it is necessary to their detoxification. It is also important in the oxidation of fatty acids, pyruvate, alpha-ketoglutaric, and acetaldehyde. It is universal in tissues and, therefore, deficiency of Vitamin B_5 is difficult to define, although some chemically sensitive patients respond to vitamin B_5 supplementation, suggesting they may, in fact, suffer from some deficiency.

Vitamin B₆ Deficiency

Vitamin B_6 catalyzes over 60 intermediary metabolism reactions. It is a cofactor in taurine metabolism and, if low, may yield extreme sensitivities to chlorines, aldehydes, alcohols, petroleum solvents, and ammonia. In chemically sensitive individuals, deficiency of vitamin B_6 could disturb the vitamin A aldehyde reactions and cause a build-up of the aldehyde with damage to the eye tissue. Gyrate atrophy, a pathologic retinal condition, has been shown to result from vitamin B_6 deficiency, which occurs in 60% of chemically sensitive individuals (see Volume I, Chapter 6).[2]

Vitamin B_6 is vitally important in the treatment of idiopathic optic neuritis and retrobulbar neuritis, both of which may be due to inhalant, food, and/or chemical sensitivity. It has also been reported to be of value in cases of diabetic proliferative retinopathy.

Vitamin C Deficiency

Much laboratory evidence suggests that ascorbic acid aids good vision and eye healing. Ascorbic acid deficiency occurs in at least 20% of chemically sensitive patients. Much animal data suggests the mechanisms for the eye changes seen in the vitamin C-deficient, chemically sensitive patient. For example, the occurrence of dehydroascorbate in the retina, subretinal fluid, and pigment epithelium indicates the presence of oxidative reaction of ascorbate on these compartments where light-induced free radicals are located.[5] If excess free radicals occur, nuclear cataracts may form with oxidative damage to the lens.[6] A sodium-dependent carrier system is involved in transport of ascorbate in primary cultures of cat retinal epithelium.[7]

The highest concentrations of semidehydroascorbate reductase (this enzyme regenerates ascorbate but is dependent on NADH) were detected in bovine retinal extracts, followed by pigment epithelium, choroid, ciliary body, and iris. No activity was detected in the ocular lens, and many facts suggest that pollutant changes can be altered in some animals and probably in the chemically sensitive individual. Retinal extracts at 5°C reduced NADH oxidation 22%, while treatment with 4mM of N-ethylmaleimide reduced it 42%.[8] Ascorbate was found to promote corneal swelling when isolated corneas were perfused without glucose.[9] Ascorbate in aqueous humor protects against myeloperoxidase and β-glucuronidase-induced oxidation. Oxidation products are then released from azurophilic granules. An inability to protect against pollutants may account for a prolonged half-life of many inflammatory mediators as well as susceptibility to *Candida* infection in some chemically sensitive patients.[10] The nonenzymatic reaction of lens crystalline with ascorbic acid may also contribute to this prolongation. Photofrin II, a commercial preparation of a hematoporphyrin derivative containing a high proportion of the active dihematoporphyrine ether, was found to convert oxygen to hydrogen peroxide and hydroxyl free radicals in the presence of ascorbate and light. In addition, a light-dependent increase in the free radical concentration was observed in hematoporphyrin derivative/ascorbate solutions with subsequent oxidation of ascorbate.[11] Ascorbic acid levels in the retina of the rat's eye decreased after operation, emphasizing that trauma causes depletion of ascorbic acid. Three months after acetone treatment, 12 animals (30%) developed cataracts, and their ascorbate levels dropped below 9 mg/100 mL after one year,[12] a situation that may be similar to the deficiency condition observed in the pollutant-sensitive chemically sensitive patient. Endotoxin-induced ocular inflammation caused a decrease in the concentration of ascorbic acid in the aqueous humor and an increase in the vitreous humor in some animals. The presence of ceruloplasmin in the aqueous humor during inflammation may contribute to the decrease of ascorbic acid,[13] since studies in the noninflamed eye of the same animal show that the aqueous humor is not the source of ascorbic acid. These animal studies provide much evidence that chronic inflammation, as seen in many chemically sensitive patients, will cause ascorbic acid deficiency and perpetuate disease.

Studies of humans provide evidence that vitamin C supplementation improves some eye problems. For example, Ishikawa et al.[14] reported the treatment of optic neuritis with a high dosage of vitamin C (2 to 10 g daily, intravenously, on an average of 29 days). The patients with unknown optic neuritis were classified into two groups according to their blood level of vitamin C. Six patients with decreased levels (<3.5 µg/mL) made up group A. Group B consisted of 5 patients within normal limits (>3.5 µg/mL). These patients were treated with high dosage (2 to 10 g) of vitamin C, and the degree of visual recovery was compared. Fifty percent of patients in group A and 37.5% of patients in group B experienced full recovery of their visual acuity. The velocity of visual recovery was also significantly greater in group A than in group B. The degree and velocity of visual recovery were not different from their previous results in patients with optic neuritis who were treated with corticosteroid.

It is known that vitamin C decreases intraocular pressure. Compared with serum, the aqueous humor of the eye is high in ascorbic acid. The administration of ascorbic acid appears to have a transient hypotonic effect on intraocular pressure in normal eyes. Its effect on glaucomatous eyes is not as well defined, particularly in the case of the rabbit. Fox et al.[15] have shown the ocular hypotensive effect (+/+) and a hereditary glaucoma (bu/bu). They have also shown that the transport of ascorbate from the serum to the aqueous is much slower in buphthalamic rabbits than in normal controls. Osmolarity is suggested as a possible hypotensive mechanism and a differential fluid transport rate between the blood and the aqueous for the different genotypes.

Vitamin E Deficiency

In rats, vitamin E has been shown to protect against oxygen-induced retinopathy. This protection is probably the result of a variety of vitamin E actions including free radical scavenger effects, inhibition of the lipid peroxide oxidant cascade, stabilization of membranes, and stabilization of enzyme effects.[16] Also, Pelle et al.[17] have shown that vitamin E protects against ultraviolent light exposure on liposomes. Again this is an antioxidant effect.

Alpha tocopherol deficiency seems to affect retina pigment epithelium. It enhances lipid peroxidation and accumulation of aging pigment lipo-fuscin. This deficiency further predisposes the chemically sensitive individual to macular degeneration. Pollutants induce absorption, and assimilation defects can then lead to alpha tocopherol deficiency.

Zinc Deficiency

Zinc deficiency is reportedly of great importance in eye pathology, as it interferes with the function of the enzyme alcohol dehydrogenase and leads to accumulation of retinaldehyde, which is toxic to the retina. In recent studies, nutritional supplementation with zinc has been found to help retard, and even reverse, early changes of macular degeneration. Zinc deficiency has been observed in 5 to 10% of chemically sensitive individuals.

Taurine Deficiency

Due to their conjugating ability and glutathione anti-free radical properties (see Volume I, Chapter 4), taurine and glutathione have been reported to be of value in maintaining the health of the retina and lens in chemically sensitive patients.[18] It appears that at least 60% of chemically sensitive patients respond to these two sulfur amino acids.

Selenium Deficiency

Selenium is necessary for glutathione peroxidase and is necessary for prevention of antioxidant injury. It is also complementary to vitamin E. In the chemically sensitive patient, free radical damge to the eye can be minimized by supplementing selenium.

MEASUREMENT OF POLLUTANT INJURY OF THE AUTONOMIC NERVOUS SYSTEM OF THE EYE

Some technology that enables us to analyze more objectively the damage to the eye caused by chemical exposure as well as the involvement of the ANS has now evolved. For example, analysis of the pupil area by means of a binocular infrared video pupillograph (iris corder: HTV TYC-831 Hamamatsu Company) was developed by Ishikawa et al.[19] This machine allows objective evaluations of the functions of the upper body area of the ANS to be ascertained and has proven efficacy in measuring autonomic dysfunction in the chemically sensitive (Figure 27.6 and 27.7). The machine is compared and calibrated with various stimulators (pilocarpine, mecholyl) and suppressors (atropine, tropicamide, cyclogyl) of the parasympathetic nervous system and stimulators (epinephrine, phenylephrine [stimulates dilator muscle]) and inhibitors (guanethidine, reserpine, alpha-methyldopa, bethanidine) of the sympathetic nervous system. These chemicals are then compared with various toxins that are stimulators (anticholinesterase drugs [eserine, cubretid, organophosphates]) and suppressors (botulinus toxin) of the parasympathetic nervous system and those that are stimulators (tyramine, which releases norepinephrine, and cocaine, which prevents the reuptake of norepinephrine) and suppressors (antihypertensive drugs such as guanethidine, reserpine, methyl dopa, and bethanidine) of the sympathetic nervous system (Figures 27.3 and 27.4).

The pupillometer uses an infrared sensitive MOS solid state camera and is able to perform quantitative and accurate measurement of the pupil area including the diameter, pupil response to light (either single or repetitive), and instantaneous measurement of the latency velocity and acceleration parameters

Figure 27.6. Iris corder for measuring autonomic dysfunction.

of the pupillary light reflex (both directly and indirectly). The data are able to be produced under an "online" system and are also stored on the disc. All calculations can be made by a built-in computer. Figure 27.8 shows how the measurements are carried out in the various areas of the pupillary response.

Analysis of the pupil area made by means of a new binocular infrared videopupillograph reveals changes following pollutant exposure (HTV Type C 831, Hamamatsu TV Company, Mamamatsu).[19] All pupil studies are carried out at constant times between 9 and 12 a.m. or 1 and 3 p.m. in order to avoid diurnal variations of the pupil (Figure 27.6).[20]

Measurements of pupil area are made with an infrared video pupillometer.[19] The video cameras in the pupillometer are a charge couple device, and their output is analyzed by a frame grabber that counts the number of pixels above a slice level (gray level) adjusted by the experimenter to discriminate between pupil and iris. The output of the frame grabber is an analog voltage, which is proportional to the pupil area (sampling rate, 60 Hz). Light-emitting diodes provide the light source (peak wavelength, 605 nm).[21]

Pupil measurements are performed in an isolated, soundless, dark room, or, as an alternative, the patients wear a mask for the 10-minute period immediately before dark adaptation occurs. At the EHC-Dallas, a chemically less-contaminated room is used to ensure no extraneous pollutant exposure. Only direct pupil response is measured. When the patient's eye position is orthophoric, the binocular light reflex is simultaneously recorded, but when the patient has a deviation, which is mainly vertical, each eye is measured separately. The stimulus of light is 50 lux with a duration of 200 m/second.

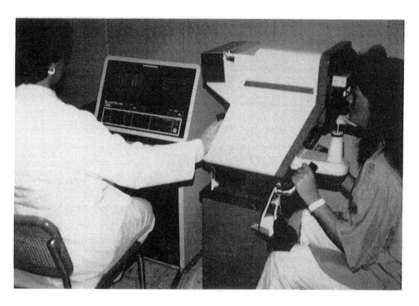

Figure 27.7. Iris corder for measuring autonomic dysfunction.

Pupillary area, latency time, amplitude of pupil constriction, half-pupil constriction time, time for maximum pupil constriction, time for pupil dilatation at 63% from maximum pupil constriction, maximum velocities of pupil constriction, and pupil dilation are determined in pupillary light reflex from the original recording and derivative and second derivative by a built-in computer

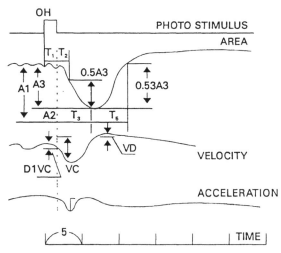

The C2515 Iriscorder uses some or all of the following twelve factors to measure Light Reflex, Alternate-Stimulus Reflex, and Near Reflex.

A1: Initial pupil area (mm²)
A2: Minimum pupil area after light stimulus (mm²)
A3: Pupil area change after light stimulus (mm²)
CR: Construction ratio A3/A1
D1: Initial diameter[1] (mm)
T1: Time from light stimulus to start of contraction[2] (m/sec)
T2: Time to half constraction (m/sec)
T3: Time to total contraction (m/sec)
T5: Time to recover to 63% of A3 after dilation from minimum state (m/sec)
VC: Maximum velocity of constriction (mm²/sec)
VD: Maximum velocity of dilation (mm²/sec)
AC: Maximum acceleration of contraction (mm²/sec²)
[1] D1 is calculated from the pupil area, assuming that the pupil is circular
[2] T1 is measured as the time from the light stimulus until the velocity of pupil contraction VC reaches 10% of the maximum velocity VCmax

Parameters of the Pupillary Light Reflex of 18 Normal Controls (Right and Left Eyes Combined)

Parameter	Mean	SD
A1	36.90	6.00
CR	0.48	0.07
T2	205.40	20.60
VC	49.60	8.60
T5	1881.80	555.50
VD	12.20	3.20

Figure 27.8. Pupillogram of light reflex and parameters of 18 normal control's pupillary light reflex (right and left eyes combined) — iris corder. (From Shirakawa, S., EHC-Dallas, 1989.)

(a detailed description of this computer has been made elsewhere).[22] At least three pupillary responses to light stimuli with an interval of 5 minutes are obtained, and they are averaged in both patients and controls.

Pupillography measurements are performed by the pupillometer throughout the sequence. There are four possible results including cholinergic, cholinolytic, sympathetic, and sympatholytic (Figure 27.3 and Table 27.3). Ishikawa[23] calibrated the pupillometer by using nondrinking, nonsmoking, noncoffee drinking, and nonmedicated controls (Table 27.4). In his initial studies, 417 Japanese males and 433 females were used for the right and left eye stimulation, respectively. In addition, 40 American controls at the EHC-Dallas confirmed the Japanese observations. Measurements were taken up to 10 times per day on 5 consecutive days. This study gave the percentage variability and reproducibility in measurements of the iris corder as well as the variation in the control population for brown and blue eyes. This machine will help quantitate pollutant injury, optic neuropathy, and visual fatigue.

Of 700 chemically sensitive patients first evaluated with the iris corder at the EHC-Dallas, 69% were found to have early pollutant injury that indicated dysfunction of the ANS (Table 27.5). This injury was then proven by inhaled or injected challenge under environmentally controlled conditions. Also, 90% of the electrically sensitive patients first evaluated with the iris corder and subsequently proven by frequency challenge under environmentally controlled conditions were identified with early pollutant injury. Additionally, the iris corder can measure the change in pupil response (thus the autonomic dysfunction) induced by a food, chemical, or biological-inhalant challenge, either oral, intradermal, or inhaled. At the EHC-Dallas, Homma et al.[24] used the iris corder in a study of 50 patients. The results of iris corder examination of chemically sensitive patients before and after double-blind inhaled ambient dose organophosphate pesticide challenge showed cholinergic changes in 20 patients. Six patients had sympatholytic changes due to ambient toluene exposure, while 22 showed either sympathetic stimulation or cholinergic suppression due to phenol exposure. Chlorine (<0.33 ppm) challenge showed nonspecific autonomic changes (Table 27.6).

The iris corder also can be used as a gauge for normalizing the ANS following environmental control treatment (avoidance of air, food, and water pollutants; avoidance of foods to which individual's have tested sensitive; intradermal injection neutralization therapy for biological inhalants, foods, and chemicals; intradermal titration therapy for biological inhalants, foods, and chemicals; and nutritional supplementation therapy) (Figure 27.9 to Figure 27.11).

Videopupillography has been used to analyze pupillary escape, which is a sign seen in optic neuropathy. Ishikawa[25] examined 30 patients with optic neuropathy thought to be due to multiple sclerosis. Pen light and intense light stimulus of about 10,000 cd/M[2] were used to evoke pupil escape phenomenon.

Ukai et al.[26] also measured visual fatigue phenomenon by the temporal change of critical flicker fusion frequency (CFF) in these multiple sclerosis

Table 27.4. Healthy Japanese and American Volunteer Controls for Pupillographic Response

	Sex	Below 10	10	20	30	40	50	60	Over 70
						Age			
A1 (mm^2)	M	41.7 ± 4.8	43.7 ± 6.6	44.5 ± 5.9	40.2 ± 5.3	34.0 ± 5.9	31.6 ± 6.7	24.5 ± 9.6	22.3 ± 6.5
	F	37.0 ± 3.9	40.9 ± 3.4	38.9 ± 6.7	38.4 ± 7.5	36.8 ± 3.1	26.5 ± 4.0	24.0 ± 7.2	17.4 ± 5.4
A3 (mm^2)	M	19.2 ± 3.3	19.9 ± 3.8	19.5 ± 2.1	17.5 ± 2.6	17.0 ± 2.1	14.5 ± 5.5	11.8 ± 3.7	11.6 ± 1.9
	F	18.3 ± 2.3	19.8 ± 1.5	19.6 ± 2.0	17.7 ± 2.8	16.8 ± 2.2	13.0 ± 1.9	12.2 ± 3.8	9.0 ± 3.4
CR (A3/A1)	M	0.46 ± 0.84	0.46 ± 0.83	0.44 ± 0.68	0.44 ± 0.44	0.50 ± 0.63	0.46 ± 0.55	0.48 ± 0.89	0.52 ± 0.96
	F	0.49 ± 0.46	0.48 ± 0.46	0.51 ± 0.57	0.48 ± 0.79	0.45 ± 0.48	0.50 ± 0.59	0.50 ± 0.80	0.52 ± 0.76
T1 (msec)	M	296 ± 17	280 ± 19	275 ± 25	286 ± 16	280 ± 25	291 ± 16	293 ± 21	312 ± 20
	F	279 ± 20	274 ± 17	273 ± 26	282 ± 18	289 ± 11	284 ± 15	295 ± 13	304 ± 25
T2 (msec)	M	194 ± 20	193 ± 16	198 ± 20	188 ± 24	182 ± 26	193 ± 26	200 ± 32	205 ± 24
	F	179 ± 13	183 ± 19	204 ± 25	189 ± 23	184 ± 18	185 ± 17	192 ± 23	191 ± 22
T3 (msec)	M	658 ± 54	670 ± 54	659 ± 45	654 ± 60	662 ± 60	667 ± 53	664 ± 87	728 ± 54
	F	638 ± 33	644 ± 37	655 ± 45	658 ± 67	654 ± 40	664 ± 51	683 ± 45	719 ± 50
T5 (msec)	M	1304 ± 252	1331 ± 189	1579 ± 349	1593 ± 370	1604 ± 342	1524 ± 425	1479 ± 356	1723 ± 502
	F	1340 ± 266	1431 ± 274	1588 ± 366	1465 ± 269	1441 ± 286	1450 ± 301	1630 ± 469	1600 ± 284
VC (mm^2/sec)	M	56.9 ± 9.2	58.2 ± 12.0	54.6 ± 9.4	51.4 ± 8.5	52.1 ± 5.6	49.8 ± 5.1	41.3 ± 9.0	38.0 ± 6.0
	F	56.1 ± 7.3	60.7 ± 5.0	55.3 ± 5.6	51.7 ± 7.4	52.3 ± 7.3	47.9 ± 5.3	43.1 ± 11.4	33.8 ± 10.5
VD (mm^2/sec)	M	17.0 ± 3.2	16.3 ± 3.0	14.8 ± 2.6	14.3 ± 2.2	13.2 ± 2.3	13.4 ± 2.5	10.9 ± 2.0	9.5 ± 2.0
	F	16.9 ± 2.2	16.0 ± 2.3	13.9 ± 2.0	13.8 ± 2.2	13.4 ± 2.2	11.4 ± 1.8	10.7 ± 3.0	8.1 ± 2.3
Healthy American volunteers		36.9 ± 6.0	0.48 ± 0.07	205.4 ± 20.6	49.6 ± 8.5	1881.8 ± 555.5	12.1 ± 3.2	40	29.7

Source: Shirakawa, S. 1988. EHC-Dallas.

Table 27.5.   Overall Results of Pupillography in 720 Patients vs. Normal Controls

	Total	Sex Male	Sex Female	Ages	Abnormal
Patient group	720[a]	231	489	8–75	69%
Comparison group	181	90	91	5–84	

[a] Each measurement was performed three times.

Source: Sujisawa, I., H. Suyama, and T. Namba. 1989–1991. EHC-Dallas.

Table 27.6.   Light Reflex Analysis by Pupillography: Double-Blind Inhaled Challenge after 4 Days Deadaptation in the ECU (50 Patients)[a]

Toxic chemical (no. of patients)	Ambient dose (ppm)	Response	Pupil
Pesticide (20)	<0.0034	Cholinergic nerve increase	A1 decrease, CR increase, VD decrease
Toluene (6)	<3.0	Sympathetic nerve decrease	CR increase, T5 increase
Phenol (22)	<0.0024	Sympathetic nerve increase Cholinergic nerve decrease	CR decrease
Chlorine (13)	<0.33 ppm	Nonspecific change	CR decrease

Notes: Increase or decrease ($p<0.05$); increase — excited; decrease — inhibited; A1 = pupil area; T5 = dilation time; CR = contraction ratio; VD = velocity of dilation.

[a] Some patients reacted to more than one chemical.

Source: Homma, K. 1993. EHC-Dallas.

patients. The results showed ten patients with a large amount of pupillary escape as manifested by CFF fatigue.

Comparative analysis between pupillary escape and visual fatigue in these same patients demonstrated an intimate relationship [r = 0.87].[26] The pen light stimulus repeated every 4 seconds in one eye also produced pupillary escape and the reduction of the amplitude of the pupillary light reflex. The rate of the reduction of the amplitude showed a tendency to decrease as visual acuity decreased. Thus, afferent pupillary defect was proven from the escape.

These two phenomena (pupillary escape and visual fatigue) seemed to be the result of a similar pathological phenomenon — optic nerve damage. Both seemed demonstrable by analyzing pupillary light reflex during pupillography. Pupillary escape sign can be an alternative test to detect optic nerve involvement in pollutant injury.

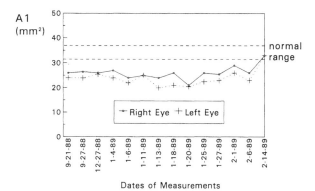

Figure 27.9. 47-year-old white female: pupillography improving with environmental control, injection therapy, and nutritional treatment. Improved with environmental control. (From EHC-Dallas. 1988.)

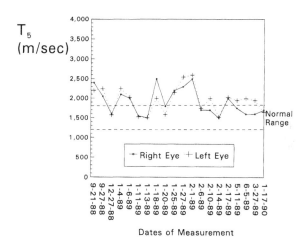

Figure 27.10. 47-year-old white female: pupillography improving with environmental control, injection therapy, and nutritional treatment. Improved with environmental control. (From EHC-Dallas. 1988.)

Ödkvist et al.[27] also have developed a procedure to study pollutant injury. They monitor eye changes that result from stimulation by a revolving chair (see Chapter 18). Ödkvist et al. first measure normal eye movements on a recorder. Then they measure the corresponding stimulus-provoked eye movements. They measure both sinusoidal swing in darkness and visual suppression. They also monitor the result of several tests including the optovestibular test, the

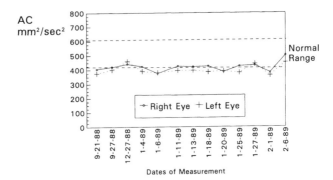

Figure 27.11. 47-year-old white female: pupillography improving with environmental control, injection therapy, and nutritional treatment. Improved with environmental control. (From EHC-Dallas. 1988.)

optokinetic test, the saccade test, and the slow pursuit movement test. They are able to evaluate pollutant injury in a large number of cases this way (Figure 27.12).

CLINICAL MANIFESTATION OF POLLUTANT DAMAGE TO THE EYE

Pollutant damage to the eye can occur in any area (Figure 27.13). The direct damage may manifest in allergic or inflammatory symptoms, but nutritional deficiencies also may be exacerbated, leaving the individual vulnerable to chemical sensitivity. The chemically sensitive individual is prone to unexplained visual fluctuations that result from pollutant exposure. All areas of the eye have been observed to be damaged in some chemically sensitive individuals. The eyelid, cornea, episclera and sclera, uveal tract, chamber, iris, lens, retina, and optic nerve will each be discussed separately.

Conjunctivitis, Lid Margins, and Adjacent Skin

Acute conjunctivitis manifests as chemosis with intense itching. It can be a result of a biological inhalant, food, or chemical exposure. Sometimes eosinophils (degranulated mast cells, basophils, plasma cells, lymphocytes, and macrophages) may be present in epithelial scraping. Other times, mononuclear leukocytes may be present in the epithelial scraping of the conjunctivitis. According to Raizman,[28] there are four types of conjunctivitis. These are vernal (occurring in the spring and fall), atopic keratoconjunctivitis, seasonal, and giant pupallomy. This classification may be false since the seasonal forms appear to be similar to the other three classes. Severe allergic conjunctivitis

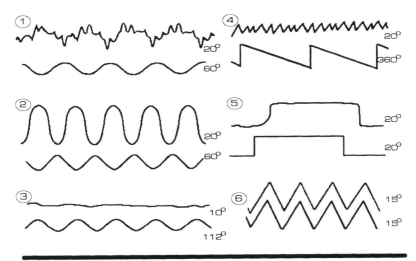

Normal eye movement recordings (upper traces in the six different vestibulo-oculomotor tests) and the corresponding stimulus provoking these eye movements (low traces).

1. Sinusoidal swing in darkness
2. Opto-vestibular test
3. Visual suppression
4. Optokinetic test
5. Saccade test
6. Slow pursuit movement test

Figure 27.12. Normal eye movement recordings (upper traces in the six different vetibulo-oculomotor tests) and the corresponding stimulus provoking these eye movements (low traces). (From Ödkvist, L. M. 1986. Personal communication.)

causes epiphora, itching of the eyes, and edema of the eyelids. At times, it is associated with sneezing, coughing, headache, dullness, and fever.

Chronic conjunctivitis gives rise to intense itching, burning, and photophobia. There may be slight edema of the conjunctiva and little congestion. The conjunctivae may show papillary proliferation, and in the long run epithelial follicles develop. Biological inhalants and foods can cause this condition as can drugs and other toxic chemicals. Organophosphate pesticides are known triggers of this type of chemical sensitivity, as are plastic contact lenses.[28]

Severe allergic conjunctivitis (vernal) in the early spring has become extremely common among the Japanese in the past 20 years. The reason for this increased incidence is a matter of dispute. The increased incidence of severe pollen-induced allergic conjunctivitis following repeated low-dose exposure to organophosphorous pesticides and organochlorine herbicides has been reported by Miyata et al.[29,30] They have postulated that repeated mucous membrane

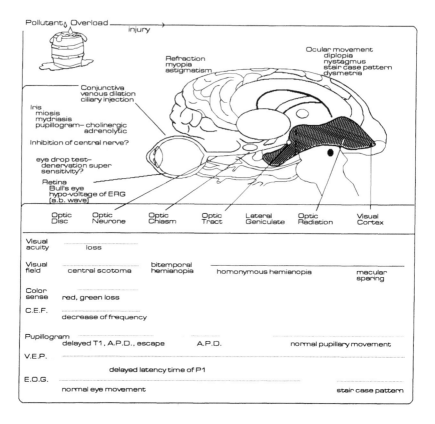

Figure 27.13. Areas of the eye susceptible to pollutant injury.

exposure to low doses of pollutants (chemicals) may cause immune deregulation and make the conjunctiva and other organs more prone to react to exposure to other incitants, which explains the increased incidences of allergic reactions in the last 20 years in Japan, where chemical exposures have also been steadily increasing. In America, according to some ophthalmologists, this type of conjunctivitis was initally thought to occur only in young boys between the ages of 5 to 10 years, but it probably occurs at any age.

Miyata et al.[29] studied guinea pigs passively immunized by an intravenous injection of antisera to Japanese cedar pollen. Thereafter, the effects due to organophosphorous pesticide (trichlorfon; DEP, Dipterex®, and fenitrothion; MEP, Sumithion®) and organochlorine herbicide (paraquat dichloride; Gramoxone®) on the conjunctival immune reaction to cedar pollen were evaluated quantitatively by measuring extravasated Evans blue of the conjunctiva.

The most sensitive allergic reaction was obtained from the groups administered with trichlorfon (from 10^{-4} to 10^{-3} mg/kg), fenitrothion (from 10^{-4} to 10^{-3} mg/kg), and paraquat dichloride (from 10^{-3} to 10^{-4} mg/kg). It was interesting that supersensitivity was observed in subjects exposed to the lower dosage, rather than the higher dosage. The effects persisted at least 2 weeks in the guinea pigs exposed to organophosphorous pesticides and at least 1 week in pigs exposed to organochlorine herbicide. These results suggest that there is a strong relationship between intake of environmental pollutants such as pesticides and herbicides and the onset of allergic conjunctivitis.

Edema of the eyelids may be associated with angioedema and urticaria in other body parts. Itchiness is commonly noted. Also periorbital edema occurs when contaminated food is eaten by chemically sensitive individuals.

Dermatitis of the eyelids is usually eczema and associated with conjunctivitis and keratitis. Eczema must be differentiated from seborrheic dermatitis and contact dermatitis. Many types of cosmetics, drugs, or anesthetics can cause contact problems as seen in the chemically sensitive patients treated at the EHC-Dallas (see Chapter 25).

Hordeolum or sty is an acute staphylococcal infection of the gland of zeis, the sebaceous gland along the side of the follicle. Underlying food and chemical triggers may often lower resistance and allow these to reoccur. Also, female hormone allergy can trigger this problem (see Volume I, Chapter 5[31]).

Chalazion and meibomitis can be triggered by food and chemical problems.

Cornea

Pollutant exposures may affect the cornea in a variety of ways. Atopic keratitis may be due to biological inhalants, foods, and/or chemicals. Mild superficial punctate lesions may occur. Deep keratitis may also occur with opacified lesions without "salmon" patches. Contact keratitis can occur from plastic contact lenses and preservatives in artificial tears. Drugs and cosmetics and other chemicals can be offenders. Interstitial keratitis also may occur. Catarrhal infiltrates and ulcers also may be seen after sensitivity reactions or chemical exposures.

Episclera and Sclera

There are two clearly differentiated types of inflammation of the sclera and episclera that may result from chemical or pollutant exposure. Episcleritis is a common, mild, and transient inflammation of the episclera. Brawny scleritis is a rare, severe, and chronic inflammation of the sclera. Scleritis near the limbus may include the cornea. Some chronic types of episcleritis are associated with rheumatoid arthritis and other collagen diseases, which can be triggered by foods and chemicals (see Chapter 23).

Uveal Tract

The uveal tract may undergo a variety of ill effects as a result of pollutant exposure. Two main types of uveitis — granulomatous and nongranulomatous

— are recognized. Granulomatous uveitis can be caused by the granuloma-producing organisms such as tuberculosis as well as those with unknown etiology such as sarcoid. Granulomatous type of uveitis may be due to micro-organisms, but it can also be due to sarcoid, talc, beryllium, wood, and silica sensitivity.[32] Sympathetic ophthalmia can also occur in one eye after trauma and then induce uveitis in the other eye (see Chapter 20). Systemic lupus erythematosis also is known to form uveitis.

Nongranulomatous uveitis is thought to be an allergic form of uveitis, since experimental evidence shows that antibodies are formed in the uveal tract. The presence of many wandering cells, particularly plasma cells, in this highly vascular tissue suggests that the tract acts as part of the reticulo-endothelial system. Also, experimentally, a sensitized eye will react and suffer uveitis if the antigen is administered systemically, and the previously unsensitized contralateral eye may develop nongranulomatous uveitis. Uveitis has been observed in atopic individuals due to allergy to pollens, animal emanations, foods, chemicals, or drugs. Bacterial allergy has also been implicated as a cause of uveitis.[33]

When compared with normal individuals, atopic individuals have a much higher percentage of cataracts, uveitis, and retinal detachment. Many atopic patients in their teens, 20s, and 30s visit the ophthalmic clinic to have a cataract operation. After the operation, the patients often are able to control develop-ment of uveitis and retinal detachment, the latter of which is often recurrent in atopic patients. Control of patients' atopic condition by environmental medi-cine specialists before and after cataract surgery is important.

Homma and Ishikawa[34] measured the blood level of benzaldehyde, a metabolite of toluene, in uveitis patients who had been exposed to toluene. They indicated that toluene might induce uveitis with mucocutaneous lesions, which was similar to Behcet's disease. This type of uveitis was the nongranulomatous type. Many cases of uveitis may be diagnosed without recognition of chemicals as triggering agents.

Anterior Chamber

The anterior chamber of either, or both, eye(s) may be involved in some cases of pollutant exposure, and as a result occular pressure may vary. Such variation in eye pressure due to toxic chemical exposure may cause glaucoma. The following is a report of a patient with glaucoma due to toxic chemical injury.

Case study. This 47-year-old white female entered the EHC-Dallas with the chief complaints of sore, dry eyes and failing vision, light sensitivity, depres-sion, headaches, and fatigue. All of these symptoms had been progressive for 16 years. The patient had been diagnosed with glaucoma and macular degen-eration. She had been on medication for both with poor results. She worked as an interior designer and was exposed to high levels of toxic substances, including xylene, at work.

This patient was found to be sensitive to biological inhalants, foods, and toxic chemicals by the various challenge tests. Her laboratory tests showed the following results: WBC = 4700/mm³ (C = 4800 to 10,000/mm³), total lymphocyte count 1551/mm³ (C = 1600 to 4200/mm³); percentage of lymphocytes = 33% (C = 21 to 49%); T lymphocytes = 86% (C = 62 to 86%); absolute T count = 1334/mm³ (C = 1260 to 2650/mm³); helper T_4 = 46% (C = 32 to 54%); helper T count = 713/mm³ (C = 670 to 1800/mm³); suppressor T_8 = 31% (C = 17 to 35%); suppressor count T_8 = 481/mm³ (C = 333 to 1070/mm³), T_4/T_8 = 1.5 (C = -1 to 2.7), B lymphocytes = 9% (C = 5 to 18%); total B count = 140/mm³ (C = 82 to 477/mm³); total complement — CH_{100} = 23% (C>70%); dichloromethane = 5.6 ppm (C<1.0 ppm); 1,1,1-trichloroethane = 3.3 ppm (C<0.3 ppm); toluene = 7.0 to 1.5 ppm (C<0.5 ppm); benzene = 2.0 ppm (C<0.10 ppm); xylene = 1.8 ppm (C<0.1 ppm); chloroform = 4.3 ppm (C<1.0 ppm); tetrachloroethylene = 2.0 ppm (C<0.3 ppm); 2 methyl pentane = 9.8 ppm, (C<0.10 ppm); 3 methyl pentane = 27 ppm (C<1.0 ppm); N-hexane = 2 ppm (C<1.0 ppm); styrene = 1.1 ppm (C<1.0 ppm); erythrocyte glutathione peroxidase = 1.9 moles NADPH/min/mg Hb (C = 4.23 to 7.234 moles NAD PH/min/mg Hb); erythrocyte superoxide dismutase = 6.51 mg Hb (C = 9.4 to 13.4 mg Hb); lipid peroxide = 6.43 nanomole/mL (C = 2.65 to 4.15 nano mole/mL); CMI = 2 of 8 positive.

This patient underwent heat depuration, injection therapy for biological inhalants, foods, and chemicals, and environmental control. She gradually cleared her symptoms. Her ocular pressure went from 40 to 15 mmHg (C = 10 to 20 mmHg) and stayed there without medication. It elevated with provocation by intravenous vitamin C, demonstrating an adverse reaction to this vitamin. She has maintained control of her glaucoma with continued environmental control and without medication for 2 years (Figure 27.14).

Endoophthalmitis phaco-anaphylacita is due to a severe allergic reaction in patients sensitized to lens protein that has entered the anterior chamber due to trauma or increased lens permeability.

Direct injury to the anterior chamber as well as injury to the ANS may occur as a result of some solvent exposures.

Iris

The iris may be involved in pollutant injury that results in miosis or mydriasis. This type of injury gives cholinergic and adrenolytic effects and will register on pupillography. The patient may develop eye pain, as was seen in several patients at the EHC-Dallas who had repeated admissions after pollutant exposure. Some chronic types of injury to the iris, which can be triggered by foods and chemicals, are associated with rheumatoid arthritis and other collagen diseases. The following is a case report of a patient with iritis and rheumatoid arthritis.

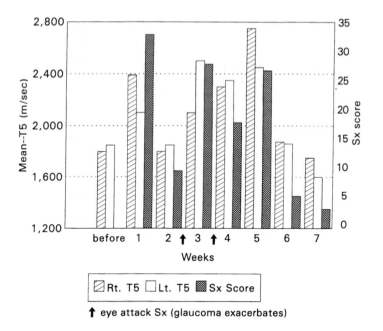

Figure 27.14. Pupillography changes due to sensitivity to vitamin C. Pupillography (T5) and symptom score: variation during heat depuration therapy.

Case study. This 22-year-old white female was admitted to the EHC-Dallas with the chief complaints of spinal arthritis, frequent episodes of iritis, tinnitus, poor memory, and muscle and joint pain. She reported that her illness began when she was 14 years old. At this time, she developed iritis and back pain. Shortly after starting college, she developed neck swelling along with stiffness and headaches. She was then diagnosed with viral encephalitis. In 1980, at age 19 years, she developed an anaerobic upper respiratory infection, and her blood test revealed toxoplasmosis. During this time, she was also diagnosed with ankylosing spondylitis with positive HLA-B 27. In 1981, she developed menorrhagia, which was corrected with birth control pills, but she eventually developed breast tumors that were thought to be linked to the birth control pills; therefore, they were stopped. She was sent to a health farm in Pennsylvania where the breast tumors dissolved with pollutant avoidance and nutritional treatment. In 1982, her physician suspected that her iritis was allergy-related. When her symptoms snowballed into numerous complaints and she was unable to gain relief, her doctor referred her to the EHC-Dallas. After admission to the ECU, she became symptom free.

Her family history was positive for hay fever, cancer, eczema, hives, diarrhea and constipation, insect allergies, drug allergies, and arthritis. At age 51 years, her father had spinal arthritis and food and chemical sensitivities. At age 50 years, her mother suffered from hay fever. She had three brothers. At age 28 years, one suffered from spondyloarthritis. At age 23 years, another

Table 27.7.   22-Year-Old White Female: Inhaled Double-Blind Challenge after 4 Days of Deadaptation and Total Load Decreased in the ECU

Incitant	Dose (ppm)	Result
Ethanol (petroleum derived)	<0.50	Blurred vision; flared iritis
Phenol	<0.0024	—
Formaldehyde	<0.20	—
Insecticide (2,4-DNP)	<0.0034	—
Chlorine	<0.33	—
Saline	—	
Saline	—	
Saline	—	

Source: EHC-Dallas. 1993.

suffered from food sensitivities and hay fever. At age 26 years, her third brother was in generally good health.

This patient was orally challenged with individual doses of a total of 22 chemically less-contaminated foods; 9 of them caused her symptoms. Her chemical sensitivities were very widespread, and she had inhalant sensitivity to dust, pollens, and molds. Double-blind inhalation challenge with ethanol (<0.50 ppm) produced crying, spaciness, headache, inability to talk, bone pain, and blurred vision. Inhaled challenges to saline placebos, formaldehyde, phenol, pesticide, and chlorine were negative. She had sensitivity to sulfa, neosynephrine, and antibiotics (Table 27.7). Abnormal test results showed a depressed T helper/suppressor ratio of 0.9 (C = 1.0 to 2.2); elevated T_8 at 43% (C = 20 to 37%); elevated IgM at 418 mg/dL (C = 56 to 352 mg/dL); and elevated IgE at 400 µ/mL (C = 0 to 180 µ/mL).

This patient was diagnosed with acute arthritis, rheumatoid HLA-B-27, fibromyositis, acute iritis and vitriitis, severe chemical hypersensitivity with atypical syncopal type episodes, immune complex disease, constipation with redundant colon, atopic disease (inhalant), multiple food hypersensitivities, and psychological factors affecting her physical condition.

This patient was placed in the ECU with nothing by mouth except spring water for 1 day. She had a total stay of 3 weeks. During this time, this patient reported irritability, headaches, and severe hunger pangs. She also reported that her joint pain lessened and her iritis and lip chapping resolved. She further reported that her headaches and tinnitus exacerbated upon provocative skin tests.

Additional response was achieved with epsom salt as a laxative to clear the gastrointestinal tract. Intravenous vitamin C, serotonin, and histamine were used to clear the symptoms produced by intradermal and oral challenges.

In accordance with determined sensitivities, immunotherapy was arranged for 24 foods, trees, grass, weeds, terpenes, cotton, fluogen, MRV, transfer factor, molds, lake algae, *Candida*, dust, dust mite, feathers, smuts, histamine, and serotonin.

An oasis that was less-polluted by the criteria established at the EHC-Dallas was created at her home. The oasis was not entirely successful in controlling her illness. This patient had to use an environmentally controlled trailer and live in a less-polluted area for a period of time in order to allow recovery to occur. The patient subsequently did well, except for three large inadvertant pollutant exposures that resulted in the activation of the iritis. These reactions cleared with avoidance, and the patient has continued to do well without medication for the past 5 years.

In one of our series of 200 inhaled double-blind challenges of chemically sensitive patients, double-blind inhaled challenges with pesticide (2,4-DNP, <0.0034 ppm), toluene (<0.1 ppm), and 1,1,1-TCE (<0.1 ppm) produced significant changes on some parameters of the iris corder light reflex test in a subset of 50 patients. Homma et al.[24] showed that such low-level exposure caused significant pupil changes in selected chemically sensitive patients (Table 27.6).

Lens

The lens is a dehydrated organ. The adult human lens contains 66% water and 33% protein. Lens dehydration is maintained by an active Na^+ ion water pump that resides within the membranes of cells in the lens epithelium and each lens fiber. When pollutants damage this Na^+ pump, lens hydration is disturbed, and this disruption leads to lens swelling, which then causes a change in vision, as is often seen in some chemically sensitive individuals.

Glutathione is synthesized in the lens, and most of it is in the reduced form GSH. Reduced glutathione is very important to the maintenance of lens transparency as it is required to maintain Na^+, K^+-ATPase transport intact. Glutathione also preserves physicochemical equilibrium of the lens proteins by maintaining high levels of reduced (SH^-) groups. It provides the lens with a major detoxifying mechanism via the enzyme glutathione peroxidase. This enzyme removes H_2O_2 and toxic lipid peroxides. Pollutants can damage this enzyme, and this damage may lead to early cataract formation. Any pollutant injury in combination with nutritional deficiency will interfere with the maintenance of lens transparency.

Once in the lens, free amino acids are incorporated into the polypeptide chain through the mRNA to form lens protein (Figure 27.15). Humans may be exposed to pollutants such as ultraviolet rays that cause changes in lens proteins with significant destruction of lens tryptophan. Some pollutants may be cataractogenic and cause early presbyopia. The lens is highly susceptible to oxidative damage by free radical (O_2^-) or its derivatives — peroxide (H_2O_2), singlet oxygen ($^1O_2$), and OH^-.

Prednisone is catabolic, and the anticholesterol drug LOBΔ is cataractogenic in rats.[35] These drugs damage the ability of the lens to maintain GSH synthesis and increase its efflux through more permeable membranes. The generalized

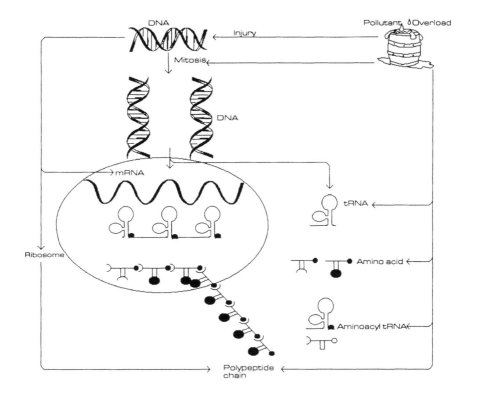

Figure 27.15. Biochemistry of protein synthesis by the lens.

disarray of lens metabolism is accompanied by ATPase loss, and the end result is a totally opaque lens (Figures 27.16 and 27.17).

Any pollutant that causes vasculitis and prolonged breakdown of the blood-aqueous barrier leads to lens vacuoles or lens opacities. Breakdown of the blood-vitreous barrier can lead to a posterior cortical or subcapsular cataract.

Trauma[36] as well as many agents, including chemicals,[37] radiation,[38] electricity,[39] and viruses[40] can induce cataracts in humans. Vitamin[41] or amino acid deficiencies[42] can also cause cataracts in humans. Cataracts may be associated with skin, CNS, and skeletal disease as well as chromosomal abnormalities.

The lens capsule surrounds the lens and maintains its structural integrity. It is made up of collagen-like glycoprotein material. It is considered a model basement membrane similar to that of the renal glomeruli and responds to pollutant injury in similar ways. Lamellar splitting of the capsule is seen in glassblowers due to pollutants or possibly infrared heat. Small cuts or rents of the capsule may result in localized opacities of the underlying lens cortex (Figure 27.16).

Cortical fiber damage has been reported with diabetes, galactosemia, and exposure to ubain (inhibits Na^+, K^+ ATPase), chlorpromazine (inhibits Ca^{2+}

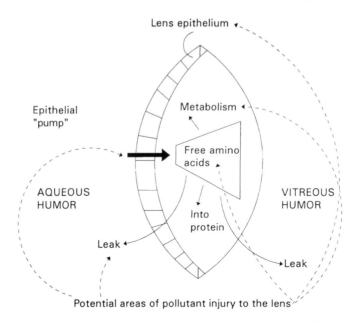

Figure 27.16. **Potential areas of pollutant injury to the lens. Amino acids are actively transported into the lens mainly by an epithelial pump, and they leak out through anterior and posterior surfaces. Free amino acids are metabolized or incorporated to form new lens fibers. Pollutant injury may disturb any of these functions.**

ATPase), and triparanol aminotriazol (inhibits catalase). The various chemically induced changes, some of which develop prior to visible cortical abnormalities, include the loss of glutathione with compensatory increase in NADPH synthesis; an increase in K^+ ion efflux; a loss of K^+ ions, inositol, and amino acids from the lens; a gain in Na^+ ions; a decrease in lens protein synthesis with a decrease in the proportion of soluble protein and an increase in insoluble protein; an increase in protein S-S groups and in Ca^{2+} ions; a decrease in the activities of most enzymes and an increase in the activities of hydrolytic enzymes; and a decrease in ATP content (Figure 27.17).

When many chemically sensitive patients are challenged with incitants, the amplitude of accommodation decreases at a young age. Physiologically, the amplitude of accommodation diminishes with increasing age. Up to the age of 8 years, the dioptric power of the eye can be raised by accommodation approximately 14D; at age 20, it drops to 11D; at age 30, it falls to 9D; and at age 50, it is less than 2D. The nearest point to which eyes accommodate so that a clear image is formed on the retina is called the "near point". In the young, this point is closest to the eye. Gradually, however, it recedes until about the age of 45. By the time the near point has become so far removed that the subject cannot read fine print, the eye is said to have become presbyopic. It is believed

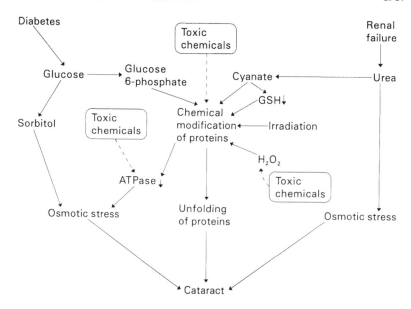

Figure 27.17.   Possible common pathways in some cataracts triggered by pollutants. Arrows link changes observed in different types of cataract. GSH = reduced glutathione.

that presbyopia is the result of sclerosis or hardening of the nucleus of the lens as well as loss of elasticity of the capsule so the lens is less strongly curved. Therefore, if presbyopia occurs in a patient of a younger age, the cause may be free radical generation by chemical incitants. If pollutants are indeed the culprit, their identification and removal may be sufficient to reverse the presbyopia, as is seen in chemically sensitive patients with formaldehyde exposures.

According to Ishikawa et al.,[43] nearsightedness and astigmatism are frequent among people who live in areas where agricultural chemicals are air-scattered. Acute organophosphate intoxication causes accommodation spasm, which is the hyperexcitation of ciliary muscles. With accommodation spasms, patients can see near, but cannot see far clearly. On the other hand, chronic organophosphate intoxication, seen in some chemically sensitive patients, causes accomodation palsy. Intoxication may cause fatigued ciliary muscle or damage the nervous pathway from the Edinger-Westphal nucleus of the oculomotor nerve nucleus, through the oculomotor nerve, short ciliary branches, ciliary ganglion, and post-ganglionic nerve to the ciliary muscle.

On double-blind inhalation challenge of chemically sensitive patients, Homma et al.[44] showed that pesticide (DDVP) and toluene produce cholinergic change on the light reflex, and 1,1,1-TCE produces tonohaptic change on light reflex. Tonohaptic change is a very rare condition in which both velocity of contraction and velocity of dilation are high. This change may be due to the excitation of the cholinergic nerve and sympathetic nerve.

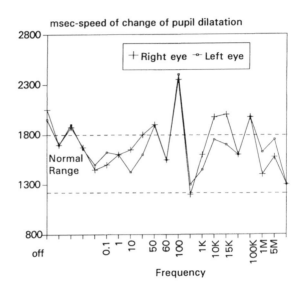

Figure 27.18.   Double-blind EMF challenge — T$_5$; 49-year-old white female studied in shielded room under environmentally controlled conditions. (From EHC-Dallas. 1990.)

It is clear from Ishikawa's and our data that accommodation is also affected by chemicals. Many chemically sensitive patients experience blurred vision immediately upon entering a contaminated area, and usually it is reversible once they leave. If blurred vision is caused by a damaged optic nerve or lens or by glaucoma, it is difficult to reverse quickly. Many complaints of chemically sensitive patients about vision can be explained by accommodation dysfunction caused by chemicals. (Sometimes the culprit is accommodation spasm, and sometimes it is accommodation palsy.) In addition, many video display terminal (VDT) workers have accommodation problems that may result from fatigue of the ciliary muscle or may be a direct effect of EMF on the ciliary muscle. The iris corder also registers changes in electromagnetically sensitive patients (Figure 27.18). Regardless of its cause, accommodation dysfunction is as sensitive as light reflex and, therefore, may be ueful in diagnosing chemically sensitive patients.

Retina

The eye requires a continuous supply of vitamin A (retinol). It has been estimated that less than 0.01% of the total quantity of vitamin A in the human body goes to the eye, where it is localized in the retina and is essential for vision. The deficiency of vitamin A causes degeneration of the retina, corneal changes in the eye, and adverse effects on the growth of the eye. Retinol, a form of vitamin A, is important to maintenance of the integrity of visual cells.

Retinoic acid, not retinol (another form of vitamin A), aids growth and keeps a person alive, but if it is subjected to light, it can be damaged, causing selective degeneration of the photoreceptor cells.

The light sensitive pigments of the human retina include the well-known rod pigment important for dim-light vision, rhodopsin, and three cone pigments responsible for color discrimination and normal daylight vision. Retinal rod pigments are abundant and more stable than cone pigments. Rhodopsin consists of an apoprotein opsin to which is attached molecule 11-*cis*-retinaldehyde or 11-*cis*-3-dehydroretinaldehyde. These may be changed in some chemically sensitive patients.

The visual cycle begins with photoisomerization of 11-*cis*-retinaldehyde bound both to opsin and to all-*trans*-retinaldehyde. This event triggers the electrophysiological response of the photoreceptor cell to light. All-*trans*-retinaldehyde is reduced to all-*trans*-retinal by a retin dehydrogenase in the outer segment of the photoreceptor membranes, and it never accumulates to any extent because of its general toxicity. The NADPH generated by the pentose cycle appears to be the exclusive reducing agent for retinaldehyde in the photoreceptor cells. Much of the retinal produced during extensive bleaching diffuses out of the photoreceptor outer segment and is esterified in the retinal pigment epithelium. Esterification appears to facilitate the retention of retinal in the eye and perhaps also protects the retina from sustained exposure to high concentrations of retinal, which is well-known to affect the stability of membranes adversely. As discussed in Volume I, Chapter 6,[2] mineral deficiency, in particular zinc deficiency in the presence of pollutants, interferes with the function of alcohol dehydrogenase and leads to accumulation of retinaldehyde, which is toxic to the retina and causes problems in visual pigment and dark adaptation and disturbances in dim-light vision (Figure 27.19).

Taurine is present in high concentrations in the retina, and light stimulates its release from rod outer segments. The role of taurine is not clear, but it may be required for intracellular osmotic regulation in photoreceptor cells. Taurine deficiency has been attributed to the degeneration of photoreceptor cells in animals.[45] Also, retinitis pigmentosa in humans is suspected to result from taurine deficiency. Taurine deficiency has been well documented in food and chemically sensitive patients.[46] It predisposes them to degenerative changes in photoreceptor cells.

Antioxidants, particularly alpha tocopherol, appear to be required to protect the pigment epithelium from the accelerated accumulation of lipofuscin, the so-called "aging-pigment". It is thought that alpha tocopherol may serve to inhibit peroxidation of polyunsaturated fatty acids and vitamin A in photoreceptors. The nutritional status of vitamins A and E is interrelated in the eye. Photoreceptor loss is enhanced in vitamin A-deficient individuals who are also vitamin E deficient. Vitamin E deficiency decreases the amount of vitamin A that can be stored in the eye. Pollutants can cause vitamin A and E deficiency affecting absorption and assimilation (see Chapter 21). Therefore, patients with food and chemical sensitivities are more prone to develop problems with dark

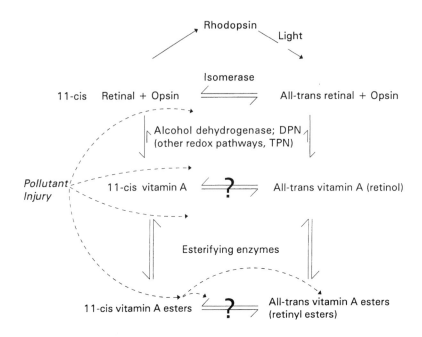

Figure 27.19.   Diagram of rhodopsin system, showing isomerization cycle. Bleaching of rhodopsin by light ends in mixture of opsin and all-*trans*-retinal. All-*trans*-retinal must be isomerized to 11-*cis* before it can regenerate rhodopsin. During regeneration, much retinal is reduced to all-*trans* vitamin A, most of which, in turn, is esterified. (From EHC-Dallas. 1994.)

adaptation as well as macular degeneration. The retina can be damaged by light (Figure 27.20).

In animal studies, Ishikawa and Miyata[47] reported retinal pigment epithelial degeneration caused by chronic organophosphate intoxication. Clinically, electroretinography (ERG) changes are sometimes seen in patients who were exposed to pesticide, and Ishikawa and Myata have identified some idiopathic retinitis pigmentosa patients who work as farmers using pesticides. Ikeda[48] reported that fluorescein angiography showed diffuse granular defects on the choroidal flow of toluene-exposed patients, demonstrating that retinochoroidal circulation can be affected by chemicals.

Retinal damage sometimes combines with optic nerve damage, so the visual field tests on toxic patients sometimes show strange changes. Rogers[49] reported the following case that emphasizes the environmental aspects of vision loss.

Case study. A 44-year-old landscape architect had developed severe vision loss 2 years prior to admission to the ECU. Within a year and a half, his vision

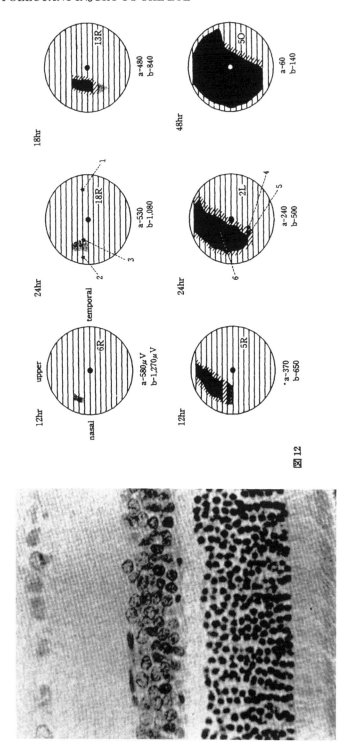

Figure 27.20.   Visual fields showing pollutant damage to the optic cells of the eye. (From EHC-Dallas. 1994.)

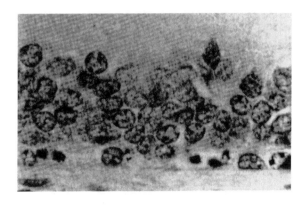

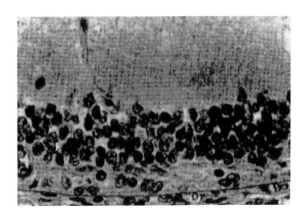

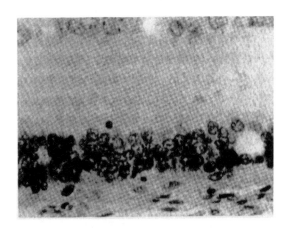

Figure 27.20. (Continued).

rapidly deteriorated to 20/200. This vision loss could not be corrected, and he could no longer drive. Also during this time, he developed a pustular rash on the soles of his feet and on the palms of his hands. He had had 20 years of recurrent sinusitis and 36 years of hay fever. He had irritable bowel, depression, headaches, and atopic dermatitis.

When treatment by his family physician failed, this patient consulted four ophthalmologists for treatment of his vision loss. He saw two allergists and two ear, nose, and throat specialists for his other problems. This patient had PE tubes. He had been previously treated for migraines and had consulted with an internist in infectious diseases. Although the dermatologist diagnosed the skin lesions as acropustulosis, and the ophthalmologists diagnosed his vision problems as retinal edema of an undetermined etiology, his symptoms went unrelieved. Earlier treatments by an allergist had failed, and two of the three allergists consulted when his vision problems developed told him his allergies were not severe enough to be treated.

A year prior to the development of his retinal edema, he had renovated his home extensively, building additional rooms, laying new carpets, and purchasing new furnishings.

Work-up revealed that he was sensitive to a variety of trees, grasses, weeds, house dust, house dust mite, and several genera of fungi. Nutrient assessments showed intracellular deficiencies of zinc, copper, folic acid, and chromium and an elevated formic acid as well as thiamine. Magnesium loading test revealed magnesium wasting and deficiency. Normal results were found on cortisol stimulation tests; triiodithronine by RIA; antithyroid antibodies; vitamins B_6, B_{12}, and A; transketolase; and RBC selenium and iron.

Environmental and dietary controls were explained to this patient, and he began a macrobiotic diet. He also was treated with injections to the titrated doses of inhalants, and his nutrient deficiencies were corrected. Within 5 months, his vision returned to normal. Subsequent testing revealed multiple foods, chemicals, and molds caused retinal swelling. In spite of marked improvement on his treatment program, he remained exceptionally sensitive to environmental pollutants. For example, opening the envelope of a scented bill was enough exposure to cause inability to see for up to 3 days. Also, ingestion of pizza left him unable to see for 3 weeks. Maintaining good environmental control and continuing on a macrobiotic diet, this patient has been fortunate to have his vision return to normal. He remains in fairly good health, although he is unable to work in his former office, which was being renovated during his recovery. Apparently, he developed multiple sensitivites to the construction materials used during this time.

Optic Nerve

Reactions to pollutant exposure can occur in the optic nerve. Before the days of antibiotics and chemotherapy, serum sickness occurred frequently, with retinal and optic nerve edema being common. One study of 75 cases of serum sickness revealed nine patients with ocular signs that occurred within 24 hours after the second dose of horse serum.[50] There were complaints of blurring and spots before the eyes with lacrimation and photophobia. Pupils were dilated and conjunctiva congested, with symptoms worsening when urticaria was at its highest. Papilledema and retinal edema were present with dilatation of the retinal vessels. This acute severe type of condition is rarely seen now, but a milder form in atopic and chemically sensitive individuals is seen. These patients complain of blurred vision associated with retinal edema and blurring of the optic disc after exposure. Chemically sensitive individuals particularly complain of blurred vision when they are exposed to chemicals in public buildings. A case of papilledema related to chemical exposure has been observed by Patel[51] and is related here.

> **Case study**. A 30-year-old white female presented with the chief complaints of blurry vision of 5 years duration. Her condition progressively worsened during the 2 months previous to her examination. Careful history revealed that this patient was consuming large amounts of diet cola. Comprehensive eye evaluation revealed she had papillodema (Figure 27.21) that was worse in her left eye. The optic nerve head (optic disc) was hyperemic with nerve swelling as evidenced by climbing retinal vessels. With elimination of colas containing aspartame for 7 months, her papilledema regressed with less swelling and hyperemia. Five years follow-up study revealed she had 20/20 vision, regressed papilledema, and reduced swelling (Figure 27.22). Unfortunately, her triggers for MS were not worked out (see Chapter 26).

Pollutant injury can occur at any area along the optic nerves, the optic tracts, the geniculate body, or the visual center in the brain. The retina can be injured by pollutants with damage occurring to the rods or cones, the peripheral choroid, or the macula. Macular degeneration also is a well-known problem that may have some relationship to pollutant injury.

Pollutant injury can occur all along the optic nerve including the disc, the optic neurons, the optic chiasm, the optic tract, the lateral geniculate body, the optic radiation, and the visual cortex.

Optic neuritis resulting from exposure to several pollutants has been documented. For example, muslin and other foreign bodies have been used to buttress vessel walls of intracranial aneurysms during surgery, causing an inflammatory reaction and thus neuritis. Substances such as toxic chemicals that are known to trigger autoimmune disease also will trigger optic nerve inflammation. A lupus-like syndrome is known to be triggered by many drugs and toxic chemicals (see Chapters 20 and 23).

Noell[52] reported a case of optic neuropathy that resulted from glue sniffing in a 25-year-old man. He had problems with blurry vision. His visual acuity and color vision were disturbed. He was treated with avoidance and anti-inflammatory drugs, which brought improvement in visual acuity and central scotoma.

Kimura and Ishikawa[53] have seen many glue sniffers and drug-abuse patients. Because glue, and many other drugs in large doses, can cause optic atrophy, these individuals are very difficult to treat successfully.

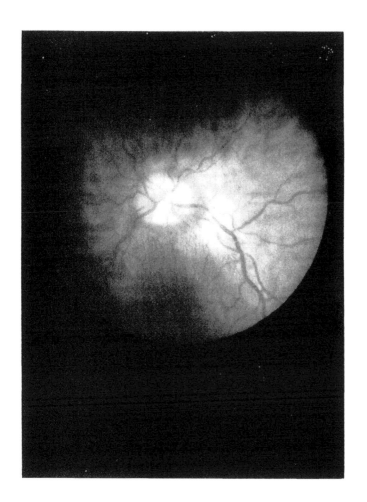

Figure 27.21. Pollutant-triggered papillodema.

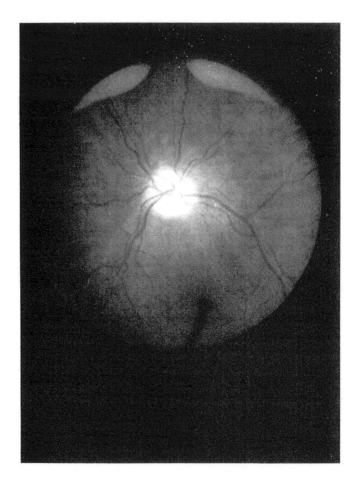

Figure 27.22. Pollutant-triggered papillodema cleared.

Homma[54] reported a case of night optic neuropathy in a 13-year-old girl. She entered the hospital, and the etiology of her eye problem remained a mystery after all exams. Her night vision was sl(–), and steroid treatment was tried. One month later, she told her physicians that she had sniffed glue 2 years before her symptoms started, and her vision loss started at the same time repainting of her junior high school commenced. Homma felt that this patient acquired chemical sensitivity to organosolvents when she was sniffing glue, and 2 years later she was in a weakened, vulnerable state when exposure to paint fumes occurred. As a result, the paint exposure caused severe vision impairment. Since this patient did not respond to steroid treatment, it was stopped, and she was kept away from chemical substances. Her vision improved to 20/20. Apparently her optic neuropathy was a symptom of chemical sensitivity, not of direct organic solvent toxicity.

In our own series, we have seen a 42-year-old Latin American female who was referred to us by her ophthalmologist. She had intractable symptoms of blurring vision to the point of episodic blindness. It was the ophthalmologist's opinion that this patient would be blind in a few months due to her optic neuritis. Her condition appeared to be intractable to all medication. She was placed in the ECU and fasted for 4 days. All her symptoms, including vision loss, disappeared. She could see well. This patient was tested on foods and biological inhalants, without problems. Inhaled double-blind challenge was negative, except for the challenge with the fumes of solder flux with which she worked daily. This challenge reproduced all of her symptoms of aches, pains, fatigue, depression, and headache in addition to transient eye pain and blindness that lasted for 45 minutes. She was instructed on the clean-up of her home environment and advised to change jobs. She did both and was asymptomatic for 5 years after treatment. Her vision returned to 20/20, and she had no more episodes of blindness. She worked daily as a clerk in a drug store and took no medications (Figure 27.23).

Pseudotumor Cereberi

Many pollutants such as fat soluble vitamin A (acutane), nicotinic acid, and tetracycline can cause pseudotumor cereberi.

Cases have been observed with headaches, visual disturbances, and proven fundoscopic changes. The following case of pseudotumor related to environmental influences is shown.

Case study. A 56-year-old white male dentist presented with severe headache, myalgia, and arthralgia. He had recently remodeled an old office. Since he was in Florida, he used a wood preservative, pentachlorophenol, to protect his wood. He was well until he moved into his complete office where he immediately developed all of his presenting symptoms. After a lengthy period

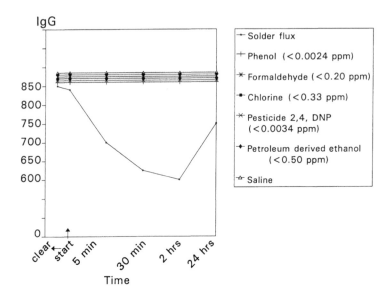

Figure 27.23.   **42-year-old Latin-American woman; pollutant injury to the eye. Symptoms reproduced on challenge included blindness, eye pain, confusion, and depression. (From EHC-Dallas. 1986.)**

of illness, medical work-up revealed pseudotumor cereberi. This patient failed to respond to treatment until he was seen at the EHC-Dallas. He was found to have many secondary sensitivities to foods and inhalants. He was severely sensitive to the ambient doses of toxic chemicals. With a massive avoidance program, injection therapy for the substances to which he was sensitive, and replenishment of his nutrient defect, he was restored to health with total remission of the pseudotumor for the 5 years posttreatment.

Absence of Eye

Every year over 120 children are born without eyes, or with tiny eyes. Until recently doctors believed that these rare conditions were due to genetic factors. This catastrophe, however, has now been linked to the fungicide benomyl. Studies on rats at the University of California have also incriminated benomyl.[55]

In one study of infants born without eyes, seven parents claimed to have been exposed to two fungicides containing elements of the chemical compound benomyl. Eyeless babies in the U.S. and Central America whose mothers were exposed to benomyl have also been reported. Further, 25 cases were reported in Britain in which no clear genetic factors were implicated, but the parents did live near fields sprayed with pesticides and fungicides. Most recently, papers published by the University of California in 1991 have

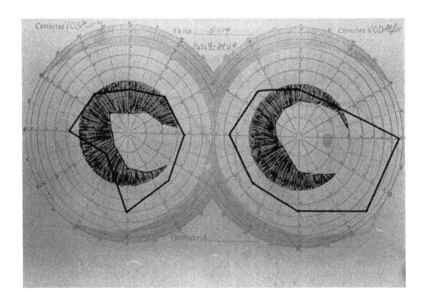

Figure 27.24. Loss of central vision; loss of peripheral vision. Treatment included avoidance of foods and chemicals to which the patient was sensitive and intradermal neutralization for foods. (Courtesy of Dor Brown, M.D. 1990. Personal communication.)

established the relationship between benomyl exposure and eye absence. Scientists showed that when pregnant rats were dosed with high levels of benomyl, nearly half the offspring had severe eye defects.[55]

VISUAL FIELDS

For the last 30 years, Brown[56] has reported numerous changes in visual fields after exposure to pollutants and foods. Figures 27.24 to 27.26 show visual field changes in some of his chemically sensitive patients due to food sensitivity. These changes returned to near normal after removal of the offending agents.

In summary, it is clear that there are many known environmental factors that can affect vision significantly, and there are many other factors that potentially may cause, or affect, chemical sensitivity in the eye. Causes of vision problems need to be sought in a meticulous manner. Many individuals who are identifed with environmentally triggerd vision deficits can be helped by definition and avoidance of environmental incitants, nutrition therapy, injection therapy for inhalants and foods, and heat depuration/physical therapy.

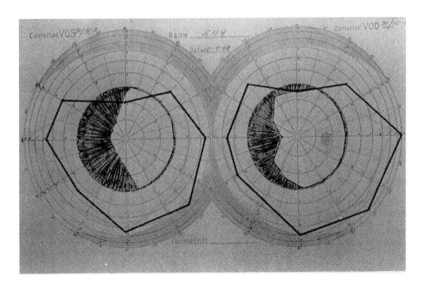

Figure 27.25.   Recovery of peripheral vision. Treatment included avoidance of foods and chemicals to which the patient was
sensitive and intradermal neutralization for foods. (Courtesy of Dor Brown, M.D. 1990. Personal communication.)

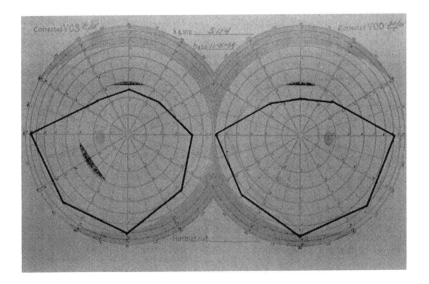

Figure 27.26.   Recovery of central vision. Treatment included avoidance
of foods and chemicals to which the patient was sensitive
and intradermal neutralization for foods. (Courtesy of
Dor Brown, M.D. 1990. Personal communication.)

REFERENCES

1. Shirakawa, S., W. J. Rea, S. Ishikawa, and A. R. Johnson. 1992. Evaluation of the autonomic nervous system response by pupillographical study in the chemically sensitive patient. *Environ. Med.* 8(4):121–127.

2. Rea, W. J. 1992. *Chemical Sensitivity, Vol. I. Mechanisms of Chemical Sensitivity.* 221. Boca Raton, FL: Lewis Publishers.

3. Rea, W. J. 1992. *Chemical Sensitivity, Vol. I. Mechanisms of Chemical Sensitivity.* 241. Boca Raton, FL: Lewis Publishers.

4. Yudkin, J. 1944. Nutritional deficiency in the pathogenesis of disease. *Br. Med. J.* 1:15.

5. Lai, Y.L., D. Fong, K. W. Lam, H. M. Wong, and A. T. Tsin. 1986. Distribution of ascorbate in the retina, subretinal fluid and pigment epithelium. *Curr. Res.* 5:933–938.

6. Lohmann, W., W. I. Schmehl, and J. Strobel. 1986. Nuclear cataract: oxidative damage to the lens. *Exp. Eye Res.,* 43:859–862.

7. Khatami, M., L. E. Stramm, and J. H. Rockey. 1986. Ascorbate transport in cultured cat retinal pigment epithelial cells. *Exp. Eye Res.* 43:607–615.

8. Khatami, M., L. E. Roel, W. Le, and J. H. Rockey. 1986. Ascorbate regeneration in bovine ocular tissues by NADH-dependent semidehydroascorbate reductase. *Exp. Eye Res.* 43:167–175.

9. Riley, M. V., C. A. Schwartz, and M. I. Peters. 1986. Interactions of ascorbate and H_2O_2: implications for in vitro studies of lens and cornea. *Curr. Eye Res.* 5:207–216.

10. Rosenbaum, J. T., E. L. Howes, Jr., and D. English. 1985. Ascorbate in aqueous humor protects against myeloperoxidase-induced oxidation. *Am. J. Pathol.* 120:244–247.

11. Buettner, G. R. and M. J. Need. 1985. Hydrogen peroxide and hydroxyl free radical production by hematoporphyrin derivative ascorbate and light. *Cancer Lett.* 25:297–304.

12. Rengstorff, R. H. and H. I. Khafagy. 1985. Cutaneous acetone depresses aqueous humor ascorbate in guinea pigs. *Arch. Toxicol.* 58:64–66.

13. McGahan, M. C. 1985. Ascorbic acid levels in aqueous and vitreous humors of the rabbit: effects of inflammation and ceruloplasmin. *Exp. Eye Res.* 41:291–298.

14. Ishikawa, S., Y. Ichibe, and M. Wakakura. 1995. Treatment of optic neuritis with high doses of vitamin C. *Acta Soc. Ophthalmol. Jpn.* In press.

15. Fox, R. R., K. W. Lam, and J. F. Coco. 1977. Effect of ascorbic acid on intraocular pressure of normal and buphthalmic rabbits. *J. Heredity* 68(3):179–183.

16. Penn, J. S. and L. A. Thum. 1989. Oxygen-induced retinopathy in the rat: role of vitamin E in defending against peroxidation. *Ann. N.Y. Acad. Sci.* 570:495–497.

17. Pelle, E., D. Maes, G. A. Padulo, E. -K. Kim, and W. P. Smith. 1989. In vitro model to assess alpha-tocopherol efficacy: protection against UV-induced lipid peroxidation. *Ann. N.Y. Acad. Sci.* 570:491–494.

18. Rea, W. J. 1992. *Chemical Sensitivity, Vol. I. Mechanisms of Chemical Sensitivity.* 47. Boca Raton, FL: Lewis Publishers.

19. Ishikawa, S., M. Naito, and K. Inabe. 1970. A new videopupillography. *Ophthalmologica* 160:248.

20. Utsumi, T., S. Ishikawa, and T. Kimura. 1976. A diurnal change of pupil as observed by bilateral infrared videopupillography (in Japanese). *Adv. Neurol. (Jpn.)* 20(5):977–989.

21. Milton, J. G., A. Longtin, T. H. Kirkham, and G. S. Francis. 1988. Irregular pupil cycling as a characteristic abnormality in patients with demyelinative-optic neuropathy. *Am. J. Ophthamol.* 105:402–407.

22. Yamazaki, A. and S. Ishikawa. 1976. Abnormal pupillary responses in myasthenia gravis. *Br. J. Ophthalmol.* 60:575–580.

23. Ishikawa, S. 1988. Personal communication.

24. Homma, K., W. J. Rea, and S. Ishikawa. 1993. Unpublished data.

25. Ishikawa, S. 1986. A new binocular videopupillography — clinical application to afferent pupillary defect. *Autonomic Nerv. Syst.* 23:242–247.

26. Ukai, K., S. Murase, and S. Ishikawa. 1982. The light reflex and visual function of patients with optic nerve disease. *Jpn. J. Clin. Ophthalmol.* 36(8):969–972.

27. Ödkvist, L. M., L. M. Berghdtz, H. Ahlfeldt, B. Andersson, C. H. Edling, and E. Strand. 1982. Otoneurological and audiological findings in workers exposed to industrial solvents. *Acta Otolaryngol. Suppl.* 386:209–251.

28. Raizman, M. B. 1992. Clinical manifestations and treatment options. *ACAI.* Symposia highlights. 12–15.

29. Miyata, M., T. Namba, and S. Ishikawa. 1990. Experimental allergic conjunctivitis due to environmental chemicals. Presented at the 8th Annual International Symposium on Man and His Environment in Health and Disease. February 22–25, 1990. Dallas, TX: The American Environmental Health Foundation.

30. Miyata, M., T. Namba, K. Horiuchi, and S. Ishikawa. 1991. Experimental allergic conjunctivitis due to environmental chemicals. *Neuro-ophthalmol. Jpn* 8(2):171–176.

31. Rea, W. J. 1992. *Chemical Sensitivity, Vol. I. Mechanisms of Chemical Sensitivity.* 155. Boca Raton, FL: Lewis Publishers.

32. Boyd, T. A. S. and T. H. Aaron. 1963. Allergy. In *Modern Ophthalmology.* Vol. II, ed. A. Sorsby. 90. Washington: Butterworths.

33. Campinchi, R., J. P. Faure, E. Bloch-Michel, and J. Haut, eds. 1973. *Uveitis: Immunologic and Allergic Phenomena.* Springfield, IL: Charles C Thomas.

34. Homma, K. and S. Ishikawa. 1992. Toluene-induced uveitis. *Environ. Med.* 8(1):9–15.

35. Hockwin, O., M. Kojima, F. Czubayko, and K. von Bergmann. 1991. Inhibition of cholesterol synthesis and cataract. *Fortsch. Ophthalmol.* 88(4):393–395.

36. Razzak, A. L. and A. Samarrai. 1991. Common risk factors in the development of cataract among Arabs residing in Kuwait. In *Distribution of Cataracts in the Population and Influencing Factors.* eds., K. Sasaki and O. Hockwin. Basel: Karger.

37. West, S., B. Munoz, E. A. Emmett, and H. R. Taylor. 1989. Cigarette smoking and risk of nuclear cataracts. *Arch. Ophthalmol.* 107(8):1166–1169.

38. Söderberg, P. G. 1988. Acute cataract in the rat after exposure to radiation in the 300 nm wavelength region. A study of the macro-, micro- and ultrastructure. *Acta Ophthalmol. (Copenh.)* 66(2):141–152.

39. Joyner, K. H. 1989. Microwave cataract and litigation: a case study. *Health Phys.* 57(4):545–549.
40. Palmer, C. G. S. and R. M. Pauli. 1988. Intrauterine varicella infection. *J. Pediatr.* 112(3):506–507.
41. Young, R. W. 1991. *Age-Related Cataract.* 88, 130, 168–169. New York: Oxford University Press.
42. Young, R. W. 1991. *Age-Related Cataract.* 81–84. New York: Oxford University Press.
43. Ishikawa, S., H. Ozawa, and M. Miyata. 1983. Abnormal standing ability in patients with organophosphate pesticide intoxication (chronic cases). *Agressologie* 24(2):143–144.
44. Homma, K., W. J. Rea, S. Ishikawa, T. Namba, A. R. Johnson, and A. Piamonte. 1993. The evaluation of the autonomic nervous system in chemical sensitivities. Presented at the Eleventh Annual International Symposium on Man and His Environment in Health and Disease. Dallas, TX: The American Environmental Health Foundation.
45. Sturman, J. A. 1990. Taurine deficiency. In *Taurine Functional Neurochemistry, Phsyiology, and Cardiology*, eds. H. Pasantes-Morales, D. L. Martin, W. Shain, and R. Martín del Río. 385–395. New York: John Wiley & Sons.
46. Pangborn, J. B. 1987. *Taurine.* Lisle, IL. Bionostics, Inc.
47. Ishikawa, S. and M. Miyata. 1980. Development of myopia following chronic organophosphate pesticide intoxication: an epidemiological and experimental study. In *Neurotoxicity of the Visual System*, eds. W. H. Merigan and B. Weis. 233–254. New York: Raven Press.
48. Ikeda, K. 1987. Visual toxicity of toluene. An experimental study. *Acta Soc. Ophthalmol. Jpn.* 91(10):903–912.
49. Rogers, S. 1993. Personal communication.
50. Theodore, F. H. and A. Schlossman. 1958. *Ocular Allergy.* Baltimore: Williams & Wilkins.
51. Patel, D. J. 1992. Personal communication.
52. Noell, W. K. 1985. Phenomena and mechanisms of retinal damage by visible light. In *The Latest Medical Book*, 1st ed. 11–22. Tokyo: Medical Education Publisher.
53. Kimura, T. and S. Ishikawa. 1976. Optico-encephano-neuropathy in glue sniffers. Proceeding of Sixth Congress of Asia-Pacific Academy of Ophthalmology Transactions. Bali, Indonesia: Airlanggo University Press. 176–181.
54. Homma, K. 1987. Personal communication.
55. McGhie, J. and M. Paduano. May 30, 1993. Babies in fungicide link. *The Observer, London,* 6.
56. Brown, D. 1983. Personal communication.

28   Children

INTRODUCTION

Throughout the first three volumes of *Chemical Sensitivity*, pollutant exposure has been shown to have wide-ranging consequences. While some individuals appear to remain unaffected or have significant time delays between an exposure, or set of exposures, and development of various sensitivities, illnesses, or end-organ diseases, others may experience immediate damage that ranges from insidious to grossly obvious effects. In any of these circumstances, the damage that does result may involve any individual or multiple organs and/or systems throughout the body. Pollutant injury may also affect any person, regardless of their age, race, or gender. In fact, pollutant-induced injury is so pervasive that not only may the unborn be damaged via its mother's exposures but also, before conception, the sperm and ova may be damaged. Dixon[1] showed, for example, that pesticide workers may have damaged sperm or ova, and Heidam[2] found that female anesthetists have problems with a higher rate of infertility and spontaneous abortions than normal populations. This increased incidence of pollutant-stimulated injury in female anesthetists is presumably due to poor quality ova, problems with its implantation, or problems with maintaining the embryo. It is clear that the sperm or ova of workers in various industries, such as those that use or develop pesticides and/or petrochemicals as well as some electronics industries, may have defects.[3,4,5]

In the history of medicine, acknowledgment of exogenous agents that can directly affect the developing fetus is recent. Prior to the thalidomide incident of toxic damage to the fetus in the late 1950s, a generally held belief was that all congenital malformations resulted from inherent genetic abnormalities. As it became clear that thalidomide exposure was the cause of damage to the developing fetus, it became equally plausible that other toxic substances might have similar effects. Eventually, the developing embryo/fetus was recognized to be more sensitive to exogenous toxic agents than the adult.

Many common agents, such as halogenated hydrocarbons, are not only toxic to the embryo/fetus, they may also alter genes or genetic expression. They can damage DNA and RNA in a variety of ways including altering cell

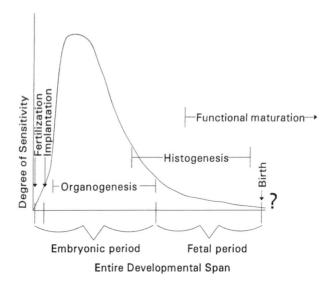

Figure 28.1. Sensitivity of the developing mammalian fetus to xenobiotic insults during gestation. The curve approximates teratogenic sensitivity at different periods in mammalian development. Although the curve does not apply specifically to any one species, it is more directly applicable to higher primates and humans in which the fetal period is prolonged and postnatal development relatively insignificant except for nervous system functions. (From Wilson, J. G. 1972. Environmental effects on development — teratology. In *Pathophysiology of Gestation.* Vol. 2, eds. N. S. Assali and C. R. Brinkman. 269–320. New York: Academic Press. With permission.)

replication, cell division, and blastogenesis. The mechanisms of pollutant-induced genetic damage have been discussed in Volume I, Chapters 3,[6] 4,[7] and 6[8] and, therefore, will not be discussed further here.

Some toxic exposures have resulted in developmental anomalies leading to fetal death, congenital malformations, functional deficits, or growth retardation in offspring. Figure 28.1 shows a curve of approximate teratogenic sensitivity, but one should be aware that only morphological changes are displayed here, while altered physiology that is often seen in the chemically sensitive individual is not. This fact is emphasized by some studies in areas of high pollution such as Katovisti, Poland, where 50% of births have anatomical genetic defects, but no metabolic defects have been recorded. However, when those so-called "normal" children without anatomic defects were examined, they were found to respond abnormally to environmental pollutants, suggesting they did have metabolic defects. Therefore, even though discussion here may be limited to anatomical teratology, the reader should realize (and this is emphasized in Volume I, Chapters 4[7] and 6[8]) that metabolic defects may well parallel, or be an alternative to, anatomical dysfunction.

The basic principles of teratology include the following: the manner in which the genotype of the ovum, sperm, zygote, and embryo/fetus interact with environmental influences, the stage of development at the time of exposure, the developmental mechanisms involved, physiological properties of the teratogen, the virulence of the environmental agent, the dose response curve of the toxic exposure, and the state of nutrition and total load of the cells involved.[9]

From preconception to birth, vulnerability occurs during various developmental periods, and ovum, sperm, zygote, embryo, and fetus are all susceptible to toxic exposure. If any of these have been chemically sensitized, exposure may increase their sensitivity. If they are undamaged but exposed to pollutants, sensitivities may develop. In either instance, the damage that results from toxic exposure depends on the properties of the chemical(s), the amount of the chemical(s), and the stage of resistance of the organism at the time of exposure.

There are periods of relative resistance and vulnerability in prenatal development (Figure 28.2). At preconception, the ova and sperm are susceptible to toxic exposure. Once conception occurs, the zygote is less susceptible, but, if it is damaged, the result is usually death and spontaneous abortion, thus protecting society from many damaged children. Then after 2 weeks of zygote resistance, susceptibility, damage, and survival lead to congenital malformations.

Congenital malformations rank as one of the main causes of perinatal mortality and postnatal morbidity in humans and occur in 2 to 3% of births in western countries. The number of births involving congenital malformations quickly rises to 15 to 18% if stillbirths and miscarriages are also calculated. This high incidence of zygotic damage occurs in some regions, suggesting an early time of high susceptibility (Table 28.1). Although functional defects have been attributed to an additional 6 to 7% of fetal abnormalities, this incidence, in light of recognition of many new metabolic defects being described today (see Volume I, Chapter 6[8]), may be too low. Nonetheless, the overall prevalence of abnormalities is still about 23 to 28% when the conservative 6 to 7% is further figured into the total number of pregnancies involving congenital malformations. These functional defects include over 2000 recognized errors of metabolism, which, along with many other factors, lead to chemical sensitivity (see Volume I, Chapter 3[6] and Chapter 24, section on abortions, for more information.)[9]

Since one purpose of this book is the prevention of pollutant injury, the remainder of this chapter examines each stage of potential pollutant damage, beginning with preconception and conception. The fetus and perinatal period are then discussed, followed by a focus on the newborn, the infant, childhood, and adolescence. At each stage of development, steps can be taken to ensure optimal health and reduce the risk of various long-term health problems associated with pollutant injury.

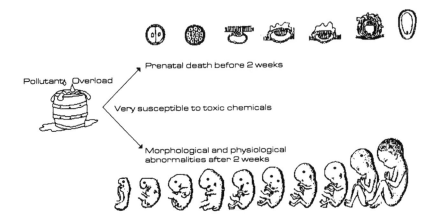

Figure 28.2. Schematic diagram of critical periods of human development.

Table 28.1.   Congenital Malformations:
The Main Cause of Perinatal
Mortality and Postnatal Morbidity —
Environmental Teratogens

Abnormality	Percent
Morphological abnormalities	2–3
Stillbirths and miscarriages	15–20.0
Functional defects in children	6–7.0
Total	23–30.0

Source: Rousseaux, C. G. and P. M. Blakely. 1991. Fetus. In *Handbook of Toxicologic Pathology*, eds. W. M. Haschek and C. G. Rousseaux. 938. San Diego: Academic Press. With permission.

PRECONCEPTION AND CONCEPTION

The best time to prevent pollutant injury is preconception. Usually, preconception prevention is impossible, however, unless informed parents are willing to adopt a good avoidance and nutrition program prior to conception in order to have an extremely healthy child. Some informed parents are independently developing such a regimen, while others are participating in organized programs such as the Foresight Program in England. Utilizing such a program means not only avoiding as many environmental pollutants in the air, food, and water as possible but also initiating a nutritional supplementation program that includes all of the antioxidants and avoidance of "junk" foods. Since most prospective parents, as well as many of their health care providers, are unaware of the environmental risks to themselves and their unborn children or the advantages of practicing good environmental control, they do not take preventive measures to resist pollutant exposure and its immediate damage or its long-term effects. To combat this present state of affairs, this chapter focuses on commonly occurring early pollutant damage, including possible harm to the mother, father, placenta, and fetus. Each facet of various stages of damage and preventive measures to counteract negative environmental influences are discussed.

Mothers

Toxic chemicals can damage the ova, causing no ovulation or poor quality ovulation or poor quality ova. This damage may result in nonunion with sperm or early abortion if union has already occurred. Stillbirth may occur at any stage of the pregnancy. Certain doses and types of toxic chemicals have been shown to go to the ovaries. Impairments induced in the ovaries have been shown to occur with exposures to substances such as pesticides (carbaryl and endosulfan)[10] and benzenes (often seen in the chemically sensitive patient).[1] These chemical exposures were shown to reduce the number of oocytes and their size as well as damage their yolk vesicles. Here the development of interfollicular spaces increases the connective tissue of the tunica albugnea and causes blood vessel dilation. Females exposed to high levels of benzene were either unable to bear children or they produced children who failed to thrive.[11] Polyaromatic hydrocarbons and alkylating agents, both of which are known to exacerbate chemical sensitivity, have experimentally induced premature ovarian failure (Figure 28.3).[1]

Toxic damage to the developing zygote can occur, affecting its quality and/or implantation. Alterations in maternal homeostasis must be severe to affect a fetus, since the needs of the fetus are usually provided at the expense of the mother. Nonetheless, initiation of congenital malformations has been associated with many maternal disorders including thyroid disease, diabetes, inborn errors of metabolism, phenylketonuria (microcephaly, mental retardation, congenital heart disease), and malnutrition. Fetal effects of these conditions are exemplified by cretinism, mental retardation, heart disease, abortion, stillbirths, neonatal deaths, and tube defects, all of which have occurred during the pregnancies of some chemically sensitive mothers. Neural tube defects can be prevented in many cases by the administration of folic acid,[12] and it follows that administration of the whole spectrum of antipollutant nutrients will prevent many congenital defects. The most recently documented damage to the fetus apparently resulting from maternal exposure to toxic wastes is anencephalia in some babies born in the Rio Grand Valley of Texas.[13]

Not all toxic exposure results in damage. During pregnancy, there is some protection against xenobiotics because of decreased intestinal absorption and

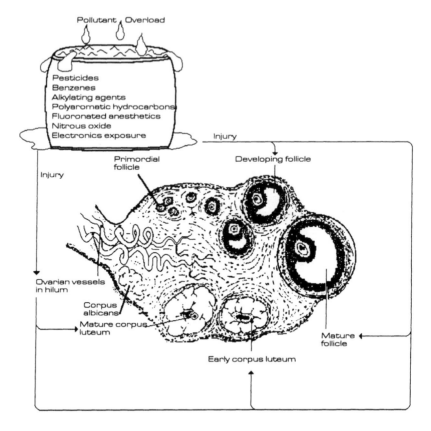

Figure 28.3. Pollutant damage to the ovum.

dilution within the mother's increased total body water and fat. However, with the progressive decrease in plasma albumin, concentrations of plasma binding capacity may be reduced allowing circulation of more toxic chemicals, which may induce or exacerbate chemical sensitivity. Since maternal enzyme activities are increased, there is a potential for reduced drug and xenobiotic metabolism because of competitive inhibition of enzymatic pathway by steroid hormones. Renal clearance of drugs and chemicals may be altered during pregnancy.

Fathers

Teratospermia (Figure 28.4), which involves induction of microscopic pathologic changes, may follow exposure to xenobiotics. For example, paternal lead poisoning can cause teratospermia that may result in reduced fertility and birth weight.[14] Direct action of certain foreign compounds, such as chlorinated pesticides, may reduce ribosomal activity, impair protein synthesis and

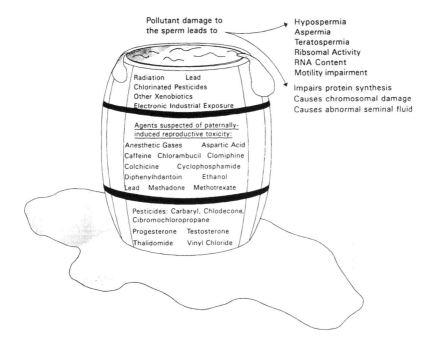

Figure 28.4. Pollutant damage to the sperm.

RNA content, produce anatomical abnormalities or chromosomal damage, and/or alter sperm motility.[15] Xenobiotics may also cause abnormal seminal fluid.[16] Hypospermia and aspermia also have been reported with exposure to some toxic chemicals, especially pesticides (see Chapter 24, section on testes, for more information). Common exposure to xenobiotics, such as to the fumes of pesticides routinely sprayed in the home, can cause this problem. One urologist saw a patient whose wife was trying to conceive. This patient's sperm count was extremely low. The low count was traced to a pesticide exposure from shoes that had been accidentally soaked in yard chemicals that the patient had been using on his lawn. After 8 months of meticulous avoidance, his sperm count returned to normal, and the couple conceived.

PLACENTA AND FETUS

The placenta and fetus are particularly vulnerable to maternal exposure to toxic chemicals. Since the detoxification systems are not totally developed, vascular dysfunction readily occurs when pollutants are introduced due to the rich blood supply of the placenta. This dysfunction can cause selective damage to various areas of the fetus.

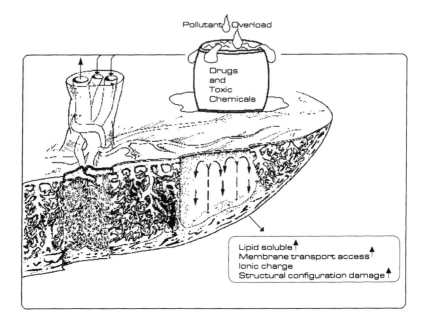

Figure 28.5. Pollutant effects upon the placenta.

Placenta

Once the zygote is implanted and the placenta develops, many changes in relation to chemical exposure are seen (Figure 28.5). Depending on their lipid solubility, molecular size, ionic charge, and structural configuration, toxic chemicals are transported across the placenta (Table 28.2). As lipid solubility of the chemicals increases, transport increases; therefore, many of the most toxic substances are able to cross the placental barrier. Those with molecular weights above 600 Da do not cross easily, but most drugs and damaging toxic chemicals are below this level. Stearic alterations may at times reduce transport. Some compounds, such as epinephrine, serotonin, prostaglandin E_2, and vasopressin, alter placental blood flow and, therefore, decrease xenobiotic placental transport.

Since the blood of the fetus is more acidic than its mother's blood, foreign compounds can be trapped more readily in the fetus than in the mother, and these compounds can then stay in the unborn for a longer time. In addition, if reactions to the chemical(s) occur in either the mother or fetus, acidosis in either or both also increases, and a vicious cycle of acidosis followed by increased trapping of compounds followed by increasing acidosis is set in motion. Trapped in this cycle, the embryo/fetus develops nutrient deficiencies, rendering it more vulnerable to additional xenobiotics. Also compromising the well-being of the fetus is the mother's exposure to certain xenobiotics, such as cadmium or trypan blue, which may diminish nutrient transport to the fetus.

Table 28.2. Pollutant Effects upon the Placenta Able To Affect Transport across the Placenta

(1) Molecular size — below 600 Da — easy to cross
(2) Lipid solubility — drugs and chemicals
(3) Ionic charge — fetal acidic type chemicals
(4) Structural configuration — stearic intolerances — reduced transport
(5) Altered blood flow with vasospasm; decreased chemical retention; increased hypoxia
(6) Nutrient transport may decrease chemical exposure
(7) Sequestration of xenobiotics impairs function — corticoids inhibit glucose and methyl mercury inhibits amino acid transport

Any of these circumstances can contribute to the development of chemical sensitivity in the fetus.

Placental function can be obstructed by sequestration of xenobiotics such as cortisone (which inhibits glucose transport) and methyl mercury (which inhibits amino acid transport). The fetal response to injury may also be a problem. With early damage, the fetus may develop cytoxicity, but in the latter half of gestation an inflammatory response may occur, which will result in chemical sensitivity if the fetus survives to birth. Alteration of glucose and amino acid metabolism can then set the placenta and embryo/fetus up for a decreased ability to detoxify other xenobiotics through the glucuronides, methionine, peptide, and sulfur conjugation detoxification systems. Then prenatal chemical sensitivity can occur. Vascular inflammation may result in local end-organ hypoxia, resulting in additional fetal damage.

Fortunately, approximately 50 to 70% of all lost conceptions occur during the first 3 weeks after conception, resulting in less strain on, and nutritional robbery of, the mother. Chromosomal abnormalities account for 60% of the abortuses. Congenital malformations can result from many factors including interference with natural cell death (polydactyly) or excess cell death, reduced cell proliferation rate (hypervitaminosis A, folic deficiency, cortisone), failed cellular interactions (actinomycin D, excess vitamin A), impeded morphogenic movements, lack of cell adhesion (EDTA), reduced biosynthesis of essential components (e.g., exposure to chloramphenicol, tetracycline, streptomycin, kanamycin, organophosphate, and carbamate), mechanical disruption due to edema or other movements of cells, and changes in intracellular pH (CO, acetazolamide, and other carbonic anhydrase inhibitors).

Fetus

There are several modifying factors that influence teratogenesis as well as metabolic defects. The first of these is the time of exposure of the organogenesis to the toxic chemical. If either the gene, embryo, or fetus are exposed to noxious substances at a critical time in the cells' life, malformation can occur. If exposure occurs after development of a fetal organ or if tissue reaches past the critical stage of development, injury may be minimal.

The second factor that modifies teratogenesis may be the ethnic background of the human involved (e.g., sickle cells in blacks and Mediterranean anemia in Greeks and Italians may predispose individuals of these races to specific toxic injury and thus early chemical sensitivity).

Organ development may exceed the threshold at which the organ may be affected by one or more genes and, thus, teratogenesis may result. If this genetically determined normal development is close to the threshold, only minor noxious stimuli may be necessary to induce abnormal development. Thus, as the number of predisposing genes increases, the severity of the teratogenic xenobiotic insult required to surpass threshold decreases. For example, cortisone, which is used by some environmentally unaware physicians to treat the inflammatory response that occurs as a result of xenobiotic overload rather than eliminating the toxic substance, damages the gene and can cause a cleft palate. This action exemplifies the multifactorial threshold concept of fetal injury. Approximately 50% of all malformations are thought to involve multifactorial inheritance. Thus, the increase in total body load of xenobiotics can be quite harmful.

Embryonic/fetal pharmacokinetics also may be involved in the modification of teratogenesis. The human liver and the adrenal cytochrome P-450 system is developed by 6 to 8 weeks, at which time detoxication of foreign compounds can occur. Since fetal detoxication functions are similar to those that occur in the adult and can result in toxic breakdown products, the presence of toxic metabolites of foreign chemicals in the fetus must also be watched for. Fetal renal excretion increases with age; therefore, as the fetus gets older, it experiences a concomitant increase of circulating toxins. Enzyme induction and inhibition can also occur in the fetus; therefore, toxic metabolites may not only occur but also certain of these may altogether stop the detoxication process, thereby allowing other toxins to accumulate and finally result in chemical sensitivity. Benzopyrene is an example of the enzyme inducer that has its toxicity increased by the induced aryl hydrocarbon hydroxylase.

Some teratogens bind preferentially to embryonic/fetal tissues, thus causing modified teratogenesis. For example, methyl mercury, lead, and thalidomide were found to bind preferentially to fetal tissue over adult tissue.

Signs and symptoms of environmental overload may be present in utero after conception. We have observed that babies who kick excessively or hiccup in utero are usually reacting to their mothers' exposures to foods, chemicals, or biological inhalants. For example, we have found that when a mother who is sensitive to wheat eats it, her fetus will become overloaded and respond with excessive activity, including excessive movement and hiccupping. Avoidance of the wheat by the mother will inhibit the baby's reaction. Also, intradermal neutralization of a mother to a specific food to which she is sensitive will often stop her fetus's hiccups and excessive activity as well as eliminate any additional symptoms. Conversely, some fetuses have been seen to be severely suppressed by chemical overload, resulting in almost no movement. At birth, these babies are sluggish and sometimes damaged.

There are also reports[17-19] of severe growth retardation, reduced head circumference, mental retardation, dysmorphogenesis, and abnormalities of facial features in babies of mothers of toxic chemical overload. This syndrome is similar to the fetal alcohol syndrome. If they survive, these babies develop chemical sensitivity.

Clinical Reports of Human Fetal Poisoning by Mercury[20]

During the past few years, three epidemics and three individual cases of mercury poisoning during pregnancy have been reported. These reports may be used by the clinician to formulate prevention guidelines for the mother who is exposed to pollution in order to have a well baby.

Minamata, Japan. Near the end of 1953, a strange nervous disorder began afflicting villagers in and around Minamata, a town in Southern Japan.[21,22] This condition was characterized by a number of neurologic manifestations including mental confusion, convulsions, and coma. About 38% of affected individuals died. Cats, crows, waterfowl, and fish also were affected. Experiments showed that the condition could be produced by feeding fish and shellfish from the bay to mice, rats, and cats.[23] These fish and shellfish contained 9 to 24 µg methyl mercury per gram. Eventually, the source of methyl mercury was traced to the effluent discharged into Minamata Bay by a nearby factory.

From 1953 to 1971, 134 cases of illness, including 78 adults and 56 children, were reported. During this period, 25 infants were born with brain damage.[24-27] From 1955 to 1959, about 6% of the children born in this area developed cerebral palsy. Other neurologic symptoms included chorea, ataxis, tremors, seizures, and mental retardation.[28] The mothers, all of whom consumed large quantities of fish, lacked the typical symptoms, although some experienced mild paresthesias. The fact that these infants had not eaten contaminated fish suggests that the neurologic symptoms resulted from toxicity in utero. On the other hand, the possibility of postnatal acquisition from breast milk cannot be ruled out.[29,30]

In 1964, an epidemic of methyl mercury poisoning similar to that in Minamata occurred in Niigata, Japan.[31,32] As in the Minamata outbreak, the source of the methyl mercury was traced to the ingestion of fish contaminated by mercury from industrial discharge. Of 46 cases, only one possible prenatal case was reported.

Iraq. In September 1971, barley and wheat grain treated with methyl mercury were distributed to Iraqi farmers. The grain was used to make bread containing about 4 mg of mercury per loaf. A widespread epidemic resulted during 1972.[29] There were 6530 hospitalized cases with 459 deaths, a mortality rate of 7%. Symptoms were similar to those of the Minamata epidemic. During this period, 31 pregnant women were hospitalized with mercury poisoning. For an undetermined reason, the mortality rate of those women in the 20 to 30 year range was 70%, whereas in women over 30, it was only 16% (2 of 12). Blood mercury levels of infants born during the epidemic were less than, or equal to, corresponding maternal values (Figure 28.6). This finding suggests poisoning via breast milk, which was found to contain 5 to 6% of the mercury concentration of maternal blood. In infants born during or after the epidemic, mercury blood values were generally greater than maternal values. In these cases, mercury was acquired in utero, during breast feeding, or both. In utero poisoning was particularly attenuated in infants less than 2 months old. Several had mercury levels greater than 25 μg/mL of blood and demonstrated evidence of severe brain damage.

U.S.S.R. In 1968, Bakulina[33] reported 10 cases of fatal poisoning among mothers who ingested grain treated with methyl mercury. Severe mental retardation was described in the babies of three of these, and decreased birth weight and muscle tone were noted in others. Since no detailed case histories were given, the extent of postnatal exposure through breast milk is unknown. However, the presence of neurologic symptoms at birth suggests prenatal intoxication. Outbreaks of poisoning by ingestion of mercury contaminated grain also occurred in Guatemala, Pakistan, and Iraq;[34,35] however, no fetal or neonatal cases were reported.

Lund, Sweden. A pregnant woman and her family, including a young infant, ate mercury-contaminated flour for several months.[36] She had no symptoms of poisoning even though she was excreting mercury up to 0.064 μg/mL of urine. She gave birth to an apparently normal, healthy infant girl who had no clinical signs of mercury poisoning. This infant's urine contained less than 0.003 μg/mL of mercury. Within a few months, however, the child developed signs of severe mental retardation. While this case does not prove that mercury ingestion during pregnancy causes mental and motor disability, it strongly suggests this effect.

King County, Washington. In 1958, a 31-year-old woman developed mercury poisoning complicated by acute renal failure following ingestion of 2.3 g of mercuric chloride in an attempt to induce abortion. Despite gastric lavage and treatment with BAL, she soon developed complete anuria. Thirteen days after the mercury ingestion, she spontaneously aborted a 3 cm fetus that appeared grossly normal. The termination of this pregnancy following intake of mercury suggests that mercuric chloride induced the abortion, but it may have resulted from the patient's general illness or therapy.

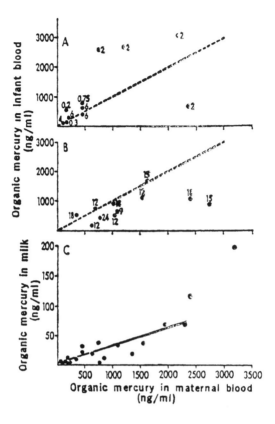

Figure 28.6. The concentrations of organic mercury in infant blood and maternal milk as a function of maternal blood concentration. (A) Blood samples from infants born during and after the epidemic and thus exposed to methylmercury both in utero and in maternal milk. (B) Blood samples from infants born prior to the epidemic and thus only exposed to methylmercury in maternal milk. (C) Maternal milk samples collected at intervals between April and July, 1972. The number adjacent to the points in A and B indicates the age of the infant in months at the time of sampling. The dashed lines in A and B are lines of identity. The line in C is the linear regression line calculated from the mercury concentrations in maternal blood that were below 2500 ng. per milliliter. (From Baker, F., S. Damlwji, L. Amin-Zaki, M. Murtadha, A. Khalidi, N. Al-Rawi, S. Tiktiti, H. Dhahir, T. Clarkson, J. Smith, and R. Doherty. 1973. Methylmercury poisoning in Iraq. An interuniversity report. *Science* 181:230. With permission.)

Alamogordo, New Mexico. A 40-year-old woman consumed mercury-contaminated pork during the third through the sixth month of her pregnancy.[37,38] Her urine mercury concentrations were 0.47 to 2.91 µg/mL during the last months of pregnancy. A sample of amniotic fluid analyzed for mercury revealed less than 0.02 µg/mL.[39] The mother's symptoms were minimal. She delivered a term male infant who appeared normal, but had some tremulous movements of the extremities that lasted for several days.[40] The neonate's urine mercury level was 2.7 µg/mL of age. An initial electroencephalogram was within normal limits for age, but by 3 months subsequent electroencephalogram showed widespread abnormal activity. At 6 months of age, the infant developed myotonic jerks and his electroencephalogram was markedly abnormal. Since this infant never nursed from the breast, these symptoms most likely resulted from toxicity acquired in utero.

The preceding cases illustrate that mercury ingestion during pregnancy may affect the fetus and be associated with mental retardation, cerebral palsy, and seizures in newborn infants. In addition, the lesions that occurred in these babies first developed when they were fetuses and became "time bombs" that were expressed during infancy.[24] Maternal exposure occurred through several sources, including contaminated fish, meat from domestic animals fed contaminated grain, direct human ingestion of mercury-treated grain, and ingestion of mercuric salts, per se.

Other environmental xenobiotic exposures can induce defects. For example, congenital abnormalities have been linked to pesticide exposure. Czeizel et al.[41] reported one series that showed that from 1989 to 1990 11 (73%) of 15 live births in an eastern European village were affected by congenital abnormalities. Five had Down syndrome (known teratogenic factors, familial inheritance, and consanguinity were excluded as causes). One each had a ventriculo-septal defect and pulmonary atresia, congenital hernia, stenosis of the left bronchus, anal atresia, choanal atresia, or cleft lip. Also, six were twins. All nine mothers had been eating fish from a pond that was found to be contaminated with the organophosphate insecticide trichlorfon. The levels in the fish were up to 100 mg/kg. Trichlorfon slowly releases dichlorvos, which is estimated to be at least 100 times more potent an anticholine-sterase agent than trichlorfon itself. Trichlorfon is known to result in low molecular weight gonads, alkylates DNA, and an increased rate of chromosomal abberations in the bone marrow of cells of Syrian hamsters and mice (Table 28.3).[42-46] The higher rates of twins may also be connected with environmental factors.[47] The cluster ceased when the chemical treatment of farmed fish was banned.

Perinatal Period

During the perinatal period, toxic chemicals can have a variety of ill effects on the developing fetus and newborn infant. For example, interference with postnatal cellular reproductive functions follows intrauterine exposure to antineoplastic agents (e.g., procarbazine), alkylating agents (busulfan), benz(a)pyrene, dimethyl benzanthracene, DDT, chlorecone, and dimethyl stilbestrol (Table 28.4). Abnormal ontogenic development of the immune system following exposure to xenobiotics may lead to immune suppression, stimulation, or deregulation followed by neoplasia or autoimmune disease with chemical sensitivity. Diphenylhydantoin, trimethdone, and primadone have been linked to mental retardation.[48] Addictive drugs such as PCP, heroin, methadone, morphine, and cocaine may produce alterations in neonatal behavior that may persist as language, attention, perception, and gross or fine motor developmental delays.[49] Methyl mercury exposure can affect the CNS (see Chapter 26). Environmentally caused "copy" of a genetically based disease may occur, resulting in inborn errors of metabolism. For example, the anatomical birth defect rate in Katovici, Poland, is 50% due to the toxic substances emitted from the steel mills. Probably the remaining 50% of births have metabolic defects.[50]

Other research demonstrating the apparent risks to the unborn that result from parents' toxic chemical exposure has emerged from the University of Arizona. Researchers here have shown that children of parents exposed to toxic chemicals are 2.5 times more likely to be born with heart abnormalities than those whose parents were not exposed. Investigators interviewed the parents of 646 children born with heart defects ranging from minor lesions to more severe defects requiring surgery. These children were conceived in Tucson, and their mothers had spent the first 3 months of pregnancy there. These parents had been exposed to a commonly used industrial solvent, and 32% of them lived or worked in a combined residential and industrial area in Tucson where trichloroethylene (TCE) contaminated the drinking water from the 1950s until 1981.[51]

Other epidemiologic studies from the past two decades have shown that serious health consequences may be associated with prolonged exposure to low levels of nitrous oxides. Some specific evidence has emerged suggesting an increase in spontaneous abortion and birth defects in the offspring of exposed women, although most studies have focused on cognitive, neurologic, hepatic, and hematopoietic side effects.[52,53]

Fetal cells are inherently susceptible to carcinogenesis.[49] The incidence of pediatric tumors commencing and ending in humans before age 15 years has been reported to be from 97.8 to 124.5/million/year. Usually, these are leukemia, lymphoma, neuroblastoma, Wilm's tumor, osteosarcomas, and/or rhabdomyosarcoma.

Testing of toxic chemicals is fraught with hazards in all types of animals and cell cultures. At best, even if a chemical is tested before it comes onto the market, safety for the individual still cannot be evaluated.[8] Tables 28.5 to 28.9 provide a partial list of chemicals known to cause prenatal damage. For further information, Shepard's[54] and Warkany's[55] books on congenital malformations catalogue teratogenic agents.

Table 28.3. Chldren Born in Rinyaszentkiraly, 1989–1990

Case[a]	Sex	Birth date[d]	Birth weight (g)	Gestational age (wk)	Congenital abnormality	Estimated date of conception[d]	Critical period[d]	Fish consumption[b]	Comment[c,d]
1	M	06/01/89	1800	36	Ventricular septal defect + pulmonary atresia	11/05/88	10–18/06/88 (days 30 to 38)	+++	Died (14/08/89)
2	M	06/01/89	2830	36	Healthy	11/05/88	—	+++	—
3	M	09/02/89	3540	40	Congenital inguinal hernia, left	18/05/88	19/11/88 to 09/01/89 (months 7 to 9)	+	Case 10 his sibling
4	M	11/04/89	2750	37	Healthy	09/08/88	—	+	Under state care
5	M	10/06/89	3450	40	Healthy	18/09/88	—	+	—
6	M	06/11/89	2650	38	Healthy	25/02/89	—	–	—
7	F	09/12/89	2980	37	Down syndrome	06/04/89	06/04/89	+	—
8	M	30/12/89	2960	38	Stenosis of left bronchus (70%)	20/04/89	24/05/89 to 04/06/89 (days 34 to 44)	+	—
9	F	13/08/90	3300	40	Anal atresia	20/11/89	25/12/89 to 01/01/90 (week 6)	+	Mother not in village throughout pregnancy
10	M	28/10/90	3300	38	Choanal atresia	03/02/90	18–25/03/90 (days 44 to 51)	+	Case 3 his sibling

11	F	01/11/90	3300	Cleft lip, left	20/02/90	26/03/90 to 10/04/90 (days 35 to 50)	++	—
12	F	21/11/90	1790	Down syndrome with endocardial cushion defect	16/04/90	16/04/90	+++	Died (21/11/90)
13	F	21/11/90	1980	Down syndrome with endocardial cushion defect	16/04/90	16/04/90	+++	Died (21/11/90)
14	M	09/12/90	2600	Down syndrome	13/04/90	13/04/90	±	Died (09/12/90)
15	M	09/12/90	2300	Robin sequence	13/04/90	07/06/90 to 14/06/90 (days 49 to 56)	±	46, XY

[a] Twins were dizygotic (1 and 2, 14 and 15) or monozygotic (12 and 13).

[b] +++ = often, in critical period certainly; ++ = often, in critical period probably; + = occasionally, in critical period probable; ± = questionable.

[c] Gypsy families for cases 1 and 2, 5, 8, 10, and 12 and 13.

[d] Day/month/year.

Source: Czeizel, A. E., E. Csaba, S. Gundy, J. Métneki, E. Nemes, A. Reis, K. Sperling, L. Tímár, G. Tusnády, and Z. Virágh. 1993. Environmental trichlorfon and cluster of congenital abnormalities. *Lancet* 341:540. With permission.

Table 28.4. Agents Producing Cytotoxicity in an Organ Analgen Followed by Malformation of that Organ

Agent	Species	Agent	Species
Cyclophosphamide	Rat	6-Aminonicotinamide	Rat
Hydroxyurea	Rat	Actinomycin D	Chick
Cytosine arabinoside	Rat	Busulfan	Rat
Cytosine arabinoside palmitate	Rat	Aminopterin	Chick
Aminothiadiazole	Rat	Thalidomide	Armadillo
6-Mercaptopurine	Rat	Urethane	Mouse
5-Fluorodeoxyuridine	Mouse Rat	5-Azacytidine	Mouse
5-Fluorouracil	Chick	5-Azauridine	Mouse
5-Fluoroortic acid	Chick	Colchicine	Hamster
Mitomycin-C	Chick	Vincristine	Hamster
Dinitrophenol	Chick	Vinblastine	Hamster
5-Fluoro-2-deoxycytidine	Mouse	Methadone	Chick
Nitrogen mustard	Mouse Chick	Ethylnitrosourea	Rat
Hadacidin	Rat	Methylnitrosourea	Rat
Ethyl alcohol	Chick	Methylbenzanthracene	Rat
5-Fluorouracil	Mouse		

Source: Scott, W. J., Jr. 1977. Cell death and reduced proliferative rate. In *Handbook of Teratology. Vol. 2. Mechanisms and Pathogenesis*, eds. J. G. Wilson and F. C. Faser. 83. New York: Plenum Press. With permission.

NEWBORN

Once the embryo/fetus has escaped, or dealt with, a series of xenobiotic exposures and reached term being healthy, he or she is still not assured an easy course through life. Instead, many pollutant exposures await.

The youngest baby treated at the EHC-Dallas was 7 days old. Our clinical experience with this child and others we have treated has given us the impression that the presence of xenobiotics in newborns can also cause problems similar to, or different from, those that occur in the fetus. For example, babies of alcoholic or drug-addicted mothers have syndromes characterized by severe growth and mental retardation, irritability, screaming, and feeding problems with failure to thrive. At the EHC-Dallas, we have seen this same type syndrome, though often to a lesser degree, occurring with exposure to many xenobiotics. Using sophisticated computerized gas chromatography and mass spectrometry, Dowty and Laseter[56] have measured cord blood in newborns and compared it to their mother's blood. They found that, although some chemicals present in the mother were absent in the newborn, many others, including

**Table 28.5. Developmental
Immunotoxicants**

Heavy metals
  Methylmercury
  Lead acetate
Hormones
  Estrogen
  Testosterone
  Cortisone
  Epinephrine
  Diethylstilbestrol
Organochlorines
  2,3,7,8-tetrachlorodibenzo-p-dioxin (TCDD)
  Polychlorinated biphenyls (PCBs)
Alkylating agents
  Busulfan
  Cyclophosphamide
Carcinogens
  Urethan
  7, 12-Dimethyl-benz(a)anthracene (DMBA)
  Methylcholanthrene (MC)
  Benz(a)pyrene
Pesticides
  Diazinon
  Carbofuran
  Chlordane
Nutrition
  Protein deficiency
Irradiation
  X-rays

Source: Roberts, D. W. and J. R. Chapman. 1981.
Concepts essential to the assessment of toxicity to the
developing immune system. In *Developmental Toxicology*, eds. C. A. Kimmel and J. Buelke-San. New
York: Raven Press. With permission.

carbon tetrachloride, chloroform, benzene, etc., were much higher in the babies
than in their mothers (Table 28.10). This bioconcentration in the newborns
suggests one of the reasons for society's overall increase in total body load with
each generation and partially explains the increase in the number and severity
of chemical sensitivities with each new generation. The following case shows
how environmental treatment with reduction of total load can decrease this
bioaccumulation.

Table 28.6.   Direct-Acting Transplacental Carcinogens

Compound	Species	Principal target organs
Sulfate and sulfonate esters		
Dimethyl sulfate	Rat	Nervous system
Diethyl sulfate	Rat	Nervous system
Methyl methanesulfonate	Rat	Nervous system
Propane sultone	Rat	Nervous system
Alkylnitrosoureas		
Methylnitrosourea	Rat	Nervous system
Ethylnitrosourea	Mouse	Lung, liver
	Mouse	Nervous system
	Rat	Nervous system
	Syrian hamster	Nervous system
	Rabbit	Kidney
	Rabbit	Nervous system
	Patas monkey	Connective tissues
n-Propylnitrosourea	Rat	Nervous system
Other acylalkylnitrosamines		
Ethylnitrosobiuret	Rat	Nervous system
Methylnitrosourethane	Rat	Nervous system, kidney, liver
Miscellaneous		
Ethylurea + sodium nitrite	Rat	Nervous system
	Syrian hamster	Nervous system
	Rabbit	Kidney
Methylazoxymethanol	Rat	Nervous system, kidney, intestine

Source: Rice, J. M. 1981. Effects of prenatal exposure to chemical carcinogens and method for their detection. In *Developmental Toxicology*, eds. C. A. Kimmel and J. Buelke-Sams. 195. New York: Raven Press. With permission.

Case study. This 23-year-old Australian mother had been ill most of her life. She eventually developed flu-like syndrome with malaise, fatigue, depression, nausea, vomiting, and weight loss. When examined at the EHC-Dallas, she was found to have extremely high blood levels of several chemicals including DDT (over 150 ppm), chlordane (over 25 ppm), heptachlor (over 20 ppm), hexachlorobenzene (over 10 ppm), and lindane (over 10 ppm). She had no relief of her symptoms until she was admitted to the EHC-Dallas. She was placed in a less-polluted porcelain house in a less-polluted outdoor area and given chemically less-contaminated food and water, intravenous and oral nutrients, and intradermal injections for her inhalant and food sensitivities. She also underwent heat depuration/physical therapy. Her health gradually improved over the next several months until most of her symptoms cleared. A decrease in her blood pesticide levels paralleled dissipation of her symptoms.

Table 28.7. Metabolism-Dependent Transplacental Carcinogens Other than Aliphatic Nitrogen-Containing Compounds

Compound	Species	Principal target organs
Polynuclear aromatic hydrocarbons		
Benzo(a)pyrene	Mouse	Lung, skin[a]
Methylcholanthrene	Mouse	Lung
7,12-Dimethylbenz(a)-anthracene	Mouse	Lung, skin[a], liver, ovary
	Rat	Nervous system, kidney
Heteroaromatic compounds and aromatic amines		
3-Hydroxyxanthine	Rat	Liver, mammary glands
Furylfuramide; AF-2	Mouse	Lung
4-Nitroquinoline-1-oxide	Mouse	Lung
o-Aminoazotoluene	Mouse	Lung
4-Dimethylaminoiazo-benzene	Mouse	Liver, lung
o-Toluidine	Mouse	Liver, lung
3,3′-Dichlorobenzidine	Mouse	Liver, lung
Natural products of plant and fungal origin		
Aflatoxin	Rat	Liver
Methylazoxymethyl-b-D-glucopyranoside (Cycasin)	Rat	Nervous system, intestine
Safrole	Mouse	Kidney
Miscellaneous		
Ethyl carbamate	Mouse	Lung, skin[a], liver, ovary
Vinyl chloride	Rat	Blood vessels, kidney, Zymbal's gland

[a] Skin tumorigenesis in mice occurred during postnatal topical application of croton oil or phorbol ester promoting agents.

Source: Rice, J. M. 1981. Effects of prenatal exposure to chemical carcinogens and method for their detection. In Developmental Toxicology, eds. C. A. Kimmel and J. Buelke-Sam. 198. New York: Raven Press. With permission.

She started feeling so well that upon a visit from her husband, she accidentally conceived a child. She elected not to terminate the pregnancy, though there was much concern about the potential harm of her chemical overload to the fetus. The pregnancy was relatively uneventful. At the baby's birth, however, blood levels were still quite high in the mother with a significant amount of transfer to the baby (Table 28.11). However, there was no bioconcentration. Although this baby appeared healthy initially, numerous feeding problems developed, suggesting that the toxic chemical load was too great for the baby's detoxification systems. With manipulation of the diet of chemically less-contaminated food and water, he has survived and done extremely well with no sequelae for the first 2 years of his life.

Table 28.8.   Etiologic Agent in Congenital Malformations
(Humans and Domestic Animals)

Agent	Effect
Alcohol	Pre- and postnatal growth retardation, MR, unusual facial features, congenital heart defects, urogenital defects, skeletal defects
Amantadine hydro-chloride (±)	Congenital heart defect, pulmonary atresia
Aminopterin	Hydrocephaly, CP, meningocele, reduced derivatives of first branchial arch
Anagyrine (lupins)	Crooked calf disease: scoliosis, arthrogryposis; CP
Anesthetics	Increased spontaneous abortions; CNS defects, musculoskeletal defects
Benztropine mesylate (±)	Left colon syndrome
Boric acid (±)	Increased risk for major malformation, especially cataracts
Bromide (±)	Short stature, small cranium, congenital heart disease
Busulfan (±)	CP, eye defects, generalized cytomegaly
Calcium carbonate	CNS defects
Carbon monoxide	CNS defects
Wild black cherry	Syrenomyelia, rudimentary external genitalia, anal atresia, blindly ending colon (pigs)
Chlorambucil (±)	Renal agenesis
Chloroquine (±)	Congenital deafness, chorioretinitis, hemihypertrophy
Cigarette smoking	Increased spontaneous abortion, prematurity, IUGR
Clomiphene (±)	Anencephaly, microcephaly
Conine (conium maculatum)	Arthrogryposis, scoliosis
Copper deficiency	"Swayback", enzootic ataxia (sheep)
Coumarin derivatives: dicumerol, warfarin	Nasal hypoplasia, calcific stippling of secondary epiphyses, hydrocephaly
Cyclopamine (Veratum californicum)	Cyclopecia, CP, cerebral defects
Cyclophosphamide (±)	Ectrodactyly, brachydactyly, flattened nasal bridge
Dextroamphetamine sulfate (±)	Atrial and ventricularseptal defect, biliary atresia, facial clefts
Diabetes	Caudal regression syndrome, CP, defects of branchial arches
Diazepam	Facial clefts, CP

Table 28.8 (Continued). Etiologic Agent in Congenital Malformations
(Humans and Domestic Animals)

Agent	Effect
Diethylstilbestrol	Hypospadias, male and female pseudohermaphrodism, vaginal adenosis
Diphenylhydantoin	CL/CP, congenital heart disease, microcephaly, hypoplasia of nails, and distal phalanges
Enovid-R (oral progestin) (±)	Female pseudohermaphrodism
17-α-Ethinyl-testosterone	Female pseudohermaphrodism
Ethionamide (±)	Congenital heart defects, spina bifida, gastrointestinal atresia
Fasting, starvation	Hydrocephaly, meningomyelocoele
Fluorine	Mottled tooth enamel
5-Fluorouracil (±)	Radial aplasia, imperforate anus, esophageal aplasia, hypoplasia of duodenum, lung, and aorta
Folic acid deficiency	Neural tube defects
Griseofulvin	CP cats
17-Hydroxyprogesterone (±)	Female pseudohermaphroditism
Hypertension	IUGR, microcephaly, patent ductus, arteriosus, hypotonia of skeletal and gut musculature
Hyperethermia	Microcephaly, microphthalmia, anencephaly, spina bifida
Hypervitaminosis A	Ectopic ureter, CP, craniofacial defects, skeletal malformations
Hypoxia (±)	Decreased birth weight, patent ductus arteriosus (may be a postnatal effect)
Imipramine (±)	Limb reduction deformities
Indomethacin (±)	Pulmonary artery changes
Insulin (±)	Fetal deaths, multiple congenital anomalies
Iodine deficiency	Endemic cretinism, hyperthyroidism
Iodine excess	Congenital goiter, hypothyroidism
Isoniazid (±)	Increased risk for malformations
Isotretionoin	Hydrocephaly, micrognathia, low-set ears, microcephaly, microphthalmia, malformed skull, ventricular septal defect
Lathyrism (*Lathyrus*)	Poorly developed muscles and connective tissue, dissecting aneurysms of aorta, spinal malformations, CP (domestic animals)
Lead	Increased stillbirth and spontaneous abortion, MR
Lithium carbonate	Epstein's anomaly

Table 28.8 (Continued). Etiologic Agent in Congenital Malformations
(Humans and Domestic Animals)

Agent	Effect
Locoweed (*Astragalus, Oxytropis*)	Arthrogryposis
Lyseric Acid (±)	Increased spontaneous abortions
Marijuana (±)	IUGR, developmental delays
Medroxyprogesterone (±)	Female pseudohermaphroditism, hypospadias
Meprobamate (±)	Congenital heart defect, increased malformation rate with no specific pattern
Mercury	Cerebral palsy, microcephaly, MR
Methallibure	Contractures of distal extremities distorted mandible and cranial bones, dysplasia of renal cortex (pigs)
Methimazole (±)	Midline defect of scalp
Methotrexate	Absence of frontal bone, premature craniosyno-stosis, rib defects, ectrodactyly
Methyltestosterone	Female pseudohermaphroditism
Myasthenia gravis	Congenital contractures
Neguvon	Congenital tremors with hypoplasia of cerebellum (pigs)
Oral contraceptives (±)	Congenital heart defects, limb reduction deformities
Oxytetracycline	Stains deciduous tooth enamel
D-Penicillamine	Lax skin, inguinal hernia, flexion contractures of knee and hip
Phenothiazine (±)	Increased malformation rate
Phenylalanine excess (material PKU)	Microcephaly, IUGR, congenital heart defects, dislocation of hips, strabismus
Phenylpropanolamine	Eye and ear defects, hypospadias
Phenobarbital (±)	Fetal hydantoin-like syndrome
Polychlorinated biphenyls	Cola-colored babies, IUGR, exophthalmus, staining of skin and gums
Pregnancy test tablets (±)	Neural tube defects, congenital heart defects
Primidone (±)	Low nasal bridge, ocular hypertelorism, pulmonic stenosis
Progesterone (±)	Hypospadias
Propylthiouracil	Congenital disorder
Quinine (±)	Congenital deafness, hydrocephaly, limb, facial, gastrointestinal and urogenital defects
Reserpine (±)	Congenital lung cysts

Table 28.8 (Continued). Etiologic Agent in Congenital Malformations (Humans and Domestic Animals)

Agent	Effect
Rheumatic disease of mother (especially systemic lupus erythematosus)	Congenital heart block
Organic solvents (±)	Neural tube defects, hydrocephaly, congenital heart defects, talipes
Streptomycin (±)	Congenital deafness
Testosterone	Female pseudohermaphroditism
Tetracycline	Staining of enamel of deciduous or permanent teeth
Thalidomide	Limb reduction anomalies, polydactyly, ear defects, facial hemangioma, esophageal or duodenal atresia, tetralogy of Fallot, renal agenesis
Tobacco stalk	Arthrogryposis (pigs)
Trimethadione	CP, cardiac defects, V-shaped eyebrows, developmental delays, low-set ears, irregular teeth
Valproic acid	Microcephaly, facial dysmorphology, congenital heart defect, neural tube defect
Virilizing tumor	Female pseudohermaphroditism
Vitamin D excess (±)	Supravalvular aortic stenosis, elfin facies, MR
X-irradiation	Microcephaly, MR, hydrocephaly, CP, hypospadias, hypoplastic genitalia, IUGR, microphthalmia, cataracts, strabismus, retinal degeneration and pigment changes, skeletal defects
Zinc deficiency	Anencephaly, achondrogenesis

Notes: MR = mental retardation; CNS = central nervous system; CL/CP = cleft lip/cleft palate; IUGR = intrauterine growth retardation; (±) = questionable association.

Source: Rousseaux, C. G. and P. M. Blakley. 1991. Fetus. In *Handbook of Toxicologic Pathology,* eds. W. M. Hascheck and C. G. Rousseaux. 959–961. San Diego: Academic Press. With permission.

It is clear now that many feeding problems and gastrointestinal upsets such as spitting up, vomiting, rumination, gastroesophagael reflux, colic, diarrhea, constipation, and malabsorption along with, skin, bronchial, ear, and behavior problems may be attributed to increases in total body pollutant loads in the parents, fetus, and newborn. When they occur, these conditions should be

Table 28.9.   Examples of Specific Defects with Known Etiology
in Mammals

System	Defect	Etiology	Species
Central nervous	Anencephaly	Methylhydrazine	Rabbit
		Clomiphene	Human
		Ethylnitrosourea	Rat
	Hydrocephalus	Chlorocyclamide-R	Rodent
		Quinine	Human
	Microencephaly	Diphenylhydantoin	Human
		Methylnitrosourea	Rat
Craniofacial	Anophthalmia/ microphthalmia	Thalidomide	Human, rabbit, monkey
	Ethylnitrosurea	5-Azauracil	Chick Rats
	Cyclopia	Veratrum californium	Ruminant
	Agnathia/ micrognathia	Alcohol	Human
		Arsenic	Mouse
		2,4 amino-5-chlorphenyl- 6-ethylpyrimidine	Rat
	Cleft lip/cleft palate	Griseofulvin	Cat
		Cigarette smoking	Human
		Diphenylhydantoin	Mouse
Cardiovascular	Atrial septal	Alcohol	Human
		Acetylcholine	Chick
		Dextroamphetamine sulfate	Mouse
	Ventricular septal	Alcohol	Human
		Trypan blue	Chick
		Dextroamphetamine sulfate	Mouse
		Thalidomide	Rabbit
	Tricuspid valve anomalies	Trypan blue	Chick
	Transposition of great vessels	Thalidomide	Rodent
Respiratory	Agenesis/aplasia	Amandine hypochloride	Human
	Hypoplasia of lungs	L-asparaginase	Rabbit
		Azetidine-2-carboxylic acid	Chick
Gastrointestinal	Diaphragmatic hernia	Thalidomide	Human
	Umbilical hernia	Benzimidazole	Chick

Table 28.9 (Continued).   Examples of Specific Defects with
Known Etiology in Mammals

System	Defect	Etiology	Species
	Omphalocoele	Colchicine	Rabbit
	Gastroschisis	Streptonigrin	Rat
Urinary	Renal agenesis	Chloramucil	Rat
		Thalidomide	Human
	Hydronephrosis	Bradykinin	Mouse
		Thalidomide	Human
	Horseshoe kidney	Chlorambucil	Rat
		Thalidomide	Human
	Ectopic ureter	Chlorambucil	Rat
	Hydroureter	Streptonigrin	Rat
		Thalidomide	Rhesus monkey
Reproductive	Anorchia/monorchia	Thalidomide	Human
	Cryptorchism	Cadmium	Rat
		Thalidomide	Human
	Intersexuality	Wild black cherries	Pig
		Androstenedione	Rat
		Envoid R	Human
Skeletal	Reduction	2-amino-1,3,4-thiadiazole	Rat
		Quinine	
	Digit Anomalies	Cyclophosphamide	Human
		Difolatan	Chick
		Aminophyllin	Rat
	Arthrogryposis	Anagyrine	Cattle
		Tobacco	Pig
		Sudan grass	Horse
	Spina Bifida	Acryflavin	Chick
	Meningocoele	Actinomycin D	Rabbit

Source: Rousseaux, C.G. and P. M. Blakley. 1991. Fetus. In *Handbook of Toxicologic Pathology*, eds. W. M. Haschek and C. G. Rousseaux. 962–963. San Diego: Academic Press. With permission.

studied for amounts of pollutants in the blood and tissues as well as levels of vitamins, minerals and amino acids. Each condition will be discussed separately for the newborn.

Gastrointestinal Tract

Any part of the gastrointestinal tract can be affected by pollutants. We will discuss the range of this problem.

Table 28.10.   Volatile Organic Constituents Identified in Human Blood of Mother and Newborn

Compound	Mother	Newborn	Compound	Mother	Newborn
Carbon tetrachloride[a]	+	>+	1,1-Dichloro-1-nitroethane	+	
Tetrachloroethylene[a]	+		Methyl isobutyl ketone	+	
Acetone	+	+	Dimethyl disulfide	+	
Dichloromethane	+	+	Toluene[a]	+	
Isobutanol	+		Xylene[a]	+	
Methylethyl ketone	+		Y-heptalactone	+	
Chloroform[a]	+	>+	Benzaldehyde	+	
Bromoform[a]	+		2-Butanone	+	
1,1,2-Trichloro-ethylene[a]	+ +		Methyl propyl ketone	+	+
Trichloroethylene[a]	+		Cyclohexanone	+	
Benzene[a]	+	>+	Acetaldehyde[a]	+	
Pentane-2-one	+		Methyl mercaptan	+	
Ethylbenzene	+		2,4,4-Trimethyl-2-pentane	+	
Styrene	+	+	Pyridine	+	
Trimethylbenzene	+	+	Dipropyl ketone	+	
Dichlorobenzene[a]	+	+	Methylfuran	+	
Dimethylethyl-benzene	+	+	BHT		+
Methylcyclopentane	+		2-Propanol	+	
			Cyclohexane	+	

[a] Compounds identified in municipal water supplies in New Orleans area. Center for Bio-organic Studies University of New Orleans, New Orleans, LA.

Source: Dowty, B. J. and J. L. Laseter. 1976. The transplacental migration and accumulation in blood of volatile organic constituents. *Pediat. Res.* 10:696–701. With permission.

Thrush

Thrush, which may be recurrent, may be seen in the newborn. This chronic *Candida* infestation may result from seeding from the mother's vaginal tract or from suppressed immune function of the newborn due to xenobiotics.

Many infants born to an untreated chemically sensitive mother will already have a food and chemical problem. As shown in the case previously presented, babies who are born prematurely or with defects (either congenital or acquired) will often have more of a tendency toward sensitivity as they grow into childhood. The presence of thrush or colic in the infant strongly suggests early food and/or chemical sensitivities.

Table 28.11. Transfer of Pesticides
across Placental Barrier Mother and
Infant after Birth with Prior
Treatment Nonbioconcentration

Pesticides	Mother (mg/L)	Newborn (mg/L)
DDE	28.8	8.5
DDD	trace	nd
DDT	1.4	nd
Heptachloroepoxide	3.3	nd
Hexachlorobenzene	1.1	nd
Lindane	trace	trace

Note: nd = nondetectable.

Source: EHC-Dallas. 1991.

Colic and Other Gastrointestinal Upset

Colic is a frequent problem in newborns. It is usually due to foods, but chemical overload may also cause colic. Both the pediatrician and generalist should search for etiologic agents when each case presents.

The causes of colic in the infant are varied. First, if a mother who is nursing her baby continues to eat food to which she is sensitive, her baby may develop colic because of prior transplacental sensitization. Second, a mother with pre-existing sensitivities may transfer these through her breast milk to her baby, which may then cause the baby to develop colic. Third, a mother's exposure to a toxic chemical may be enough to affect her milk by its presence and cause colic in her baby. More frequently, environmental sensitivity in children develops when a child is being weaned off breast milk early while simultaneously being introduced to formula and solid foods. This child may become rapidly sensitive to each new substance introduced. Cow's milk is the most common offender in the newborn. However, we have observed some colic due to other causes including foods, dust, pollen, and/or xenobiotics.

Spitting up, vomiting, and excessive drooling frequently reflect an infant's intolerance to formulas or its sensitivity to mother's milk, assuming, of course, that a mother's feeding technique is adequate. Foods that most commonly sensitize infants are milk, corn, wheat, soy, and chocolate. The following case report emphasizes the futility of not looking for environmental causes as opposed to the ease with which problems from exposure may be resolved once these pollutants have been identified and appropriate intervention started.

Case study. A 1-month-old infant was seen by his pediatrician for recurrent vomiting since birth. This baby's vomiting had become so severe that anything he ingested resulted in vomiting and regurgitation. Extensive work-up revealed massive esophageal reflux, but no other metabolic problems. This baby had a feeding gastrostomy placed and was fed entirely with artificial feedings for the next 2 years. He presented at the EHC-Dallas never having been able to eat anything in his life without vomiting. He was worked up with oral and intradermal challenge and found to be sensitive to biological inhalants, foods, and chemicals. He also was very sensitive to his gastrostomy feedings, which also triggered vomiting. This patient was placed on a rotary diet of organic foods and was given injections for foods and biological inhalants, as well as having his total body pollutant load reduced, and did well with total cessation of vomiting. He has continued to thrive without a need for the gastrostomy for 2 years. Needless to say, if this baby's physician had been aware that pollutant overload was causing the baby's health problems, he could have prevented much suffering for this infant and his family.

Constipation and Diarrhea

Constipation and diarrhea are also seen in some sensitive newborns and can result from formula, related enzyme deficiency, xenobiotic overload, or mild pancreatic insufficiency (see Chapter 21).

Skin

Skin problems from pollutant injury are not uncommon in the newborn, who may develop erythematous rashes, boils, or eczema. These conditions are often due to pollutant overload, which results in malfunction of the immune and nonimmune detoxification systems with secondary food and biological inhalant sensitivity causing vascular dysfunction and inflammation.

Seborrheaic Dermatitis

Seborrheaic dermatitis, which is extensive and spreading, should have its etiology sought. Often, its causes will be found in food and chemical overload.

Recurrent monilial diaper dermatitis may also occur in newborns who are overloaded with toxic chemicals, including antibiotics.

Eczema

Eczema may have a food and chemical etiology. It may also be due to biological inhalants. Many times exacerbation is seen in winter and summer. Most cases of eczema seen at the EHC-Dallas in this group of newborns were easily solved by dietary manipulation and food neutralization injections.

Perianal Dermatitis

Perianal dermatitis may be due to fungus, food, and/ or chemicals.

Respiratory System

Chronic nasal congestion, recurrent cough, and respiratory infections may have an etiology in a suppressed immune system due to food and/or chemical overload. Cows' milk is the most common cause.

Cystic Fibrosis

Often children with cystic fibrosis have recurrent pulmonary and bronchial infections. These children can be spared some bronchial damage and recurrent infections by eliminating incitants from their environment and by manipulating their diet. For further discussion of cystic fibrosis, see Chapter 18.

Endocrine System

Mild to severe endocrine abnormalities may occur as a result of food or chemical sensitivities. Pérez-Comas[57] and Sáenz de Rodriguez et al.[58] presented a large series of babies with excess estrogen who developed adrenogenital syndrome from dietary xenobiotic exposure (see Chapter 24, section on ovary).

The clinical appearance of a chemically overloaded or food sensitive child can vary. However, often seen is a baby with dark circles around the eyes, flushed cheeks, a protuberant abdomen, thin legs, and a generally scrawny appearance. Also, symptoms may include irritability, screaming, unwillingness to be cuddled, and an inability to pacify.

We have seen chemically sensitive newborns who were extremely difficult to treat. In addition to many who have food sensitivity, a few will be sensitive to dust and other inhalants. Overlooking this aspect of their sensitivities will lead to failure of treatment.

The effects of pollutants on children are widespread and poorly appreciated by the general medical community. These pollutant effects may be very subtle and wide ranging. They may involve hyperactivity or dullness or sluggishness, or they may result in severely incapacitating problems involving whole systems such as cardiovascular, gastrointestinal, dermatological, respiratory, genitourinary, auditory, or visual.

INFANTS

The 2-month to 2-year-old child can be strongly influenced by environmental overload. In particular, the malfunction that results from this overload may involve gastrointestinal upsets, recurrent respiratory infections, and/or hyperactivity. At the EHC-Dallas, we have observed and treated more than 50 infants between 2 months and 1 year of age who had their problems cleared by avoidance of foods, chemicals, and biological inhalants and who had symptoms reproduced by subsequent challenge. Many times an infant does well while being fed mother's milk. However, once the infant is introduced to solid foods or different formulas, problems that may affect the aforementioned systems appear.

We have observed one family whose experience critically illustrates the point of how desensitizing breast milk can be. A mother, 37-years-old and severely allergic to pollen, dust, and molds, had an IgE of 10,004 IU/mL. She required injection therapy for inhalants and food up to 150 times the starting dose, daily for 7 years. She also required 1 unit of transfer factor two times per week. Her older son also required similar doses in order to eat and breathe without symptoms. Strict rotary diet of chemically less-contaminated foods was also needed. She continued this therapy through the gestation and birth of a set of normal fraternal twins. The children were entirely asymptomatic until age 8 months when, at the onset of the spring pollen season, they developed rhinorrhea. The mother found that if she breast fed the babies every 5 hours she could completely inhibit the runny nose and irritability of each child. If she failed to nurse within this time, the children would develop symptoms. Throughout the remainder of the pollen season, the children stayed symptom free as long as their mother maintained her 5-hour nursing schedule. Now 4 years old, these children appear to have acquired immunity. They require neither medication nor injections, and they remain symptom free.

Gastrointestinal System

Constipation or diarrhea, gas, bloating, and cramps may occur in some infants. Again, these symptoms may be the result of dietary or toxic chemical overload. One series of Puerto Rican patients seen at the EHC-Dallas were those who did not have adreno-genital syndrome but had developed an inability to hold down foods, failure to thrive, and diarrhea. These infants usually had to have intravenous hyperalimentation to survive. When triggering agents were sought, they again appeared to be food contaminants. The infants had a widespread food sensitivity that necessitated food injection therapy and a long rotary diet (7 days). In addition, good environmental control at home was used in order to reduce the total body load and allow a decrease of the total load of the food handling mechanism. All of these infants were found to be extremely sensitive to xenobiotics upon challenge testing. Studies of their local environment and diet in Puerto Rico found them to be highly contaminated with toxic chemicals that apparently triggered these children's sensitivities.

Skin

Eczema and dry scaly skin are the major dermatological problems in infants. We have been able to define food and chemical triggers such as milk, yeast, and formaldehyde in these children. We have treated a series of 50 infants and children who cleared their symptoms with avoidance and reproduced them with challenge. They cleared without medications. The following case is an extreme example.

> **Case study.** This 1-year-old white male entered the ECU covered entirely with eczema (Figure 28.7). He had to have his hands in sock gloves to prevent his skin tearing completely open. This child was on high doses of prednisone. He was found to react to all biological inhalants, foods, and chemicals. He was placed on a rotary diet of chemically less-contaminated foods and water, while those substances that caused the worst reactions were altogether avoided. He was given injections for foods and biological inhalants. The gas was taken out of his home, and the carpet was removed from his bedroom. He was weaned off his steroids, and his eczema gradually cleared. Five-year follow-up showed the child to be totally asymptomatic. He was not taking medications or injections. He still used environmental control and a rotary diet to maintain his good health.

Respiratory System

As in newborns, chronic nasal congestion may occur in infants. It may be isolated or accompanied by postnasal drip, chronic cough and wheezing, and reactive airway disease.

Each of these entities may have their origin in biological inhalant, food, or chemical sensitivity.

The Heiner syndrome of recurrent pulmonary infiltrates is an example of food sensitivity causing pulmonary problems. Usually, this syndrome is due to milk sensitivity, but in selected cases, pulmonary infiltrates can also be due to other toxic substances.

Until 2 to 3 years of age, children often develop recurrent otitis media and middle ear problems. Most of the time, in the experience of environmental medicine specialists, these problems can be explained by the presence of food sensitivities or chemical overload. Waikman,[59] Rapp,[60] and Boris[61] have reported cases that were food and chemically triggered. They were able to stop these recurrent infections by avoidance with rotary diets and injections for the foods to which the children were sensitive. The use of tubes in the ears dropped markedly in these children.

Neurology and Behavior

Some infants are excessively active, constantly moving, and accident prone. Often they have temper tantrums and/or bite. Frequently, these behaviors are responses to food overload, but chemical intolerance should not be overlooked as a causative factor. Such a child's behavior is not always on an even keel, and fluctuations of mood are a frequent occurrence. The following case illustrates environmentally triggered neurological problems.

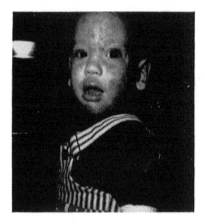 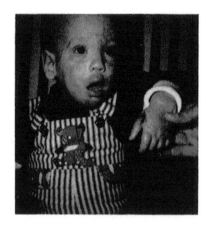

Figure 28.7. Severe eczema triggered by inhalants, foods, and chemicals.

Case study. A 3-month-old girl was referred to the ECU for evaluation of persistent diarrhea, failure to thrive, apathy, and poor weight gain. She had experienced nasal congestion and conjunctivitis, persistent vomiting, diarrhea, and colic since 1 month of age. Changing her formula had not remedied her problems at all. She became apathetic and lethargic to the point that she would lie in her crib, stare, and not move. Neurologists suggested she was suffering from an arrested brain function. Prognosis was dismal.

This patient was placed in an environmentally controlled room constructed of porcelain. She was tested intradermally for foods, pollen, dust, and molds. She was also challenged orally with chemically less-contaminated foods and tested against chemicals in an airtight, steel and glass exposure booth. This child was diagnosed with multiple sensitivities, including apple juice, apricot juice, carrot juice, black cherry juice, bananas, papaya, sweet potato, potato, peaches, moo-soy, beets, rice, squash, goat's milk, and cow's milk. She was also sensitive to formaldehyde, phenol, and chlorine.

She was started on a program of antigen injections, chemically less-contaminated foods, and environmental modification. On this routine her vomiting, diarrhea, and colic resolved. She became alert and lost all signs of neurological defects. She started sleeping through the night, and she rapidly gained weight. Follow-up at 3 years of age showed that she was eating and sleeping well, gaining weight, and that she had been able to discontinue antigen injections. She had continued to follow a rotation diet. To date, she is totally asymptomatic, and she has normal brain function. This patient has progressed remarkably well.

It is clear that many infants can be made ill depending upon their exposure to different incitants. Care should be taken to evaluate total load in these patients, since cases of sensitivity to biological inhalants as well as foods and chemicals have been seen.

CHILDREN

From the ages of 2 to 12 years, children may have a wide range of problems. Older children appear to develop many problems relating to recurrent respiratory infections and/or behavior problems. However, some have neurological, cardiovascular, gastrointestinal, and renal ailments.

Respiratory System

In early childhood, chronic ear infections leading to tubes in the ears are often seen. These infections are similar to those described in previous sections. Large series of these cases have been reported to be triggered by foods and some chemicals.[62,63] Few children should have tubes or tonsillectomy without first having an environmental work-up.

Bronchitis and Asthma

Childhood asthma appears to be on the rise. The diagnosis here is no different than with adults (see Chapter 18). We have seen many children whose bronchitis and asthma were triggered by food and chemical problems and whose pediatricians and pulmonologists prescribed steroids without ever looking for triggering agents. In our opinion, the use of steroids without a proper environmental evaluation is condemnable.

The overall death rate for children with asthma has increased over the last 20 years.[64] Many authors seem to be puzzled by this fact. However, when one considers that the etiology is seldom looked for, that triggering agents are rarely defined and eliminated, and that synthetic chemicals are used to treat what turns out to be a chemically triggered problem, it is not surprising that an increase in death rate has occurred.

In one series of 30 children (22 males and 8 females) with recurrent "intractable" asthma seen at the EHC-Dallas, multiple triggering agents were found. These children ranged from the ages of 2 to 12 years. All had been placed on aminophyllin and cortisone compounds as well as bronchodilators without any effort to evaluate environmental triggers. All had been considered failures of therapy by their families. IGEs ranged from 0 to 1906 2 IU/mL with 67% being above control. No gammaglobulin deficiencies were found. Each child was carefully placed on a diagnostic program using withdrawal and oral, intradermal, and inhaled challenge. Meticulous definition of triggering agents was done, measuring peak flows (when the child was large enough to cooperate with the flow testing) before and after challenge. An avoidance and injection program for foods, biological inhalants, and chemicals was designed and instituted. These children were weaned off their medications and became free of asthma for the first time. Long-term follow-up showed the chemically sensitive patients treated at the EHC-Dallas remained free of asthma with extremely low morbidity (10 minor episodes of wheezing after massive exposure requiring extra injections to the substances to which they were sensitive and no mortality). Table 28.12 shows asthma and its triggering agents.

Another group of children have persistent, recurrent bronchitis usually, but not always, accompanied by recurrent rhinitis and sinusitis. These children are constantly on antibiotics. Although we have seen hundreds of these patients, one series of eight will be presented in order to illustrate several points.

These patients (4 males and 4 females) ranged in age from 1 to 11 years. IgEs ranged from 0 to 141 IU/mL with 38% being above control ranges, while the rest were normal. One had measureable low IgA, while the majority of those measured had impaired phagocytosis. These patients were systematically diagnosed using oral, inhaled, and intradermal challenge under environmentally controlled conditions. Rotary diets that eliminated the severe offenders were instituted as was injection therapy for foods, biological inhalants, and chemicals. An oasis was created in each patient's home. These children were followed for 5 years with very little recurrence of their infections. Three children had two recurrences and two had one recurrence. These all occurred after inadvertant exposures to their triggering agents and cleared without the use of long-term antibiotics. Table 28.13 shows the types of triggering agents.

Musculoskeletal System

Chronic fatigue with muscle and joint aches is seen in some children (see Chapter 23). After infectious diseases are ruled out, a host of environmental triggers remain. Many children complain of legs aching with myofascial pains. Often these pains are dismissed as growing pains when, in fact, they are symptoms of pollutant overload with food and chemical sensitivity. If pressed, the older child recognizes that the pain is episodic and usually unrelated to exercise. This hangover pain is like all pain that occurs after vigorous exercise. Careful work-up reveals food and chemical triggers.

Gastrointestinal System

Some older pediatric patients may develop abdominal pain, which may or may not be abdominal angina. Frequently these patients have peripheral vasospasm, and probably most of the abdominal pain is spastic in nature. Some patients even develop anaphylaxis along with the abdominal angina, as is described in the following case.

Case study. A 6-year-old child was evaluated in the ECU after a life-threatening anaphylactic reaction as well as recurring intractable abdominal pain following ingestion of a small portion of a peanut. At the time, she was also in the presence of a dog to which she was known to be sensitive. Her main complaint through the previous years was recurrent abdominal pain.

Table 28.12.   Data of 30 Asthmatic Children

Positive skin testing		High toxic chemical in blood	
Antigen	**No.**	**Chemical**	**No.**
Inhalant:		Toluene	1
Molds	28	Trimethylbenzenes	1
Algae	1	1,1,1-Trichloroethane	1
Dust	21	2 Methyl pentane	2
Dust mite	21	3 Methyl pentane	1
Danders	16	n-Hexane	2
T.O.E.	16	**Treatment**	
Candida	17		
Fluogen	22	**Immuno-therapy**	**Rotation**	**Adrenal**
MRV	23	**(antigen)**	**diet**	**steroids**
Pollens	28	30	30	4
Terpenes	21		
Cotton	15		
Smuts	5		
Ants	1		
Chemicals:			
Cigarette smoke	6		
Women's cologne	1		
Men's cologne	2		
Ethanol	2		
Formaldehyde	14		
Newsprint	1		
Orris root	9		
Diesel	1		
Glycerine	3		
Chlorine	1		
Food:			
(Range = 3 to 57 kinds; mean = 17 kinds)	30		

Notes: Sex = 22 M and 8 F. Age range = 1 to 12 years; average, 6 to 7 years. Normal IgE = 10 patients (33%); high IgE = 20 patients (67%); IgE range = 0 to 1906 IU/mL.

Source: EHC-Dallas. 1993.

This child had a life-long history of multiple allergies, rhinitis, incapacitating abdominal pain with vomiting and dizziness, eczema, and urticaria in addition to anaphylaxis. As a smaller child, she had undergone extensive medical evaluation with a variety of specialists without relief of any symptoms. She also had had an endoscopy, with a bowel biopsy for the recurrent abdominal pain. She was thought to have lactase deficiency. From the age of 6 weeks, she was a screaming infant with colic and profuse diarrhea. She had multiple

Table 28.13. Data of Eight Patients with Bronchitis

Associated diseases	No.	Positive skin testing				Toxic chemical in blood compound	No.	Treatment	No.
		Antigen	No.	Immunoglobulin	No.				
Allergic rhinosinusitis	2	Inhalant		IgE — High	3	HCB	1	L-glutathione	1
Asthma	1	Molds	6	IgE — Normal	3	α-BHC	1	Antibiotic	1
Eczema	1	Dust	6	(Range: 0 to 141 IU/mL)		β-BHC	2	Vitamin C	2
Irritable bowel syndrome	1	Dust mite	6	IgA — low	1	Oxychlordane	1	Oxycard	1
Inhalant sensitivity	6	Danders	3			trans-Nonachlor	1	Antigen	7
Food sensitivity	7	Tree pollen	6			DDT	1	Elimination diet	1
		Grass pollen	6			Penta-chlorophenol	1	Rotation diet	5
		Weed pollen	6					Zephron chlorine	2
		Fluogen	5					Nilstat	1
		MRV	5						
		Candida	5						
		T.O.E.	5						
		Insect	1						
		Smuts	1						
		Cotton	1						
		Terpenes	3						
		Chemical							
		Formaldehyde	4						
		Orris root	1						
		Foods	7						
		(Range: 3 to 18 kinds; (mean: 10 kinds)							

Notes: Sex = 4 M and 4 F. Age range = 1 to 11 years; mean age = 4 years.

Source: EHC-Dallas. 1994.

food intolerance; therefore, changing her formula made very little difference. She continued to be a difficult infant. Her problems peaked when she developed laryngeal edema.

At age 3 years, she received a course of prednisone for a severe widespread rash, even though food triggering was not evaluated by proper environmental methods. At one point, her nutrition was so poor and food intolerance so troublesome that she needed hyperalimentation for 3 weeks. Additionally, she had, at that time, laryngeal edema with asthma after eating a cookie. She was evaluated at a prominent medical school, but given no hope or help.

When this patient came to the EHC-Dallas, she was placed in an environmentally controlled porcelain room and fasted on spring water for $2^1/_2$ days. During this time, she experienced withdrawal symptoms of nausea, vomiting, muscle pains, nightmares, joint pains, and headaches, all of which cleared after fasting. With oral and intradermal challenges, she reacted to 36 foods including apples, oranges, potatoes, carrots, chicken, lettuce, cane sugar, turkey, salmon, oats, and rice. She reacted to dog dander. She also reacted with abdominal pain, coughing, sleepiness, and rash to intradermal peanut challenge. Double-blind, inhaled chemical challenge revealed reactions to <0.50 ppm petroleum derived ethanol and <0.20 ppm formaldehyde. Placebos of saline were negative, as was challenge to <0.33 ppm chlorine, <0.002 ppm phenol, and <0.0034 ppm of insecticide (2,4-DNP). Open challenge to natural gas (open flame) reproduced symptoms (Table 28.14). In the hospital, investigation of her symptoms showed that her WBC was elevated to 14,700/mm^3 (C = 5000 to 10,000/mm^3) with an IgM that was elevated at 292 mg/dL (C = 60 to 280 mg/dL) and a depressed IgA of 88 mg/dL (C = 90 to 450 mg/dL).

This child was started on a comprehensive program of rotary diet, environmental avoidance of offending agents, antigen injection therapy for inhalants and food, and nutritional support. The family removed their gas heat and carpeting, creating an oasis in the child's bedroom. Her mother reported a dramatic change in her health within 2 weeks of these changes. Having been treated with dietary control and antigen injections, she was doing very well 2 years later, the time of her follow-up examination. She showed good food tolerance, including tolerance to peanuts. She was also able to tolerate dogs.

This patient remains environmentally sensitive, but now she has only rare urticarial rashes. Further, she has shown significant improvement with 9 years absence from the life-threatening anaphylactic reactions and abdominal angina that had previously troubled her. She has become a typical teenager. Injections for foods and inhalants, which she takes once a week, are the only treatment she continues.

Behavior

At the EHC-Dallas, we see many children with behavior and learning problems. We have seen over 400 who were having trouble learning in school. Some had attention deficits, inability to concentrate, and hypo- and hyperactivity. Frequently these difficulties may be triggered by food and chemical

**Table 28.14.   Double-Blind Inhaled Challenge after 4
Days Total Load Reduction and Deadaptation
in the ECU (6-Year-Old White Female)**

Challenge	Amount (ppm)	Result[a]
Ethanol-petroleum derived	<0.50	+
Formaldehyde	<0.20	+
Chlorine	<0.33	−
Phenol	<0.0034	−
Insecticide (2,4-DNP)	<0.0024	−
Saline	—	−
Natural gas[b]	open flame	+

[a] Reproduced her admitting symptoms.
[b] Single-blind.

Source: EHC-Dallas. 1992.

**Table 28.15.   8-Year-Old White Male after
4 Days Deadaptation with Total Load
Decreased in the ECU — Double-Blind
Inhaled Challenge Reaction**

Chemical	Dose (ppm)	Reaction
Pesticide (2,4-DNP)	<0.0034	+[a]
Formaldehyde	<0.2	−
Ethanol	<0.5	−
Chlorine	<0.33	−
Phenol	<0.002	−
Saline		−

[a] Severe paranoia for 48 hours.

Source: EHC-Dallas. 1991.

sensitivities (Table 28.15). The following case is an example of environmentally triggered behavior and learning dysfunction.

Case study. This 8-year-old white male entered the second grade in school with no problems. He suddenly developed paranoia and an inability to learn. He had to drop out of school because of his paranoia and an inability to concentrate. The family and school authorities were ready to send this boy to the state mental hospital before he was worked up at the EHC-Dallas. Here he was placed in an environmentally controlled room and fasted for 2 days. His symptoms cleared including his paranoia. He was able to concentrate for the first time in many months, and he loved to study his school work and solve problems. Biological inhalants appeared to be minor problems. The child was given inhaled double-blind challenges of the ambient doses of some toxic

Table 28.16. Improvement of Immune Parameters
After Treatment

Immune parameters		Before treatment 6/18/86	After treatment 6/9/87	Normal range
WBC		5700	5100	4800–10,000/mm^3
Lymphocyte	(%)	30	39	21–49%
	C	1710	1989	1600–4200/mm^3
T_{11}	(%)	64	79	62–86%
	C	1094	1571	1260–2650/mm^3
T_4	(%)	35	42	32–54%
	C	599	835	670–1800/mm^3
T_8	(%)	28	28	17–35%
	C	479	557	333–1070/mm^3
T_4/T_8		1.3	1.5	1–2.7
B Lymphocyte	(%)	14	18	5–18%
	C	239	358	82–477/mm^3

Source: EHC-Dallas. 1993.

chemicals. The odor of 1 sniff of <0.0034 ppm pesticide (2,4-DNP) caused severe paranoia and an inability to concentrate for 48 hours. Investigation revealed that the school had been sprayed with insecticide the day the child became paranoid. He was treated with avoidance of xenobiotic exposures in the air, food, and water, injection neutralization of foods and inhalants, and nutritional supplementation. His immune parameters improved (Table 28.16). However, he had to have home-bound teaching for 2 years before he could return to school.

Since children have a wider range environment than infants, their potential for exposure to triggering agents is increased. However, the common ones, such as milk, corn, wheat, and formaldehyde, prevail.

As the child grows, school environments become a problem. Many children do well until they start nursery school, where they are not only exposed to new and higher levels of bacteria and viruses but also bombarded by pollutants emanating from a variety of sources. School workers, teachers, and students themselves pollute their environment when they use such products as heavily scented cosmetics (phenols), deodorants, hair sprays, or nail polish. Also, cleaners such as scented soap, toxic cleaning solvents and disinfectants, and scented waxes and polishes (phenols) pose problems. Products made of particle board or that use formaldehyde-based wood finishes or glue, such as portable school buildings, desks, tables, chairs, shelves, or paneling, contribute to school pollution, as do synthetic drapes and carpeting, room finishes and paints, and flooring. The heating/cooling system in a school can further exacerbate pollution problems if it becomes contaminated with dust and mold.

Also, inadequate ventilation to reduce pollutants generated indoors increases the indoor pollution load. Many products such as art supplies including crayons, magic markers, ink pens, paints and varnishes, glue, paste, clay, and putty are insidious contaminants used directly in the learning process. Chalkboards, paper products, and newsprint are also offenders. A miscellany of other sources of contamination in the school environment include lead-lined water coolers, book depositories, indoor plants and animals, pesticides, and electromagnetic fields set up by copy machines, computers, video players, and televisions. Some products used in building insulation may be offenders as are, finally, outdoor industrial emissions and auto exhaust that find their way indoors. Tight-building construction intended to facilitate energy efficiency also complicates the school pollution picture.

Air pollution in the schools can cause severe problems. It can retard learning and also cause sickness, such as recurrent infections. Although not a child, the 16-year-old described in the following case study illustrates the problems with air pollution in the schools.

> **Case study.** This healthy 16-year-old high school student developed dizziness, ear infection, malaise, inability to learn, and bronchitis. The roof of her school had been undergoing repair, and warnings were given to the students to move all of their cars because the sealant being used for repairs could corrode the paint, although students were assured that the sealant would not be harmful to their health. When this patient was seen at the EHC-Dallas following her exposure to the sealant, her blood levels revealed toluene (2 ppb) which had not been present on her previous health exams. Her reaction was consistent with known possible effects of the sealant that was used on the roof. She stayed away from school until the toluene there had dissipated, her symptoms had disappeared, and her blood had become toluene-free. Then she returned to school without further problems.

Anaphylaxis

Recurrent and fatal anaphylaxis, especially to foods and bee stings, has been reported in some children. This devastating complication of biological inhalant, food, and chemical sensitivity should be searched for and prevented if at all possible. For the last 25 years, environmental medicine physicians have reported that this entity can be treated and prevented by using pollutant avoidance plus injection therapy.

Brown[65] and Lee[66] have had much success at treating and preventing anaphylaxis in children. Over the last 20 years, occasional reports of a child dying from anaphylaxis resulting from peanut butter hidden in a food have occurred. The American Academy of Allergy has steadfastly denied that food injection therapy is helpful in treating food sensitivity, and this position has resulted in the unnecessary deaths of some children with the hidden food-anaphylaxis problem. Recently, Oppenheimer et al.[67] at the National Jewish Hospital in Denver reported the death of a patient on whom they were performing "scientific" studies of a "new technique" for food injection therapy. Our group at the EHC-Dallas and other environmental medicine speacialists object to the needless generation of mortality and morbidity in a quest for injection therapy for foods, when a proven technique such as intradermal injection neutralization used in thousands of cases during the last 25 years is already available and safe.

Case study. A 6-year-old white female entered the ECU after several near fatal episodes of anaphylaxis requiring administration of epinephrine and local hospitalization. Each episode occurred after she smelled the odor of peanut butter at a public place, such as kindergarten or church. Her family related that once the anaphylactic episode occurred, this patient would then develop a respiratory infection that often progressed into bronchopneumonia requiring antibiotics and 3 weeks of confinement at home or in the hospital before it resolved. It appeared that if precise definition was not done, this child would one day suffer a fatal episode. The child was placed in the ECU and fasted for 2 days until her symptoms were eliminated. She was then challenged orally, intradermally, and by inhalation until a definitive program of elimination of triggering agents could be found. Injection therapy for biological inhalants, foods, and chemicals was started. It was noted that when the patient was being tested, another child who had eaten peanut butter for lunch came into the testing room. The odor on the child's breath triggered severe angioedema and laryngeal edema. In spite of that super-sensitivity, this child was able to be neutralized by the intradermal injection method for peanut butter. Following neutralization, she was usually able to be in a room with the odor of peanut butter without anaphylaxis. Further environmental history revealed that, prior to this patient becoming ill, her house was treated wtih chlordane. Also, blood tests revealed the presence of chlordane (0.6 ppm — C<0.1 ppm), and subsequent air analysis revealed chlordane in the child's bedroom. The bedroom was cleaned and all areas sealed until no trace of chlordane remained.

Skin testing revealed this child was secondarily sensitive to the following biological inhalants: molds, dust, dust mites, animal danders, weeds, tree and grass pollens, and 29 foods including peanuts. She was sensitive to the following chemicals: newsprint, phenol, chlorine, formaldehyde, diesel, car exhaust, perfume, formaldehyde, orris root, and terpenes of pine, sage, trees, cut grass, and cedar.

As her environment was cleaned up and made free of dust and chemicals, this child became less fragile, and her episodes of anaphylaxis diminished and then ceased. She, of course, continues her injections for peanuts and these other substances on a biweekly basis.

This case exemplifies the basic interaction of toxic chemicals (e.g., chlordane) with the body's immune and nonimmune system in creating conditions that cause severe anaphylaxis (see Volume I, Chapters 4[7] and 5[68].)

We have studied a larger series of children at the EHC-Dallas who have had anaphylaxis usually generated by toxic chemicals but also triggered by foods and inhalants. Each child's anaphylaxis had multiple triggers of biological inhalants, food, and toxic and nontoxic chemicals. Each child's treatment necessitated a reduction in total body load by a rigid avoidance program of indoor and outdoor environmental pollutants, food and water contaminants, and food and biological inhalant triggers; injection therapy for biological inhalants, foods, and some food-contaminating chemicals; and nutrient supplementation of vitamins, minerals, and amino acids to which they were deficient. Long-term follow-up of up to 15 years showed these patients not only were free of anaphylaxis but also of most of their sensitivity problems.

ADOLESCENT

Because of the individual emotional instability sometimes brought on by puberty, the physician may encounter difficulty discerning problems in an adolescent that occur as a part of a child's development vs. problems that result from pollutant exposures. Thus, diagnosing and treating adolescents can be particularly problematic (see Chapter 24). Alert to the possibility of the compounding effects of hormonal changes and environmental influences in the adolescent, however, the astute physician can still proceed to evaluate an individual through the usual techniques and principles outlined throughout this book. The systems that may be affected in the adolescent are the same as in any other age group, including the gastrointestinal tract, the respiratory system, the neurological system, and/or the skin. Also, some adolescents may have involvement of the genitourinary and the cardiovascular systems.

Musculoskeletal System

Musculoskeletal aches, pains, and fatigue are uncommon in children. However, when they do occur, pollutant injury should be searched for and treated if found. Chronic fatigue is occurring more frequently in youths, and usually the cause can be found by studying their past history for pollutant exposures and the subsequent effects of these pollutants on their well-being.

Apparently a certain segment of patients with juvenile rheumatoid arthritis can also be helped by environmental control methods. The following case study provides an example.

> **Case study**. A 15-year-old, white male student was admitted to the ECU with the chief complaints of swelling of the joints. Three years previously, he had had rheumatic fever, which seemed to be the start of his problem, with most of his joints becoming swollen at this time. He saw a rheumatologist without gaining relief. He developed recurrent sinusitis and was seen by an allergist for this illness, again without relief. His eyes watered frequently, and exposure to cigarette smoke or mist sprays produced a headache. Car exhaust would make his nose run and become congested. Nylon rugs also seemed to affect him, and he noted that candle shops clogged his sinuses and would

make him sneeze. Generally, he felt better outdoors, especially in the country. He noted that if he got enough cortisone, all joint swelling would disappear. Aspirin would temporarily loosen his joints. As a child, he had had a tonsillectomy for recurrent tonsillitis.

His physical examination was normal except for his swollen joints, which appeared to have permanent changes of swelling and deformity in all the phalangeal joints of the hands and the feet. This was classic symptomatology for juvenile rheumatoid arthritis.

Abnormal lab tests showed elevated biotin at 591 pg/mL (C = 200 to 500 pg/mL), elevated vitamin E at 1.29 mg/dL (C = 0.6 to 1.2 mg/dL), depressed vitamin B_1 at 21 ng/mL (C = 25 to 75 ng/mL), and elevated magnesium at 2.19 mEq/L (C = 1.3 to 2.1 mEq/L).

He had detectable blood levels of 14 toxic chemicals, the most common being: DDE at 0.82 ppb, heptachlor epoxide at 0.2 ppb, alpha-BHC at 0.2 ppb, xylene at 2.0 ppb, dichloromethane at 0.4 ppb, chloroform at 4.0 ppb, bromoform at 0.5 ppb, trichloroethane at 1.0 ppb, trichloroethylene at 0.6 ppb, and tetrachloroethylene at 2.0 ppb. These are inordinately high levels and an abnormal total number of chemicals for an individual of this age. The average number is less than two.

His intradermal food and chemical challenges showed he had sensitivity to 18 foods, 9 inhalants, histamine, orris root, and cigarette smoke as well as mix respiratory vaccine (MRV). Chicken and pork gave him headaches.

In 1976, this patient was admitted to the ECU. He was placed in an environmentally controlled room. During his hospitalization, he felt much better, and his symptoms cleared with fasting. He was treated with antigen immunotherapy injections, a rotary diversified diet, and environmental control. He became progressively better on this routine. Fourteen year follow-up showed he had finished medical school and was pursuing a rheumatology fellowship without medications.

Gastrointestinal System

Some adolescents develop gastrointestinal upset, irritable bowel syndrome, failure to thrive, ulcers, regional enteritis, or ulcerative colitis. The following case study is an example.

Case study. An 11-year-old girl was admitted to the ECU with severe complaints of tonsillitis, pharyngitis, lethargy, difficulty hearing, a rash on her left foot, diarrhea, confusion, and sneezing. She was approximately 9 years old when her symptoms began. Her growth had slowed, and she had stopped gaining weight. She had continuous sore throats, colds, and bronchitis. Symptoms seemed to be most severe during the afternoons in the early spring and were frequently associated with fever and/or signs of infection. She became ill and developed respiratory symptoms of a rather frequent nature. She did become short of breath at times. She had hiccups most of her life at rather frequent intervals. She also had some stomach distention, abdominal cramping, and passing of intestinal flatus. She had diarrhea at times and occasional constipation. Sudden headaches, which could last for hours, occurred on the vertex. These headaches interfered with her normal mental function. She was out of school for 3 years because of these symptoms.

She had been exposed to inhalants and chemicals that were present in her classroom and home. Specifically, odors from printing inks, solvents, and oils were brought into her home environment by her father, who worked on cars and in a print shop.

She had a family history of hay fever, epilepsy, migraine, nervousness, vertigo, heart disease, indigestion, diarrhea, bronchitis, insect allergy, and drug allergy as well as arthritis.

This patient's general physical exam was normal, except that her nasal mucus membranes and tonsils were a little redder than normal. Her left foot had an erythematus nonpapular rash.

Diagnostic chemical challenge was conducted in a steel and glass airtight booth using a double-blind procedure after 4 days of deadaptation in the ECU. Inhaled chemicals that triggered symptoms were natural gas, chlorine at <0.33 ppm, and phenol at <0.002 ppm, whereas formaldehyde at <0.20 ppm and saline placebos did not produce symptoms. Her skin testing revealed that she had sensitivity to 35 foods. Oral challenge results showed she had sensitivity to 30 foods.

Abnormal blood analysis revealed elevated strontium at 31 µg/dL (C = 1 to 18 µg/dL), elevated zinc at 199 µg/dL (C = 100 to 160 µg/dL), depressed ALT/SGPT at 8 µ/L (C = 11 to 43 µ/L), elevated inorganic phosphorus at 5.1 ng/dL (C = 2.5 to 4.5 ng/dL), elevated chloride at 111 mEq/L (C = 98 to 106 mEq/L), depressed glucose at 60 mg/dL (C = 65 to 115 mg/dL), depressed creatinine at 0.6 mg/dL (C = 0.7 to 1.4 mg/dL), elevated A/G ratio at 2.4 (C = 1.2 to 2.2), slightly depressed T cells at 57% (C = 60 to 80%), and the absolute number of all T lymphocytes was normal at 1610/mm^3 (C>1000/ mm^3). B cells were depressed at 11% (C = 20 to 40%), and the absolute number of B cells was also depressed at 305/mm^3 (C>500/mm^3). She had elevated vitamin B$_6$ at 142 ng/mL (C = 30 to 80 ng/mL) and elevated biotin at 660 pg/mL. Her 24-hour urine analysis revealed that calcium was depressed at 3.3 mEq/24 hours (C = 50 to 400 mEq/24 hours); sodium was low at 40 mEq/24 hours (C = 50 to 225 mEq/24 hours), and phosphorous was low at 0.2 g/24 hours (C = 0.9 to 1.3 g/24 hours).

Her serum pesticide screening test showed β-BHC, DDT, dieldrin, heptachlor epoxide, hexachlorobenzene, and PCBs were low at <1 ppb, and aldrin, 2-BHC, 5-BHC, and chlordane, DDE, endosulfan I, II, endrin, heptachlor, and mirex were nondetectable (<0.05 ppb). In our experience at the EHC-Dallas, this number of detectable pesticides in an individual is extremely high and always results in disease.

She was diagnosed with cephalgia, malaise, irritable bowel syndrome, rhinitis, food and chemical sensitivities, and inhalant sensitivities.

During her hospital course, this patient was placed in an environmentally controlled room. She also received detoxification therapy. This patient fasted for 5 days, and her symptoms cleared during the fast, with all symptoms disappearing. She also underwent immunotherapy. After discharge, recommendation was made for her to continue a rotation diet of chemically less-contaminated foods, to drink only spring water, to keep a daily diet diary, to remove all inciting chemicals from her home, and to follow a treatment plan of immunotherapy. She steadily improved; she gained 35 pounds in 1985, and in 1986 she went back to school and tolerated it well. Her problems with mental function resolved. She became much more intelligent than she had been before her illness. Even though she was out of school for 3 years, she jumped from fifth to eighth grade, and she won many academic honors every year. She hopes to become a doctor in the future and has now graduated from high school with honors.

This case illustrates that environmentally controlled surroundings and detoxification therapy with immunotherapy can be beneficial for patients with chemical sensitivities. Because her mother was incapacitated by a severe chemical exposure at the same time this patient became ill, we believe her immune deregulation was probably also triggered by toxic exposure. A second case illustrates the severity of environmentally triggered problems associated with diabetes.

Case study. This 11-year-old white male presented at the EHC-Dallas with polydyspia, polyurea, chronic diarrhea, and wasting. He was well until age 5 years when he developed chronic diarrhea with 20 stools per day. Small bowel biopsy resulted in a diagnosis of celiac disease. This patient did not respond to a gluten-free diet. For the next 6 years, he continued to have diarrhea with 8 to 10 watery bowel movements, which usually floated. He had no nausea, vomiting, melena, or hematemasis. His mother found that many substances in the environment exacerbated his diarrhea. This child was diagnosed as having juvenile diabetes, which was controlled with diet and insulin. Even though his diabetes was controlled, his diarrhea peristed. Physical exam revealed an emaciated white male with distended abdomen.

This child was placed in the ECU, and after 2 days his diarrhea stopped. Intradermal antigen challenge revaled sensitivity to molds, weeds, grass, terpenes, algae, foods, chemicals, molds, wheat, chicken, Brewer's yeast, beef, and cow's milk. Corn and cabbage reproduced his gastrointestinal upset.

This patient was placed on food injection every 4 days, a rotary diversified diet during which he avoided the foods that triggered his diarrhea, and 3 units of humalin.

Laboratory tests revealed high blood sugar up to 700 mg/100 mL before dietary and insulin control was instituted. IGE was 1509 IU/mL. Stool analysis was high for Pseudomonus Aerugenosa, and serum protein was low at 5.5 g/dL (normal = 6 to 8 g/dL).

This patient was discharged to his home where, 15 minutes after entering, he developed fulminant diarrhea and loss of control of his blood sugar. Since these conditions persisted for several days, he was transferred to a chemically less-contaminated porcelain ECU, and his diarrhea and blood sugar came under control within 24 hours. This scenario was repeated 4 times over the next month, until it was clear that something in the house triggered the exacerbations. The contamination was traced to pesticides.

This boy has lived in less-polluted housing during the last 8 years and done extremely well. He is robust, eats well, and lives a vigorous life. He does notice that if he stays too long in an area containing pesticides, he has gastrointestinal upset and his blood sugar increases. The same happens if his mold shot contains the wrong dose.

Cardiovascular System

Frank congenital anomalies will not be discussed since they are presented well in many textbooks. However, there exists a group of children who have problems with arrhythmias and/or cardiac failures. These problems appear to be environmentally induced and congenital.

Case study. One child seen at the EHC-Dallas had incapacitating supraventricular arrhythmias. There had been a gas leak at his home, and he apparently had become sensitized to gas fumes. Some foods and most toxic chemicals triggered his problem. After gas, phenol, formaldehyde, insecticide, cigarette smoke, alcohol, and chlorine challenges, the Holter monitor showed sinus arrhythmia and sinus bradycardia and bradyrhythmia with PVCs. Sinus tachycardia was at 170 per minute, and runs of ventricular tachycardia were 150 per minute. He had a flushed face and tired, weak legs. After treatment that included the removal of gas from his home and implementation of a rotary diet, his health improved. He has done well for the past 10 years.

Case study. Another boy seen at the EHC-Dallas developed unexplained atrial fibrillation, ventricular tachycardia, and heart failure. This 15-year-old white male was admitted to the ECU with the chief complaint of atrial fibrillation, ventricular tachycardia, angina, and hard heart beating. He developed tachycardia, hives, and fatigue from a second treatment dose of amoxicillin, which was initially prescribed for a sore throat. He was never well after this reaction. When he was exposed to chemicals in the school environment, his heart rate increased, and he developed chest pain. He had, therefore, been tutored at home for 7 years previous to this admission due to continuing, severe chest pain. He was hospitalized as a result of this persistent chest pain. In 1985, he underwent heart catheterization and an echocardiogram and was diagnosed as having a cardiomyopathy. He was found to be quite sensitive to pollens, dusts, molds, chemicals, and food. He experienced anaphylactic-type reactions that were linked to the odors of leather and plastic. He also experienced occasional hives. He had also fractured three ribs 1 year previously, while doing stunts on his bicycle. He had been hospitalized three times — 1976, 1980, and 1985 — for severe ventricular arrhythmias. He had had a cystoscope at age 3 years for dysuria, but no cause was found.

His family history was positive for hay fever, cancer, asthma, hives, migraines, nervousness, heart disease, indigestion, diarrhea, hypertension, insect allergies, arthritis, and drug allergies.

Physical examination was normal. His heart showed normal rhythm without murmur. Chest X-ray showed an enlarged heart with all chambers involved. Left heart endiastolic pressure was elevated, and the ejection fraction was reduced to 25.

He was placed in an environmentally controlled room with nothing by mouth except spring water for 3 days. During this time he experienced weakness and a slight headache. At the time of his discharge, he felt very well, although he related that his chest pain would exacerbate on an intermittent basis depending on the provocative testing.

Intradermal antigen challenge showed that he had sensitivities to five foods, tree mix, two grasses, weeds mix, three terpenes, including grass terpene, ragweed terpene, Mountain Cedar terpene, cotton, fluogen, MRV, five molds, lake algae, TOE, *Candida*, dust, dust mite, two smuts, histamine, serotonin, and six chemicals — orris root, ethanol, formaldehyde, newsprint, perfume, and phenol.

Chemical inhalant testing was done in a steel and glass airtight booth by double-blind method (Table 28.17). The findings were negative for placebo, chlorine, ethanol, and formaldehyde. They were positive for phenol (<0.002 ppm) and insecticides (<0.0034 ppm).

The laboratory data revealed that plasma sodium was high at 148 mEq (C = 133 to 144 mEq). Iron was elevated at 1.92 ppm (C = 0.5 to 1.5 ppm), phosphorus was low at 95 ppm (C = 107 to 120 ppm), and sulfur was high at 1119 ppm (C = 610 to 950 ppm). RBI minerals showed high magnesium at 50 ppm (C = 39 to 49 ppm), low copper at 0.611 ppm (C = 0.641 to 0.863 ppm), low chromium at 0.09 ppm (C = 0.119 to 0.209 ppm), and low selenium at 0.145 ppm (C = 0.153 to 0.246 ppm).

His immune parameters improved after treatment (Table 28.16). His blood toxic chemical analysis showed DDE at 0.5 ppm and HCB at 0.2 ppm. His EKG findings were normal sinus rhythm and normal EKG. His exercise test showed no ST change at 75% MPHR.

This patient was sent home on an environmental control program. He created an oasis in his bedroom, ate chemically less-contaminated foods, and drank chemically less-contaminated water. He also took injections for biological inhalants, foods, and chemicals. He was on a vitamin and mineral program. As this patient gradually cleared his symptoms, his heart improved and he became a state champion bicycle rider.

Peripheral vascular dysfunction can be devastating, causing strokes and limb loss. Often vascular spasm can occur. One case involved a young girl who developed gangrene in her left foot after she was exposed to pesticides (see Volume II, Chapter 13[69]).

Case study. A 4-year-old white female developed a history of sensitivities to some foods. She suddenly developed a fever of unknown origin and cyanosis of the left foot. The cyanosis progressed and ascended the left leg. She had a decreased left femoral pulse. She was found to have severe inflammation of the arteries and veins with microvascular clotting. She was placed on intravenous antibiotics and steroids at a children's hospital in Florida and then transferred to the EHC-Dallas. The antibiotics and steroids were continued briefly, and intravenous vitamin C (7.5 g/day) was started. As can be seen in Figures 3 and 4 of Volume II, Chapter 13,[70] this patient had gangrene of her

**Table 28.17.   Inhaled Double-Blind Challenge
of 15-Year-Old White Male Cardiomyopathy
with Ventricular Arrhythmia and Heart Failure —
4 Days of Deadaptation in the ECU
with Decreased Total Load**

Challenge	Amount (ppm)	Result[a]
Phenol	<0.005	+
Insecticide - 2,4-DNP	<0.0034	+
Chlorine	<0.33	−
Formaldehyde	<0.20	−
Ethanol-petroleum derived	<0.50	−
Saline		−

[a] Signs reproduced were multifocal PVCs and ventricular tachycardia.

Source: EHC-Dallas. 1987.

toes and left foot. She was treated for inhalant, food, and chemical sensitivity by intravenous nutrients and subcutaneous injections of foods and biological inhalants in the controlled environment. The gangrene gradually slowed, but she ended up losing the end of two toes. The exposure was traced to this child's crawling in a chemically treated lawn. She apparently inhaled the chemical fumes and then developed vascular spasm, vasculitis, clotting, and finally gangrene. A similar chemical exposure occurred again 3 years later when this child was playing on a lawn that a neighbor had treated with toxic chemicals. She developed a cyanotic foot and yellow color to the skin. She was immediately transferred to the environmental unit and given intravenous vitamin C (7.5 g/day). She rapidly cleared her system without sequalae. Chemical testing revealed she was sensitive to phenol and formaldehyde, cigarette smoke, and pesticides. Deficiencies in the intracellular minerals, iron, and sulfur were found. Blood volatile organic chemicals revealed toluene, 0.5 ppm (C<0.5 ppm); dichloromethane, 10.4 ppm (C<0.1 ppm); 1,1,1-trichloromethane, 4 ppm (C<0.5 ppm); trichloroethylene, 1.3 ppm (C<0.5 ppm); oxychlorodane, 0.7 ppm (C<0.3 ppm); transnonachlor, 1 ppm (C<0.3 ppm); DDE, 12.2 ppm (C<0.3 ppm); pentachlorophenol, 7.2 ppm (C<0.5 ppm); and *n*-hexane, 5.5 ppm (C<1 ppm). It was clear that this patient was overloaded with toxic chemicals that caused sympathetic nervous system dysfunction and vascular deregulation.

With the low sulfur and iron, this patient would have impaired glutathione, methionine, and taurine conjugation as well as a less-than optimally functioning cytochrome P-450 system. She now is in good health, using a massive avoidance and nutrient supplementation program and taking her food injections every 4 days.

Genitourinary System

At the EHC-Dallas, we have seen many children who experience bedwetting or recurrent cystitis due to food and/or chemicals. See Volume II, Chapter 14,[71] and Chapter 22 for examples and discussion.

Dermatological System — Acne

Many children in their teen-age years develop acne. This may be due to hormonal imbalance. However, many cases are triggered by food and chemical exposures.

DIAGNOSIS AND TREATMENT

Diagnosing and treating children with environmental triggers involve no special problems. In fact, children are generally much easier to diagnose, and, once diagnosed, they respond rapidly and with fewer problems to treatment. They can be diagnosed and treated by the environmental approach used with adults.

Children under one year of age may be fasted 1 day if necessary. However, unless the case is severe, fasting in a child under 1 year is usually unnecessary. Instead, substitution of different formulas, especially those made of less-contaminated food such as lamb or beef base with spring water, banana base with spring water, oat base with spring water, etc. can be tried. Older children are fasted from 1 to 4 days until their signs and symptoms improve. Usually, however, most newborns, infants, and children with environmental problems do not have to be fasted in order to clear their symptoms. Rotary diets usually suffice. Biological inhalant problems can be worked out with techniques similar to those used with adults, such as serial dilution endpoint titration or the symptom neutralization intradermal injection techniques. (See Volume IV, chapters on diagnosing and treating patients with chemical sensitivities.)

Food sensitivity is worked out in a manner similar to that used with adults. Fasting, rotary elimination diets, intradermal injection, and oral challenge testing may all be used (see Chapter 34).

Chemical sensitivity is worked out in a manner similar to that used with adults via inhaled, oral, and intradermal challenge. See Volume IV, Chapters 31, 34, and 36.

The policy at the EHC-Dallas is to counsel parents before conception. Our feeling is that the optimum time to treat newborns and infants is preconception. If a mother and father are healthy or even if they have chemical sensitivities for which they are being treated, they can have babies who have excellent health.

The principles of reduction of total body load and good nutrition seem to apply when children who are born to once severely sensitive parents are studied. These children appear to be much healthier than their parents, and they require little treatment. We now have a large number of "ECO babies". Levels of cord blood pollutants are lower in these babies than the blood levels of their mothers who have been previously detoxified. This finding sharply contrasts with Dowty and Laseter's study[56] that showed that toxic volatile organic hydrocarbons bioconcentrate in the fetus. Of course, Dowty and Laseter studied people who paid no attention to chemical overload and were presumably more contaminated than our sample parents and their infants. Our studies suggest that the reduction of total body load and better delivery of nutrients to expectant mothers results in improved detoxification systems and, thus, healthier babies.

Over 200 infants and children with problems related to the gastrointestinal, genitourinary, cardiac, vascular, or musculoskeletal systems have been taken through the hospital environmental control process of diagnosis and treatment at the EHC-Dallas without problems. We have diagnosed pollutant triggers and treated children with intractable asthma, seizures, cardiac arrhythmias and cardiomyopathy, diarrhea, vomiting, arthritis, diabetes, dysuria and cystitis, bedwetting, and autism. Intradermal and sublingual tests confirming diagnosis and treatment have been performed without problems on over 1000 outpatient children.

In one study, we compared oral and intradermal challenge on 21 patients. A total of 1176 oral and 1176 intradermal challenges were performed (Table 28.18). Three hundred and forty-five (29%) oral food challenges were positive, and 394 (33%) intradermal challenges were positive (see Volume IV, Chapter 36). There was an 88% correlation between a positive oral and a positive intradermal food challenge. It is difficult to say if the small percent more positive by intradermal challenge were false positives, since some patients had such severe sensitivity that they reacted to most substances tested on their skin.

Of the 200 hospitalized children we treated, a cross-section sample (30) of them were prospectively studied in a defined series. This group ranged in age from 1 week to 16 years and included 20 males and 10 females. The gender of the children with apparent chemical sensitivity was the direct opposite of the adult population, where females present more commonly than males. The ratio of male to female again equalizes as end-stage disease occurs in older populations. Then mortality is the same for males and females, although some years later for the female.

These infants and children had numerous toxic chemicals in their blood including 4.2 pesticides per patient (Table 28.19). This finding contrasted to that found in outpatient, allergic children, who usually had one or two. Ninety-one percent of the 12 patients studied had DDE, while another 40% contained dieldrin. Aromatics and aliphatics were also found in these children (Table 28.12).

Table 28.18.   Comparison of Food Sensitivity by Oral and
Intradermal Challenge in 21 ECU Hospitalized Children
after 2 to 4 Days Fasting and Reduced Total Load

	Oral challenge	Intradermal challenge
No. of positive foods	345	394
Average reactions per patient	16	19

Note: No. of challenge tests = 1176.

Source: EHC-Dallas. 1992.

Table 28.19.   The Frequency of Chlorinated Pesticides
in 12 Hospitalized Children

Pesticides	No. of patients	No. of positive	Frequency (%)
Aldrin	12	2	16.
α-BHC	12	2	16.7
β-BHC	12	2	16.7
DDD	12	0	0.0
DDE	12	11	91.7
Dieldrin	12	5	41.7
Endosulfan I	12	1	8.3
Heptachlor	12	2	16.7
Heptachlor epoxide	12	9	75.0
HCB	12	11	91.7
trans-Nonachlor	12	5	41.7

Source: EHC-Dallas. 1992.

Hemaglobins (Table 28.20) were low in 38% of the patients, with WBC being abnormal in 28% and eosinophil counts being abnormal in 57%. SMACs are shown in Table 28.21.

Immune parameters (Tables 28.22 to 28.24) including IGE and Total T and B lymphocytes and complements were performed on the sickest children. Their IGE was abnormal in 50%; IGG was abnormal in 25%, and IGA was abnormal in 22%. Serum complements were tested in nine patients. Eight had some abnormality.

T and B lymphocytes were measured in 18 children. All had some parameter abnormality, while many had multiple parameters that were abnormal.

Vitamin levels (Table 28.25) were abnormally low in about the same ratio as adults, with 60% having low B_6 and B_2. Serum and intracellular mineral levels were measured in nine patients. Many abnormalities were seen, with levels of Mg, Cr, Si, Se, Zn, Ca, and K predominating (Tables 28.26 and 28.27). Testing for chemicals by double-blind inhalant (9) and intradermal (30) challenge was accomplished (Table 28.28). One hundred percent of those skin tested and 75% of those who were challenged by inhalation reacted. Upon

Table 28.20.   The Frequency of Abnormal Hematology Analysis
in 29 Hospitalized Children

Hematology analysis	No. of patients tested	No. above normal	No. below normal	No. abnormal	(%)
WBC	29	7	1	8	27.6
RBC	29	0	6	6	20.7
HGB	29	0	11	11	37.9
HCT	29	2	9	11	37.9
MCV	29	0	2	2	6.9
MCH	25	0	3	3	12.0
MCHC	25	0	1	1	4.0
Seg	25	0	4	4	16.0
Band	11	1	0	1	9.1
Eos (%)	15	3	0	3	20.0
Eos C	14	4	4	8	57.1
Lym (%)	27	7	4	11	40.7
Mono (%)	23	4	0	4	17.4

Source: EHC-Dallas. 1992.

Table 28.21.   The Frequency of Abnormal SMAC Chemistry Analysis
in 29 Hospitalized Children

SMAC chemistry analysis	No. of patients tested	No. above normal	No. below normal	No. abnormal	(%)
Glucose	29	2	11	13	44.8
Uric acid	29	1	2	3	10.3
BUN	29	3	0	3	10.3
CO_2	16	0	3	3	18.8
Total protein	29	0	4	4	13.8
Alkaline phosphatase	29	14	0	14	48.3
SGOT	27	6	0	6	22.2
SGPT	21	0	0	0	0.0
LDH	29	10	1	11	37.9
CPK	16	1	0	1	6.3
Triglyceride	22	3	0	3	13.6
Cholesterol	29	2	4	6	20.7

Source: EHC-Dallas. 1992.

Table 28.22.   The Frequency of Abnormal Immune
Antibodies in 24 Hospitalized Children

Immune antibodies	No. of patients tested	No. above normal	No. below normal	No. abnormal	(%)
IgA	18	0	4	4	22.2
IgE	24	12	0	12	50.0
IgG	20	1	4	5	25.0
IgM	18	0	0	0	0.0

Source: EHC-Dallas. 1992.

Table 28.23.   The Frequency of Abnormal Complements
in 9 Hospitalized Children

Complement	No. of patients tested	No. above normal	No. below normal	No. abnormal	(%)
CH_{100}	9	0	4	4	44.4
C_1	7	2	0	2	28.6
$C1q$	7	0	0	0	0.0
C_2	7	0	0	0	0.0
C_3	9	0	1	1	11.1
C_4	9	0	3	3	33.3
C_5	7	2	0	2	28.6

Source: EHC-Dallas. 1992.

intradermal challenge to orris root, 77% of the patients reacted. This type of challenge has proven to be a good, nonspecific test for chemical sensitivity. Seventy percent tested by intradermal challenge reacted to cigarette smoke.

These findings emphasize the seriousness of the problem "passive smoking" poses for environmentally sensitive children who are exposed to cigarette smoke at home. We insist that there should be no smoking in the homes and schools of chemically sensitive children. Sixty-two percent of subjects in our study reacted to newsprint, which may give some insight into those children who experience learning problems and to the increasing number of new cases of sensitivity seen in each new generation of children.

Sixty percent of our subjects were also sensitive to petroleum-derived ethanol. With this substance being both ubiquitous in the environment and insidious in its effects, exposure is quite common, and the prevalence of sensitivity to this chemical found in our study emphasizes how it can both spark symptoms in the chemically sensitive individual and cause an increase

Table 28.24.   The Frequency of Abnormal Immune Parameters in 18 Hospitalized Children

Immune parameters	No. of patients tested	No. above normal	No. below normal	No. abnormal	(%)
WBC	18	4	1	5	27.8
Lym (%)	15	6	3	9	60.0
Lym C	15	1	3	4	26.7
T_{11} (%)	17	0	4	4	23.5
T_{11} C	17	1	5	6	35.3
T_4 (%)	11	1	1	2	18.2
T_4 C	11	0	2	2	18.2
T_8 (%)	11	1	1	2	18.2
T_8 C	11	2	2	4	36.4
T_4/T_8	11	1	2	3	27.3
Bly (%)	17	5	0	5	29.4
Bly C	17	5	0	5	29.4

Source: EHC-Dallas. 1992.

Table 28.25   The Frequency of Abnormal Vitamins and Vitamin-Related Enzymes in 11 Hospitalized Children

Vitamins related enzymes	No. of patients tested	No. above normal	No. below normal	No. abnormal	(%)
EGPT (B_6)	10	6	0	6	60.0
ETK (B_1)	11	1	0	1	9.1
EGR (B_2)	10	3	3	6	60.0
FIGLU (Folate)	9	1	0	1	11.1
KRYPTO (B_6)	7	0	0	0	0.0
XA (B_6)	8	0	0	0	0.0
KYN (B_6)	8	1	0	1	12.5
MMA (B_{12})	8	1	0	1	12.5
1 NMN (B_3)	8	2	2	4	50.0
Lipid peroxide	2	2	0	2	11.0
ESOD	1	0	1	1	100.0
Vitamin C/WBC	9	0	3	3	33.3
Vitamin D_3	9	1	2	3	33.3
Vitamin A	8	1	3	4	50.0
β-Carotene	8	0	2	2	25.0

Notes: EGPT = erythrocyte glutathione pyruvate transaminase. ETK = erythrocyte transketolase. EGR = erythrocyte glutathione reductase. FIGLU = formimin oglutamic acid. KRYPTO = kryptopyrole. XA = xanthurenic acid. KYN = kynurenic acid. MMA = methylmalonic acid. 1 NMN = 1-N-methylno cotinamide. ESOD = erythrocyte superoxide dismatase.

Source: EHC-Dallas. 1992.

Table 28.26.   The Frequency of Abnormal RBC Minerals
in 9 Hospitalized Children

RBC minerals	No. of patients tested	No. above 2 SD	No. below 2 SD	No. abnormal	(%)
Ca	8	1	0	1	12.5
Mg	9	6	1	7	77.8
Na	6	0	0	0	0.0
K	6	3	0	3	50.0
Cu	9	2	1	3	33.3
An	9	1	2	3	33.3
Fe	9	0	2	2	22.2
Mn	8	0	0	0	0.0
Cr	9	0	6	6	66.7
P	6	1	0	0	16.7
Se	8	0	2	2	16.7
Si	6	0	2	2	33.3
S	6	0	2	2	33.3
Ba	6	1	0	1	16.7

Source: EHC-Dallas. 1992.

Table 28.27.   The Frequency of Abnormal Plasma Minerals
in 10 Hospitalized Children

RBC minerals	No. of patients tested	No. above 2 SD	No. below 2 SD	No. abnormal	(%)
Ca	10	0	3	3	30.0
Mg	10	0	2	2	20.0
Na	6	3	1	4	66.7
K	6	1	0	1	16.7
Cu	10	1	5	6	60.0
Zn	10	2	3	5	50.0
Fe	10	4	3	7	70.0
Mn	10	0	0	0	0.0
Cr	10	0	0	0	0.0
P	10	6	1	7	70.0
Se	10	2	1	3	30.0
Si	6	0	0	0	0.0
S	6	6	0	6	100.0
Ba	6	1	0	1	16.7

Source: EHC-Dallas. 1992.

Table 28.28.   Intradermal and Inhalant Challenge of Chemicals
in 30 Hospitalized Children

Chemicals	Intradermal challenge (n = 30)			Inhalant challenge (double-blind) (n = 9)		
	No. tested	No. positive	(%)	No. tested	No. positive	(%)
Cigarette smoke	27	20	74.1	—	—	—
Orris root	27	21	77.8	—	—	—
Ethanol	30	18	60.0	9	5	55.6
Formaldehyde	13	13	100.0	8	6	75.0
Newsprint	24	15	62.5	—	—	—
Perfume	27	13	48.1	—	—	—
Phenol	10	2	20.0	8	6	66.7
Chlorine	6	1	16.7	9	5	55.6
Pesticides	—	—	—	7	4	57.1
Placebo	—	—	—	20	0	0.0

Source: EHC-Dallas. 1992.

in the number of new cases of sensitivity that develop annually. Perfumes also were a problem in 48% of the children. Certainly, this sensitivity can cause nasal and respiratory problems, as well as interfere with learning.

Terpene sensitivity was prevalent in 100% of 28 children studied (Table 28.29). Eighty-nine percent were sensitive to grass terpene. Ragweed and pine and Mountain Cedar terpenes followed close behind. Terpene sensitivity was found to be a problem indoors as well as outside. Sensitive children exposed to pine floors, pine wood cabinets, cedar closets, toys, and beds made of pine often develop severe problems that never seem to clear as they are constantly around these substances.

For long-term treatment, each child had to have some degree of environmentally controlled living quarters and chemically less-contaminated food and water. The more complex patients had to have nutrient supplementation, intradermal injections, and heat depuration/physical therapy.

EFFECTS OF CONTAMINATED FOOD AND AIR ON CHILDREN

Concern about health effects from exposure to pesticides in the food of children is growing as is evidenced by recent public outcry and the National Academy of Sciences[72] study on the pesticides in foods consumed by children. Clearly, our chapter on pesticides and our work with many children whose symptoms were triggered by exposure to pesticides indicates we too are concerned. Many of the pesticides applied to food crops in this country are

Table 28.29.   Terpenes Intradermal Test
in 28 Hospitalized Children

Terpenes	No. of patients	No. positive	(%)
Pine	28	17	60.7
Tree	28	12	42.9
Grass	28	25	89.3
Ragweed	28	18	64.3
Mountain Cedar	27	16	59.2
Mesquite	7	3	42.9
Sage	2	1	50.0

Source: EHC-Dallas. 1992.

present in foods that children eat. Figure 28.8 and Table 28.30 show the percentage of positive detections for 6 pesticides in various foods for samples larger than 25. As can be seen, up to 50% of the samples contain pesticide with a range of 0 to 50 (see Volume II, Chapter 8[73]). They are common foods that babies frequently eat. Table 28.31 shows six foods. Among the 18 foods most frequently consumed by infants, pesticide residue was detected in every food and often more than one pesticide was identified. Some studies by Elkins in 1989[74] reported data on the effects of food processing on pesticide residues. In most instances, washing the food was shown to reduce residues. Blanching reduced them even further, and the canning process led to still greater decreases. Elkins' data indicated that in tomatoes and green beans subjected to each of these three steps, the levels of malathion reduced 99%. However, some pesticide was still present. Carbonyl concentrations decreased 99% to 73%. Levels of parathion were reduced 66% in spinach and 10% in frozen broccoli. Some processing actually increases pesticide levels such as the case of ETU, which increased 94.5% in frozen turnip greens as a result of maneb degradation during cooking in a sauce pan. These facts again emphasize that no matter how food is treated and prepared, if it contains any pesticide, the potential for problems exists.

Millions of children in the U.S. receive up to 35% of their entire lifetime dose of some carcinogenic pesticides by the age of 5 years. This pattern is most evident for pesticides used on foods heavily consumed in the first years of life, such as the fungicides captan (35% of lifetime risk by age 5 years) and benonyl (29%) and the insecticide dicotol (32%). Infants and children are routinely exposed to combinations of two or three (in rare cases as many as eight) pesticides. In an analysis of 4500 samples of fruits and vegetables taken from supermarket warehouses from 1990 to 1992, 201 more pesticides were found on 52% of oranges sampled, 44% on apples sampled, and from 1/4 to 1/3 of samples of cherries, peaches, strawberries, celery, pears, and grapes. In addition, the FDA found 108 different pesticides in just 22 fruits and vegetables; 42 different pesticides were detected on tomatoes, 38 on strawberries, and 34 on apples.[75] By the average child's first birthday, the combined cancer risk

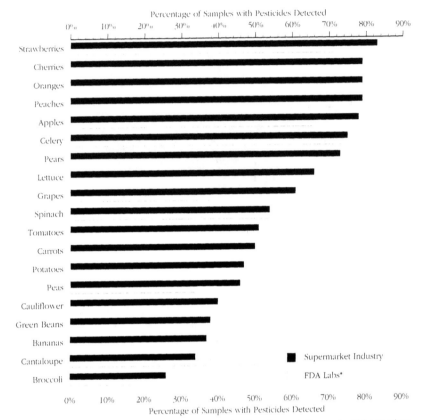

A Comparison of Supermarket Warehouse
Data and FDA Labs*

Figure 28.8. **Percent of fruits and vegetables heavily consumed by young children with pesticides detected: a comparison of supermarket warehouse data and FDA Labs. (From Wiles, R. and C. Campbell. 1993.** *Pesticides in Children's Food.* **Washington D.C.: Environmental Working Group. With permission.)**

from just 18 pesticides on 20 foods exceeds the EPA's lifetime level of acceptable risk of one-in-one-million additional cancers. Lifetime risks from these same pesticides in foods is slightly more than ten times the one-in-a-million standard. Certainly ingestion of these toxic chemicals might trigger chemical sensitivity and may be a significant factor in the upsweep in chemical

Table 28.30.   The Percentage of Positive Detections for Six Pesticides
in Various Foods for Samples Larger than 25

Chemical	Foods	No. in sample	Positive samples No.	Positive samples (%)
Ethylene-bis-dithiocarbamate	Peas	124	49	40
	Bananas	100	4	4
Benomyl	Apples	134	35	26
	Peaches	26	13	50
	Bananas	72	8	11
Captan	Apples	978	125	13
	Peaches	574	48	8
	Pears	431	18	4
	Carrots	390	1	0.3
	Peas	705	2	0.3
	Beans	776	3	0.4
Chlorpyrifos	Apples	968	116	12
	Peaches	581	14	2
	Pears	454	6	1
	Carrots	402	2	0.5
	Bananas	347	16	5
	Peas	751	9	1
	Beans	834	23	3
Dimethoate	Apples	959	10	1
	Peaches	561	2	0.4
	Pears	455	10	2
	Bananas	347	2	0.6
	Peas	720	70	10
	Beans	805	43	5
Parathion-methyl	Apples	957	37	4
	Peaches	570	4	0.7
	Beans	809	5	0.6
	Peas	732	56	0.7

Source: National Research Council. 1993. *Pesticides in the Diets of Infants and Children*. 254. Washington D.C.: National Academy Press. With permission.

sensitivity in children. Millions of children drink water from midwestern rivers and reservoirs that are contaminated with carcinogenic herbicides. These are known to trigger chemical sensitivity. It is clear from these and our studies at the EHC-Dallas that children would have better health if they ate organically grown food (see Volume II, Chapters 7[73] and 8[76]).

We have seen many problems in children from homes that have undergone indoor extermination and spraying with toxic chemicals. The greatest hazard posed by these chemicals appears to be to infants who are crawling or sleeping on carpet that has been sprayed or treated with toxic chemicals. Because of

Table 28.31. Six Foods Among the 18 Most Frequently Consumed by Infants and the Tolerances and Residue Levels of 6 Pesticides Detected in Them

| | | Residue level, ppm | | |
| | | Mean detected | Maximum detected | EPA tolerance |
Food	Pesticide			
Apples, fresh	Ethylenebisdithio-carbamate (EBDC)	0.0937	0.7700	7.0
	Benomyl	0.1635	2.6400	7.0
	Captan	0.0352	3.4000	25.0
	Chlorpyrifos	0.0064	0.9000	1.5
	Dimethoate	0.0006	0.1900	2.0
	Parathion-methyl	0.00176	0.2600	1.0
Peaches, fresh	EBDC	0.0154	0.2000	10.0
	Benomyl	0.2095	1.1100	15.0
	Captan	0.1329	9.6900	50.0
	Chlorpyrifos	0.0013	0.1300	0.05
	Dimethoate	0.0001	0.0875	NT
	Parathion-methyl	0.0024	0.0900	1.0
Pears, fresh	EBDC	0	0	7.0
	Benomyl	0.0691	1.5900	7.0
	Captan	0.0155	1.0000	25.0
	Chlorpyrifos	0.0003	0.0400	0.05
	Dimethoate	0.0024	0.3850	2.0
	Parathion-methyl	0	0	1.0
Carrots	EBDC	0	0	7.0
	Benomyl	0	0	0.2
	Captan	0.0021	0.8300	2.0
	Chlorpyrifos	0.0001	0.0200	NT
	Dimethoate	0	0	NT
	Parathion-methyl	0	0.0100	1.0
Peas, succulent, garden	EBDC	2.5964	23.0000	7.0
	Benomyl	0.1000	0.4100	NT
	Captan	0	0	2.0
	Chlorpyrifos	0.0017	0.4800	NT
	Dimethoate	0.0341	2.8775	2.0
	Parathion-methyl	0.0009	0.5500	1.0
Beans, succulent, green	EBDC	0.1042	2.0000	10.0

Table 28.31 (Continued). Six Foods Among the 18 Most Frequently
Consumed by Infants and the Tolerances and Residue Levels
of 6 Pesticides Detected in Them

Food	Pesticide	Residue level, ppm		
		Mean detected	Maximum detected	EPA tolerance
	Benomyl	0	0	2.0
	Captan	0.0028	1.0000	NT
	Chlorpyrifos	0.0027	0.5650	0.05
	Dimethoate	0.0120	2.3450	2.0
	Parathion-methyl	0.0001	0.7000	1.0

Note: NT = no tolerance level has been established by EPA.

Source: Based on FDA Surveillance Data, 1988–1989, unpublished. National Research Council. 1993. *Pesticides in the Diets of Infants and Children*. 256, Washington D.C.: National Academy Press. With permission.

these contaminations, we have always recommended nonpesticided and non-toxic chemical treatments for houses.

Fenske et al.[77] report that home pesticide applications may pose health risks for occupants, particularly infants and children. A study of 20 adult residents exposed to dichlorvos following crack-and-crevice treatments found a slight decrease in group mean serum cholinesterase, with three residents reporting a feeling of illness 24 hours post-applications[78] (Table 28.32). Six of 21 children admitted to Arkansas Children's Hospital for organophosphate intoxication were judged to have been exposed following insecticide spraying inside the home.[79] A review of 37 hospitalized pesticide poisonings among infants and children at the Children's Medical Center in Dallas, Texas, revealed that five cases were due to pesticide absorption following spraying or fogging inside residences.[80] California State agencies reviewed 44 illness reports in 1984 associated with residential pesticide use,[81] and subsequently conducted a hazard assessment of acute intoxication potential among infants following indoor pesticide treatments.[82] Concern also exists regarding the chronic health effects of chldren's exposure to pesticides[83] (Table 28.33).

Residential monitoring studies have focused primarily on ambient air concentrations of pesticides following several types of applications, including termiticide applications,[84–88] DDVP resin strips hung in households,[89–92] and crack-and-crevice or baseboard treatments.[93-98] Recently, the U.S. Environmental Protection Agency initiated a major study of pesticide air concentrations in homes.[99] Limited studies of two additional application methods have been conducted. Broadcast applications[100,101] involve the handspraying of an aqueous mixture directly onto target surfaces (e.g., rugs, floors, furniture),

Table 28.32. Chlorpyrifos Residues Available by Wipe Sampling: Effects of Time and Ventilation

		Ventilated rooms			Nonventilated rooms	
Sample time (hour)	N	Mean residue[a] ($\mu g/cm^2$)	Percent initial deposit[c]	N	Mean residue[b] ($\mu g/cm^2$)	Percent initial deposit
<0.5 Immediate	6	1.69	12.4	3	3.90	28.6
0.5 to 6.5 Day 1	35[d]	0.69	5.0	18	1.56	11.4
24 Day 2	6	0.28	2.0	3	0.48	3.5

[a,b] All ventilated room values significantly lower than nonventilated room values (ANOVA: Student-Neuman-Keuis test; $p<0.01$).
[c] Deposition following application = 13.6 $\mu g/cm^2$.
[d] 25.2 $\mu g/cm^2$ value excluded as outlier.

Source: Fenske, R. A., K. G. Black, K. P. Elkner, C.-L. Lee, M. M. Methner, and R. Soto. 1990. Potential exposure and health risks of infants following indoor residential pesticide applications. *Am.. J. Public Health* 80(6):689–693. With permission.

Table 28.33. Total Estimated Absorbed Chlorpyrifos Dose for Infants: Day of Application (Day 1) and Day Following Application (Day 2)

Room condition	Dermal dose (mg/kg)	Percent total	Respiratory dose (mg/kg)	Percent total	Total dose (mg/kg)	Percent of NOEL[a]
Day 1						
Ventilated	0.052	69	0.023	31	0.075	250
Nonventilated	0.120	76	0.038	24	0.158	527
Day 2						
Ventilated	0.022	58	0.016	42	0.038	127
Nonventilated	0.037	67	0.018	33	0.055	183

[a] Human no observable effect level = mg/kg/day.

Source: Fenske, R. A., K. G. Black, K. P. Elkner, C.-L. Lee, M. M. Methner, and R. Soto. 1990. Potential exposure and health risks of infants following residential pesticide applications. *Am. J. Public Health* 80(6):689–693. With permission.

while total aerosol release applications (foggers)[102-104] involve activation of a pressurized canister in an unoccupied room, with the pesticide aerosol covering all surfaces (walls, floor, ceiling). Both of these methods produce residues on surfaces to be contacted by occupants. Flea treatment by broadcast application was selected for study here because of the potential for both high air concentrations and substantial dermal exposure. A preliminary risk assessment has been conducted for crawling infants (9 to 10 months of age).

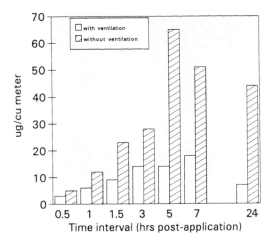

Figure 28.9. Chlorpyrifos air concentrations in the sitting adult breathing zone (100 cm). (From Fenske, R. A., K. G. Black, K. P.
Elkner, C.-L. Lee, M. M. Methner, and R. Soto. 1990. Potential exposure and health risks of infants following indoor
residential pesticide applications. *Am. J. Public Health*
80(6):689–693. With permission.)

Fenske et al.[77] measured air and surface chlorpyrifos residues for 24 hours
following a 0.5% Dursban® broadcast application for fleas inside a residence.
Two of the three treated rooms were ventilated following application. Maximum air concentrations were measured 3 to 7 hours postapplication. Peak
concentrations in the infant breathing zone were 94 µg/m³ in the nonventilated
room and 61 µg/m³ in the sitting adult breathing zone. Concentrations of
approximately 30 µg/m³ were detected in the infant breathing zone 24 hours
postapplication. Surface residues available through wipe sampling were 0.7 to
1.6 µg/cm³ of carpet on the day of application and 0.3 to 0.5 µg/cm³ 24 hours
postapplication. Estimated total absorbed doses for infants were 0.08 to 0.16
mg/kg on the day of application and 0.04 to 0.06 mg/kg the day following
application, with dermal absorption representing approximately 68% of the
totals. These doses are 1.2 to 5.2 times the human tolerated dose (Figures 28.9
and 28.10).

In summary, toxic chemicals have been shown to affect the sperm, ovum,
zygote, embryo, fetus, newborn, infant, child, and adolescent in many ways.
Therefore, in order to enhance health and prevent disease, attention to environmental causes of illness is now necessary. Again, massive avoidance of pollutants in air, food, and water; injection therapy for the secondary sensitivities to
pollens, molds, foods, and some chemicals; and nutrient supplementation and
replacement are necessary for proper prevention and treatment of chemical
sensitivity in the young. Heat depuration/physical therapy and immune modulators may also be beneficial.

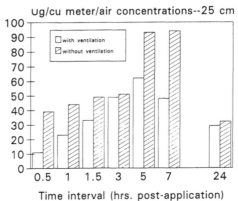

Figure 28.10.   Chlorpyrifos air concentrations in the infant breathing zone (25 cm). (From Fenske, R. A., K. G. Black, K. P. Elkner, C.-L. Lee, M. M. Methner, and R. Soto. 1990. Potential exposure and health risks of infants following indoor residential pesticide applications. *Am. J. Public Health* 80(6):689–693. With permission.)

REFERENCES

1. Dixon, R. L. 1986. Toxic responses of the reproductive system. In *Casarett and Doull's Toxicology: The Basic Science of Poisons*, 3rd ed., eds. C. D. Klaassen, M. O. Amdur, and J. Doull. 432. New York: Macmillan.

2. Heidam, L. Z. 1984. Spontaneous abortions among dental assistants, factory workers, painters, and gardening workers: a follow-up study. *J. Epidemiol. Community Health* 38(2):149–155.

3. Sager, D. B. 1983. Effect of postnatal exposure to polychlorinated biphenyls on adult male reproductive function. *Environ. Res.* 31:76–94.

4. Sager, D., D. Girard, and D. Nelson. 1991. Early postnatal exposure to PCBs: sperm function in rats. *Environ. Toxicol. Chem.* 10:737–746.

5. Barlow, S. M. and F. M. Sullivan. 1982. *Reproductive Hazards of Industrial Chemicals*. London: Academic Press.

6. Rea, W. J. 1992. *Chemical Sensitivity. Vol. I. Mechanisms of Chemical Sensitivity*. 17. Boca Raton, FL: Lewis Publishers.

7. Rea, W. J. 1992. *Chemical Sensitivity. Vol. I. Mechanisms of Chemical Sensitivity*. 47. Boca Raton, FL: Lewis Publishers.

8. Rea, W. J. 1992. *Chemical Sensitivity. Vol. I. Mechanisms of Chemical Sensitivity*. 221. Boca Raton, FL: Lewis Publishers.

9. Rousseaux, C. G. and P. M. Blakley. 1991. Fetus. In *Handbook of Toxicologic Pathology,* eds. W. M. Haschek and C. G. Rousseaux. 937. San Diego: Academic Press.

10. Hayes, W. J. 1982. *Pesticides Studied in Man.* 253, 444. Baltimore: Williams & Wilkins.

11. Hamilton, A. and H. L. Hardy. 1974. *Industrial Toxicology,* 3rd ed. Acton, England: Publishing Sciences.

12. Rousseaux, C. G. and P. M. Blakley. 1991. Fetus. In *Handbook of Toxicologic Pathology,* eds. W. M. Haschek and C. G. Rousseaux. 959. San Diego: Academic Press.

13. U.S. Environmental Protection Agency. 1994. Lower Rio Grande Valley. Environmental Monitoring Study. Report to the Community on the Pilot Project. Dallas, TX.

14. Lancranjan, I., H. I. Papescu, O. Gavanescu, I. Klepsch, and M. Serbanescu. 1975. Reproductive ability of workmen occupationally exposed to lead. *Arch. Environ. Health* 30:396–401.

15. Crease, D. M. and P. M. D. Foster. 1991. Male reproductive system. In *Handbook of Toxicologic Pathology,* eds. W. M. Haschek and C. G. Rousseaux, 829. San Diego: Academic Press.

16. Dixon, R. L. 1986. Toxic responses of the reproductive system. In *Casarett and Doull's Toxicology: The Basic Science of Poisons,* 3rd ed., eds. C. D. Klaassen, M. O. Amdur, and J. Doull. 434. New York: Macmillan.

17. Cripps, D. J., H. A. Peters, A. Gocmen, and I. Dogramici. 1984. Porphyria turcica due to hexachlorobenzene: a 20 to 30 year follow-up study on 204 patients. *Br. J. Dermatol.* 11(4):413–422.

18. Rosati, P., G. Noia, M. Conte, M. De Santis, and S. Mancuso. 1989. Drug abuse in pregnancy: fetal growth and malformations. *Panminerva Med.* 31(2):71–75.

19. Thomas, D. C., D. B. Petitti, M. Goldhaber, S. H. Swan, E. B. Rappaport, and I. Hertz-Picciotto. 1992. Reproductive outcomes in relation to malathion spraying in the San Francisco Bay Area, 1981–1982. Epidemiology 3(1):32–39.

20. Koos, B. J. and L. D. Longo. 1976. Mercury toxicity in the pregnant woman, fetus, and newborn infant. *Am. J. Obstet. Gynecol.* 126(3):390–406.

21. Kurland, L., S. Faro, and H. Siedler. 1960. Minamata disease. The outbreak of a neurologic disorder in Minamata, Japan, and its relationship to the ingestion of seafood contaminated by mercuric compounds. *World Neurol.* 1:370.

22. Takeuchi, T., T. Kambara, and N. Morikawa. 1950. Pathologic observations of the Minamata disease. *Acta Pathol. Jpn.* 9:769.

23. Birke, G., A. Johnels, L. Plantin, B. Syostrand, S. Skerfving, and T. Westermark. 1972. Studies on humans exposed to methyl mercury through fish consumption. *Arch. Environ. Health* 25:77.

24. Matsumoto, H., G. Koya, and T. Takeuchi. 1964. Fetal Minamata disease. A neuropathological study of two cases of intrauterine intoxication by a methyl mercury compound. *J. Neuropathol. Exp. Neurol.* 24:563.

25. Moriyama, H. 1967. A study on the congenital Minamata disease. *J. Kumamoto Med. Soc.* 41:506.

26. Murakami, U. 1972. Embryo-fetotoxic effect of some organic mercury compounds. *Ann. Rep. Res. Inst. Environ. Med. Nagoya Univ.* 19:61–68.

27. Tatetsu, S. and M. Harada. 1968. Mental deficiency resulting from mercury intoxication in the prenatal period. *Adv. Neurol. Sci.* (Tokyo) 12:181.

28. Nelson, N. (Chairman), Expert Committee. 1971. Hazards of mercury. Special report to the secretary's pesticide advisory committee, Department of Health, Education, and Welfare. *Environ. Res.* 4:1.

29. Bakir, F., S. Damluji, L. Amin-Zaki, M. Murtadha, A. Khalidi, N. Al-Rawi, S. Tikriti, H. Dhahir, T. Clarkson, J. Smith, and R. Doherty. 1973. Methyl mercury poisoning in Iraq. An interuniversity report. *Science* 181:230.

30. Yang, M., J. Wang, J. Garcia, E. Post, and K. Lei. 1973. Mammary transfer of ^{203}Hg from mothers to brains of nursing rats. *Proc. Soc. Exp. Biol. Med.* 142:723.

31. Takizawa, Y. 1970. Studies on the Niigata episode of Minamata disease outbreak. *Acta Med. Biol. (Niigata)* 17:293.

32. Takizawa, Y., T. Kosaka, and R. Sugai. 1972. Studies on the cause of the Niigata episode of the Minamata disease outbreak. *Acta Med. Biol.* 19:193.

33. Bakulina, A. V. 1968. The effect of a subacute ethylmercury coated grain poisoning on the progeny. *Sov. Med.* 31:60.

34. Haq, I. U. 1963. Agrosan poisoning in man. *Br. Med. J.* 1:1579.

35. Ordónez, J. V., J. A. Carillo, C. M. Miranda, and J. L. Gale. 1971. Epidemiological study of an illness in the Guatemala highlands believed to be an encephalitis. *Bol. Of. Sanit. Panam.* 172:65.

36. Engleson, G. and T. Herner. 1952. Alkyl mercury poisoning. *Acta Paediatr. Scand.* 41:289.

37. Pierce, P., J. Thompson, W. Likosky, L. Nickey, W. Barthel, and A. Hinman. 1972. Alkyl mercury poisoning in humans. Report of an outbreak. *JAMA* 220:1439.

38. Snyder, R. D. 1971. Congenital mercury poisoning. *N. Engl. J. Med.* 18:1014.

39. Curley, A., V. Sedlak, E. Girling, R. Hawk, W. Barthel, P. Pierce, and W. Likosky. 1971. Organic mercury identified as the cause of poisonings in humans and hogs. *Science* 172:65.

40. Skerfving, S. 1971. Organic mercury compounds — relation between exposure and effects. In *Mercury in the Environment. A Toxicological and Epidemiological Appraisal, PB-205–000,* eds. L. Friberg and J. Vostal. National Technical Information Service. U.S. Department of Commerce, 8–1 to 8–44. Washington, D.C.: U.S. Government Printing Office.

41. Czeizel, A. E., C. Elek, S. Gundy, J. Mëtneki, E. Nemes, A. Reis, K. Sperling, L. Tímár, G. Tusnády, and Z. Vigágh. 1993. Environmental trichlorfon and cluster of congenital abnormalities. *Lancet* 341:539–542.

42. Dzwonkowska, A. and H. Hubner. 1986. Induction of chromosomal aberrations in the Syrian hamster by insecticides tested in vivo. *Arch. Toxicol.* 58:152–156.

43. Degraeve, N., M. C. Chollet, and J. Moutschen. 1984. Evaluation of the mutagenic potential of four commercial mixtures of insecticides. *Food Chem. Toxicol.* 22:683–687.

44. Degraeve, N., M. C. Chollet, and J. Moutschen. 1984. Cytogenetic and genetic effects of subchronic treatments with organophosphorus insecticides. *Arch. Toxicol.* 56:66–67.

45. Degraeve, N., M. C. Chollet, and J. Moutschen. 1984. Cytogenetic effects induced by organophosphorus pesticides in mouse spermatocytes. *Toxicol. Lett.* 21:315–319.

46. Moutschen-Dahmen, J., M. Moutschen-Dahmen, and N. Degraeve. 1981. Metrifonate and dischlorvos: cytogenetic investigations. *Acta Pharmacol. Toxicol. (Copenh.)* 49(Suppl. 5):29–39.

47. Staples, R. E. and E. H. Goulding. 1979. Dipterex teratogenicity in the rat, hamster and mouse when given by gavage. *Envir. Health Perspect.* 30:105–113.

48. Rousseaux, C. G. and P. M. Blakley. 1991. Fetus. In *Handbook of Toxicologic Pathology,* eds. W. M. Haschek and C. G. Rousseaux. 954. San Diego: Academic Press.

49. Rousseaux, C. G. and P. M. Blakley. 1991. Fetus. In *Handbook of Toxicologic Pathology,* eds. W. M. Haschek and C. G. Rousseaux. 955. San Diego: Academic Press.

50. Guminska, M. June 6–7, 1991. Air pollution and health effects in the region of Cracow. Presented at Conference of the Prophylactic Role of Clean Environment in Health Preservation. Cracow, Poland.

51. Childhood heart defect linked to TCE solvent. Winter 1985. *Toxic News Digest.* 2.

52. Cohen, E. N., M. L. Wu, C. E. Whitcher, J. B. Brodsky, H. C. Gift, W. Greensie, T. W. Jones, and E. J. Driscoll. 1980. Occupational disease in dentistry and chronic exposure to trace anesthetic gases. *J. Am. Dent. Assoc.* 101:21–31.

53. Baird, P. A. 1992. Occupational exposure to nitrous oxide — not a laughing matter. *N. Engl. J. Med.* 327(14):1026–1027.

54. Shepard, T. H. 1983. *Catalogue of Teratogenic Agents,* 4th ed. Baltimore: Johns Hopkins University Press.

55. Warkany, J. 1971. *Congenital Malformations: Notes and Comments.* Chicago: Year Book Medical Publishers.

56. Dowty, B. J. and J. L. Laseter. 1976. The transplacental migration and accumulation in blood of volatile organic constituents. *Pediatr. Res.* 10:696–701.

57. Pérez-Comas, A. 1982. Precocious sexual development in Puerto Rico. *Lancet* 1:1299–1300.

58. Sáenz de Rodriguez, C. A., A. M. Bongiovanni, and L. Conde de Borrego. 1985. An epidemic of precocious sexual development in Puerto Rican children. *J. Pediatr.* 107:393–396.

59. Waikman, F. 1986. Personal communication.

60. Rapp, D. 1986. Personal communication.

61. Boris, M. 1986. Personal communication.

62. Lecks, H. I. 1963. Allergy of the middle ear. In *The Allergic Child,* ed. F. Speer. 289–298. New York: Haeber Medical Division.

63. Pelikan, Z. 1987. Rhinitis and secretory otitis media: a possible role of food allergy. In *Food Allergy and Intolerance,* eds. J. Brostoff and S. J. Challacombe. 467. London: Ballière Tindall.

64. So, S. Y., M. M. T. Ng, M. S. M. Ip, and W. K. Lam. 1990. Rising asthma mortality in young males in Hong Kong 1976–85. *Respir. Med.* 84:457–461.

65. Brown, D. 1980–1993. Personal communication.

66. Lee, C. 1974–1978. Personal communication.

67. Oppenheimer, J. J., H. S. Nelson, S. A. Bock, F. Christensen, and Y. M. Leung. 1992. Treatment of peanut allergy with rush immunotherapy. *J. Allergy Clin. Immunol.* 90(2):256–262.

68. Rea, W. J. 1992. *Chemical Sensitivity. Vol. I. Mechanisms of Chemical Sensitivity.* 155. Boca Raton, FL: Lewis Publishers.

69. Rea, W. J. 1994. *Chemical Sensitivity. Vol. II. Sources of Total Body Load.* 837. Boca Raton, FL: Lewis Publishers.

70. Rea, W. J. 1994. *Chemical Sensitivity. Vol. II. Sources of Total Body Load.* 843–844. Boca Raton, FL: Lewis Publishers.

71. Rea, W. J. 1994. *Chemical Sensitivity. Vol. II. Sources of Total Body Load.* 941. Boca Raton, FL: Lewis Publishers.

72. National Research Council. 1993. Pesticides in the diets of infants and children. Washington, D.C.: National Academy Press.

73. Rea, W. J. 1994. *Chemical Sensitivity. Vol. II. Sources of Total Body Load.* 579. Boca Raton, FL: Lewis Publishers.

74. Elkins, E. R. 1989. Effect of commercial processing on pesticide residues in selected fruits and vegetables. *J. Assoc. Off. Anal. Chem.* 72(3):535.

75. Environmental Working Group. 1993. *Pesticides in Children's Food,* 1. Washington D.C.: Agricultural Pollution Prevention Project.

76. Rea, W. J. 1994. *Chemical Sensitivity. Vol. II. Sources of Total Body Load.* 535. Boca Raton, FL: Lewis Publishers.

77. Fenske, R. A., K. G. Black, K. P. Elkner, C.-L. Lee, M. M. Methner, and R. Soto. 1990. Potential exposure and health risks of infants following indoor residential pesticide applications. *Am. J. Public Health* 80(6):689–693.

78. Gold, R. E., T. Holeslaw, D. Tupy, and J. B. Ballard. 1984. Dermal and respiratory exposure to applicators and occupants of residences treated with dichlorvos (DDVP). *J. Econ. Entomol.* 77:430–436.

79. Woody, R. C. 1984. The clinical spectrum of pediatric organophosphate intoxications. *Neurotoxicology* 5:75.

80. Zwiener, R. J. and C. M. Ginsburg. 1988. Organophosphate and carbamate poisoning in infants and children. *Pediatrics* 81:121–126.

81. Knaak, J. B., J. Schreider, and P. Berteau. 1987. Hazard assessment of indoor use of chlorpyrifos, dichlorvos, propoxur and other organophosphates and N-methyl carbamates. Worker Health and Safety Branch Report No. HS-1423. Sacramento, CA: California Dept. of Food and Agriculture.

82. Berteau, P. E., J. B. Knaak, and D. C. Mengle. 1989. Insecticide absorption from indoor surfaces: hazard assessment and regulatory requirements. In *Biological Monitoring for Pesticides Exposure*, eds. R. G. M. Wang, C. A. Franklin, R. C. Honeycutt, and J. C. Reinert. ACS Symposium Series 382. Washington, D.C.: American Chemical Society.

83. Lowengart, R. A., J. M. Peters, C. Cicioni, J. Buckley, L. Bernstein, S. Preston-Martin, and E. Rappaport. 1987. Childhood leukemia and parents' occupational and home exposures. *JNCI* 79:39–46.

84. Livingston, J. M. and C. R. Jones. 1981. Living area contamination by chlordane used for termite treatment. *Bull. Environ. Contam. Toxicol.* 27:406.

85. Wright, C. G. and R. B. Leidy. 1982. Chlordane and heptachlor in the ambient air of houses treated for termites. *Bull. Environ. Contam. Toxicol.* 20:617.

86. Wright, C. G., R. B. Leidy, H. E. Dupree, and T. J. Sheets. 1985. Subterranean termite control: chlordane residues in soil surrounding and air within houses. In *Dermal Exposure Related to Pesticide Use,* eds. R. C. Honeycutt, G. Zweig, and N. N. Ragsdale. ACS Symposium Series 273. Washington, D.C.: American Chemical Society.

87. Moye, H. A. and M. H. Malagodi. 1987. Levels of airborne chlordane and chlorpyrifos in two plenum houses: Saranex S-15 as a vapor barrier. *Bull. Environ. Contam. Toxicol.* 39:533–540.

88. Anderson, D. J. and R. A. Hites. 1988. Chlorinated pesticides in indoor air. *Environ. Sci. Technol.* 22:717–720.

89. Wright, C. G., R. B. Leidy, and H. E. Dupress. 1988. Chlorpyrifos in the ambient air of houses treated for termites. *Bull. Environ. Contam. Toxicol.* 40:561–568.

90. Gillet, J. W., J. R. Harr, F. T. Linstrom, D. A. Mount, A. D. St. Clair, and L. J. Weber. 1972. Evaluation of human health hazards on use of dichlorvos (DDVP), especially in resin strips. *Residue Rev.* 44:115–159.

91. Gillet, J. W., J. R. Harr, A. D. St. Clair, and L. J. Weber. 1972. Comments on the distinction between hazards and safety in evaluation of human health hazards on use of dichlorvos, especially in resin strips. *Residue Rev.* 44:161–184.

92. Leary, J. S., W. T. Keane, C. Fontenot, E. S. Feictmeir, D. Schultz, D. A. Koos, L. Hirsch, E. M. Lavor, C. C. Roan, and C. H. Hine. 1974. Safety evaluation in the home of polyvinyl chloride resin strip containing dichlorvos (DDVP). *Arch. Environ. Health* 29:308–314.

93. Wright, C. G. and M. D. Jackson. 1971. Propoxur, chlordane and dizinon on porcelain china saucers after kitchen cabinet spraying. *J. Econ. Entomol.* 64:457.

94. Wright, C. G. and M. D. Jackson. 1975. Insecticide residues in non-target areas of rooms after two methods of crack and crevice application. *Bull. Environ. Contam. Toxicol.* 13:123.

95. Wright, C. G. and M. D. Jackson. 1976. Insecticides movement following application to crevices in rooms. *Arch. Environ. Contam. Toxicol.* 4:492.

96. Wright, C. G. and R. B. Leidy. 1978. Chlorpyrifos residues in air after application to crevices in rooms. *Bull. Environ. Contam. Toxicol.* 19:340.

97. Wright, C. G., R. B. Leidy, and H. E. Dupree. 1981. Insecticides in the ambient air of rooms following their application for control of pests. *Bull. Environ. Contam. Toxicol.* 20:548.

98. Wright, C. G., R. B. Leidy, and H. E. Dupree. 1984. Chlorpyrifos and diazinon detection on surfaces in dormitory rooms. *Bull. Environ. Contam. Toxicol.* 31:259–264.

99. Hsu, J. P., H. G. Wheeler, D. E. Camann, H. J. Schattenberg, R. G. Lewis, and A. E. Bond. 1988. Analytical methods for detection of nonoccupational exposure to pesticides. *J. Chromatogr. Sci.* 26:181–189.

100. Maddy, K. T., J. Lowe, and N. Saini. 1985. Indoor air concentration of ethylene glycol monoethyl ether following application of propétamphos insecticide emulsifiable concentrate. Report No. HS-1264. Sacramento, CA: Worker Health and Safety Unit, California Department of Food and Agriculture.

101. Naffziger, D. H., R. J. Sprenkel, and M. P. Mattler. 1985. Indoor environmental monitoring of Dursban L. O. following broadcast application. In *Down to Earth,* No. 41. Midland, MI: Dow Chemical Co.

102. Maddy, K. T., S. Edmiston, and A. S. Fredrickson. 1981. Monitoring residues of DDVP in room air and on a horizontal surface following use of a room fogger. Report No. HS-897. Sacramento, CA: Worker Health and Safety Unit, California Department of Food and Agriculture.

103. Maddy, K. T., S. Edmiston, and E. Ochi. 1984. Dissipation of DDVP and propoxur following the release of an indoor fogger — a preliminary study. Report No. HS-1259, Sacramento, CA: Worker Health and Safety Unit, California Department of Food and Agriculture.

104. Kreiger, R., J. Ross, T. Thongsinthusak, and H. Fong. 1988. Measuring potential dermal transfer of surface pesticide residue generated from indoor fogger use. Sacramento, CA: Worker Health and Safety Branch Report, California Department of Food and Agriculture.

Index

A

Abortion, spontaneous, 1641
Acetaldehyde, 1199
Acidophil cells, 1599
Acne, 1130
Acrolein, 1199
Adaptation, 1107
Adrenal gland
  insufficiency, 1622
  medulla, 1599
Adrenogenital syndrome,
  1628
Air conditioner lung, 1253
Albumin, 1146, 1203
Alcohol, 1135
Aldehyde, 1215
Alkali
  phosphatase, 1482
  salts, 1481
Alkylating agents, 1229
Allergy
  alveolitis, 1252
  conjunctivitis, 1907
  reaction, 1908
Alloplastic, 1182
Alpha
  receptors, 1738
  tocopherol, 1919
Alphachloratose, 1148
Alveola
  capillary block, 1256
  ducts, 1219
Alzheimer's disease, 1763
American Academy of
    Allergy, 1976
Amikacin, 1135
Amino acids, 1223
Amygdala, 1841

Anabolic steroid, 1648
Analgesics, 1135, 1749
Anaphylaxis, 1976
Anaerobic glycolysis, 1331
Anesthetics, 1758
Angiotensin receptors,
  1851
Animal danders, 1126
Anion gap, 1146
Anoxia, 1785, 1794
Antepartum, 1638
Antibiosis, 1419
Anticholinergics, 1433
Anti-DNA levels, 1577
Anxiety, 1614
Aorta
  -autonomic
      neuroendocrine cells,
      1612
  bodies, 1209
Apple, 1126
Aqueous fluid, 1887
Arachnoiditis, 1784
Argyrophilic cells, 1613
Aroma
  amine carcinogens,
    1521
  hydrocarbons, 1804
  volatile, 1362
Arrhythmias, 1130
  supraventricular, 1982
Arterial blood gases,
  1261
Arterial occlusion, 1339
Arteriosclerosis, 1324
Arthralgia, 1584
Arthritic, 1547
Arthrogenic diet, low,
  1552

Artificial breast implants,
  1272
Asbestos, 1214
Ascorbic acid, 1896
Asthma, 1243
Atherosclerosis, 1386
Atopic keratitis, 1908
Atrial fibrillation, 1392
Atrophic rhinitis, 1116
Attention deficit, 1785
Autonomic nervous system,
  1300
Audiogram, 1140
Auditory system
  brainstem response,
    1136
  hallucinations, 1846
Auto workers, 1802
Autoantibodies, 1174
Autoimmune disease,
  1577
Autonomic nervous system,
    1303, 1410, 1605
  deregulation, 1738
  reflex pathways, 1734
Autosomal recessive gene,
  1211
Axonal degeneration, 1211

B

B lymphocytes, 1525
Baby-milk powder, 1652
Baker's yeast, 1151
Barbiturates, 1150
Barley, 1946
Baroreceptors, 1301
Bedwetting, 1523
Beets, 1151

Behavior
functions, 1834
responses, 1795
Bell's palsy, 1133
Benzaldehyde, 1910
Benzopyrene, 1464
Benzphetamine,
N-demethylase,
1204
Benzyl alcohol, 1784
Beriberi, 1894
Bile pigments, 1446
Binocular infrared, 1899
Biochemical individuality,
1195
Bioflavonoid, 1658, 1662
phloretin, 1660
Biological inhalants, 1125
Bipolarity, 1195
Bismuth, 1806
Bladder, 1517
spasm, 1613
Bleomycin, 1135, 1583
Blindness, 1885
Blood
pressure, low, 1130
vessels, 1212
Bond, reduction, 1423
Bone pain, 1597
Brain, 1110
Branching nerve terminal,
1535
Breast, 1267
cancer, 1269, 1272
implant, 1275
capsule, 1279
milk, 1966
Bromobenzene, 1509
Bronchi, 1210
bronchitis, 1130
mucosa, 1213
Bronchiectasis, 1204
Bronchiomeric group, 1610
Bronchoconstrictor response,
1211
Bronchospasm, 1747
Broncoconstrictor, 1200
Bruises, spontaneous, 1155
Buffering capacity, 1480
Bumetanide, 1135
Butylated hydroxy toluene,
1445

C

Cadmium, 1321
Calcium, 1321
calcification, 1274
Cancer, 1111, 1784
Candida, 1167
Cane sugar, 1151
Capsaicin, 1213
Car exhaust, 1178
Carbohydrate, 1447
Carbon
dioxide, 1209
disulfide, 1356
monoxide, 1202
Carcinogenic pesticides,
1993
Cardiac arrhythmias, 1379
Cardiomyopathy, 1320
Cardiovascular system,
1379
Carcinogenic potencies, 1112
Carotid sinus reflex, 1308
Carotodynia, 1105
Carpal tunnel syndrome,
1549
Cataracts, 1910
Catecholamines, 1608
Caudal epithalamus, 1603
Caustic soda, 1260
Celiac disease, 1435
Cement pipes, 1233
Cephalgia, 1980
Cerebellar atazia, 1764
Cerebellar inhibition, 1154
Cerebral cortex, 1785
Chalazion, 1908
Chemical
challenge, 1525
inhalant testing, 1983
messengers, 1425
transmitter systems, 1847
yellows, 1301
Chenopodium, oil of, 1135
Chest wall syndrome, 1231
Chloramphenicol, 1135
Chlorinated hydrocarbson,
volatile, 1360
Chlorination
organic solvents, 1505
pesticides, 1654
phenols, 1807

Chlorine, 1119
Chloroform, 1465
Chloropromazine, 1841
Chloroquine, 1135
Chlorpyrifos, 1999
Chocolate, 1342
Cholecystitis, 1470
Cholecystokinin, 1480
Cholesterol, 1624
Cholinolytic, 1214
Chromaffin cells, 1617
Chromium, 1542
Chromosomal abnormalities,
1943
Cigarette, 1122
Cilia, 1111
transport, 1213
Cis-platinum, 1135
Citric acid, 1496
Claudication, 1316
Cleaning material, 1196
Clitoris, 1632
Clonidine, 1616
Cobalt, 1214
salts, 1463
Codfish, 1393
Coenzyme
A, 1451
FAD, 1894
Coke dust, 1110
Cold, 1130
Colitis, 1442
Collagen-like glycoprotein,
1915
Comatose, 1475
Commensalism, 1419
Commercial foods, 1381
Competitive inhibition, 1419
Conchal cartilage, 1183
Conductive hearing loss,
1136
Congenital malformations,
1949
Conjunctivitis, 1906
Constipation, 1435
Contractility, 1537
Cord, 1128
Corn, 1137
Cornea, 1908
Cortical fiber, 1915
Cortical response, 1840
Costochondral grafts, 1183

Co-trimoxazule, 1567
Cotton, 1831
Cough, 1201
Cow's milk, 1965
Cranial neuralgia, 1179
Cyanosis, 1983
Cyclophosphomide, 1218
Cysteine, 1504
Cystic fibrosis, 1252
Cystic mastitis, 1271
Cystitis, 1525
Cytoarchitectural, 1606
Cytokine, 1115
Cytologic, 1810

D

Dehydration, 1839
Dehydroxylation, 1423
Deiodinated metabolites, 1659
Dental inflammation, 1429
Deranged mineral metabolism, 1356
Dermatitis, 1908
Desert Storm veterans, 1583
Detoxification systems, 1941
Detrusor, 1518
contractions, 1517
Diabetes, 1476
Diarrhea, 1135
Dichlorvos, 1997
Diethyldithiocarbamate, 1762
Diethylpropion hydrochloride, 1583
Dihydrosteptomycin, 1135
Diisocyanates, 1244
Dimethylcarbamyl chloride, 1112
Dioxide, 1214
Dioxin, 1503
Dipalmitoyl phosphatidylcholine, 1226
Diplopia, 1894
Distention, 1128
Dizygotic twins, 1211
Dizziness, 1173
Dopamine system, 1847
Drainage tubes, 1136
Dryness
mouth, 1121

throat, 1201
Dysmenorrhea, 1635
Dyspareunia, 1640
Dysphagia, 1431

E

Eczema, 1964
Edema, cyclic, idiopathic, 1622
Ejaculate, 1648
Electromyography, 1133
Emaciation, 1468
Embryo, pharmakinetics, 1944
Emotion
instability, 1978
states, 1848
Emphysema, 1204
Encephalin-endorphin system, 1848
Endocrine system, 1965
effects, 1336
function, 1473
pancreatic dysfunction, 1475
Endoophthalmitis phaco-anaphylacita, 1911
End-organ damage, irreversible, 1105
Endothelium, 1751
cells, 1501
Enlarged heart, 1982
Enolase enzyme, 1607
Enteric plexuses, 1410
Enterochromaffin-like cells, 1427
Eosinophilic fasciitis, 1548
Eosinophils, 1245
Epigastric pain, 1462
Episcleritis, 1908
Epithelium, 1213
hepatic cells, 1446
inflammation, 1116
Equilibrium disturbances, 1152
Erectile impotence, 1525
Erosion, 1182
Erythromycin, 1135
Estrogen, 1268
Ethacrynic acid, 1135
Ethmoid, 1146

Eustachian tube, 1106
Excitatory amino acids, 1753
Exocrine, 1425
buffering system, 1481
dysfunction, 1252
enzymes, 1478
pancreas, 1478
Eye
absence of, 1928
irritation, 1201
movements, stimulus-provoked, 1905

F

Fabrics, 1109
Faciculus gracilis, 1211
Farmer's lung disease, 1254
Fasciculi, 1107
Fasting, 1431
Fat
acids, 1620
degeneration, 1322
Fatigue, visual, 1902
Fetus, 1937
cells, 1949
Fever, 1366
Fibrillar mucoprotein, 1751
Fibromyalagia, 1547
Fibronectin, 1217
Fibrosis, 1584
Fields, visual, 1929
Figs, 1475
Finnish rayon, 1818
Fixed-named disease, 1209
Flood chamber, 1460
Fluorine, 1215
Fluorocarbons, 1324
Follicular carcinoma, 1183
Food
additives, 1322
challenge, 1248
cravings, 1496, 1634
injection, 1121
neutralization injections, 1964
sensitivity, 1252, 1322
tests, 1264
Formaldehyde, 1111
Free amino acids, 1914
Free radicals, 1222

Frontal lobe, 1730
Fruits, 1552
Fulminant diarrhea, 1982
Fumigants, 1509
Fungi, 1420
  benomyl, 1928

G

Gallbladder, 1470
Ganglion nodosom, 1612
Gastric dysfunction, 1433
Gastritis, 1431
Gastrointestinal tract, 1130,
  1416
  upset, 1979
Genetic factors, 1211
Genitourinary system, 1495
  paraganglion cells, 1613
Glands, 1212
Glomerular membrane,
  1501
Glomerulonephritis, 1506
Glossophanyrgeal IX, 1209
Glucagon, 1474
Glucaric acid, 1450
Glucocorticoid, 1623
Gluconeogenesis, 1619
Glucosidases, 1424
Glue sniffing, 1924
Glutaraldehyde, 1792
Glutathione, 1914, 1983
  nephrotoxicity, 1503
  peroxidase, 1898
Glycerol
  esterified, 1452
  test, 1173
Glycogen, 1450
Glycolytic zone, 1332
Glycosidases, 1423
Goblet cell hyperplasia, 1111
Goiter, 1656
Goodpasture's syndrome,
  1371
Gram-negative bacteria,
  1664
Grand mal, 1785, 1789
Grasses, 1151, 1178, 1522
Grip strength, 1553
Growth
  hormone, 1601
  retardation, 1945

Guanethidenes, 1889
Gut flora, 1424

H

Halogen
  alkanes, 1324
  hydrocarbons, 1388, 1935
Halothane, 1210
Harrell-Butler
    Neuropsychological
    Screen
Hay fever, 1980
Head raising, 1842
Headache, 1177
  traction, 1179
Hearing, 1979
  discrimination, 1169
Heart, 1300, 1331
  failure, 1982
Heartburn, 1431
Heart-lung transplants, 1611
Heat, depuration, 1118,
    1148, 1163
Heiner syndrome, 1967
Hemolytic serum
    complements, 1381
Hemoptysis, 1260, 1261
Hemorrhagic trigonitis, 1522
Hepatic system
  glycogen, 1448
  microsomal enzyme
    inducers, 1660
Hepatocarcinogens, 1464
Hepatocyte, 1464
  hypertrophies, 1461
Hepatotoxicity, 1461
Heptachlorepoxide, 1765
Heroin, 1762
Herpes simplex virus, 1132
Hexadine, 1135
Hexose monophosphate
    shunt, 1450
Hiccup, 1944
High blood sugar, 1982
Hippocampus, 1845
Histamine, 1203
Hormones, 1466
Human placenta, 1663
Huntington's Chorea, 1778
Hydrocephalus, 1784
Hydrogen ions, 1209

Hydroxylators, 1766
Hypofunction, 1470, 1621
Hypoglycemia, 1150
Hypothalamus, 1110
  medial, 1839
  pituitary axis, 1601
Hypothyroidism, 1665

I

Icca syndrome, 1579
Ifosfamide, 1506
Imidazole agents, 1645
Immune system, 1325
  defense system, 1169
  dysfunction, 1156
  parameters, 1987
Immunochemicals, 1743
Immunoglobulins, 1384
Immunosuppressive agent,
  1510
Immunotherapy, 1134
Inattention, 1868
Incontinence, 1523
Indecisiveness, 1831
Industrial solvents, 1150
Infants, 1965
Infarction, 1813
Inferior meatus, 1106
Inflammation, 1115
Infrared, 1899
Inhalant
  extract subcutaneous,
    1136
  sensitivities, 1176
  skin testing, 1136
Inhibition, 1727
Injection therapy, 1276,
    1527, 1867, 1911,
    1999
Inner ear, 1150
Inorganic chemicals, toxic,
    1110
Insecticide, 1127, 1664
Insulin-producing B-cell,
    1473
Integrative brain functions,
    1829
Intense itching, 1907
Intercostal, 1209
Intermediary metabolism
    reactions, 1895

Intermediate
   Psychoneurological
      Exam, 1855
Interstitial fibrosis, 1214
Interstitial-cell stimulating
   hormone, 1600
Intestinal system
   malignancy, 1445
   microflora, 1414
   peptides, 1428
Intra-adrenal
   paragangliomas,
      1617
Intracellular edema, 1621
Intracellular potassium, 1770
Intractable sinusitis, 1118
Intradermal antigen
   challenge, 1463
   injection testing, 1241
Intradermal challenge, 1151,
   1175, 1178, 1864
Intradermal neutralization,
   1128, 1169
Intradermal provocation,
   1754
Intradermal test, 1136, 1166
Intradermal titration therapy,
   1902
Intradermal wheals, 1135
Intraocular pressure, 1887
Intrathoracic paragangliomas,
   1614
Intravagal neuroendocrine
   cells, 1612
Intrinsic muscles, 1887
Iodides, 1579
Iodine uptake, 1604
Iris, 1911
   corder, 1118, 1728, 1889
Iritis, 1914
Irritant, corrosives, 1197
Isobutylacetate, 1117
Isothiocyanates, 1656

J

Japanese cedar pollen, 1908
Jet, fuel, 1154, 1178

K

Kinin, 1203, 1390

L

Labrynthinitis, 1132
Lactation, 1270
Lactose, 1203
Lactovegetarian diet, 1564
Lactulose, 1424
Larynx, 1128
   dysfunction, 1211
Lavage, 1198
Lead, 1135
   encephalopathy, 1825
   poisoning, 1829
   salts, 1834
Learning, 1160
Leather, 1110
Lecithin, 1767
Lectins, 1485
Legumes, 1128
Lens, 1914
Lesions, premalignant, 1271
Leukocyte, 1115
   adhesiveness of, 1216
Leukoplakia, 1430
Leukotriene B4, 1216
Limbic cortex, 1842
Lindane, 1625
Lingula, 1249
Lipid, 1452
   peroxidation, 1505
Lipotropes, 1453
Lipoxygenase prostaglandin
   system, 1749
Liver, 1434, 1446
   cancer, 1465
   detoxifying processes,
      1461
   vascular system, 1459
Lung, biopsies, 1253
Lymphadenopathy, 1182
Lymphatic system, 1106,1124
   drainage, 1567
Lymphocyte, 1497
Lysosome formation,
   spontaneous, 1333

M

Ménière's disease, 1105,
   1117, 1169, 1170,
   1171, 1173, 1174,
   1175, 1176

Macrobiotic diet, 1923
Macrophages, 1214, 1746
Magnesium, 1543, 1788
   sulfate, 1615
Malabsorption, 1433
Maladaptive response, 1318
Malaise, 1384
Masking, 1756
Mast cells, 1169
Maxillary sinus, left, 1146
Mediators, 1198, 1213
Medulla oblongata, 1209,
   1735
Melanocyte, 1746
Membrane
   barrier, 1195
   structure, 1327
Memory, 1830
   loss, 1793
Menarche, 1627
Menopause, 1646
Mental retardation, 1946
Mercuric chloride, 1946
Mercury-contaminated flour,
   1946
Mesothelioma, 1235
Mesothelium, 1640
Metaplasia, 1115
Methyl
   chloroform, 1150
   isocyanate, 1641
   mercury, 1509, 1785
   testosterone, 1654
Microchemistry, 1754
Microcirculation, 1326
Microsomes, 1204
Micturition, 1518
Middle ear disease, 1134
Migraine, 1177, 1779
   classic, 1179
Milk, 1172, 1210, 1646,
   1661, 1975
Millet, 1662
Minamata Bay, mercury
   disaster, 1859
Mineral, 1321
   excess, 1459
Minimum change nephrosis,
   1510, 1514
Miosis, 1911
Mitochondrial, 1207
Mold, 1117, 1126

Monoamines, 1607
Mononuclear cell infiltrates, 1116
Monooxygenase, 1204
Mood, 1820
  altering drugs, 1795
Motor
  conduction velocities, 1146
  defects, 1118
  neurons, lower, 1132
  response, 1158, 1826
Mouth, 1410
  ulcers, 1429, 1430
Movement, coordination test, 1159
Mucus, 1197, 1214, 1251, 1412
  intestinal, 1412
  layers, 1107
Multiendocrinopathy, 1597
Multifocal PVCs, 1391
Multiple behavioral effects, 1809
Multiple sclerosis, 1769
Muscle
  contractions, 1408
  cramps, 1539, 1540
  weakness, 1620
Musculoskeletal system, 1157, 1275
Mustard gas, 1218
Mutualism, 1416
Myalgia, 1539
Mycotoxin, 1310
Myelinated nerve fibers, large, 1533
Myesthenia gravis, 1536, 1784
Myocardial ischemia, 1761
Myoneural junction, 1408

N

Narcotics, 1210
Nasal mucosa, 1112
Nasal stuffiness, 1176
Nasal vault, 1107
Natural gas, 1118
Nausea, 1151
Navy beans, 1486
Nearsightedness, 1917

Neck, stiff, 1107
Neoplasia, 1464
Nephrosclerosis, 1506
Nerve
  biopsy, 1817
  cells, 1407
  conduction velocity changes, 1153
  supply, 1106
Nervous system, 1727
Neural crest, 1605
Neural pathways, 1154
Neuralgia, 1134
Neurocardiovascular, 1117
Neuroendocrine system, 1425
  cell neoplasms, 1611
  stimulation, 1614
Neurogenic mechanism, 1747
Neurogenic vascular responses, 1740
Neuromechanism, 1206
Neuron, 1211
  atrophy, 1822
  death, 1755
Neuropeptides, 1743
Neuroresponses, 1207
Neurotoxicity, 1813
  disorders, 1758
Neurotoxins, 1150
Neurotransmitters, 1736
Neurovascular responses, 1195
Neutrophils, 1217
Newborn, 1952
Nickel, 1110
  sensitivity, 1323
Nigral iron, 1766
Nitrogen dioxide, 1259
Nitrous oxide, 1214, 1949
  sulfurs, 1215
Nodal points, 1851
Noncardiac pain, 1612
Nongranulomatous uveitis, 1910
Nonimmune mechanisms, 1325
Nonmigraine vascular headaches, 1179
Nonneoplastic endocrine cell hyperplasia, 1426

Nonsteroidal antiinflammatory drugs, 1561
Norepinephrine, 1409
  secretory neurons, 1847
Nose
  airway resistance, 1115
  burning, 1115
  epithelium, 1112, 1113
  irritation, 1198, 1201
Noxious odor, 1110
Nucleus
  soliaris, 1208, 1209
  tuberis, 1731
Nutrients, 1323
  deficiency, 1866, 1891
  fuel deficiency, 1107
Nystagmus, 1148

O

Oats, 1135
Occlusion, 1182
Occular damage, 1894
Occular pressure, 1910
Occupational chemical exposure, 1110
Occupational history, 1819
Ocular-motor changes, 1153
Oculo-motor nerve, 1886
Odor
  hallucination, 1825
  sensitivity, 1109
Oil mist, 1110
Olfaction, 1108
  bulb, 1853
  nerve, 1107
  tract, 1110
Oligodendrial cells, 1746
Oligodendrocyte, 1752
Oligopeptide, 1606
Onion, 1210, 1662
Opiates, 1852
Optic system, nerve, 1885, 1923, 1924
  damage, 1920
  degeneration, 1843
Oral challenge, 1169, 1242
Oral nutrient supplementation, 1118
Organ ischemia, 1348

Organic solvents, 1154
Organochloride pesticides, 1522
Organochlorine, 1339, 1472, 1864, 1889
Organophosphate, 1763, 1764, 1778
  seizures, 1789
Orris root, 1261
Osteoarthritis, 1182
Osteoporosis, 1150
Otalgia, 1179
Otitis media, 1130
Otosclerosis, 1150
Outgassers, 1197
Ovarian dysfunction, 1630
Oxidants, 1202
Oxidative degradation, 1454
Oxychlordane, 1166
Oxytocin, 1600
Ozone, 1199

P

Pacemaker, 1310
Paget's disease, 1150
Pain, 1182
Palpitations, 1308
Pancreas, 1470
  pancreatectomy, 1475
  pancreatic polypeptide, 1481
Panic disorder, 1806
Papillary tumors, 1219
Papilledema, 1924
Paraganglia, 1618
Paragangliomas, 1613
Parahormone, 1624
Paranoia, 1974
Parasympathetic system, 1303, 1519, 1888
Parathormones, 1624
Parkinson's disease, 1847
Parquat, 1221
Paynaud's phenomenon, 1155
Peaches, 1134
Peak flow, 1262
Pedal edema, 1468
Pelvic peritoneum, 1639
Penile prostheses, 1525

Pentose phosphate pathway, 1223
Peptide
  biosynthesis, 1612
  hormone, 1481
Peptide-containing nerves, 1474
Perceptual/motor problems, 1834
Perfume, 1109, 1128, 1522
Perianal dermatitis, 1964
Periaqueductal gray, 1850
Periorbital edema, 1121, 1780
Perivascular lymphocytic infiltrates, 1119
Personality
  test, 1834
  traits, 1868
Petechiae, 1155
Phagocytosis, 1214
Phenol, 1119
Phenoxy herbicide, 1646
Pheochromocytomas, 1613
Phlebitis, 1371, 1373
Phospholipids, 1455, 1456
Photoisomerization, 1919
Phrenic nerve, 1209
Phthalates, 1464
Physiologic adaptation, 1207
Phytoestrogens, 1651
Pilocarpine, 1888
Pinocytotic vesicles, 1750
Pituitary gland, 1599
Plasma
  amino acid, 1251
  colloid osmotic pressure, 1502
  renin, 1356
Pleural effusion, 1235
Pneumotactic center, 1209
Pollen, 1264
Polyarthritis, 1580
Polychlorinated biphenyls, 1638, 1652
Polydyspia, 1982
Polymonuclear leukocytes, 1369
Polymorphism, 1200
Polyneuropathy, 1826
Polypeptide chain, 1849
Polyphenol, 1662

Polyprotein gene expression, 1608
Polyps, 1116, 1117
Porcelain steel house, 1783
Pork, 1138
Porphyria, 1462
Positional nystagmus, 1148
Positive skin whealing, 1170
Post-polio syndrome, 1573
Postural proteinuria, 1510
Posturography, 1156
Postvagotomy-cell lypoplasia, 1427
Potasium bromate, 1135
Potato, 1137
Poultry, 1629
Powder burn, 1640
Prednisone, 1914
Premenstrual syndrome, 1634, 1636
Progesterone, 1630, 1645
Progestin, 1631
Prolactin, 1268
Prostatitis, 1497
Protein, 1270, 1457
  -sparing effect, 1602
Proteolysis, 1659
Prototype anaerobes, 1420
Provocative food test, 1169
Pseudo tumor cereberi, 1927
Pseudo-stratified epithelium, 1111
Psychomotor
  dysfunction, 1868
  speed, 1792
Psychoneurological responses, 1804
Ptyalin, 1412
Pulmonary emphysema, 1260
Pulmonary fibrosis, 1204
Pupil
  escape, 1904
  light reflex, 1904
  measurements, 1899
Pupillometer, 1898
Pure tone audiometry, 1136

Q

Quantitiative electroencephalogram, 1147

R

Raynaud's disease, 1106,
  1182, 1339, 1351,
  1549, 1584
Red blood cell magnesium,
  1542
Redundancy of structure, 1810
Relapsing steroid-responsive
  MCD, 1511
Renal, system, nervous, 1498
Renal system
  ailments, 1968
  artery stenosis, 1352
  failure, 1515
Reproductive cycles, 1602
Reproductive system, 1626
Resistance, stage of, 1937
Resorcinol, 1659
Respiratory system, 1420,
  1967, 1968
  center, 1207, 1830
  dysfunction, 1210
  failure, 1207, 1261
  tree, lower, 1213
  wall, lower, 1195
Restlessness, 1807
Retina, 1918
  pigment epithelium, 1897,
  1920
Rhabdosphincter, 1517
Rhesus monkeys, 1639
Rheumatic fever, 1391
Rheumatoid ankylosing
  spondylitis, 1564
Rheumatoid arthritis, 1551,
  1553
Rhinitis, 1116
  seasonal, 1176
  transient, 1125
Rhinorrea, 1203
Riboflavin deficiency, 1894
Ribonuclease, 1479
Rod pigment, 1919
Root canal, 1179
Rotary diet, 1435
Rotatory response, 1150
Russian thistle pollen, 1138

S

Saliva, 1121, 1604
Salivary gland, 1116
  malfunction, 1116

Sclerosis, 1182
Seborrheaic dermatitis,
  1964
Secretin, 1481
Secretory proteins, 1426
Seine-flu vaccine, 1770
Selenium deficiency, 1898
Selye, 1205
Semen, 1653
Semidehydroascorbate
  reductase, 1896
Sensitivities, 1105
Sensory system
  damage, 1832
  irritants, 1197
  organization, 1161,
  1168
Serotonin, 1603
  -secreting neurons, 1848
  secretory, 1427
Serous otitis media, 1132
Sertoli cell number, 1653
Serum total complement,
  1138
Sex
  excitability, 1626
  maturity, 1604
Shock, 1261, 1622
Shortness of breath, 1216
Short-term memory loss,
  1146
Silica, 1215, 1582
Silicone breast implant
  patients, 1281,
  1284
Sinus, 1130
  sinusitis, 1119
Sleep, apnea, 1210
Smooth muscles, 1408
  organs, 1388
Sneezing, 1110, 1176
Sodium, 1346, 1409, 1538,
  1983
  chloride, 1354
  nitrate, 1550
Solitary nucleus, 1209
Solvent, 1111, 1163, 1210,
  1213, 1582, 1728,
  1779, 1864, 1949
Somatostatin-producing
  D-cell hyperplasia,
  1428
Somesthetic messages,
  1536

Soy, 1170, 1210
Spasm, 1316, 1442
Spermatocytes, 1218
Spermicidal foams, 1647
Sphenopalatine ganglion,
  1108
Sphincterotomy, 1485
Spine
  arthritis, 1912
  cord anterior horn cells,
  1207
Spino-pontine-spinal reflex,
  1518
Spirometry, 1202
Spitting up, 1963
Spreading phenomenon,
  1195, 1833, 1869
Spring waters, 1386
Squamous cell, 1235
  carcinoma, 1218
  epithelium, stratified,
  1111
Staphylococcus aureus, 1228
Startle reflexes, 1822
Sternoclavicular grafts,
  1183
Steroid
  -dependent asthmatics,
  1239
  hormones, 1641
Stillbirths, 1937
Stridor, 1128
Strokes, 1983
Styrene, 1111
Substantia nigra, 1763
Sulfoxidation, 1769
Sulfur, 1214
  dioxide, 1226
Superoxides, 1322
Suppression, visual, 1155
Sway-referenced support,
  1162
Sweat glands, 1737
Sweet potato, 1262
Swelling index, 1552
Symbol formation, 1829
Sympathetic system
  fibers, 1214
  nerves, 1470
  trunks, 1735
Synergistic relationships,
  1419
Synthetic fabrics, 1372
Synthetic hepatotoxins, 1464

T

T lymphocytes, 1129
Tachycardia, 1308
Tamoxifen, 1521
Taurine, 1919
   deficiency, 1898
Teratogenesis, 1944
Teratospermia, 1939
Terpene, 1151, 1178
   sensitivity, 1778
Testosterone, 1496
Tetrachlorotheylene, 1210
Thalamus, 1268
Thalidomide, 1935
Thalmus, 1209
Thelarche, premature,
   1627
Thiamine, deficiency, 1150
Throat, 1124
Throiditis, 1664
Thromboembolisms, 1631
Thrush, 1962
Thyroid, 1654
   C-cells, 1612
   hormones, 1848
Thyroidectomy, 1183
Tinnitus, 1105, 1177
T-lymphocytes, 1123
Tobacco smoke, 1196
Toluene, 1166
   diisocynate, 1244
Total body load, 1122, 1197,
   1978
Total mercury levels, 1641
Touch, 1825
Toxicity, neuropathy, 1761
Trachea, 1214
   bronchial tree, 1213
Tracheostomy, 1263
Transamination, 1458
Transmitters, 1154
Transplacental sensitization,
   1963
Trauma, 1134
Trees, 1128
Tremors, 1830
Trichloroethane, 1197,
   1582
Trichloroethylene, 1150
Trichomonacide, 1423
Trigeminal nerve, 1886

Triggering agents, 1122
Trochlear nerve, 1886
Tryptophan metabolism,
   1582
T-suppressor cells, 1668
Tube
   osmotic clearance, 1498
   osmotic pressure, 1502
Tuberculosis, 1249
Turbinates, 1106
Twitching, 1542
Tympanogram, 1138

U

Ureaformaldehyde foam
   insulation, 1113
Urticaria, 1184
Uterus, 1482

V

Vagal paraganglioma, 1614
Vagal reflex, 1481
Vagina, 1646
   discharge, 1647
Vagus nerve, 1208, 1210
Vascular system, 1299, 1344
   injury, 1299
   permeability, 1206
   spasm, 1106, 1351, 1484
Vasculitides, 1336
Vasculitis, 1336, 1638
Vasoconstriction, 1307
Vasomotor rhinitis, 1203
Vasospasm, 1609
Ventral neural crest, 1610
Ventricular arrhythmia,
   1311, 1320, 1389,
   1391
Vertigo, 1145
Vessel, vasculitis, large,
   1339, 1345, 1348
Vestibular system
   abnormalities, 1156
   nystagmus, 1155
   oculomotor, 1154
Vestibule-oculomotor
   exposure, 1154
Videopupillography, 1902
Vinyl chloride, 1504
Viomycin, 1135

Viral encephalitis, 1569
Visceral-autonomic
   paraganglia, 1612,
   1613
Vision, 1822, 1885, 1920
Visual-motor performance,
   1803
Vitamin
   A, 1891, 1893
   B_1, 1894
   B_2, 1894
   B_3, 1895
   B_5, 1895
   B_6, 1895
   C, 1630, 1895
   E, 1897, 1979
   levels of, 1987
Vitronectin receptors, 1217
Voluntary nervous systems,
   1727
Vomiting, 1524
Vulva, 1646
Vulvitis, 1647

W

Wastewater effluents, 1661
Weather changes, 1318
Wechsler Adult Intelligence
   Scale-Revised, 1868
Wegener's granulomatosis,
   1368
Wood
   cutter, 1252
   dust, 1110

X

Xenobiotics, sequestration
   of, 1943
Xylene, 1110

Y

Yeast, 1126

Z

Zinc, 1251, 1897, 1980
Zygote, 1942
Zymogen, 1478
   granules, 1479

Printed and bound by CPI Group (UK) Ltd, Croydon, CR0 4YY

28/10/2024

01780093-0001